Eicosanoids and Other Bioactive Lipids in Cancer, Inflammation, and Radiation Injury, 4

ADVANCES IN EXPERIMENTAL MEDICINE AND BIOLOGY

Recent Volumes in this Series

Volume 460
MELATONIN AFTER FOUR DECADES
Edited by James Olcese

Volume 461
CYTOKINES, STRESS, AND DEPRESSION
Edited by Robert Dantzer, Emmanuele Wollman, and Raz Yirmiya

Volume 462
ADVANCES IN BLADDER RESEARCH
Edited by Laurence S. Baskin and Simon W. Hayward

Volume 463
ENZYMOLOGY AND MOLECULAR BIOLOGY OF CARBONYL METABOLISM 7
Edited by Henry Weiner, Edmund Maser, David W. Crabb, and Ronald Lindahl

Volume 464
CHEMICALS VIA HIGHER PLANT BIOENGINEERING
Edited by Fereidoon Shahidi, Paul Kolodziejczyk, John R. Whitaker,
Agustin Lopez Munguia, and Glenn Fuller

Volume 465
CANCER GENE THERAPY: Past Achievements and Future Challenges
Edited by Nagy A. Habib

Volume 466
CURRENT VIEWS OF FATTY ACID OXIDATION AND KETOGENESIS:
From Organelles to Point Mutations
Edited by P. A. Quant and S. Eaton

Volume 467
TRYPTOPHAN, SEROTONIN, AND MELATONIN: Basic Aspects and Applications
Edited by Gerald Heuther, Walter Kochen, Thomas J. Simat, and Hans Steinhart

Volume 468
THE FUNCTIONAL ROLE OF GLIAL CELLS IN HEALTH AND DISEASE:
Dialogue between Glia and Neurons
Edited by Rebecca Matsas and M. Tsacopoulos

Volume 469
EICOSANOIDS AND OTHER BIOACTIVE LIPIDS IN CANCER, INFLAMMATION, AND
RADIATION INJURY, 4
Edited by Kenneth V. Honn, Lawrence J. Marnett, and Santosh Nigam

Eicosanoids and Other Bioactive Lipids in Cancer, Inflammation, and Radiation Injury, 4

Edited by

Kenneth V. Honn
Wayne State University School of Medicine
Detroit, Michigan

Lawrence J. Marnett
Vanderbilt University School of Medicine
Nashville, Tennessee

Santosh Nigam
Free University Berlin
Berlin, Germany

and

Edward A. Dennis
University of California at San Diego
San Diego, California

Springer Science+Business Media, LLC

Library of Congress Cataloging-in-Publication Data

Eicosanoids and other bioactive lipids in cancer, inflammation, and radiation injury 4/
edited by Kenneth V. Honn, Lawrence J. Marnett, Santosh Nigam.
    p. cm. -- (Advances in experimental medicine and biology ; v.469)
    Includes bibliographical references and index.
    ISBN 978-0-306-46138-5   ISBN 978-1-4615-4793-8 (eBook)
    DOI 10.1007/978-1-4615-4793-8
    1. Eicosanoids--physiological effect--Congresses. 2.
Eicosanoids--Pathophysiology--Congresses. 3. Lipids--Pathophysiology--Congresses. 4.
Cancer--Pathophysiology--Congresses. 5. Inflammation--Mediators--Congresses. 6.
Radiation injuries--Pathophysiology--Congresses. I. Series. II. Marnett, Lawrence J. III.
Nigam, S. K. (Santosh K.) IV. International Conference on Eicosanoids and Other
Bioactive Lipids in Cancer, Inflammation, and Related Diseases (5th : 1997 : La Jolla,
San Diego, Calif.)
    [DNLM: 1. Eicosanoids--genetics--Congresses. 2.
Eicosanoids--metabolism--Congresses. 3. Gene Expression Regulation--Congresses. 4.
Inflammation--physiopathology--Congresses. 5. Neoplasms--metabolism--Congresses. 6.
Radiation Injuries--metabolism--Congresses. W1 AD559 v.469 1999]
QP752.E53 E362 1999
612'.01577 21 --dc21

                                                                    99-040840

ISBN 978-0-306-46138-5

© 1999 Springer Science+Business Media New York
Originally published by Kluwer Academic / Plenum Publishers , New York in 1999

10 9 8 7 6 5 4 3 2 1

A C.I.P. record for this book is available from the Library of Congress

PREFACE

This volume constitutes the proceedings of the 5th International Conference on Eicosanoids and Other Bioactive Lipids in Cancer Inflammation and related diseases, which was held at the Sheraton Grand Torrey Pines in La Jolla, California, in September 1997. This conference built on four previous meetings which were held in Detroit 1989, Berlin 1991, Georgetown 1993, and Hong Kong 1995.

Since their discovery over sixty years ago, eicosanoids have come to represent a diverse family of bioactive lipid modulators, including prostaglandins, thromboxanes, leukotrines, lipoxins, isoprostanes, hepoxilins, hydroxy acids, epoxy, and hydropoxy fatty acids.

The conference contained presentations regarding the regulation of eicosanoid enzymes and, in particular, cyclooxygenases, lipoxygenases, and phospholipases. In addition, recent evidence over the last seven years has led to the identification of a number of receptors for these bioactive lipids. The new field of isoprostanes was represented. It has become increasingly evident that eicosanoids play a critical role in signal transduction, both in normal cells and in pathological processes. These aspects were discussed in relation to cellular events, such as apoptosis, angiogenesis, and cancer prevention and treatment.

The editors are deeply indebted to our major contributors, which include: Radiation Oncology Research & Development Corporation, Detroit; Monsanto Corporation; Merck Frosst—Quebec; ONO Pharmaceuticals Co., Ltd.; and Searle Pharmaceuticals. Additional generous contributions were received from 14 other corporations, institutes, etc., which are acknowledged elsewhere in this volume. We are grateful to the invited speakers and participants, and for the submission of manuscripts.

As with all conferences, much of the work falls to members of the local organizing committee. We are especially grateful to Dr. Edward Dennis, the University of California, San Diego, and to Hiroko Vogt, for their excellent organizatoin of the meeting site and social programs. We are also indebted to Ms. Terri Larrew, Kathy Pastorelli, and Celeste Riley for their help during the conference. Finally, special acknowledgment to Ms. Terri Larrew for her help in preparing this volume.

Kenneth V. Honn
Wayne State University
Detroit, Michigan

Lawrence J. Marnett
Vanderbilt University
Nashville, Tennessee

Santosh Nigam
Free University Berlin
Berlin, Germany

ACKNOWLEDGMENTS

The Organizing Committee along with the participants of this Symposium gratefully acknowledge the primary support provided by the following organizations:

> Merck Frosst—Quebec
> ONO Pharmaceutical Co., Ltd.
> Radiation Oncology Research & Development Center, Detroit
> Searle Pharmaceuticals

Additional support was provided by the following organizations:

> Avanti Polar Lipids, Inc.
> BIOMOL Research Laboratories, Inc.
> Cayman Chemical
> Everlight Chemical Industrial Corporation
> Glaxo Wellcome Inc.
> Japan Tobacco Inc., Central Pharmaceutical Research Institute
> Merck & Co. Inc.
> Nu-Check Prep, Inc.
> Oxford Biomedical
> Parke-Davis Pharmaceutical Research, Division of Warner Lambert Company
> Shionogi & Co., Ltd.
> SmithKline Beecham Pharmaceuticals
> The Council for Tobacco Research
> Topcro

The International Organization Committee wishes to thank the Local Organizing Committee and the International Advisory Committee for their help in planning this conference.

COMMITTEES

Local Organizing Committee

Dr. Edward A. Dennis and staff (U.C.S.D.)
Ms. Hiroko Vogt (Intermedd Inc.)

International Advisory Committee

K. Abe (Aoba-ku Sendai)
A. Bennett (London)
P. Borgeat (Quebec)
A.R. Brash (Nashville)
W.C. Chang (Taiwan)
K. Crowshaw (Seattle)
L.J. de Asua (Beunos Aires)
T.E. Eling (Research Triangle Park)
G. Fitzgerald (Philadelphia)
M.L. Foegh (Washington, D.C.)
A.W. Ford-Hutchinson (Quebec)
M. Fukushima (Osaka)
C. Funk (Philadelphia)
E. Goetzl (San Francisco)
R.R. Gorman (Kalamazoo)
P.V. Haluskha (Charleston)
S. Hammarstrom (Linkoping)
Y. Hannun (Durham)
M. Hughes-Fulford (San Francisco)
R. Jones (Hong Kong)
J. Kock (South Africa)

H. Kuhn (Berlin)
W. Lands (Bethesda)
S. Narumiya (Kyoto)
S. Nishimura (Tokyo)
C.R. Pace-Asciak (Toronto)
R. Phipps (Rochester)
S. Prescott (Salt Lake City)
A. Raz (Detroit)
B.S. Reddy (Valhalla)
C.C. Reddy (University Park)
L.J. Roberts (Nashville)
M. Schneider (Berlin)
C.N. Serhan (Boston)
T. Shimizu (Tokyo)
E. Sigal (Palo Alto)
W. Smith (Lansing)
F. Snyder (Lansing)
J. Vanderhoek (Washington, D.C.)
T.J. Williams (London)
S. Yamamoto (Tokushima)

CONTENTS

REGULATION OF EICOSANOID ENZYMES

1. Function and Regulation of Prostaglandin Synthase 2 3
   Harvey R. Herschman, Weilin Xie, and Srinivasta Reddy

2. Regulation of Cyclooxygenase-2 by the Activated p38 MAPK Signaling Pathway 9
   Zhonghong Guan, ShaAvhree Y. Buckman, Lisa D. Springer, and Aubrey R. Morrison

3. Fatty Acid Cyclooxygenase Induction and Prostaglandin D Synthesis in a Human
   Megakaryoblastic Cell Line CMK Differentiated by Phorbol Ester 17
   Shozo Yamamoto, Natsuo Ueda, Ishtiaq Mahmud, Hiroko Yamaguchi, Rieko Yamashita,
   Kei Yamamoto, Kazunori Ishimura, Yoshihiro Urade, Yoshihide Kanaoka, and Osamu Hayaishi

4. Eicosanoid Metabolism in Human Platelets is Modified by Albumin 23
   Marina Dadaian and Pär Westlund

5. Diverse Functional Coupling or Prostanoid Biosynthetic Enzymes in Various Cell Types ... 29
   Ichiro Kudo and Makoto Murakami

6. Regulation of Prostaglandin, Leukotriene, and Platelet-activating Factor Metabolism
   in Mast Cells 37
   Makoto Murakami, Kinji Tada, Ko-ichi Nakajima, and Ichiro Kudo

7. Lipid Peroxides and Neuronal Plasticity 43
   Makoto Nishiyama, Nobuaki Hori, Takashi Watanabe, Tomokatsu Hori, Katsuhiko Ogata,
   Keiichi Watanabe, Eiichi Maru, and Takao Shimizu

8. Secretion of Lipocalin-type Prostaglandin D Synthase (β-trace) from Human Heart
   to Plasma During Coronary Circulation 49
   Yoshihiro Urade, Yutaka Eguchi, Naomi Eguchi, Yoshiyuki Kijima, Yasuhiko Matsu-ura,
   Hiroshi Oda, Kousuke Seiki, and Osamu Hayaishi

9. Lipocalin-type Prostaglandin D Synthase (β-trace) Binds Non-Substrate Lipophilic Ligands   55
   Carsten T. Beuckmann, Yoshihiro Urade, and Osamu Hayaishi

10. Features of Mammalian Lipoxygenases 61
    Bernd J. Thiele, Mario Berger, Holger Thiele, Antje Huth, and Iris Reimann

11. Identification and Characterization of an Enhancer Sequence in the Promoter Region
    of Human 15-Lipoxygenase (15-LO) Gene 67
    Uddhav Kelavkar, Susheng Wang, Angel Montero, and Kamal Badr

12. Cytokine Induced Regulation of 15-Lipoxygenase and Phospholipid Hydroperoxide
    Glutathione Peroxidase in Human Cells 75
    Kerstin Schnurr, Roland Brinckmann, and Hartmut Kühn

13. Investigation of a Second 15S-Lipoxygenase in Humans and Its Expression in Epithelial
    Tissues 83
    Alan R. Brash, Mitsuo Jisaka, William E. Boeglin, Min S. Chang, Diane S. Keeney,
    Lillian B. Nanney, Susan Kasper, Robert J. Matusik, Sandra J. Olson, and Scott B. Shappell

14. Sequence Determinants for the Positional Specificity of Mammalian and Plant
    Lipoxygenases 91
    S. Borngräber, R-J. Kuban, and Hartmut Kühn

15. X-Ray Absorption Studies into the Iron Ligand Sphere of Plant and Animal Lipoxygenases   99
    Hartmut Kühn, R. Kuban, M. Walther, and Gerrit A. Veldink

16. Transcriptional and Posttranscriptional Regulation of 5-Lipoxygenase mRNA Expression
    in the Human Monocytic Cell Line Mono Mac 6 by Transforming Growth Factor-β
    and 1,25-Dihydroxyvitamin D_3 105
    Damaris Harle, Olof Radmark, Bengt Samuelsson, and Dieter Steinhilber

CYCLOOXYGENASES

17. Comparison of Prostaglandin H Synthase-1 and -2 Structural Stabilities 115
    Guishan Xiao, Wei Chen, and Richard J. Kulmacz

18. Modulation of the Expression of the Cyclooxygenase 1 and 2 Genes in Rat Mammary
    Glands: Role of Hormonal Status and Dietary Fat 119
    Alaa F. Badawi, Ahmed El-Sohemy, Laurie L. Stephen, Amit K. Ghoshal,
    and Michael C. Archer

19. Regulation of Cerebrovascular Cyclooxygenase-2 by Pro- and Anti-Inflammation Cytokines   125
    Steven A. Moore, Elizabeth Yoder, Gretchen Rich, MacKenzie Hilfers, and Jeffrey Albright

20. Visualization and Quantitation of Cyclooxygenase-1 and -2 Activity by Digital Fluorescence
    Microscopy 131
    Richard L. Ornberg and Alane T. Koki

21. Discovery of a New Class of Selective Cyclooxygenase-2 (COX-2) Inhibitor that Covalently
    Modifies the Isozyme 139
    Amit S. Kalgutkar, Brenda C. Crews, Scott W. Rowlinson, Carlos Garner,
    and Lawrence J. Marnett

22. Mitigation of Arthritis by High-Dose Administration of a COX-2 Inhibitor in the Collagen-
    Induced Arthritis Model in the Mouse 145
    Mark G. Obukowicz and Richard L. Ornberg

23. Indomethacin Inhibition of Pristane Plasmacytomagenesis in Genetically Susceptible Inbred
    Mice 151
    Michael Potter

24. Role of COX-2 Inhibition on the Formation and Healing of Gastric Ulcers Induced
    by Indomethacin in the Rat 157
    Nuria Godessart, Carolina Salcedo, Andres G. Fernandez, and José M. Palacios

25. Regulation of *in vivo* Prostaglandin Biosynthesis by Glutathion 165
Alon Margalit, Peter C. Isakson, and Scott D. Hauser

26. Different Effects of Reactive Nitrogen Intermediates on Prostaglandin E_2 Synthesis
in Cultured Rat Microglia and Raw 264.7 Cells 169
Cecilia Guastadisegni, Luisa Minghetti, Alessia Nicolini, Elisabetta Polazzi, Maria Balduzzi,
and Giulio Levi

PHOSPHOLIPASES/PAF

27. Activation of cPLA$_2$ in Vascular Smooth Muscle 177
Edward F. LaBelle and Erzsebet Polyak

28. Role of Type IIA Secretory Phospholipase A$_2$ in Arachidonic Acid Metabolism 183
Hiroshi Kuwata, Hisashi Sawada, Makoto Murakami, and Ichiro Kudo

29. Phospholipase A$_2$ is Involved in Chemotaxis of Human Leukocytes 189
Ulrich Tibes, Markus Hinder, Werner Scheuer, Walter-Gunar Friebe, Stephan Schramm,
and Beate Kaiser

30. Suppression of Acute Experimental Inflammation by Antisense Oligonucleotides Targeting
Secretory Phospholipase A$_2$ (sPLA$_2$). *In vitro* and *in vivo* experiments 199
Ulrich Tibes, Sigrid P. Rohr, Werner Scheuer, Elke Amandi-Burgermeister, and Anette Litters

31. Comparison of Recombinant Types IIA, V, and IIC Phospholipase A$_2$S, The Three Related
Mammalian Secretory Phospholipase A$_2$ Isozymes 209
Satoko Shimbara, Makoto Murakami, Terumi Kambe, and Ichiro Kudo

32. Respective Roles of the 14 KDA and 85 KDA Phospholipase A$_2$ Enzymes in Human
Monocyte Eicosanoid Formation 215
Lisa A. Marshall, Brian Bolognese, and Amy Roshak

33. Modulation of Long-term Potentiation in the CA1 Area of Rat Hippocampus by Platelet-
Activating Factor 221
Kunio Kato

RECEPTORS

34. Ppars: Nuclear Receptors for Fatty Acids, Eicosanoids, and Xenobiotics 229
Pallavi R. Devchand, Annemieke Ijpenberg, Béatrice Devesvergne, and Walter Wahli

35. Molecular Cloning and Characterization of Leukotriene B$_4$ Receptor 237
Takashi Izumi, Takehiko Yokomizo, Toshio Igarashi, and Takao Shimizu

36. Determinants of Receptor Subtype Specificity in the LPA-like Lipid Mediator Family 245
G. Tigyi, D.J. Fischer, K. Liliom, Z. Guo, T. Virag, C. Sun, D.D. Miller,
K. Murakami-Murofushi, s. Kobayashi, and J.R. Erickson

37. Subunits and Cellular Occurrence of The 12(S)-HETE Binding Complex 253
Helena Herbertsson, Tobias Kühme, and Sven Hammarstrom

38. A Subfamily of G Protein-Coupled Cellular Receptors for Lysophospholipids and
Lysosphingolipids 259
Edward J. Goetzel and Songzhu An

39. Evidence for Involvement of Prostaglandin I_2 as a Major Nociceptive Mediator in Acetic
Acid-Induced Writhing Reaction: A Study Using IP-Receptor Distrupted Mice 265
Sachiko Oh-Ishi, Akinori Ueno, Hideki Matsumoto, Takahiko Murata, Fumitaka Ushikubi,
and Shuh Narumiya

40. Local Anesthetic Effects on TXA_2 Receptor Mediated Platelet Aggregation Using Quenched
Flow Aggregometry 269
Christian W. Hönemann, Bernard Lo, Jo S. Errera, Renata Polanowska-Grabowska,
Adrean RL Gear, and Marcel Duriex

41. Volatile and Local Anesthetics Interfere with Thromboxane A_2 Receptors Recombinantly
Expressed in *Xenopus* Oocytes 277
Christian W. Hönemann, John A. Arledge, Tobias Podranski, Hugo Van Aken,
and Marcel Durieux

LEUKOTRIENES AND LIPOXINS

42. Aspirin-Triggered 15-Epi-Lipoxin A_4 and Novel Lipoxin B_4 Stable Analogs Inhibit
Neutrophil-Mediated Changes in Vascular Permeability 287
Charles N. Serhan, Tomoko Takano, Clary B. Clish, Karsten Gronert, and Nicos Petasis

43. Cleavage of Leukotriene D_4 in Mice with Targeted Disruption of a Membrane-Bound
Dipeptidase Gene 295
Geetha M. Habib and Michael W. Lieberman

44. γ-Glutamyl Leukotrienase Clevage of Leukotriene $_4$ 301
Michael W. Lieberman, Jefry E. Shields, Yvonne Will, Donald J. Reed, and Bing J. Carter

45. LTE_4 Blood Levels in Infants with Congenital Heart Lesions 307
Jean Pierre Gascard and Charles Brink

46. Evaluation of the Pharmacological Activity of the Pure Crsteinyl-Leukotriene Receptor
Antagonists CGP 45715A (Irakukast) and CGP 57698 in Human Airways 313
Valerie Capra, Saula Ravasi, Manlio Bolla, Serena Viappiani, Silvia Pagliardini,
P. Angelo Belloni, Maurizio Mezzetti, G. Carlo Folco, Simonetta Nicosia, and G. Enrico Rovati

47. Evidence for a Carbocation Intermediate in the Enzymatic Transformation of Leukotriene
A_4 into Leukotriene B_4 319
Martina Andberg, Mats Hamberg, and Jesper Z. Haeggstrom

48. A Random Rapid Equilibrium Mechanism for Leukotriene C_4 Synthase 327
Namrata Gupta, Michael J. Gresser, and Anthony W. Ford-Hutchinson

ISOPROSTANES

49. Formation of Reactive Products of the Isoprostane Pathway: Isolevuglandins and
Cyclopentenone Isoprostanes 335
L. Jackson Roberts, II, Cynthis J. Brame, Yan Chen, Jason D. Morrow, and Robert G. Salomon

50. Formation of Novel Isoprostane-like Compounds from Docosahexaenoic Acid 343
Jason D. Morrow, Andrew R., Tapper, William E. Zackert, James Yang, Stephanie C. Sanchez,
Thomas J. Montine, and L. Jackson Roberts

LYSOPHOSPHATIDIC ACID (LPA) AND LPA RECEPTORS

51. cDNA Cloning, Expression, and Chromosomal Localization of Two Human
Lysophosphatidic Acid Acyltransferases 351
Christin Eberhardt, Patrick W. Gray, and Larry W. Tjoelker

52. The First Cloned and Identified Lysophospholipid (LP) Receptor Gene, VZG-1:
Implications for Related Receptors and the Nervous System 357
Jerold Chun

SIGNAL TRANSDUCTION

53. Involvement of Protein Kinase C, P38 Map Kinase and ERK in Arachidonic Acid
Stimulated Superoxide Production in Human Neutrophils 365
Charles S.T. Hii, Zhi H. Huang, Andrea Bilney, Kathryn Stacey, Andrew W. Murray,
Deborah A. Rathjen, and Antonio Ferrante

54. 13(S)-HPODE Augments Epidermal Growth Factor Signal Transduction by Attentuating
EGF Receptor Dephosphorylation 371
Wayne C. Glasgow, Putai Hui, Shiranthi Jayawickreme, Julie Angerman-Steward,
Bing-Bing Han, and Thomas E. Eling

55. Role of Lipoxygenase in the Regulation of Glucose Transport in Aortic Vascular Cells 377
Shlomo Sasson, Ana Davarashvili, and Reuven Reich

56. Functions of the p21-Activated Protein Kinases (PAKS) in Neutrophils and Their
Regulation by Complex Lipids 385
Dwight Robinson, RiYun Huang, Jian P. Lian, Alex Toker, and John A. Badwey

57. Inhibition of Neurofibromin and p120 GTPase Activating Protein (GAP) by Dietary Fatty
Acids 391
Joung H. Lee, Jyoti A. Harwalker, Sophia S. Bryant, Vidyokhaya Sundaram, Richard Jove,
and Mladen Golubic

BIOACTIVE LIPIDS IN INFLAMMATION

58. Spinal Synthesis and Release of Prostanoids after Peripheral Injury and Inflammation 401
David M. Dirig and Tony L. Yaksh

59. Prostaglandin E2 as a Modulator of Lymphocyte Mediated Inflammatory and Humoral
Responses 409
Kuljeet Kaur, Sarah O. Harris, Josue Padilla, Beth A. Graf, and Richard P. Phipps

60. Lipid Mediators Stimulate Reactive Oxygen Species Formation in Immortalized Human
Keratinocytes 413
Rachel Goldman, Sandra Moshonov, and Uriel Zor

61. Nitric Oxide Synthesis is Increased in Periodontitis 419
Michael Matejka, Christian Ulm, Lukas Partyka, Martin Ulm, and Helmut Sinzinger

62. Augmentation Effects of High Glucose on Endotoxin-Induced Nitric Oxide Production
in Murine Macrophages 425
Soo H. Lee, Hyun Goo Woo, Ji Y. Kim, and Chang-Hyun Moon

63. 1,2-diacylglycerol Hydroperoxide Induces the Generation and Release of Superoxide Anion
from Human Polymorphonuclear Leukocytes 431
Yorihiro Yamamoto, Yasuhiro Kambayashi, Takashi Ito, Keiichi Watanabe, and Minoru Nakano

64. Kinetic Evaluation of Endogenous Leukotriene B4 and E4 Acute Activation of
Inflammatory Cells in the Rabbit 437
Antonio Celardo, Giuseppe Dell'Elba, Stefano Manarini, Virgilio Evangelista,
Giovanni de Gaetano, and Chiara Cerletti

65. Lipopolysaccharide- and Liposome-Encapsulated MTP-PE-Induced Formation of
Eicosanoids, Nitric Oxide and Tumor Necrosis Factor-α in Macrophages 443
P. Dieter, U. Hempel, B. Malessa, E. Fitzke, T.A. Tran-Thi, J. MacLouf, C. Creminon,
Y. Kanaoka, and Y. Urade.

66. LTA4 Hydrolase Expression During Glomerular Inflammation: Correlation of
Immunohistochemical Localization with Cytokine Regulation 449
Angel Montero, Susumu Uda, Karen A. Munger, and Kamal F. Badr

67. 12-Lipoxygenase Products Increase Monocyte: Endothelial Interactions 455
C. Hedrick, M. Kim, R. Natarajan, and J. Nadler

LINOLEIC AND LINOLENIC ACIDS

68. Linoleic Acid Metabolites in Health and Disease 463
Michael R. Buchanan

69. Toxicity of Linoleic Acid Metabolites 471
Jessica F. Greene and Bruce D. Hammock

70. Dramatic Increase of Linoleic Acid Peroxidation Products by Aging, Atherosclerosis,
and Rheumatoid Arthritis 479
Wolfgang Jira and Gerhard Spiteller

71. Modulation of Atherogenesis by Dietary Gamma-Linolenic Acid 485
Yang-Yi Fan, Kenneth S. Ramos, and Robert S. Chapkin

72. The Selective Cytotoxicity of γ-Linolenic Acid (GLA) is Associated with Increased
Oxidative Stress 493
Sujata Vartak, Mike E.C. Robbins, and Arthur A. Spector

73. γ-Linolenic Acid (GLA)-Mediated Cytotoxicity in Human Prostate Cancer Cells 499
Mike Robbins, Kamran Ali, Ryan McCaw, Josalyn Olsen, Sujata Vartak, and David Lubaroff

74. Five-Lipoxygenase Inhibitors Reduce Panc-1 Survival: Synergism of MK886 with Gamma
Linolenic Acid 505
J.E. Harris, W.A. Alrefai, J. Meng, and K. M. Anderson

ANANDAMIDE AND OTHER BIOACTIVE LIPIDS

75. Enzymological and Molecular Biological Studies on Anandamide Amidohydrolase 513
    Natsuo Ueda, Kazuhisa Katayama, Yuko Kurahashi, Mitsujiro Suzuki, Hiroshi Suzuki,
    Shozo Yamamoto, Itsuo Katoh, Vincenzo Di Marzo, and Luciano De Petrocellis

76. A Pathway for the Biosynthesis of Fatty Acid Amides 519
    Kathleen A. Merkler, Laura E. Baumgart, Jodi L. DiBlassio, Uta Glufke, Lawrence King III,
    Kimberly Ritenour-Rodgers, John C. Vederas, Benjamin J. Wilcox, and David J. Merkler

77. Investigation of Structural Analogs of Prostaglandin Amides for Binding to and Activation
    of CB_1 and CB_2 Cannabinoid Receptors in Rat Brain and Human Tonsils 527
    Barbara A. Bergland, Daniel L. Boring, and Allyn C. Howlett

78. Hepoxilin A_3 is Metabolized into its ω-Hydroxy Metabolite by Human Neutrophils 535
    Cecil R. Pace-Asciak, Denis Reynaud, Olga Rounova, Peter Demin, and Kazimir K. Pivnitsky

79. Oxidation of Arachidonate Containing Glycerophospholipids in Intact Red Blood Cells
    and Red Blood Cell Membranes with Tert-Butylhydroperoxide 539
    Tatsuji Nakamura, Lisa Hall, and Robert C. Murphy

80. Cannabinoid Modulation of Neuronal Activity in Adult Rat Hippocampus 547
    Paul Schweitzer, George R. Siggins, and Samuel G. Madamba

APOPTOSIS

81. Cyclooxygenase-independent Induction of $P21^{WAF-1/CIP1}$, Apoptosis and Differentiation
    by L-745,337 and Salicylate in HT-29 Colon Cancer Cells 555
    Paola Patrignani, Giovanna Santini, Maria G. Sciulli, Rosanna Marinacci, Ornella Fusco,
    Luciana Spoletini, Clara Natoli, Antonio Procopio, and Jacques Maclouf

82. Modulation of Cellular Proliferation and Induction of Apoptosis in a Human Lymphoma
    Cell Line after Treatment with Selective Lipoxygenase Inhibitors 563
    Kenning M. Anderson, Frank G. Ondrey, and Jules E. Harris

83. Mechanism of Peroxynitrite Induced Apoptosis in HL-60 Cells 569
    King-Teh Lin, Ji-Yan Xue, and Patrick Y.-K. Wong

84. Central Role of Arachidonate 5-Lipoxygenase in the Regulation of Cell Growth
    and Apoptosis in Human Prostate Cancer Cells 577
    Jagadananda Ghosh and Charles E. Myers

85. Role of Autocrine Motility Factor in a 12-Lipoxygenase Dependent Anti-Apoptotic Pathway 583
    Keqin Tang, Daotai Nie, and Kenneth V. Honn

CANCER PREVENTION AND TREATMENT

86. The Possible Involvement of 15-Lipoxygenase/Leukocyte Type 12-Lipoxygenase
    in Colorectal Carcinogenesis 593
    Hideki Kamitani, Mark Geller, and Thomas Eling

87. Induction of Leukotriene B$_4$ Metabolism by Cancer Chemopreventive Agents 599
Thomas Primiano, Thomas W. Kensler, Michael A. Trush, and Thomas R. Sutter

88. Intestinal Tumor Load in the Min/+ Mouse Model is Not Correlated with Eicosanoid
Biosynthesis 607
Jay Whelan, Chun-Hung Chiu, and Michael F. McEntee

89. 12-Lipoxygenase Expression in Human Melanoma Cell Lines 617
Joszef Timar, Erzsebet Raso, Kenneth V. Honn, and Wolfgang Hagmann

90. Platelet-Type 12-Lipoxygenase Regulates Angiogenesis in Human Prostate Carcinoma 623
Daotai Nie, Gilda G. Hillman, Timothy Geddes, Keqin Tang, Christopher Pierson,
David J. Grignon, and Kenneth V. Honn

91. Prostaglandin H Synthase and Lipoxygenase Mediated Activation of Xenobiotics
in Platelets 631
Kyoung-Min Lim, Joo-Young Lee, Jung-Sun Kim, and Jin-Ho Chung

92. Characterization of Two Spliced Variants of Human Phosphatidic Acid Phosphatase cDNAS
that are Differentially Expressed in Normal and Tumor Cells 639
David W. Leung, Christopher K. Tompkins, and Thayer White

93. Eicosapentaenoic Acid Alters Manganese Superoxide Dismutase Immunoreactive Protein
Levels in Normal But not Malignant Central Nervous System Derived Cells 647
Geoffrey D. Girnum, Larry W. Oberley, Steven A. Moore, and Mike E.C. Robbins

94. Gamma Radiation and Release of Norepinephrine in the Hippocampus 655
Sathasiva B. Kandasamy

BIOACTIVE LIPIDS IN NON-MAMMALIAN CELLS

95. (3R)-Hydroxy-Oxylipins—A Novel Family of Oxygenated Polyenoic Fatty Acids of Fungal
Origin 663
Santosh Nigam, Tankred Schewe, and J. Lodewyk F. Kock

96. Eicosanoids in the Brain of Warm- and Cold-Acclimated Bullfrogs 669
Ceil A. Herman and Georgia Luczy-Bachman

97. Production of 3-Hydroxy Fatty Acids by the Yeast *Dipodascopsis uninucleata*. Biological
Implications 675
J. Lodewyk F. Kock, Pierre Vener, Alfred Botha, Dennis J. Coetzee, Pieter W.J. vanWyk,
Dandre P. Smith, Tankred Schewe, and Santosh Nigam

98. Catalytic Properties of Linoleate Diol Synthase of the Fungas *Gaeumannomyces graminis*:
A Comparison with PGH Synthases 679
Ernst H. Oliw, Chao Su, and Margareta Sahlin

CONTRIBUTORS 687

INDEX 693

Eicosanoids and Other Bioactive Lipids in Cancer, Inflammation, and Radiation Injury, 4

REGULATION OF EICOSANOID ENZYMES

1

FUNCTION AND REGULATION OF PROSTAGLANDIN SYNTHASE 2

Harvey, R. Herschman, Weilin Xie, and Srinivasta Reddy

Departments of Biological Chemistry and Molecular and Medical Pharmacology
UCLA School of Medicine
Molecular Biology Institute
611 Circle Drive East
Los Angeles, CA 90025-1570

INTRODUCTION

The prostanoids (prostaglandins, prostacyclins, thromboxanes) are one subset of arachidonic-acid derived lipid mediators of inflammation. The leukotrienes constitute the second major class of arachidonic-acid derived lipid mediators. A variety of inflammatory signals (including exogenous mediators such as endotoxin/lipopoly-saccharide and endogenous mediators such as inflammatory cytokines) stimulate cellular phospholipase activities. As a result, arachidonic acid is released from phospholipid membrane stores. This free arachidonic acid is converted either (i) to leukotrienes via the lipoxygenase pathway(s), or (ii) to prostanoids. The key enzyme in prostanoid biosynthesis is prostaglandin synthase (PGS), also known as cyclooxygenase (COX). PGS/COX converts free arachidonate first to prostaglandin G_2 (PGG$_2$) by a cyclooxygenase reaction, and then --in a second enzymatic step-- to prostaglandin H_2 (PGH$_2$) by a hydroperoxidase reaction. PGH$_2$ is then converted to the various prostaglandins (PGE$_2$, PGD$_2$, PGF$_{2a}$, etc), prostacyclin, (PGI$_2$) or thromboxane by cell-type restricted synthetases that use PGH$_2$ as substrate. The various prostanoids have a wide range of positive and negative immuno-modulatory effects on lymphoid, myeloid, and endothelial cells. New pharmacologic agents that modulate the activity of enzymes involved in prostanoid synthesis or the expression of the genes encoding these enzymes will provide improved directions in the management of acute and chronic inflammatory diseases.

IDENTIFICATION OF PROSTAGLANDIN SYNTHASE 2

In the late 1980's we were using differential cloning techniques to identify genes induced in quiescent, non-dividing fibroblasts by mitogenic stimulation[1,2]. One cDNA that we cloned, TIS10, when sequenced[3], demonstrated substantial homology to the prostaglandin synthase gene first cloned in 1988. When expressed, the TIS10 protein had both the cyclooxygenase and hydroperoxidase activities of a prostaglandin synthase (PGS). We now refer to this gene as the prostaglandin synthase 2, or PGS2 gene. Our laboratory was the first to clone the cDNA for what was clearly a distinct, inducible PGS gene[3], to demonstrate its enzyme activity[4], to clone the gene for PGS2[4], and to demonstrate the inducibility of the PGS2 promoter in a reporter gene transfection assay[4]. The chicken gene was cloned at about

Eicosanoids and Other Bioactive Lipids in Cancer, Inflammation, and Radiation Injury, 4
Edited by Honn *et al.*, Kluwer Academic / Plenum Publishers, New York, 1999.

3

the same time[5], and two other labs subsequently reported cloning of the murine PGS2 cDNA as a growth factor or oncogene induced gene[6,7]. Using sequence information from the murine gene, the human[8,9] and rat[10] genes were later cloned. The discovery of PGS2 immediately suggested (i) that regulation of prostanoid synthesis might be more complex than previously thought, and (ii) that PGS2 might be a new a target for pharmacologic intervention in inflammatory responses.

PROSTAGLANDIN SYNTHASE 2 IS INDUCED IN A VARIETY OF CELL TYPES, IN RESPONSE TO A WIDE RANGE OF LIGANDS AND INDUCERS.

PGS2 was originally discovered as a mitogen and oncogene inducible message in murine fibroblasts[3,6,7]. However, in the six years since these initial reports, induction of the PGS2 gene has been described in macrophages, monocytes, epithelial cells, endothelial cells, smooth muscle cells, synovial microvascular cells, granulosa cells, uterine stromal cells, amnion cells, decidual cells, calvariae, osteoblasts, mesangial cells, neurons, and mast cells. In addition to oncogenes, polypeptide hormones (PDGF, EGF, TNFa, endothelin, FSH, LH, thrombin, INFg, TGFb, FGF, parathyroid hormone, IL-1a, IL-1b, kit ligand, BDNF, NT-3), pharmacologic agents (phorbol esters, forskolin) prostaglandins (PGF2a, PGE1, PGE_2), LPS/endotoxin, neurotransmitters (norepinephrine, serotonin), depolarization, and aggregation of IgE receptors are able to induce induction of PGS2 gene expression in appropriate target cells[11,12]. Expression of the PGS2 gene seems likely to play a critical role in a variety of cellular responses to many different stimuli.

PROSTAGLANDIN SYNTHASE 2 PLAYS A MAJOR ROLE IN PRODUCTION OF PROSTANOIDS IN INFLAMMATORY REACTIONS, AND IS THE PRIMARY TARGET OF MOST NON-STEROIDAL ANTI-INFLAMMATORY DRUGS.

Studies in animal models demonstrated that, *in vivo*, acute and chronic inflammatory responses are accompanied by PGS2 induction. Aspirin, ibuprofen, and the majority of other non-steroidal anti-inflammatory drugs (NSAIDs) exert their major anti-inflammatory effects by inhibiting the action of prostaglandin synthase[13]. The major side effects of NSAIDs are gastric ulcers and nephrotoxicity. With the discovery of PGS2, and the demonstration that it is induced by inflammatory stimuli, the pharmaceutical industry entered into a major commitment to drug discovery programs whose endpoint is the identification of specific inhibitors of PGS2 that will not inhibit PGS1. The assumption driving this research is that a PGS2-specific inhibitor that does not inhibit PGS1 will be --like aspirin-- anti-pyretic, analgesic, and anti-inflammatory, but will not be ulcerogenic or nephrotoxic. Data from several animal models have been cited to suggest that new compounds that preferentially inhibit PGS2 are effective as anti-inflammatory agents, but are neither ulcerogenic nor nephrotoxic[13]. The suppression of inflammation by aspirin and other NSAIDs which inhibit PGS demonstrates the importance of PGS2-mediated prostaglandin synthesis in disease. Understanding the role of PGS2 in arachidonic acid utilization and prostanoid production will provide insights for drug development and disease management in allergic inflammation.

ARACHIDONIC ACID RELEASED FROM ENDOGENOUS PHOSPHOLIPIDS IS PREFERENTIALLY UTILIZED BY INDUCED PGS2, AND IS NOT AVAILABLE TO CONSTITUTIVE PGS1.

The conventional view of ligand-induced prostanoid synthesis has, until recently, been that the rate-limiting step for prostaglandin production has been the ligand-stimulated activation of a phospholipase(s) to release arachidonic acid from membrane phospholipids. According to this model, constitutive prostaglandin synthase is available for conversion of free arachidonate, released from membrane phospholipids, to PGG_2/PGH_2; the availability of free arachidonate is limiting. This model raises a paradox: why should cells make PGS2 if PLA$_2$ activation is the rate-limiting step in prostaglandin synthesis?

We found that preventing ligand-induced synthesis of PGS2 by cycloheximide[14], PGS2 antisense RNA[15], or blocking the activity of PGS2 with a PGS2-specific inhibitor that has no effect on PGS1[11], all prevent mitogen-induced PGE$_2$ synthesis in fibroblasts and endotoxin-induced PGE$_2$ synthesis in macrophages. Prostaglandin is not produced in response to these stimuli, despite the release of arachidonic acid from membranes following ligand stimulation, and despite the presence of active PGS1 enzyme in the cells (demonstrated by the ability of these cells to use exogenous arachidonate as substrate for prostaglandin synthesis). Our data suggest that arachidonic acid released from membranes by phospholipase activation in response to ligand stimulation is not available to constitutive PGS1, but *is* available to PGS2, synthesized in response to the ligand stimulation. Recent studies have suggested that PGS1 and PGS2 may occupy overlapping but distinct subcellular locations in fibroblasts -- PGS1 is present in the endoplasmic reticulum while PGS2 can be found in both the ER and the nuclear envelope[16]. In many instances the rate-limiting step for prostaglandin production is the induction of PGS2 enzyme activity. Moreover, arachidonic acid released from phospholipids following ligand stimulation is --in some way-- sequestered from PGS1 and metabolically channeled to PGS2 in many cell types. These data suggest that previously unknown regulatory mechanisms for prostaglandin production contribute to normal and pathologic states, and provide attractive targets for intervention in immune diseases.

WE CHARACTERIZED THE REGULATORY REGIONS OF THE PGS2 GENE, THE SIGNAL TRANSDUCTION PATHWAYS, AND THE TRANSCRIPTION FACTORS THAT MEDIATE INDUCTION OF PGS2 GENE EXPRESSION BY THE V-SRC ONCOGENE AND BY PLATELET DERIVED GROWTH FACTOR (PDGF).

We cloned and sequenced the 5' regulatory region of the murine PGS2 gene[4]. Using deletion analysis, with chimeric luciferase reporter constructs, we identified the region of the PGS2 promoter required for *v-src*[17] and PDGF[18] induction. Mutational analysis identified the CRE present at nucleotides -56 to -48 from the transcription start site as the critical cis-acting element regulating both *v-src*[17] and PDGF[18] induction. Using cotransfection assays with plasmids encoding dominant-negative and/or constitutively activated forms of the Ras, MAP kinase, and MEKK families, we demonstrated (i) that the key transcription factor mediating both *v-src* and PDGF induction of PGS2 gene expression via the CRE of the PGS2 promoter is the Jun protein, and (ii) that both *v-src* and PDGF induction are mediated by the Ras/MEKK1/JNK/Jun signal transduction pathway[17,18]. Our studies resulted in identification of the pathway that mediates PGS2 gene expression as a result of activation of the PDGF receptor by ligand[18] and of the pathway that mediates *v-src* oncogene induction[17]. Other investigators have implicated an NF-IL6 response element[19] and an NFkB element[20] of the PGS2 promoter in response to other ligands. In preliminary studies, we have demonstrated induction of PGS2-luciferase reporter constructs in macrophages by endotoxin treatment, and in mast cells by aggregation of IgE receptors. Identification and characterization of induction pathways of PGS2 gene expression by inflammatory stimuli will provide alternative targets for regulation of this gene in acute and chronic inflammatory diseases.

SYNTHESIS OF PGD$_2$ IN ACTIVATED MAST CELLS OCCURS IN TWO PHASES, IN CONTRAST TO PROSTAGLANDIN SYNTHESIS IN LIGAND-STIMULATED FIBROBLASTS, MACROPHAGES, AND OTHER CELLS.

We found that mast cells activated by aggregation of their IgE receptors produce prostaglandin D$_2$ (PGD$_2$) in two distinct phases[21]. The early phase of PGD$_2$ synthesis is complete within 10-15 minutes. The delayed phase begins at about 1-2 hours after activation, and is completed within 6-8 hours. Using PGS1 and PGS2 inhibitors, we demonstrated that the early phase of PGD$_2$ synthesis is completely dependent on PGS1, and does not require PGS2. In contrast, the late phase of PGD$_2$ synthesis is dependent on PGS2, and does not require PGS1[21]. Murakami *et al*[22] confirmed our conclusion that PGS1 and PGS2 mediate distinct phases of prostaglandin production in activated mast cells. We suspected that the early phase of PGD$_2$ synthesis in activated mast cells is due to the action of a secreted phospholipase A2 (sPLA$_2$), released from mast cells following activation, and that the late phase results from activation of cPLA$_2$. Using a series of pharmacologic reagents we demonstrated[23] that, in activated mast cells, (i) secreted sPLA$_2$ mediates the release of arachidonic acid for early, PGS1-dependent PGD$_2$ synthesis, (ii) secreted sPLA$_2$ does not play a role in the late, PGS2-dependent PGD$_2$ synthesis, (iii) cytoplasmic PLA$_2$ mediates the release of arachidonic acid for late, PGS2-dependent PGD$_2$ synthesis, and (iv) a cytoplasmic PLA$_2$-dependent step precedes secretory PLA$_2$ activation and is necessary for optimal PGD$_2$ production by the secretory PLA$_2$/PGS1-dependent early pathway (see Figure 1).

Bingham *et al*[24] also demonstrated that sPLA$_2$ is required for PGD$_2$ synthesis following mast cell activation. However, they suggest that sPLA$_2$ is required for the *late* phase of PGD$_2$ synthesis, and is not required for the early phase of PGD$_2$ production. Their conclusions are, therefor, just the opposite of ours. These results demonstrate the need for additional pharmacologic and genetic experiments to clarify the pathways that couple phospholipase isoforms to prostaglandin synthase isoforms in mast cells and in other cells. Type IIa sPLA$_2$ has been thought to be the isoform that mediates PGD$_2$ release in activated macrophages[25]. Recently, both we[26] and Bingham *et al*[24] found that mast cells from mice homozygous for disruption of the *PLA2G2A* gene encoding type IIa sPLA$_2$ produce PGD$_2$ normally. We find that type V sPLA$_2$, previously thought to be expressed only in heart, is likely to mediate PGD$_2$ production in activated mast cells[26]. Regardless of the experimental resolution of the conditions under which sPLA$_2$ is coupled to PGS1 or PGS2, the description of two distinct pathways for PGD$_2$ synthesis in activated mast cells identifies possible roles for this prostanoid in early and late phases of airway responses to allergens. Moreover, the identification of type V sPLA$_2$ as a mediator of PGD$_2$ production suggests a new target for pharmacologic intervention.

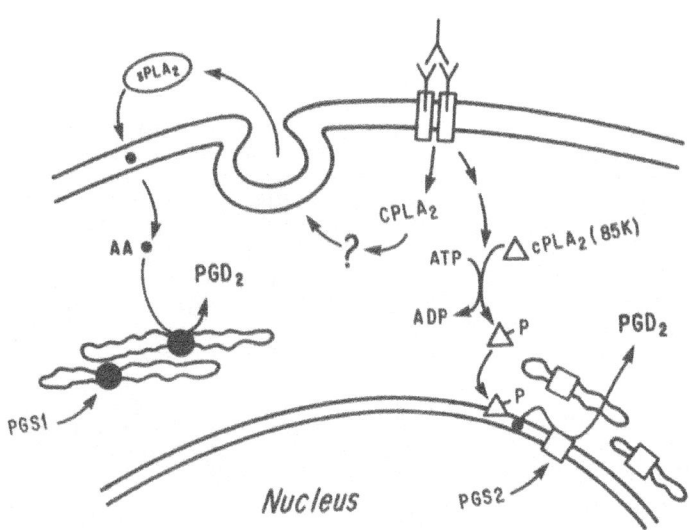

Figure 1. A model for prostaglandin D2 production in mast cells activated by aggregation of IgE receptors. Arachidonic acid released from one membrane pool is available to PGS2, but not to PGS1. Arachidonic acid released by sPLA$_2$ activity is available to constitutive PGS1.

SPLA₂ RELEASED FROM ACTIVATED MAST CELLS INITIATES TRANSCELLULAR PROSTAGLANDIN SYNTHESIS.

We speculated that sPLA$_2$, released from mast cells in response to aggregation of IgE receptors, might be able to activate prostaglandin synthesis in distal bystander cells via a PGS1-mediated pathway. Because mast cells produce PGD$_2$ and fibroblasts produce PGE$_2$, we could use mixed cell cultures to investigate the ability of mast cell-derived sPLA$_2$ to induce "transcellular" prostaglandin synthesis. Using mixed mast cell/fibroblast cell cultures, recombinant sPLA$_2$ enzyme, sPLA$_2$ inhibitors, and supernatants from activated mast cells, we demonstrated that sPLA$_2$ released from activated mast cells could induce rapid, PGS1-dependent PGE$_2$ synthesis in fibroblasts. We prelabeled the membranes of the mast cells and the fibroblasts with radioactive arachidonic acid, to determine whether the arachidonic acid that is converted to PGE$_2$ during "transcellular" prostaglandin production is derived from the membranes of the mast cells or from the membranes of the fibroblasts. Radioactivity is incorporated into PGE$_2$ when the membranes of the fibroblasts are labeled with radioactive arachidonic acid, but not when the membranes of the mast cells contain radioactive arachidonate; "transcellular" synthesis uses arachidonic acid released from the fibroblast membranes in response to mast cell-derived sPLA$_2$ and does not utilize arachidonate released from the mast cell[27]. We conclude that fibroblasts have the capacity for early phase prostaglandin synthesis, from a distinct arachidonate pool, if exposed to sPLA$_2$. Prior to our studies, transcellular prostaglandin synthesis had not been unequivocally described. Demonstration of transcellular prostaglandin synthesis, dependent on PGS1 in the distal cell, suggests that (i) PGS2-specific inhibitors may not be effective in all inflammatory conditions, and (ii) transcellular prostaglandin synthesis may be a target for intervention in inflammatory diseases.

REFERENCES

1. Herschman, H.R. Primary response genes induced by growth factors and tumor promoters, In *Annual Review of Biochemistry, vol. 60* (Richardson, C.C., Abelson, J.N., Meister, A., and Walsh, C.T., eds.) Annual Reviews, Inc., Palo Alto, California, pp. 281-319 (1991).
2. Herschman, H.R., Kujubu, D.A., Fletcher, B.S., Ma, Q-F., Varnum, B.C., Gilbert, R.S., and Reddy, S.T. "The TIS genes, primary response genes induced by growth factors and tumor promoters." in *Progress in Nucleic Acids Research and Molecular Biology, vol. 47* (Cohen, W., and Moldave, K., eds.) Academic Press, Orlando, Florida, pp. 113-148 (1994).
3. Kujubu, D.A., Fletcher, B.S., Varnum, B.C., Lim, R.W., and Herschman, H.R. TIS10, a phorbol ester tumor promoter-inducible mRNA from Swiss 3T3 cells, encodes a novel prostaglandin synthase/cyclooxygenase homologue. *J. Biol. Chem.* 266:12866-12872 (1991).
4. Fletcher, B.S., Kujubu, D.A., Perrin, D.M., and Herschman, H.R. Structure of the Mitogen-Inducible TIS10 gene and demonstration that the TIS10 encoded protein is a functional prostaglandin G/H synthase. *J. Biol. Chem.* 267:4338-4344 (1992).
5. Xie, W., Chipman, J.G., Robertson, D.L., Erikson, R.L., and Simmons, D.L. Expression of a mitogen-responsive gene encoding prostaglandin synthase is regulated by mRNA splicing. *Proc. Natl. Acad. Sci.* 88:2692-2696 (1991).
6. O'Banion, M.K., Sadowski, H.B., Winn, V., and Young, D.A. A serum-and glucocorticoid-regulated 4-kilobase mRNA encodes a cyclooxygenase-related protein. *J. Biol. Chem.* 266: 23261-23267 (1991).
7. Ryseck, R.P., Raynoschek, C., Macdonald-Bravo, H., Dorfman, K., Mattei, M., and Bravo, R. Identification of an immediate early gene, phs-B, whose protein product has prostaglandin synthase/cyclooxygenase activity. *Cell Growth Differ.* 3:443-450 (1992).
8. Hla, T., and Neilson, K. Human cyclooxygenase-2 cDNA. *Proc. Natl. Acad. Sci.* 89:7384-7385 (1992).
9. Jones, D.A., Carlton, D.P., McIntyre, T.M., Zimmerman, G.A., and Prescott, S.M. Molecular cloning of human prostaglandin endoperoxide synthase type II and demonstration of expression in response to cytokines. *J. Biol. Chem.* 268:9049-9054 (1993).

10. DuBois, R.N., Tsujii, M., Bishop, P., Awad, J.A., Makita, K., and Lanahan, A. Cloning and characterization of a growth factor-inducible cyclooxygenase gene from rat intestinal epithelial cells. *Am. J. Physiol.* 266:G822-G827 (1994).

11. Herschman, H.R., Xie, W., and Reddy, S. Inflammation, reproduction, cancer, and all that.... The regulation and role of the inducible prostaglandin synthase. *BioEssays* 17:1031-1037 (1995).

12. Herschman, H.R. Prostaglandin synthase 2. *Biochim. Biophys. Acta* 1299:125-140 (1996).

13. Vane, J. Towards a better aspirin. *Nature* 367:215-216 (1994).

14. Kujubu, D.A., and Herschman, H.R. Dexamethasone inhibits mitogen iduction of the TIS10 prostaglandin synthase/cyclooxygenase gene. *J. Biol. Chem.* 267:7991-7994 (1992).

15. Reddy, S.T., and Herschman, H.R. Ligand-induced prostaglandin synthesis requires expression of the TIS10/PGS-2 prostaglandin synthase gene in murine fibroblasts and macrophages. *J. Biol. Chem.* 269:15473-15480 (1994).

16. Morita, I., Schindler, M., Regier, M.K., Otto, J.C., Hori, T., DeWitt, D.L., and Smith, W.L. Different intracellular locations for prostaglandin endoperoxide H synthases-1 and -2. *J. Biol. Chem.* 270:10902-10908 (1995).

17. Xie, W., and Herschman, H.R. *v-src* induces prostaglandin synthase 2 gene expression by activation of the c-Jun N-terminal kinase and the c-Jun transcription factor. *J. Biol. Chem.* 270:27622-27628 (1995).

18. Xie, W., and Herschman, H.R. Transcriptional regulation of prostaglandin synthase 2 gene expression by platelet derived growth factor and serum. *J. Biol Chem.* 271:31742-31748 (1996).

19. Inoue, H., Yokoyama, C., Hara, S., Tone, Y., and Tanabe, T. Transcriptional regulation of man prostaglandin-endoperoxide synthase-2 gene by lipopolysaccharide and phorbol ester in vascular endothelial cells. Involvement of both nuclear factor for interleukin-6 expression site and cAMP response element. *J. Biol. Chem.* 270: 24965-24971 (1995).

20. Yamamoto, K., Arakawa, T., Ueda, N., and Yamamoto, S. Transcriptional roles of nuclear factor kappa B and nuclear factor-interleukin-6 in the tumor necrosis factor alpha-dependent induction of cyclooxygenase-2 in MC3T3-E1 cells. *J. Biol. Chem.* 270:31315-31320 (1995).

21. Kawata, R., Reddy, S.T., Wolner, B., and Herschman, H.R. Prostaglandin synthase 1 and prostaglandin synthase 2 both participate in activation-induced PGD_2 production in mast cells. *J. Immunol.* 155:818-825 (1995).

22. Murakami, M., Bingham, C.O. 3rd, Matsumoto, R., Austen, K.F., and Arm, J.P. IgE-dependent activation of cytokine-primed mouse cultured mast cells induces a delayed phase of prostaglandin D2 generation via prostaglandin endoperoxide synthase-2. *J. Immunol.* 155:4445-4453 (1995).

23. Reddy, S.T., and Herschman, H.R. Prostaglandin synthase-1 and prostaglandin synthase-2 are coupled to distinct phospholipases for the generation of prostaglandin D_2 in activated mast cells. *J. Biol Chem.* 72:3231-3237 (1997)

24. Bingham, C.O. 3rd, Murakami, M., Fujishima, H., Hunt, J.E., Austen, K.F., and Arm, J. A heparin-sensitive phospholipase A2 and prostaglandin endoperoxide synthase-2 are functionally linked in the delayed phase of prostaglandin D2 generation in mouse bone marrow-derived mast cells. *J. Biol. Chem.* 271:25936-25944 (1996).

25. Fonteh, A.N., Bass, D.A., Marshall, L.A., Seeds, M., Samet, J.M., and Chilton, F.H. Evidence that secretory phospholipase A_2 plays a role in arachidonic acid release and eicosanoid biosynthesis by mast cells. *J. Immunol.* 152:5438-5446 (1994).

26. Reddy, S.T, Winstead, M.V., Tischfield, J.A., and Herschman, **H.R.** Analysis of the secretory phospholipase A_2 that mediates prostaglandin production in mast cells. *J. Biol. Chem.* 272:13591-13596 (1997).

27. Reddy, S.T., and Herschman, H.R. Transcellular prostaglandin production following mast cell activation is mediated by proximal secretory phospholipase A_2 and distal prostaglandin synthase 1. *J. Biol. Chem.* 271:186-191 (1996).

2

REGULATION OF CYCLOOXYGENASE-2 BY THE ACTIVATED p38 MAPK SIGNALING PATHWAY

Zhonghong Guan, ShaAvhree Y. Buckman, Lisa D. Springer and Aubrey R. Morrison

Departments of Medicine and Molecular Biology and Pharmacology, Washington University School of Medicine, St. Louis, Missouri 63110

INTRODUCTION

Cyclooxygenase (Cox) is a ubiquitous enzyme involved in various inflammatory processes. Two isoforms of Cox have been identified, Cox-1 and Cox-2. Cox-1 is constitutively expressed in most tissues and mediates physiologic responses such as regulation of renal sodium and water reabsorption, vascular homeostasis, and cytoprotection of the stomach. By comparison Cox-2 is primarily considered an inducible immediate-early gene whose synthesis can be upregulated by mitogenic or inflammatory stimuli including: tumor promoters [1], IL-1β [2], endotoxins [3], growth factors [4], and serum[5]. The pathophysiological role of Cox has been the topic of much interest. Cox is the main therapeutic target for non-steroidal anti-inflammatory drugs (NSAIDs) which exhibit their antipyretic, analgesic, and anti-inflammatory effects in humans via inhibition of prostaglandin biosynthesis [6]. NSAIDS have been effective in the reduction of inflammatory symptoms in carrageenan-induced rat paw inflammation models [7] and in the reduced incidence of colon cancer [8, 9].

Recently, a novel class of cytokine-suppressive anti-inflammatory drugs (CSAIDs) have been shown to act as inhibitors of endotoxin stimulated TNF and IL-1 induction [10]. These CSAIDs have been shown to inhibit the catalytic activity of p38 MAPK. These recent findings serve to potentially link prostaglandin biosynthetic pathways with activation of the mitogen-activated protein kinase (MAPK) signaling cascade. In mammalian cells, several different subfamilies of MAPK have been identified including the extracellular signal-regulated kinases (ERK 1 and 2), stress activated protein kinases (SAPK α, β, γ), also called c-Jun-N-terminal kinases (JNK 1 and 2), the p38 MAPKs (p38 MAPK α, β1, β2, γ, δ) and ERK 3,5 and 7. These kinases are activated by distinct upstream MAPK kinases (MKKs). [11] Once phosphorylated, these MAPKs then activate their specific substrates on serine and/or threonine residues to produce their effects on downstream targets. Recent work has demonstrated that both SAPK/JNK and p38 MAPK cascades are the important signaling mechanisms underlying the inflammatory processes.

Recent work has linked activation of the prostaglandin generative pathway with the MAPK pathway. Lin et al. have demonstrated that the activation of cytosolic phospholipase A_2 (cPLA$_2$) is mediated by MAPK [12]. Furthermore, Kramer et al. have demonstrated that thrombin activates both ERK and p38 MAPK in human platelets which may utilize cPLA$_2$ as a downstream target [13]. We have shown that IL-1β stimulation of renal mesangial cells mediates prostaglandin E_2 production and Cox-2 expression with concomitant activation of p38 MAPK and SAPK mediated signaling pathways [14, 15].

In this report, we first demonstrate that inhibition of p38 MAPK results in the down-regulation of IL-1β mediated Cox-2 expression and PGE$_2$ production, suggesting a role for the activation of p38 MAPK in the regulation of prostaglandin synthesis. Previous work by Templeton et al. has shown that MEKK1 directly activates the SAPK pathway [16]. The availability of this model system encouraged exploration of possible SAPK mediated effects on prostaglandin production suggested by our previous cytokine stimulation data. In the current study, we also investigate whether the activation of MEKK1 modulates Cox-2 expression and PGE$_2$ production and analyze the intermediate kinases which may be involved in this regulation.

RESULTS AND DISCUSSION

IL-1β Induces p38 MAPK Activation and Phosphorylation. To investigate whether the p38 MAPK pathway is involved in IL-1β signal transduction, we examined whether the activation of p38 MAPK was induced by IL-1β in glomerular mesangial cells. Cells were treated with 100 U/ml of IL-1β, and an immunocomplex assay using myelin basic protein (MBP) as the substrate was performed to directly measure p38 activity. We found that IL-1β rapidly and transiently activated p38 MAPK. The increase of p38 activity induced by IL-1β was detected within 1 minute of exposure to IL-1β, reached a peak in 5-10 minutes and declined by 60 minutes towards baseline. Western blot analysis using the anti-phospho-specific p38 MAPK also showed that IL-1β quickly increased phosphorylation of p38 MAPK in mesangial cells. The time course of IL-1β induced phosphorylation of p38 MAPK was highly consistent with that of p38 MAPK activation. Together, these results establish that the p38 MAPK pathway is rapidly activated by the proinflammatory cytokine IL-1β in rat mesangial cells and that phosphorylation of Tyr 182 is associated with the activation of p38 MAPK in mesangial cells stimulated by IL-1β. We therefore believe that the activation of p38 MAPK may mediate IL-1β signal amplification and modulation that results in later events such as PGE$_2$ biosynthesis.

The Activation of p38 MAPK Is Required for the Induction of Cox-2 Expression Stimulated by IL-1β. Recently, a group of pyridinyl imidazole compounds have been identified which bind and inhibit p38 MAPK activity. SC68376, kindly provided to us by G.D. Searle, is a pyridinyl oxazole, and has similar properties. We first examined the specificity of this compound on its ability to inhibit the MAP kinase family of enzymes. We found that SC68376, directly inhibits p38 MAPK catalytic activity (but did not affect JNK or ERK activity). Thus, this compound is a potentially useful pharmacological tool to explore the physiological and pathophysiological function of p38 MAPK. To determine whether the activation of p38 MAPK is involved the induction of Cox-2 expression by IL-1β, we studied the effects of SC68376 on IL-1β-induced Cox-2 expression in renal mesangial cells. SC68376, in the range 0.1-100 μM, dose-dependently inhibited IL-1β-induced PGE$_2$ production, Cox-2 protein and Cox-2 mRNA expression. These observations indicate that selective inhibition of p38 MAPK suppresses IL-1β-induced Cox-2 expression and PGE$_2$ production, suggesting that IL-1β-induced PGE$_2$ synthesis is up-regulated by the activation of p38 MAPK pathway. The activation of p38 MAPK is necessary for the induction of Cox-2 expression by IL-1β.

MEKK1 Activation Induces Cox-2 Expression and PGE$_2$ Biosynthesis. We also examined whether activation of MEKK1 expression induces prostaglandin biosynthesis. A MEKK1 inducible system was utilized in which NIH 3T3 cells were stably transfected with a constitutively active truncated mutant of MEKK1 (the C-terminal 324 amino acids, _MEKK1) under the control of an IPTG inducible promoter. Comparisons of untreated MEKK1 3T3 cells with cells induced by IPTG demonstrate an approximately 4-fold enhancement in PGE$_2$ production. We further examined whether the enhancement in PGE$_2$ production was correlated with changes in the expression of Cox. Induction of MEKK1 by IPTG resulted in a significant increase in Cox-2 (but not Cox-1) protein expression.(Figure 1)These data clearly demonstrate that MEKK1 promotes Cox-2 expression and PGE$_2$ production.

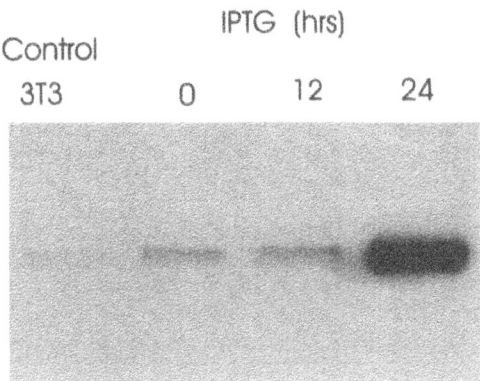

Figure 1. Activation of MEKK1 induces Cox-2 protein expression. NIH 3T3 cells or MEKK1 inducible NIH 3T3 cells were stimulated with or without 1 mM IPTG for the indicated time periods and Cox-2 expression was detected by western blot analysis using an anti-Cox-2 antibody. Densitometric analysis was performed and results were indicated above.

MEKK1 Induces SAPK and p38 MAPK Activity. To understand the signal transduction mechanisms responsible for mediating induction of Cox-2 expression in MEKK1 3T3 cells, in-gel kinase assays using His-c-jun (1-79) as a substrate were performed to measure SAPK activity. The data shows that two bands corresponding to the molecular weights of p46 SAPK and p54/55 SAPK were maximally induced in MEKK1 3T3 cells by 24 hours after IPTG induction. Similarly, western blot analysis using an anti-phospho specific JNK which recognizes both phosphorylated p46 SAPK and p54/55 SAPK further illustrated that MEKK-1 activation could increase phosphorylation of both p46 SAPK and p54/55 SAPK respectively. These above results demonstrate that JNK/SAPK is activated as a consequence of MEKK1 signaling.

Several investigators have indicated that MEKK1 functions as an upstream activator of the SAPK pathway *in vivo*. In our current study, IPTG-induced MEKK1 3T3 cells were assayed for p38 MAPK activity using an immune-complex kinase assay with GST-ATF-2 (1-96) and MBP as substrates. We found that p38 MAPK activity increased after MEKK1 induction. This increase in activity was further correlated with western blot data illustrating increased phosphorylation of p38 MAPK following IPTG stimulation. Together, these results suggest a role for MEKK1 as an activator of both the SAPK pathway and p38 MAPK.

p38 MAPK Mediates MEKK 1-Induced COX-2 Expression. To analyze whether p38 MAPK signaling pathways mediate PGE_2 production and cyclooxygenase expression induced by MEKK1 activation, we studied the effects of SC68376, an inhibitor of p38 MAPK..

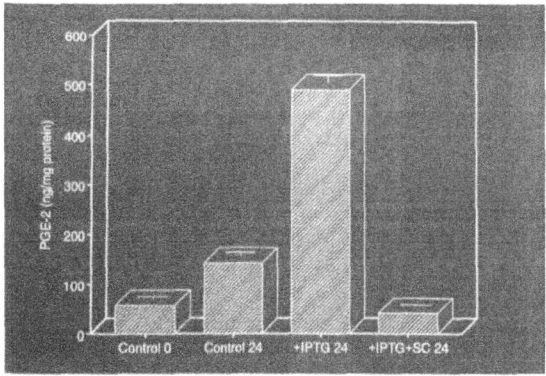

Figure 2. p38 MAPK inhibitor inhibits ΔMEKK1 induced PGE₂ production. MEKK1 NIH3T3 cells were treated with or without 1 mM IPTG in the presence or absence of 10 μM of SC68376 for 24 hours. PGE₂ in the culture media was measured by ELISA.

When MEKK1 3T3 cells were treated with 1mM IPTG and 10μM inhibitor for 24 hours, this completely eliminated Cox-2 expression and concomitant PGE$_2$ production normally induced by MEKK1 activation. (Figure 2) The above observations confirm that the activation of p38 MAPK provides a crucial signaling mechanism which promotes Cox-2 expression and prostaglandin biosynthesis.

MEKK1→SEK/MKK4 is a Possible Upstream Activator of p38 MAPK. Previous data has suggested that SEK/MKK4 is the immediate upstream kinase of SAPK/JNK [17] whereas MKK3 and MKK6 are immediately upstream of p38 MAPK. We therefore investigated whether this proposed signaling cascade was responsible for activation of p38 MAPK in MEKK1 inducible cells. Surprisingly, IPTG induction of MEKK1 expression induced only phosphorylation (assessed by western blot assay using an anti-phospho-MKK4 antibody) and activation (assessed by immunocomplex kinase assay using GST-p38 MAPK as substrate) of SEK/MKK4. However, phosphorylation (assessed by western blot assay using an anti-phospho-MKK3/MKK6 antibody) and activation of MKK3 and MKK6 (assessed by immunocomplex kinase assay using GST-p38 MAPK as substrate) were not induced by IPTG stimulation. As a positive control, anisomycin did activate and phosphorylate MKK3/MKK6 in NIH 3T3 cells. These findings are suggestive of a signaling mechanism for activation of p38 MAPK involving phosphorylation of SEK/MKK4 induced by MEKK1. To further demonstrate our hypothesis that SEK1/MKK4 may mediate this signaling mechanism, SEK1-WT (wild type), SEK1-ED (constitutively active form) and SEK1-AL (kinase dead form) were stably transfected in NIH 3T3 cells. Overexpression of the constitutively active form of SEK1 results in increased phosphorylation of JNK/SAPK and p38 MAPK, whereas the kinase negative mutation, SEK1-AL, inhibits phosphorylation of both aforementioned kinases. This evidence suggests that SEK1/MKK4 is upstream of both the JNK/SAPK and p38 MAPK pathways. Furthermore, our experiments also demonstrate that activated SEK1 induces Cox-2 expression and PGE$_2$ production whereas dominant negative SEK1 inhibits this response. SEK1-induced Cox-2 expression and PGE$_2$ production are completely inhibited by the p38 MAPK inhibitor. Together, these results illustrate the ability of SEK1/MKK4 to mediate MEKK1-induced Cox-2 expression and PGE$_2$ production and serve as an upstream kinase modulating p38 MAPK activation. We therefore believe that the MEKK1→SEK1/MKK4→p38 MAPK cascade is an important upstream signaling mechanism promoting Cox-2 expression and PGE$_2$ production.

CONCLUSION

The MAPK cascade is believed to function as an important regulator of prostaglandin biosynthesis. IL-1β induces activation of JNK/SAPK and p38 MAPK with concomitant upregulation of Cox-2 expression and PGE$_2$ synthesis. Inhibition of p38 MAPK results in the down-regulation of IL-1-mediated Cox-2 expression and PGE$_2$ synthesis, suggesting a critical role for p38 MAPK in the modulation of prostaglandin biosynthesis. Evidence presented in this paper demonstrates that overexpression of MEKK1 activates both JNK/SAPK as well as p38 MAPK. Furthermore, overexpression of the activated form of MEKK1 induces increased prostaglandin biosynthesis with enhanced Cox-2 protein expression. Our findings suggest a novel signaling pathway by which MEKK1→SEK/MKK4→p38 MAPK→→→COX-2. This pathway provides further evidence for the role of the p38 MAPK pathway in the mediation of prostaglandin biosynthesis and the modulation of inflammatory response.

ACKNOWLEDGMENT:

This work was supported by U.S. Public Health Awards DK 38111 and DK 50606.

REFERENCES

1. Kujubu, D. A., B. S. Fletcher, B. C. Varnum, R. W. Lim, and H. R. Herschman. 1991. TIS10, a phorbol ester tumor promoter-inducible mRNA from Swiss 3T3 cells, encodes a novel prostaglandin synthase/cyclooxygenase homologue. *Journal of Biological Chemistry.* 266 (20):12866-72.

2. Maier, J. A., T. Hla, and T. Maciag. 1990. Cyclooxygenase is an immediate-early gene induced by interleukin-1 in human endothelial cells. *Journal of Biological Chemistry.* 265 (19):10805-8.

3. Xie, W., D. L. Robertson, and D. L. Simmons. 1992. Mitogen-inducible prostaglandin G/H synthase: A new target for nonsteroidal anti-inflammatory drugs. *Drug Development Research.* 25:249-265.

4. Habenicht, A. J., M. Goerig, J. Grulich, D. Rothe, R. Gronwald, U. Loth, G. Schettler, B. Kommerell, and R. Ross. 1985. Human platelet-derived growth factor stimulates prostaglandin synthesis by activation and by rapid de novo synthesis of cyclooxygenase. *Journal of Clinical Investigation.* 75 (4):1381-7.

5. O'Banion, M. K., V. Winn, and D. A. Young. 1992. cDNA cloning and functional activity of a glucocorticoid-regulated inflammatory cyclooxygenase. *Proceedings of the National Academy of Sciences of the United States of America.* 89:4888-4892.

6. Smith, W. L., E. A. Meade, and D. L. DeWitt. 1994. Interactions of PGH synthase isozymes-1 and -2 with NSAIDs. *Annals of the New York Academy of Sciences.* 744:50-7.

7. Seibert, K., Y. Zhang, K. Leahy, S. Hauser, J. Masferrer, W. Perkins, L. Lee, and P. Isakson. 1994. Pharmacological and biochemical demonstration of the role of cyclooxygenase 2 in inflammation and pain. *Proc. Natl. Acad. Sci. USA.* 91:12013-12017.

8. Marnett, L. J. 1992. Aspirin and the potential role of prostaglandins in colon cancer. [Review]. *Cancer Research.* 52 (20):5575-89.

9. Rao, C. V., A. Rivenson, B. Simi, E. Zang, G. Kelloff, V. Steele, and B. S. Reddy. 1995. Chemoprevention of colon carcinogenesis by sulindac, a nonsteroidal anti-inflammatory agent. *Cancer Research.* 55 (7):1464-72.

10. Lee, J. C., J. T. Laydon, P. C. McDonnell, T. F. Gallagher, S. Kumar, D. Green, D. McNulty, M. J. Blumenthal, J. R. Heys, S. W. Landvatter, and et al. 1994. A protein kinase involved in the regulation of inflammatory cytokine biosynthesis. *Nature.* 372 (6508):739-46.

11. Kyriakis, J. M., P. Banerjee, E. Nikolakaki, T. Dai, E. A. Rubie, M. F. Ahmad, J. Avruch, and J. R. Woodgett. 1994. The stress-activated protein kinase subfamily of c-Jun kinases. *Nature.* 369 (6476):156-60.

12. Lin, L. L., M. Wartmann, A. Y. Lin, J. L. Knopf, A. Seth, and R. J. Davis. 1993. cPla2 is phosphorylated and activated by Map kinase. *Cell.* 72 (2):269-78.

13. Kramer, R. M., E. F. Roberts, S. L. Um, A. G. Borsch-Haubold, S. P. Watson, M. J. Fisher, and J. A. Jakubowski. 1996. p38 mitogen-activated protein kinase phosphorylates cytosolic phospholipase A2 (cPla2) in thrombin-stimulated platelets. Evidence that proline-directed phosphorylation is not required for mobilization of arachidonic acid by cPla2. *Journal of Biological Chemistry.* 271 (44):27723-9.

14. Guan, Z. H., T. Tetsuka, L. D. Baier, and A. R. Morrison. 1996. Interleukin-1-Beta Activates C-Jun NH_2-Terminal Kinase Subgroup Of Mitogen-Activated Protein Kinases In Mesangial Cells. *American Journal of Physiology - Renal Fluid & Electrolyte Physiology.* 39 (4):F 634-F 641.

15. Guan, Z. H., L. D. Baier, and A. R. Morrison. 1997. p38 Mitogen-Activated Protein Kinase Down-Regulates Nitric Oxide and Up-Regulates Prostaglandin E(2) Biosynthesis Stimulated By Interleukin-1-Beta. *Journal of Biological Chemistry.* 272 (12):8083-8089.

16. Yan, M., T. Dai, J. C. Deak, J. M. Kyriakis, L. I. Zon, J. R. Woodgett, and D. J. Templeton. 1994. Activation of stress-activated protein kinase by MEKK1 phosphorylation of its activator SEK1. *Nature.* 372 (6508):798-800.

17. Derijard, B., J. Raingeaud, T. Barrett, I. H. Wu, J. Han, R. J. Ulevitch, and R. J. Davis. 1995. Independent human MAP-kinase signal transduction pathways defined by MEK and MKK isoforms. *Science.* 267 (5198):682-685.

3

FATTY ACID CYCLOOXYGENASE INDUCTION AND PROSTAGLANDIN D
SYNTHESIS IN A HUMAN MEGAKARYOBLASTIC CELL LINE CMK
DIFFERENTIATED BY PHORBOL ESTER

Shozo Yamamoto[1], Natsuo Ueda[1], Ishtiaq Mahmud[1],
Hiroko Yamaguchi[1], Rieko Yamashita[1], Kei Yamamoto[1],
Kazunori Ishimura[2], Yoshihiro Urade[3], Yoshihide Kanaoka[3],
and Osamu Hayaishi[3]

Departments of [1]Biochemistry and [2]Anatomy, Tokushima University,
School of Medicine, Tokushima 770-8503, Japan
[3]Osaka Bioscience Institute, Osaka 565-0874, Japan

INTRODUCTION

The biosynthesis of prostaglandin (PG) and thromboxane (TX) is initiated by a bifunctional enzyme with oxygenase and peroxidase activities, and PGH_2 is produced from arachidonic acid via PGG_2[1]. This bifunctional enzyme (prostaglandin endoperoxide synthase, E.C. 1.14.99.1) will be referred to briefly as cyclooxygenase in this chapter. There are two cyclooxygenase isozymes (COX-1 and COX-2)[2], and COX-1 is generally known to be present in human platelets and involved in the synthesis of proaggregatory and vasoconstrictive TXA_2. Since platelets have no nucleus and can not synthesize mRNA, the COX-1 protein is derived presumably from megakaryocytes which are precursor cells of platelets.

INDUCTION OF CYCLOOXYGENASE-1

A human megakaryoblastic cell line CMK was established by Takeyuki Sato and others from peripheral blood of an acute megakaryoblastic leukemia patient with Down's syndrome. By treatment with 12-O-tetradecanoylphorbol 13-acetate (TPA) or dimethyl sulfoxide the cells differentiate to mature megakaryocyte-like cells, expressing glycoprotein IIb/IIIa, and give rise to platelets[3,4]. It was reported earlier that the cells produced TXA_2, implicating the presence of cyclooxygenase[5].

CMK cells were cultured for 4 days in the presence of 0.1 μM TPA[6]. After sonication of the cells thus cultured, the lysate was incubated with ^{14}C-labeled arachidonic acid, and the products were separated by thin layer chromatography. As shown in Figure 1, we observed many radioactive bands corresponding to the cyclooxygenase metabolites of arachidonic acid. In sharp contrast, arachidonic acid was essentially unchanged with the lysate of the cells which were not treated by TPA. By incubation with whole cells treated with TPA, PGD_2 and TXB_2 were found. This product profile will be discussed later.

Eicosanoids and Other Bioactive Lipids in Cancer, Inflammation, and Radiation Injury, 4
Edited by Honn *et al.*, Kluwer Academic / Plenum Publishers, New York, 1999.

17

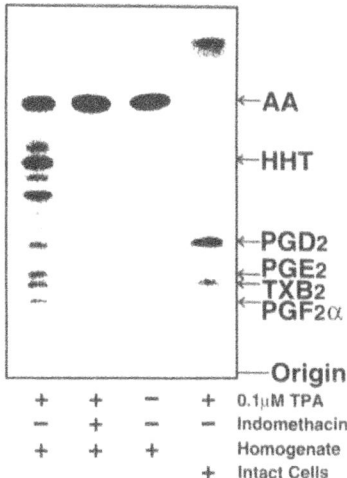

Figure 1. Arachidonate metabolites in CMK cells. The cells were cultured for 4 days in the presence or absence of 0.1 μM TPA. The cell lysate was incubated with ^{14}C-arachidonic acid, and the ethereal extract was analyzed by thin layer chromatography. The whole cells were also allowed to react with arachidonic acid.

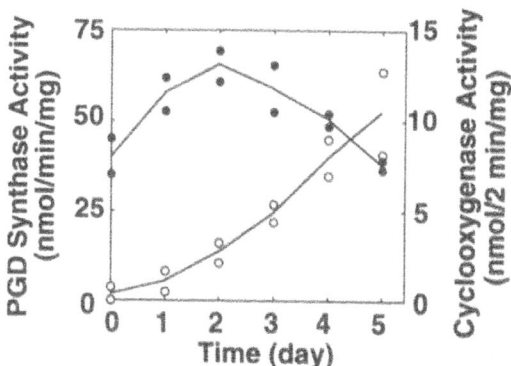

Figure 2. Time-dependent effect of phorbol ester on the activities of cyclooxygenase and PGD synthase in CMK cells. The cells were cultured in the presence of 0.1 μM TPA. At various time intervals the cells were harvested, and the homogenate was incubated with ^{14}C-arachidonic acid for the cyclooxygenase assay (open circles) or with ^{14}C-PGH2 for the PGD synthase assay (closed circles).

As shown in Figure 2, when TPA was added to the cells, the cyclooxygenase activity appeared after a lag phase of several hours and continued to increase for several days[6]. Dimethyl sulfoxide also brought about a similar increase in the cyclooxygenase activity. We attempted to find out whether the increased enzyme activity was attributed to COX-1 or COX-2. The cyclooxygenase activity of CMK cells was not affected by NS-398 (a COX-2-specific inhibitor) up to 10 μM. Upon Western blotting using an antibody against human platelet COX-1 which did not cross-react with human COX-2, the COX-1 protein increased time-dependently for up to 4 days. In contrast, when an antibody specific for human COX-2 was used, we could hardly detect a band comigrating with the COX-2. The cyclooxygenase activity of CMK cells was found in the microsomal fraction upon differential centrifugation. The enzyme was solubilized from the microsome by the use of Tween-20, and the solubi-lized enzyme was incubated with anti-COX-1 antibody. As the amount of the antibody was increased, the enzyme activity was recovered in the pellet, and the activity in the supernatant decreased. Furthermore, the Northern blotting showed a time-dependent increase of COX-1 mRNA in a time course consistent with the changes of the enzyme activity and protein mentioned above. The COX-induction by TPA was not affected by anti-inflammatory steroid. The finding is consistent with the reported nature of COX-1 induction. On the basis of these findings we identified most of the TPA-induced cyclooxygenase in CMK cells as COX-1 rather than COX-2. Northern blotting at the initial phase demonstrated a slight transient induction of COX-2 mRNA around 3 h after the addition of TPA, but the cyclo-oxygenase activity was almost undetectable at that time. The significance of this transient COX-2 mRNA induction is unknown.

The COX-1 induction in CMK cells was also demonstrated immunocytochemically[6]. TPA-treated cells were subjected to immunoelectron microscopy with anti-COX-1 antibody. The dark deposits of immunoperoxidase reaction were found in nuclear envelope and endoplasmic reticulum. When Sato and others established CMK cells, the so called platelet peroxidase (PPO) was localized in nuclear envelope and endoplasmic reticulum of the TPA-treated cells, and the density of the PPO activity was increased by the addition of TPA[4]. We would identify the PPO as COX-1 with prostaglandin hydroperoxidase activity[1].

PROSTAGLANDIN D SYNTHASE

As mentioned above (Figure 1), PGD2 was produced together with TXB2 upon incubation of CMK cells with arachidonic acid. Previously it was a debatable subject whether or not platelets can synthesize PGD2 in addition to TXA2. Watanabe and others investigated the PGD2 production by human platelet-rich plasma which many other investigators had observed, and they demonstrated the conversion of PGH2 to PGD2 by platelet-poor plasma but not by washed platelets. The PGD2 synthesizing protein was purified and identified as serum albumin[7]. Indeed, as presented in Figure 3, when ^{14}C-labeled PGH2 was incubated with the cytosol of human platelets, there was no PGD2 production above the level of non-enzymatic degradation of PGH2[8]. However, an efficient conversion of PGH2 to PGD2 was observed with the cytosol fraction of the CMK cell homogenate, but not with the particulate fraction. The cytosol of CMK cells produced PGD2 in the presence of 1 mM glutathione, but not in its absence. The heat-denatured cytosol was inactive. The rate of PGD2 synthesis increased depending on the concentration of PGH2, however its Km value was as high as 200 μM. The enzyme activity was also dependent on the concentration of glutathione with a high Km of about 1 mM. In the standard assays we

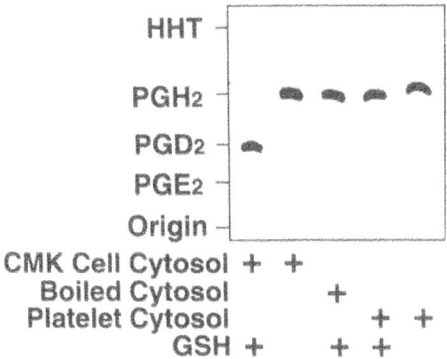

Figure 3. Incubation of PGH2 with the cytosol fractions of platelets and CMK cells. [14]C- PGH2 at 40 μM was incubated with the cytosol fraction of CMK cells or human platelets in the presence of 1 mM glutathione or in its absence. The ethereal extract was subjected to silica gel thin layer chromatography. GSH, glutathione; HHT, 12-hydroxy-heptadecatrienoic acid.

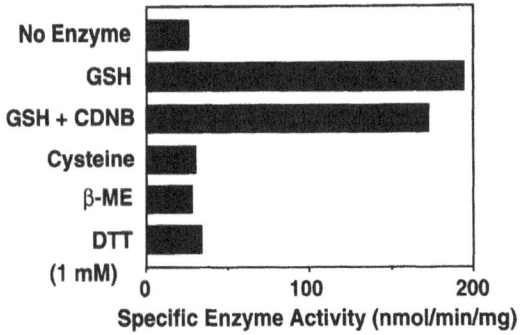

Figure 4. Selective requirement of glutathione for the PGD synthase activity of CMK cell. GSH, glutathione; CDNB, 1-chloro-2,4-dinitrobenzene; β-ME, β-mercaptoethanol; DTT, dithiothreitol.

used PGH2 and glutathione at subsaturating concentrations of 40 μM and 1 mM, respec-tively.

As has been extensively investigated by the group of Osaka Bioscience Institute, two isoforms of PGD synthase are known[9]. The lipocalin type, which is also mentioned as the brain type, is activated by various sulfhydryl compounds, whereas the hematopoietic type or the spleen type requires glutathione selectively. The two isozymes are distinguished by inhibition by 1-chloro-2,4-dinitrobenzene. The lipocalin type is found in brain and cerebro-spinal fluid, and the hematopoietic type is found in peripheral tissues such as spleen, mast cells and skin. PGD synthase of CMK cells required selectively glutathione, and was not inhibited by 1-chloro-2,4-dinitrobenzene (Figure 4). Therefore, the enzyme was charac-terized to be a hematopoietic type [8].

In view of such a requirement of glutathione, we attempted to purify the enzyme by affinity chromatography[8]. The cytosol of CMK cells was applied to a glutathione-Sepharose 4B column. A bulk of proteins passed through the column, and PGD synthase was eluted with a significant delay without the addition of glutathione, resulting in a high purification up to a specific enzyme activity of about 300 μmol/min/mg protein. Active fractions showed 26-kDa bands stained with silver nitrate. With rabbit antiserum against human hematopoietic PGD synthase, the purified enzyme gave a positive band, but not with antiserum against human lipocalin type enzyme. PGD synthase activity was followed during differentiation of CMK cells (Figure 2). As described above, when the cells were cultured in the presence of 0.1 μM TPA, the COX-1 activity was almost undetectable at first, and then increased time-dependently. In contrast, PGD synthase activity was considerably high at day 0, increased only about 2 fold at day 2, and then decreased to the basal level. Western blotting showed the change in the amount of enzyme protein in parallel with the enzyme activity. As examined by Northern blotting, two bands around 1.8 kb and 1.0 kb were observed, and the most intense bands were observed at day 1.

DISCUSSION

When CMK cells were cultured in the presence of TPA, the COX-1 activity was almost undetectable at first, and then increased time-dependently. In contrast, the CMK cells had a considerable activity of PGD synthase which increased only about 2 fold at day 2 upon the addition of TPA, and then decreased to the basal level. It should be noted that the COX-1 activity was always rate-limiting. Thus, PGD synthase, which was not found in human, porcine and rat platelets, was found at a high basal level in CMK cells. Since the PGD synthase of CMK cells required selectively glutathione, and was not inhibited by 1-chloro-2,4-dinitrobenzene, the enzyme was identified to be of the hematopoietic type. This identification was confirmed by Western and Northern blottings. Similar findings were also reported with CMK cells and other megakaryoblastic cell lines by Suzuki et al.[10].

The PGD synthase thus found in CMK cells does not appear in platelets which are derived from megakaryocytes. Since CMK cells are leukemia cells, such a high expression of PGD synthase may not be a physiological event. Alternatively, at a certain step of mega-karyoblastic differentiation PGD2 may be produced and further transformed to PGJ2 and its derivatives[11] which are known to inhibit cell growth and may regulate cell proliferation. Therefore, it would be of interest to examine a possible transient expression of PGD synthase in native megakaryoblasts and megakaryocytes. The induced COX-1 is to be incorporated into platelets as a constitutive enzyme. We are investigating the COX-1 and COX-2 induction in megakaryocyte differentiation stimulated by physiological factors like thrombopoietin.

REFERENCES

1.   S.Yamamoto, Enzymes in the arachidonic acid cascade, in: *Prostaglandins and Related Substances* , C.R.Pace-Asciak, and E.Granström, eds., Elsevier Science Publishers, pp.171-202 (1983).
2.   H.R.Herschman, Prostaglandin synthase 2, *Biochim. Biophys. Acta* 1299:125 (1996).
3.   N.Komatsu, T.Suda, M.Moroi, N.Tokuyama, Y.Sakata, M.Okada, T.Nishida, Y.Hirai, T.Sato, A.Fuse, and Y.Miura, Growth and differentiation of a human megakaryoblastic cell line, CMK, *Blood* 74:42 (1989).
4.   T.Sato, A.Fuse, M.Eguchi, Y.Hayashi, R.Ryo, M.Adachi, Y.Kishimoto, M. Teramura, H.Mizoguchi, Y.Shima, I.Komori, S.Sunami, Y.Okimoto, and H.Nakajima, Establishment of a human leukaemic cell line (CMK) with megakaryocytic characteristics from a Down's syndrome patient with acute megakaryoblastic leukaemia, *Br. J. Haematol.* 72:184 (1989).
5.   R.Ryo, A.Yoshida, M.Adachi, W.Sugano, M.Yasunaga, N.Yoneda, N.Yamaguchi, And T. Sato, Cytosolic calcium mobilization and thromboxane synthesis in a human megakaryocytic leukemia cell, *Exp. Hematol.* 18:271 (1990).
6.   N.Ueda, R.Yamashita, S.Yamamoto, and K.Ishimura, Induction of cyclooxygenase-1 in a human megakaryoblastic cell line (CMK) differentiated by phorbol ester, *Biochim. Biophys. Acta* 1344:103 (1997).
7.   T.Watanabe, S.Narumiya, T.Shimizu, and O.Hayaishi, Characterization of the biosynthetic pathway of prostaglandin D2 in human platelet-rich plasma, *J. Biol. Chem.* 257: 14847 (1982).
8.   I.Mahmud, N.Ueda, H.Yamaguchi, R.Yamashita, S.Yamamoto, Y.Kanaoka, Y.Urade, and O.Hayaishi, Prostaglandin D synthase in human megakaryoblastic cells, *J. Biol. Chem.* 272:28263 (1997).
9.   Y. Urade, K. Watanabe, and O.Hayaishi, Prostaglandin D, E, and F synthases, *J. Lipid Mediators and Cell Signalling* 12:257 (1995).
10.   T.Suzuki, K.Watanabe, Y.Kanaoka, T.Sato, and O.Hayaishi, Induction of hematopoietic prostaglandin D synthase in human megakaryocytic cells by phorbol ester, *Biochem. Biophys. Res. Commun.* 241:288 (1997).
11.   M.Negishi, T.Koizumi, and A.Ichikawa, Biological actions of Δ^{12}-prostaglandin J2, *J. Lipid Mediators and Cell Signalling* 12:443 (1995).

4

EICOSANOID METABOLISM IN HUMAN PLATELETS IS MODIFIED BY ALBUMIN

Marina Dadaian and Pär Westlund

Department of Woman and Child Health
Division of Reproductive Endocrinology
Karolinska Hospital, Bldg. L5
S-171 76 Stockholm, Sweden

INTRODUCTION

Activation of human platelets leads to release of arachidonic acid from cellular phospholipids, the step which is believed to be the rate limiting for the metabolism of this fatty acid. Once released arachidonic acid can be metabolised by two platelet enzymes. Cyclooxygenase produces endoperoxides with the subsequent formation of thromboxane A_2 (TXA_2). Another enzyme, 12-lipoxygenase, catalyses conversion of arachidonic acid to 12-hydroperoxyeicosatetraenoic acid (12-HPETE) which is rapidly reduced to 12-hydroxyeicosatetraenoic acid (12-HETE)[1]. The physiological significance of the latter pathway is poorly understood to date. Previously we have shown that albumin prevents metabolism of 12-HETE by polymorphonuclear leukocytes *in vitro* and protects it from metabolism by blood cells *in vivo*[2,3]. In the present study we have investigated the influence of albumin on the activity of 12-lipoxygenase in the presence of phospholipase A_2 and C inhibitors. In addition, we studied the influence of 15-hydroxyeicosatetraenoic acid (15-HETE), which has been reported to be a selective platelet 12-lipoxygenase inhibitor[4], and 5-hydroxyeico-satetraenoic acid (5-HETE), which is a substrate for 12-lipoxygenase, on the endogenous 12-HETE production in the presence of albumin.

METHODS

Platelets were prepared from buffy coat as described[2]. Washed platelets were treated with aspirin (3 mM, 30 min incubation at room temperature) and finally suspended in phos-phate buffered saline. Then the platelets ($1,5$-1×10^9 cells/ml) in a volume of 0.5 ml were preincubated for 5 min at 37°C in the presence or absence of albumin (600 μM). Some samples were also preincubated with the phospholipase A_2 inhibitor, arachidonyl trifluoro-methyl ketone (AACOCF$_3$, 60 µM), or phospholipase C inhibitor, U73122, a steroidal amine (20 µM), or with both inhibitors. The reactions were started by addition of 15-HETE or 5-HETE (50 µM) and the calcium ionophore A23187 (3 µM) simultaneously. In all the incubations the fatty acids were added in a mixture with the corresponding labelled compound, [^{3}H$_8$]15-HETE (specific activity 180 mCi/mmol) and [^{3}H$_8$]5-HETE (specific activity 200 mCi/mmol). The incubation time varied from 5 min to 60 min and the reactions were stopped by the addition of 3 volumes of ethanol. The precipitated proteins were removed by centrifugation and the samples were extracted using Sep Pak C$_{18}$ cartridges. 12-HETE detection and quantitation was done using RP-HPLC with a diode array detector and a Nucleosil C$_{18}$ column by comparison of the peak heights with the corresponding calibration curve. The data are presented as mean±standard error and the statistics were performed using Kruskal-Wallis analysis of variance on ranks.

RESULTS AND DISCUSSION

Effect of 5- and 15-HETE on the formation of 12-HETE in the presence of albumin

It is known that 5-HETE is a substrate for platelet 12-lipoxygenase and can be metabolised by this enzyme with the production of 5(S),12(S)-dihydroxyeicosatetraenoic acid (5(S),12(S)-DHETE)[5]. In order to understand how this reaction would affect 12-HETE production from endogenous arachidonic acid in the presence of albumin, platelets were incubated with 50 µM 5-HETE for 60 min in the presence or absence of albumin.

The amount of 12-HETE produced during these incubations did not vary significantly from the control levels, neither addition of albumin influenced these figures thus suggesting that 5-HETE as well as albumin had no effect on 12-HETE formation during long time incubations within the used concentration range (Figure 1).

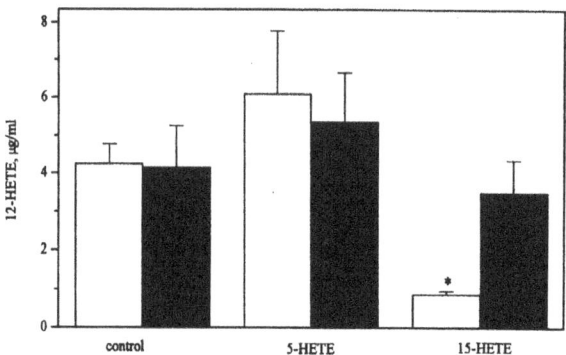

Figure 1. Effect of albumin on endogenous 12-HETE production in the presence of 5-HETE and 15-HETE. The amount of 12-HETE is expressed as µg/ml of platelet suspension (1.2x10^9 cells/ml) and values are mean of four experiments. Open bars - without albumin, filled bars - with albumin. * - P<0.05 compared to all other groups.

In contrast, during incubation of 50 μM 15-HETE with platelets under the same conditions in the absence of albumin the level of 12-HETE produced was reduced dramatically in comparison to control incubations (Figure 1). The presence of albumin in the incubation medium restored the ability of the platelet suspension to produce 12-HETE comparable to the control incubations (Figure 1). 15-HETE was found to be a selective inhibitor for platelet 12-lipoxygenase[4], however, the mechanism of its action was not defined. These results suggest that albumin abolishes the inhibitory effect of 15-HETE on 12-HETE production by binding with the compound. We have shown previously using affinity chromatography with Blue Sepharose, that albumin indeed binds with 5- and 15-HETE to the same extend[6].

Platelet 12-lipoxygenase is able to metabolise 15-HETE with the production of 14,15-dihydroxyeicosatetraenoic acid (14,15-DHETE)[7]. We have shown previously that albumin sequesters 5-HETE from the metabolism by platelet 12-lipoxygenase but in sharp contrast, enhances the conversion of 15-HETE to 14,15-DHETE 4 to 6 fold during 60 min incubations with stimulated platelets[6]. Taken together our results show that albumin, on one hand, enhances metabolism of 15-HETE by 12-lipoxygenase and on the other hand abolishes the inhibitory effect of 15-HETE on the same 12-lipoxygenase. This effect seems to be contradictory. This contradiction might be explained by the fact that platelet 12-lipoxygenase is activated by 12-HPETE[8]. The kinetic of the 12-lipoxygenase catalysed reaction is characterised by the slow initial velocity followed by the increased rate of catalysis[9]. It seems that the hydroperoxide at position 12 (12-HPETE) is essential for this activation of the enzyme. In the case of 15-HETE the formation of 14,15-DHETE is not accompanied by the formation of the hydroperoxy group at position 12, and the mechanism of the reaction is thought to involve radical migration to C-14 after initial C-10 hydrogen atom abstraction and subsequent intermolecular reaction with oxygen[7]. Possibly due to this the catalysis rate of 12-lipoxygenase in the presence of 15-HETE is reduced. In comparison, the mechanism of 5(S),12(S)-DHETE formation involves the addition of oxygen to give a 12(S) hydroperoxide. This might be the possible explanation for the fact that 15-HETE, but not 5-HETE possesses inhibitory action on 12-lipoxygenase. However, albumin binding of 15-HETE probably increases the availability of endogenous arachidonic acid for 12-lipoxygenase enough to initiate and elevate the rate of the reaction.

Albumin effect on 12-lipoxygenase activity in the presence of phospholipase A$_2$ and C inhibitors

In order to study the kinetics of 12-lipoxygenase reaction with different susbstrates in intact platelets we decided to block arachidonic acid release from the phospholipids. Two types of inhibitors were chosen for this purpose, the arachidonic acid analogue AACOCF$_3$, a phospholipase A$_2$ inhibitor, and a phospholipase C inhibitor, U73122, both of which have been shown to inhibit arachidonic acid release in platelets[10,11].

Platelets were preincubated with AACOCF$_3$ or U73122, or with both inhibitors in the presence or absence of albumin for 5 min prior to initiation of the reactions by addition of calcium ionophore A23187. After 5 min the reactions were stopped and analysed for the formation of 12-HETE. When albumin was present in the incubations the incubation time was prolonged to 30 min.

The formation of 12-HETE, as expected, was decreased when the inhibitors were present in the incubation medium (Figure 2, panel a). In the presence of AACOCF$_3$ the amount of 12-HETE formed was decreased by 50%, and U73122 was more efficient, decreasing 12-HETE formation by about 90%.The same result was obtained when both inhibitors were present in the incubations - the formation of 12-HETE was decreased by about 90%. This shows that probably activation of the phospholipase C cascade reactions contributes more to arachidonic acid release during platelet stimulation by calcium ionophore rather than activation of phospholipase A$_2$. On the other hand we can not rule out the possibility of the direct influence of U73122 on platelet 12-lipoxygenase activity, although it has been shown that U73122 does not appear to inhibit the uptake of exogenous arachidonic acid or its conversion to TXA$_2$[11]. In contrast, AACOCF$_3$ was shown to affect the cyclooxygenase pathway in human platelets, but did not inhibit platelet 12-lipoxygenase[10].

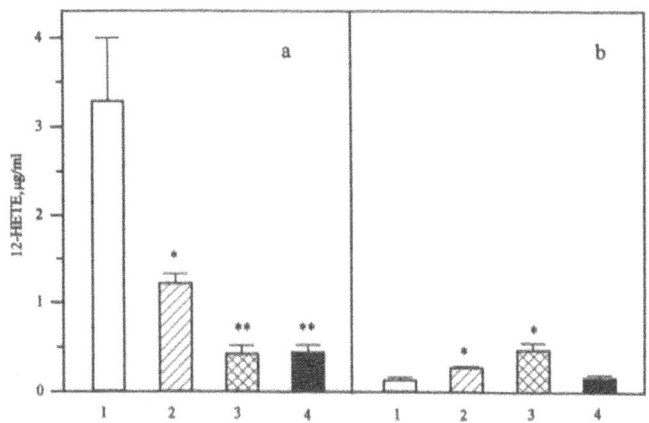

Figure 2. Influence of phospholipase A$_2$ and C inhibitors on the production of 12-HETE by human platelets. The platelets were incubated in the presence of calcium ionophore A23187 for 5 min (panel a) or in the presence of albumin together with A23187 for 30 min (panel b). The amount of 12-HETE is expressed as μg/ml of platelet suspension (1.2×10^9 cells/ml) and values are mean of three experiments. 1- control, 2 - addition of AACOCF$_3$, 3 - addition of U73122, 4 - addition of both inhibitors. * - P<0.05 compared to control; ** - P<0.01 compared to control.

During 30 min incubation albumin dramatically reduced the endogenous production of 12-HETE when no inhibitors were added to the incubations (Figure 2, panel b). However, in the presence of albumin and the inhibitors no additional inhibition could be recorded. Moreover, the amount of the formed 12-HETE was enhanced in the presence of the inhibitors, especially pronounced when the platelets were preincubated with phospholipase C inhibitor U73122 (Figure 2, panel b).

These results suggest that albumin probably binds both inhibitors as well as arachidonic acid and these compounds are competing for the same binding site. The binding of the inhibitors with albumin provides more arachidonic acid for the production of 12-HETE, since albumin by sequestering them abolishes the inhibitory effect on phospholipases. Since it seems that phospholipase C is more important in providing substrate for 12-lipoxygenase, binding with this inhibitor leads to enhanced production of 12-HETE in comparison to AACOCF$_3$.

CONCLUSIONS

Our results show that albumin has important impact on the regulation of 12-lipoxygenase in human platelets. The ability of this protein to efficiently bind different fatty acids as well as other structurally related compounds leads to a regulation of the availability of these compounds for the metabolising enzyme. Since the 12-lipoxygenase reaction once initiated proceeds continuously, until all available substrate is consumed, albumin seems to be an important regulator for this enzyme.

REFERENCES

1. M. Hamberg and B. Samuelsson, Prostaglandin endoperoxides. Novel transformations of arachidonic acid in human platelets, Proc. Natl. Acad. Sci. USA 71:3400 (1974).
2. M. Dadaian, E. Granström, and P. Westlund, Albumin prevents metabolism of 12-hydroxyeicosatetraenoic acid by leukocytes in vitro, Biochim. Biophys. Acta 1303:154 (1996).
3. M. Dadaian, E. Granström, and P. Westlund, 12-Hydroxyeicosatetraenoic acid is a long lived substance in the rabbit circulation, Prostaglandins and Other Lipid Mediators 53:00 (1998).
4. J.Y. Vanderhoek, R.W. Bryant, and J.M. Bailey, 15-Hydroxy-5,8,11,13-eicosatetraenoic acid. A potent and selective inhibitor of platelet lipoxygenase, J. Biol. Chem. 255:5996 (1980).
5. A.J. Marcus, M.J. Broekman, L.B. Safier, H.L.Ullman, and N. Islam, Formation of leukotrienes and other hydroxy acids during platelet-neutrophil interactions in vitro, Biochem. Biophys. Res. Commun. 109:130 (1982).
6. M.Dadaian and P. Westlund, Different effects of albumin on the metabolism of hydroxyeicosatetraenoic acids in human platelets, in: Proceedings of the Fourth International Congress on Essential Fatty Acids and Eicosanoids, AOCS Press, in press (1998).
7. R.L. Maas and A.R. Brash, Evidence for a lipoxygenase mechanism in the biosynthesis of epoxide and dihydroxy leukotrienes from 15(S)-hydroperoxyeicosatetraenoic acid by human platelets and porcine leukocytes, Proc. Natl. Acad. Sci. USA 80:2884 (1983).
8. C. Edenius, S. Tornhamre, and J.Å. Lindgren, Stimulation of lipoxin synthesis from leukotriene A$_4$ by endogenously formed 12-hydroperoxyeicosatetraenoic acid in activated human platelets, Biochim.Biophys. Acta 1210:361 (1994).
9. M.I. Siegel, R.T. McConnell, S.L. Abrahams, N.A. Porter, and P. Cuatrecasas, Regulation of arachidonate metabolism via lipoxygenase and cyclooxygenase by 12-hydroperoxyeicosatetraenoic acid, the product of human platelet lipoxygenase, Biochem. Biophys. Res. Commun. 89:1273 (1979).
10. D. Riendeau, J. Guay, P.K. Weech, F. Laliberte, J. Yergey, c. Li, S. Desmarais, H.Perrier, S. Liu, D. Nicoll-Griffith, and I.P. Street, Arachidonyl trifluoromethyl ketone, a potent inhibitor of 85-kDa phopholipase A$_2$ blocks production of arachidonate and 12- hydroxyeicosatetraenoic acid by calcium ionophore-challenged platelets, J. Biol. Chem. 269:15619 (1994).
11. J.E. Bleasdale, G.L. Bundy, S. Bunting, F.A. Fitzpatrick, R.M. Huff, F.F. Sun, and J.E.Pike, Inhibition of posphilipase C dependent processes by U-73,122, in: Advances in Prostaglandin, Thromboxane, and Leukotriene Research, Vol. 19, B. Samuelsson et al.ed., Raven Press, Ltd., New York (1989).

5

DIVERSE FUNCTIONAL COUPLING OF PROSTANOID BIOSYNTHETIC ENZYMES IN VARIOUS CELL TYPES

Ichiro Kudo and Makoto Murakami

Department of Health Chemistry
School of Pharmaceutical Sciences
Showa University
1-5-8 Hatanodai, Shinagawa-ku
Tokyo 142, Japan

INTRODUCTION

In mammalian cells, prostaglandin (PG)-biosynthetic pathways utilizing endogenous arachidonic acid (AA) are subdivided into three distinct phases, which show different kinetics and may recruit different sets of biosynthetic enzymes. The *constitutive immediate response*, which occurs within several minutes after stimuli causing a rapid and transient increase in cytoplasmic Ca^{2+}, is regulated by post-translational activation of constitutively expressed enzymes. The *delayed response*, which proceeds gradually for several hours after proinflammatory stimuli, requires *de novo* synthesis of particular biosynthetic enzymes. The *induced immediate response*, which is elicted by Ca^{2+}-mobilizing stimuli after priming by proinflammatory stimuli, reflects the combination of the above two responses, and involves both constitutive and inducible enzymes. Here we introduce the differential coupling of the initial [phospholipase A_2 (PLA_2)], intermediate [cyclooxygenase (COX)], and terminal [terminal PG synthase (tPGS)] biosynthetic enzymes in each phase using rat peritoneal macrophages, mouse osteoblasts and other cell types as model systems.

DIFFERENTIAL TXA$_2$ AND PGE$_2$ BIOSYNTHESIS IN RAT PERITONEAL MACROPHAGES

We examined the functional linkage of enzymes regulating the initial, intermediate and terminal steps of PG synthesis in rat peritoneal macrophages treated with lipopolysaccharide (LPS) and/or A23187 (Figure 1A).[1,2] Quiescent cells stimulated with A23187 produced thromboxane (TX) A_2 in marked preference to PGE_2 within 30 min (*constitutive immediate response*). Pharmacological studies revealed that this immediate response was mediated by preexisting cytosolic PLA_2 ($cPLA_2$), COX-1, and TX synthase (TXS). Although type IIA $sPLA_2$ ($sPLA_2$-IIA)

Eicosanoids and Other Bioactive Lipids in Cancer, Inflammation, and Radiation Injury, 4
Edited by Honn *et al.*, Kluwer Academic / Plenum Publishers, New York, 1999.

29

was abundantly expressed in rat macrophages, it was not utilized in A23187-induced TXA$_2$ generation. Other sPLA$_2$ isozymes were minimally expressed in rat peritoneal macropahges.

Cells treated with LPS predominantly produced PGE$_2$ during culture for 3-24 h with no appreciable TXA$_2$ generation (*delayed response*) (Figure 1B). Delayed PGE$_2$ generation was accompanied by increased expression of cPLA$_2$ and *de novo* induction of COX-2. In the delayed response, cPLA$_2$ and sPLA$_2$-IIA functioned cooperatively with inducible COX-2, which was in turn coupled with PGE$_2$ synthase (PGES). Measurement of the conversion of exogenous PGH$_2$ to each prostanoid in cell lysates revealed a LPS-dependent increase in PGES activity that was reduced by pretreatment with actinomycin D or cycloheximide, implying *de novo* induction of this critical terminal enzyme. COX-1 was non-functional in the delayed response, even though it was constitutively expressed in the cell.

The preferred coupling of the two inducible enzymes, COX-2 and PGES, was confirmed by the ability of LPS-treated cells to convert exogenous AA to PGE$_2$ optimally when both enzymes were simultaneously induced. Furthermore, cells primed for 12 h with LPS and then stimulated for 30 min with A23187 produced PGE$_2$ in marked preference to TXB$_2$ from endogenous AA (*priming response* or *induced immediate response*) (Figure 1C). In this response, induced cPLA$_2$, COX-2 and PGES were functionally linked.

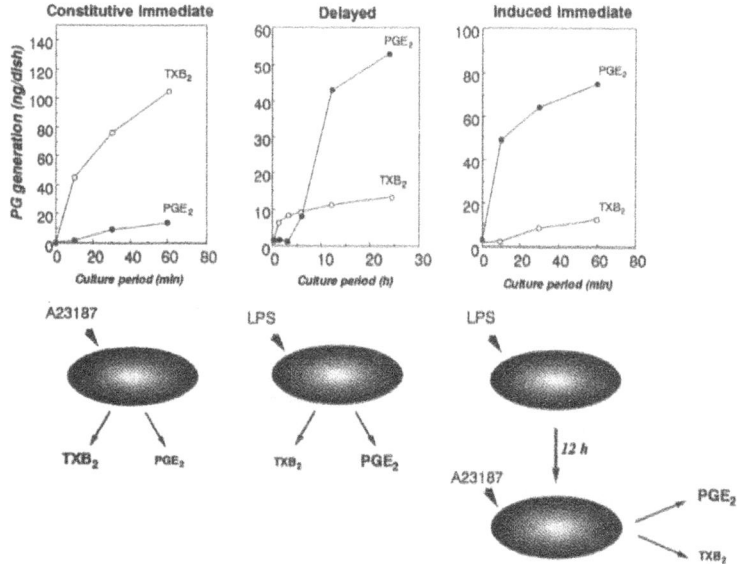

Figure 1. Three phases of the PG-biosynthetic pathway in rat peritoneal macrophages. (A) A23187 stimulation of cells induced rapid TX generation in marked preference to PGE$_2$ (*constitutive immediate response*). (B) LPS-stimulated cells produced PGE$_2$ predominantly over 3-24 h (*delayed response*). (C) Cells primed for 12 h with LPS and then treated with A23187 rapidly produced PGE$_2$ in preference to TX (*induced immediate response*).

Taken together, these results suggest that distinct PG-biosynthetic enzymes display segregated functional coupling following different transmembrane stimuli even when enzymes that catalyze similar reactions *in vitro* coexist in the same cell (Figure 2).

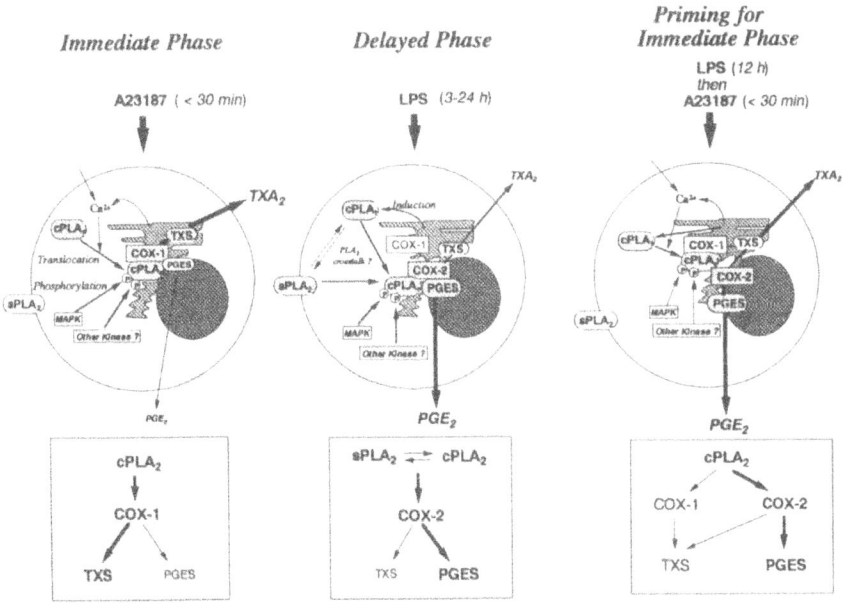

Figure 2. Schematic model of the PG-biosynthetic pathway in rat peritoneal macrophages. The A23187-induced constitutive immediate response is mediated by cPLA₂/COX-1/TXS, the LPS-initiated delayed response by cPLA₂ and sPLA₂/COX-2/PGES, and the LPS-primed, A23187-induced immediate response by cPLA₂/COX-2/PGES.

PGE₂ BIOSYNTHESIS IN MOUSE OSTEOBLASTS

We used the osteoblastic cell line MC3T3-E1, which originates from a C57BL/6J mouse that is genetically sPLA₂-IIA deficient,[3] to study the sPLA₂-IIA-independent route of the PG biosynthetic pathway. Kinetic and pharmacological studies showed that *delayed* PGE₂ generation by this cell line in response to IL-1β and TNFα was dependent upon cPLA₂ and COX-2, both of which were inducible.[4] Induced expression of these two enzymes was reduced by cPLA₂ or COX-2 inhibitors and was restored by adding exogenous AA or PGE₂, indicating that PGE₂ produced by these cells acted as an autocrine amplifier of delayed PGE$_2$ generation through enhanced cPLA$_2$ and COX-2 expression.

Exogenous addition or enforced expression of sPLA$_2$-IIA significantly increased IL-1β/TNFα-initiated PGE$_2$ generation in MC3T3-E1 cells. PGE$_2$ production augmented by sPLA$_2$-IIA introduction was accompanied by increased expression of both cPLA$_2$ and COX-2 and was suppressed by inhibitors of these enzymes. These results revealed specific crosstalk between the two PLA$_2$ enzymes and COX-2 during delayed PGE$_2$ biosynthesis by a sPLA$_2$-IIA-deficient cell line, in which cPLA$_2$ is responsible for initiating COX-2-dependent delayed PGE$_2$ generation and sPLA$_2$-IIA, if introduced, enhances PGE$_2$ generation by increasing cPLA$_2$ and COX-2 expression via endogenous PGE$_2$ (Figure 3).

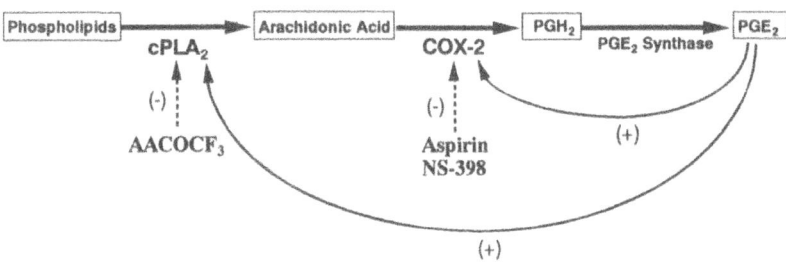

Figure 3. PGE$_2$ biosynthesis in MC3T3-E1 osteoblastic cells. Cytokine stimulation causes the induction of cPLA$_2$ and COX-2 expression. PGE$_2$ positively regulates cPLA$_2$ and COX-2 expression, thereby revealing an autocrine amplification loop via the end product PGE$_2$.

In contrast to MC3T3-E1 cells, Ddy mouse primary calvaria osteoblasts did not display delayed PGE$_2$ generation in response to IL-1α, even though there was marked induction of COX-2.[5] Exogenously added AA was readily converted into PGE$_2$ via COX-2 in IL-1α-primed, but not control, cells, indicating that the cellular AA-releasing step is rate-limiting in this situation. However, cPLA$_2$ is responsible for the initiation of the induced immediate response, which is elicited by the secondary Ca^{2+}-inducing agents. Thus, addition of platelet-derived growth factor (PDGF) to IL-1α-primed cells markedly increased the phosphorylation state of cPLA$_2$, which then released AA that was rapidly metabolized to PGE$_2$ via COX-2. Thus, this *priming response* exemplifies the two-step regulation of PGE$_2$ synthesis; COX-2 induction by IL-1α and cPLA$_2$ activation by PDGF, which mobilizes intracellular Ca^{2+} and activates the MAP kinase pathway.

SUMMARY

As also detailed in our accompanying papers in this issue, recent studies have revealed functional crosstalk and segregation between PLA$_2$s, COXs, and terminal PG synthases in various cells (Table I).[1-14] Among the PLA$_2$s, cPLA$_2$ is required for all three responses, and sPLA$_2$-IIA augments the delayed response in preference to the immediate response. sPLA$_2$-IIA associates with proteoglycans on the surface of stimulus-primed cells to exert its functions. COX-1 is utilized only in the immediate response and COX-2 is a prerequisite for the delayed response. The induced immediate response is often mediated by COX-2 rather than by COX-1, especially when the end product is PGE$_2$. In addition to segregated utilization of these enzymes, significant crosstalk and/or synergism between them, which is often cell type specific, is also obvious. For instance, sPLA$_2$ acts as an enhancer of COX-2 expression in rat mast cells,[6] functional cPLA$_2$ is required for sPLA$_2$ induction in rat fibroblasts,[7] and sPLA$_2$ augments cPLA$_2$ and COX-2 expression in mouse osteoblasts via endogenous PGE$_2$.[4] Moreover, differential coupling between COXs and downstream terminal PG synthases is also evident in macrophages, in which COX-1 and COX-2 are preferentially coupled with TXS and PGES, respectively.[1] Thus, different PG-biosynthetic enzymes, acting on different cellular AA pools at different locations and being regulated by separate but interacting mechanisms, confer on the system great versatility in ensuring that both immediate and delayed AA-derived mediators are efficiently generated during cellular responses.

ACKNOWLEDGMENTS

We thank Drs. S. Oh-ishi (Kitasato University, Tokyo, Japan) and T. Suda (Showa University, Tokyo, Japan) for their helpful suggestions.

Table 1. Functional coupling of PG-biosynthetic enzymes in various cells.

Cells	Stimuli	PLA$_2$	COX	tPGS	Refs.
A. *Constitutive immediate response*					
Rat macrophages	A23187	cPLA$_2$	COX-1	TXS	1
Mouse BMMC[1]	IgE/Ag, KL	cPLA$_2$	COX-1	hPGDS	8, 9
Rat CTMC[2]	IgE/Ag, KL, NGF[6]	cPLA$_2$	COX-1	hPGDS	6
MMC-34[3]	IgE/Ag	cPLA$_2$ + sPLA$_2$-V	COX-1	hPGDS	10
Rat fibroblasts[4]	A23187	cPLA$_2$	COX-1	PGES	7
Human Platelets	A23187, throbmin	cPLA$_2$	COX-1	TXS	11
HUVEC[5]	thrombin	cPLA$_2$	COX-1	PGIS	12

Cells	Stimuli	PLA₂	COX	tPGS	Refs.
B. *Delayed response*					
Rat macrophages	LPS	cPLA₂ + sPLA₂-IIA	COX-2	PGES	1
Mouse BMMC	IgE/Ag, KL + IL-1, IL-10	cPLA₂ + sPLA₂(IIA?)⁷	COX-2	hPGDS	8, 9, 13
Rat CTMC	NGF	cPLA₂ + sPLA₂-IIA	COX-2	hPGDS	6
MMC-34	IgE/Ag	cPLA₂	COX-2	hPGDS	10
Rat fibroblasts	IL-1, TNF	cPLA₂ + sPLA₂-IIA	COX-2	PGES	7
Mouse osteoblasts	IL-1, TNF	cPLA₂	COX-2	PGES	4
HUVEC	TNF	cPLA₂ + sPLA₂(IIA?)⁷	COX-2	PGIS	12
C. *Induced immediate (Priming) response*					
Rat macrophages	LPS then A23187	cPLA₂	COX-2	PGES	1
Mouse BMMC	KL then IgE/Ag	cPLA₂	COX-1	hPGDS	14
Mouse osteoblasts	IL-1 then PDGF	cPLA₂	COX-2	PGES	5

[1] BALB/cJ-derived mouse IL-3-dependent bone marrow-derived mast cells.

[2] Connective tissue mast cells.

[3] Mouse IL-3-independent mast cell line.

[4] 3Y1 cells.

[5] Human umbilical vein endothelial cells.

[6] Nerve growth factor.

[7] Sensitive to sPLA₂-IIA inhibitors, but the precise sPLA₂ isozyme involved is uncertain.

REFERENCES

1. H. Naraba, M. Murakami, H. Matsumoto, S. Shimbara, A. Ueno, I. Kudo, and S. Oh-ishi. Segregated coupling of phospholipases A₂, cyclooxygenases, and terminal prostanoid synthases in different phases of prostanoid biosynthesis in rat peritoneal macrophages. *J. Immunol.* 160: 2974 (1998)

2. H. Matsumoto, H. Naraba, M. Murakami, I. Kudo, I., K. Yamaki, A. Ueno, and S. Oh-ishi. Concordant induction of prostaglandin E₂ synthase with cyclooxygenase-2 leads to preferred production of prostaglandin E₂ to thromboxane and prostaglandin D₂ in lipopolysaccharide-stimulated rat peritoneal macrophages. *Biochem. Biophys. Res. Commun.* 230: 110 (1997)

3. M. MacPhee, K.P. Chepenik, R. Liddell, K.K. Nelson, L.D. Siracusa, and A.M. Buchberg. The secretory phospholipase A₂ gene is a candidate for the *Mom1* locus, a major modifier of *Apc^min*-induced intestinal neoplasia. *Cell* 81: 957 (1995)

4. M. Murakami, H. Kuwata, H., Y. Amakasu, S. Shimbara, Y. Nakatani, G. Atsumi, and I. Kudo. Prostaglandin E₂ amplifies cytosolic phospholipase A₂ and cyclooxygenase-2-dependent delayed prostaglandin E₂ generation in mouse osteoblastic cells: enhancement by secretory phospholipase A₂. *J. Biol. Chem.* 272: 19891 (1997)

5. Q-R. Chen, C. Miyaura, S. Higashi, M. Murakami, I. Kudo, S. Saito, T. Hiraide, Y. Shibasaki, and T. Suda. Activation of cytosolic phospholipase A₂ by platelet-derived growth factor is essential for cyclooxygenase-2-

dependent prostaglandin E_2 synthesis in mouse osteoblasts cultured with interleukin-1. *J. Biol. Chem.* 272: 5952 (1997)

6. M. Murakami, K. Tada, K. Nakajima, and I. Kudo. Cyclooxygenase-2-dependent delayed prostaglandin D_2 generation is initiated by nerve growth factor in rat peritoneal mast cells: its augmentation by extracellular type II secretory phospholipase A_2. *J. Immunol.* 159: 439 (1997)

7. H. Kuwata, Y. Nakatani, M. Murakami, and I. Kudo. Cytosolic phospholipase A_2 is required for cytokine-induced expression of type IIA secretory phospholipase A_2 that mediates optimal cyclooxygenase-2-dependent delayed prostaglandin E_2 generation in rat 3Y1 fibroblasts. *J. Biol. Chem.* 273: 1733 (1998)

8. M. Murakami, R. Matsumoto, K. F. Austen, and J. P. Arm. Prostaglandin endoperoxide synthase-1 and -2 couple to different transmembrane stimuli to generate prostaglandin D_2 in mouse bone marrow-derived mast cells. *J. Biol. Chem.* 269: 22269 (1994)

9. M. Murakami, C. O. Bingham, R. Matsumoto, K. F. Austen, and J. P. Arm. IgE-dependent activation of cytokine-primed mouse cultured mast cells induces a delayed phase of prostaglandin D_2 generation via prostaglandin endoperoxide synthase-2. *J. Immunol.* 155: 4445 (1995)

10. S. T. Reddy, M. V. Winstead, J. A. Tischfield, and H. R. Herschman. Analysis of the secretory phospholipase A_2 that mediates prostaglandin production in mast cells. *J. Biol. Chem.* 272: 13591 (1997)

11. F. Bartoli, H.-K. Lin, F. Ghomashchi, M. H. Gelb, M. K., Jain, and R. Apitz-Castro. Tight binding inhibitors of 85-kDa phospholipase A_2 but not 14-kDa phospholipase A_2 inhibit release of free arachidonate in thrombin-stimulated human platelets. *J. Biol. Chem.* 269: 15625 (1994)

12. M. Murakami, I. Kudo, and K. Inoue. Molecular nature of phospholipase A_2 involved in prostaglandin I_2 synthesis in human umbilical vein endothelial cells: possible participation of cytosolic and extracellular type II phospholipases A_2. *J. Biol. Chem.* 268: 839 (1993)

13. C. O. Bingham, M. Murakami, H. Fujishima, J. E. Hunt, K. F. Austen, and J. P. Arm. A heparin-sensitive phospholipase A_2 and prostaglandin endoperoxide synthase-2 are functionally linked in the delayed phase of prostaglandin D_2 generation in mouse bone marrow-derived mast cells. *J. Biol. Chem.* 271: 25936 (1996)

14. M. Murakami, R. Matsumoto, Y. Urade, K.F. Austen, and J.P. Arm. c-*Kit* ligand mediates increased expression of cytosolic phospholipase A_2, prostaglandin endoperoxide synthase-1, and hematopoietic prostaglandin D_2 synthase and increased IgE-dependent prostaglandin D_2 generation in immature mouse mast cells. *J. Biol. Chem.* 270, 3239 (1995)

6

REGULATION OF PROSTAGLANDIN, LEUKOTRIENE, AND PLATELET-ACTIVATING FACTOR METABOLISM IN MAST CELLS

Makoto Murakami, Kinji Tada,
Ko-ichi Nakajima, and Ichiro Kudo

Department of Health Chemistry
School of Pharmaceutical Sciences
Showa University
1-5-8 Hatanodai, Shinagawa-ku
Tokyo 142, Japan

INTRODUCTION

There are at least two distinct populations of mast cells: connective tissue mast cells (CTMC) and mucosal mast cells (MMC). CTMC contain heparin proteoglycan, CTMC-specific proteases and large amounts of histamine, and respond to Fc_eRI-dependent activation with preferred generation of the cyclooxygenase (COX) pathway product, prostaglandin (PG) D_2. MMC contain chondroitin sulfate proteoglycan, MMC-specific proteases, and less histamine than CTMC, and generate leukotrime (LT) C_4 via the 5-lipoxygenase (5-LO) pathway in preference to PGD_2 following Fc_eRI-dependent activation. Mouse bone marrow-derived mast cells (BMMC) developed in interleukin (IL)-3, a progenitor population of mast cells, resemble MMC in terms of their granule contents and preferred Fc_eRI-dependent LTC_4 generation, but express mast cell proteases different from those expressed in CTMC and MMC. Coculture of BMMC with 3T3 fibroblasts results in morphological and functional development toward a more mature CTMC-like phenotype. Many of these alterations are supported by the stromal cytokine, c-kit ligand (KL). Another fibroblast-derived cytokine, nerve growth factor (NGF), and several hematopoietic cytokines, such as IL-3, IL-4, IL-9 and IL-10, either alone or in combination, modulate the proliferation and maturation of mast cells in vitro. Recent advances in molecular biology as well as culture techniques have enabled us to identify the regulatory mechanisms of the lipid mediator metabolism in mast cells. In this paper, we describe the regulation of the biphasic PGD_2 biosynthetic pathways in activated mast cells, developmentally segregated regulation of the PGD_2 and LTC_4 pathways in lineage-related committed mast cell progenitors, and an intriguing anti-inflammatory action of mast cells, degradation of platelet-activating factor (PAF).

Eicosanoids and Other Bioactive Lipids in Cancer, Inflammation, and Radiation Injury, 4
Edited by Honn *et al.*, Kluwer Academic / Plenum Publishers, New York, 1999.

37

BIPHASIC PGD₂ BIOSYNTHESIS IN MOUSE CULTURED MAST CELLS

When BALB/cJ mouse BMMC were activated with IgE and antigen (IgE/Ag) or KL in the presence of the two accessory cytokines IL-10 and IL-1b, two sequential phases of PGD₂ generation were elicited; the first phase occurred by 1 h and the second phase occurred from 2 to 10 h after BMMC activation.[1,2] In the absence of these accessory cytokines, only the immediate phase of PGD₂ generation, which was accompanied by LTC₄ synthesis and the secretion of pre-formed mediators stored in granules, was elicited. The delayed phase of PGD₂ generation was accompanied by the de novo induction of COX-2 expression. The immediate phase of PGD₂ generation was completely abrogated by the irreversible inhibition of pre-exisiting COX-1 by aspirin pretreatment, whereas the delayed phase of PGD₂ generation was almost undetectable in the presence of the COX-2 inhibitor NS-398, indicating that the immediate and delayed phases of PGD₂ generation are dependent upon COX-1 and COX-2, respectively. IgE/Ag, KL and IL-10 each initiated and stabilized COX-2 mRNA expression, leading to an increase in the expression of its protein, whereas IL-1b contributed to stabilizing the COX-2 protein without affecting its mRNA level.

Induction of COX-2 in BMMC derived from BALB/cJ mice was accompanied by the increased expression of type IIA secretory phospholipase A₂ (sPLA₂-IIA), but not cytosolic PLA₂ (cPLA₂).[3] Delayed PGD₂ generation was suppressed markedly by an antibody raised against recombinant murine sPLA₂-IIA or by treatment with heparin, which prevents sPLA₂-IIA association with the cell surface. Dexamethasone, which attenuated the induction of sPLA₂-IIA and COX-2 expression markedly, suppressed the delayed phase of arachidonic acid (AA) release and PGD₂ generation. These results support the hypothesis that sPLA₂-IIA might be a critical enzyme that is coupled with COX-2-dependent PGD₂ generation.

However, BMMC derived from C57BL/6J mice, which are genetically deficient in sPLA₂-IIA, display normal PGD₂ generation.[4,5] Lysates of C57BL/6J-derived BMMC contained a sPLA₂-like activity, which exhibited column chromatographic profiles similar to those of sPLA₂-IIA. These results imply that a sPLA₂ similar to, but distinct from, sPLA₂-IIA is present in BMMC and compensates for the sPLA₂-IIA deficiency in C57BL/6J-derived BMMC. RT-PCR analysis revealed that the sPLA₂-IIA-related enzyme, sPLA₂-V, was also expressed in BMMC derived from BALB/cJ or C57BL/6J and RBL-2H3 cells (Figure 1). The transcript for another sPLA₂ subtype, sPLA₂-IIC, was also detected in BALB/cJ BMMC and RBL-2H3 cells, but not in C57BL/6J BMMC.

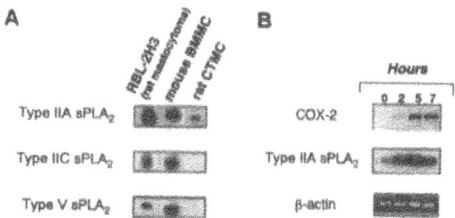

Figure 1. (A) Expression of sPLA₂ isozymes in BALB/cJ mouse BMMC, rat mastocytoma RBL-2H3, and rat serosal CTMC. (B) Coordinated induction of sPLA₂-IIA and COX-2 in BALB/cJ BMMC after stimulation for the indicated periods with the cytokine triad, KL + IL-1 + IL-10. Expression of sPLA₂s and COX-2 was assessed by RT-PCR/Southern blotting and immunoblotting, respectively. β-Actin was used as a control, the expression of which did not change during the culture period.

Both the immediate and delayed phases of PGD_2 generation are suppressed efficiently by $cPLA_2$ inhibitors,[5] demonstrating the essential role of $cPLA_2$ in both phases. In the immediate response, $cPLA_2$ is transiently phosphorylated by MAP kinase and translocates from the cytosol to the perinuclear envelope in a cytoplasmic Ca^{2+}-dependent manner.[6] The role of $cPLA_2$ in the immediate phase of eicosanoid synthesis by mast cells is further supported by recent studies of $cPLA_2$ knock-out mice in which the acute allergic reaction was dramatically mitigated.[7] The sensitivity of delayed PGD_2 generation to both $cPLA_2$ and $sPLA_2$ inhibitors suggests that both enzymes function cooperatively in the delayed phase of PGD_2 generation.

Reddy et al.[5] showed, in studies using mouse mast cell line MMC-34, that $sPLA_2$-V is involved in the immediate, but not the delayed phase of PGD_2 generation. In contrast to our results,[4] they did not detect $sPLA_2$s-IIA and -IIC expression in their mast cells.[5] It is not clear whether this difference between our results and theirs is due to mast cell heterogeneity, different experimental systems, or differences in the $sPLA_2$ isozymes utilized. In our preliminary studies, $sPLA_2$-V was constitutively expressed, whereas $sPLA_2$-IIA was inducible, in BALB/cJ mouse BMMC. It is notable that our anti-$sPLA_2$-IIA antibody, used in our current studies, did not crossreact with $sPLA_2$-V appreciably, indicating its strict specificity to $sPLA_2$-IIA. Therefore, suppression of delayed PGD_2 generation by BALB/cJ mouse BMMC by this antibody is likely to be due to the neutralization of $sPLA_2$-IIA, and not to cross-inhibition of $sPLA_2$-V.

BIPHASIC PGD_2 BIOSYNTHESIS IN RAT SEROSAL MAST CELLS

Although biphasic PGD_2 generation in the immature mast cell population is thus being established, it remains to be elucidated whether the delayed phase of PGD_2 generation is specific to BMMC or also occurs in mature mast cell phenotypes. When rat serosal CTMC were stimulated with NGF, the immediate PGD_2 generation, which occurred within 10 min in the presence of the activation cofactor lysophosphatidylserine (lysoPS), was followed by delayed PGD_2 generation that occurred between 2 and 24 h, reaching levels as high as ~50 ng and ~300 ng/10^6 cells in the absence or presence of lysoPS, respectively.[8] This delayed PGD_2 generation was accompanied by de novo induction of COX-2, and was completely suppressed by the COX-2-inhibitor NS-398, irrespective of the coexistence of constitutive COX-1. Although the expression of $cPLA_2$ and $sPLA_2$-IIA did not change appreciably over 24 h of stimulation with NGF, delayed PGD_2 generation was suppressed almost completely by inhibitors of these PLA_2s, indicating that both PLA_2s are required for the COX-2-dependent PGD_2 biosynthetic pathway in rat CTMC. Unlike mouse BMMC, expression of $sPLA_2$-IIC and -V was barely detectable in rat CTMC (Figure 1), implying again that the expression profile of $sPLA_2$ family members is mast cell phenotype-specific, as is the case for mast cell-specific proteases. Notably, addition of exogenous $sPLA_2$-IIA, at concentrations comparable to those detectable in inflammatory exudates, to NGF-stimulated CTMC significantly augmented delayed, but not immediate, PGD_2 generation, and this augmentative effect was mediated by enhanced COX-2 expression as well as increased release of AA by $sPLA_2$-IIA. Furthermore, treating CTMC with the $sPLA_2$-IIA inhibitor significantly reduced NGF-induced COX-2 expression, suggesting that endogenous $sPLA_2$-IIA is involved in COX-2-induction in an autocrine manner. We conclude, therefore, that delayed PGD_2 generation in CTMC is initiated by the functional coupling of $cPLA_2$ and COX-2, and that $sPLA_2$-IIA contributes to the enhancement of this response in two ways; supplying AA and increasing COX-2 expression further (Figure 2).

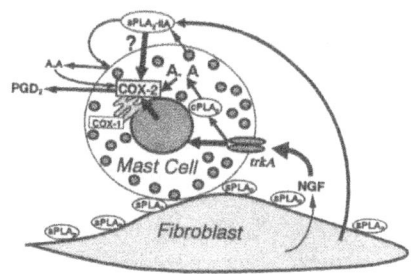

Figure 2. Regulation of delayed PGD$_2$ generation in rat serosal CTMC. CTMC are activated by fibroblast-derived NGF via its tyrosine kinase receptor, c-trkA. AA released by the action of cPLA$_2$ is supplied to the de novo induced COX-2 to be metabolized to PGD$_2$. sPLA$_2$-IIA released from CTMC contributes to supplying AA as well as enhancing COX-2 expression further. It is at present unclear whether the effect of sPLA$_2$-IIA occurs through lipid messengers or through the sPLA$_2$ receptor-dependent process.

LEUKOTRIENE C$_4$ BIOSYNTHESIS IN MAST CELLS

The formation of LTC$_4$ in mast cells is initiated by Ca^{2+}-mobilizing stimuli, such as Ca^{2+} ionophore, IgE/Ag and KL, with the release of AA by the action of cPLA$_2$.[6] The released AA then binds to an integral perinuclear membrane protein, 5-lipoxygenase-activating protein (FLAP), and is presented to 5-LO, which, like cPLA$_2$, undergoes the Ca^{2+}-dependent translocation from the cytosol to the perinuclear envelope.[9] 5-LO metabolizes AA in two sequential steps to form 5-HPETE and then LTA$_4$. The integral perinuclear protein LTC$_4$ synthase (LTCS), which belongs to the same novel gene family as FLAP, conjugates LTA$_4$ with reduced glutathione (GSH) to form the cysteinyl LT, LTC$_4$.[10] Minimal production of 5-LO metabolites occurs in the delayed response, probably because Ca^{2+} signaling, which is a prerequisite for 5-LO activation, is limited during this period.[6]

DEVELOPMENT OF THE CYCLOOXYGENASE AND 5-LIPOXYGENASE PATHWAYS DURING MATURATION OF MAST CELLS

When IL-3-dependent mouse BMMC were cultured with KL, with IL-3, IL-9 or IL-10 acting as accessory cytokines, subsequent IgE-dependent immediate production of PGD$_2$ increased several fold above the level produced by the cells maintained in IL-3 alone.[11] The increased IgE-dependent immediate PGD$_2$ generation was apparent after 1 day of culture and reached a maximum after 2-4 days. In contrast, IgE-dependent LTC$_4$ production increased only modestly, revealing that the activity of KL is specific to the COX pathway. Increased IgE-dependent immediate PGD$_2$ generation was accompanied by increased expression of cPLA$_2$, COX-1, and PGD$_2$ synthase (PGDS), without changes in the expression of 5-LO, FLAP and LTCS. IL-3, IL-9 and IL-10 each enhanced COX-1 expression further. IL-4 counter-regulated the induction of cPLA$_2$ by KL, thereby inhibiting this priming event.[12] These findings demonstrate that KL up-regulates the expression of the constitutive enzymes cPLA$_2$, COX-1 and PGDS, leading to selective development of the COX pathway.

Polycationic mast cell activators, such as compound 48/80 and substance P, activate CTMC, but not other mast cell phenotypes, by interacting directly with the G_i family of trimeric GTP-binding proteins. BMMC developed in IL-3, which lack responsiveness to these compounds, underwent maturation toward a phenotype that responded to them after only 4-6 days of coculture with fibroblasts in concert with recombinant KL.[13] These cells exhibited biphasic PGD_2 generation in response to substance P in a G_i-dependent manner, with COX-1 and COX-2 functioning in the immediate and delayed phases, respectively.

To study cytokine regulation of the development of the 5-LO pathway, BMMC that minimally expressed each protein in the pathway were established using a novel culture system, lacking IL-3.[14] BMMC developed from mouse bone marrow cells with KL + IL-10 expressed minimal $cPLA_2$, 5-LO, FLAP and LTCS and produced only a limited amount of LTC_4 in response to IgE/Ag. When these BMMC were cultured for an additional 2-5 weeks with IL-3, the IgE-dependent LTC_4 generation increased more than 20-fold and was accompanied by increased expression of $cPLA_2$, 5-LO, FLAP and LTCS. That IL-3 up-regulates the expression of each protein in the 5-LO pathway contrasts with the fact that KL up-regulates the expression of the enzymes involved in the COX pathway. This finding is compatible with the fact that IL-3 has been implicated in the in vivo development of MMC that predominantly generate LTC_4.

PLATELET-ACTIVATING FACTOR METABOLISM IN MAST CELLS

IgE-dependent and -independent activation of BMMC elicited rapid and transient production of PAF, which reached a maximal level by 2-5 min and was then degraded rapidly, returning to baseline levels by 10-20 min.[15] Inactivation of PAF was preceded by the release of PAF-acetylhydrolase (PAF-AH) activity, which plateaued after 3-5 min and paralleled the exocytosis. PAF-AH release from rat serosal CTMC was also observed. Immunochemical and molecular biological studies revealed that the PAF-AH released from activated mast cells was identical to the plasma-type isoform. In support of the autocrine action of exocytosed PAF-AH, adding exogenous recombinant plasma-type PAF-AH markedly reduced PAF accumulation in activated BMMC. Furthermore, culture of BMMC with the cytokine triad KL + IL-10 + IL-1 for more than 24 h led to an increase in plasma-type PAF-AH expression, accompanied by a reduction in stimulus-initiated PAF production. Collectively, these results suggest that plasma-type PAF-AH released from activated mast cells sequesters proinflammatory PAF produced by these cells, thereby revealing an intriguing anti-inflammatory aspect of mast cells (Figure 3).

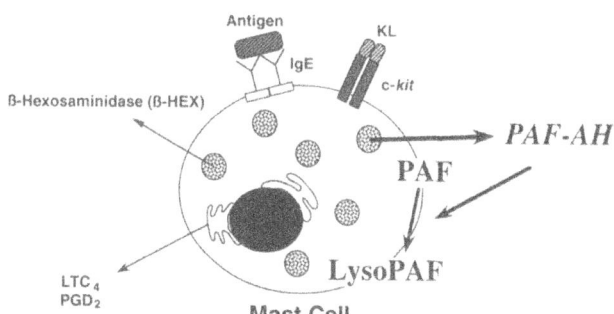

Figure 3. Regulation of PAF metabolism in mast cells. Mast cells rapidly produce PAF in response to various stimuli, such as IgE/Ag and KL. Plasma-type PAF-AH, released from activated mast cells via exocytosis, degrades membrane-associated PAF to lysoPAF. Thus, PAF accumulation in activated mast cells is only transient.

REFERENCES

1. M. Murakami, R., Matsumoto, K.F. Austen, and J.P. Arm. Prostaglandin endoperoxide synthase-1 and -2 couple to different transmembrane stimuli to generate prostaglandin D_2 in mouse bone marrow-derived mast cells. *J. Biol. Chem.* 269:22269 (1994)

2. M. Murakami, C.O. Binham, R. Matsumoto, K.F. Austen, and J.P. Arm. IgE-dependent activation of cytokine-primed mouse cultured mast cells induces a delayed phase of prostaglandin D_2,generation via prostaglandin endoperoxide synthase-2. *J. Immunol.* 155:4445 (1995)

3. C.O. Bingham, M. Murakami, H. Fujishima, J.E. Hunt, K.F. Austen, and J.P. Arm. A heparin-sensitive phospholipase A_2 and prostaglandin endoperoxide synthase-2 are functionally linked in the delayed phase of prostaglandin D_2 generation in mouse bone marrow-derived mast cells. *J. Biol. Chem.* 271:25936 (1996)

4. M. Murakami, K. Tada, S. Shimbara, T. Kambe, H. Sawada and I.Kudo. Detection of secretory phospholipase A_2s related but not identical to type IIA isozyme in culture mast cells. *FEBS Lett.* 413:249 (1997)

5. S.T. Reddy, M.V. Winstead, J.A. Tischfield, and H.R. Herschman. Analysis of the secretory phospholipase A_2 that mediates prostaglandin production in mast cells. *J. Biol. Chem.* 272:13591 (1997)

6. M. Murakami, F.K., Austen, K.F., and J.P. Arm. The immediate phase of c-*kit* ligand stimulation of mouse bone marrow-derived mast cells elicits rapid leukotriene C_4 generation through posttranslational activation of cytosolic phospholipase A_2 and 5-lipoxygenase. *J. Exp. Med.* 182:197 (1995)

7. N. Uozume, K. Kume, T. Nagase, N. Nakatanim S. Ishii, F. Tashiro, Y. Komagata, K. Maki, K. Ikuta, Y.Ouchi, J. Miyazaki, and T. Shimizu. Role of cytosolic phospholipase A_2 in allergic response and parturition. *Nature* 390: 618 (1997)

8. M. Murakami, K. Tada, K. Nakajima, and I. Kudo. Cyclooxygenase-2-dependent delayed prostaglandin D_2 generation is initiated by nerve growth factor in rat peritoneal mast cells: its augmentation by extracellular type II secretory phopholipase A_2. *J. Immuno.* 159:439 (1997)

9. R. Malaviya, R. Malaviya, and B.A. Jakschik. Reversible translocation of 5-lipoxygenase in mast cells upon IgE/antigen stimulation. *J. Biol. Chem.* 268:4939 (1993)

10. B.K. Lam, J.F. Penrose, K.Xu, M.H. Baldasaro, and K.F. Austen. Site-directed muragenesis of human leukotriene C_4 synthase. *J. Biol. Chem.* 268: 4939 (1993)

11. M. Murakami, R. Matsumoto, Y. Urade, K.F. Austen, and J.P. Arm. c-*Kit* ligand mediates increased expression of cytosolic phopholipase A_2, prostaglandin endoperoxide synthase-1, and hematopoietic prostaglandin D_2 synthase and increased IgE-dependent prostaglandin D_2 generation in immature mouse mast cells. *J. Biol. Chem.* 270, 3239 (1995)

12. M. Murakami, J.F. Penrose, Y. Urade, K.F. Austen, and J.P. Arm. Interleukin-4 suppresses c-*kit* Ligand-induced expression of cytosolic phospholipase A_2 and prostaglandin endoperoxide synthase-2 and their roles in separate pathways of eicosanoid synthesis in mouse bone marrow-derived mast cells. *Proc. Natl. Acad. Sci. USA* 92: 6107 (1995)

13. T. Ogasawara, M. Murakami, T. Suzuki-Nishimura, U.K. Uchida, and I. Kudo. Mouse bone marrow-derived mast cells undergo exocytosis, prostanoid generation and cytokine expression in response to G protein-activating polybasic compounds after coculture with fibroblasts in the presence of c-*kit* ligand. *J. Immunol.* 158:393 (1997)

14. M.Murakami, K.F. Austen, C.O. Bingham, D.S. Friend, J.F.Penrose, and J.P. Arm. Interleukin-3 regulates development of the 5-lipoxygenase/leukotriene C_4 pathway in mouse mast cells. *J. Biol. Chem.* 270:22653 (1995)

15. K. Nakajima, M. Murakami, R. Yanoshita, Y. Samejima, K. Karasawa, M. Setaka, S. Nojima, and I. Kudo. Activated mast cells release extracellular type platelet-activating factor acetylhydrolase that contributes to autocrine inactivation of platelet-activating factor. *J. Biol. Chem.* 272:19708 (1997)

7

LIPID PEROXIDES AND NEURONAL PLASTICITY

Makoto Nishiyama[1], Nobuaki Hori[2], Takashi Watanabe[1], Tomokatsu Hori[1], Katsuhiko Ogata[3], Keiichi Watanabe[3], Eiichi Maru[4], and Takao Shimizu[5]

[1]Division of Neurosurgery, Tottori University, Faculty of Medicine, Yonago 683
[2]Department of Pharmacology, Kyushu University, Faculty of Dentistry, Fukuoka 812
[3]Department of Pathology, Tokai University, School of Medicine, Isehara 257-11
[4]Department of Physiology, Nippon Medical School, Tokyo 113
[5]Department of Biochemistry, Faculty of Medicine, University of Tokyo, Tokyo 113

INTRODUCTION

Long-term potentiation (LTP) of synaptic transmission is a well defined form of neuronal plasticity. The induction of LTP at perforant path-dentate granule cell and Schaffer collateral-CA1 pyramidal cell synapses of the hippocampus is known to require influx of Ca^{2+} through postsynaptic N-methyl-D-aspartate (NMDA) receptors, while the expression of LTP has been proposed to be mediated at least in part through presynaptic[1]. Epileptic seizures which are associated with massive increases of intracellular cation concentration may also account for the persistent increases in efficacy of excitatory synaptic transmission in the hippocampus[2]. This seizure-induced synaptic potentiation (SIP) and LTP have some features in common such as the requirement of NMDA receptor activation[1, 3] and the consequent increase in neurotransmitter release[4, 5]. These findings led to the concept of a retrograde messenger by which the presynaptic terminals are informed that the postsynaptic sites have been activated. Arachidonic acid (AA) has been recognized as the potential candidate for a retrograde messenger[1].

Here, we have addressed the involvement of lipid peroxides, especially lipoxygenase metabolites, in neuronal plasticity.

[³H]12-HPETE BINDING ASSAY

Membrane fraction of intact hippocampal formations were prepared from three mongrel dogs, as described previously[6, 7]. [³H]-labeled and unlabeled 12S-hydro(pero)xy-eicosatetraenoic acid (12S-H(P)ETE) were also prepared as described previously[8]. The standard assay mixture (70 μl) containing [³H]12-HPETE (2.3 nM, 50,000 dpm) in the presence or absence of 10 μM nonradiolabeled 12-HPETE and membrane fractions (54 μg of protein) in 50 mM Tris-HCl (pH 7.3), 10 mM $MgCl_2$ were incubated at 4°C for 30 min.

Eicosanoids and Other Bioactive Lipids in Cancer, Inflammation, and Radiation Injury, 4
Edited by Honn *et al.*, Kluwer Academic / Plenum Publishers, New York, 1999.

43

HPETE binding. The specific binding activity was 118.4±19.7 fmol/ mg protein (mean of the late-phase peak±SD, n=5). The late phase of [³H]12-HPETE binding, but not the early phase, was eliminated by the pretreatment with 1 μM GTPγS. However, both phases of [³H]12-HPETE binding were abolished in the Mg^{2+}-free reaction mixture, suggesting that both binding phases require energy. Contrarily this membrane fraction did not show any significant specific binding to [³H]12-HETE.

These results suggested that the membrane fraction of canine hippocampus possessed at least two types of [³H]12-HPETE binding sites. Since the part of specific [³H]12-HPETE binding was sensitive to GTPγS, it may occur at GTP binding protein-coupled receptors.

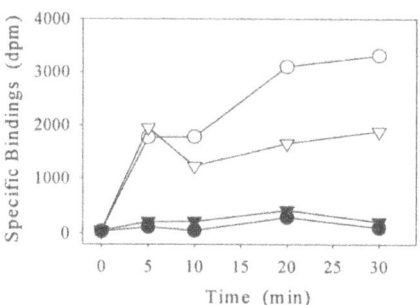

Figure 1. Membrane fraction of the canine hippocampus showed the specific binding to [³H]12-HPETE (○) but not to [³H]12-HETE (●). [³H]12-HPETE binding to membrane fraction in 1 μM GTPγS pretreatment (▽) and Mg^{2+} free (▼) conditions. Plots show the means of quadruplicate samples.

MEMBRANE TRANSLOCATION OF 12-LIPOXYGENASE ASSOCIATED WITH THE EXPRESSION OF SEIZURE-INDUCED POTENTIATION

SIP at the perforant path-dentate granule cell synapse in canine hippocampus, following tetanic stimulation (400 Hz, 1 mA, 50 μs for 1 sec; 12-15 trains every 2 sec), was examined *in vivo*. SIP was successfully induced in 9 out of 11 cases, producing a significant potentiation of the field excitatory postsynaptic potential (fEPSP) (277±120%, mean±SD). Tetanic stimulation induced hippocampal seizure in all performed cases, but failed to induce long-lasting synaptic potentiation in 2 out of 11 cases.

Immunostaining with immunoaffinity-purified polyclonal antibodies against 12-lipoxygenase (12-LOX)[9] was performed in dentate granule cells from control animals in which tetanic stimulation failed to induce SIP and successfully SIP-induced granule cells (3 hours following tetanic stimulation). In control animals where tetanic stimulation failed to induce SIP, uniform 12-LOX immunofluorescence was observed throughout the cytoplasm of granule cells which was similar to unstimulated control animals (Fig. 2A)[9]. In SIP-induced animals, the distribution of 12-LOX immunofluorescence was drastically changed from a uniform to a punctuated pattern 3 hours after tetanic stimulation (Fig. 2B).

Like phospholipases C[10] and A_2[11], 12-LOX translocates from the cytosol to the plasma membrane, following the raise in $[Ca^{2+}]_i$[12, 13]. In the present study, 12-LOX also translocates to the plasma membrane in association with the expression of SIP. This translocation may be a critical step to increase its enzyme activity, since an inhibitory activity against 12-LOX is present in the cytoplasm[14)] and arachidonic acid (AA) and other poly-unsaturated fatty acids are exclusively located in cellular membrane.

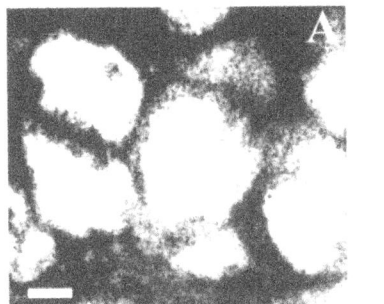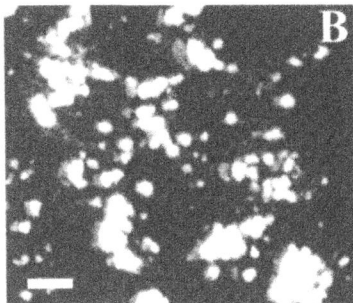

Figure 2. Membrane translocation of arachidonate 12-lipoxygenase (12-LOX) was induced in dentate granule cells, associated with seizure-induced synaptic potentiation (SIP). (A) Seizure-induced dentate granule cells not associated with a synaptic potentiation. 12-LOX was diffusely localized in the cytoplasm of cell bodies and apical dendrites. (B) SIP-induced dentate granule cells (3 hours following tetanic stimulation). 12-LOX translocated to the membrane, displaying as a finely particulated pattern. Bars: (A) and (B)= 5 μm.

PRODUCTION OF EICOSANOIDS DURING THE INDUCTION OF LTP IN RAT HIPPOCAMPAL CA1.

In rat hippocampal slices, LTP was induced in CA1 region by the tetanic stimulation (100 Hz, 1sec). Krebs-Ringer solution perfusing LTP-induced slices was collected every 5 min on ice. Eicosanoids were immediately extracted with diethyl ether as described previously[15)]. After evaporation of the organic solvent, the extract was analyzed by straight-phase HPLC on a silica column (Nucleosil 50-5) with the solvent system: n-hexane/ n-isopropanol/ acetic acid (99: 1: 0.01, v/v, 2 ml/ min), as the mobile phase.

As shown in Fig. 3, three peaks of conjugated dienes (retention time: 7.2, 9.2, 13.0 min) of a hundred nanomolars to one micromolar were detected in the perfusing buffer medium.

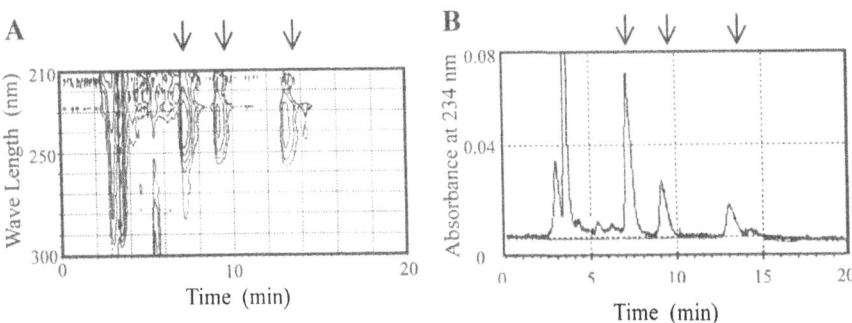

Figure 3. Straight-phase HPLC analysis of the perfusing bath medium of LTP-induced slices. Contour Image (A) and Monitoring at 234 nm (B) of the bath medium extract showing 3 peaks of conjugated dienes (arrows).

12-HPETE ENHANCED SPIKE POTENTIATION IN CA1 AREA OF THE RAT HIPPOCAMPAL SLICES.

Bath application of 100 nM 12-HPETE and 500 nM AA for 5 min caused the significant enhancement of spike potentiation (SP) induced by tetanic stimulation (100 Hz, 1sec) in CA1 area of the rat hippocampal slices (262.1±43.2% and 210.3±20.8% compared with control, 144.3±12.0%, respectively) (mean±SE, n=6, Fig. 4).

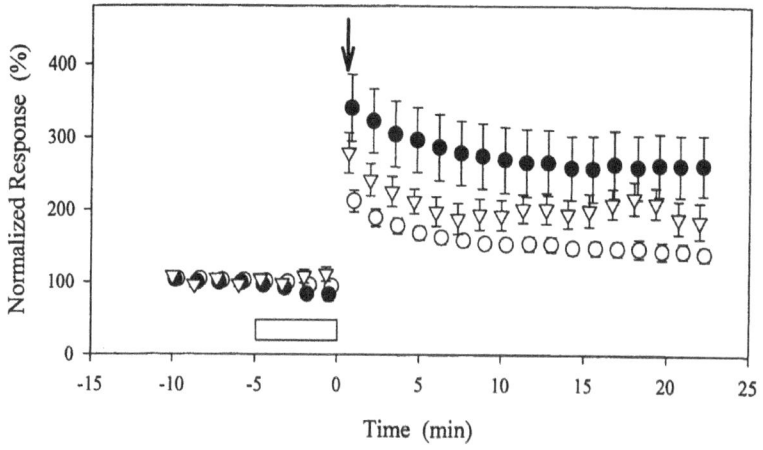

Figure 4. Spike potentiation in CA1 was enhanced by 100 nM 12-HPETE (●) and 500 nM AA (∇). Control (○). The arrow indicates the time of tetanic stimulation applied. Open bar denotes eicosanoids application.

Furthermore, 2-amino-5-phosphono-valeric acid (APV), a NMDA channel blocker, failed to inhibit the SP induced by tetanic stimulation in 12-HPETE (100 nM)-treated slices (130.7±7.7%, mean± SE, n=7, Fig. 5A). However, the tetanic stimulation induced spike depression in AA (100 nM)-treated slices in the presence of 10 μM APV (63.7±11.5%, mean±SE, n=5, Fig. 5B). Pretreatment with 10 nM 12-HPETE also caused spike depression where the tetanic stimulation was applied in the presence of APV (data not shown).
Considering that 5-10% of AA converts to 12-HPETE in the brain tissues[15], it is consistent that 12-HPETE can replace the effects of AA as shown above. 12-HPETE is recognized as a K^+ channel (S-channel) opener in the nervous system[16]. However, 100 nM 12-HPETE did not affect the impedance of postsynaptic cell membrane (unpublished data). Furthermore, 100 nM 12-HPETE did not affect a postsynaptic responsiveness to γ-aminobutylic acid ionotophoretically applied or evoked postsynaptic Cl⁻ currents but decreased that to quisqualate and NMDA (unpublished data). Thus, 12-HPETE neither showed any activating effect on the postsynaptic site nor decreased the inhibition. Since 100 nM 12-HPETE decreased the paired-pulse facilitation ratio, 12-HPETE is suggested to act on presynaptic site as a diffusible messenger, inducing the increase of neurotransmitter release.

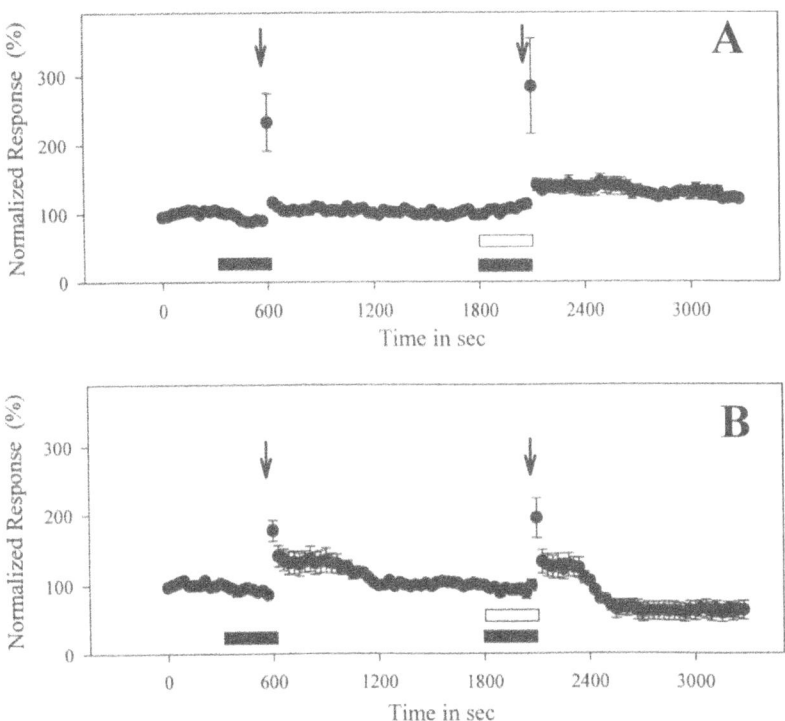

Figure 5. Spike potentiation was successfully induced by tetanic stimulation in the presence of NMDA channel blocker by 12-HPETE. Treated with 100 nM 12-HPETE (A) and 100 nM AA (B). The arrow indicates the time of tetanic stimulation applied. Open and closed bars denote applications of eicosanoids and APV, respectively.

In this study, we demonstrated that membrane translocation of 12-LOX coincided with the long-lasting synaptic potentiation induced by tetanic stimulations in canine hippocampus. We have found that 12-HPETE but not AA plays a significant role in the induction of synaptic and also neuronal plasticities which might mediate through the receptor-operated manner, since bath-application of 100 nM 12-HPETE can induce LTP even in the presence of APV[17] and glibenclamide, potent eicosanoid receptor blocker[8, 18, 19, 20], can abolish this synaptic potentiation (unpublished data).

REFERENCE

1) T.V.P. Bliss and G.L. Collingridge, A synaptic model of memory: long-term potentiation in the hippocampus, *Nature* 361:31 (1993).

2) R.M. Douglas and G.V. Goddard, Long-term potentiation of the perforant path-granule cell synapse in the rat hippocampus, *Brain Res*. 86: 205 (1975).

3) M.E. Gilbert and C.M. Mack, The NMDA antagonist, MK-801, suppresses long-term potentiation, kindling, and kindling-induced potentiation in the perforant path of the unanesthetized rat, *Brain Res*. 519: 89 (1990).

4) A.C. Dolphin, M.L. Errington, and T.V.P. Bliss, Long-term potentiation of the perforant path *in vivo* is associated with increased glutamate release, *Nature* 297: 496 (1982).

5) P.A. Jarvie, T.C. Logan, C. Geula, and J.T. Slevin, Entorhinal kindling permanently enhances Ca^{2+}-dependent L-glutamate release in regio inferior of rat hippocampus, *Brain Res*. 508: 188 (1990).

6) M. Nishiyama, Participation of arachidonate 12-lipoxygenase metabolites in the neuronal transmission, *Natl. Def. Med. J*. 42: 279 (1995).

7) T. Watanabe, T. Shimizu, I. Miki, C. Sakanaka, Z. Honda, Y. Seyama, T. Teramoto, T. Matsushima, M. Ui, and K. Kurokawa, Characterization of the guinea pig lung membrane Leukotriene D_4 receptor solubilized in an active form, *J. Biol. Chem*. 265: 21237 (1990).

8) M. Nishiyama, H. Okamoto, T. Watanabe, T. Hori, T. Sasaki, T. Kirino, and T. Shimizu, Endothelium is required for 12-hydroperoxyeicosatetraenoic acid-induced vasoconstriction, *Eur. J. Pharmacol*. 341 57 (1998).

9) M. Nishiyama, T. Watanabe, N. Ueda, H. Tsukamoto, and K. Watanabe, Arachidonate 12-lipoxygenase is localized in neurons, glial cells, and endothelial cells of the canine brain, *J. Histochem. Cytochem*. 41: 111 (1993).

10) U.H. Kim, H.S. Kim, and S.G. Rhee, Epidermal growth factor and platelet-derived growth factor promote translocation of phospholipase C-gamma from cytosol to membrane, *FEBS Lett*. 270: 33 (1990).

11) J.D. Clark, L.L. Lin, R.W. Kriz, C.S. Ramesha, L.A. Sultzman, A.Y. Lin, N. Milona, and J.L. Knopf, A novel arachidonic acid-selective cytosolic PLA_2 contains a Ca^{2+}-dependent translocation domain with homology to PKC and GAP, *Cell* 65:1043 (1991).

12) A. Baba, S. Sakuma, H. Okamoto, T. Inoue, and H. Iwata, Calcium induces membrane translocation of 12-lipoxygenase in rat platelets, *J. Biol. Chem*. 264: 15790 (1989).

13) W. Hagmann, D. Kagawa, C. Renaud, and K.V. Honn, Activity and protein distribution of 12-lipoxygenase in HEL cells: induction of membrane-association by phorbol ester TPA, modulation of activity by glutathione and 13-HPODE, and Ca^{2+}-dependent translocation to membrane, *Prostaglandins* 46: 471 (1993).

14) M. Nishiyama and T. Watanabe, Localization and characterization of arachidonate 12-lipoxygenase in brain tissues, *Adv. in Neurol. Scis*. 36: 933 (1992).

15) M. Nishiyama, H. Okamoto, T. Watanabe, T. Hori, T. Hada, N. Ueda, S. Yamamoto, H. Tsukamoto, K. Watanabe, and T. Kirino, Localization of arachidonate 12-lipoxygenase in canine brain tissues, *J. Neurochem*. 58: 1395 (1992).

16) D. Piomelli, A. Volterra, N. Dale, S.A. Siegelbaum, E.R. Kandel, J.H. Schwartz, and F. Belardetti, Lipoxygenase metabolites of arachidonic acid as second messengers for presynaptic inhibition of Aplysia sensory cells, *Nature* 328: 38 (1987).

17) M. Nishiyama, N. Hori, T. Watanabe, T. Hori, K. Suzuki, E. Maru, and T. Shimizu, 12-Hydroperoxy-eicosatetraenoic acid induced the long-term potentiation in rat hippocampal CA1, *Soc. Neurosci. Abstr*. 22: 1459 (1996).

18) T.M. Cocks, S.J. King, and J.A. Angus, Glibenclamide is a competitive antagonist of thromboxane A_2 receptor in dog coronary *in vitro*, *Br. J. Pharmacol*. 100: 375 (1990).

19) H. Zhang, N. Stockbridge, B. Weir, C. Krueger, and D. Cook, Glibenclamide relaxes vascular smooth muscle constriction produced by prostaglandin$F_{2\alpha}$, *Eur. J. Pharmacol*. 195: 27 (1991)

20) M. Nishiyama, H. Hashitani, H. Fukuta, Y. Yamamoto, and H. Suzuki, Potassium channels activated in the endothelium-dependent hyperpolarization in coronary artery of the guinea-pig, *J. Physiol. (Lond.)* 510.2: 455(1998)

8

SECRETION OF LIPOCALIN-TYPE PROSTAGLANDIN D SYNTHASE (β-TRACE) FROM HUMAN HEART TO PLASMA DURING CORONARY CIRCULATION

Yoshihiro Urade,[1,2] Yutaka Eguchi,[3] Naomi Eguchi,[2,4,5] Yoshiyuki Kijima,[6] Yasuhiko Matsu-ura,[6] Hiroshi Oda,[7] Kousuke Seiki,[7] and Osamu Hayaishi[1]

[1]Department of Molecular Behavioral Biology, Osaka Bioscience Institute
Suita, Osaka 565-0874, Japan

[2]CREST, Japan Science and Technology Corporation
c/o Osaka Bioscience Institute, Suita, Osaka 565-0874, Japan

[3]Intensive Care Unit, Shiga University of Medical Science
Otsu, Shiga 520-2121, Japan

[4]PRESTO, Japan Science and Technology Corporation
c/o Osaka Bioscience Institute, Suita, Osaka 565-0874, Japan

[5]Department of Morphological Brain Science, Faculty of Medicine,
Kyoto University, Kyoto 606-8315, Japan;

[6]Cardiovascular Division, Ishinkai Yao General Hospital
Yao, Osaka 581-0036, Japan;

[7]Central Research Institute, Maruha Corporation
Tsukuba, Ibaragi 300-4247, Japan

INTRODUCTION

Prostaglandin (PG) D_2 is actively formed in a variety of tissues and cells,[1] and is involved in many physiological events; e.g., it regulates sleep and ocular pressure, prevents platelet aggregation, and induces vasodilation and bronchoconstriction.[2,3] Two distinct types of PGD synthase (PGDS), which catalyzes the isomerization of PGH_2 to PGD_2, have been isolated and characterized:[4] one is glutathione-independent, the lipocalin-type PGDS (L-PGDS);[5] and the other is glutathione-requiring, the hematopoietic PGDS.[6] L-PGDS is responsible for biosynthesis of PGD_2 in the central nervous system and male genital organs of various mammals, and is secreted into the cerebrospinal fluid and seminal plasma, respectively, as "β-trace".[7,8] In this study, we found that mRNA for human L-PGDS was most intensely expressed in the heart among various tissues examined and that the L-PGDS-like

Eicosanoids and Other Bioactive Lipids in Cancer, Inflammation, and Radiation Injury, 4
Edited by Honn *et al.*, Kluwer Academic / Plenum Publishers, New York, 1999.

49

immunoreactivity is localized in myocardial cells, atrial endocardial cells, and the synthetic state of smooth muscle cells in the arteriosclerotic plaques. We also demonstrated that the enzyme, β-trace, is secreted into the plasma of the coronary circulation of angina patients.[9]

RESULTS AND DISCUSSION

DOMINANT EXPRESSION OF L-PGDS mRNA IN HUMAN HEART

By Northern blot analysis with poly-$(A)^+$RNA obtained from various human tissues (Figure 1), the mRNA for L-PGDS was detected most intensely in the heart, moderately in the brain, and very weakly in the placenta, lung, liver, skeletal muscle, kidney, and pancreas. The expression of the mRNA in human heart was much higher than that in any other tissues including the brain, whereas the expression was not detected in rat heart.[10] These results indicate that the gene expression of L-PGDS is regulated in a highly species-specific manner, similar to the case of DP receptor, a prostanoid receptor for PGD_2.[11-13]

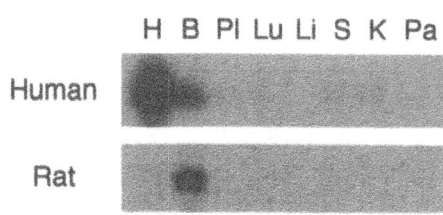

Figure 1. Tissue specificity of gene expression of L-PGDS (β–trace). Poly-$(A)^+$RNA (2 μg/lane) of various human tissues (upper lane) and total RNA (20 μg/lane) from rat tissues (lower lane) were analyzed by Northern blot assay. H, heart; B, brain; Pl, placenta; Lu, lung; Li, liver; S, skeletal muscle; K, kidney; Pa, pancreas.

CELLULAR LOCALIZATION OF L-PGDS-LIKE IMMUNOREACTIVITY IN .HUMAN HEART

The cellular localization of L-PGDS was examined by immunostaining of human autopsy specimens with monoclonal antibodies against human L-PGDS.[14] Myocardial cells were positively stained, in which the immunoreactivity decreased markedly after extensive cardiopulmonary resuscitation, suggesting that L-PGDS is actively produced in beating myocardial cells and that its intracellular concentration decreases after a decrease in contraction.

The immunoreactivity was also detected in atrial endocardial cells but not in endothelial cells of the coronary artery, although both types of cells were immunoreactive with anti-von Willebrand factor antibody. In the arteriosclerotic specimens, the L-PGDS-immunoreactivity was localized in the synthetic state of smooth muscle cells in the intimal plaques and accumulated in the region of hyaline degeneration (Figure 2). However, smooth muscle cells in the contractile state were negative for L-PGDS.

Two different monoclonal and polyclonal antibodies against human L-PGDS showed essentially identical immunostaining profiles. When IgGs obtained from non-immunized animals or the polyclonal antibodies preabsorbed with excess amounts of the purified enzyme were used instead of the primary antibody, no positive immunostaining was detected.

These results suggest that L-PGDS is a useful marker for identification of the functional or differentiation stages of myocardial, endocardial, and smooth muscle cells.

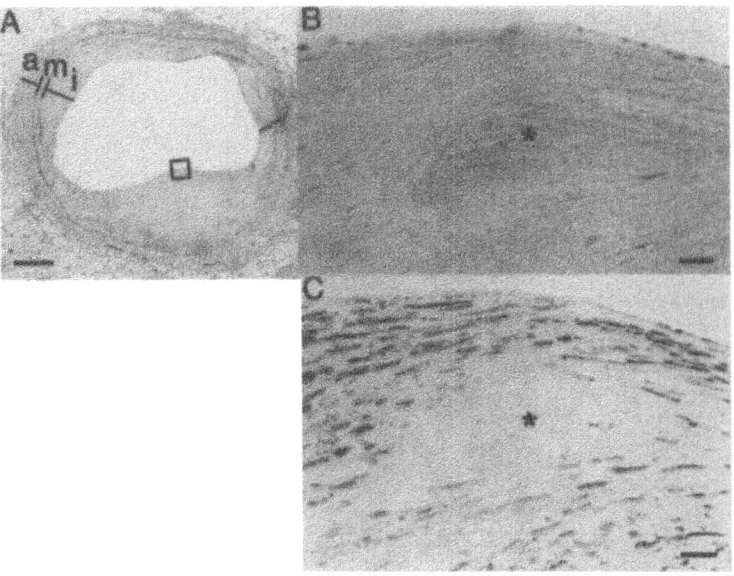

Figure 2. Immunohistochemical demonstration of L-PGDS (β-trace) in the advanced arteriosclerotic left coronary artery (50% stenosis). The serial sections were immunostained with monoclonal antibodies against L-PGDS (A, B) and alpha-smooth muscle actin (C). (B) A high-magnification view of the thickened intima (squared in A). The immunoreactivity accumulates in a region of hyaline degeneration (*). Bar = 500 μm for A and 20 μm for B and C. a, Adventitia; m, media; i, intima.

SECRETION OF L-PGDS INTO PLASMA DURING CORONARY CIRCULATION

L-PGDS in the brain, eye, and male genital organ is secreted into the cerebrospinal fluid,[7,8] interphotoreceptor matrix,[15] aqueous and vitreous humors,[16] and seminal plasma,[17,18] respectively. Since the enzyme may also be secreted from human heart into the plasma, we determined the plasma concentration of L-PGDS before and after coronary circulation. The plasma was collected from the orifice of the left coronary artery and great cardiac vein during coronary angiography for clinical diagnosis (Figure 3). The patients were classified into two groups, i.e., patients with stable angina and normal subjects. There were no statistical differences between these two groups in terms of age, past history of hypertension, diabetes mellitus, habit of smoking tobacco, and serum levels of total cholesterol, triglyceride, glutamic-oxaloacetic transaminase, lactate dehydrogenase, and creatine kinase.

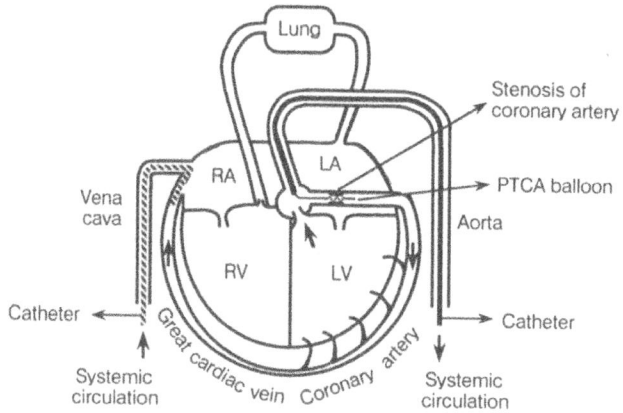

Figure 3. Schematic representation of positions of the inserted catheter used for sampling plasma.

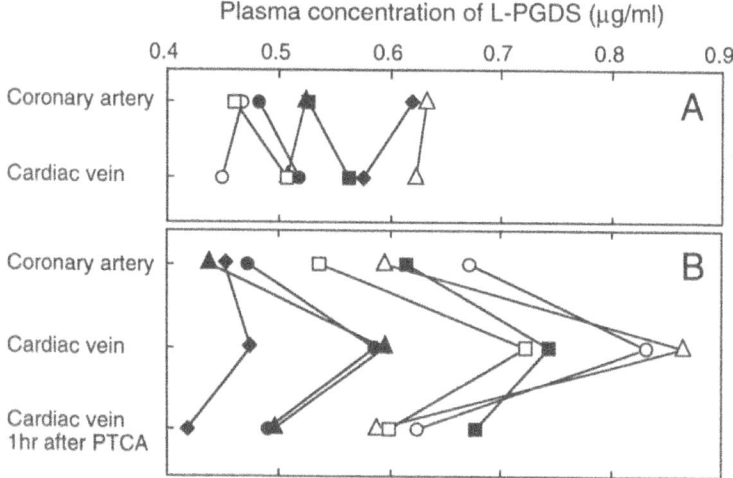

Figure 4. The plasma concentration of L-PGDS (β–trace) in the cardiac vein and coronary artery of normal subjects without stenosis (A) or of patients with stable angina (B). PTCA: percutaneous transluminal coronary angioplasty.

As shown in Figure 4, the plasma concentration of L-PGDS in the cardiac vein (0.69 ± 0.05 μg/ml) of patients with stable angina (n = 7) was significantly (p < 0.01) higher than the concentration in the coronary artery (0.55 ± 0.03 μg/ml), indicating that L-PGDS is accumulated in the plasma during the coronary circulation of these patients. In normal subjects without stenosis (n = 7), such veno-arterial difference in the plasma concentration of L-PGDS was not observed. After percutaneous transluminal coronary angioplasty of the patients, the L-PGDS level in the cardiac vein decreased significantly (p < 0.01) to 0.61 ± 0.05 μg/ml at 20 min later and reached to a level similar to the arterial level within 1 hr (0.56 ± 0.03 μg/ml).

These results indicate that occurrence of atherosclerotic plaques of the coronary artery causes the veno-arterial difference in the plasma concentration of L-PGDS. PGD_2 may function to protect against platelet aggregation in atherosclerotic blood vessels as does PGI_2, although its antiaggregatory potency is 3- to 10-fold weaker than that of PGI_2.[19]

ACKNOWLEDGMENTS

This work was supported in part by grants-in-aid from the Scientific Research Program of the Ministry of Education, Science, and Culture of Japan (07558108 and 07457033), and by grants from the Sankyou Foundation of Life Science, Japan Foundation for Applied Enzymology, and The Cell Science Research Foundation. We are grateful to Drs. K. Hashimura and Y. Kato, Ishinkai Yao General Hospital, and D. Irikura, Osaka Bioscience Institute, for their skillful technical assistance.

REFERENCES

1. M. Ujihara, Y. Urade, N. Eguchi, H. Hayashi, K. Ikai, and O. Hayaishi, Prostaglandin D_2 formation and characterization of its synthetases in various tissues of adult rats, *Arch. Biochem. Biophys.* 260:521 (1988).
2. S. Ito, S. Narumiya, and O. Hayaishi, Prostaglandin D_2: a biochemical perspective, *Prostaglandins Leukotrienes Essent. Fatty Acids* 37:219 (1989).
3. M. Negishi, Y. Sugimoto, and A. Ichikawa, Prostanoid receptors and their biological actions, *Prog. Lipid Res.* 32:417 (1993).
4. Y. Urade, K. Watanabe, and O. Hayaishi, Prostaglandin D, E, and F synthases, *J. Lipid Mediator Cell Signalling* 12:257 (1995).
5. H. Toh, H. Kubodera, N. Nakajima, T. Sekiya, N. Eguchi, T. Tanaka, Y. Urade, and O. Hayaishi, Glutathione-independent prostaglandin D synthase as a lead molecule for designing new functional proteins, *Protein Engineering* 9:1067 (1996).
6. Y. Kanaoka, H. Ago, E. Inagaki, T. Nanayama, M. Miyano, R. Kikuno, Y. Fujii, N. Eguchi, H. Toh, Y. Urade, and O. Hayaishi, Cloning and crystal structure of hematopoietic prostaglandin D synthase. *Cell* 90:1085 (1997).
7. A. Hoffmann, H.S. Conradt, G. Gross, M. Nimtz, F. Lottspeich, and U. Wurster, Purification and chemical characterization of β-trace proteins from human cerebrospinal fluid; its identification as prostaglandin D synthase, *J. Neurochem.* 61:451 (1993).
8. K. Watanabe, Y. Urade, M. Mäder, C. Murphy, and O. Hayaishi, Identification of β-trace as prostaglandin D synthase, *Biochem. Biophys. Res. Commun.* 203:1110 (1994).

9. Y. Eguchi, N. Eguchi, H. Oda, K. Seiki, Y. Kijima, Y. Matsu-ura, Y. Urade, and O. Hayaishi, Expression of lipocalin-type prostaglandin D synthase (β-trace) in human heart and its accumulation in the coronary circulation of angina patients, *Proc. Natl. Acad. Sci. USA.* 94:14689 (1997).
10. Y. Urade, A. Nagata, Y. Suzuki, Y. Fujii, and O. Hayaishi, Primary structure of rat brain prostaglandin D synthetase deduced from cDNA sequence, *J. Biol. Chem.* 264:1041 (1989).
11. M. Hirata, A. Kakizuka, M. Aizawa, F. Ushikubi, and S. Narumiya, Molecular characterization of a mouse prostaglandin D receptor and functional expression of the cloned gene, *Proc. Natl. Acad. Sci. USA.* 91:11192 (1994).
12. Y. Boie, N. Sawyer, D.M. Slipetz, K.M. Metters, and M. Abamovitz, Molecular cloning and characterization of the human prostanoid DP receptor, *J. Biol. Chem.* 270:18910 (1995).
13. D.Y. Gerashchenko, C.T. Beuckmann, Y. Kanaoka, N. Eguchi, W.C. Gordon, Y. Urade, N.G. Bazan, and

O. Hayaishi, Dominant expression of rat prostanoid DP receptor mRNA in leptomeninges, inner segments of photoreceptor cells, iris epithelium, and ciliary processes. *J. Neurochem.*: in press (1998).

14. H. Oda, N. Eguchi, Y. Urade, and O. Hayaishi, Quantitative sandwich immunosorbent assay of human secretory prostaglandin D synthase (β-trace), *Proc. Japan Acad.* 72:108 (1996).

15. C.T. Beuckmann, W.C. Gordon, Y. Kanaoka, N. Eguchi, V.L. Marcheselli, D.Y. Gerashchenko, Y. Urade, O. Hayaishi, and N.G. Bazan, Lipocalin-type prostaglandin D synthase (β-trace) is located in pigmented epithelial cells of rat retina and accumulates within interphotoreceptor matrix, *J. Neurosci.* 16:6119 (1996).

16. D.Y. Gerashchenko, C.T. Beuckmann, V.L. Marcheselli, W.C. Gordon, Y. Kanaoka, N. Eguchi, Y. Urade, O. Hayaishi, and N.G. Bazan, Localization of lipocalin-type prostanoid D synthase (β-trace) in iris, ciliary body, and eye fluids, *Invest. Ophthalmol. Vis. Sci.* 39:198 (1998).

17. Y. Tokugawa, I. Kunishige, Y. Kubota, K. Shimoya, T. Nobunaga, T. Kimura, F. Saji, Y. Murata, N. Eguchi, H. Oda, Y. Urade, and O. Hayaishi, Lipocalin-type prostaglandin D synthase in human male reproductive organs and seminal plasma, *Biol. Reprod.* 58:600 (1998).

18. R.L. Gerena, D. Irikura, Y. Urade, N. Eguchi, D.A. Chapman, and G.J. Killian, Identification of a fertility-associated protein in bull seminal plasma as lipocalin-type prostaglandin D synthase, *Biol. Reprod.* 58:826 (1998).

19. B.J.R. Whittle, S. Moncada, and J.R. Vane, Comparison of the effects of prostacyclin (PGI$_2$), prostaglandin E$_1$ and D$_2$ on platelet aggregation in different species, *Prostaglandins* 16:373 (1978).

9

LIPOCALIN-TYPE PROSTAGLANDIN D SYNTHASE (β-TRACE) BINDS NON-SUBSTRATE LIPOPHILIC LIGANDS

Carsten T. Beuckmann,[1] Yoshihiro Urade,[1,2] and Osamu Hayaishi[1]*

[1]Department of Molecular Behavioral Biology
Osaka Bioscience Institute
6-2-4 Furuedai, Suita, Osaka 565-0874, Japan
[2]CREST, Japan Science and Technology Corporation
c/o Osaka Bioscience Institute
6-2-4 Furuedai, Suita, Osaka 565-0874, Japan

INTRODUCTION

The second most abundant protein in human cerebrospinal fluid after albumin, named β-trace[1], has been recently identified to be lipocalin-type prostaglandin (PG) D synthase ((5Z,13E)-(15S)-9α,11α-epidioxy-15-hydroxyprosta-5,13-dienoate-D-isomerase, EC 5.3.99.2)[2]. This enzyme catalyzes the conversion of PGH_2 to PGD_2 and is localized to the central nervous system[3,4] and male genital organs of mammals[5].

The PGD synthase has been shown to be a member of the lipocalin superfamily, a group of small secretory proteins which bind and transport a large variety of small lipophilic biomolecules[6]. Among the lipocalins, PGD synthase is the only known enzyme, due to the presence of a unique Cys^{65} residue, which is well conserved among lipocalin-type PGD synthases of various species but has not been found in other lipocalins.

It was recently demonstrated by Tanaka et al.[7] that PGD synthase binds retinoids in vitro with affinities comparable to those of already known extracellular retinoid-binding proteins, suggesting the possibility that β-trace binds other small lipophilic molecules, as well as its natural substrate PGH_2.

Eicosanoids and Other Bioactive Lipids in Cancer, Inflammation, and Radiation Injury, 4
Edited by Honn *et al.*, Kluwer Academic / Plenum Publishers, New York, 1999.

In this report, we examined the ability of lipocalin-type PGD synthase to bind other lipophilic ligands such as bilirubin and L-thyroxine (T4) by fluorescence quenching measurement and circular dichroism (CD) spectroscopy and found that PGD synthase binds both lipophilic molecules.

EXPERIMENTAL PROCEDURES

Materials

Bilirubin and T4 were purchased from Sigma and used without further purification. Bis(2-methoxyethyl)ether was obtained from Tokyo Kasei (Tokyo, Japan) and dimethylsulfoxide from WAKO (Tokyo, Japan). [1-^{14}C]PGH$_2$ was prepared from [1-^{14}C]arachidonic acid (2.20 GBq/mmol; Du Pont-New England Nuclear, Boston, MA) as described[3]. All other chemicals were from Sigma unless otherwise specified.

Recombinant Rat Lipocalin-type PGD Synthase

The Cys[65]-Ala substituted PGD synthase was used in this study[8]. It was expressed using a glutathione transferase-fusion protein system. Briefly, the mutated cDNA was inserted into the EcoRI-BamHI site of pGEX-2T (Pharmacia Biotech, Tokyo, Japan) and expressed in E. coli DH5α. The fusion protein was bound to glutathione-Sepharose (Pharmacia Biotech) and incubated with thrombin to release the mutated PGD synthase. The recombinant PGD synthase was further purified by Mono-S column chromatography to homogeneity as judged by polyacrylamide gel electrophoresis with subsequent silver staining.

Stock Solutions of Lipophilic Ligands

Bilirubin was dissolved in dimethylsulfoxide to give a 2 mM stock solution. T4 was dissolved in bis(2-methoxyethyl)ether/ethanol (1:1) to give a stock solution of 4 mM. The concentrations of these lipophilic ligands were determined spectroscopically with ε_{453} in chloroform for bilirubin = 61,700 M^{-1}·cm^{-1} and ε_{325} at pH 11 for T4 = 6,180 M^{-1}·cm^{-1}. Stock solutions were stored in the dark at -20°C.

Fluorescence Quenching Assays

The lipophilic substances (10 µl) were incubated at 20°C for 60 min with 1.5 µM recombinant rat PGD synthase in 990 µl of 5 mM Tris/HCl, pH 8.0. After incubation, the intrinsic tryptophan fluorescence was measured in a RF-5000 spectrofluorophotometer (Shimadzu, Kyoto, Japan) with an excitation wavelength at 282 nm and an emission wavelength at 338 nm.

CD spectra

PGD synthase (80 μM) was incubated at $10^{\circ}C$ with 20 μM T4 in 1 ml of PBS containing 1% (v/v) bis(2-methoxyethyl)ether/ethanol (1:1). In case of bilirubin, the protein (final concentration 30 μM) was added to the ligand (30 μM) in PBS with 1% (v/v) dimethylsulfoxide. CD spectra were recorded in a J-600 spectropolarimeter (JASCO, Tokyo, Japan) at $10^{\circ}C$. The spectra were measured ten times with a band width of 1 nm and a resolution of 1 nm at a scan speed of 200 nm/min.

RESULTS

Fluorescence Quenching of PGD Synthase

Several members of the lipocalin family, such as epididymal retinoic acid-binding protein[9], β–lactoglobulin[10], and plasma retinol-binding protein[11] show quenching of their intrinsic tryptophan fluorescence by binding of their respective ligands. The fluorescence quenching was also observed for PGD synthase by binding of retinoic acid or retinal[7]. We therefore examined the effect of bilirubin and T4 on the tryptophan fluorescence of PGD synthase. Both lipophilic ligands quenched fluorescence of the protein in a concentration-dependent manner (Figure 1).

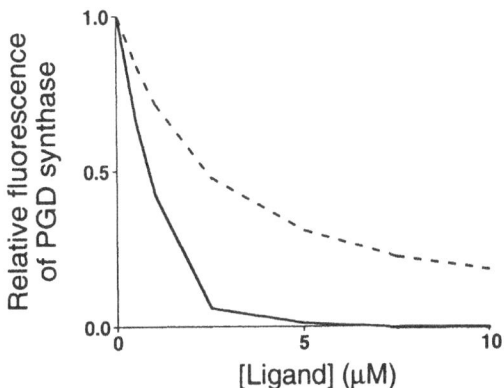

Figure 1. Fluorescence quenching of PGD synthase caused by binding of bilirubin (solid line) and T4 (dashed line). The protein (1.5 μM) was incubated with various concentrations of lipophilic ligands and fluorescence was measured.

The concentrations for quenching 50% of initial fluorescence intensity were calculated to be 790 nM for bilirubin and 2.4 μM for T4.

Bilirubin showed a strong interaction with the intrinsic tryptophan residue of PGD synthase, resulting in a complete disappearance of its fluorescence in the presence of excess amounts. In contrast T4 showed a weaker binding to PGD synthase, giving 20% of initial fluorescence under the condition of ligand excess.

Induced CD Spectra of Ligands Bound to PGD Synthase

Bilirubin, T4, and PGD synthase did not show any CD signal in a wavelength range above 300 nm. However, after incubation of ligands with protein, CD spectra of the ligands appeared in a range of absorption wavelength for each ligand with peaks at 422 nm (-), 472 nm (+), and 523 nm (-) for bilirubin and 328 nm (+) for T4, respectively, indicating the ligands are bound to PGD synthase (Figure 2).

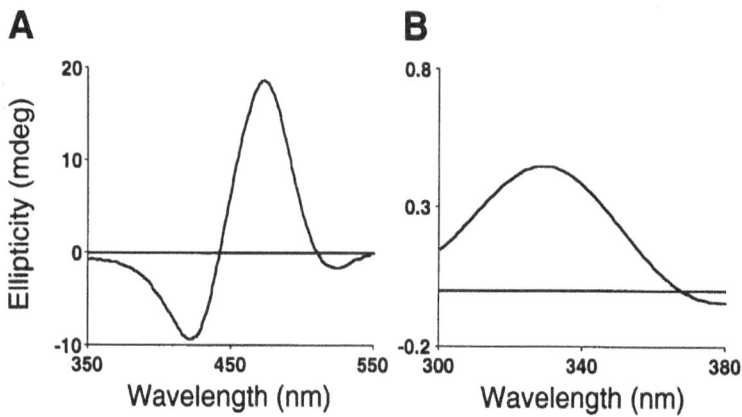

Figure 2. CD spectra of lipophilic ligands induced by PGD synthase. Bilirubin (A) and T4 (B) showed CD signals only in the presence of the protein.

DISCUSSION

In this study, we demonstrated by fluorescence quenching assay (Figure 1) and CD spectroscopy (Figure 2) that lipocalin-type PGD synthase, β-trace, binds bilirubin and T4. Although intracellular PGD synthase acts as an enzyme synthesizing PGD_2 because the substrate PGH_2 is readily supplied within cells, extracellular PGD synthase secreted into the cerebrospinal fluid and other body fluids could have a different function rather than an enzyme. This report, together with our previous report of retinoid-binding by the enzyme, clearly demonstrates that lipocalin-type PGD synthase may bind and transport these small lipophilic molecules.

The tertiary structures of several members of the lipocalin family, such as rat epididymal retinoic acid-binding protein[9] and bovine β–lactoglobulin[10], have been determined. These lipocalins possess a β–barrel structure with an internal ligand binding cavity in its center. A tryptophan residue is located at the bottom of this hydrophobic cavity[12] and its fluorescence is considered to be quenched upon binding of a ligand[9,11]. Homology modeling of lipocalin-type PGD synthase revealed that the enzyme comprised a typical lipocalin fold with a hydrophobic cavity in which a conserved Trp residue was found at the bottom[12].

The intensity of fluorescence quenching evoked by a lipophilic ligand is thought to inversely correlate to the distance between the ligand and the Trp residue. Therefore, we assume that bilirubin electronically interacts much better with the Trp residue than T4 does, because the fluorescence quenching evoked by bilirubin is much stronger than T4. It may be due to orientational reasons, i.e., the aromatic π-electron system of bilirubin may be closer or better orientated to the Trp residue than is the aromatic system of T4.

ACKNOWLEDGMENTS

This work was supported in part by an International Fellowship Grant from Takeda Science Foundation and Science and Technology Agency, Japan (C.T.B.), grants-in-aid for Scientific Research Program of the Ministry of Education, Science, and Culture of Japan (07558108 and 07457033) (Y.U.), and grants from the Sankyou Foundation of Life Science, Japan Foundation for Applied Enzymology, and The Cell Science Research Foundation (Y.U.). We thank Mr. Daisuke Irikura for preparing the expression vector for PGD synthase.

FOOTNOTES

*To whom correspondence should be addressed: Tel.: 81-6-872-4812
FAX: 81-6-872-4818.
The abbreviations used are: PG, prostaglandin; T4, L-thyroxine; CD, circular dichroism.

REFERENCES

1. J. Clausen, Proteins in normal cerebrospinal fluid not found in serum, *Proc. Soc. Exp. Biol. Med.* 107:170 (1961).
2. A. Hoffmann, H.S. Conradt, G. Gross, M. Nimtz, F. Lottspeich, and U. Wurster, Purification and chemical characterization of β-trace protein from human cerebrospinal fluid: its identification as prostaglandin D synthase, *J. Neurochem.* 61:451 (1993).

3. Y. Urade, N. Fujimoto, and O. Hayaishi, Purification and characterization of rat brain prostaglandin D synthetase, *J. Biol. Chem.* 260:12410 (1985).

4. C.T. Beuckmann, W.C. Gordon, Y. Kanaoka, N. Eguchi, V.L. Marcheselli, D.Y. Gerashchenko, Y. Urade, O. Hayaishi, and N.G. Bazan, Lipocalin-type prostaglandin D synthase (β-trace) is located in pigment epithelial cells of rat retina and accumulates within interphotoreceptor matrix, *J. Neurosci.* 16:6119 (1996).

5. Y. Tokugawa, I. Kunishige, Y. Kubota, K. Shimoya, T. Nobunaga, T. Kimura, F. Saji, Y. Murata, N. Eguchi, H. Oda, Y. Urade, and O. Hayaishi, Lipocalin-type prostaglandin D synthase in human male reproductive organs and seminal plasma, *Biol. Reprod.* 58:600 (1998).

6. D. R. Flower, The lipocalin protein family: structure and function, *Biochem. J.* 318:1 (1996).

7. T. Tanaka, Y. Urade, H. Kimura, N. Eguchi, A. Nishikawa, and O. Hayaishi, Lipocalin-type prostaglandin D synthase (β-trace) is a newly recognized type of retinoid transporter, *J. Biol. Chem.* 272:15789 (1997).

8. Y. Urade, T. Tanaka, N. Eguchi, M. Kikuchi, H. Kimura, H. Toh, and O. Hayaishi, Structural and functional significance of cysteine residues of glutathione-independent prostaglandin D synthase. Identification of Cys[65] as an essential thiol, *J. Biol. Chem.* 270:1422 (1995).

9. M.E Newcomer, R.S. Pappas, and D.E. Ong, X-ray chrystallographic identification of a protein-binding site for both all-*trans* and 9-*cis*-retinoic acid, *Proc. Natl. Acad. Sci. USA* 90:9223 (1993).

10. Y. Cho, C.A. Batt, and L. Sawyer, Probing the retinol-binding site of bovine beta-lactoglobulin, *J. Biol. Chem.* 269:11102 (1994).

11. G. Zanotti, G. Malpeli, and R. Berni, The interaction of N-ethyl retinamide with plasma retinol-binding protein (RBP) and the crystal structure of the retinoid-RBP complex at 1.9-Å resolution, *J. Biol. Chem.* 268:24873 (1993).

12. H. Toh, H. Kubodera, N. Nakajima, T. Sekiya, N. Eguchi, T. Tanaka, Y. Urade,, and O. Hayaishi, Glutahthione-independent prostaglandin D synthase as a lead molecule for designing new functional proteins, *Protein Engineering* 9:1067 (1996).

10

FEATURES OF MAMMALIAN LIPOXYGENASES

Bernd-J. Thiele, Mario Berger, Holger Thiele, Antje Huth and Iris Reimann

Institute of Biochemistry, University Clinics Charité, Humboldt-University Berlin, Hessische Str. 3-4, D-10115 Berlin, Germany

INTRODUCTION

Lipoxygenases (LOXs) are enzymes which oxygenate polyenoic fatty acids containing 1,4-pentadiene structures to their corresponding hydroperoxy derivatives (Yamamoto, 1992). They are distributed over the whole animal and plant kingdoms.

According to the currently used nomenclature LOXs are classified with respect to their positional specificity of arachidonic acid oxygenation into 5-, 8-, 12- and 15-LOXs. 5-LOXs which introduce molecular oxygen at C-5 in arachidonic acid are involved in the biosynthesis of leukotrienes which are important mediators of anaphylactic and inflammatory processes (Samuelsson et al., 1987; Lewis et al., 1990). In contrast, the biological functions of 8-, 12- and 15-LOXs are less well investigated (Kühn, 1996). There is evidence for the implication of 15-LOX in red cell maturation (Rapoport, 1990) and atherogenesis (Kühn and Chan, 1997), and of 12- and 8-LOX in tumor metastasis (Honn et al. 1994) and skin carcinogenesis (Fürstenberger et al. 1991).

The expression of LOXs is regulated at the transcriptional (O'Prey and Harrison, 1995) and at the translational level (Thiele et al., 1982). In particular, the role of regulatory proteins which interact with the 3'-untranslated region of 15-LOX mRNA has been studied in detail in recent years (Ostareck-Lederer et al., 1994; Ostareck et al., 1997).

This article focusses exclusively on mammalian LOXs with emphasis on our results on the structure and regulation of expression of rabbit leukocyte-type 12- and 15-LOX.

RESULTS AND DISCUSSION

Classification of Mammalian Lipoxgenases

Since the first publication of a sequence of a mammalian LOX, a partial sequence of the rabbit reticulocyte 15-LOX (Thiele et al. 1987), the list of cloned LOXs is steadily growing. This list encompasses currently at least 16 enzymes of seven different mammalian species. LOXs are classified traditionally according to their positional selectivity in the oxygenation of arachidonic acid. In table 1, which summarizes numerical values of sequence homologies between mammalian LOXs at the amino acid level, they are arranged according to their positional specificity into 5-, 8-, 12- and 15-LOXs. The oxygenation of polyenoic fatty acids occurs in a stereospecific way. All mammalian LOXs isolated so far have proven to be (S)-lipoxygenases. LOXs with (R) specificity are known from lower animal phyla like the 12(R)-LOX from the coral *P. homomalla* (Brash et al. 1997).

Table 1. Sequence similarities of mammalian lipoxygenases

Percent Similarity

	1	2	3	4	5	6	7	8	9	10	11	12	13	14	15	16		
1		91.9	95.5	96.4	40.0	34.7	37.0	37.1	35.9	35.9	35.9	36.0	34.1	35.3	38.5	34.1	1	r.5-LOX(J03960)
2	7.6		93.2	93.3	40.7	35.8	39.2	38.6	37.1	37.1	37.6	37.3	35.6	36.6	40.1	35.9	2	h.5-LOX(J03571)
3	4.3	6.4		95.8	40.4	34.9	38.2	37.9	38.2	26.5	38.5	36.7	34.4	35.5	39.2	34.7	3	ha.5-LOX(U43333)
4	3.1	6.8	3.7		40.4	35.2	37.7	37.7	37.0	36.8	36.8	37.0	35.0	36.3	39.8	38.0	4	m.5-LOX(I42198)
5	56.9	56.4	56.3	56.7		32.3	35.8	35.6	34.8	34.8	34.8	34.8	33.8	34.6	77.5	33.8	5	m.e8-LOX(U93277)
6	61.2	60.2	61.4	60.9	62.9		59.2	59.5	65.6	59.1	65.4	59.4	61.8	66.2	33.2	61.6	6	m.e12-LOX(U24181)
7	59.4	57.8	58.6	58.9	61.8	40.5		84.6	64.3	56.6	65.2	56.7	60.8	65.4	35.6	60.9	7	n.p12-LOX(M36792)
8	59.1	58.0	58.7	58.9	61.0	40.0	15.4		83.0	57.0	63.3	57.2	59.4	63.1	34.4	69.4	8	m.p12-LOX(U04334)
9	59.3	58.6	59.3	58.6	61.5	34.0	35.0	36.1		71.9	88.7	72.1	78.4	86.0	36.3	78.3	9	b.l12-LOX(M62516)
10	59.8	58.9	59.5	59.1	61.6	40.6	42.7	42.1	28.1		70.4	98.5	71.3	72.8	34.7	71.2	10	m.l12-LOX(I34570)
11	59.6	58.3	59.2	59.1	61.6	34.1	34.1	35.8	11.3	29.6		70.6	79.2	86.0	35.1	79.3	11	p.l12-LOX(M31417)
12	59.4	58.5	59.2	58.8	61.6	40.3	42.6	42.0	27.9	0.5	29.4		71.5	73.3	34.8	71.3	12	r.l12-LOX(I06040)
13	61.9	60.6	61.8	61.2	63.0	37.8	38.4	39.6	21.6	28.7	20.8	28.5		80.8	34.1	99.1	13	rb.l12-LOX(Z97654)
14	59.8	59.0	59.9	59.2	61.6	33.7	34.5	36.3	13.6	26.7	13.6	26.3	18.7		34.9	81.0	14	h.15-LOX(M23892)
15	57.9	56.8	57.2	57.1	21.9	63.5	61.7	61.4	60.0	61.7	60.9	61.7	62.4	61.2		34.4	15	h.15-LOX2(U78294)
16	61.9	60.3	61.5	61.2	62.8	37.9	38.2	39.6	21.7	28.8	20.7	28.7	0.9	18.6	62.1		16	rb.15-LOX(M27214)
	1	2	3	4	5	6	7	8	9	10	11	12	13	14	15	16		

Percent Divergence

For the alignment the "MegAlign" programme of the Macintosh "DNA Star" package was used.
For each sequence the EMBL database accession number is shown.

Sequence alignments as performed in table 1 demonstrate that a classification with respect to positional specificity has become ambiguous. This is evident particularly in the group of the 12-LOXs. They are historically classified according to the tissues from which they primarily have been isolated into leukocyte-type, platelet-type and epidermal 12-LOXs. Homologies between the members of this group are however in the range of only about 60%. The other extreme is the rabbit leukocyte-type 12-LOX which has been cloned by us (see below). It is much more similar to the reticulocyte-type 15-LOX of the same species (99% homology) than to 12-LOXs of other tissue types. The newly discovered human 15-LOX2 is more related to the mouse epidermal 8-LOX than to the human reticulocyte-type 15-LOX1. These examples may demonstrate that a more comprehensive classification scheme based on sequence homology may be introduced. Such a new system should be oriented in gene structures rather than in enzymatic properties.

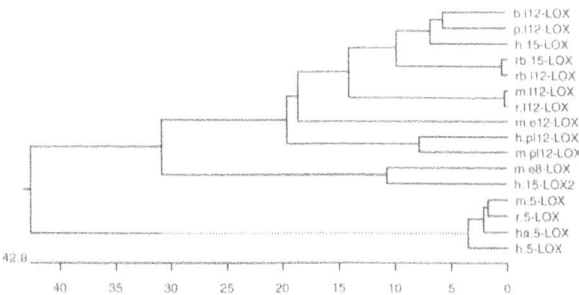

Figure 1. Phylogenetic tree of mammalian lipoxygenases. It was constructed using the Macintosh programme "MegAlign" with sequences from the EMBL database under accession numbers as shown in table 1. The abbrevations for lipoxygenases of different species are as follows: b. bovine, p. porcine, h. human, ha. hamster, r. rat and rb. rabbit. The tissue specificities refer to l - leukocyte-type, e - epidermal-type and pl - platelet-type. Evolutionary distances are given in million years.

Figure 1 presents a phylogenetic tree based on sequence alignments of the available mammalian LOX sequences. The difficulty with that kind of presentation is that two different relations are tried to compare. It affects the diversification of an ancient LOX-gene into different independent lines of LOX-genes and evolutionary distances between modern species. This problem can be overcome when LOX-genes of only one species are compared. In mouse five independent LOX-genes have been described up to now.

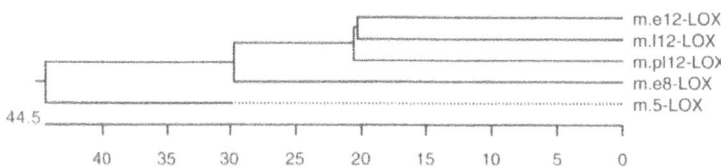

Figure 2. Phylogenetic tree of mouse lipoxygenases. For construction and abbrevations see legend to fig. 1.

In figure 2 the phylogenetic tree of mouse-LOXs is shown. From this currently most comprehensive collection of LOX sequences of one mammalian species a classification could be suggested which organizes this gene family with respect to the number of independent LOX-genes into at least five classes. All LOXs cloned so far fit in such a scheme. Class I: 5-LOXs; class II: a group which currently involves the mouse epidermal 8-LOX and the human epidermal 15-LOX2; class III: the platelet-type 12-LOXs, classIV: a group which has only one member at present, the mouse epidermal 12-LOX; and class V: the highly related leukocyte-type 12- and reticulocyte type 15-LOXs.

As we have shown (Berger et al. 1998, see next paragraph), in rabbits exist two independent genes for the reticulocyte-type 15- and the leukocyte-type 12-LOX. If one assumes a similar situation in mouse, in this species the number of mammalian LOX-genes would be six. Nevertheless, due to the high degree of homology between these two genes they should be handled as one class (V).

Co-Expression of Leukocyte-Type 12-LOX and Reticulocyte-Type 15-LOX in one Mammalian Species

The question whether a reticulocyte-type 15-LOX and a leukocyte-type 12-LOX coexist in one species has been a matter of discussion for several years. The lack in convincing evidence for a simultaneous expression of a leukocyte-type 12-LOX and a reticulocyte-type 15-LOX in one species supported the assumption that the former enzyme may constitute the functional equivalent of the reticulocyte 15-LOX in various species.

Although several authors reported the simultaneous formation of 12- and 15-LOX products in different tissues (for review see Kühn, 1996), the enzymes responsible for this product formation have not been well characterized. In most cases, the question whether 12-HETE formation was due to the action of a platelet-type or a leukocyte-type 12-LOX has not been carefully addressed. Recent quantitative PCR studies using sub-type specific primer combinations suggested the expression of a leukocyte-type 12-LOX in man (Kim et al., 1995). In order to obtain experimental evidence for the simultaneous expression of a reticulocyte-type 15-LOX gene and a leukocyte-type 12-LOX gene in rabbits we rescreened a reticulocyte cDNA library and isolated a full length cDNA clone which coded for a leukocyte type 12-lipoxygenase (Thiele et al. 1998). The data indicate for the first time the existence of two separate genes coding for a reticulocyte-type 15- and a leukocyte-type 12-LOX in one animal species which are expressed in a tissue-specific way. Since these enzymes have been implicated in cell development and atherogenesis this co-expression may be of physiological relevance.

The predicted amino acid sequence of this enzyme shared a high degree (99%) of identity with the reticulocyte-type 15-lipoxygenase (table 1). Among the six amino acids different in both enzymes a phe-leu exchange was detected at position 353. Recently, site-directed mutagenesis studies have revealed that this amino acid exchange converts a 15-lipoxygenase to a 12-lipoxygenase (Borngräber et al. 1996). Other enzymatic properties too charcterized this novel enzyme as a leukocyte-type 12-LOX. To exclude that this cDNA sequence represents a simple polymorphism a search for the leukocyte-type 12-LOX coding gene was started. Genomic PCR fragments were generated using 12-LOX specific primers and characterized by sequencing. A summarizing scheme of the structure of the 3'-terminal third of the novel rabbit leukocyte-type 12-LOX gene in comparison to the highly related 15-LOX gene is shown in figure 3.

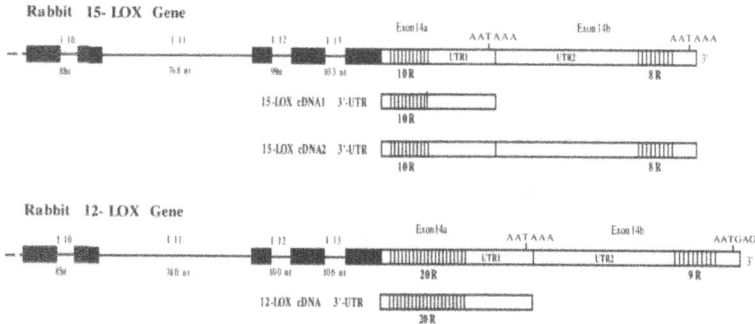

Figure 3. Characterization of the 3'-terminal structures of the genes and cDNAs coding for the rabbit leukocyte-type 12-LOX and reticulocyte-type 15-LOX.

64

The intron/exon organization is identical to all other LOX-genes analyzed so far. Introns were about 90% homologous to each other. A remarkable difference has been detected in the structure of the repetitive differentiation control element (DICE) in the 3'-untranslated region of the mRNA which is responsible for translational control (Ostareck-Lederer et al. 1994). In contrast to the DICE-motif of the 15-LOX mRNA which consists of a 10-fold repeat of a 19 nucleotide basic CU-rich element, a very similar 20-fold repeat is found in the 12-LOX mRNA (Berger et al. 1998).

Posttranscriptional Control of 15-LOX Synthesis

15-LOX synthesis is an interesting subject for the investigation of postranscriptional control of gene expression. In early stages of red blood cell development (erythroblasts) 15-LOX mRNA is transcribed, but stored as a translationally inactive mRNP complex until it is activated at the reticulocyte level. As we have shown previously this translational repression is mediated by the interaction of mRNA binding proteins (LOX-BP) with the repetitive DICE-motif in the 3'-untranslated region of 15-LOX mRNA (Ostareck-Lederer et al. 1994, Ostareck et al. 1997). This type of translational control is found to be associated with the 15-LOX mRNA1 which is exclusively expressed in erythroid tissues (short 3'-UTR, fig. 3).

In nonerthroid tissues a synthesis of 15-LOX can be induced, for instance by anemic stress. In these tissues the message for LOX synthesis however is 15-LOX mRNA2 which contains an about 1000 nucleotides longer 3'-UTR (Thiele et al., Nucl. Acids Res., submitted for publication). In its 3'-UTR2 (exon 14b) a novel 8-fold repetitive element (8R) has been found which is related but not identical to the 10-fold repeat (10R) in 3'-UTR1 (exon14a). The protein binding properties of 8R differ remarkably from those of 10R. Proteins which bind to the 3'-UTR2 element 8R have been purified by RNA affinity chromatography and analyzed in cell-free translation. The results are summarized in figure 4.

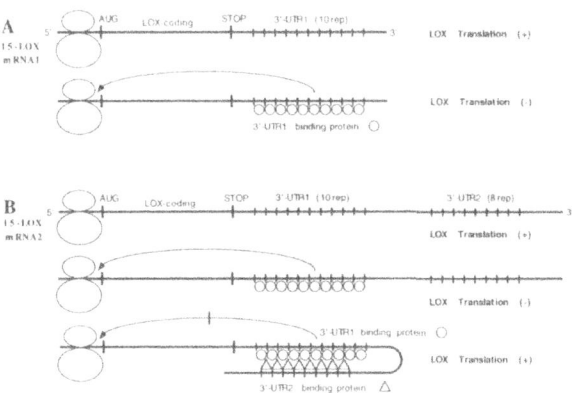

Figure 4. Schematic representation of translational control of 15-LOX synthesis. A) in reticulocytes, B) in nonerythroid tissues. The mRNAs are not drawn to scale.

Figure 4 gives an idea of the complex posttranscriptional regulation of 15-LOX. In reticulocytes 3'-UTR1 binding proteins affect initiation of translation with 15-LOX mRNA1 by an unknown mechanism (figure 4A). In nonerythroid tissues translational inhibition can be abolished by binding of 3'-UTR2 binding proteins to the 8R UTR2 element and secondary protein/protein interaction. This is schematically shown in figure 4B.

Currently we are investigating the molecular details of the proposed mechanisms and the impact on the physiology of 12- and 15-LOX expression.

REFERENCES

Berger, M., Schwarz, K., Thiele, H., Reimann, I., Huth, A., Borngräber, S., Kühn, H. and Thiele, B.J., 1998, Simultaneous expression of leukocyte-type 12-lipoxygenase and reticulocyte-type 15-lipoxygenase in rabbits. *J Mol. Biol.*, in press.

Borngräber, S., Kuban, R.-J., Anton, M. and Kühn, H., 1996, Phenylalanine 353 is a primary determinant for the positional specificity of mammalian 15-lipoxygenase. *J. Mol. Biol.* 264:1145.

Brash, A.R., Boeglin,W.E., Chang, M.S. and Shieh, B.H., 1996, Purification and molecular cloning of an 8R-lipoxygenase from the coral Plexaura homomalla reveal the related primary structures of R-and S-lipoxygenases. *J. Biol. Chem.* 271:20949.

Fürstenberger, G., Hagedorn, H., Jacobi, T., Besemfelder, E., Stephan,M., Lehmann, W.-D. and Marks, F., 1991, Characterization of an 8-lipoxygenase activity induced by the phorbol ester tumor promoter 12-O-tetradecanoylphorbol-13-acetate in mouse skin in vivo. *J. Biol. Chem.* 266:15738.

Honn, K.V., Tang, D.G., Gao, X., Butovitch, I.A., Liu, B., Timar, J. and Hagman, W., 1994, 12-lipoxygenase and 12(S)-HETE: role in cancer metastasis. *Cancer Metastasis Rev.* 13:365.

Kim, J.A., Gu, J.L., Natarajan, R., Berliner, J.A. and Nadler, J.L., 1995, A leukocyte type of 12-lipoxygenase is expressed in human vascular and mononuclear cells. *Arterioscler. Thromb. Vasc. Biol.* 15:942.

Kühn, H., 1996, Biosynthesis, metabolization and biological importance of the primary 15-lipoxygenase metabolites 15-hydro(pero)xy-5Z, 8Z, 11Z, 13E-eicosatetraenoic acid and 13-hydro(pero)xy-9Z, 11E-octadecadienoic acid. *Prog. Lipid Res.* 35:203.

Kühn, H. and Chan, L., 1997, The role of 15-lipoxygenase in atherogenesis: pro- and antiatherogenic actions. *Curr. Opinion Lipidol.* 8:111.

Lewis, R.A., Austen, K.F. and Soberman, R.J., 1990, Leukotrienes and other products of the 5-lipoxygenase pathway. Biochemistry and relation to pathobiology in human diseases. *New Engl.J. Med.* 323:645.

O'Prey, J. and Harrison, P., 1995, Tissue specific regulation of the rabbit 15-lipoxygenase gene in erythroid cells by a transcription silencer. *Nucleic Acid Res.* 23:3664.

Ostareck-Lederer, A., Ostareck, D.H., Standart, N. and Thiele, B.J., 1994, Translation of 15-lipoxygenase mRNA is inhibited by a protein that binds to repeated sequence in the untranslated region. *EMBO J.* 13:1476.

Ostareck, D.H., Ostareck-Lederer, A., Wilm, M., Thiele, B.J., Mann, M. and Hentze, M.W., 1997, mRNA silencing in erythroid differentiation: hnRNP K and hnRNP E1 regulate 15-lipoxygenase mRNA translation from the 3' end. *Cell*, 89:597.

Rapoport, S.M., Schewe, T. and Thiele, B.J., 1990, Maturational breakdown of mitochondria and other organelles in reticulocytes. In: *Blood Cell Biochemistry*, Vol.1 "Erythroid Cells", J.R.Harris Ed.), Plenum Press, New York 1990, p.151.

Samuelsson, B., Dahlen, S.E., Lindgren, J.A., Rouzer, C.A. and Serhan, N.J., 1987, Leukotrienes and lipoxins: structures, biosynthesis, and biological effects. *Science*, 237:1171.

Thiele, B.J., Andree, H., Höhne,M. and Rapoport, S.M., 1982, Lipoxygenase mRNA in rabbit reticulocytes. Its isolation, characterization and translational repression. *Eur. J. Biochem.*, 129:133.

Thiele, B.J., Fleming, J., Kasturi, K., O'Prey, J., Black, E., Chester, Rapoport, S.M. and Harrison P.R., 1987, Cloning of a rabbit erythroid-cell-specific lipoxygenase mRNA. *Gene* 57:111.

Yamamoto, S., 1992, Mammalian lipoxygenases: molecular structures and functions. *Biochim. Biophys.Acta*, 1128:117.

11

IDENTIFICATION AND CHARACTERIZATION OF AN ENHANCER SEQUENCE IN
THE PROMOTER REGION OF HUMAN 15-LIPOXYGENASE (*15-LO*) GENE

Uddhav Kelavkar , Susheng Wang , Angel Montero and Kamal Badr
Renal Division, Emory University, and Center for Glomerulonephritis, Veterans Affairs
Medical Center, Atlanta, GA 30322.

INTRODUCTION:

Lipoxygenases are lipid-peroxidating enzymes that are implicated in the pathogenesis of a variety of inflammatory disorders [1-3], membrane remodeling [4-6] and atheroma formation [7-9]. Formation of 15-(S)-hydroxyeicosatetraenoic acid [15-(S)-HETE] and lipoxin (LX) A4 in human leukocytes, mediated by *15-LO* dependent catalysis of arachidonic acid, likely represents a component of endogenous pro- and anti-inflammatory influences that ultimately regulate the extent and severity of inflammatory reactions [1-4, 10].

15-(S)-HETE and LXA4 have been proposed as endogenous anti-inflammatory molecules that suppress white cell chemotaxis, adherence, and activation and to specifically antagonize the functional responses of proinflammatory 5-LO derivatives, the leukotrienes [1-4, 10]. The role and functions of leukotriene C4 (LTC4), leukotriene D4 (LTD4) and leukotriene B4 (LTB4) have been described earlier [1-3]. 15-(S)-HETE decreases LTB4 generation in leukocytes [11], antagonizes neutrophil chemotaxis by LTB4 [12] and suppresses leukocyte activation in response to activators [11, 13]. LXA4 exerts similar effects as it inhibits natural killer cytotoxicity, attenuates LTB4-induced chemotaxis, and decreases leukocyte-endothelial cell adhesion [1, 10]. The effects of 15-(S)-HETE on cellular activation and of LXA4 to trigger selective responses have been discussed [11, 13, 14].

Lymphokines secreted by helper T-lymphocytes (TH) also modulate the inflammatory response through their actions on monocytes. Evidences in the recent past have established that lymphokines derived from TH-lymphocyte populations differentially regulate lipoxygenase enzymatic pathways [15-21]. IL-4 was the only cytokine known to induce *15-LO* in human monocytes/macrophages [22, 23] until recently. Nassar *et al.*, [24] have demonstrated that IL-13, but not IL-10, induces *15-LO* mRNA and protein synthesis in human blood monocytes leading to enhanced production of 15-HETE. This was inhibited by IFN- ɣ, as described in IL-4 induced monocytes [22]. Though *12-* and *15-LO* from human and *15-LO* from rabbit have been cloned and sequenced [25-27] and a human *5-LO* promoter [28] characterized, little is known about the 5' promoter region of human *15-LO*.

The regulation of genes encoding for the enzymes involved in arachidonate oxygenation pathways is a recently recognized aspect of cytokine biology. It is particularly intriguing in view of the apparent "logic" in the pattern of specificity between cytokines such as IL-13 which exert anti-inflammatory actions in macrophages and their respective target enzymes in the lipoxygenase pathway, namely *15-LO*. It is possible that this aspect of cytokine biology involves mechanisms for transcriptional regulations which are distinct from those of Stat proteins, [29-32] which have been implicated in cytokine modulation of phenomena such as cell surface receptor expression and immunoglobulin class switching during allergic-type reactions [32].

To identify the cis-acting element that is responsible for the cell-specific expression of the 15-lipoxygenase gene, we analyzed the promoter activity of its 5'-flanking region by transient expression assays. The fusion genes were constructed by inserting the 5'-flanking region of the *15-LO* gene upstream from the firefly luciferase gene and were introduced into HeLa cells. We thus found an enhancer element, located between -2.0 and -0.8 kilobase pairs (kb) upstream from the transcription initiation site, that enhances the transient expression of the luciferase reporter gene in HeLa cells. Using the fusion genes containing putative enhancer elements under the control of the heterologous simian virus 40 promoter, we identified the IL-13 induced cell-specific (in monocytes) enhancer of approximately 500 base pairs (bp) between -1.2 Kb and -790 bp and further localized the core sequence to a 41 bp region. This element was then confirmed to direct the HeLa cell-specific expression of the reporter gene under the *15-LO* gene promoter. We thus propose that this core element is responsible for the IL-13 induced cell-specific expression of the human *15-LO* gene. In this paper, we describe the characterization of the 5' flanking enhancer region of human *15-LO* for the first time.

MATERIALS AND METHODS:

5' Rapid Amplification of cDNA Ends (5' RACE): 5' RACE was carried out using the Gibco-BRL 5' RACE System Kit (Cat.No 18374-025). The gene-specific antisense primer was 5'-GCCATATTCAGAATTAACCCGT-3' (436 to 458 bp of *15-LO* cDNA)(GSP1), and the nested gene-specific primer was 5'-GGTAGTTCCACCTTGAGTTCTGTCT-3' (136 to 158 bp of *15-LO* cDNA) (GSP2). Poly(A)$^+$ RNA prepared using RNAzol method [24], from the human peripheral monocytes induced by IL-13, was reverse transcribed using the primer GSP1. The anchor-ligated cDNA was then PCR amplified using the anchor primer and an internal primer, GSP2. PCR amplified, products were purified from the agarose gel and subcloned into a plasmid vector pCRTMII using the TA Cloning Kit (Invitrogen) and sequenced by dideoxynucleotide chain-termination method as described by Sambrook et al. [33].

Isolation of Genomic Clones: A 185 bp (5' RACE product) insert was excised by *Eco*RI from the plasmid as the probe for screening a human genomic library constructed in Lambda FIX II vector. Dr. Cam Patterson, Harvard School of Public Health, Massachusetts, USA kindly provided this library. The probe was labeled with [α^{32} P]dCTP with Oligolabeling Kit (Pharmacia Biotech). Plaques were transferred onto nylon immobilization membranes (Schleicher and Schuell) and hybridization was carried out at 42°C for 16-20 h in standard hybridization solution. Final Washing of the membranes was carried out in 2X SSC containing 0.2% SDS at 65°C for 30 min. The membranes were then exposed to X-ray film. Secondary screening was performed to identify and confirm positive plaques. The phage DNA was extracted by the method described by Sambrook et al. [33].

Southern blot analysis, subcloning and sequencing of 5' flank region of the gene: The DNA extracted from the positive plaques were digested with restriction enzyme *Not*I and electrophoresed on 0.4% agarose gel. An insert of 18 kb was purified from the agarose gel using QIAGEN gel extraction kit. This fragment was then further digested with *Pst*I, *Sac*I, *Bam*HI, *Hind*III, *Kpn*I and *Eco*RI respectively; resolved by electrophoresis on 1% agarose gel and was blotted on a nylon membrane. The hybridization was carried out using the $[\alpha$ ^{32} P]dCTP labeled 185 bp fragment of the cDNA end as the probe with a standard procedure [33]. The fragments that hybridized with the probe were gel purified and subcloned into the pBluescript II SK+ plasmid vector (Stratagene). The DNA sequence was analyzed by automated sequencer (Applied Biosystems 377), using fluorescent methodology.

Standardization of transfection efficiencies: HeLa transfection efficiencies were standardized to the quantity of luciferase in pGL3 vector present in the transfected cells. To do this, the concentration of DNA in one sample was determined and the volume containing 3µg plasmid calculated. This volume from each sample was blotted onto a nylon membrane (Schleicher and Schuell) and probed with a random prime labeled ($[\alpha^{32}$P]dCTP) 2.5 kb (*Hind*III-*Cla*I) luciferase fragment from pGL3. Hybridization was carried out at 62°C for 16-20 h in standard hybridization solution. Final washing of the membrane was carried out in 0.2X SSC containing 0.2% SDS at 65°C for 30 min. and then exposed to X-ray film. The resulting autoradiogram was scanned on an ultrascan XL laser densitometer (LKB, Bromma).

Subcloning DNA segments into the reporter plasmid, transfection and luciferase assay: The 5' flank sequence of the gene was PCR amplified using different primers. The products were subcloned into a luciferase reporter vector pGL3 (Promega). The integrity of all constructs was determined by restriction enzyme site mapping and all constructs produced by PCR were sequenced using fluorescent methodology.

HeLa cells were cultured in Dulbecco's modified Eagle's medium (DMEM), supplemented with 10% fetal calf serum (FCS). Cells were grown to 80-90% confluency, trypsinized, washed once with phosphate buffered saline (PBS) and resuspended (3.75×10^7 cells/ml) in ice cold PBS. Transient transfections were carried out using a luciferase reporter construct (a promoter fragment cloned into pGL3; Promega). For each experiment, equimolar quantities of the various promoter constructs were used and the total mass of DNA was standardized for transfections using pGL3 DNA.

All experiments were performed in either duplicates or triplicates. Three µg DNA was transfected by lipofection with LipofectAmine reagent into HeLa cells. In 100µl of serum-free medium (-FCS), DNA-liposome complex was allowed to form at RT for 45 min. Then 0.8 ml of DMEM (-FCS) medium was added, gently mixed, and the complex overlayed onto the HeLa cells and incubated at 37°C for 5 h and then on in complete medium (+FCS) for 24 h. Cells were washed twice with PBS and harvested into 400 µl reporter lysis buffer (Promega). The lysate was clarified by centrifugation, and the supernatant transferred to a new tube. Luciferase activity of the supernatant (20µl) was assayed using Promega's luciferase assay system (100 µl) and a luminometer (Analytical Luminescence Laboratories).

Extraction of nuclear proteins and gel mobility shift assay: Nuclear extracts from HeLa, A549, Thp1 cells and human peripheral monocytes were prepared according to the method decribed by Dignam et al. [34]. Human peripheral blood monocytes were isolated as decribed by Nassar et al. [24]. Protein concentrations in nuclear extracts were determined by protein assay kit (Bio-Rad). Gel shift assay was performed as described by Austin et al. [35]. Binding reaction were conducted using the radiolabeled DNA fragment from 15-LO gene 5' flanking -1092 to -1051 bp

```
-2667  ggtacatttt tcacctatcc gattggctaa ggtaaggtta tgaacacacc attggcaaga gtgaggaaga atgggcattc tcatacacag tccactggtg
-2567  ggagtctaaa tcagtacaa ctttctacaga gggcaattga gcagtttctg cctaaattcg atatgctttt ggtccagaaa ttctatctcc agggttttgt
-2467  tctacagaga tactttcata cgagaaatgc tgtatatgcg aggtttattt ctttcagcac tttttcatag ctacaaaaga ttactaataa cctgtgtcca
-2367  ctacaagaga ctagtagaat ttatggcgtt acagcaaatg gcagccttag caaagaacaa ggaactcttt atgcatggat atggaatgac ctcctgttgc
-2267  actatggaaa aacgagatgc agaataaggc atgctacaat gtgcagaaaa agacaaagat aggcgtgtat ttgctcagta atattagtgc acataagaga
-2167  ctggtaactg tgattgctgg gagtggagca actgggtggc ttggggccaa gaggaggga gaaactagtga atatcacttt gtaccttttg aattttgaat
-2067  cacatgaatg tacttcccag taaacaagca gggccggacg tgatggctca tggctgtaa cccagcactt ggggaggccg aggcaggtgg atcacttgag
-1967  gtcaggagtt caagaccagc ctggccaaca tggtgaaacc ccatctcgac tgaaaataca aaaagtagcc gggcatgttg gtgggtgcct gtaatcccag
-1867  ctactctgga ggctgaggca ggagaatcgc ttgaacccgg caggtggagg ttgcagtgag ccgagatcac accactgtac tctagcccgg gcaacagagc
-1767  aagactccat ctcattaaaa aaaaaaacac caaaaacaaa aaacaagcaa accagaaaat atgcatcaca gacatatttt cctggacttc atttgtgtta
-1667  ccaggctgat tgtaatccat gccatgaagt ttatgttagt attccccttt tgtttacatt gcctgataac ttgtttccct ttgtctcccc caagggaagc
-1567  attcatttat aatgaatttt tcagctcagt gtgtttcagc aaacacttat tgacacccac aaacctgaca ctgtgccgtc tccaggggaa acaaaagtga
-1467  gagtcctgct gggtgtgtag tccctgccta ggacgggctc acattagtgg gggatacaca catacaaatg ctcacactgc agtattaact gtgcaacagt
-1367  agaagtgtgc gcaaggagaa aagtgttaga aaggaaagct gggccgggcg ccgtggctca cacctgtaat tccagcactt tgggagactg aggtgggtgg
-1267  atcacaaggt caggagttca agaccagcgt ggccaacatg gagaaacccc gtctctacta aaaatacaaa aattatccag gcatggtggc aggcgcctgt

-1167  aatcccagct actcggaagg ctgaggcaga gaattgcttg agcccgggag gcagaggttg ccgtgagccg agatcgtgcc attgcactcc accctgcatt

-1067  ccagggtgac agagtgagac tctatatctc aaaaaaaaaa aaaggaaagc taattcactc tggtggggtg gtggggggtg aaagtaccgg gtaagacttt

 -957  cctgagaaac cggagtgaac tagggtttag aagcatgaat agatgataga ctaatccagg agcaattctt ttgagagaac agcaggtggg gttctgggga
 -867  acacatggat ccagcccaaa aagaaactgg agacgaatta gggtttagaa gcatgaccag actaacccag aagtgag
```

Figure 1: Sequence of the 5' flanking promoter region of the human *15-LO* (Accession No. AF 030294). Nucleotides are numbered from the translational start codon (+1). The location of the enhancer sequence is shown by an open box.

(2 X 10^4 cpm, 40 fmol for each reaction). Whenever added, 3 pmol competitor (unlabeled oligonucleotide) was used.

Computer assisted sequence analysis: Sequencing results were analyzed by using DNASIS [36] and BLAST search [37].

RESULTS:

DNA sequence of the 5'-flanking region of the *15-LO* gene: To gain information relating to potential target sequences in the promoter of the *15-LO* gene, a 2.8 kb complete sequence was obtained out of which, 2 kb fragment (Fig.1) is of significant relevance to this paper [38].

Identification of the enhancer region of the *15-LO* gene: To identify enhancer DNA sequence regulating *15-LO* gene expression, the region from -1290 to -790 bp; (+1 is adenine in the ATG start codon), was amplified by PCR from the 3 kb fragment previously cloned in LOPB-5. This fragment was then purified and restricted with *Kpn*I and *Hind*III and cloned into a plasmid pGL3 containing the luciferase reporter gene. The resulting plasmid was then transiently transfected into the HeLa cells. The luciferase activity assay demonstrated that the 500 bp DNA fragment clearly had enhancer activity in the cells tested.

To further determine the site of the minimal *15-LO* enhancer region, a series of deletion mutants were constructed. The sequence of the primers were as follows:

KF-4 5'TTAAAGGTACCAGGGTGACAGAGTG 3'
KF-5 5' TTAAAGGTACCATGGACTTCATTTGT 3'
KF-7 5' TACTAGGTACCTGTAATTCCAGCACT 3'
KF-8 5' TACTAGGTACCAGTAAACAAGCAGGGC 3'
XHR 2 5' TACAGTCTCGAGCTCCTGGATTAGTCTA 3'
XHR 3 5'TACAGTCTCGAGGTTATCGAACAATGTAA 3'
XHR 4 5' TACAGTCTCGAGTCCTGACCTTGTGATC 3'
XHR-5 5'TACAGTCTCGAGTCTGGTCATGCTTCTA 3'

The different PCR fragments obtained and their respective sizes (mentioned in the parenthesis) were:

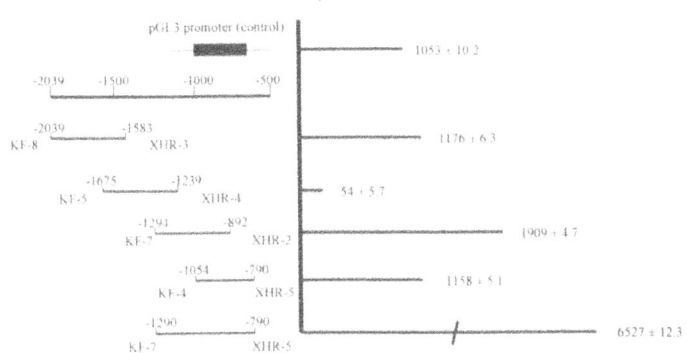

Figure 2: Enhancer activity of the different regions encompassing the 5' flanking region of human *15-LO*. The histogram shows the average +/- SEM activity derived from duplicate or triplicate assays.

1	KF8 and XHR 3	-2039 to -1585 (454 bp)
2	KF5 and XHR 4	-1675 to -1239 (436 bp)
3	KF7 and XHR 2	-1290 to -892 (402 bp)
4	KF4 and XHR 5	-1054 to -790 (264 bp)
5	KF7 and XHR5	-1290 to -790 (500 bp)

These fragments were then purified and restricted with *Kpn*I and *Hind*III and individually cloned into a plasmid pGL3 containing the luciferase reporter gene. The resulting plasmids were then individually transiently transfected into the HeLa cells. The activities of the different segments of 5' *15-LO* promoter/enhancer are shown in Fig.2. As can be seen, the enhancer activity is found associated with fragment from -1290 to -750 nt. Deletion of 100 bp from the 3' end or from 5' end led to a 100% loss in activity. Therefore, it appears that the enhancer lies within the -1290 to -750 bp of the 5'-flanking region. These observations were further confirmed

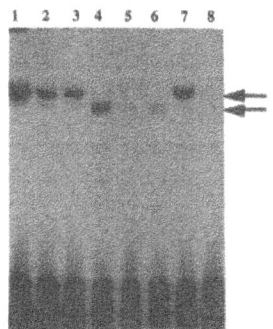

Figure 3: Mobility shift DNA binding analysis of nuclear protein extracts from human peripheral blood monocytes, HeLa, Thp1 and A549 cells. Binding reaction were conducted using the radiolabeled 41 bp DNA fragment from 15-LO gene 5' flanking (-1092 to 1051 bp), [2 x10⁴ cpm, 40 fm for each reaction]. Whenever added, 3 pmol competitor (unlabeled DNA fragment) was used. Lane 1: Probe plus nuclear extracts from HeLa cells. Lane 2: A549 cells (a) Lane 3: (a) Plus probe as a competitor. Lane 4: IL-23 induced monocytes (b). Lane 5: (b) Plus probe as a competitor. Lane 6: Control monocytes (uninduced). Lane 7: Probe plus nuclear extracts from Thp 1 cells. Lane 8: Probe alone. Arrows indicate specific bands.

by mobility shift assays and localized the core sequence to a 41 bp probe (-1092 to –1051 bp). The sequence is as follows:

GTGCCATTGCACTCCACCCTGCATTCCAGGGTGACAGAGTG
CACGGTAACGTGAGGTGGGACGTAAGGTCCCACTGTCTCAC

Specific interaction of human _15-LO_ promoter with nuclear proteins from human peripheral blood monocytes induced by IL-13: The DNA fragment (41 bp) of the promoter (-1092 to – 1051 bp) DNA was [α^{32} P]dCTP labeled at one end by a fill-in reaction and incubated with nuclear proteins extracts prepared from A549, Thp1 and the human peripheral blood monocytes induced and uninduced (control) by IL-13. (Fig. 3). Gel shift indicated that the enhancer element was responsive only in the presence of IL13 whereas a nonspecific or as yet to be identified factor bound; from the extracts in the A549, HeLa and Thp1 cells respectively.

DISCUSSION:

We have reported the isolation, mapping and characterization of 1056 bp 5' flanking sequence for the human _15-LO_ gene [38] . The organization of the 5' flanking region of the human 15-lipoxygenase gene is schematically represented in Figure 4. We identified conserved domains that are all G+C rich, which averages 65% over its length and may act as potential targets for transcriptional factor binding sites. The promoter region was found to have putative Sp1, Ap2 binding sites and CATTC and TTATT sequences, but lacked 100% consensus sequence match with TATA and CCAAT sequences.

Chimeric genes consisting of various regions of the promoter were generated to determine the enhancer element. Luciferase activity from transfected HeLa cells was found to be associated with the fragment, -1290 to -790 nt (Fig. 2). The human fragment was active in initiating transcription of the reporter gene in HeLa cells. Further, gel shift assays with nuclear protein extracts from several tissues (Fig.3) indicated that the enhancer element was responsive only in the presence of IL13 whereas a nonspecific or as yet to be identified factor bound from the nuclear extracts of A549, HeLa and Thp1 cells respectively. It has been shown earlier that G+C rich TATA-less promoters of widely expressed genes usually contain several transcription initiation sites spread over a fairly large region, rather than at a single base position [39]. It strongly suggests that the enhancer element is specific for IL13 and may play an important role in _15-LO_ expression control and pattern.

Genes with high G+C content in promoters are also considered to express proteins with a "house keeping" function in the cell. Furthermore, several genes devoid of TATA or CCAAT motifs in their promoter regions have been shown to encode proteins that are highly regulated [40, 41]. _15-LO_, although expressed in various tissues, also appears to be specifically regulated in different cell types. Taken together, these observations suggest that although _15-LO_ is expressed widely,

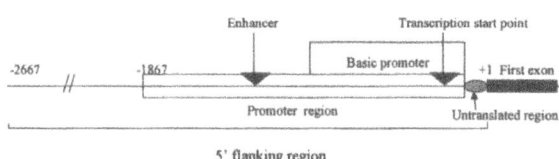

Figure 4: Organization of the 5'flanking region of the human 15-lipoxygenase gene.

there is a possibility of developmental and tissue specific regulation of expression, as described [2]. Thus, there is only one promoter region at the 5'-end of the *15-LO* gene [38], and the control of widely expressed full-length transcript may be regulated via tissue specific transcription factors binding to as yet uncharacterized elements or enhancers. The *15-LO* promoter seems to be a single basal element that is active in all tissues with specialized expression controlled by cell-specific transcription factors.

Identification of the elements that regulate the expression of *15-LO* in monocytes and their role in gene induction by IL-13 may provide insight into the potential for up-regulating *15-LO* as a therapeutic strategy in inflammatory disorders [1-3]. Initially, we identified regions, namely DP1 and DP2, that bind to IL-13 induced nuclear binding protein(s) [38]. Notably, these sequences are distinct from the ones that bind Stat proteins, TTCTAAGAA or TTCCCAGAA [30, 32], also induced by IL-4/IL-13 in B-lymphocytes, which suggests a novel monocyte-specific mechanism(s) for regulation of *15-LO* by IL-13.

Though similar gene organization between human *15-LO* and other species such as rabbit *15-LO*, the different cell specificities for expression of the various lipoxygenases indicate that different regulatory DNA elements have been added to the genes in this family after gene divergence. The close chromosomal localization of *15-LO* [38] to that reported previously for *12-LO* [27, 42], is consistent with a common evolutionary origin. In fact in mice, rats, and other species, the two enzymes "merge" into a single protein (*12/15-LO*), which performs both 12 and 15 lipoxygenations [1, 6, 43].

ACKNOWLEDGMENT:

National Institutes of Health (NIH) grant # 2RO1DK43883 to K.B supported this work.

REFERENCES:

1. B. Samuelsson, SE. Dahln, J.A. Lindgren, C.A. Rouzer, and C.N. Serhan, *Science* 7:1171 (1987)
2. A. Ford-Hutchinson, *Eicosanoids* 4: 65-74 (1991)
3. K.F. Badr, *Kidney Int.* 442: S101-S108 (1992)
4. S. M. Rapoport and T. Schewe, *Biochim. Biophys. Acta.* 864, 471-495 (1986)
5. H. Kuhn, J. Belkner, R. Wiesner and A.R. Brash, *J. Biol. Chem.* 265: 18351-18361 (1990)
6. J.J. Murray and A.R. Brash, *Arch. Biochem. Biophys.* 265: 514-523 (1988)
7. C.P. Sparrow, S. Parthasarathy and D.J. Steinberg, *Lipid Res.* 29: 745-753 (1988)
8. S. Yla-Herttuala, M.E. Rosenfeld, S. Parthasarathy, C.K. Glass, E. *U.S.A.* 87: 6959-6963 (1991)
9. S. Yla-Herttuala, M.E. Rosenfeld, S. Parthasarathy, E. Sigal, T. Sarkioja, J.L. Witztum and D.J. Steinberg, *J. Clin. Invest.* 87: 1146-1152 (1991)
10. C.N. Serhan, *Biochim. Biophy. Acta.* 12: 1-25 (1994)
11. M.E. Brezinski and C.N. Serhan, *Proc.Acad. Natl. Sci.U.S.A.* 87: 6248-6252 (1990)
12. D.B. Fischer, J.W. Christman and K.F. Badr, *Kidney Int.* 41: 1155-1160 (1992)
13. A.B. Legrand, J.A. Lawson, B.O. Meyrick, I.A. Blair and J.A. Oates, *J. Biol. Chem.* 266: 7570-7577 (1991)
14. S. Fiore, J.F. Maddox, H. Daniel Perez and C.N. Serhan, *J. Exp. Med.* 180: 253-260 (1994)
15. T.R. Mosmann and R.L. Coffman, *Annu. Rev. Immunol.* 7: 145-173 (1989)
16. F. Powrie and R.L. Coffman, *Immunol. Today.* 14: 270-274 (1993)
17. A. Minty, P. Chalon, J-M. Derocq, X. Dumont, I-C. Guillemot, M. Kaghad, C. Labit, P. Leplatois, P. Liauzun, B. Miloux, C. Minty, P. Casellas, G. Loison, J. Lupker, S. Shire, P. Ferrara and D. Caput, *Nature* 362: 248-250 (1993)

18. A.N.J. McKenzie, J.A. Culpepper, R. De Waal Malefyt, F. Briere, J. Punnonen, G. Aversa, A. Sato, W. Dang, B.G. Cocks, S. Menon, J.E. De Vries, J. Banchereau and G. Zurawski, *Proc. Natl. Acad. Sci. U.S.A.* 90: 3735-3739 (1993)

19. R. De Waal Malefyt, C.G. Figdor, R. Huijbens, S. Mohan-Paterson, B. Bennett, J. Culpepper, W. Dang, G. Zurawski and J.E. De Vries, *J. Immunol.* 151: 6370-6381 (1993)

20. H.L. Wong, G.L. Costa, M.T. Lotze and S.M. Wahl, *J. Exp. Med.* 177: 775-781 (1993)

21. D.F. Fiorentino, A. Zlotnik, T.R. Mosmann, M.H. Howard and A. O'Garra, *J. Immunol.* 147: 3815-3822 (1991)

22. D.J. Conrad, H. Kuhn, M. Mulkins, E. Highland and E. Sigal, *Proc. Acad. Sci. U.S.A.* 89: 217-221 (1992)

23. B.D. Levy, M. Romano, H.A. Chapman, J.J. Reilly, J. Drazen and C.N. Serhan, *J. Clin. Invest.* 92: 1572-1579 (1993)

24. G.M. Nassar, J.M. Morrow, L.J. Roberts II, F.G. Lakkis and K.F. Badr, *J. Biol. Chem.* 269: 27631-27634 (1994)

25. J. Fleming, B.J. Thiele, J. Chester, J. O'Prey, S. Janetzki, A. Aitken, I.A. Anton, S.M. Rapoport and P.R. Harrison, *Gene* 79: 181-188 (1989)

26. E. Sigal, C.S. Craik, E. Highland, D. Grunberger, L.L. Costello, R.A.F. Dixon and J.A. Nadel, *Biochem. Bioph. Res. Comm.* 157: 457-464 (1989)

27. T. Yoshimoto, T. Arakawa, T. Hada, S. Yamamoto and E. Takahashi, *J. Biol. Chem.* 267: 24805-24809 (1992)

28. S. Hoshiko, O. Raadmark and B. Samuelsson, *Proc. Natl. Acad. Sci. U.S.A.* 87: 9073-9077 (1990)

29. I. Köhler and E.P. Rieber, *Eur. J. Immunol.* 23: 3066-3071 (1993)

30. J. Hou, U. Schindler, W.J. Henzel, T.C. Ho, M. Brasseur and S.L. McKnight, *Science.* 265: 1701-1706 (1994)

31. I. Köhler, P. Alliger, A. Minty, D. Caput, P. Ferrara, B.H. Neugebauer, G. Rank and E.P. Rieber, *FEBS. Lett.* 345: 187-192 (1994)

32. T. Taniguchi, *Science.* 268: 251-255 (1995)

33. J. Sambrook, E.F. Fritsch and T. Maniatis, *Molecular cloning: A laboratory manual, 2nd Ed., Cold Spring Harbor Laboratory Press,* Plainview, New York (1989)

34. J.D. Dignam, R.M. Labovitz and R.G. Roeder, *Nucleic Acids Res.* 11: 1475-1489 (1983)

35. G.E. Austin, W.G. Zhao, W. Zhang, E.D Austin, H.W. Findlev and J. Murtagh, *Leukemia.* 9: 848-857 (1995)

36. DNA sequence input and analysis software system for PC-DOS: Ver VIII C0410-604 (1991)

37. S. Altschul, W. Gish, M. Webb, E. Myers and D.J. Lipman, *J. Mol. Biol.* 215: 403-410 . (1990)

38. U. Kelavkar, S. Wang, A. Montero, J. Murtagh, K.Shah and K. Badr, *Mol. Biol.Reports,* 25: 1-10 (1998)

39. S.T. Smale and D. Baltimore, *Cell* 57: 103-113 (1989)

40. K.L. Luskey, *Mol. Cell. Biol.* 7: 1881-1899 (1987)

41. Y. Li, S. Camp, T.L. Rachinsky, C. Bongiorno and P. Taylor, *J. Biol. Chem.* 268: 3563-3572 (1993)

42. C. Funk, L. Funk, G. FitzGerald and B. Samuelsson, *Proc. Natl. Acad. Sci. U.S.A.* 89: 3962-3966 (1992)

43. T. Katoh, F. Lakkis, N. Makita and K. Badr, *Kidney Int.* 46: 341-349 (1994)

12

CYTOKINE INDUCED REGULATION OF 15-LIPOXYGENASE AND PHOSPHOLIPID HYDROPEROXIDE GLUTATHIONE PEROXIDASE IN MAMMALIAN CELLS

Kerstin Schnurr, Roland Brinckmann and Hartmut Kühn

Institute of Biochemistry, University Clinics Charité, Humboldt University, Hessische Str. 3-4, 10115 Berlin, Germany

INTRODUCTION

Lipid peroxidation has been implicated in the pathogenesis of various diseases such as cancer, atherosclerosis, and neurodegenarative disorders (Louvel et al., 1997). Under resting conditions the intracellular peroxide level of mammalian cells is rather low (Weitzel and Wendel, 1993) and this may be due to the high reducing capacity of most mammalian cells. However, during cell activation the cellular peroxide tone may be increased leading to membrane damage and even cell death. The cellular peroxide tone is a steady state concentration which depends *inter alia* on the expression of pro- and anti-oxidative enzymes. With respect to lipid peroxidation 15-lipoxygenases (15-LO's) and the phospholipid hydroperoxide glutathione peroxidase (PH-GPx) constitute opposite enzymes in the cellular metabolism of hydroperoxy ester lipids.

15-LO's are lipid peroxidizing enzymes capable of oxygenating complex ester lipids and even lipid-protein assemblies such as biomembranes and lipoproteins (Kühn, 1996). These hydroperoxy lipids may subsequently induce radical chain reactions and thus, may exhibit cytotoxic effects if not reduced to the corresponding hydroxy derivatives. In most resting mammalian cells lipid hydroperoxides are rapidly reduced. It is generally accepted that glutathione peroxidases may be involved in the reduction of free and esterified lipid hydroperoxides. Glutathione peroxidases constitute a family of selenoperoxidases (Burk and Hill, 1993) and several isoenzymes have been described (Ursini et al., 1995). Among them, only the PH-GPx is capable of reducing hydroperoxy fatty acids which are esterified to biomembranes or lipoproteins. We have studied the interaction of 15-LO and PH-GPx in reconstituted *in vitro* systems (Schnurr et al., 1996) and found that pre-incubation of biomembranes with PH-GPx completely prevented their susceptibility towards oxygenation by 15-LO's. When both 15-LO and PH-GPx were added simultaneously to the membrane preparation a partial inhibition of membrane lipid oxygenation was observed. These data suggested that PH-GPx down-regulates the enzymatic activity of the 15-LO.

The expression of 12/15-LO's in several mammalian cells is up-regulated by the interleukins-4 and -13 (Nassar et al., 1994; Conrad et al., 1992; Levy et al., 1993; Brinckmann et al., 1996, Cornicelli et al., 1996). Considering the inverse functional relation between 15-LO and PH-GPx one may assume that cytokine induced up-regulation of 15-LO expression may be accompanied by a down-regulation of PH-GPx. Unfortunately, no data are currently available as to the regulation of PH-GPx by interleukins. This lack of information prompted as to study the impact of interleukins-4 and -13 on the expression of PH-GPx *in vitro* and *in vivo*. The data obtained indicate that 12/15-LO's and PH-GPx are inversely regulated by these interleukins in human lung carcinoma cells (A549) and in human peripheral monocytes. The regulatory processes involved may create a oxidizing environment inside the cells. These findings may be of patho-physiological importance for disorders which are associated with up-regulation of oxidative metabolic events.

MATERIALS AND METHODS

The chemicals used were from the following sources: NADPH, glutathione, glutathione reductase, sodium selenite, penicillin-streptomycin solution, immunochemicals from Sigma (Deisenhofen, Germany); trypsin/versene from BioWhittaker (Heidelberg, Germany); DMEM from Gibco BRL (Eggenstein, Germany); recombinant human IL-4 and IL-13 from PBH (Hannover, Germany). All solvents used were of HPLC grade and purchased from J.T.Baker (Deventer, The Netherlands).

For immunoblotting and immunocytochemical stainings a polyclonal anti-rabbit-15-LO antibody (FPLC purified IgG fraction of a guinea pig) was used. This antibody was tested to strongly cross-react with the human 15-LO, the murine leukocyte-type 12-LO and with the porcine leukocyte-type 12-LO but not with the human platelet-type 12-LO and with the human 5-LO. For PH-GPx staining a peptide derived polyclonal antibody was used which was raised in rabbits against a central peptide of the porcine PH-GPx which exhibited low homology to other GPx isoforms.

Human lung carcinoma cells (A549) which were purchased from A.T.C.C. (Rockville, MD) were maintained in Dulbecco´s modified Eagle´s medium (DMEM) containing 10% human AB-serum, antibiotics (100 units/ml penicillin and 100 μg/ml streptomycin) and sodium selenite (50 nM). Human peripheral monocytes were isolated from buffy coats of healthy volunteers by density gradient centrifugation and adherence to plastic dishes (Conrad et al., 1992). They were cultured in RPMI 1640 medium supplemented with 10% human AB-serum, antibiotics and sodium selenite for 5 days. The cell viability was usually higher than 98% as indicated by Trypan exclusion. After plating the cells at a density of about 2×10^6 cells/Petri dish they were cultured for 24 h in the absence of exogenous cytokines and then IL-4 or IL-13 (670 pM) were added.

For measurements of the cellular PH-GPx activity the cells were trypsinized from the culture dishes, washed and resuspended in sodium chloride/phosphate solution (137 mM NaCl, 2.7 mM KCl, 8.1 mM Na_2HPO_4, 1.5 mM KH_2PO_4), pH7.4. The cells were sonicated on ice (15 times for 0.5 s at 40 W with a microtip sonifier; Braun; Melsungen, Germany). The homogenate was centrifuged for 5 min at 20,000 x g and 4° C, and the supernatant was assayed for PH-GPx activity. The assay mixture was composed of 1.2 ml 0.1 M Tris-HCl (pH 7.4) containing 5 mM EDTA, 0.1% Triton X-100, 0.2 mM NADPH, 3 mM glutathione and 1 U of glutathione reductase. As source of PH-GPx the 20,000 x g supernatant of the cell homogenate (1 mg protein/ml) was added. This mixture was pre-incubated at 37°C for 5 min and then the reaction was started by addition of 25 µM dilinoleoyl phosphatidylcholine hydroperoxide as substrate. The decrease in absorbance at 340 nm reflecting the consumption of NADPH was followed at 37°C with a Shimadzu UV-2100 spectrophotometer. For measurement of the LO activity, cells were washed and resuspended in PBS at a concentration of 1 x 10^6 to 1 x 10^7 cells/ml. After addition of linoleic acid as substrate (100 µM final concentration) the formation of specific 15-LO products was quantified by RP-HPLC as measure for the cellular 15-LO activity (Brinckmann et al., 1996).

Transgenic mice which overexpress IL-4 under the control of MHC I regulatory elements (Erb et al., 1994) were kindly provided by A. Schimpl (Würzburg, Germany). Animals were killed by diethyl ether inhalation, the organs were removed and were shock-frozen in liquid nitrogen. After thawing the organs (about 100-500 mg wet weight) were homogenized on ice in 6 ml of sodium chloride/phosphate solution, pH 7.4 using an Ultraturax microhomogenizer. The homogenates were centrifuged at 20,000 x g and aliquots of the 20,000 x g supernatant was used for PH-GPx activity assays in the standard spectrophotometric assay (see above). Arachidonic acid oxygenase activity was assayed as described in reference (Brinckmann et al., 1996) with the exception that frozen pieces of the transgenic tissues were used instead of A549 cells.

RESULTS

Human lung carcinoma cells A549 do not express the 15-LO as indicated by immunohistochemical staining, quantitative RT-PCR and activity assays. However, when the cells were cultured for 3 days in the presence of IL-4 the 15-LO was expressed (Brinckmann et al, 1996). In contrast, the PH-GPx is expressed in resting A549 cells as indicated by an immunoreactive protein band of about 20 kDa which cross reacted with a our anti-human PH-GPx antibody (Fig. 1).

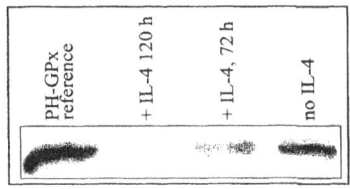

Figure 1. Immunoblots of PH-GPx in resting, and IL-4 treated A549 cells. As reference the recombinant human PH-GPx expressed in the baculovirus/insect cell system was used. Before applying to SDS gel electrophoresis the enzyme was purified by FPLC on a MONO-S column.

After culturing the cells for 3 days in the presence of recombinant human IL-4 the intensity of the immunoreactive band was markedly impaired suggesting a decrease in the cellular PH-GPx concentration (Fig. 1). After 5 days of culturing in the presence of IL-4 the immunoreactive band disappeared almost completely. A similar decrease in PH-GPx expression was observed in immunocytochemical staining and when the PH-GPx mRNA was quantified by competitive RT-PCR (data not shown).

In order to find out whether this regulatory phenomenon is also detectable on the level of enzyme activity 15-LO and PH-GPx activity assays were carried out. It can be seen from Table 1 that IL-4 treatment led to a strong increase in the 15-LO activity. In resting A549 cells no 15-LO activity was detectable. The small amounts of hydroxy fatty acids found originated from non-enzymatic oxidation processes as indicated by analysis of their chemical structure. However, when the cells were treated with IL-4 a time dependent increase in the formation of hydroxy fatty acids was observed. HPLC and GC/MS analysis of the chemical structure indicated 13S-HODE(Z,E) as major oxygenation product when exogenous linoleic acid was used as substrate. Other hydroxy linoleic acid isomers such as 9-HODE(E,Z) or the E,E-isomers were only detected in trace amounts. These data are consistent with a 15-LO origin of the 13S-HODE(Z,E) formed. In contrast to the increase in cellular 15-LO activity, PH-GPx activity was impaired during IL-4 treatment of A549 cells (Table 1) suggesting a IL-4 induced down-regulation of enzyme expression. A similar inverse regulation of 15-LO and PH-GPx was observed for IL-4 and IL-13 treatment of human peripheral monocytes.

Table 1. Inverse regulation of the cellular activities of 15-LO and PH-GPx in A549 cells upon IL-4 treatment

conditions	15-LO activity (μg 13-HODE/Mill cells)	PH-GPx activity (ΔA_{340}/min x mg protein)
no IL-4	<0.01	0.074
plus IL-4 3 days	0.6	0.03
plus IL-4 5 days	2.43	0.02

In order to find out whether these regulatory effects may also occur *in vivo* we assayed the arachidonate oxygenase activity and the PH-GPx activity in various tissues of mice which systemically overexpress IL-4. Here we found that in heart (Fig. 2), kidney, liver and spleen the arachidonic acid oxygenase activity was higher in the tissues of transgenic mice when compared with inbred controls of identical genetic background. No significant differences were observed in the liver. PH-GPx activity was impaired in lung, spleen and liver of the transgenic animals, but no differences were seen in heart and kidney. Thus, the spleen was the only organ in which we observed an inverse regulation of 15-LO and PH-GPx. In other tissues either the 15-LO was up-regulated while PH-GPx was not affected or PH-GPx was down-regulated and 15-LO expression was not influenced. These data suggest tissue specific regulation mechanisms.

DISCUSSION

IL-4 and IL-13 are capable of inducing the expression of 15-LO's in various mammalian cells when the cells were cultured *in vitro* in the presence of these cytokines. Here we showed for the first time that in A549 human lung carcinoma cells and in human peripheral monocytes the up-regulation of 15-LO is accompanied by a down-regulation of PH-GPx expression. Interestingly, the regulatory effects of IL-4 are not restricted to *in vitro* incubations but they also occur *in vivo*. In transgenic mice which systemically overexpress IL-4 the arachidonic acid oxygenase activity is up-regulated in various tissues. This is the first experimental evidence that IL-4 induced 15-LO expression does actually occur *in vivo*.

Up-regulation of 15-LO in A549 cells and peripheral monocytes requires cytokine concentrations which are at least two orders of magnitude higher than the IL-4 plasma concentrations measured in humans. It should, however, be stressed that the plasma level may not reflect the IL-4 concentrations in other parts of the body. In fact, when the IL-4 plasma concentration was assayed in the transgenic mice which overexpress this cytokine (Erb et al., 1994) no significant differences to the inbred control animals were observed. However, clear cut IL-4 related effects on T-cell subset distribution and on B-cell hyperreactivity (Erb et al., 1994) have been reported in these animals. These data suggest that increased plasma concentrations may not be required for IL-4 induced cell physiological effects.

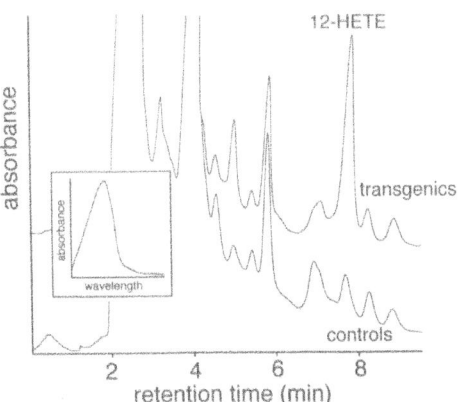

Figure 2. Increased arachidonic acid oxygenase activity in the heart of transgenic mice overexpressing IL-4 systemically. Inset: uv-spectrum of 12-HETE.

The intracellular mechanism of IL-4 and/or IL-13 induced up-regulation of 12/15-LO and of the concomitant down-regulation of the PH-GPx remains to be investigated. In the transgenic mice there appears to be no strong coupling of up- and down-regulation of both enzymes. Only in spleen an up-regulation of 12/15-LO and a concomitant down-regulation of PH-GPx was observed. The observed time course of up-regulation of 15-LO expression in monocytes (Conrad et al., 1992) and A549 cells (Brinckmann et al., 1996) suggests that the 12/15-LO gene may not belong to the immediate early genes turned on by this cytokine. More work is needed in the future to characterize the signal transduction cascade leading to the up-regulation of 12/15-LO's and down-regulation of PH-GPx. For the up-regulation of the 15-LO in human monocytes preliminary data may suggest the involvement of a new transcription factor (Kkelavkar et al., 1997). Whether this protein may also occur in other cells and whether it is involved in the down-regulation of PH-GPx remains to be investigated.

IL-4 and IL-13 are pleiotropic cytokines which have a broad range of biological activities (Paul, 1991). IL-4 plays a critical role in the pathogenesis of allergic disease since it is necessary for the isotype switching of B cells to produce Ig E (Kuhn et al., 1991). At this point of research it is rather difficult to speculate whether the up-regulation of the 15-LO and the concomitant down-regulation of the PH-GPx in A549 cells and in the spleen of transgenic mice may be related to any of the known IL-4 effects. Experiments with specific LO inhibitors and/or studies with transgenic animals which lack or overexpress either 12/15-LO's and/or PH-GPx are required in order to sort out which role both enzymes may play in the cytokine induced metabolic network. 12/15-LO deficient mice have already been created (Sun and Funk, 1996). When these animals would be cross bred with IL-4 overexpressing mice (Erb et al., 1997) more detailed information on the role of 15-LO in the IL-4 induced signal transduction may be obtained.

REFERENCES

Brinckmann, R., M. Topp, I. Zalan, D. Heydeck, P. Ludwig, H. Kühn, W.E. Berdel, and A.J.R. Habenicht, 1996, The regulation of 15-lipoxygenase expression in lung epithelial cells by interleukin 4. *Biochem. J.* 318: 305.

Burk, R.F., and K.E. Hill, 1993, Regulation of selenoproteins. *Annu. Rev. Nutr.* 13:65.

Conrad, D.J., H. Kühn, M. Mulkins, E. Highland, and E. Sigal. 1992. Specific inflammatory cytokines regulate the expression of human monocyte 15-lipoxygenase. *Proc. Natl. Acad. Sci. USA* 89:217.

Cornicelli, J.A., K. Welch, B. Auerbach, S.J. Feinmark, and A. Daugherty. 1996. Mouse peritoneal macrophages contain abundant omega-6 lipoxygenase activity that is independent of interleukin-4. *Arterioscler. Thromb. Vasc. Biol.* 16:1488-1494.

Erb, K.J., B. Ruger, M. von-Brevern, B. Ryffel, A. Schimpl, and K. Rivett, 1997, Constitutive expression of interleukin (IL)-4 in vivo causes autoimmune-type disorders in mice. *J. Exp. Med.* 185:329.

Erb, K.J., T. Holtschke, K. Muth, I. Horak, and A. Schimpl, 1994, T cell subset distribution and B cell hyperreactivity in mice expressing interleukin-4 under the control of major histocompatibility omplex class I regulatory sequences. *Eur. J. Immunol.* 24:1143.

Kkelavkar, U., S. Wang, A. Montero, and K.F. Badr, 1997, Identification and characterization of promoter and enhancer sequences of human 15-lipoxygenase (15-LO) gene. *5th International Conference of Eicosanoids & other bioactive Lipids in Cancer, Inflammation and releated Diseases* LaJolla,CA.

Kühn, H., 1996, Biosynthesis, metabolization and biological importance of the primary 15-lipoxygenase metabolites 15-hydro(pero)xy-5Z,8Z,11Z,13E-eicosatetraenoic acid and 13-hydro(pero)xy-9Z,11E-octadecadienoic acid. *Prog. Lipid Res.* 35:203.

Kuhn, R., K. Rajewsky, and W. Muller, 1991, Generation and analysis of interleukin-4 deficient mice. *Science* 254:707.

Levy, B.D., M. Romano, H.A. Chapman, J.J. Reilly, J. Drazen, and C.N. Serhan, 1993, Human alveolar macrophages have 15-lipoxygenase and generate 15(S)-hydroxy-5,8,11-cis-13 trans-eicosatetraenoic acid and lipoxins. *J. Clin. Invest.* 92:1572.

Louvel, E., J. Hugon, and A. Doble, 1997, Therapeutic advances in amyotrophic lateral sclerosis. *Trends Pharmacol. Sci.* 18:196.

Nassar, G.M., J.D. Morrow, L.J. Roberts II, F.G. Lakkis, and K.F. Badr, 1994, Induction of 15-lipoxygenase by interleukin-13 in human blood monocytes. *J. Biol. Chem.* 269:27631.

Paul, W.E., 1991, Interleukin-4: a prototypic immunoregulatory lymphokine. *Blood* 77:1859.

Schnurr, K., Belkner, J., Ursini, F., Schewe, T., Kühn, H., 1996, The selenoenzyme phospholipid hydroperoxide glutathione peroxidase controls the activity of the 15-lipoxygenase with complex substrates and preserves the specificity of the oxygenation products. *J. Biol. Chem.* 271: 4653.

Sun, D., and C.D. Funk, 1996, Disruption of 12/15-lipoxygenase expression in peritoneal macrophages. Enhanced utilization of the 5-lipoxygenase pathway and diminished oxidation of low density lipoprotein. *J. Biol. Chem.* 271:24055.

Ursini, F., M. Maiorino, R. Brigelius-Flohé, K.D. Aumann, A. Roveri, D. Schomburg, and L. Flohé, 1995, Diversity of glutathione peroxidases. *Meth. Enzymol.* 252:38.

Weitzel, F., and A. Wendel, 1993, Selenoenzymes regulate the activity of leukocyte 5-lipoxygenase via the peroxide tone. *J. Biol. Chem.* 268:6288.

13

INVESTIGATION OF A SECOND 15S-LIPOXYGENASE IN HUMANS AND ITS EXPRESSION IN EPITHELIAL TISSUES

Alan R. Brash[1], Mitsuo Jisaka[1,7], William E. Boeglin[1], Min S. Chang[2], Diane S. Keeney[3], Lillian B. Nanney[4], Susan Kasper[5], Robert J. Matusik[5], Sandra J. Olson[6], and Scott B. Shappell[6]

[1]Departments of Pharmacology, Ophthalmology[2], Biochemistry[3], Cell Biology[4], Urology[5], and Pathology[6]
Vanderbilt University Medical Center
Nashville, TN 37232-6602

[7]Present address:
Mitsuo Jisaka, PhD
Department of Life Science and Biotechnology
National Shimane University
1060, Matsue, Shimane 690
Japan

INTRODUCTION

Three Human Lipoxygenases, 1975-1977

The first reports of a lipoxygenase in animal cells, the 12S-lipoxygenase of platelets, were published in 1974-75 (Hamberg and Samuelsson, 1974; Nugteren, 1975). At nearly the same time, Rapoport and colleagues recognized the existence of the 15S-lipoxygenase in reticulocytes (Schewe et al., 1975), and discovery of the leukocyte 5S-lipoxygenase was reported in the following year (Borgeat et al., 1976). Subsequent metabolism studies and molecular analyses indicated the occurrence of the same enzymes in selected other tissues; for example, both the 12S- and 15S-lipoxygenases were expressed also in skin (Nugteren and Kivits, 1987; Hussain et al., 1994; Takahashi et al., 1993). For the past two decades it had appeared that the three enzymes could account for all known lipoxygenase activities in man.

Indications of the potential existence of additional mammalian lipoxygenases came from reports of formation of other HETE metabolites. Most notably, Fürstenberger, Marks and colleagues described an enzyme in phorbol ester-treated mouse skin that forms 8-HETE (Gschwendt et al., 1986). Subsequent studies confirmed the enzyme is an 8S-lipoxygenase (Hughes and Brash, 1991; Fürstenberger et al., 1991). This enzyme represented the fourth known lipoxygenase type in the mouse, the others being the leukocyte 5S-lipoxygenase, the platelet-type of 12S-lipoxygenase, and the leukocyte-type of 12S-lipoxygenase (generally considered the homologue of the 15S-lipoxygenase of human and rabbit reticulocytes, macrophages and eosinophils) (Chen et al, 1994, 1995). In 1996, a third type of 12S-lipoxygenase was cloned from mouse skin, and this was named the epidermal 12-lipoxygenase (eLox) (Funk et al., 1996).

DISCOVERY OF A SECOND 15S-LIPOXYGENASE IN HUMANS

A complex mixture of eicosanoids is reported in human skin, and these include hydroxy fatty acids of both the R and S stereoconfigurations (Hammarström et al., 1975; Camp, 1982; Woollard, 1986; Baer et al., 1991; Baer and Green, 1993). In searching for an enzyme that might account for the formation of 12R-HETE, we carried out a homology-based RT-PCR strategy to screen for novel lipoxygenase(s) (Brash et al, 1997). The cDNA was prepared from human hair follicles as this represented an easily acquired source of human skin cell types for preparation of RNA. Using a nested PCR protocol we detected many clones of the known 12S- and 15S-lipoxygenases, and additionally, we identified clones of a previously unrecognized lipoxygenase transcript (Brash et al., 1997).

The complete cDNA sequence of the new lipoxygenase was obtained by 5' and 3' RACE, and a clone corresponding to the open reading frame was prepared by PCR. The cDNA encodes a protein of 676 amino acids with a calculated molecular weight of 76 kD. The amino acid sequence has approximately 35-45% identity to the known human 5S-, 12S- and 15S-lipoxygenases. When the cDNA was expressed in HEK 293 cells, the enzyme converted [^{14}C]arachidonic acid to 15S-HPETE (Figure 1). Thus, the enzyme represents the second 15S-lipoxygenase in humans.

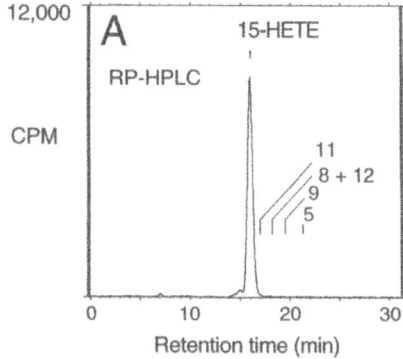

Figure 1. Reversed-phase HPLC analysis of [^{14}C]arachidonic acid metabolism by the second type of human 15S-lipoxygenase

The cDNA was expressed in HEK 293 cells and the cell sonicate incubated with [^{14}C]arachidonic acid. [^{14}C]15-HETE is the only significant HETE product. Retention times of the other HETEs (11, 8+12, 9, 5) are indicated. The small peak on the front shoulder of the [^{14}C]15-HETE co-chromatographs with 15-ketoeicosatetraenoic acid. (From Brash et al, 1997, with permission).

There are distinct differences in the properties and expression of the two human 15S-lipoxygenases. The two enzymes are certainly not functional homologues, and the two scarcely qualify as isozymes, as their catalytic activities differ very significantly. Three distinctively different characteristics of each enzyme are highlighted here: substrate specificity, regiospecificity, and tissue expression.

Substrate Specificity: Preference for C18.2ω6 versus C20.4ω6

As noted many years ago, the original reticulocyte type of 15S-lipoxygenase (designated here as 15-Lox-1) uses linoleic acid in preference to arachidonic acid as substrate (Soberman et al., 1985). In contrast, the new enzyme cloned from skin (15-Lox-2) has a distinct preference for arachidonic acid. Linoleic acid is also metabolized by 15-Lox-2, but at about one third the efficiency of arachidonate (Brash et al., 1997). It is well known that 15-Lox-1 is able to oxygenate phospholipids and other esterified polyunsaturated fatty acids. This additional property related to substrate specificity awaits study with 15-Lox-2.

Specificity of Oxygenation of Arachidonic Acid

The two enzymes differ significantly in the regiospecificity of their oxygenation of arachidonic acid. 15-Lox-1 catalyzes mainly 15S-oxygenation, but 12S-HPETE accounts for about 10-20% of the products (Bryant et al., 1982). The stereo- and regio- fidelity of 15-Lox-1 are also degraded considerably at normal physiological temperatures (as opposed to room temperature or below) (Murray and Brash, 1988), and by other factors such as the concentration and nature of the substrate (Kühn et al., 1990; Belkner et al., 1991). In comparison, 15-Lox-2 catalyzes almost exclusive 15S-oxygenation of arachidonic acid (Figure 1). There is no significant proportion of 12S-HPETE or other by-products. The conclusion is that 15-Lox-2 is an enzyme designed for the specific conversion of arachidonic acid to 15-H(P)ETE. In contrast, 15-Lox-1 is comparatively non-specific in its regio- and stereospecificity. This enzyme appears to be designed for the catalysis of non-selective oxygenations, releasing free radical intermediates that can induce co-oxidation and co-oxygenation reactions on unsaturated fatty acids and other components of the cell. This property is central to the proposed roles of 15-Lox-1 in the maturation of reticulocytes (structural modification of red cell membranes) or in the initiation of atherosclerosis by promoting the oxidation of LDL.

Tissue Localization

As noted earlier, expression of 15-Lox-1 is associated with reticulocytes, macrophages, eosinophils and additionally with certain epithelia including trachea and skin. So far, 15-Lox-2 has not been found in any blood cells. As determined by Northern analysis, 15-Lox-2 is also *not* expressed in heart, brain, placenta, liver, skeletal muscle, kidney, pancreas, spleen, thymus, testis, ovary, small intestine, colon, retina and iris (Brash et al., 1997). Expression of the mRNA was detected in lung, prostate, and cornea. Together with skin, this gives four known sites of expression of 15-Lox-2, all epithelia, or tissues having a major epithelial cell component.

CHARACTERIZATION OF A MURINE HOMOLOGUE OF 15-LOX-2

The Structural Homologue of 15-Lox-2 in the Mouse: the Phorbol Ester-Inducible 8S-Lipoxygenase

Using a slight modification of the homology-based PCR strategy with cDNA from phorbol ester-treated mouse skin as template, we cloned a novel mouse lipoxygenase transcript (Jisaka et al., 1997). This enzyme was difficult to express in HEK 293 cells, but was successfully and consistently expressed in vaccinia virus-infected Hela cells (Jisaka et al., 1997). The enzyme converts arachidonic acid almost exclusively to 8S-HPETE. In DNA and protein sequence, this mouse 8S-lipoxygenase shares about 78% identity to the human 15-Lox-2. Indeed, the two enzymes share a very unusual amino acid substitution in the position of a putative fifth amino acid ligand to the active site iron (Minor et al., 1996). Both enzymes have a serine in place of the usual histidine or asparagine at amino acid positions 558/557 (8-Lox, 15-Lox-2 respectively).

Expression of the Mouse 8S-Lipoxygenase in Skin

The reaction of mouse skin to phorbol ester treatment is very strain-dependent (Fischer et al., 1988). Using mice that react well to TPA, we demonstrated a strong induction in mRNA, protein, and 8S-lipoxygenase enzymatic activity (Figure 2). Immunohistochemical analyses of the

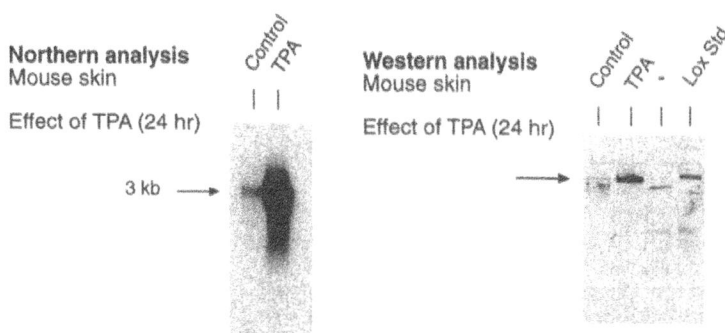

Figure 2. Effect of phorbol ester on expression of the mouse skin 8S-lipoxygenase: Northern and Western analyses (In part from Jisaka et al, 1997, with permission).

8-lipoxygenase expression were conducted in mouse back skin and tail skin (the latter having a thicker epidermis where differences in expression level throughout the cell layers are more readily discerned). The 8S-lipoxygenase is expressed in the outer, most differentiated cells of the epidermis (the stratum granulosum). Results of *in situ* hybridization of the mRNA confirmed the epidermal expression and clearly revealed also expression in the differentiating cells of the hair follicle. Phorbol ester induction of 8S-lipoxygenase activity is associated with hyperplasia of the 8S-lipoxygenase positive cells. Thus, the increase in 8-HETE formation is accounted for by induction of enzyme in the stratum granulosum, as well as by an increase in the number of 8-lipoxygenase-positive cells (Jisaka et al., 1997).

Other Sites of Expression of the Mouse 8S-Lipoxygenase

Northern analysis using stringent conditions and a 8S-lipoxygenase-specific probe revealed hybridization also to transcripts in mouse brain (Jisaka et al., 1997), the significance of which requires further evaluation. In the many published studies of arachidonic acid metabolism in brain tissue, there are no reports of the specific synthesis of 8-HETE.

HUMAN 15-LOX-2 AND MOUSE 8-LOX: DIFFFERENCES IN EXPRESSION

Expression of 15-Lox-2 in Human Prostate

We have initiated a study of 15-Lox-2 expression in the human prostate. We have detected formation of 15-HETE by an enzyme activity assay (HPLC analysis of the conversion of [^{14}C]arachidonic acid to product), and we detected 15-Lox-2 in several samples of normal prostate by Western and Northern analyses. The initial indications from immunohistochemical analyses are of selective expression of 15-Lox-2 in the prostate epithelial cells. The aim now is to study of changes in expression in hyperproliferative conditions and cancer of the prostate. Potentially, there could be some association between 15-Lox-2 expression and epithelial hyperplasia, as there appears to be for the mouse 8S-lipoxygenase in skin.

Human 15-Lox-2 and Mouse 8-Lox: Functional Homologues?

While the two enzymes we cloned appear to be structurally related, it remains to be established if the two are functional homologues. There appear to be species differences in the expression patterns of the two enzymes. For example, 8S-lipoxygenase activity was not detected in normal mouse prostate (Jisaka et al., 1997).

The end product of the mouse enzyme, 8S-HETE, is one of the recognized ligands of the peroxisome proliferator activated receptor, PPAR-α (Yu et al., 1995). It is worth emphasizing that that mouse is the only animal, and mouse skin is the only tissue, in which the specific synthesis of 8S-HETE has been clearly demonstrated. If the activity of 8S-HETE on PPAR-α has physiological relevance, and if the mouse 8S-lipoxygenase and 15-Lox-2 are functional homologues, then one would predict that the human equivalent of 8S-HETE should be 15S-HETE. At present, however, there are no published data to support this supposition on the bioactivity of 15-HETE in man.

ACKNOWLEDGMENTS

This work was supported by NIH grant GM-53638. DSK acknowledges support by NIH grant KO1 AR02002.We acknowledge also the contributions of the Vanderbilt Skin Disease Research Center (5P30 AR41943) and the DNA Sequencing Core laboratory of the Vanderbilt Cancer Center (1P30 CA68485).

REFERENCES

Baer, A.N., Costello, P.B., and Green, F.A. (1991) Stereospecificity of the products of the fatty acid oxygenases derived from psoriatic scales. *J. Lipid Research* 32:341.

Baer A.N., and Green F.A. (1993) Fatty acid oxygenase activity of human hair roots. *J. Lipid Research* 34:1505.

Belkner, J., Wiesner, R., Kühn, H., and Lankin, V. Z. (1991) The oxygenation of cholesterol esters by the reticulocyte lipoxygenase. *FEBS Lett.* 279:110.

Borgeat, P., Hamberg, M., and Samuelsson, B. (1976) Transformation of arachidonic acid and homo-γ-linolenic acid by rabbit polymorphonuclear leukocytes. Monohydroxy acids from novel lipoxygenases. *J. Biol. Chem.* 251:7816.

Brash, A. R., Boeglin, W. E., and Chang, M. S. (1997) Discovery of a second 15S-lipoxygenase in humans. *Proc. Natl. Acad. Sci. USA* 94:6148.

Bryant, R. W., Bailey, J. M., Schewe, T., and Rapoport, S. M. (1982) Positional specificity of a reticulocyte lipoxygenase. Conversion of arachidonic acid to 15S-hydroperoxy-eicosatetraenoic acid. *J. Biol. Chem.* 257:6050.

Camp, R. D. R., Mallet, A. I., Woollard, P. M., Brain, S. D., Kobza-Black, A., and Greaves, M. W. (1983) The identification of hydroxy fatty acids in psoriatic skin. *Prostaglandins* 26:431.

Chen, X-S., Kurre, U., Jenkins, N. A., Copeland, N. G., and Funk, C. D. (1994) cDNA cloning, expression, mutagenesis of C-terminal isoleucine, genomic structure, and chromosomal localizations of murine 12-lipoxygenases. *J. Biol. Chem.* 269:13979.

Chen, X-S., Naumann, T. A., Kurre, U., Jenkins, N. A., Copelend, N. G., and Funk, C. D. (1995) cDNA cloning, expression, mutagenesis, intracellular localization, and gene chromosomal assignment of mouse 5-lipoxygenase. *J. Biol. Chem.* 270:17993.

Fischer, S.M., Baldwin, J.K., Jasheway, D.W., Patrick, K.E., and Cameron, G.S. (1988) Phorbol ester induction of 8-Lipoxygenase in inbred SENCAR (SSIN) but not C57BL/6J mice correlated with hyperplasia, edema, and oxidant generation but not ornithine decarboxylase induction. *Cancer Res.* 48:658.

Funk, C.D., Keeney, D.S., Oliw, E. H., Boeglin, W.E., and Brash, A.R. (1996) Functional expression and cellular localization of a mouse epidermal lipoxygenase. *J. Biol. Chem.* 271:23338.

Fürstenberger, G., Hagedorn, H., Jacobi, T., Besemfelder, E., Stephan, M., Lehmann, W. D., and Marks, F. (1991) Characterization of an 8-lipoxygenase activity induced by the phorbol ester tumor promoter 12-O-tetradecanoylphorbol-13-acetate in mouse skin *in vivo*. *J. Biol. Chem.* 266:15738.

Gschwendt, M., Fürstenberger, G., Kittstein, W., Besemfelder, E., Hull, W. E., Hagedorn, H., Opferkuch, H. J., and Marks, F. (1986) Generation of the arachidonic acid metabolite 8-HETE by extracts of mouse skin treated with phorbol erter *in vivo*; identification by ^{1}H-n.m.r. and GC-MS spectroscopy. *Carcinogenesis* 7:449.

Hamberg. M., and Samuelsson, B. (1974) Prostaglandin endoperoxides. Novel transformations of arachidonic acid in human platelets. *Proc. Natl. Acad. Sci. USA* 71:3400.

Hughes, M. A., and Brash, A. R. (1991) Investigation of the mechanism of biosynthesis of 8-hydroxyeicosatetraenoic acid in mouse skin. *Biochim. Biophys. Acta* 1081:347.

Hammarström, S., Hamberg, M., Samuelsson, B., Duell, E. A., Stawiski, M., and Voorhees, J. J. (1975) Increased concentrations of nonesterified arachidonic acid, 12L-hydroxy-5,8,10,14-eicosatetraenoic acid, prostaglandin E_2, and prostaglandin $F_{2\alpha}$ in epidermis of psoriasis. *Proc. Nat. Acad. Sci. USA* 72:5130.

Hussain, H., Shornick, L. P., Shannon, V. R., Wilson, J. D., Funk, C. D., Pentland, A. P. and Holtzman, M. J. (1994) Epidermis contains platelet-type 12-lipoxygenase that is overexpressed in germinal layer keratinocytes in psoriasis. *Am. J. Physiol.* 266:C243.

Jisaka, M., Kim, R. B., Nanney, L. B., Boeglin, W. E., and Brash, A. R. (1997) Molecular cloning and functional expression of a phorbol ester-inducible 8S-lipoxygenase from mouse skin. *J. Biol. Chem.* 272:24410.

Kühn, H., Belkner, J., Wiesner, R., and Brash, A.R. (1990) Oxygenation of biological membranes by the pure reticulocyte lipoxygenase. *J. Biol. Chem.* 265:18351.

Minor, W., Steczko, J., Stec, B., Otwinowski, Z., Bolin, J.T., Walter, R., and Axelrod, B. (1993) Crystal structure of the soybean lipoxygenase L-1 at 1.4Å resolution. *Biochemistry* 35:10687.

Murray, J. J., and Brash, A.R. (1988) Rabbit reticulocyte lipoxygenase catalyzes specific 12(S) and 15(S) oxygenation of arachidonyl-phosphatidylcholine. *Arch. Biochem. Biophys.* 265:514.

Nugteren, D. H. (1975) Arachidonate lipoxygenase in blood platelets.*Biochim. Biophys. Acta* 380:299.

Nugteren, D. H., and Kivits, G. A. A. (1987) Conversion of linoleic acid and arachidonic acid by skin epidermal lipoxygenases. *Biochim. Biophys. Acta* 921:135.

Schewe, T., Halangk, W., Hiebsch, Ch., and Rapoport, S. M. (1975) A lipoxygenase in rabbit reticulocytes which attacks phospholipids and intact mitochondria. *FEBS Lett.* 60:149.

Soberman, R. J., Harper, T. W., Betteridge, D., Lewis, R. J., and Austen, K. F. (1985) Characterization and separation of the arachidonic acid 5-lipoxygenase and linoleic acid ω-6 lipoxygenase (arachidonic acid 15-lipoxygenase) of human polymorphonuclear leukocytes. *J. Biol. Chem.* 260:4508.

Takahashi, Y., Reddy, G. R., Ueda, N., Yamamoto, S. and Arase, S. (1993) Arachidonate 12-lipoxygenase of platelet-type in human epidermal cells. *J. Biol. Chem.* 268:16443.

Woollard, P.M. (1986) Stereochemical difference between 12-hydroxyeicosatetraenoic acid in platelets and psoriatic lesions. *Biochem. Biophys. Res. Commun.* 136:169.

Yu, K., Bayona, W., Kallen, C. B., Harding, H. P., Ravera, C. P., McMahon, G., Brown, M., and Lazar, M. A. (1995) Differential activation of peroxisome proliferator-activated receptors by eicosanoids. *J. Biol. Chem.* 270:23975.

14

SEQUENCE DETERMINANTS FOR THE POSITIONAL SPECIFICITY OF MAMMALIAN AND PLANT LIPOXYGENASES

S. Borngräber, R.-J. Kuban and H. Kühn

Institute of Biochemistry, University Clinics Charité, Humboldt University, Hessische Str. 3-4, 10115 Berlin, Germany

INTRODUCTION

Lipoxygenases (LO's), which catalyze the conversion of polyunsaturated fatty acids to their corresponding hydroperoxy derivatives are widely distributed in plants and animals. They contain one gram-atom non-heme iron per mole enzyme and this iron appears to be involved in the rate limiting step of the oxygenase reaction, the initial hydrogen removal. In mammalian cells, mainly linoleic acid and arachidonic acid serve as substrates for the LO reaction whereas in plants linolenic acid may constitute the major LO substrate. According to the currently used nomenclature, LO's are categorized with respect to their positional specificity of arachidonic acid oxygenation into 5-, 8-, 11-, 12- and 15-LO's. Several studies have been carried out in the past on the structural reasons for the positional specificity of LO's. Comparing the sequences of various mammalian 12- and 15-LO's, Sloane and colleagues identified conserved differences in the primary structure of these enzymes. When the amino acids ile418 and met419 of the human 15-LO were mutated to their counterparts present in the human platelet-type 12-LO (ala and val), this enzyme was converted to a 12-lipoxygenating species (Sloane et al., 1991). It was hypothesized that these residues form the bottom of the substrate binding pocket so that their bulkiness and the geometry of their side-chains may determine the substrate alignment at the active site. Later on, these findings were confirmed for the human platelet 12-LO (Chen and Funk, 1993) and for the porcine leukocyte-type 12-LO (Suzuki et al., 1994). However, similar attempts to alter the positional specificity of the rat leukocyte-type 12-LO failed (Watanabe and Haeggstrom, 1993; Hada et al., 1994) and the authors suggested that additional „structural features" may be of importance for the positional specificity of these LO isoforms. To obtain more detailed information on these „structural features", we constructed chimeric LO species combining different parts of the cDNA of various LO isoforms and identified amino acid 353 as sequence determinant for the positional specificity of 12- and 15-LO's (Borngräber et al., 1996). These data were used to establish a structural model for the alignment of substrate fatty acids at the active site of mammalian 12- and 15-LO's.

Here we report more recent mutagenesis studies, which confirm the importance of residue 353 in mammalian 12- and 15-LO's for the oxygenation specificity of these enzymes. Furthermore, we investigated, whether the amino acids important for the specificity of mammalian LO's are also involved in determining substrate alignment at the active site of plant enzymes.

MATERIALS AND METHODS

The chemicals used were from the following sources: arachidonic acid (5Z,8Z,11Z,14Z-eicosatetraenoic acid) and sodium borohydride from Serva (Heidelberg, Germany), ampicillin from Gibco (Eggenstein, Germany), isopropyl-ß-D-thiogalactopyranoside (IPTG) from Sigma-Aldrich (Deisenhofen, Germany), HPLC solvents from Merck (Darmstadt, Germany). Restriction enzymes were purchased from New England Biolabs (Schwalbach, Germany). Phage T4 ligase, Pwo-polymerase and sequencing kits were obtained from Boehringer Mannheim (Mannheim, Germany) and the *Escherichia coli* strain HB 101 was purchased from Invitrogen (San Diego, CA). Oligonucleotide synthesis was carried out by TiB-Molbiol (Berlin, Germany).

For bacterial expression of the wild-type 15- and 12-LO's as well as of mutant enzyme species, the following cDNA constructs basing on the plasmid pKK 233-2 (Pharmacia, Upsalla, Sweden) were used: rabpKK-15LOX containing the coding region of the rabbit reticulocyte-type 15-LO cDNA (Kühn et al., 1993), rabpKK-12LOX containing the coding region of the rabbit leukocyte-type 12-LO cDNA (Berger et al., 1998). Site-directed mutagenesis was performed by PCR-overlap extension technique using mismatching synthetic oligonucleotides and all constructs were sequenced.

Enzyme expression and activity assays were performed as previously described (Borngräber et al., 1996). Briefly, bacterial clones were cultured at 37° C in 1 ml of LB medium containing 0.1 mg/ml ampicillin to an optical density at 600 nm of at least 0.5. Lipoxygenase expression was induced by addition of IPTG (1 mM final concentration). After 12 hours of incubation, bacteria were spun down, washed and resuspended in phosphate buffered saline. For activity assays, 0.1 mM arachidonic acid (final concentration) was added, the suspension was sonicated and incubated for 10 minutes at 37° C. The hydroperoxy fatty acids formed were reduced to their more stable hydroxy derivatives by addition of sodium borohydride. After lipid extraction (two times with 1 ml of chloroform/2-propanole, 1:1, v/v), the solvent was evaporated and the residue was reconstituted in 0.3 ml of methanol/water (2:1, v/v). Aliquots of this solution were subjected to HPLC analysis. RP-HPLC was carried out on a Nucleosil C-18 column (Macherey/Nagel, KS-system, 250 x 4 mm) using a solvent system of methanol/water/acetic acid (85:15:0.1; v:v:v) and a flow rate of 1 ml/min. Absorbance at 235 nm was monitored. 12- and 15-HETE were identified by coinjections with authentic standards and by their uv-spectroscopic properties.

RESULTS

Previous mutagenesis studies indicated the amino acid at position 353 as primary determinant for the positional specificity of 12- and 15-LO's (Borngräber et al., 1996). Depending on the side-chain geometry of this residue and on the amino acids at positions 418 and 419 (Sloane et al., 1991), arachidonic acid is aligned with presenting either the doubly allylic methylene C-10 (12-lipoxygenation) or C-13 (15-lipoxygenation) to the enzyme bound iron. When phe353 of the rabbit 15-LO was replaced with leu, the rat and murine leukocyte-type 12-LO counterparts, the positional specificity of the rabbit enzyme was altered. We found that arachidonic acid was converted to 15-H(P)ETE and 12-H(P)ETE in a ratio of about 1:3 (Table 1). A similar product pattern has been reported for arachidonic acid oxygenation by the wild-type murine 12-LO (Sloane et al., 1995). These data are consistent with our model suggesting that substitution of a small residue at position 353 for the rather bulky phe may approach the doubly allylic C-10 of arachidonic acid to the non-heme iron leading to an increased 12-lipoxygenation irrespective of the chemical structure of residues 418 and 419.

Table 1. Positional specificity of lipoxygenase mutants

Lipoxygenase mutant	relative product formation	
	12-H(P)ETE	15-H(P)ETE
rab15-LO[1]	1	15
rab15-LO-phe353leu	3	1
rab12-LO[2]	3	1
rab12-LO-leu353phe	1	15

[1]rabbit reticulocyte-type 15-LO, [2]rabbit leukocyte-type 12-LO

Recently, a novel rabbit leukocyte-type 12-LO was cloned (Berger et al., 1998), which shares a 99% amino acid identity with the well-characterized rabbit 15-LO. In contrast to the reticulocyte-type 15-LO, the amino acid residue at position 353 of this enzyme is a rather small leu. As major product of arachidonic acid oxygenation 12-H(P)ETE was identified (Table 1) although positions 418 and 419 are occupied by the space-filling amino acids ile and met (Fig. 1). Mutation of leu353 to phe resulted in an enzyme species, which converted arachidonic acid predominantly to 15-H(P)ETE (Table 1). In fact, the 15-H(P)ETE/12-H(P)ETE-ratio of the mutant enzyme was comparable to that of wild-type rabbit 15-LO (Table 1). It should be stressed that the mutant enzymes listed in Table 1 were expressed in similar amounts as the corresponding wild-type species as indicated by immunoblotting (data not shown).

Summarizing our mutagenesis data on amino acid 353 of mammalian 12- and 15-LO's, one may conclude that this residue constitutes an important sequence determinant for the positional specificity of these enzymes. However, amino acid sequence alignments indicated that the soybean LO-1 exhibiting arachidonic acid 15-LO activity contains a fairly small ser at the corresponding position (Fig. 1), which should lead to arachidonate 12-lipoxygenation.

	352	353		417	418	419
human reticulocyte-type 15-LO	D	F		Q	I	M
rabbit reticulocyte-type 15-LO	D	F		Q	I	M
rabbit leukocyte-type 12-LO	D	L		Q	I	M
murine leukocyte-type 12-LO	D	L		K	V	M
soybean 15-LO (isoenzyme 1)	D	S		T	T	F
	491	492		556	557	558

Figure 1. Sequence alignment of selected regions of various lipoxygenases. [1]numbering for mammalian enzymes, [2]numbering for the soybean enzyme

On the other hand, these alignments also show that in the soybean LO-1 the amino acid, which corresponds to residues 419 of mammalian enzymes, is a rather bulky phe (Fig. 1). We supposed that phe558 may shield ser492 from interacting with the methyl terminus of the substrate fatty acid and thus, could favor 15-lipoxygenation. To test this hypothesis, additional mutagenesis studies were carried out. First, we mutated phe353 of the rabbit 15-LO to ser and expected a predominant 12-lipoxygenation of arachidonic acid. However, the resulting mutant was enzymatically inactive (Table 2). In a second step, the double mutant phe353ser+met419phe was created. Unfortunately, this mutant was also inactive, although it was expressed at a similar level as the wild-type enzyme (immunoblot analysis, data not shown). As possible reason for the loss in enzyme activity, one can discuss that the introduction of a polar OH-group at position 353 may disturb the enzyme-substrate interaction. In fact, when phe353 of the wild-type rabbit enzyme was mutated to tyr, an inactive enzyme species was expressed (Table 2). These results support the hypothesis that introduction of a polar OH-group at position 353 of rabbit 15-LO may influence enzyme/substrate interaction.

Table 2. Activity of lipoxygenase mutants

Lipoxygenase mutant	H(P)ETE formation (µg/ml culture x min)
rab15-LO	1.1
rab15-LO-phe353ser	<0.1
rab15-LO-phe353ser+met419phe	<0.1
rab15-LO-phe353tyr	<0.1

For this reason, the catalytically active phe353leu mutant of rabbit 15-LO, which exhibits mainly 12-LO activity, was used to test whether a bulky residue at at position 419 may shield amino acid 353 from interacting with the substrate fatty acid. Replacement of met419 with phe in this 12-lipoxygenating mutant resulted in an enzyme species, which converted arachidonic acid to 15- and 12-H(P)ETE in a ratio of about 3:1 (Table 3).

Table 3. Positional specificity of rabbit 15-LO mutants

lipoxygenase species	product formation (%)	
	12-H(P)ETE	15-H(P)ETE
rab15-LO	6	94
rab15-LO-phe353leu	75	25
rab15-LO-phe353leu+met419phe	25	75
rab15-LO-phe353leu+met419trp	8	92

This experiment displays the mutual interplay of residues 353 and 419. It may be concluded that the bulkiness of phe419 decreases the volume of the substrate binding pocket, so that arachidonic acid alignment required for 12-lipoxygenation is sterically hindered. Instead, C-13 of the substrate may be localized closer to the non-heme iron leading to a predominant 15-lipoxygenation (Fig. 2). In order to increase the share of 15-H(P)ETE, we introduced an even more bulky trp at position 419. HPLC analysis of that mutant revealed an almost exclusive formation of 15-H(P)ETE (Table 3). It may be hypothesized that trp with its rigidity and its space filling side chain narrows the substrate binding cavity that much, that predominantly arachidonate 15-lipoxygenation occurs.

DISCUSSION

The positional specificity is a property of LO's for which the structural reasons appear to be rather complex. There is no doubt that the alignment of the substrate fatty acid at the active site, in particular its spatial relation to the non-heme iron, is crucial for the positional specificity of the oxygenase reaction but there are no direct experimental data characterizing the substrate orientation for any of the LO isoforms. Although the atom

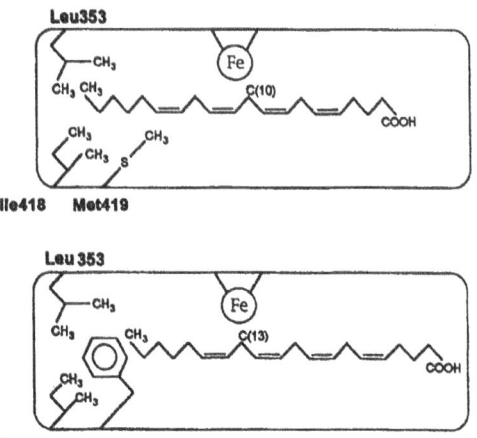

FIGURE 2. Impact of side chain geometry of amino acids 353, 418 and 419 on substrate binding in rabbit reticulocyte-type 15-LO mutants

coordinates for the soybean LO's -1 and -3 (Boyington et al., 1993, Minor et al., 1996, Skrzypczak-Jankun et al., 1997) as well as for a rabbit reticulocyte-type 15-LO/inhibitor complex (Gillmor et al., 1997) are available, no X-ray data on LO/fatty acid complexes have been collected. Initial mutagenesis studies and multiple sequence alignments suggested that ile418 and met419 may constitute important primary determinants for the specificity of 12- and 15-LO's. Later on, we showed that this was only the case for a certain sub-set of mammalian 12- and 15-LO's and that for other LO isoforms amino acid residue 353 is of higher impact. The data presented here indicate that for plant LO's the situation appears to be even more complex and that the rules established for mammalian enzymes may not explain the specificity of plant LO's. Moreover, so far only the positional specificity with arachidonic acid as substrate has been investigated but no information is available as to the impact of the critical amino acids on the oxygenation of other substrates such as linoleic acid, linolenic acid, 15-HETE, phospholipids, cholesterol esters, biomembranes and lipoproteins.

Furthermore, we do not know anything about the sequence determinants for the oxygenation specificity of 5-LO's. When ile418 and met419 of the rabbit reticulocyte 15-LO were mutated to ala and asn (these amino acids align with ile418 and met419 when the rabbit reticulocyte type-15-LO and the human 5-LO are compared) an inactive mutant resulted. The major question yet to be solved in this respect is whether or not arachidonic acid shows an inverse „head to tail" orientation in the case of 5-LOs. The possibility of an inverse orientaion has been concluded from the stereochemistry of the LO reaction (Kühn et al., 1986; Lehman, 1994) but recent structural studies suggested a similar substrate orientation for 15-, 12-, and 5-LO's (Gillmor et al., 1997). According to this hypothesis, the volume of the substrate binding pocket may be crucial for positional specificity. It was calculated that the substrate binding cavity of 5-LO's is 20 % larger than that of 15-LO's. If these size differences are the major reasons for the oxygenation specificity, it should be possible to convert a 15- to 5-LO by mutating critical space filling amino acids to smaller ones. However, at this point of research, such critical amino acids remain to be identified.

REFERENCES

1. Berger, M., Schwarz, K., Thiele, H., Borngräber, S., Kühn, H. and Thiele, B.J. ,1998, Simultaneous expression of leukocyte-type 12-lipoxygenase and reticulocyte-type 15-lipoxygenase in rabbits. *J. Mol. Biol.* (in press).
2. Boyington, J.C., Gaffney, B.J. and Amzel, L.M., 1993, The three-dimensional structure of an arachidonic acid 15-lipoxygenase. *Science* 260: 1482.
3. Borngräber, S., Kuban, R.J., Anton, M. and Kühn, H., 1996, Phenylalanine 353 is a primary determinant for the positional specificity of mammalian 15-lipoxygenases. *J. Mol. Biol.* 264: 1145.
4. Chen, X.S. and Funk, C.D., 1993, Structure-function properties of human platelet 12-lipoxygenase: chimeric enzyme and in vitro mutagenesis studies. *FASEB J.* 7: 694.
5. Gillmor, S., Villasenor, A., Fletterick, R., Sigal, E. and Browner, M., 1997, The structure of mammalian 15-lipoxygenase reveals similarity to the lipases and the determinants of substrate specificity. *Nature Struct. Biol.* 4: 1003.

6. Hada, T., Hagiya, H., Suzuki, H., Arakawa, T., Nakamura, M., Matsuda, S., Yoshimoto, T., Yamamoto, S., Azekawa, T., Morita, Y., Ishimura, K. and Kim, H.Y., 1994, Arachidonate 12-lipoxygenase of rat pineal glands: catalytic properties and primary structure deduced from its cDNA. *Biochim. Biophys. Acta* 1211: 221.
7. Kühn, H. Schewe, T and Rapoport, S.M., 1986, The stereochemistry of the reaction of lipoxygenases and their metabolites. A proposed nomenclature for lipoxygenases and related enzymes. *Adv. Enzymology* 58: 273.
8. Kühn, H., Thiele, B.J., Ostareck-Lederer, A., Stender, H., Suzuki, H., Yoshimoto, T. and Yamamoto, S., 1993, Bacterial expression, purification and partial characterization of recombinant rabbit reticulocyte 15-lipoxygenase. *Biochim. Biophys. Acta* 1168: 73.
9. Lehmann, W.D., 1994, Regio-and stereoselectivity of the dioxygenation reaction catalyzed by (S)-type lipoxygenases or by the cyclooxygenase activity of prostaglandin H synthase. *Free Rad. Biol. Med.* 16: 241.
10. Minor, W., Steczko, J., Stec, B., Otwinowski, Z., Bolin, J.T., Walter, R., and Axelrod, B., 1996, Crystal structure of soybean lipoxygenase L-1 at 1.4 Å resolution. *Biochemistry* 35: 10687.
11. Skrzypczak-Jankun, E., Amzel, L.M., Kroa, B.A. and Funk, M.O., 1997, Structure of soybean lipoxygenase L3 and a comparison with its L1 isoenzyme. *Proteins* 29: 15.
12. Sloane, D.L., Leung, R., Craik, C.S. and Sigal, E., 1991, A primary determinant for lipoxygenase positional specificity. *Nature* 354: 149.
13. Sloane, D.L., Leung, R., Barnett, J., Craik, C.S. and Sigal, E., 1995, Conversion of human 15-lipoxygenase to an efficient 12-lipoxygenase: the side-chain geometry of amino acids 417 and 418 determines positional specificity. *Protein Engineering* 8, 275.
14. Suzuki, H., Kishimoto, K., Yoshimoto, T., Yamamoto, S., Kanai, F., Ebina, Y., Miyatake, A. & Tanabe, T., 1994, Site-directed mutagenesis studies on the iron binding domain and the determinant for the substrate oxygenation site of porcine leukocyte arachidonate 12-lipoxygenase. *Biochim. Biophys. Acta* 1210: 308.
15. Watanabe, T. and Haeggström, J.Z., 1993, Rat 12-lipoxygenase: mutations of amino acids implicated in the positional specificity of 15- and 12-lipoxygenases. *Biochem. Biophys. Res. Commun.* 192: 1023.

15

X-RAY ABSORPTION STUDIES INTO THE IRON LIGAND SPHERE OF PLANT AND ANIMAL LIPOXYGENASES

Hartmut Kühn[1], R. Kuban[1], M. Walther[1] and Gerrit A. Veldink[2]

[1]Institute of Biochemistry, University Clinics Charité, Humboldt University, Hessische Str. 3-4, 10115 Berlin, Germany.
[2]Bijvoet Center for Biomolecular Research, Department of Bio-organic Chemistry, Utrecht University, Padualaan 8, NL-3584 CH Utrecht, The Netherlands

INTRODUCTION

Lipoxygenases (LO's) constitute a family of lipid peroxidizing enzymes which are classified according to their positional specificity of arachidonic acid oxygenation (Kuhn, 1996). In plants LO's are known for a long time. Some 60 years ago a lipoxidase was identified in dry soybeans. This enzyme and other plant LO's were later on purified and characterized with respect to their protein chemical and enzymatic properties (Siedow, 1991). For a long time it was believed that LO's are restricted to the plant kingdom but along with the advances in eicosanoid research LO's have been discovered in the animal kingdom. In 1975 an arachidonate 15-LO was detected in rabbit reticulocytes which is high level expressed in these immature red blood cells. Since then a variety of LO's isoforms have been described in plants and animals and our knowledge on the enzymology and molecular biology of LO's has significantly increased during the past 15 years. However, direct structural information on LO's is still somewhat limited. LO's contain one gram-atom non-heme iron per mole enzyme. In 1993 two independent low resolution crystal structures for the soybean LO-1 were published (Boyington et al., 1993, Minor et al., 1993). Later on high resolution crystal structures were reported for two soybean LO isoenzymes (Minor et al., 1996; Skrzypczak-Jankun et al., 1997). More recently, the crystal structure of a rabbit reticulocyte 15-LOX/inhibitor complex was solved (Gillmor et al., 1997). In addition to these crystallographic studies there is a number of spectroscopic investigations aimed to determine the geometry of the iron ligand sphere. Investigations of the magnetic susceptibility (Petersson et al., 1985) suggested that the ferrous active site ground state is high spin with S=2. Measurements of the X-ray absorption (van der Heijdt et al., 1992; Scarrow, 1994), Mossbauer spectroscopy (Dunham, et al., 1990) and experiments on the magnetic circular dichroism (Whittaker and Solomon, 1988) predicted a distorted octahedral iron ligand sphere for plant LO's. Studies of the circular dichroism as well as evaluation of the 1s-3d transition pre-edge signal in the X-ray absorption spectra suggested that plant and mammalian 15-LO's in their non-activated ground state may contain 6-coordinate ferrous iron (Pavlosky et al., 1995). Site directed mutagenesis studies suggested 4 histidines and the C-terminal isoleucine as iron liganding amino acids of various mammalian LO's and the sixth ligand was supposed to be a water molecule (Kuban et al., 1998). In the soybean LO-1 one iron liganding histidine is substituted for by an asparagine, and here again a water oxygen was identified as sixth iron ligand. In a recent review article new insights into the structure/function relationship of LO's are provided (Solomon, 1997).

For this study we investigated the iron ligand sphere of a mammalian and a plant LO by X-ray absorption studies under strictly comparable conditions and found that both enzymes contain ferrous six coordinate iron and six nitrogen and/or oxygen atoms function as direct iron ligands which surround the central atom in a distance of 0.19 nm to 0.32 nm.

MATERIAL AND METHODS

The soybean LO-1 and the rabbit reticulocyte 15-LO were prepared to apparent homogeneity as described before (Petersson et al., 1985; Belkner et al., 1993). After the final preparation step the buffer was exchanged to a 50 mM Bis-Tris-buffer, pH 6.8 and the enzymes was concentrated using a collodion bag micro-concentrator (Sartorius, FRG). Precipitates formed during the concentration procedure were spun down and the clear enzyme solution was transferred to a home-made plastic cell (1 mm of thickness) the measuring windows of which were covered by a Kapton membrane. Then the preparation was shock frozen in liquid nitrogen.

X-ray absorption measurements were carried out at the EMBL (European Molecular Biology Laboratory) X-ray absorption spectrometer at HASYLAB (c/o Deutsches Elektronen Synchrotron, Hamburg). The X-ray absorption spectra were recorded by monitoring the X-ray fluorescence of the samples with a 13 element Germanium solid-state detector. A series of 22-55 spectra were taken for each sample at 20° K. The spectra were recorded in the range of 6900-7900 eV with steps of 0.3 eV in the near edge region. No evident damage of the protein sample occurred during the exposure to the X-ray beam as far as can be judged from the identity of the first and the last spectrum collected. Evaluation of the absorption spectra was carried out with the Excurve Program (version 92, CCLRC Daresbury Lab., Warrington, UK). A more detailed description of the measuring procedure and on the data evaluation is given in the reference (Kuban et al., 1998).

RESULTS

Metalloproteins have characteristic X-ray absorption spectra and the shape of the spectra depends on the metal atom and its ligand sphere. The near edge region of the X-ray absorption spectra (Fig. 1) provides information on the chemical nature of the metal atom (e.g. copper, iron, magnesium etc.), on the valency status (e.g. ferrous or ferric iron) and on the number of metal ligands (four-, five- or six-coordinate). From the extended region of the absorption spectra the chemical nature of the direct metal ligands (nitrogen, sulfur etc.) and the binding distances can be extracted. Unfortunately, oxygens and nitrogens cannot be differentiated as liganding atoms by X-ray absorption spectroscopy. In Fig. 1 a typical X-ray absorption spectrum of the rabbit reticulocyte 15-LO is shown.

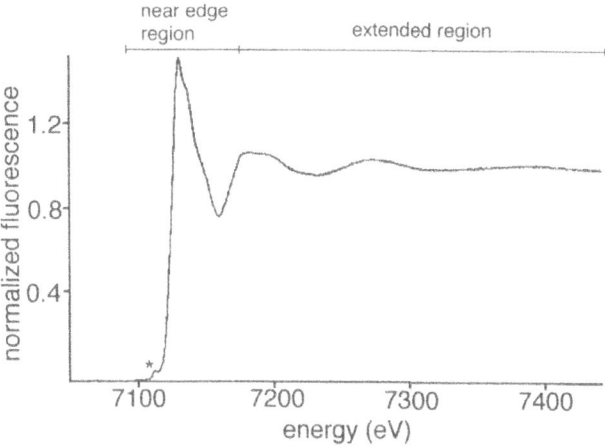

Figure 1. X-ray absorption spectrum of the rabbit 15-LO. * indicates the 1s-3d transition signal which provides information on the number of iron ligands

Evaluating this spectrum the following structural conclusions can be drawn: i) The edge position of the spectrum indicated the presence of ferrous iron in the rabbit 15-LO. For the soybean LO-1 the absorption spectrum also indicated the presence of ferrous iron (not shown). ii) The intensity of the pre-edge 1s-3d transition signal suggested a six-coordinate iron for both, the rabbit and the soybean LO's. iii) 6 nitrogens and/or oxygens may constitute the direct iron ligands in both enzymes and the binding distances of these ligands vary between 0.195 nm and 0.304 nm for the soybean enzyme and between 0.213 nm and 0.293 nm for the rabbit LO.

In order to assign the binding distances extracted from the X-ray absorption spectra to specific amino acids molecular modeling was carried out. The crystallographic data of the soybean LO-1 identified the ε-nitrogen atoms of His499, 504, 690, the δ1 oxygen of Asn694, a carboxylic oxygen of the C-terminal Ile839 and a water molecule constitute the direct iron ligands (Minor et al., 1996). For the rabbit reticulocyte 15-LO His360, His365, His540, His544 and the C-terminal Ile662 have been suggested as protein iron ligands (Gillmor et al., 1997). In both enzymes the water oxygen is about 0.25 nm distant from the central iron (0.247 nm for soybean LO-1, 0.242 nm for the rabbit LO). The results of our modeling studies are summarized in Fig. 2.

DISCUSSION

Our X-ray absorption data suggest that plant and mammalian LO's contain 6-coordinate ferrous iron and the ligand sphere can be described as distorted bipyramid (octahedron). In the soybean LO-1 His690 is located diagonal to the water molecule, His499 diagonal to the C-terminal Ile839 and His504 diagonal to Asn694 (Fig. 3). The iron ligands His499 and His504 are parts of a long helix (leading helix) which spans almost the entire enzyme molecule (Minor et al., 1996). It should be noted that due to the binding forces of the central iron the structure of this leading helix is disturbed between the two iron ligands. His690 and Asn694 are parts of another helix. Here again, the helical structure is disturbed between the iron ligands. For the rabbit 15-LO, His540 which aligns with

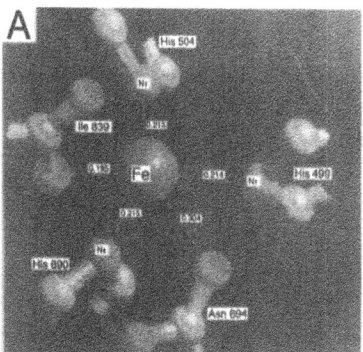 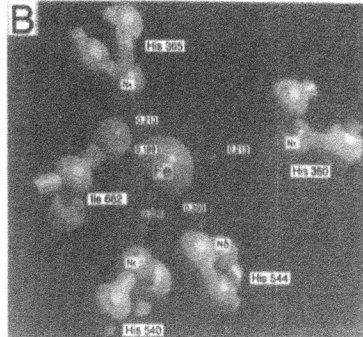

Figure 2. Binding distances of the protein iron ligands of LO's. A) soybean LO-1, B) rabbit reticulocyte 15-LO. The binding distances are given in nm. The water molecules are not included in this figure.

His690 of the soybean LO-1 is located diagonal to the water molecule. His360 which aligns with His499 of the soybean enzyme is located diagonal to Ile662 and His544 is localized across to His544. The two iron ligands His360 and His365 are constituents of the long spanning leading helix which also contains Phe353. This amino acid was recently identified as primary determinant for the positional specificity of 12- and 15-LO's (Borngräber et al., 1996).

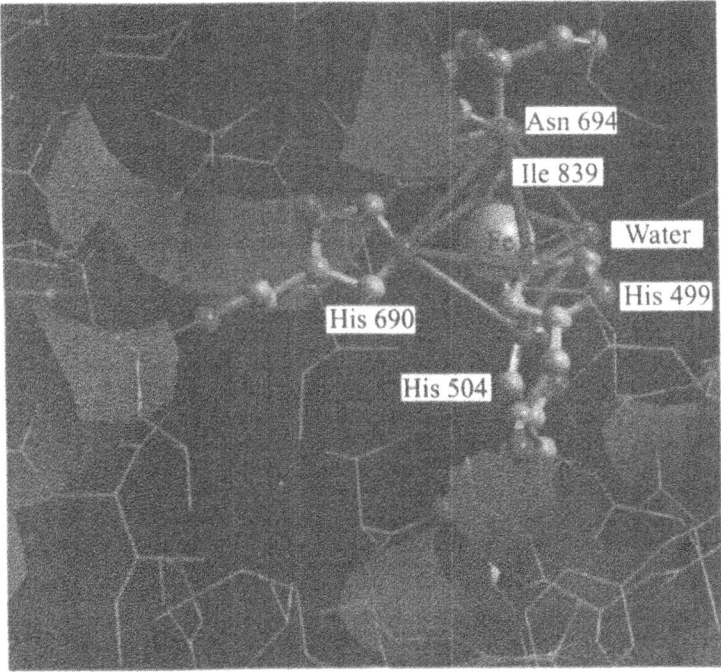

Figure 3. Iron ligand sphere of the soybean LO-1

At this time it is not entirely clear how the fatty acids are bound at the active site. It might be possible that the substrates are aligned along the leading helix and that it may displace the water from the sixth ligand position (Minor et al., 1996; Borngräber et al., 1996). In fact, such a displacement of the water molecule was shown for a LOX inhibitor in the crystallographic studies of the rabbit 15-LOX/inhibitor complex (Gillmor et al., 1997).

ACKNOWLEDGMENT

The authors wish to thank Dr. H. Nolting and Dr. V.A. Solé from the EMBL Outstation, c/o Deutsches Elektronen Synchrotron DESY for their support in taking and interpreting the X-ray absorption spectra. The technical assistance of Mrs. M. Anton and Mrs. C. Gerth at the preparation of the rabbit reticulocyte 15-lipoxygenase enzyme is also acknowledged.

REFERENCES

Belkner, J., Wiesner, R., Rathman, J., Barnett, J., Sigal, E., and Kühn, H., 1993, Oxygenation of lipoproteins by mammalian lipoxygenases. *Eur. J. Biochem.* 213: 251.

Borngräber, S., Kuban, R.J., Anton, M. and and Kühn, H., 1996, Phenylalanine 353 is a primary determinant for the positional specificity of mammalian 15-lipoxygenases. J. Mol. Biol. 264: 1145.

Boyington, J.C., Gaffney, B.J. and Amzel, L.M., 1993, The three-dimensional structure of an arachidonic acid 15-lipoxygenase. *Science* 260: 1482.

Dunham, W.R., Carroll, R.T., Thompson, J.F., Sands R.H., and Funk, M.O., 1990, The initial characterization of the iron environment in lipoxygenase by Mossbauer spectroscopy. *Eur. J. Biochem.* 190: 611.

Gillmor, S., Villasenor, A., Fletterick, R., Sigal, E. and Browner, M., 1997, The structure of mammalian 15-lipoxygenase reveals similarity to the lipases and the determinants of substrate specificity. *Nature Struct. Biol.* 4: 1003.

Kuban, R.J., Wiesner, R., Rathman, J., Veldink, G., Nolting, H., Solé, V.A. and Kühn, H., 1998, The iron ligand sphere geometry of mammalian 15-lipoxygenases. *Biochem. J.* (in press).

Kuhn, H., 1996, Biosynthesis, metabolization and biological importance of the primary 15-lipoxygenase metabolites 15-hydro(pero)xy-5Z,8Z,11Z,13E-eicosatetraenoic acid and 13-hydro(pero)xy-9Z,11E-octadecadienoic acid. *Prog. Lipid Res.* 35: 203.

Minor, W., Steczko, J., Bolin, J.T., Otwinowski, Z., and Axelrod, B. ,1993, Crystallographic determination of the active site iron and its ligands in soybean lipoxygenase L-1. *Biochemistry* 32: 6320.

Minor, W., Steczko, J., Stec, B., Otwinowski, Z., Bolin, J.T., Walter, R., and Axelrod, B., 1996, Crystal structure of soybean lipoxygenase L-1 at 1.4 Å resolution. *Biochemistry* 35: 10687.

Pavlosky, M.A., Zhang, Y., Westre, T.E., Gan, Q.F., Pavel, E., Campochiaro, C., Hedman, B., Hodgson, K.O., and Solomon, E.I., 1995, Near infrared circular dichroism, magnetic circular dichroism, and X-ray absorption spectral comparison of the non-heme ferrous active sites of plant and mammalian 15-lipoxygenases. *J. Am. Chem. Soc.* 117: 4316.

Petersson, L., Slappendel, S., and Vliegenthart, J.F.G., 1985, Magnetic susceptibility of soybean lipoxygenase-1. Implications for the symmetry of the iron environment and the possible coordination of dioxygen to Fe^{2+}. *Biochim. Biophys. Acta* 828:81.

Scarrow, R.C., Trimitsis, M.G., Buck, C.P., Grove, G.N., Cowling, R.A., and Nelson, M.J., 1994, X-ray absorption spectroscopy of the iron site in soybean lipoxygenase-1: changes in coordination upon oxidation or addition of methanol. *Biochemistry* 33: 15023.

Skrzypczak-Jankun, E., Amzel, L.M., Kroa, B.A. and Funk, M.O., 1997, Structure of soybean lipoxygenase L3 and a comparison with its L1 isoenzyme. *Proteins* 29: 15.

Siedow, J.N., 1991, Plant lipoxygenases: structure and function. *Annu. Rev. Plant Physiol. Plant Mol. Biol.* 42: 145.

Solomon, E.I., 1997, New insights from spectroscopy into the structure/function relationship of lipoxygenases, *Chem. Biol.* 4:795.

van der Heijdt, L.M., Feiters, M.C., Navaratnam, S. Nolting, H.F., Hermes, C. Veldink, G.A., and Vliegenthart, J.F.G., 1992, X-ray absorption spectroscopy of soybean lipoxygenase-1. *Eur. J. Biochem.* 207: 793.

Whittaker, J.W. and Solomon, E.I., 1988, Spectroscopic studies on ferrous non heme iron active site: magnetic circular dichroism of mononuclear Fe sites in superoxide dismutase and lipoxygenase. *J. Am. Chem. Soc.* 110; 5329.

16

TRANSCRIPTIONAL AND POSTTRANSCRIPTIONAL REGULATION OF
5-LIPOXYGENASE mRNA EXPRESSION
IN THE HUMAN MONOCYTIC CELL LINE MONO MAC 6
BY TRANSFORMING GROWTH FACTOR-ß AND 1,25-DIHYDROXYVITAMIN D₃

Damaris Härle[1], Olof Rådmark[2], Bengt Samuelsson[2] and Dieter Steinhilber[1]

[1]Institute of Pharmaceutical Chemistry, Johann Wolfgang Goethe-University
Frankfurt, Marie-Curie-Str. 9, D-60439 Frankfurt, Germany
[2]Karolinska Institute, Division MBB, S-17177 Stockholm, Sweden

INTRODUCTION

5-lipoxygenase (5-LO) is expressed in myeloid cells like granulocytes, monocytes macrophages and in B-lymphocytes. In addition, 5-LO expression was reported for neuronal tissues and keratinocytes. In myeloid cells, expression of the 5-LO pathway (5-LO) is upregulated during cell maturation. It has been shown by various authors that differentiation of HL-60 cells is accompanied by an upregulation of 5-LO mRNA, protein, and activity (Bennett et al., 1993; Kargman and Rouzer, 1989). Depending on the differentiation inducer, 3-6-fold differences in 5-LO mRNA levels were observed for HL-60 cells (Bennett et al., 1993; Steinhilber et al., 1993a). The combination of 1,25-dihydroxyvitamin D₃ (calcitriol) and transforming growth factor-ß (TGFß), either alone or in combination with dimethylsulfoxide, was the most prominent inducer of the 5-LO pathway, leading to cellular protein expression and activity comparable to (or higher than) what is found for normal human granulocytes (Brungs et al., 1994; Steinhilber et al., 1993b).
Prominent induction of the 5-LO pathway was also observed during the in-vitro differentiation of the human monocytic cell line Mono Mac 6. There was a more than 500-fold induction of 5-LO activity and an over 128-fold induction of protein expression when the cells were differentiated with TGFß plus calcitriol. These changes in protein expression and activity were accompanied by a 64-fold induction of 5-LO mRNA (Brungs et al., 1995). Thus, 5-LO belongs to the family of

genes which are upregulated by calcitriol or the combination of calcitriol and TGFß like CD14 or alkaline phosphatase. However, among the considerable number of calcitriol regulated genes only a few among them (e.g. 24-hydroxylase) show a strong response to calcitriol on the transcriptional level. On the other hand, for many genes like calbindin$_{28k}$ or CD14 only modest effects of calcitriol on gene transcription were found which do not account for the changes in the respective mRNA levels (Christakos *et al.*, 1996; Zhang *et al.*, 1994). Therefore, it was of interest to study the transcriptional and posttranscriptional effects of TGFß and calcitriol on 5-LO gene expression in Mono Mac 6 cells.

METHODS

Cell culture

Mono Mac 6 cell culture was performed in RPMI 1640 medium with various supplements as described before (Brungs *et al.*, 1995).

Nuclear run-off analysis

Mono Mac 6 cells were cultured with TGFß (1 ng/ml) and calcitriol (50 nM) for the indicated times. Then, nuclei were prepared by density centrifugation through a sucrose cushion and the in-vitro transcription was performed for one hour at RT as described (Merscher *et al.*, 1994). Total nuclear RNA was extracted by the guanidinium isothiocyanate method and specific labelling of the RNA was checked as described in the Boehringer Mannheim DIG user's guide. The nuclear RNA was hybridized to pT7T3-LX12A1:B1 plasmid (7 µg) which covers intron 1 (position 179-2468), pT7T3-GAPDH (4 µg, positive control) and pT7T3-vector (4 µg, negative control) immobilized on a Nylon membrane (Boehringer Mannheim, F.R.G.). Hybridization, washing and detection was performed as described in the Boehringer Mannheim DIG user's guide.

RT-PCR analysis

Determination of 5-LO RNA expression by reverse transcription and PCR analysis was performed as described before in detail (Härle *et al.*, 1997). In brief, total cellular RNA was isolated by the guanidinium thiocyanate method and reverse transcribed either by random hexamer priming or d(T)$_{12-18}$-priming into cDNA in a total volume of 40 µl. PCR was carried out in a total volume of 50 µl containing 2 µl of reverse transcription reaction mixture, 1x PCR buffer, 0.2 mM dNTP and 0.5 µM each of 5'- and 3'- primers. The annealing temperature was individually adjusted to the primers used in order to avoid the generation of unwanted side products during the PCR reaction (Härle *et al.*, 1998). Also, cycle numbers were minimized for every series of experiments in order to keep the PCR reaction in the exponential phase and to avoid saturation effects. Three different 5-LO RNA species (primary transcripts, immature and mature 5-LO RNA) were analyzed.

5-LO primary transcripts were analyzed using random primed cDNA and PCR primers corresponding to intron 1 of the 5-LO gene. 5-LO transcripts reaching exon 14, that have been polyadenylated but not necessarily spliced, have been determined by priming of the reverse transcriptions with $d(T)_{12-18}$, followed by PCR with primers corresponding to exon 14. 5-LO transcripts reaching exon 14, that have been spliced but not necessarily polyadenylated, have been determined by priming of the reverse transcription reactions with random primers, followed by PCR with primers spanning the last intron. Finally, mature 5-LO mRNA has been determined by priming of the reverse transcription reactions with $d(T)_{12-18}$, followed by PCR with primers spanning the last intron.

RESULTS AND DISCUSSION

Previous studies revealed that there is an up to 64-fold induction of 5-LO mRNA after treatment of Mono Mac 6 cells with TGFß and calcitriol. Therefore, it was of interest to study the contribution of transcriptional and posttranscriptional effects on 5-LO mRNA induction by these two agents. There was no significant difference in 5-LO mRNA half life between differentiated and undifferentiated cells (data not shown).

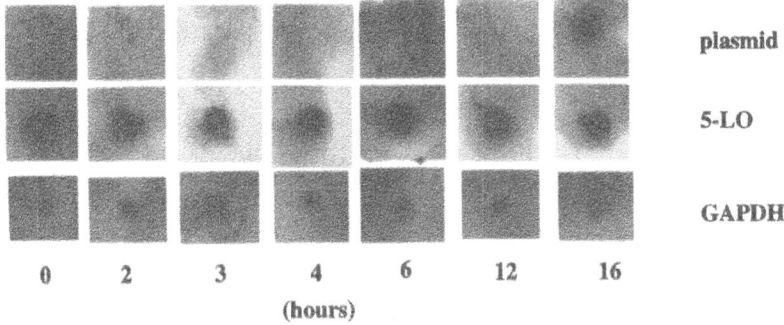

Figure 1. Nuclear run-off analysis of 5-LO gene transcription. Mono Mac 6 cells were grown for the indicated times in the presence of TGFß (1 ng/ml) and calcitriol (50 nM). Then, nuclei were prepared and nuclear run-off analysis was performed as described above.

As can be seen from fig. 1, no significant increase in 5-LO transcripts could be detected by the nuclear run-off analysis.

In order to confirm these results, an RT-PCR method was developed to directly analyze changes in 5-LO primary transcripts. Total RNA was isolated from Mono Mac 6 cells and transcribed into cDNA by random hexamer priming. 5-LO primary transcripts were analyzed by PCR using intron-1-specific primers. Using this method, a 4-5-fold increase in 5-LO primary transcripts was detected which could be inhibited by cycloheximide (50 µM). Possible explanations for the discrepancy between RT-PCR analysis and nuclear run-off could be that the nuclear run-off method is either not sensitive enough to detect small changes in 5-LO transcription, or that the increased transcription caused by calcitriol is obliterated by the nuclear run-off analysis procedure. Possibly, the effect of a diffusible ligand (calcitriol) is lost during preparation of nuclei and incubation with labelled nucleotides (totally 2-3 h).

No significant increase in 5-LO transcription was found with either TGFß or calcitriol alone. Thus, in order to check whether the TGFß or calcitriol effect depends on protein synthesis, the cells were cultured for 1 day either in the presence of TGFß or calcitriol. Then, the corresponding second inducer (calcitriol or TGFß, respectively) was added either alone or together with cycloheximide. About 4-fold increases in 5-LO transcription were obtained upon addition of calcitriol or TGFß as second inducer. Cycloheximide (50 µM) inhibited the TGFß effect but not the calcitriol effect. Obviously, calcitriol directly stimulates 5-LO transcription in the presence of TGFß-induced proteins.

Since the effects of TGFß and calcitriol on 5-LO transcription were too small to explain the dramatic increase in mature 5-LO mRNA by TGFß/calcitriol, posttranscriptional effects were analyzed. The human 5-LO gene spans more than 82 kb and consists of 14 exons. Thus, we speculated that the effect of the differentiation inducers could be of different magnitude in various parts of the 5-LO gene which could be due to putative transcription arrest sites. In order to detect different effects of TGFß and calcitriol on the induction of 5-LO RNA at the 5'- and 3'-end of the 5-LO gene, the cells were cultured with TGFß (1 ng/ml) and calcitriol (50 nM) for the indicated times and the relative changes in 5-LO RNA species were analyzed by RT-PCR for every exon as reported previously (Härle et al., 1998), (fig. 2).

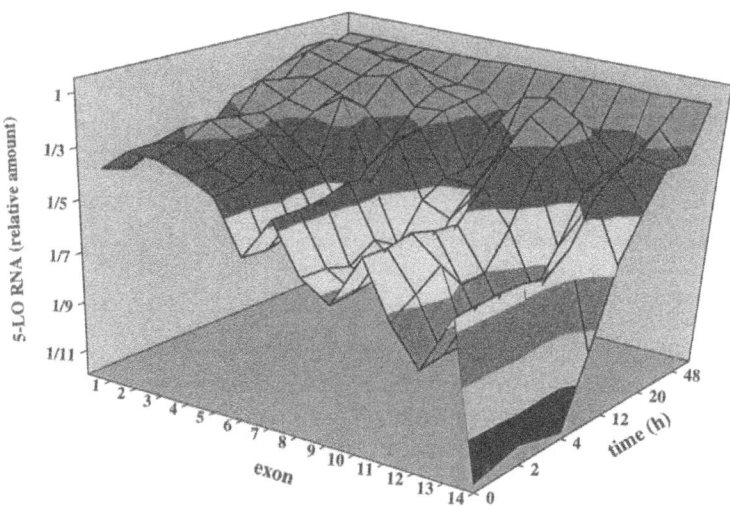

Figure 2. Exon-specific analysis of 5-LO RNA induction. Mono Mac 6 cells were grown for the indicated times in the presence of TGFß (1 ng/ml) and calcitriol (50 nM). Then, RNA was isolated, transcribed into cDNA by random hexamer priming and analyzed by PCR using primers corresponding to every 5-LO exon. The results are expressed as relative 5-LO RNA amounts compared to the 48 h sample. The 5-LO RNA amount of the 48h sample was set to 1.

It can be observed that the relative levels of 5-LO RNA, measured at each exon, are practically constant up to 4 hours after addition of TGFß and calcitriol (fig. 2). Then, the 5-LO RNA amounts increase, particularly during 8-12 hours. With PCR primers corresponding to exon 1, a maximal 3.6-fold induction in 5-LO mRNA was observed after TGFß and calcitriol treatment of Mono Mac 6 cells, and it can be seen that the relative increments at exons 1-5 were similar. Interestingly, from exon 5 to exon 14, the relative induction of 5-LO RNA increased from 3.6 to 12-fold. The data clearly indicate that the TGFß/calcitriol effects on 5-LO RNA levels increase from the 5' to the 3'-end of the gene. The most reasonable explanation for this phenomenon is that calcitriol and TGFß induce 5-LO transcript elongation.

In order to check whether TGFß/calcitriol also affect 5-LO RNA maturation, the effects of both agents on the levels of immature and mature 5-LO RNA species were investigated. When RT-PCR analysis was performed in a way which detected both mature and immature 5-LO RNA species (see Methods), 12-15-fold increases in 5-LO RNA by calcitriol/TGFß were found. However, specific detection of mature 5-LO mRNA gave a 42-fold induction (table 1). This suggests that calcitriol and TGFß affect 5-LO RNA maturation. Interestingly, the induction of mature 5-LO RNA by both, calcitriol and TGFß, could be inhibited by cycloheximide (data not shown) which suggests that both hormones obviously induce other proteins involved in 5-LO RNA processing.

Taken together, our data suggest that calcitriol and TGFß induce 5-LO mRNA in Mono Mac 6 cells through several mechanisms including enhanced gene transcription, transcript elongation and mRNA maturation.

Table 1. Induction of the 5-lipoxygenase pathway by TGFß and calcitriol in Mono Mac 6 cells.

Parameter	induction by TGFß and calcitriol	time required for maximal effect
5-LO primary transcripts	5-fold	8 h
5-LO total RNA (exon1)	4-fold	12-16 h
5-LO total RNA (exon 14)	12-fold	16 h
5-LO mRNA	42-64-fold	24 h
5-LO protein	> 128-fold	48 h
5-LO activity	> 500-fold	72 h

The effects of TGFß/calcitriol on various 5-LO parameters are summarized in table 1. There is an early onset in 5-LO transcription with subsequent upregulation of mature 5-LO mRNA after 24 h,

5-LO protein expression (after 48h) and finally 5-LO activity after 72 h. Interestingly, similar slow responses with respect to calcitriol (or calcitriol and TGFß) were also found for other genes like calbindin-D_{28k} or CD14 (Theofan et al., 1986; Zhang et al., 1994).

Recently, it was shown that the induction of alkaline phosphatase by retinoic acid is due to the stabilization of newly synthesized transcripts (Zhou et al., 1994). Moreover, a study on the cleavage-polyadenylation specificity factor CPSF suggested a link between transcription initiation, RNA elongation by RNA-polymerase II and processing of the 3'end of RNA (Dantonel et al., 1997). It might be possible that many vitamin D-regulated genes with modest calcitriol effects on gene transcription are regulated in a similar way as 5-LO and that upregulation of the RNA processing of particular genes like alkaline phosphatase, CD14 or 5-LO is linked with cellular differentiation processes (Morikawa et al., 1990; Testa et al., 1993).

REFERENCES

Bennett, C.F., Chiang, M.-Y., Monia, B.P. and Crooke, S.T., 1993, Regulation of 5-lipoxygenase and 5-lipoxygenase-activating protein expression in HL-60 cells, Biochem. J. 289:33.

Brungs, M., Rådmark, O., Samuelsson, B. and Steinhilber, D., 1994, On the induction of 5-lipoxygenase expression and activity in HL-60 cells: effects of vitamin D3, retinoic acid, DMSO and TGFß, Biochem. Biophys. Res. Commun. 205:1572.

Brungs, M., Rådmark, O., Samuelsson, B. and Steinhilber, D., 1995, Sequential induction of 5-lipoxygenase gene expression and activity in Mono Mac 6 cells by transforming growth factor-beta and 1,25-dihydroxyvitamin D3, Proc Natl Acad Sci USA 92:107.

Christakos, S., Raval-Pandya, M., Werny, R.P. and Yang, W., 1996, Genomic mechanisms involved in the pleiotropic actions of 1,25-dihydroxyvitamin D3, Biochem. J. 316:361.

Dantonel, J.-C., Murthy, K.G.K., Manley, J.L. and Tora, L., 1997, Transcription factor TFIID recruits factor CPSF for formation of 3'end of mRNA, Science 389:399.

Härle, D., Rådmark, O., Samuelsson, B. and Steinhilber, D., 1998, Calcitriol and transforming growth factor-beta upregulate 5-lipoxygenase mRNA expression by increasing gene transcription and mRNA maturation, Eur. J.Biochem. submitted.

Kargman, S. and Rouzer, C.A., 1989, Studies on the regulation, biosynthesis, and activation of 5-lipoxygenase in differentiated HL60 cells, J. Biol. Chem 264:13313.

Merscher, S., Hanselmann, R., Welter, C. and Dooley, S., 1994, Nuclear run-off transcription analysis using chemiluminescent detection, BioTechniques 16:1024.

Morikawa, M., Harada, N., Soma, G. and Yoshida, T., 1990, Transforming growth factor-beta 1 modulates the effect of 1-alpha, 25-dihydroxyvitamin D3 on leukemic cells, In Vitro Cell Dev Biol 26:682.

Steinhilber, D., Hoshiko, S., Grunewald, J., Rådmark, O. and Samuelsson, B., 1993a, Serum factors regulate 5-lipoxygenase activity in maturating HL-60 cells, Biochim. Biophys. Acta 1178:1.

Steinhilber, D., Rådmark, O. and Samuelsson, B., 1993b, Transforming growth factor beta upregulates 5-lipoxygenase activity during myeloid cell maturation, Proc. Natl. Acad. Sci. USA 90:5984.

Testa, U., Masciulli, R., Tritarelli, E., Pustorino, R., Mariani, G., Martucci, R., Barberi, T., Camagna, A., Valtieri, M. and Peschle, C., 1993, Transforming growth factor-beta potentiates vitamin-D3 induced terminal monocytic differentiation of human leukemic cell lines, J Immunol 150:2418.

Theofan, G., Nguyen, A.P. and Norman, A.W., 1986, Regulation of calbindin-D28K gene expression by 1,25-dihydroxyvitamin D3 is correleated to receptor occupancy, J. Biol. Chem. 261:16943.

Zhang, D.E., Hetherington, C.J., Gonzalez, D.A., Chen, H.M. and Tenen, D.G., 1994, Regulation of CD14 expression during monocytic differentiation induced with 1α,25-dihydroxyvitamin D3, *J Immunol* 153:3276.

Zhou, H., Manji, S.S., Findlay, D.M., Martin, T.J., Heath, J.K. and Ng, K.W., 1994, Novel action of retinoic acid. Stabilization of newly synthesized alkaline phosphatase transcripts, *J. Biol. Chem.* 269:22433.

CYCLOOXYGENASES

17

COMPARISON OF PROSTAGLANDIN H SYNTHASE-1 AND -2 STRUCTURAL STABILITIES

Guishan Xiao, Wei Chen, and Richard J. Kulmacz

Division of Hematology, Department of Internal Medicine
University of Texas Health Science Center at Houston
Houston, TX 77030

INTRODUCTION

Two isoforms of prostaglandin H synthase (PGHS) are known, with PGHS-1 believed to be a housekeeping enzyme and PGHS-2 assigned roles in cell proliferation and inflammation[1]. Both isoforms have two catalytic activities: oxygenation of arachidonate to form PGG_2 (cyclooxygenase), and reduction of PGG_2 to form PGH_2 (peroxidase). The overall amino acid identity between the human PGHS isoforms is about 60%, with much higher conservation in residues required for catalysis[2]. Crystallographic results have shown PGHS-1 and PGHS-2 to have very similar folding patterns[3,4,5].

In spite of these similarities, the two PGHS isoforms have remarkably distinct selectivities for cyclooxygenase inhibitors[1]. The different interaction with inhibitors has been attributed partly to the larger cyclooxygenase pocket in PGHS-2 made possible by having valine at position 509, instead of the isoleucine at the corresponding position in PGHS-1[4,5]. However, this single side chain substitution cannot completely account for the inhibitor specificity differences[6,7,8], and it seems likely that dynamic aspects of the isoform structures may contribute to the discrimination among inhibitors. To begin comparing the dynamic structural properties of the two isoforms, we examined their resistance to denaturation by guanidinium hydrochloride (GdmHCl)[9].

CONTROLLED DENATURATION OF PGHS-1 AND -2

The basic experimental approach was to pre-equilibrate the PGHS-1 and -2 apoenzymes with 0-6 M GdmHCl concentrations for 30 min at room temperature before assessing the extent of structural or functional disruption. Several assays were used. The secondary structure of both isoforms is predominantly alpha helix[3,4,5], and this was monitored from the intensity of the trough at 222 nm in the circular dichroism (CD) spectrum. Each isoform has several tryptophan residues buried in the globular portion of the protein[3,4,5]. The increased exposure of these tryptophans to the aqueous environment upon unfolding was followed from the fluorescence at 338 nm (excitation at 282 nm). Effects on quaternary structure were assessed by gel permeation chromatography and by cross-linking with glutaraldehyde. Peroxidase and cyclooxygenase assays were used to monitor the active

Eicosanoids and Other Bioactive Lipids in Cancer, Inflammation, and Radiation Injury, 4
Edited by Honn *et al.*, Kluwer Academic / Plenum Publishers, New York, 1999.

115

site structures. The integrity of the cyclooxygenase site was also quantitated from the fluorescence quenching observed upon formation of a tight complex with indomethacin or flurbiprofen[10]. The results are summarized in Figure 1.

Active site function was disrupted at lower GdmHCl levels than the overall secondary, tertiary, or quaternary structure for both isoforms (compare the EC_{50} levels indicated by vertical bars in Figure 1). This indicates that the active sites are particularly unstable regions, consistent with observations with other enzymes[11]. For each isoform, peroxidase and cyclooxygenase active site functions were lost at almost identical denaturant levels, suggesting the possibility of structural coupling between the two catalytic regions.

The two isoforms differed significantly in their susceptibility to denaturation. Active site functions and overall secondary and tertiary structure were lost at considerably lower GdmHCl levels for PGHS-2 than for PGHS-1 (Figure 1), pointing to a generally lower structural stability in the PGHS-2 subunits. On the other hand, the midpoint for dissociation into monomers was observed at similar denaturant levels for the two isoforms, suggesting

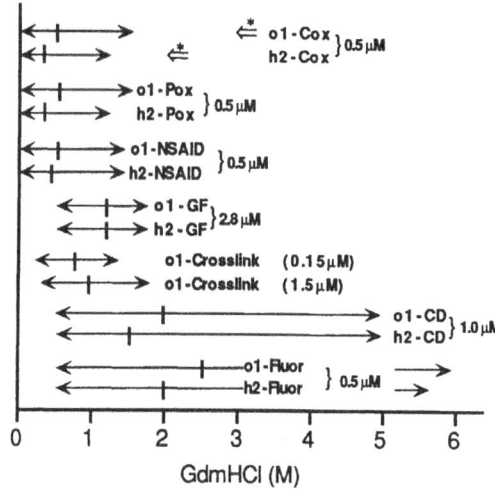

Figure 1. Summary of chemical denaturation results with ovine PGHS-1 (o1) and human PGHS-2 (h2) (from data in Ref. 9). Cyclooxygenase activity (Cox), peroxidase activity (Pox), indomethacin and flurbiprofen binding activity (NSAID), dimeric structure measured by gel filtration (GF) and glutaraldehyde cross-linking (Crosslink), alpha helix content (CD), and tryptophan sequestration (Fluor) were assessed after 30 min preincubation of the apoenzymes with the indicated GdmHCl levels. Horizontal arrows represent the ranges of denaturant concentration required for changes in each parameter; vertical bars show denaturant concentrations for half-maximal effect (EC_{50} values). Protein concentrations are shown at right. Open arrows marked with asterisks indicate the upper denaturant level at which loss of cyclooxygenase activity was fully reversible upon dilution.

comparable monomer-monomer interaction stability in PGHS-1 and -2. Dimer dissociation measured by cross-linking (only for PGHS-1) was more extensive at lower protein concentration, as expected for an equilibrium between dimer and monomers. Disruption of quaternary structure occurred at denaturant levels above those needed for loss of most function (Figure 1), indicating that inactive dimers can accumulate. Loss of function and secondary and tertiary structure were sensitive to the protein concentration, showing that subunit structure is affected by the oligomeric state in both isoforms[12]. Overall, the present observations suggest that the unfolding pathways of PGHS-1 and -2 include at least two intermediates between native dimers and unfolded monomers, as schematized in Figure 2.

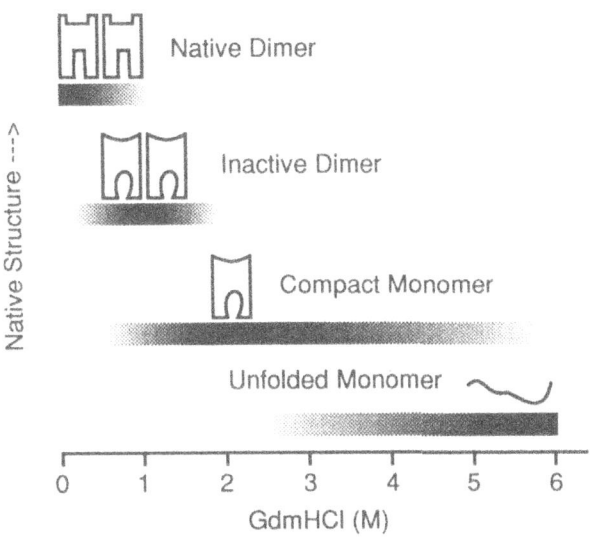

Figure 2. Hypothetical distribution of PGHS-1 and -2 structural species as a function of denaturant level.

The denaturation studies revealed a surprising fundamental difference in active site folding stability between PGHS-1 and -2. The challenge now is to identify the sequence differences which lead to the distinct folding stabilities, and to evaluate the relationship between active site stability and active site function in the two isoforms.

Acknowledgment

This work was supported by a grant (GM 52170) from the National Institutes of Health.

REFERENCES

1. H.R. Herschman, Prostaglandin synthase 2, Biochim. Biophys. Acta 1299:125 (1996).
2. T. Hla and K. Neilson, Human cyclooxygenase-2 cDNA, Proc. Natl. Acad. Sci. U.S.A. 89:7384 (1992).
3. D. Picot, P.J. Loll, and R,M, Garavito, The X-ray crystal structure of the membrane protein prostaglandin H2 synthase-1, Nature 367:243 (1994).
4. C. Luong, A. Miller, J. Barnett, J. Chow, C. Ramesha, and M.F. Browner, Flexibility of the NSAID binding site in the structure of human cyclooxygenase-2, Nature Struct. Biol. 3:927 (1996).
5. R.G. Kurumbail, A.M. Stevens, J.K. Gierse, J.J. McDonald, R.A. Stegeman, J.Y. Pak, D. Gildehaus, J.M. Miyashiro, T.D. Penning, K. Seibert, P.C. Isakson, and W.C. Stallings, Structural basis for selective inhibition of cyclooxygenase-2 by anti-inflammatory agents, Nature 384:644 (1996).

6. J.K. Gierse, J.J. McDonald, S.D. Hauser, S.H. Rangwala, C.M. Koboldt, and K. Seibert, A single amino acid difference between cyclooxygenase-1 (COX-1) and -2 (COX-2) reverses the selectivity of COX-2 specific inhibitors, J. Biol. Chem. 271:15810 (1996).

7. Q. Guo, L.-H. Wang, K.-H. Ruan, and R.J. Kulmacz, Role of Val509 in time-dependent inhibition of human prostaglandin H synthase-2 cyclooxygenase activity by isoform-selective agents, J. Biol. Chem. 271:19134 (1996).

8. E. Wong, C. Bayly, H.L. Waterman, D. Riendeau, and J.A. Mancini, Conversion of prostaglandin G/H synthase-1 into an enzyme sensitive to PGHS-2-selective inhibitors by a double His513 -- Arg and Ile523 -- Val mutation, J. Biol. Chem. 272:9280 (1997).

9. G. Xiao, W. Chen, and R.J. Kulmacz, Comparison of structural stabilities of prostaglandin H synthase-1 and -2, J. Biol. Chem., in press (1998).

10. V. Houtzager, M. Ouellet, J.P. Falgueyret, L.A. Passmore, C. Bayly, and M.D. Percival, Inhibitor-induced changes in the intrinsic fluorescence of human cyclooxygenase-2, Biochemistry 35:10974 (1996).

11. C.-L. Tsou, Conformational flexibility of enzyme active sites, Science 262:380 (1993).

12. K.E. Neet and D.E. Timm, Conformational stability of dimeric proteins: quantitative studies by equilibrium denaturation, Protein Sci. 3:2167 (1994).

18

MODULATION OF THE EXPRESSION OF THE CYCLOOXYGENASE 1 AND 2 GENES IN RAT MAMMARY GLANDS: ROLE OF HORMONAL STATUS AND DIETARY FAT

Alaa F. Badawi, Ahmed El-Sohemy, Laurie L. Stephen, Amit K. Ghoshal, and Michael C. Archer

Laboratory of Pharmacogenetics, Clinical Pharmacology, Atlantic Veterinary College, 550 University Avenue Charlottetown, Prince Edward Island CANADA C1A 4P3

INTRODUCTION

A role for ovarian and pituitary hormones in human breast cancer development has been suggested by observations such as the protective effects of a late age of menarche, an early age of menopause and first pregnancy and bilateral oopherectomy before the age 40[1,2]. Similarly, completion of pregnancy and lactation or ovariectomy prior to carcinogen administration protects against mammary carcinogenesis in the rat[3,4]. Administration of a mammary carcinogen shortly before pregnancy enhances mammary tumor formation during the period of pregnancy and lactation compared to age-matched virgin controls[5,6]. Furthermore, treatment of rats with tamoxifen[7,8] or 2-bromo-α-ergocryptine[8], suppressers of estrogen and prolactin secretion respectively, at the time of, or shortly after carcinogen exposure, significantly inhibits tumorigenesis.

A number of mechanisms have been suggested to explain the association between hormonal status and breast cancer. Among these, particular attention has been paid to their effects on prostaglandin (PG) synthesis. Several lines of evidence suggest that PGs play a role in breast cancer development. Important among these are: 1) malignant breast tumors and tumor cell lines from humans and rodents contain elevated levels of PGs compared to the normal tissues from which they arise[9,10], 2) inhibiting the key regulatory enzyme in PG synthesis, prostaglandin H-synthase/cyclooxygenase (COX), by non-steroidal anti-inflammatory agents (NSAIDs) hinders the development of chemically-induced as well as transplantable mammary cancer[11,12]. NSAIDs can inhibit both COX isoenzymes, the constitutive COX-1 and the inducible COX-2 isoforms.

There is some evidence linking hormonal effects on mammary carcinogenesis to PG synthesis. There is a loss of PG-receptors during progression of rat mammary tumors from hormonal dependence to autonomy of growth[13]. Addition of growth hormones to rat mammary cell lines results in a significant increase in the release of PGs[14]. The PGE_2 content of 7,12-dimethylbenz[α]anthracene (DMBA)-induced mammary tumors correlates with growth responses to alterations in endogenous estrogen and prolactin[15]. Finally, elevated levels of PGE_2 have been associated with aggressive, metastatic, estrogen receptor-negative tumors[16], although other reports show either that estrogen receptor-negative tumors produce lower levels of PGs[9] or that both hormone-dependent and autonomous mammary tumors produce the same amounts of PGs[13].

Dietary fats are known to modulate mammary gland carcinogenesis in experimental animals[17,18] and may have similar effects in human breast cancer development[19]. Several studies in rodents have shown that vegetable oils rich in n-6 polyunsaturated fatty acids (PUFAs) promote mammary carcinogenesis whereas similar levels of marine oils rich in the n-3 PUFAs, appear to inhibit[17,20].

Evidence linking the effects of dietary fat on mammary carcinogenesis to PG synthesis derives mainly from the observation that NSAIDs inhibit the promoting effects of diets rich in n-6 PUFAs[21]. Furthermore, the n-6 PUFA linoleic acid (LA) is a precursor of the PG 2 series that are known to be mitogenic for normal human and rodent mammary epithelial cells[22]. On the other hand the n-3 PUFAs eicosapentaenoic acid (EPA) and docosahexaenoic acid (DHA), are precursors of PG 3 series that lack such a mitogenic effect.

In order to provide further information on the possible association between PGs and breast cancer development, we have recently been investigating the levels of PG synthesis in the mammary glands of rats that have different levels of susceptibility to mammary gland carcinogenesis. This susceptibility was either associated with various hormonal conditions or with feeding different levels and types of dietary fatty acids.

EFFECT OF HORMONAL STATUS ON PG SYNTHESIS

The expression of the *COX-1* and *COX-2* genes and levels of PG synthesis was examined in the mammary glands of rats that have different levels of susceptibility to mammary gland carcinogenesis associated with pregnancy, lactation, post-lactational involution, and ovariectomy in order to understand the possible association between hormone effects on breast cancer development and PGs. Female Sprague-Dawley rats were mated with syngeneic males to yield groups (6-8 animals/group) of pregnant, lactating, and post-lactational animals with a group of virgin controls. The animals were sacrificed on day 7 of pregnancy, day 7 of lactation, or day 15 after ceasing lactation. At the time of sacrifice all animals were 100 ± 5 days of age. In a second experiment, a group of 35 day-old animals was ovariectomized, acclimatized for 7 days, then administered 17β-estradiol and progesterone, s.c. (50 mg and 5 mg, respectively, in 10% ethanol in olive oil) daily for 10 consecutive days. The animals were sacrificed one day following the last treatment. The controls were either age-matched ovariectomized or sham-operated animals that were administered only the vehicle. Mammary gland mRNA was examined for *COX-1* and *COX-2* expression.

COX-1 mRNA, measured by Northern blot analysis[23], was similar in virgin, lactating, pregnant, and post-lactational animals of the same age (Figure 1a). Ovariectomized animals exhibited significantly lower levels of COX-1 mRNA (~40%) compared to the sham-operated controls or the ovariectomized animals treated with estradiol and progesterone (Figure 1b). COX-2 mRNA, measured by RT-PCR[23], was detectable only in the mammary glands of lactating animals and ovariectomized animals administered estradiol and progesterone (Figure 2).

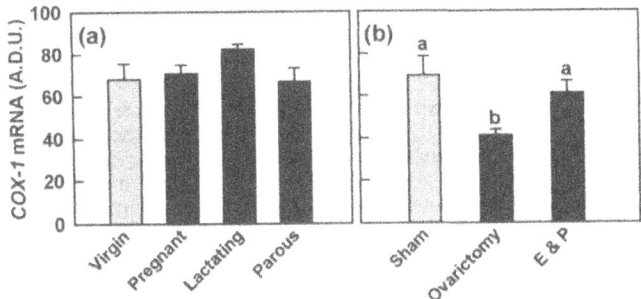

Figure 1. Effect of hormonal status on the levels of COX-1 mRNA in rat mammary glands measured by Northern blot analysis and quantified as arbitrary density units (A.D.U.) by measuring the band intensity relative to β-actin. (a) 100±5 day-old animals and (b) 52 ± 5 day-old animals. Data are means ± S.D. for 6-8 animals. Groups with different letters are significantly different from each other ($p<0.05$) using Student's t- test.

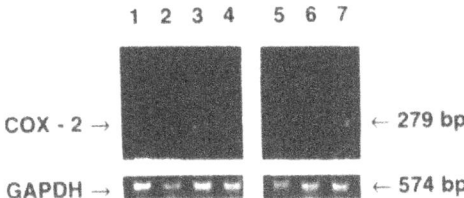

Figure 2. Effect of hormonal status on the expression of COX-2 mRNA in rat mammary glands measured by RT-PCR. Single stranded cDNA was reverse transcribed from total RNA from the mammary glands and used for PCR amplification with *COX-2-* (upper panel) and GAPDH- (lower panel) specific primers. Lanes are: 1, virgin; 2, pregnant; 3, lactating; 4, post-lactational; 5, sham-operated; and 6, ovariectomized animals; and 7; ovariectomized animals administered estradiol and progesterone. Groups 1-4 are 100±5 day-old while groups 5-7 are 52±5 day-old.

COX enzymatic activities, determined in the mammary gland microsomes by measuring the conversion rate of [1-^{14}C]-arachadonic acid to prostanoids[23,24], showed that lactating animals had a significantly higher activity compared to virgin (~40%), pregnant (~30%), or post-lactational animals (~40%). Ovariectomized animals had significantly lower COX enzymatic activity compared to the sham operated animals. Significant induction of COX activity, however, was observed in ovariectomized animals administered estradiol and progesterone. These changes in COX enzymatic activity were paralleled by similar changes in the mammary gland PGE$_2$ content, measured in the tissue homogenate by enzyme immunoassay[23].

Our results show that the level of PG synthesis in the mammary gland correlates with its susceptibility to cancer development. Increased synthesis of PGs, particularly series 2, can stimulate the proliferation of mammary epithelial cells and may thereby stimulate the proliferation of initiated and/or preneoplastic cells. It is particularly interesting that *COX-2* expression is induced during lactation and after treatment with estrogen and progesterone, the hormonal conditions that are known to enhance tumor formation[25,26]. These results suggest that the effect of hormones on the genesis of mammary cancer in the rat may be mediated, at least in part, by their effects on *COX-2* expression and PG synthesis.

EFFECTS OF DIETARY FATTY ACIDS ON PG SYNTHESIS

In order to provide information on the mechanism by which dietary fats modulate breast cancer development, we examined the effect of feeding rats different levels of n-3 and n-6 PUFAs on the expression of the *COX-1* and *COX-2* genes and PG synthesis in mammary glands. Female Sprague-Dawley rats (38-43 days old) were acclimatized for 7 days on an AIN-93G standard reference diet. The animals were then randomized into four groups (5/group) and were fed diets containing either low- (7%; LF) or high- (21%; HF) fat diets that were rich in either n-6 PUFAs (Safflower oil, S) or n-3 PUFAs (Menhaden oil, M) for 3 weeks.

COX-1 mRNA levels were approximately the same in groups fed diets containing either level of Menhaden oil, but were significantly increased by ~30% in the LFS and HFS groups (Figure 3a). Transcripts of the inducible *COX-2* gene were not detectable in the n-3 PUFAs fed groups, but this gene was expressed in animals fed either level of n-6 PUFAs (Figure 3b) and in the HFS group was associated with increased levels of COX enzymatic activity and production of PGE_2.

n-3 PUFAs have been shown to inhibit the growth of human breast cancer cells[27]. These effects were attributed to the production of series 3 PGs, rather than series 2, and also to the ability of metabolites of the n-3 PUFAs to inhibit the metabolism of n-6 PUFAs at various stages in the AA cascade. The findings of this study suggest that the inhibitory effect of n-3 PUFAs on mammary carcinogenesis might be due to their modulating effects on COX enzymatic activity. In contrast, dietary n-6 PUFAs upregulate *COX-2* and, to some extent, *COX-1* expression. There was a concomitant increase in COX enzyme activity and PG synthesis in the mammary glands of rats fed high levels of n-6 PUFAs[23]. These observations may explain, at least in part, the promoting effects of dietary n-6 PUFAs on mammary carcinogenesis.

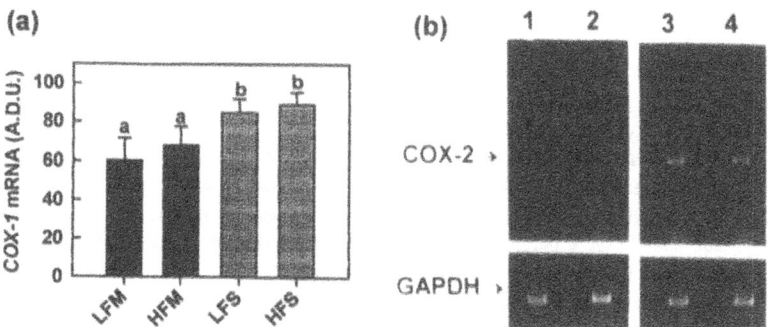

Figure 3. Expression of *COX-1* and *COX-2* genes in the mammary glands of rats fed diets containing different levels of Menhaden or Safflower oil. (a) Levels of COX-1 mRNA measured by Northern blot analysis and quantified as arbitrary density units (A.D.U.) by measuring the band intensity relative to β-actin. Data are means ± S.D. of 5 animals. Groups with different letters are significantly different from each other ($p<0.05$) using Student's *t*- test. (b) RT-PCR analysis of COX-2 mRNA. Single stranded cDNA was reverse transcribed from total RNA from the mammary glands and used for PCR amplification with *COX-2*- (upper panel) and GAPDH- (lower panel) specific primers. Lanes are: 1, LFM; 2. HFM; 3, LFS; and 4, HFS.

CONCLUSION

Our results show that the level of PG synthesis in the mammary gland correlates with its susceptibility to cancer development. Increased PG synthesis could promote cancer development in at least two ways. First, via its peroxidase activity, cyclooxygenase can contribute to the activation of chemical carcinogens[28]. Second, elevated PG biosynthesis, particularly series 2, can stimulate the proliferation of mammary epithelial cells and may thereby stimulate the proliferation of initiated and/or preneoplastic cells. It is particularly interesting that *COX-2* expression is induced under hormonal and dietary conditions that are known to enhance tumor formation. *COX-2* overexpression in stably transfected rat intestinal epithelial cells has been shown to enhance growth and/or decrease loss of initiated or preneoplastic cells[29]. Therefore, our results suggest that the effect of hormones and dietary fat in the genesis of mammary cancer in the rat may be mediated, at least in part, by their effects on *COX-2* expression and PG synthesis.

REFERENCES

1. J.L. Kelsely, M.D. Gammon, and E.M. John, Reproductive factors and breast cancer. *Epidemiol. Rev.* 15:36 (1993).
2. J.R. Harris, M.E. Lippman, U. Veronesi, and W. Willett, Breast cancer. *N. Eng. J. Med.* 327:319 (1993).
3. I.H. Russo, and J. Russo, Mammary gland neoplasia in long-term rodent studies. *Environ. Heallth. Prespect.* 104:938 (1996).
4. C.W. Welsch, Host factors affecting the growth of carcinogen-induced rat mammary carcinomas: A review and tribute to Charles Brenton Huggins. *Cancer Res.* 45:3415 (1985).
5. G.M. McCormick, and R.C. Moon, Effect of pregnancy and lactation on growth of mammary tumors induced by 7,12-dimethylbenz(a)anthracine (DMBA). *Br. J. Cancer.* 19:160 (1965).
6. S. Sakamoto, Y. Imamura, S. Sassa, and R. Okamoto, DMBA-induced mammary tumor and hormone environment in the rat during pregnancy, postpartum, and long term lactation. *Toxicol. Lett.* 4:237 (1979).
7. M.M. Gottardis, and V.C. Jordan, Antitumor actions of ketoxifene and tamoxifen in the *N*-nitrosomethyl-urea-induced rat mammary carcinoma model. *Cancer Res.* 47:4020 (1987).
8. C.W. Welsch, M. Goodrich-Smith, C.K. Brown, D. Mackie, and D. Johnson, 2-Bromo-α-ergocryptin (CB-154) and tamoxifen (ICI 46,474) induced supression of the genesis of mammary carcinomas in female rats treated with 7,12-dimethylbenzanthracene (DMBA): a comparison. *Oncol.* 39:88 (1982).
9. R.A. Karmali, S. Welt, H.T. Thaler, and F. Lefevre, Prostaglandins in breast cancer: relationship to disease stage and hormone status. *Br. J. Cancer.* 48:689 (1983).
10. W.C. Tan, O.S. Privett, and M.E. Goldyne, Studies of prostaglandins in rat mammary tumors induced by 7,12-dimethylbenz(a)anthracene. *Cancer Res.* 34:3229 (1974).
11. C.A. Carter, R.J. Milholland, W. Shea, and M.M. Ip, Effect of the prostaglandin synthase inhibitor indomethacin on 7,12-dimethylbenz(a)anthracene-induced mammary tumorigenesis in rats fed different levels of fat. *Cancer Res.* 43:3559 (1983).
12. G.M. Kollmorgen, M.M. King, S.D. Kosanke, and C. Do, Influence of dietary fat and indomethacin on the growth of transplantable mammary tumors in rats. *Cancer Res.* 43:4714 (1983).
13. H. Abou-Issa, and J.P. Minton, Loss of prostaglandin E receptors during progression of rat mammary tumors from hormonal dependence to autonomy. *J. Natl. Cancer Inst.* 76:101 (1986).
14. P.S. Rudland, A.C.T. Davies, and S.W. Tsao, Rat mammary preadipocytes in culture produce a trophic agent for mammary epithelia-prostaglandin E_2. *J. Cell Physiol.* 120:364 (1984).
15. M.K. Foecking, W.E. Kibbey, H. Abou-Issa, R.H. Matthews, and J.P. Minton, Hormone dependence of 7,12-dimethylbenz[a]anthracene-induced mammary tumor growth: correlation with prostaglandin E_2 content. *J. Natl. Cancer Inst.* 69:443 (1982).

16. X-H. Liu, and D.P. Rose, Differential expression and regulation of cyclooxygenase-1 and -2 in two human breast cancer cell lines. *Cancer Res.* 56:5125 (1996).

17. L.A. Cohen. (1986) Dietary fat and mammary cancer. in: *Diet, Nutrition, and Cancer: A Critical Evaluation.* B.S. Reddy and L.A. Cohen, eds., CRC Press, Inc, Boca Raton, FL (1986).

18. U. Mohr, and J.P. Lewkowski, The effect of diet on tumor development in animals. *Expl. Pathol.* S12:5 (1989).

19. C.P.J. Caygill, A. Charlett, and M.J. Hill, Fat, fish, fish oil, and cancer. *Br. J. Cancer.* 74:159 (1996).

20. M. Craig-Schmidt, M.T. White, P. Teer, J. Johnson, and H.W. Lane, Menhaden, coconut, and corn oils and mammary tumor incidence in BALB/c virgin female mice treated with DMBA. *Nutr. Cancer.* 20:99 (1993).

21. D.C. Cunningham, L.Y. Harrison, and T.D. Shultz, Proliferative responses of normal human mammary and MCF-7 breast cancer cells to linoleic acid, conjugated linoleic acid and eicosanoid synthesis 32. inhibitors in culture. *Anticancer Res.* 17:197 (1997).

22. D.F. Horrobin, The role of essential fatty acids and prostaglandins in breast cancer. in: *Diet, Nutrition, and Cancer: A Critical Evaluation.* B.S. Reddy, and L.A. Cohen, eds. CRC Press, Inc, Boca Raton, FL (1986).

23. A.F. Badawi, A. El-Sohemy, L.L. Stephen, A.K. Ghoshal, and M.C. Archer, The effect of dietary n-3 and n-6 polyunsaturated fatty acids on the expression of cyclooxygenase 1 and 2 and levels of p21[ras] in rat mammary glands. *Carcinogenesis.* in press (1998).

24. A.F. Badawi and M.C. Archer, Effect of hormonal status on the expression of the cyclooxygenase 1 and 2 genes and prostaglandin synthesis in rat mammary glands. *Proc. Am. Assoc. Cancer Res.* 39:590 (1998).

25. C.W. Welsch, Interaction of estrogen and prolactin in spontaneous mammary tumorigenesis in mouse. *J. Toxicol. Environ. Health.* 1:161 (1976).

26. C.W. Welsch, J.W. Jenkins, and J. Meites, Increased incidence of mammary tumors in the female rat grafted with multiple pituitaries. *Cancer Res.* 30:1024 (1970).

27. J.M. Connolly, and D.P. Rose, Effects of fatty acids on invasion through reconstituted basement membrane ('Matrigel') by human breast cancer cell line. *Cancer Lett.* 75:137 (1993).

28. T.E. Eling, and J.F. Curtis, Xenobiotic metabolism by prostaglandin *H* synthase. *Pharmac. Ther.* 53:261 (1992).

29. M. Tsuji, and R.N. DuBois, Alteration in cellular adhesion and apoptosis in epithelial cells overexpressing prostaglandin endoperoxide synthase 2. *Cell.* 83:493 (1995).

19

REGULATION OF CEREBROVASCULAR CYCLOOXYGENASE-2 BY PRO- AND ANTI-INFLAMMATORY CYTOKINES

Steven A. Moore, Elizabeth Yoder, Gretchen Rich, MacKenzie Hilfers, and Jeffrey Albright

Department of Pathology
The University of Iowa
Iowa City, IA 52242

INTRODUCTION

Arachidonic acid metabolism is important in regulating the cerebral circulation and its response to CNS injury [1,2], a response that often involves vascular cell proliferation, inflammatory cell infiltrates, and glial cell reactions. Induction and regulation of cyclooxygenase-2 (COX-2) may be important in these processes. Since a number of pro- and anti-inflammatory cytokines, including TGFβ1, TNFα, and IL-1β, are produced during the course of CNS reaction to injury [3,4,5], we investigated the regulation of cerebrovascular COX-2 expression by these agents. In addition to cytokine production, the release of polyunsaturated fatty acids (PUFA), particularly arachidonic acid (20:4n-6) and docosahexaenoic acid (22:6n-3), is a common early reaction to CNS injury [6,7]. Therefore, the potential interactions of PUFA and proinflammatory cytokines in the regulation of COX-2 were also investigated.

METHODS

Mouse-derived cerebral microvascular endothelial and smooth muscle cultures were utilized between passages 5 and 20 [8,9]. Cells were incubated up to 12 h with pro- and anti-inflammatory cytokines (all murine recombinant cytokines from R&D) prior to isolating total RNA or total protein for northern or western blot analysis, respectively. The COX-1 and COX-2 probes were purchased from Oxford Biochemical and the COX-2 antibody from Cayman Chemical. The COX-1 antibody was a gift from Dr. William Smith, Michigan State University. Blotting was carried out as described previously [10,11].

Eicosanoids and Other Bioactive Lipids in Cancer, Inflammation, and Radiation Injury, 4
Edited by Honn *et al.*, Kluwer Academic / Plenum Publishers, New York, 1999.

125

RESULTS

COX-2 Induction by Proinflammatory Cytokines

Incubation of cerebral microvascular endothelium or smooth muscle with interleukin-1β (IL-1β) or tumor necrosis factor α (TNFα) causes 3 to 10 fold increases in basal and stimulated prostaglandin (PG) production in both cell types (data not shown). The increased PG production is blocked almost entirely by the selective COX-2 inhibitor NS-398 [12]. Levels of COX-2 mRNA and protein are rapidly, transiently, and greatly induced by these same proinflammatory cytokines (western blots not shown). Peak protein and mRNA levels are seen by 6h and 3h, respectively, after cytokine exposure. Examples of the time course of COX-2 induction are shown in Figure 1. In contrast to COX-2, levels of COX-1 protein and mRNA change very little, however, a small induction of COX-1 mRNA was noted in vascular smooth muscle cells following exposure to TNFα (see lower right panel of Figure 1). The induction of COX-2 is dose dependent (maximal effect produced by 500 pg/ml IL-1β or 100 ng/ml TNFα) and is prevented by 1 μM dexamethasone (data not shown).

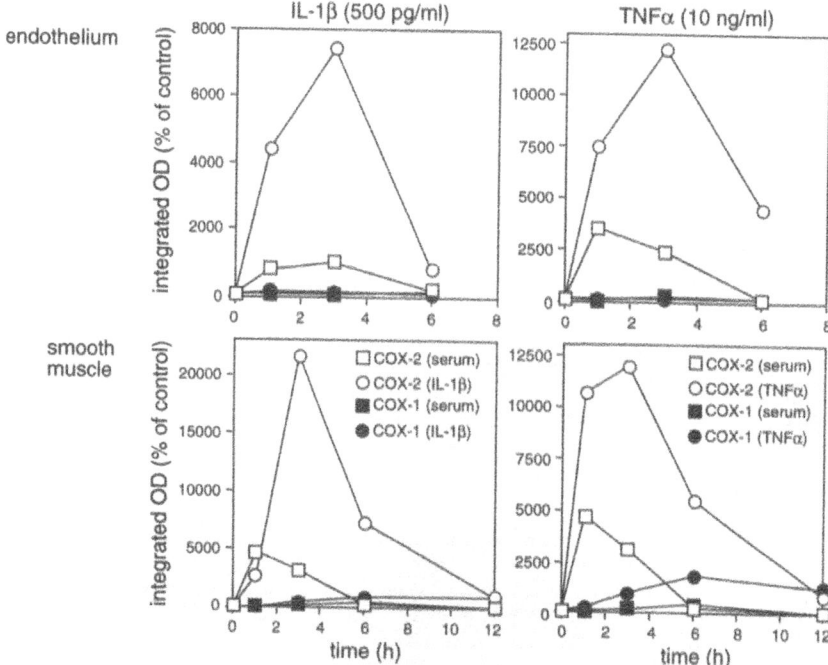

Figure 1. Northern blot analysis of the time course of COX-1 and COX-2 induction in cerebral endothelium and smooth muscle by proinflammatory cytokines. The culture medium was changed at time zero and cells were incubated up to 12 h in routine culture medium containing 10% serum or 10% serum plus cytokines. IL-1β was utilized at a concentration of 500 pg/ml and TNFα at a concentration of 10 ng/ml. These graphs represent quantitative densitometry data derived from the northern blots.

Polyunsaturated Fatty Acids Enhance COX-2 Induction

In order to determine if exposure to PUFA might alter the induction of COX-2 by proinflammatory cytokines, cerebral endothelial cells were incubated up to 96 h with 60 μM arachidonic or docosahexaenic acid. Cells grown under routine conditions express very little COX-2 mRNA, while both IL-1β and TNFα induce readily detectable amounts of COX-2 (Figure 2). Both short term (3h) and long-term (96h) incubation with PUFA (either arachidonate or docosahexaenoate) also induce COX-2. Surprisingly, the combination of PUFA and proinflammatory cytokines results in much greater COX-2 induction than either PUFA or cytokine alone. This enhanced induction occurs whether the cells are preincubated with PUFA for 96h prior to the addition of cytokines or whether the PUFA and cytokines are added simultaneously. Although the data is not shown here, similar inductions of COX-2 occur with the combinations of arachidonate with TNFα and docosahexaenoate with IL-1β. Similar, but smaller inductions of COX-2 protein are also stimulated by these PUFA/cytokine combinations (data not shown).

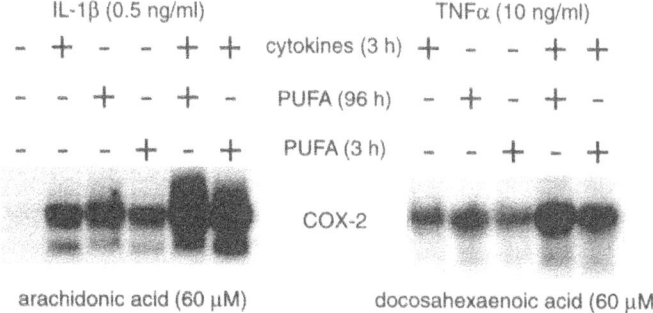

Figure 2. Northern blot analysis of the induction of COX-2 in cerebral endothelium by PUFA and proinflammatory cytokines. Cultures that received neither PUFA nor cytokine expressed very low levels of COX-2. Cultures exposed to cytokines alone, PUFA alone, or cytokines in combination with PUFA are induced to express high levels of COX-2. The combinations of PUFA and cytokines were carried out by either preincubating cells with PUFA for 96h prior to the addition of cytokine or by simultaneously adding PUFA and cytokine to the cells.

COX-2 Induction is Blunted by the Anti-inflammatory Cytokine TGFβ₁

The anti-inflammatory cytokine, transforming growth factor β_1 (TGFβ_1 at 1 ng/ml) is able to blunt TNFα induction of cerebrovascular smooth muscle COX-2 by about 50%, especially if it is present 6 to 12h prior to the addition of TNFα (Figure 3, left side). The mechanism by which TGFβ_1 blunts the TNFα COX-2 response is at least partially due to its effects on COX-2 mRNA half life (Figure 3, right side). This was demonstrated by adding 4 μg/ml actinomycin D to cultures that had previously been incubated 3h with TNFα alone or with TNFα following a 6h preincubation with TGFβ_1. COX-2 mRNA levels were analyzed over a 6h time course following the addition of actinomycin D. TGFβ_1 significantly shortened the COX-2 mRNA half-life.

DISCUSSION

These data indicate that COX-2 expression in cerebrovascular cells, both endothelium and vascular smooth muscle, are regulated by pro- and anti-inflammatory cytokines. Both IL-1β and TNFα rapidly induce large, transient increases in COX-2 expression. The time course and degree of induction are quite similar in both cell types. These two cytokines are prominent proinflammatory cytokines produced at the site of many types of brain pathology [3,4,5]. Sources of these cytokines are likely to include glia (astrocytes and microglia) and infiltrating leukocytes [5]. Proinflammatory cytokines such as IL-1β and TNFα also are likely to mediate some of the CNS changes that occur during systemic inflammatory or infectious diseases [13].

In additional work pertenent to modeling brain injury states, the proinflammatory cytokine induction of cerebrovascular cell COX-2 is greatly enhanced by PUFA, especially the major n-6 and n-3 fatty acids found in brain, arachidonic acid (20:n-6) and docosahexaenoic acid (22:6n-3). Each of these PUFAs can be released in large amounts from glycerolipid stores (mainly membrane phospholipid) during the reaction to CNS

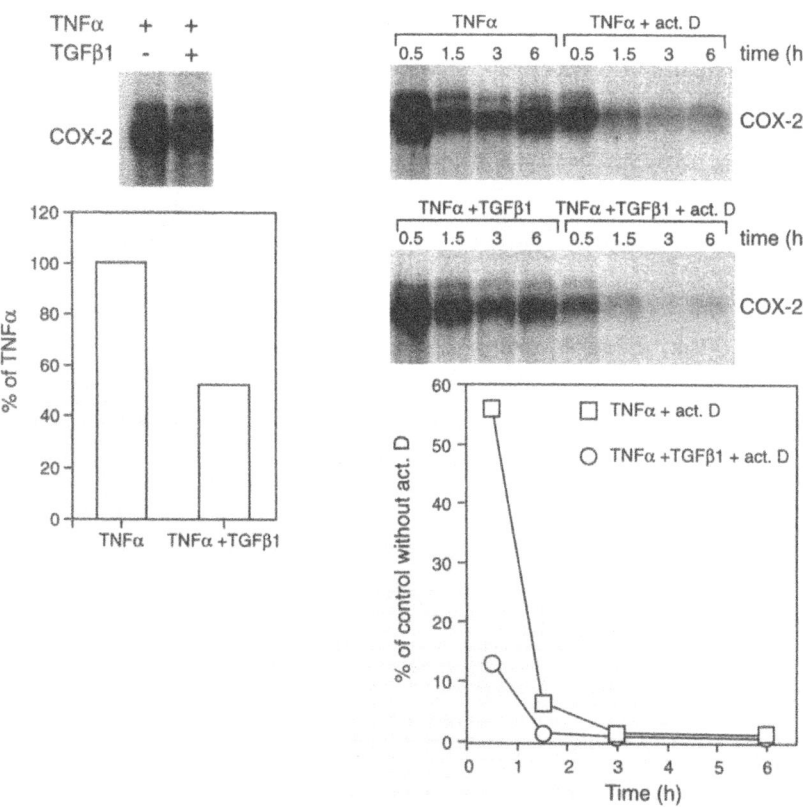

Figure 3. Northern blot analysis of the effects of TGFβ₁ on COX-2 induction by TNFα. The left side of the figure depicts COX-2 expression in cerebrovascular smooth muscle following a 3h exposure to 10 ng/ml TNFα with or without a 6h preincubation with 1 ng/ml TGFβ₁. Densitometry quantitation of the blot reveals about a 50% reduction in COX-2 mRNA in the TGFβ₁-treated cells. The right side of the figure is a COX-2 mRNA half-life study where cerebral smooth muscle cells were exposed 3h to 10 ng/ml TNFα with or without a 6h preincubation with 1 ng/ml TGFβ₁. Actinomycin D (4 μg/ml) was then added to half the cultures and COX-2 mRNA assayed over the ensuing 6h. Quantitative densitometry shows that the half-life of COX-2 mRNA was substantially shortened by TGFβ₁.

injuries [6,7]. This is probably best documented in brain trauma, brain ischemia, and seizures. Since proinflammatory cytokines are also upregulated in these pathologic states [3,4,5], it is important to understand that genes such as COX-2 may be super-induced by the combined action of PUFA and cytokines. Massive release of PUFA may, therefore, exacerbate the proinflammatory environment of these types of brain lesions.

Anti-inflammatory cytokines are also produced in the setting of CNS pathology [3,4,5] and their synthesis often lags behind the induction of proinflammatory cytokines. One of the major anti-inflammatory cytokines produced is $TGF\beta_1$. Like the proinflammatory cytokines, $TGF\beta_1$ is likely made by a combination of glia and infiltrating leukocytes [5]. In cerebrovascular cells, $TGF\beta_1$ can blunt the COX-2 inducting effect of IL-1β and TNFα by approximately 50%. At least one mechanism involved in this inhibitory effect is the shortening of COX-2 mRNA half-life.

In summary, the data suggest that cerebrovascular PG production may be modified by regulation of endothelial or smooth muscle cell COX-2 in disease states such as trauma, stroke, infection, or neoplasia where cytokines such as IL-1β, TNFα, and $TGF\beta_1$ are produced [3,4,5,14]. Thus, the cerebrovascular reaction to CNS pathology may in part be dependent upon COX-2 expression within the cerebral vessel wall itself.

ACKNOWLEDGMENTS

This work was supported by NIH grants NS-24621 and NS-09858 and by and American Heart Association Grant-in-Aid 96 50661N.

REFERENCES

1. Leffler CW and Busija DW. Arachidonic acid metabolites and perinatal cerebral hemodynamics. *Sem Perinatol* 11:31-42, 1987.
2. Wahl M, Unterberg A, Baethmann A, and Schilling L. Mediators of blood-brain barrier dysfunction and formation of vasogenic brain edema. *J Cerbral Blood Flow Metab* 8:621, 1988.
3. Feuerstein GZ, Liu T, and Barone FC. Cytokines, inflammation, and brain injury: role of tumor necrosis factor-α. *Cerebrovasc Brain Metabol Rev* 6:341-360, 1994.
4. Schobitz B, de Kloet ER, and Holsboer F. Gene expression and function of interleukin 1, interleukin 6, and tumor necrosis factor in the brain. *Prog Neurobiol* 44:397-432, 1994.
5. Eng, L. F., R. S. Ghirnikar, and Y. L. Lee. 1996. Inflammation in EAE: Role of chemokine/cytokine expression by resident and infiltrating cells. *Neurochem. Res.* 21:511-525.
6. Bazan NG, Birkle DL, Tang W and Reddy TS. The accumulation of free arachidonic acid, diacylglycerols, prostaglandins, and lipoxygenase reaction products in the brain during experimental epilepsy. *Adv Neurol* 44:879-902, 1986.
7. Chan PH and Fishman PA. The role of arachidonic acid in vasogenic brain edema. *Fed Proc* 43:210-213, 1984.
8. Moore SA, Yoder E, and Spector AA. Role of the blood-brain barrier in the formation of long-chain ω-3 and ω-6 fatty acids from essential fatty acid precursors. *J Neurochem* 55:391-402, 1990.
9. Moore SA, Strauch AR, Yoder EL, Rubenstein P and Hart MN. Cerebral microvascular smooth muscle in tissue culture. *In Vitro* 20:512-520, 1984.
10. Rich G, Yoder EJ, Prokuski L, and Moore SA. Prostaglandin production in cerebral microvascular smooth muscle is serum dependent. *Am J Physiol* 270 (Cell Physiol 39):C1379-C1387, 1996.
11. Rich G, Yoder EJ, and Moore SA. Regulation of prostaglandin H synthase-2 expression in cerebromicrovascular smooth muscle by serum and EGF. *J Cell Physiol*, accepted.
12. Futaki N, Takahashi S, Yokoyama M, Arai I, Higuchi S, Otomo S. NS-398, a new anti-inflammatory agent, selectively inhibits prostaglandin G/H synthase/cyclooxygenase (COX-2) activity in vitro. *Prostaglandins* 1994; 47:55-59.
13. Cao C, Matsumuta K, Yamagata K, Watanabe Y. Induction by lipopolysaccharide of cycloxygenase-2 mRNA in rat brain: its possible role in the febrile response. *Brain Res* 697:187-196, 1995.
14. Tada M, De Tribolet N. Recent advances in immunobiology of brain tumors. *J Neurooncol* 1993; 17:261-271.

20

VISUALIZATION AND QUANTITATION OF CYCLOOXYGENASE-1 AND -2 ACTIVITY BY DIGITAL FLUORESCENCE MICROSCOPY.

Richard L. Ornberg and Alane T. Koki

Analytical Sciences Center
Science and Technology
Monsanto Company
St. Louis, MO 63198

INTRODUCTION

Cyclooxygenase (COX), also known as prostaglandin endoperoxide synthase, catalyzes the initial step in the conversion of arachidonic acid to prostaglandins and thromboxanes. Two distinct forms of the enzyme, COX-1 and COX-2 are known to be encoded by separate genes and are expressed in different normal and diseased tissues (Pairet and Engelhardt, 1996; Williams and DuBois, 1996). COX-1 is expressed in a wide variety of normal cells and tissues and produces prostaglandins that mediate homeostatic function. COX-2 is virtually undetectable under normal physiological conditions but is induced in a variety of cells at sites of inflammation (Masferrer, 1995), cancer (Koki et.al., 1998), and in the central nervous system (Kaufman et.al., 1996). Despite the similar catalytic activity and high amino acid sequence homology of the two isoforms, selective inhibitors have been synthesized for COX-2 and these are expected to provide a new therapy for arthritis and pain (Lipsky, 1998), inflammation (Masferrer 1995) and cancer (Kawamori T, et.al., 1997)

Much of what is known about the enzymology of PHGS has been gained from isolated enzyme preparations from various tissues and transfected cells. Recent work suggests that the functional role of COX-1 and COX-2 in normal and pathophysiological conditions may be dependent upon their subcellular location (Reddy and Herschman, 1994). To address how intracellular compartmentalization and the local cellular environment may modulate COX activity, we have developed a method that enables the kinetic study and

Eicosanoids and Other Bioactive Lipids in Cancer, Inflammation, and Radiation Injury, 4
Edited by Honn *et al.*, Kluwer Academic / Plenum Publishers, New York, 1999.

quantitation of enzyme activity in situ. COX has two catalytic activities, a cyclooxygenase which bis oxygenates arachidonate at the C-9 and C-11 positions to make prostaglandin G2 and a peroxidase which reduces the endoperoxide to form prostaglandin H2 and free radicals. In this paper we describe a method of capturing arachidonate induced, COX-derived oxygen radicals in mouse fibroblasts with an intracellular fluorescent specific reporter substrate 5-(and-6)-carboxyl 2',7'dichlorohydro-fluorescein diacetate (CDCF-DA). Using low light level digital time lapse fluorescence microscopy, we can visualize and measure COX activity as a fluorescence intensity increase in cells in response to exogenous arachidonic acid. We demonstrate that the fluorescent shifts are appropriately sensitive to selective PHGS inhibitors and that the observed activity coincides with immunolocalization of COX protein. These results suggest this assay may be an excellent tool to study the functional significance of PGH-synthase subcellular compartmentalization, and to determine the local physiological parameters that mediate prostaglandin levels in specific intracellular locations.

EXPERIMENTAL PROCEDURES:

Cell Culture and Fluorescence Microscopy

NIH 3T3 fibroblast were grown to 80% confluence in Dulbecco's modified Eagles medium (DMEM) containing 10% fetal calf serum. Cells were passed in 0.25% trypsin in EDTA for 5 min, plated at approximately 4×10^5 cells per well in untreated double well Lab-Tek chamber slides and cultured for 48 hr in DMEM containing 0.2% fetal calf serum. COX-2 expression was induced by culturing cells in DMEM containing 10% fetal calf serum for 2 hr prior to study. For immunohistology, cells were fixed in prechilled methanol, permeabilized in 0.2% saponin in phosphate buffered saline, and labeled with isoform specific rabbit polyclonal antibodies synthesized in-house, #566 (COX-1) and #539 (COX-2). Specific label was visualized with a goat anti-rabbit antibody conjugated to Cy3 (Jackson Labs) and photographed by digital fluorescence microscopy.

The reactive oxygen reporter dye, CDCF-DA, (Molecular Probes, Inc.) was used to detect reactive oxygen radicals produced by cyclooxygenases in response to exogenously applied arachidonic acid. Prior to visualization, cells were rinsed twice in phosphate buffered saline containing $CaCl_2$ and $MgCl_2$ at 0.1 mg/ml followed by incubation in 10 uM CDCF-DA in PBS for 15 min at $37°C$ in the dark. Excess dye was washed off with PBS. Arachidonic acid from a 6.6 mM sodium salt stock solution was diluted to the 2x the final desired concentration in PBS and an equal volume was added to the cell culture well to initiate COX activity. Cyclooxygenase inhibitors were dissolved to 10 mM in DMSO and added to cells at the appropriate concentration in PBS 30 min prior to visualization. The final concentration of DMSO was less than 0.1%.

Cell fluorescence was imaged and measured with a Nikon inverted fluorescence light microscope and workstation-based digital imaging system (Inovision Corp., Durham, NC). Under computer control, a field of cells were exposed to excitation light from a 75W xenon lamp and images were collected at 658 x 550 pixels by 4096 gray levels with a cooled low light level CCD camera (Photometrix Inc. model PXL). Excitation and emission wavelengths were determined by a Nikon B2_E filter cube. Typically, images from 50 or 100 msec exposures were acquired every 5 - 10 sec beginning at 50 sec prior to and 3 - 10 min after adding an equal volume of arachidonic acid in PBS. The exposure intensity and time were set such that the background fluorescence of the extracellular solution was at approximately 200 units and the basal cell intensity was 300 - 400 units. Excitation light intensity was measured to be 40 - 60 uW/cm^2-sec such that each image delivered a total radiation dose of less than 0.6 uW/cm^2. Raw images were corrected for uneven illumination by dividing with a flatfield image and multiplying by the mean intensity of the flatfield image. The flatfield image was obtained from a uniform region of uranium glass prior to each experiment. The corrected images consisting of a time lapse series of a field of cells were stored on disc for subsequent analysis of cellular fluorescence. Using the Isee image analysis software (Inovision Corp., Durham, NC), 30 cells from each time lapse series were selected at random and the maximum intensity of each cell was measured and stored at every time point. The mean and standard deviation of these maximum intensities was calculated and graphed as the difference of the mean intensity minus the background versus time using Microsoft Excel spreadsheet.

RESULTS

COX-1 activity in serum starved cells appeared as a diffuse rise in fluorescence through out the entire cytosolic compartment of the cell after adding 10 uM AA, Figure 1A. Serum induced COX-2 expression resulted in a more robust fluorescence increase following arachidonate, Figure 1B. To specifically visualize COX-2 activity, serum induced cells were treated with a selective COX-1 inhibitor, SC-560. The resultant fluorescence intensity in these cells appeared to localize over the nuclear region suggesting that COX-2 may be preferentially concentrated in or near the nucleus, Figure 1C. Immunocytochemical localization of COX-1 and COX-2 coincided with the activity described above, Figure 2. COX-1 was found through out the cytoplasm including the fine pseudopodia of all cells. COX-2 was concentrated to the perinuclear region and nucleus with little or no stain in the fine pseudopodia.

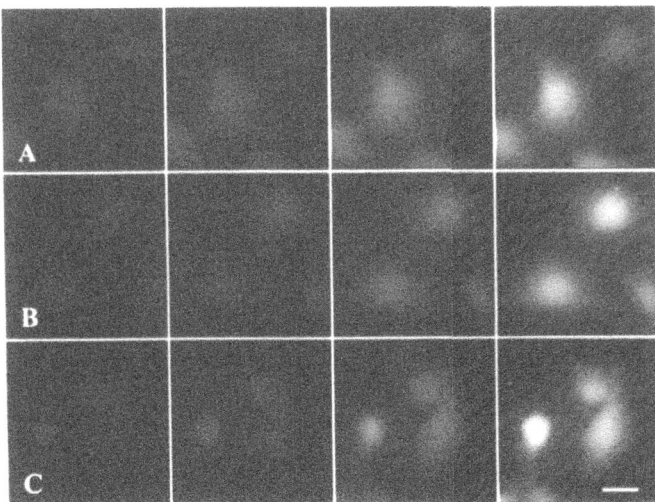

Figure 1. Time-lapse fluorescence light microscopy images of COX activity in NIH 3T3 cells after (A) no serum stimulation, (B) after 2 hr serum treatment and (C) 2 hr serum treatment in presence of a COX-1 inhibitor, SC-560. Images from left to right are at 0, 100, 200, and 300 sec after adding 10 uM arachidonic acid. Bar = 5 micron.

Quantitative measurements of these activities from a one typical set of experiments are shown in Figure 3. In serum starved cells, fluorescence induced by 100 nM to 10 uM (10 uM data shown) resulted in a gradual curvilinear rise with a delay of 10 - 60 sec, depending on the concentration of arachidonate. Preincubation of serum starved cells with a non-selective COX inhibitor, indomethacin (10 uM) or a selective COX-1 inhibitor, SC-560, (100 nM) resulted in a dose dependent inhibition of AA induced fluorescence. As expected a COX-2 selective inhibitor, SC-125 had no effect on quiescent cell activity, (data not shown). This observation confirms previous work that COX-1 is the predominant form of COX activity in quiescent fibroblasts.

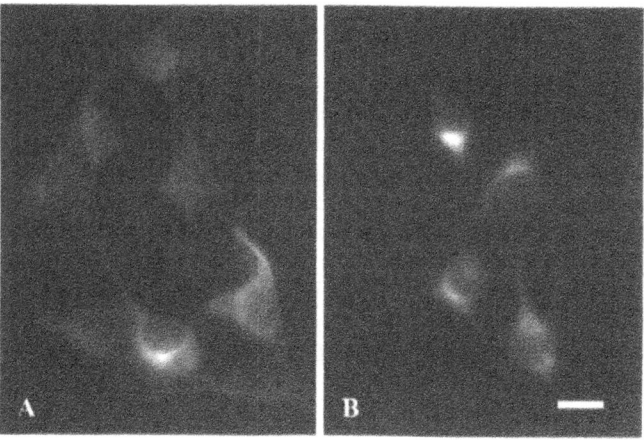

Figure 2. Fluorescence Immunocytochemical labeling of COX-1 (A) and COX-2 (B) in 2 hr serum-treated 3T3 fibroblasts. COX-1 is found through out the cytoplasm. COX-2 is intensely localized to the perinuclear region and nucleus. Bar = 20 micron.

The fluorescence increase observed in serum stimulated cells was 30 to 50 % higher than in quiescent cells depending on the time after AA. Treatment with the COX-1 inhibitor resulted in a linear fluorescence increase which was 30 to 40% less than no treatment. Treatment with a selective COX-2 inhibitor, SC-125 (10 nM) also diminished the overall response by approximately 30% at 2 min after arachidonate and typically resulted in a slightly more curvilinear fluorescence increase resembling that of COX-1 activity in quiescent cells. All arachidonate induced fluorescence increases were blocked by more than 90% in cells pre-treated for 15 min. with the non-selective COX inhibitor, 10 µM indomethacin.

DISCUSSION

The oxygen radical reporter, CDCF-DA has been used to visualize and measure oxidative activity in a variety of cells and organs by spectrofluorometry, flow cytometry and fluorescence light microscopy including cyclooxygenase activity (Morita, et.al.). However, few reports have used CDCF -DA to follow reactive kinetic as described here. CDCF-DA is photolabile and requires reduced light illumination that is synchronized with a low light level digital camera for multiple time lapse image collections. Hence, in our hands, confocal imaging failed in that the intense laser excitation photoxidized the dye to produce brightly fluorescent cells in the absence of arachidonate after only one image. Although this feature proved useful for assessing dye loading, great care was taken to insure that fluorescence changes observed occurred only after arachidonate administration.

Although a number of peroxidase activities are capable of oxidizing CDCF, we believe the signals we have measured are cyclooxygenase-derived for the following reasons. We observed no change in fluorescence in the absence of exogenous arachidonic acid over a period of ten min. The arachidonate induced fluorescence increase was selectively inhibited by non-steroidal ant-inflammatory drugs. Finally, the observed pattern of fluorescence in the presence of the isoform specific inhibitors correlates well with immunohistochemical staining for COX-1 and COX-2, suggesting the measured peroxidase activity in this system are primarily COX dependent.

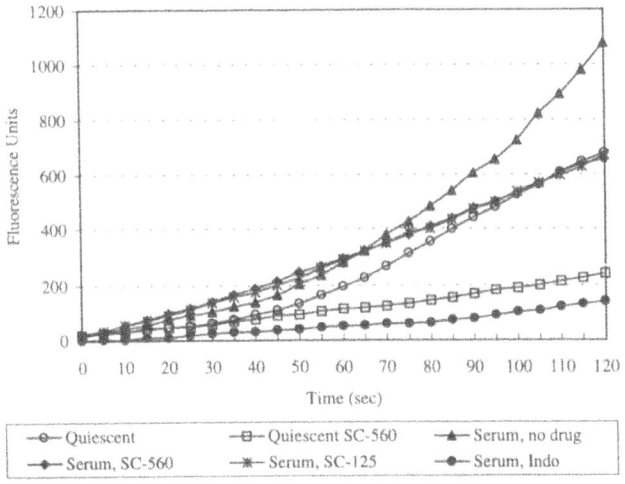

Figure 3. COX activity in quiescent and serum stimulated NIH 3T3 cells as measured by quantitative image analysis of the average fluorescent intensity of thirty cells following 10 uM arachidonate. Curves for quiescent cells treated with 10 uM indomethacin (Indo) were identical to that shown for serum stimulated cells.

The cyclooxygenase activities we have observed in situ are strikingly different. COX-1 has a curvilinear activity after a variable delay whereas COX-2 activity is essentially linear. Although the exact shape of these curves is effected by AA concentration, the observed

135

differences hold for AA concentrations from 100 nm to 10 uM. It is clear that this difference is related to substrate presentation to the enzyme. However, preliminary experiments not shown here with peroxide generating regents in situ have been shown to accelerate the COX kinetics similar to that describe for in vitro enzyme (Kulmacz and Ho-Wang, 1995) suggesting that the physiological state of the cell plays an important role in COX function. We expect that the methodology for measuring in situ COX activity as we described here will play a significant role in understanding COX function.

BIBILIOGRAPHY

Williams, C.S. and DuBois, R.N., 1996, Prostaglandin endoperoxide synthase: why two isoforms?, A. J. Physiol. 270:G393.

Pairet M. and Engelhardt, G., 1996, Distinct isoforms (COX-1 and COX-2) of cyclooxygenase: possible physiological and therapeutic implications, Fundam. Clin. Pharmacol. 10:1.

Seibert, K., Zhang, Y., Leahy, K., Hauser, S., Masferrer, J., Perkins, W., Lee, L., and Isakson, P., 1994, Pharmacological and biochemical demonstration of the role of cyclooxygenase 2 in inflammation and pain, Proc. Nat. Acad. Sci. USA. 91:12013-12017.

Koki, A.T., Edwards, D.A., Dannenberg, A.J., Ornberg, R.L., Zweifel, B.S., Meyer, D.M., Altorki, N., Masferrer, J.L., 1998, Cyclooxygenase-2 is expressed in animal and human cancers, Proc. Amer. Assoc. Cancer Res. 39:150.

Kaufman, W.E., Worley, P.F., Pegg, J., Bremer, M., Isakson, P., 1996, COX-2, a synpatically induced enzyme, is expressed by excitatory neurons at postsynaptic sites in rat cerebral cortex, Proc. Nat. Acad. Sci. USA 93: 2317.

Lipsky, P.E., Isakson, P.C., 1997, Outcome of specific COX-2 inhibition in rheumatoid arthritis, J. Rheumatol. 24: 9-14.

Masferrer, J.L., Zweifel, B.S., Colburn, S.M., Ornberg, R.L., Salvemini, D., Isakson, P., Seibert, K., 1995, The role of cyclooxygenase in inflammation, Am. J. Therapeutics 2:1.

Kawamori, T., Rao, C.V., Seibert, K., Reddy, B.S., 1998 Chemopreventive activity of celecoxib, a specific cyclooxygenase-2 inhibitor, against colon carcinogenesis, Cancer Res. 58: 409-12

Reddy, S.T. and Herschman, H.R., 1994, Ligand-induced prostaglandin synthesis requires expression of the TIS10/PGS-2 prostaglandin synthase gene in murine fibroblast and macrophages., J. Biol. Chem. 269:15473.

Morita, I., Schindler, M., Regier, M.K., Otto, J.C., Hori, T., DeWitt, D.L., Smith, W.L., 1995, Different intracellular locations for prostaglandin endoperoxide H synthase-1 and -2, J. Biol. Chem. 270:10902.

Kulmacz, R.J. and Wang, L., 1995, Comparison of hydroperoxide initiator requirements for the cyclooxygenase activities of prostaglandin H synthase -1 and -2, J. Biol. Chem. 270: 24019.

21

DISCOVERY OF A NEW CLASS OF SELECTIVE CYCLOOXYGENASE-2 (COX-2) INHIBITOR THAT COVALENTLY MODIFIES THE ISOZYME

Amit S. Kalgutkar,[1] Brenda C. Crews,[1] Scott W. Rowlinson,[1] Carlos Garner,[2] and Lawrence J. Marnett[1,2]

A. B. Hancock, Jr., Memorial Laboratory for Cancer Research,
Departments of [1]Biochemistry and [2]Chemistry
Center in Molecular Toxicology
Vanderbilt University School of Medicine
Nashville, Tennessee 37232-0146

INTRODUCTION

Prostaglandin endoperoxide synthase (PGHS) catalyzes the first two steps of the arachidonic acid cascade leading to prostaglandin and thromboxane biosynthesis.[1] Two distinct PGHS isozymes exist in mammals.[2] PGHS-1 is constitutively expressed and involved in the synthesis of cytoprotective prostaglandins in the stomach and thromboxanes in platelets, whereas PGHS-2 is inducible and is a major contributor to prostaglandin biosynthesis in inflammatory cells and in the brain.[3] The differential tissue distribution of the isozymes provides a basis for the development of selective PGHS-2 or COX-2 inhibitors as antiinflammatory and analgesic agents without the gastrointestinal and vascular liabilities that plague all currently available NSAIDs. Most COX-2 inhibitors developed to date, including the diarylheterocycles[4] and the acidic sulfonamides[5], inhibit the enzyme by binding tightly but noncovalently at the cyclooxygenase active site.[6]

Our efforts focused on the development of a selective covalent modifier of COX-2. Aspirin is the only NSAID which covalently modifies COX-1 and COX-2 by acetylating an active site serine residue.[7] Replacement of the carboxylate moiety of aspirin with an alkylmercapto group led to the synthesis of a series of 2-acetoxyphenylalkyl sulfides as selective COX-2 inhibitors. The most potent inhibitor in this series, 2-acetoxyphenylhept-2-ynyl sulfide (APHS) was 60 times more reactive against COX-2 than aspirin and 100 times more selective for its inhibition. Tryptic digestion and peptide mapping studies confirmed that APHS acetylated the same serine as aspirin but site-directed mutagenesis studies suggested that the basis for selectivity results from novel protein-inhibitor interactions. This paper summarizes our efforts on the development of the first, selective covalent inactivators of COX-2 and places the results in the perspective of our current understanding of the molecular and structural basis of selective COX-2 inhibition.

Eicosanoids and Other Bioactive Lipids in Cancer, Inflammation, and Radiation Injury, 4
Edited by Honn *et al.*, Kluwer Academic / Plenum Publishers, New York, 1999.

MATERIALS AND METHODS

Detailed synthetic procedures of the compounds included in this study will be reported elsewhere. Purification of COX-1 from sheep seminal vesicles, preparation of apoCOX-1, and inhibition assays utilizing thin layer chromatography (TLC) are detailed in previous publications from this laboratory.[8] Tryptic digestion, peptide mapping, liquid chromatography/mass spectrometry (LC/MS) methodologies and, *in vitro* inhibition of COX-2 in intact cells (RAW264.7 macrophages and HCA-7 colon cancer cells) have been described in detail by Kalgutkar *et al.*[9] Samples of human COX-2 were a gift from Jim Gierse (Monsanto). Murine COX-2 mutants were produced using the Strategene QuickChangeTM site-directed mutagenesis kit (La Jolla, CA). The wild-type and mutant mCOX-2 proteins were produced in SF-9 insect cells by means of the pVL1393 expression vector (Pharmingen, San Diego, CA).

RESULTS

SAR studies on 2-acetoxyphenylalkyl sulfides as selective COX-2 inhibitors

Inhibitory potencies and selectivities were determined against purified human COX-2 and ovine COX-1. Sulfide (1) was identified as an early lead compound that exhibited moderate inhibitory potency and selectivity for COX-2 (Figure 1). Under these conditions, aspirin displayed selective COX-1 inhibition (IC_{50} (COX-2) ~ 62.5 µM; IC_{50} (COX-1) 12.5 µM). That sulfone 2 did not inhibit either isozyme activity is a result which is opposite to the inhibitory profile displayed by the diarylheterocycles in which sulfones display selective COX-2 inhibition and sulfides are selective COX-1 inhibitors.[10] Increments in chain length of the S-alkyl group in 1 afforded compounds with significant increases in COX-2 potencies, the COX-1 isozyme was also inhibited. The most potent inhibitor in this series, 2-acetoxyphenylheptyl sulfide (3) was only 3 times more selective for COX-2 (see Figure 1). Replacement of the sulfur in 3 with other heteroatoms such as Se, $N(CH_3)$, CH_2 and O or oxidation of the sulfide to the corresponding sulfoxide or sulfone afforded compounds that were inferior in potency towards either isozyme (IC_{50} COX-1 or COX-2 > 10 µM). Incorporation of a triple bond in the 2 position of the heptyl side chain of 3 led to the synthesis of APHS, the most potent and selective COX-2 inhibitor in the series (see Figure 1). Kinetic analysis revealed that APHS was a time- and dose-dependent inhibitor of COX-2 (k_{inact}/K_i ~ 0.18 $min^{-1} \cdot µM^{-1}$). As in the saturated alkyl series, the hept-2-ynyl chain was optimum for potency and selectivity. Furthermore, 2-hydroxyphenylhept-2-ynyl sulfide (11), the hydrolysis product of APHS was inactive against either isozyme.

Pharmacology of APHS as a selective COX-2 inhibitor

In addition to the *in vitro* inhibition of purified COX-2, APHS inhibited COX-2 activity in RAW264.7 macrophages stimulated with lipopolysaccharide and γ-interferon with an IC_{50} of 0.12 µM. In parallel experiments, aspirin displayed an IC_{50} value of 100 µM for inhibition of COX-2 activity in macrophages. Comparison of the effect of APHS on the growth in soft agar of HCA-7 colon cancer cells, which constitutively express high levels of COX-2 and are sensitive to COX-2 inhibitors[11], with its effect on HCT-15 colon cancer cells, which do not express COX-2 and are resistant to the effect of COX-2 inhibitors was also examined. HCA-7 cells were sensitive to growth inhibition by APHS (IC_{50} ~ 2 µM) whereas HCT-15 were insensitive. This IC_{50} value is lower than that observed for the inhibition of the growth of HCA-7 cells by the diarylheterocycle, SC-58125.[11]

Selective covalent modification of COX-2 by APHS and identification of the acetylated amino acid residue

The degree of incorporation of the [^{14}C-acetyl] moiety into COX-2 and COX-1, following reaction with [^{14}C-acetyl]-APHS, correlated well with the relative inhibitory potency against the

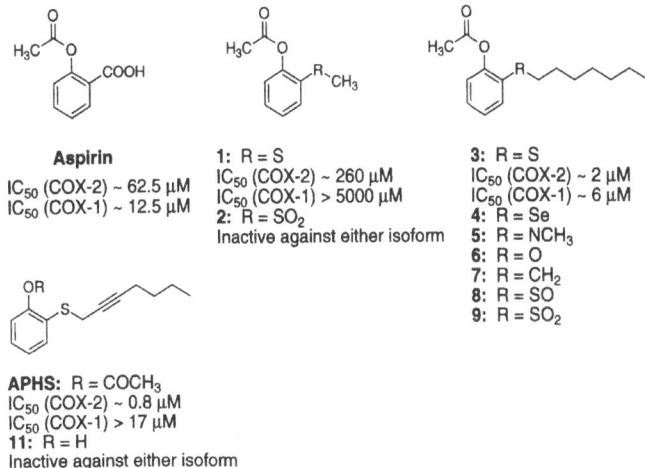

Aspirin
IC$_{50}$ (COX-2) ~ 62.5 μM
IC$_{50}$ (COX-1) ~ 12.5 μM

1: R = S
IC$_{50}$ (COX-2) ~ 260 μM
IC$_{50}$ (COX-1) > 5000 μM
2: R = SO$_2$
Inactive against either isoform

3: R = S
IC$_{50}$ (COX-2) ~ 2 μM
IC$_{50}$ (COX-1) ~ 6 μM
4: R = Se
5: R = NCH$_3$
6: R = O
7: R = CH$_2$
8: R = SO
9: R = SO$_2$

APHS: R = COCH$_3$
IC$_{50}$ (COX-2) ~ 0.8 μM
IC$_{50}$ (COX-1) > 17 μM
11: R = H
Inactive against either isoform

Figure 1. SAR studies on 2-acetoxyphenylalkyl sulfides as selective COX-2 inhibitors.

two enzymes (ratio of [^{14}C-acetyl] incorporated into COX-2 vs COX-1 = 15.4). Tryptic digestion and peptide mapping of acetylated COX-2 indicated that the [^{14}C-acetyl] moiety was incorporated in a single peptide that included the serine acetylated by aspirin. Mass spectrometric analysis identified the structure as an acetylated tripeptide S-L-K (Figure 3), which was confirmed by comparison with authentic standards of the tripeptide. This peptide is present in the COX-2 sequence at positions 516-518 and contains the serine residue acetylated by aspirin.

Structural basis for the selective covalent modification of COX-2 by APHS

The interaction of APHS with three different site-directed murine COX-2 mutants was examined. The Arg106Gln and the Tyr341Ala mutants which are important determinants in the binding of arachidonic acid and all of the carboxylate-containing NSAIDs[13], were more sensitive to inhibition by APHS. In contrast, these mutants were resistant to inhibition by carboxylate-containing NSAIDs including aspirin. Furthermore, the side pocket triple mutant Val509Ile:Arg499His:Val420Ile which is resistant to inhibition by the diarylheterocycles[14] was potently inhibited by APHS. Thus, the selectivity of inhibition of APHS appears to result from its binding at a previously uncharacterized region of the protein.

CONCLUSION

This study reveals that selective, covalent modifiers of COX-2 can be designed that may provide the therapeutic benefits associated with aspirin without the deleterious ulcerogenic side effects, which limits aspirin's use in long term therapy. Inhibition of COX-2 activity by APHS, in LPS- and γ-interferon-treated inflammatory cells and its selectivity in attenuating growth of COX-2 expressing colon cancer cells indicate that 2-acetoxyphenylalkyl sulfides may not only serve as nonulcerogenic antiinflammatory agents, but may also find potential applications as cancer chemopreventive agents.

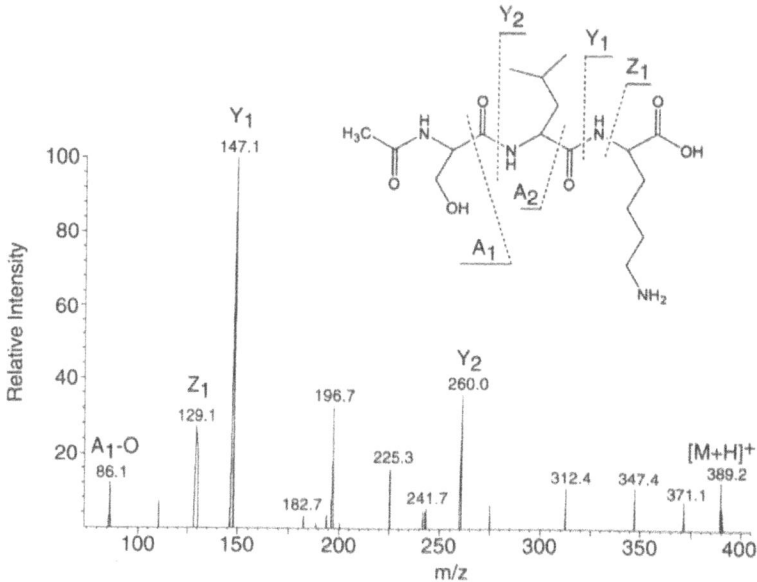

Figure 2. Mass spectrum of the molecular ion of the acetylated peptide isolated from COX-2 treated with [¹⁴C-acetyl]-APHS.

ACKNOWLEDGMENT

We thank Jim Gierse for samples of recombinant human COX-2. This study was supported by a research grant and center grants from the National Institutes of Health (CA47479, ES00267, and CA68485). Scott W. Rowlinson is a recipient of an Australian National Health and Medical Research Council C.J. Martin Fellowship.

REFERENCES

1. W. L. Smith, R. M. Garavito, and D. L. DeWitt, Prostaglandin endoperoxide H synthases (cyclooxygenases)-1 and -2, *J. Biol. Chem.* 271:327 (1996).
2. H. R. Herschman, Prostaglandin synthase 2, *Biochim. Biophys. Acta Lipids Lipid Metab.* 1299:125 (1996).
3. T. Hla, and K. Neilson, Human cyclooxygenase-2 cDNA, *Proc. Natl. Acad. Sci. U.S.A.* 89:7384 (1992).
4. T. D. Penning, J. J. Talley, S. R. Bertenshaw, J. S. Carter, P. W. Collins, S. Docter, M. J. Graneto, L. F. Lee, J. W. Malecha, J. M. Miyashiro, R. S. Rogers, D. J. Rogier, S. S. Yu, G. D. Anderson, E. G. Burton, J. Nita Cogburn, S. A. Gregory, C. M. Koboldt, W. E. Perkins, K. Seibert, A. W. Veenhuizen, Y. Y. Zhang, and P. C. Isakson, Synthesis and biological evaluation of the 1,5-diarylpyrazole class of cyclooxygenase-2 inhibitors: Identification of 4-[5-(4-methylphenyl)-3-(trifluoromethyl)-1H-pyrazol-1-yl]benzenesulfonamide (SC- 58635, celecoxib), *J. Med. Chem.* 40:1347 (1997).
5. C.-S. Li, W. C. Black, C.-C. Chan, A. W. Ford-Hutchinson, J.-Y. Gauthier, R. Gordon, D. Guay, S. Kargman, C. K. Lau, J. Mancini, N. Quimet, P. Roy, P. Vickers, E. Wong, R. N. Young, R. Zamboni, and P. Prasit, Cyclooxygenase-2 inhibitors. Synthesis and pharmacological activities of 5-methanesulfonamido-1-indanone derivatives, *J. Med. Chem.* 38:4897 (1995).

6. R. A. Copeland, J. M. Williams, J. Giannaras, S. Nurnberg, M. Covington, D. Pinto, S. Pick, and J. M. Trzaskos, Mechanism of selective inhibition of the inducible isoform of prostaglandin G/H synthase, *Proc. Natl. Acad. Sci. U.S.A.* 91:11202 (1994).

7. G. J. Roth, N. Stanford, and P. W. Majerus, Acetylation of prostaglandin synthase by aspirin *Proc. Natl. Acad. Sci. U.S.A.* 72:3073 (1975).

8. A. S. Kalgutkar, B. C. Crews, and L. J. Marnett, Kinetics of the interaction of nonsteroidal antiinflammatory drugs with prostaglandin endoperoxide synthase-1 studied by limited proteolysis, *Biochemistry* 35:9076 (1996).

9. A. S. Kalgutkar, B. C. Crews, S. W. Rowlinson, C. Garner, K. Seibert, and L. J. Marnett, Aspirin-like molecules that covalently inactivate cyclooxygenase-2, *Science* (1998) In Press.

10. C. A. Lanzo, J. M. Beechem, J. Talley, and L. J. Marnett, Investigation of the binding of isoform-selective inhibitors to prostaglandin endoperoxide synthases using fluorescence spectroscopy, *Biochemistry* 37:217 (1998).

11. H. Sheng, J. Shao, S. C. Kirkland, P. Isakson, R. J. Cofey, J. Morrow, R. D. Beauchamp, R. N. DuBois, Inhibition of human colon cancer cell growth by selective inhibition of cyclooxygenase-2, *J. Clin. Invest.* 99:2254 (1997).

12. J. L. Masferrer, B.S. Zweifel, P. T. Manning, S. D. Hauser, K. M. Leahy, W. G. Smith, P. C. Isakson, and K. Seibert, Selective inhibition of inducible cyclooxygenase 2 *in vivo* is antiinflammatory and nonulcerogenic, *Proc. Natl. Acad. Sci. U.S.A.* 91:3228 (1994).

13. D. K. Bhattacharyya, M. Lecomte, C. J. Rieke, R. M. Garavito, and W. L. Smith, Involvement of arginine 120, glutamate 524, and tyrosine 355 in the binding of arachidonate and 2-phenylpropionic acid inhibitors to the cyclooxygenase active site of ovine prostaglandin endoperoxide H synthase-1, *J. Biol. Chem.* 271:2179 (1996).

14. S. W. Rowlinson and L. J. Marnett, Unpublished results.

22

MITIGATION OF ARTHRITIS BY HIGH-DOSE ADMINISTRATION OF A COX-2 INHIBITOR IN THE COLLAGEN-INDUCED ARTHRITIS MODEL IN THE MOUSE

Mark G. Obukowicz[1] and Richard L. Ornberg[2]

[1]Discovery Pharmacology, G.D. Searle
[2]Monsanto Corporate Research
800 N. Lindbergh Blvd.
St. Louis, MO 63167 USA

INTRODUCTION

Non-steroidal anti-inflammatory drugs (NSAIDs), including selective cyclooxygenase (COX)-2 inhibitors, improve many parameters of arthritis in the adjuvant-induced arthritis model in the rat (Anderson et al., 1996). However, results with NSAIDs in the collagen-induced arthritis model in the mouse (Wooley, 1988) have been equivocal, possibly due to a mechanism that is prostaglandin-independent and because the level of dosing is limited due to gastrointestinal (GI) toxicity (Griswold et al., 1988; Phadke et al., 1985; Smith et al., 1990). With the advent of selective COX-2 inhibitors, it was possible to evaluate the efficacy of high-level dosing of a COX-2 inhibitor in the context of a GI-sparing background. The specific objectives of this study were to evaluate whether SC-046, a selective COX-2 inhibitor, mitigates the incidence and/or severity of arthritis and, possibly, is disease-modifying in the collagen-induced arthritis model in the mouse.

MATERIALS AND METHODS

SC-046 is a selective and potent human COX-2 inhibitor *in vitro* and is efficacious in the adjuvant-induced arthritis (AIA), carrageenan-induced paw edema, and carrageenan-induced air models of inflammation (Table 1). The GI-sparing properties of SC-046 in the mouse (>200 mpk/day) allowed it to be evaluated at high-dose levels for efficacy in the collagen-induced arthritis model in the mouse. SC-046 was prepared as a suspension in vehicle (0.5 methylcellulose + 0.025% Tween-20) by wet milling (Retsch, model MM2 wet miller). SC-046 or vehicle was administered by gavage (ig) as a 0.1 ml bolus dose twice per day (bid) at doses of 3, 10, 30, or 100 mpk.

DBA-1 male mice were purchased at approximately ten weeks of age and fed a standard chow diet. Mice were divided into seven groups (n=10/group), sensitized with collagen (chick type II) + complete Freund's adjuvant on day 0, and then boosted with collagen on day 21 (Wooley, 1988). Dosing with SC-046 was initiated on day 0 (0 [vehicle], 3, 10, 30, or 100 mpk, bid) or day 21 (0 [vehicle] or 30 mpk, bid). Arthritis was scored visually (incidence and severity), starting on day 26 and ending on day 76, the experimental endpoint. SC-046 was quantified in plasma at approximate peak (2 hr post-gavage) and trough (20 hr post-gavage) times by solid-phase (C18 SepPak) extraction followed by HPLC quantification. The anti-collagen antibody titer in plasma was determined by ELISA (Wooley, 1988) on day 35.

Eicosanoids and Other Bioactive Lipids in Cancer, Inflammation, and Radiation Injury, 4
Edited by Honn *et al.*, Kluwer Academic / Plenum Publishers, New York, 1999.

145

Table 1. Structure and anti-inflammatory properties of SC-046.

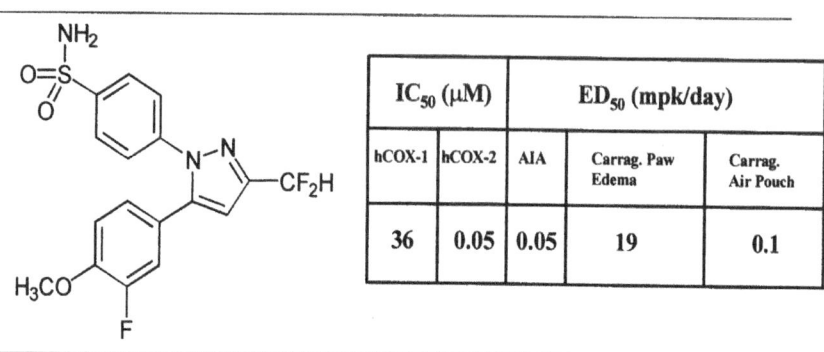

IC$_{50}$ (μM)		ED$_{50}$ (mpk/day)		
hCOX-1	hCOX-2	AIA	Carrag. Paw Edema	Carrag. Air Pouch
36	0.05	0.05	19	0.1

Front and hind paws of every mouse were prepared for joint histological analysis by light microscopy at day 76, the experimental endpoint. Paws were fixed in Optiprobe® (Oncor, Inc.), decalcified, embedded in paraffin, sectioned, and stained with hematoxylin and eosin. Every joint was evaluated for arthritic damage in a blinded manner and scored subjectively for 1) the number of joints affected, 2) synovial hyperplasia, 3) cell infiltration, and 4) overall joint disease ([cell infiltration + synovial hyperplasia] x arthritis incidence [%] per group).

RESULTS

Prophylactic dosing (day 0) with SC-046 mitigated incidence (Fig. 1A) and severity (Fig. 2A) of arthritis as well as histological scores of joint pathology (Table 2) dose-dependently. In corroboration with the visual scoring results, SC-046 mediated dose-dependent improvements in the histological assessment of joint pathology, *viz*, joint disease score, number of joints affected, synovial hyperplasia, articular cartilage damage, and cell infiltration (Table 2). Comparisons of approximate peak (Cmax) and trough plasma levels showed that SC-046 was cleared quickly at all doses (Fig. 3). Cmax plasma levels of SC-046 were correlated directly with dose (Fig. 3) and inversely with severity (Fig. 2A). At the highest dose tested, 100 mpk, there was no detectable arthritis, visually or histologically, in any of the paw joints of the mice in the group (n=10). At

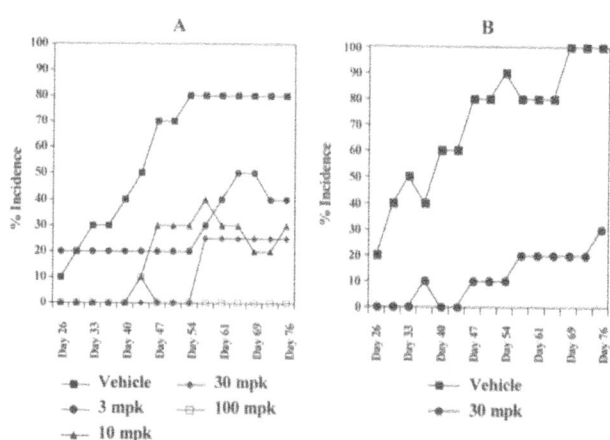

Figure 1. Effect of SC-046 on the incidence of arthritis. Dosing was initiated on day 0 **(A)** or day 21 **(B)**.

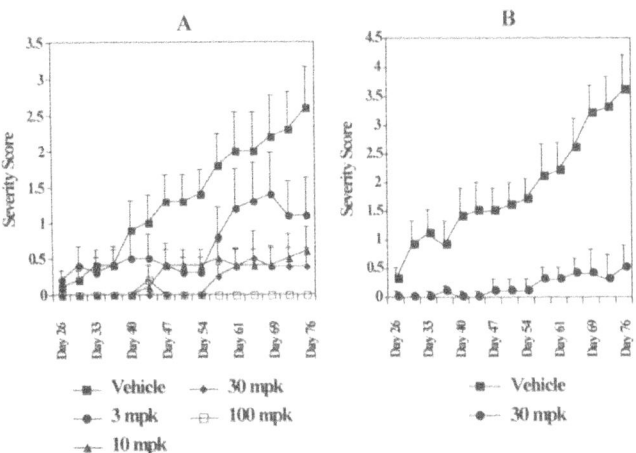

Figure 2. Effect of SC-046 on the severity of arthritis. Dosing was initiated on day 0   (**A**) or day 21 (**B**).

this highest level of dosing, the approximate Cmax was 10 μg/ml (25 μM) (Fig. 3), a concentration close to the *in vitro* IC_{50} value (36 μM) for COX-1 inhibition (Table 1). At an intermediate dose of SC-046 (30 mpk), incidence (Fig. 1A and B), severity (Fig. 2A and B), Cmax plasma levels (data not shown), and joint pathology scores (Table 2) were nearly identical, regardless of whether dosing was initiated coincident with collagen sensitization (day 0) or collagen boosting (day 21). For comparison, the onset of arthritis was first observed visually at day 26 (Fig. 1 and 2).

Table 2. Histological assessment of joint pathology.

Treatment	Joint Disease Score	# of Joints Affected	Synovial Hyperplasia	Cell Infiltration	Comments[1]
SC-046 100mpk	1.00	0	-	-	Normal
SC-046 30mpk	2.13	1	+	+	Mild
SC-046 10mpk	2.00	1	+	+/-	Mild
SC-046 3mpk	2.40	2	++	++	More Severe
Vehicle	3.60	>4	++++	++++	Most Severe

[1]Normal = essentially unaffected
Mild = disruption of articular cartilage, but no bone involvement
Severe = articular cartilage destroyed with bone involvement

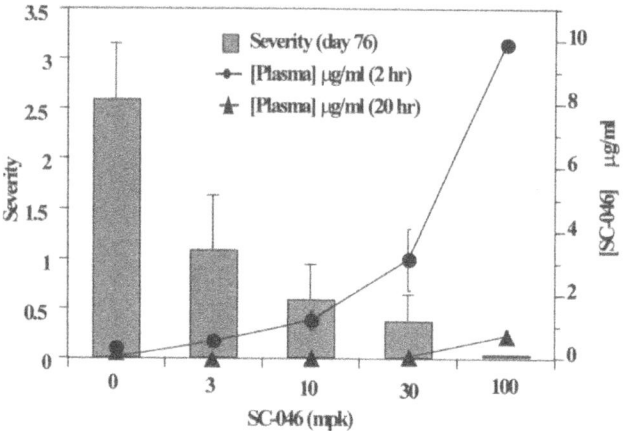

Figure 3. Severity vs. plasma level of SC-046.

The anti-collagen antibody titer was quantified in naïve mice and in mice sensitized with collagen and then dosed, beginning or day 0 or day 21, with vehicle or SC-046 (Fig. 4). No anti-collagen antibody was detected in naïve mice, whereas a relatively high anti-collagen antibody titer was detected in collagen-sensitized mice dosed with vehicle, beginning on day 0 or day 21. SC-046 caused a dose-dependent decrease in the anti-collagen antibody titer when dosing was initiated on day 0. Decreases in the anti-collagen antibody titer were not affected if dosing with an intermediate dose of SC-046 (30 mpk) was initiated at day 0 or day 21; the anti-collagen antibody titer was reduced approximately 50% compared to the respective vehicle control groups.

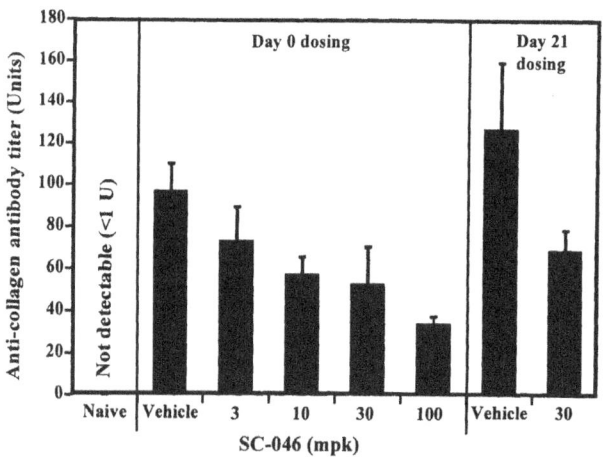

Figure 4. Effect of SC-046 on the anti-collagen antibody titer.

Light micrographs of representative joints from the paws of naïve mice and collagen-sensitized mice dosed, beginning on day 0, with vehicle (bid) or SC-046 (100 mpk, bid) for 76 days are shown for comparison (Fig. 5A-C). No damage was evident in the joint of a paw of a naïve mouse (Fig. 5A), whereas severe damage was present in the joint of a paw of a mouse sensitized with collagen and dosed with vehicle (Fig. 5B). Joint damage consisted of synovial inflammation and disruption of normal articular cartilage structure. In addition, there was macrophage and neutrophil infiltration in the synovial lining and surrounding tissue and signs of synovial hyperplasia. Synovial hyperplasia was evidenced by thickening of the synovial epithelial lining that was accompanied by multiple cell layers in the lining, neovascularization, and mitotic synoviocytes. In contrast, the joint of a paw of a mouse sensitized with collagen and dosed with SC-046 (Fig. 5C) was nearly identical in appearance to the joint of a paw of a naïve mouse (Fig. 5A); no joint pathology was evident.

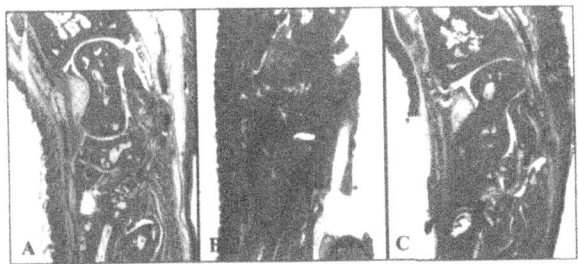

Figure 5. Light micrographs of representative joints from the paws of naïve mice (**A**) and collagen-sensitized mice dosed with vehicle (**B**) or SC-046 (**C**) for 76 days.

DISCUSSION

The selective COX-2 inhibitor, SC-046, mitigated visual incidence and severity and histopathology scores of arthritis in a dose-dependent manner. The mitigation of incidence and severity was correlated with the approximate Cmax plasma level of SC-046. At the highest dose tested, 100 mpk, bid, the plasma Cmax was similar to the IC_{50} value for COX-1 inhibition. SC-046 caused decreases in the anti-collagen antibody titer at doses >10 mpk, bid, whether dosing was initiated prophylactically on day 0 or on day 21, just before the onset of visible arthritis on day 26. These results suggest that the mechanism of action may, in part, be due to the immunosuppressive properties of SC-046 at high doses. However, the data do not support immune modulation by SC-046 during the initiation phase of the immune response. Combined COX-2/COX-1 inhibition could be responsible for the complete mitigation of arthritis at high doses, either by partial immunosuppression or by yet unknown mechanisms beyond inhibiton of prostaglandin synthesis.

REFERENCES

Anderson, G.D., Hauser, S.D., McGarity, K.L., Bremer, M.E., Isakson, P.C., and Gregory, S.A., 1996, Selective inhibition of cyclooxygnease (COX)-2 reverses inflammation and expression of COX-2 and interleukin 6 in rat adjuvant arthritis, *J. Clin. Invest.* 97:2672-2679.

Griswold, D.E., Hillegass, L.M., Meunier, P.C., DiMartino, M.J., and Hanna, N., 1988, Effect of inhibitors of eicosanoid metabolsim in murine collagen-induced arthritis, *Arthrit. Rheumat.* 31:1406-1412.

Smith, S.R., Pennline, K.J., Bober, L.A., Kenworthy-Bott, L.P., Pellerito, F., DaFonseca, M., and Terminelli, C., 1990, Inhibition of collagen II-induced arthritis in mice - a comparison of the effects of SCH 24937, immunosuppressants and nonsteroidal antiinflammatory drugs on the clinical expression of disease, *Int. J. Immunopharmac.* 12:165-173.

Wooley, P.H., 1988, Collagen-induced arthritis in the mouse, in: *Methods in Enzymology*, 162:361-373, G. DiSabato, ed., Academic Press, Inc., New York.

150

23

INDOMETHACIN INHIBITION OF PRISTANE PLASMACYTOMAGENESIS IN GENETICALLY SUSCEPTIBLE INBRED MICE

Michael Potter

Laboratory of Genetics
National Cancer Institute
National Institutes of Health
Bethesda, MD 20892

INTRODUCTION

In the peritoneal plasmacytomagenesis (PCTG) model system, plasmacytomas (PCTs) are induced when different kinds of metabolically inert substances (paraffin oils, silicones, polycarbonate solid discs) are introduced into the peritoneal cavities of genetically susceptible BALB/cAn mice. The PCTs develop morphologically in the chronic inflammatory tissue that forms in peritoneal connective tissues in response to these agents. The most extensively studied agent is pristane (2,6,10,14-tetramethylpentadecane)[1,2]. This induction system has been useful in identifying steps and biological determinants in plasma cell neoplasia. Important features of pristane PCTG in the mouse are: 1) the remarkable dependence on the BALB/cAn genotype and, specifically, the chromosome-4 PCT susceptibility genes *Pctr1* and Pctr2[3,4]; 2) the consistent (>95%) appearance of chromosomal translocations t(6;15)or t(12;15) that are associated with rearrangement of Immunoglobulin (Ig) genes and c-*myc* and result in the deregulation of *c-myc* transcription[5,6]; 3) the influence of natural antigenic stimulation [Specific Pathogen Free BALB/cAnPt mice that have a very restricted exposure to environmental antigens are resistant to PCT induction by pristane[7]; 4) the dependence upon the special vascular and immunological properties of the peritoneal space[8]; and 5) the dependence for growth and progression on the reactive chronic inflammatory microenvironments and cytokines, e.g., IL-6[9-11].

CHRONIC INFLAMMATORY ENVIRONMENT INDUCED BY PARAFFIN OILS (PRISTANE)

The initial engagement of the oil droplets by macrophages and neutrophils probably requires some form of natural emulsification process that coats the oil droplet with material permitting adherence by macrophages and neutrophils. Large droplets are often surrounded by many cells, while microdroplets are ingested into individual macrophages. Complexes of oil droplets and inflammatory cells soon adhere to peritoneal surfaces. The mesentery appears to be the important but not exclusive site for oil granuloma (OG) formation. Here the deposition of the oil/macrophage complexes causes the underlying blood vessels to form new channels to vascularize the developing granuloma, and the mesothelium expands and covers the newly formed fixed tissue[8]. B lymphocytes can gain access to the OG from the mesenteric circulation, and also B cells that reside in the peritoneal space and connective tissues can directly be incorporated into the developing OG. Some of the B cells appear to mature and become plasma cells; rarely, others form foci of proliferating plasma cells[4]. The PCTs appear to develop from cells in these proliferative foci.

The OG may affect the process of PCT development at different stages of B cell development, first by exposing circulating B lymphocytes to a microenvironment that contains many cells involved in phagocytosis and, hence, generating large amounts of reactive oxygen species (ROS). Second, the OG is a tissue that is permissive for the maturation of some B cells into plasma cells. An intuitive hypothesis is that oncogenic mutations occur at the B lymphocyte level because plasma cells are usually terminally differentiated cell types that drop out of cell cycling and/or die by apoptosis. Thus, normal plasma cells are unlikely targets for developing a series of oncogenic mutations. Evidence to support this is the finding that the consistent chromosomal translocation t(12;15) that activates the c-myc protooncogene can be detected by PCR in lymphoid tissues of normal mice[12].

INDOMETHACIN IS A POTENT INHIBITOR OF PRISTANE-INDUCED PCTG

Initial studies of indomethacin (INDO) inhibition utilized a single 1.0 ml dose of pristane[13]. This produced PCTs in only 40% of the BALB/c mice and was utilized to provide a greater chance for INDO to act. INDO treatment begun before and 60 days after the injection of pristane was highly effective in preventing PCT development. INDO treatment did not alter the transplantability and growth of primary PCTs in pristane-conditioned mice, indicating that INDO was not directly cytotoxic to PCT cells at the concentrations attained by ingesting drinking water containing 20 μg/ml[13,14]. A second series of experiments was carried out in BALB/cAn.DBA/2-Idh1-Pep3 (C.D2I/P) congenic mice which are hypersusceptible to PCT induction[14]. Sixty-five to 76% of C.D2-I/P mice given three 0.5ml injections on days 0, 60, and 120 develop PCTs by 300 days[4,14]. We again found that continuous INDO treament was inhibitory; however, when the INDO treament was begun 50 to 70 days after the first injection of pristane, the incidence of PCTs was reduced from 76% in controls to 25% in the mice begun on day 50 and to 30.2% in the mice begun on day 70. These results suggested that by days 50 and 70 after pristane was begun, some developing PCT cells had progressed to an INDO refractory state. INDO did not appear to affect the formation of T(12;15) *c-myc* activating chromosomal translocations. A plausible hypothesis is that INDO is inhibiting the early establishment of B/plasma cell precursors carrying translocations in the chronic inflammatory microenvironment.

INDO also effectively inhibits the formation of plastic disc induced PCTs[14,15]. The reactive tissues in these mice are very different from those injected with oil. One reactive tissue is a patchy fibroplastic reaction on peritoneal surfaces. Nodules containing connective tissues and blood vessels which resemble the polyp-like structures that form in the OG were also seen. The omentum adheres to the plastic disc usually by a stalk of connective tissue containing blood vessels. The movement of the disc can often disrupt this connection, resulting in vascular necrosis of the tissues attached to the disc. PCTs again appear to develop in these reactive tissues.

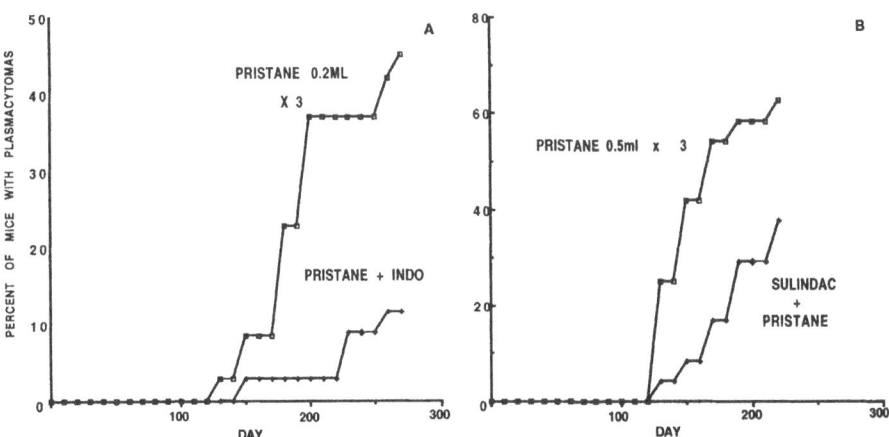

Figure 1. Plasmacytoma induction studies in C.D$_2$ I/P mice showing inhibitory effects of INDO (A) and Salindac (B). The conditions of the experiments differed. In panel A the mice were give 0.2 ml pristane x 3 on days 0, 60 and 120 and treated with a low dose of INDO (10 µg.ml) in the drinking water (d/w). In panel B the mice were given 0.5 ml pristane x 3 and low dose of Salindac in the d/w 31 µg/ml.

Mechanisms of Action of Indomethacin and Other NSAIDS that are Relevant to PCT Development

Indomethacin is a lipophilic molecule that integrates into the plasma membrane where it potentially can interfere with membrane-associated proteins[16]. INDO irreversibly binds and inactivates

PHS (prostaglandin H synthases), i.e., the cyclooxygenases COX-1 and COX-2 at low concentrations *in vitro*, e.g., 7.4 and 9.7 x 10^{-7} M^{17}. INDO binds to or near the arachidonic acid binding site, thus blocking the access of the major substrate arachidonic acid (AA) to the site[18,19], resulting in the loss of PG biosynthesis and of PGE_2, PGD_2, PGI_2, PGF_{2a} and $TBXA_2$ in the chronic inflammatory tissues. INDO has been shown to activate at least one intracellular receptor, the peroxisome proliferator-activated receptor (PPARγ)[20]. Three general mechanisms of INDO inhibition applicable to the B/plasma cell system are discussed below. A single mechanism for how non-steroidal anti-inflammatory agents (NSAIDS) inhibit PCT development is not yet defined, and we are currently presented with multiple explanations. In a chronic inflammatory process such as the OG, INDO could potentially exert its inhibitory actions on different inflammatory cells (neutrophils, monocytes and macrophages, fibroblasts), and T lymphocytes and not in the B cell lineage itself. If this is true, the effector molecules produced by non-B cells control neoplastic development in B/plasma cells.

Inhibition of Prostaglandin Synthesis

The extrordinary sensitivity of the PHS enzymes to the irreversible inhibition by INDO favors the hypothesis that the loss of PG synthesis and the availability of one of the PHS products plays a central role in PCTG. One explanation is that one of the PGs, probably PGE_2, produced by the inflammatory cells in the OG is an important survival/proliferation factor for developing PCT cells, yet there is little if any direct evidence to support this. Rather, the evidence supports the notion that PGE_2 does appear to stimulate IL-6 formation in macrophages[9] and, possibly, also IGF-1[21] which are growth factors for B cells. PGE_2 also inhibits T cell development and IL-2 formation [see[14] for discussion]. The role of T cells in PCTG is not clearly defined, probably because there are multiple actions of different T cell subsets at different stages of PCT development. It has been found in pristane-treated mice that there is a 3-5 fold increase in $CD4^+$ T cells in the OG and peritoneal fluid 50-100 days after pristane injection and that this increase is inhibited by INDO[22]. The findings to date are consistent with the hypothesis that INDO works by inhibiting PGHS and limiting the availability of PGE_2 or another PHS product. A mechanism, however, is still not clear.

Inhibition of Reactive Oxygen Species

PMNLs and monocyte/macrophages are present in great abundance in the peritoneal space and become the chief components of the OG. These cells interact with pristane and phagocytose or attempt to phagocytose larger oil globules. This potentially can lead to the formation of ROS that induce DNA damage to nearby B cells[23,24]. Lymphocytes are particularly sensitive to developing DNA damage from H_2O_2 as they have a low catalase content[24]. The coincubation of PMA stimulated neutrophils and B cell target cells (LPS blasts or PCT cells) has been shown to cause DNA strand breaks in B target cells. This induction of DNA damage was not inhibited by 50μM INDO[23]. An important action of INDO and other NSAIDs, however, is the inhibition of superoxide generation by neutrophil NADPH oxidase (EC 1.6.99.6)[25-27]. In human neutrophils INDO at 180μM reduces by 50% the formation of superoxide anion by neutrophil NADPH oxidase[25], while INDO at 10μM gave a 50% inhibition of Guinea pig macrophages[27]. These results may reflect species differences. The precise biochemical mechanism has not been established, but INDO appears to interfere with the activation and assembly of the multicomponent NADPH oxidase complex in both the human cell and cell free systems[25]. A direct demonstration that neutrophils activated by pristane can induce DNA damage in B cells *in vivo* has not been achieved.

NSAIDS have also been shown *in vitro* to scavenge certain ROSs[28]. Additionally, PGH peroxidase function may also yield peroxyl radicals from natural substrates, and these ROSs as pointed out by Marnett are relatively long lived[29]. These ROSs could be generated in neutrophils, monocytes/macrophages or fibroblasts and transmitted to B cells. The intracellular origin of peroxyl radicals in the membranes of B cells themselves is not clearly defined as little is known about PHS in B cells. Raising the oxidant levels in B cells could lead to mutagenesis or other changes in enzyme function that are dependent upon REDOX potentials. An example is IκB, which is sensitive to oxidant levels[30] and requires the presence of oxidants for activation. Interestingly, it has been shown that paraffin oil immunological adjuvant (incomplete Freund's adjuvants) enhancement of immune responses can be inhibited by antioxidants such as N-acetylcysteine[31]. This type of inhibition could be important in two ways in PCTG: it would reduce the load of potentially mutagenic oxidants, but it would also reduce the oxidation and denaturation of proteins and cell particles. Possibly, one of the by-products of the chronic inflammatory process is to generate immunogenic autoantigens from naturally occurring molecules and particulates (e.g., nucleosomes, RNP particles, membranes).

In summary, indirect evidence supports the possibililty that ROS generated by OG cells are potential DNA-damaging agents for B/plasma cells.

DIRECT ACTION OF INDO IN COLONOCYTES SUGGESTS NEW POSSIBILITIES

NSAIDS, notably sulindac and indomethacin, have been shown both clinically and experimentally to inhibit the growth of adenomatous polyps, the precursor stage of colon carcinomas[17,32]. These NSAIDs can also inhibit proliferation of colon carcinoma cells *in vitro* in situations where the target cells lack PHS enzymes[33]. This in addition to other evidence indicates that the NSAIDS can exert growth inhibitory effects through a non-PHS dependent pathway. Several mechanisms have been advanced. In colonocyte cell lines it has been shown that sulindac sulfide and 300μM INDO induce striking increases in AA, the activation of sphyingomyelinase, and the formation of ceramide, a lipid second messenger that turns on the apoptotic process and apoptosis[34]. The loss of the apoptotic potential in colonocytes has been postulated to be responsible for polyp formation[34]. Whether INDO produces apoptosis at some stage in B cell development remains to be demonstrated. Nonetheless, this finding presents the exciting possibility that INDO could have a direct action that is independent of PHS.

REFERENCES

1. Anderson, P. N. & Potter, M. *Nature* 222, 994-995 (1969).
2. Potter, M. & Wax, J. S. *J. Natl. Cancer Inst.* 71, 391-395 (1983).

3. Mock, B. A., Krall, M. M. & Dosik, J. K. *Proc. Natl. Acad. Sci. USA* 90, 9499-9503 (1993).

4. Potter, M., Mushinski, E. B., Wax, J. S., Hartley, J. & Mock, B. A. *Cancer Res.* 54, 969-975 (1994).

5. Mushinski, J. F. in *Cellular Oncogene Activation* (ed Klein, G.) 181-222 (Marcel Dekker, Inc. New York, 1988).

6. Cory, S. *Adv. Cancer Res.* 47, 189-211 (1986).

7. Byrd, L. G., McDonald, A. H., Gold, L. G. & Potter, M. *J. Immunol.* 147, 3632-3637 (1991).

8. Anderson, A. O., Wax, J. S. & Potter, M. *Curr. Top. Microbiol. Immunol.* 122, 242-253 (1985).

9. Hinson, R. M., Williams, J. A. & Shacter, E. *Proc. Natl. Acad. Sci. USA* 93, 4885-4890 (1996).

10. Nordan, R. P. & Potter, M. *Science* 233, 566-569 (1986).

11. Lattanzio, G., Libert, C., Aquilina, M., et al. *Am.J.Pathol.* 151, 689-696 (1997).

12. Muller, J. R., Jones, G. M., Janz, S. & Potter, M. *Blood* 89, 291-296 (1997).

13. Potter, M., Wax, J. S., Anderson, A. O. & Nordan, R. P. *J. Exp. Med.* 161, 996-1012 (1985).

14. Potter, M., Wax, J. & Jones, G. M. *Blood* 90, 260-269 (1997).

15. Merwin, R. M. & Redmon, L. W. *J. Natl. Cancer Inst.* 31, 990-1007 (1963).

16. Abramson, S. B. & Weissmann, G. *Arthritis Rheum.* 32, 1-9 (1989).

17. Levy, G. N. *FASEB J.* 11, 234-247 (1997).

18. Loll, P. J. in *New targets in inflammation. Inhibitors of COX-2 or adhesion molecules* (eds Bazan, N., Botting, J. & Vane, J.) 13-21 (Kluwer Academic Publishers, Dordrecht, 1996).

19. Picot, D., Loll, P. J. & Garavito, R. M. *Nature* 367, 243-249 (1994).

20. Lehmann, J. M., Lenhard, J. M., Oliver, B. B., Ringold, G. M. & Kliewer, S. A. *J. Biol. Chem.* 272, 3406-3410 (1997).

21. Arkins, S., Rebeiz, N., Biragyn, A., Reese, D. L. & Kelley, K. W. *Endocrinology* 133, 2334-2343 (1993).

22. McDonald, A. H. & Degrassi, A. *Cell.Immunol.* 146, 157-170 (1993).

23. Shacter, E., Beecham, E. J., Covey, J. M., Kohn, K. W. & Potter, M. *Carcinogenesis* 9, 2297-2304 (1988).

24. Schraufstatter, I., Hyslop, P. A., Jackson, J. H. & Cochrane, C. G. *J.Clin.Invest.* 82, 1040-1050 (1988).

25. Umeki, S. *Biochem.Pharmacol.* 40, 559-564 (1990).

26. Smith, R. J. & Iden, S. S. *Biochem.Pharmacol.* 29, 2389-2395 (1980).

27. Oyanagui, Y. *Biochem.Pharmacol.* 25, 1473-1480 (1976).

28. Aruoma, O. I. & Halliwell, B. *Xenobiotica* 18, 459-470 (1988).

29. Marnett, L. J. *Environ. Health Perspect.* 88, 5-12 (1990).

30. Schreck, R. & Baeuerle, P. A. *Trends Cell Biol.* 1, 39-42 (1991).

31. Di Gianni, P., Minnucci, F. S., Alves Rosa, M. F., Vulcano, M. & Isturiz, M. A. *Scand J. Immunol.* 43, 413-420 (1996).

32. Marnett, L. J. *Cancer Res.* 52, 5575-5589 (1992).

33. Hanif, R., Pittas, A., Feng, Y., et al. *Biochem.Pharmacol.* 52, 237-245 (1996).

34. Chan, T. A., Morin, P. J., Vogelstein, B. & Kinzler, K. W. *Proc. Natl. Acad. Sci. U. S. A.* 95, 681-686 (1998).

24

ROLE OF COX-2 INHIBITION ON THE FORMATION AND HEALING OF GASTRIC ULCERS INDUCED BY INDOMETHACIN IN THE RAT

Nuria Godessart, Carolina Salcedo, Andrés G. Fernández, and José M. Palacios.

Research Center,
Almirall Prodesfarma,
Cardener 68-74.
08024 Barcelona (Spain).

INTRODUCTION

The role that the two forms of cyclooxygenase (COX) plays in basal conditions and during the induction and the repair process of gastric lesions is not well understood. Prostaglandins derived from COX-1 are believed to be essential for the maintenance of gastric homeosthasis but, although expressed constitutively, COX-1 can be induced in stomach under certain circumstances[1]. The expression of COX-2 in normal gastric mucosa remains still controversial but a role for constitutive COX-2 products on the gastroprotective effects evoked by mild irritants such as ethanol has been suggested[2].

There is good evidence that selective COX-2 inhibitors do not induce gastric injury *per se* but they may interfere with the healing of preexisting ulcers. COX-2 is induced during the course of ulcer formation in humans[3] and in experimentally-induced ulcers in rodents [4-6]. The involvement of COX-2 on gastric cell replication has also been postulated[6].

Using the selective COX-2 inhibitors L-745.337 and NS-398, several groups have showed a delay in the healing of ulcers induced experimentally by cryoprobe in rats[4] and acetic acid-induced ulcers in mice [5]. Nevertheless, other authors have not found any effect of L-745,337 on the healing of acetic acid-induced ulcers in the rat[7] or have suggested that this effect may depends on the particular compound[8].

To our knowledge, there is no information concerning the effect of COX-2 in the course and healing of NSAID-induced ulcers. Only recently, Elliott et al[9] have reported that aspirin and indomethacin induce the expression of COX-2 in rat stomach mucosa and that this effect seems to be specific for NSAIDs because it is not observed when ulcers are induced by 40% ethanol. Our aim has been to study the effect of a selective COX-2 inhibitor on the formation and healing of gastric ulcers induced by a single administration of indomethacin in the rat. Other NSAIDs such as diclofenac, ketorolac and aspirin have also been tested.

Eicosanoids and Other Bioactive Lipids in Cancer, Inflammation, and Radiation Injury, 4
Edited by Honn *et al.*, Kluwer Academic / Plenum Publishers, New York, 1999.

MATERIAL AND METHODS

Animals and Drugs

Male Wistar rats (Interfauna UK Ltd.) weighing about 140-160 g were used. They were maintained on a 12:12 hour light-dark cycle (lights on 7:00 a.m.) at room temperature ($22\pm1°C$). For all experiments, rats were fasted for 18 h prior to the experiment with free access to drinking water.

NS-398 (Taisho Pharmaceuticals), diclofenac and ketorolac have been synthesized by Almirall Prodesfarma Medicinal Chemistry Department. Indomethacin and aspirin have been supplied by Sigma. All drugs were administered orally in a volume of 10 ml/kg in a vehicle composed of 0.1% Tween 80 and 0.5% methylcellulose in distilled water.

Experimental Procedures

Effect of NS-398 Preadministration on NSAID-induced Ulcers. NS-398 (10 mg/kg) or vehicle were administered by oral route and 30 min later, the animals received indomethacin at doses ranging from 1 to 30 mg/kg (7-10 rats/group). Animals were sacrificed 5 h after the NSAID dosage. The stomach of each rat was removed, opened and gently washed. The macroscopic severity of the erosions was assessed using a parametric scale[10], evaluating the number and size of the ulcers in the glandular stomach. The score was performed by an observer unaware of the treatment. In another series of experiments, diclofenac 50 mg/kg, ketorolac 50 mg/kg or aspirin 250 mg/kg were used instead of indomethacin.

Effect of Time of NS-398 Administration on Indomethacin-induced Ulcers. Six groups of 7 animals each received indomethacin (20 mg/kg) by the oral route at t = 0 . Two groups of animals previously received NS-398 (10 mg/kg) or vehicle 30 min before, whereas the other four groups received the treatment or vehicle 1 h or 2.5 h after indomethacin administration. All animals were sacrificed 5 h after indomethacin dosage. The stomachs were processed and ulcers assessed as described before.

Effect of COX-2 Inhibition on Healing of Gastric Ulcers. Indomethacin (20 mg/kg) was administered by oral route to groups of 7-10 animals. Five hours later, a group of animals was sacrificed and the stomachs processed as indicated, to obtain a lesion index at this point (5 h score). The other groups received vehicle or NS-398 and were mantained in fed conditions for the following 19h. Twenty four hours after indomethacin administration, stomachs were opened and lesion index assessed (24 h score). Alternatively, diclofenac 50 mg/kg, ketorolac 50 mg/kg or aspirin 250 mg/kg were used instead of indomethacin.

Determination of PGE_2 Levels in Gastric Mucosa. Samples of gastric mucosa were obtained according to the method of Whittle et al.[11]. Briefly, stomachs (7 rats/group) were opened along the small curvature, gently washed and strips of gastric mucosa (150-200 mg) freed from underlying muscle were obtained. The strips were minced by scissors in 1 ml of Tris-HCl buffer (pH 8.4), vortexed for 1 min at room temperature and centrifuged for 30 sec in a microfuge. Supernatants were freezed in liquid nitrogen and PGE_2 levels measured by means of an ELISA kit (Cayman).

Statistics. A non-parametric test (Mann-Whitney) was used to perform statistical comparisons between lesion indexes in NS-398-treated and vehicle-treated groups in each assay. T test was used to compare PGE_2 values among groups. A value of $p<0.05$ was considered to be significant.

RESULTS AND DISCUSSION

The effect of NS-398 preadministration on gastric ulcers induced by indomethacin is shown in Figure 1. The treatment with the selective COX-2 inhibitor had no effect on the severity of ulcers induced by indomethacin at doses ranging from subulcerogenic to highly ulcerogenic thus suggesting that COX-2 inhibition is not involved in the formation or exacerbation of ulcers induced by this NSAID.

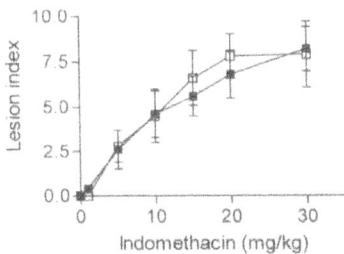

Figure 1. Effect of preadministration of 10 mg/kg NS-398 (■) or vehicle (□) on the formation of gastric ulcers induced by increasing doses of indomethacin.

At 20 mg/kg of indomethacin, no effect of COX-2 inhibition on ulcer index was observed irrespective of the time of NS-398 administration (Table 1), suggesting that COX-2, if induced during the process, is not crucial for the ulcers to develop in these experimental conditions.

Table 1. Effect of time of NS-398 administration on indomethacin-induced ulcers in rat stomach.

Treatment			Ulcer index
			5 h
0 h	1 h	2.5 h	
Indomethacin	None	None	6.4±1.3
Indomethacin	NS-398	None	5.2±2.0
Indomethacin	Vehicle	None	6.6±2.4
Indomethacin	None	NS-398	6.2±2.0
Indomethacin	None	Vehicle	7.8±2.2

To study the effect of COX-2 in ulcer healing we first determined how was the curative process affected by the experimental conditions. As shown in Table 2, the healing of gastric ulcers induced by indomethacin becomes almost complete in 24 h when animals are fed after the acute stage (5 h after indomethacin dosage). Rats mantained in fasting conditions did not significantly improved the ulcer index observed at 5 h, indicating that repair was not allowed. Feeding of rats from 5 to 24 h was the experimental protocol chosen for further experiments.

Table 2. Effect of experimental conditions on the healing of ulcers induced by indomethacin.

Treatment	5 h Score	24 h Score	
		Fasted (5-24 h)	Fed (5-24 h)
Vehicle	0.4 ± 0.3	0.4 ± 0.3	0.3 ± 0.3
Indomethacin	7.1 ± 1.3	5.2 ± 1.0	1.1 ± 0.4 *

*$p < 0.05$, Mann-Whitney *vs.* 5h score

Figure 2 shows that the healing of indomethacin-induced ulcers was not inhibited when animals were treated with NS-398 at the acute stage. Only indomethacin at 10 mg/kg is able to block the repair of ulcers in feeding conditions, suggesting a role for COX-1.

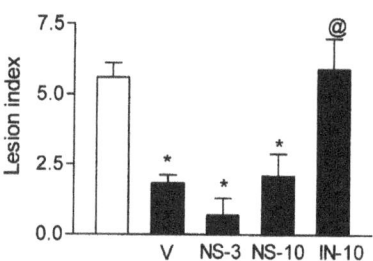

Figure 2. Effect of vehicle (V), NS-398 (NS) and indomethacin (IN) on the healing of ulcers induced by 20 mg/kg indomethacin. (White bars: 5 h score, black bars: 24 h score). *$p < 0.05$ vs 5h score, @ $p < 0.05$ vs vehicle 24 h score.

The contents of PGE_2 in rat stomachs from the different experimental conditions were measured. The levels of PGE_2 were not affected by the administration of NS-398 in any case. The presence of food stimulated the synthesis of this prostaglandin in vehicle-treated rats.

Table 3. Effect of NS-398 at10 mg/kg on the gastric PGE_2 levels (pg/100 mg mucosa) in the ulcers at 5 and 24 h after indomethacin at 5 and 24h.

Treatment		5h Score	24h Score	
			Fasted 5-24h	Fed 5-24h
Vehicle		5670 ± 919	4646 ± 1440	11099 ± 3549
NS-398		5905 ± 485		
Indom. + NS-398		92 ± 6**		
Indomethacin		83 ± 17**		
Indomethacin	Vehicle	330 ± 125*	627 ± 67*	
	NS-398	329 ± 194*	546 ± 105*	

* $p < 0.05$** $p < 0.005$ *vs* vehicle at 5h, Ttest. Basal PGE_2 levels: 10.400-16.000.

Only a partial recovery of PGE_2 levels was seen at 24 h in indomethacin-treated rats in feeding conditions when compared with the vehicle-treated group, despite those animals had shown a significant amelioration in the ulcer index. This apparent discrepancy between gastric healing and PGE_2 levels could be explained by the involvement of other prostaglandins different form PGE_2 in this process or by a prostaglandin-independent, direct effect of COX-2 (i.e., targeting directly to the nucleus of gastric cells).

To discard that our results were due to the particular characteristics of indomethacin (high COX-2 inhibitory activity, pharmacokinetics,...), other NSAIDs were used in similar experiments as inducers of gastric lesions. The effect of NS-398 in ulcers evoked by aspirin, diclofenac or ketorolac was tested. As shown in Figure 3, no significant differences in ulcer index were obtained between vehicle- and NS-398-treated rats.

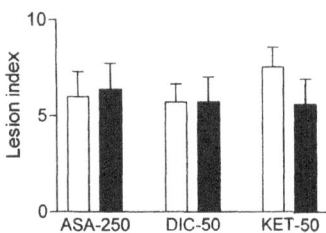

Figure 3. Effect of COX-2 inhibition on gastric ulcers induced by aspirin (ASA), diclofenac (DIC) and ketorolac (KET). (Rats were pretreated with vehicle- white bars - or with 10 mg/kg NS-398 -black bars- before NSAIDs administration).

The effect of COX-2 inhibition on the healing of ulcers induced by the different NSAIDs is shown in Table 4.

Table 4. Effect of NS-398 on ulcer index of rat stomachs at 24h after administration of several NSAIDs.

Treatment t = 0	Dose (mg/kg)	Ulcer index 5 h Score	Administration at t=5 h	Ulcer index 24 h Score
Vehicle		0.16 ± 0.16	Vehicle	0
			NS-398	0
		6 ± 1.4		
Aspirin	250		Vehicle	0 ± 0 **
		8 ± 2	NS-398	0 ± 0 **
Diclofenac	50	6.3 ± 0.8	Vehicle	0.2 ± 0.1**
			NS-398	0.3 ± 0.1**
Ketorolac	50		Vehicle	0.5 ± 0.3**
			NS-398	0.3 ± 0.2**

** $p < 0.005$ Mann-Whitney test *vs* 5h score.

No blockade in the healing process was observed when animals were treated with 10 mg/kg NS-398 at the acute stage of ulcerogenesis following NSAIDs administration and kept in fed conditions. The "mechanical" effect of food probably causes the release of arachidonic acid that can be used by COX-1 to restore prostaglandin levels[12]. COX-1 activity blocked by NSAIDs can be restored by new enzyme synthesis (aspirin) or by clearance of the inhibitor (faster with diclofenac because of its short half life, and slower with drugs with a long half life such as indomethacin or ketorolac).

CONCLUSIONS

Our results suggest that COX-2 is not involved in the development of gastric ulcers induced by NSAIDs. Healing of these ulcers is not affected by treatment with a COX-2 selective inhibitor at doses that it impairs healing in other ulcer models [5,6] probably indicating that in different models, different repair mechanisms are involved. Our results do not exclude a role for COX-2 in ulcer healing but they strongly indicate that in our conditions, a functional COX-1 is needed for the repair process to occur.

Acknowledgments. We thank S. Petit and N.Torán for excellent technical support.

REFERENCES

1.J.G.P. Ferraz, A. Sharkey, B.H. Reuter, S. Asfaha, A.W. Tigley, M.L. Brown, W. McNight, and J.L. Wallace, Induction of cyclooxygenase 1 and 2 in the rat stomach during endotoxemia; Role of resistance to damage. *Gastroenterology* 113: 195-204, (1997).
2. B. Gretzer, N. Lambrecht, N. Ehrlich, M. Chantrain, and B.M. Peskar, Prostaglandins involved in mild irritant evoked gastroprotection are derived via the cyclooxygenase 2 pathway. *Gastroenterology* 112: A133 (1997).

3. A. Tarnawski, J. Kidao, S. Jaafar, R. Itano, R. Pai, H. Gergely, and I.J. Sarfeh, Expression and localization of cyclooxygenase-2 in normal and ulcerated gastric mucosa, and in cultured human endothelial cells. *Gastroenterology* 112 : A309 (1997).

4. A. Schmassmann, P. Netzer, B. Flogerzi, and F. Halter, Role of cyclooxygenase-1 and -2 in gastric ulcer healing in rats.*Gastroenterology* 112 : A283 (1997).

5. H. Mizuno, Ch. Sakamoto, K. Matsuda, K. Wada, T. Uchida, H. Noguchi, T. Akamatsu, and M. Kasuga, Induction of cyclooxygenase 2 in gastric mucosal lesions and its inhibition by the specific antagonist delays healing in mice. *Gastroenterology* 112: 387-397 (1997).

6. S.Tsuji, W.H. Sun, H. Sawaoka, M. Tsujii, E.S. Gunawan, H. Murata, K. Nagano, and S.A. Kawano, A specific inhibitor for cyclooxygenase-2 delays gastric ulcer healing. *Gastroenterology* 112 : A316 (1997).

7. A.Guglietta, C.A. Lesch, E.R. Kraus, R. Dyer, D. Schrier, and B. Sanchez, Effect of PGHS-2 inhibitors and other antinflammatory drugs on healing of acetic acid-induced ulcer in rats. *Gastroenterology* 112 : A135 (1997).

8. S. Elliott, W. McKnight, G. Cirino, and J.L. Wallace, A nitric oxide-releasing nonsteroidal anti-inflammatory drug accelerates gastric ulcer healing in rats. *Gastroenterology* 109: 524-530 (1995).

9. N.M. Davies, K.A. Sharkey, S. Asfaha, W.K. MacNaughton, and J.L. Wallace, NSAIDs cause a rapid and selective upregulation of cyclooxygenase-2 expression in the rat stomach. *Gastroenterology* 112: 98 (1997).

10. R. Cosen and P. Mazure, Restrain-induced ulcer with feeding: A test for pharmacological studies, in *Experimental Ulcer*. Th.G.W.Gheorghiu, ed., Witzostrock, Baden-Baden, (1975).

11. B.J.R. Whittle, G.A. Higgs, K.E. Eakins, S. Moncada, and J.R.Vane, Selective inhibition of prostaglandin production in inflammatory exudates and gastric mucosa. *Nature* 284: 271-273 (1980).

12. A. Robert, Prostaglandins and digestive diseases, in *Advances in Prostaglandins and Thromboxane Research* 8:1533-1539. B.Samuelsson, P.W.Ramwell and R.Paoletti, ed., Raven Press, New York (1980).

25

REGULATION OF *IN VIVO* PROSTAGLANDIN BIOSYNTHESIS BY GLUTATHIONE

Alon Margalit[1], Scott D. Hauser[2] and Peter C. Isakson[2]

[1]The Institute for Applied Biochemistry
Ben-Gurion University of the Negev
P.O. Box 653   Beer-Sheva 84105
Israel

[2]Searle Research and Development
700 Chesterfield Parkway North
St. Louis, MO 63198

INTRODUCTION

Monosodium urate crystals are a naturally occurring pro-inflammatory agent associated with the manifestations of the rheumatic disorder, gout. This disease is characterized by occasional periods of severe pain and inflammatory swelling that, if untreated, resolves spontaneously within a few days[1]. Although urate crystals are known to be the etiologic agent for gout, and their accumulation seems to be the major pathophysiological event leading to inflammatory episodes, additional factors influence the inflammatory response. Urate crystal deposition in joints may occur without associated inflammation[2] and inflammatory attacks may remit although urate crystals are still present in the joints[3]. Prostaglandins (PGs) appear to play a prominent role in the onset of gout attacks because non-steroidal anti-inflammatory drugs (NSAIDs), which inhibit PG formation, are considered the drugs of choice for the treatment of acute attacks[4]. Using a mouse model of urate crystal-induced peritonitis, we have found that while urate crystals triggered an initial burst of eicosanoid products, at later times additional arachidonic acid metabolism was attenuated in a fashion independent of substrate concentration[5]. The present study further investigate this phenomena and demonstrate that urate crystal administration resulted in five fold elevation of the intracellular and extracellular pools of reduced glutathione (GSH) and that the elevated GSH attenuated PG formation by shifting the cyclooxygenase (COX) products profile from PG to 12-hydroxyheptadecatrienoic acid (12-HHT).

RESULTS AND DISCUSSION

A major factor limiting PG biosynthesis is free AA[6, 7]. In addition, PG production is also dependent on *de novo* COX synthesis[8] and on regulation by cellular cofactors such as hydroperoxide[9-11], peroxynitrite[12] and GSH[13,14].

In previous work, we demonstrated that urate crystals attenuated peritoneal eicosanoid production independent of AA levels [5]. To study the effect of urate crystal administration on the expression of COX-1 and COX-2, total RNA was harvested from peritoneal lavage cells collected at various times after urate crystal injection, and the steady-state mRNA levels for COX-1 and COX-2 were measured by nuclease protection. COX-1 mRNA was readily detectable in lavage cells from untreated animals, but declined rapidly after urate crystal treatment, while COX-2 mRNA, which was undetectable in untreated control animals, was markedly induced. GAPDH expression appeared to decrease initially at 1 hr, but returned to control levels by 6 hours. Western blot analysis of COX-1 and COX-2 protein using isoform specific antibodies showed a corresponding decrease in COX-1 and increase in COX-2 protein levels. The rapid decline in both COX-1 and GAPDH mRNA levels that occurred *in vivo,* but were not reproduced upon *in vitro* treatment of lavage, suggests that urate crystal treatment induces a switch in the peritoneal-cell population and that infiltrating mononuclear cells express less COX-1 than the residential peritoneal macrophages. Such a change in peritoneal cell population from peritoneal macrophages to blood monocytes was reported earlier after i.p. administration of *Listeria monocytogenes* [15]. However, the *de novo* synthesis of COX-2, which follows a time-course similar to that seen in other inflammation models[16,17] indicates that the attenuation of PG production observed at 4 and 6 hours after urate crystal administration is not dependent on enzyme levels.

GSH is a major cellular factor controlling PG biosynthesis. In the presence of glutathione-s-peroxidase (GSSPx), GSH reduced the hydroperoxide tone and thereby inhibited COX activity[13]. Alternatively, GSH may also have a direct regulatory effect on COX activity. Under *in vitro* conditions concentrations of GSH in the mM range shifted the product profile of COX-1 and COX-2 from PGE_2 to 12-hydroxyheptadecatrienoic acid (12-HHT).

To investigate a possible role for GSH in the inhibition of PG biosynthesis by urate crystals, mice were injected first with either urate crystals or zymosan, followed by the injection of exogenous AA at different time intervals. Lavage fluid was collected 10 minutes after AA administration, and analyzed for AA metabolites by electrospray and tandem mass spectrometry. Consistent with previous results, the metabolism of AA to 6-keto $PGF_{1\alpha}$, PGE_2 and 12-HETE was reduced by 90, 65 and 80 % respectively following urate crystal injection. In contrast, levels of 12-HHT were markedly elevated after AA administration in the presence of urate crystals. Pretreatment with zymosan resulted in a moderate reduction of 6-keto $PGF_{1\alpha}$, and an elevation of PGE_2. Zymosan did not significantly affect 12-HETE and 12-HHT production.

To determine whether administration of urate crystals alters peritoneal GSH levels, lavages were collected as a function of time after urate crystal injection and analyzed for extracellular and intracellular GSH. Both the intracellular and the extracellular pools of GSH, were elevated approximately 5 fold in response to urate crystal administration. Zymosan induced a 2 fold elevation of extracellular GSH but did not affect intracellular GSH levels. The calculated intracellular GSH concentration based on mean cell volume of 470 fL[18] was 4.5 mM for untreated cells and 20 mM after urate crystal administration.

These values are within the physiological range of GSH[19, 20]. To verify that these levels of GSH can alter AA metabolism in peritoneal macrophages, lavage cells were harvested from untreated mice and exposed to 10 µg /ml AA in the presence of exogenous GSH. Concentrations of 10 and 20 mM GSH significantly inhibited PGE$_2$, 6-keto PGF$_{1a}$ while augmenting 12-HHT production. GSH also inhibited the production of 12-HETE.

GSH biosynthesis can be inhibited by L-buthionine-(SR)-sulfoximine (BSO)[21]. Pretreatment of the mice with BSO partially restored conversion of AA to PGs and 12-HETE at the expense of 12-HHT. In the same experiment, BSO failed to completely inhibit the extracellular GSH elevation induced by urate crystals. This was probably due to an alternative GSH recycling pathway which involves ascorbic acid[22, 23].

In addition to the effects urate crystals exert on PG biosynthesis, we have also observed a reduction of the *in vivo* formation of 12-HETE[5]. Likewise, direct treatment with GSH inhibited the formation of 12-HETE in isolated peritoneal macrophages. Interestingly, GSH has also been shown to negatively regulate 12-lipoxygenase activity[24, 25]. Inhibition of a second, independent enzymatic pathway in this model further support the conclusion that, in response to urate crystals, the peritoneal concentration of GSH is elevated to levels that are capable of altering the enzymatic activity of enzymes that metabolize AA.

In summary, our study provides evidence that GSH can attenuate PG biosynthesis *in vivo*. This conclusion was based on the following observations: a) urate crystal administration caused a sustained shift in the eicosanoid profile from PG to 12-HHT b) urate crystal administration elevated both the acellular and extracellular GSH levels in peritoneal lavages c) physiological concentrations of GSH (19, 20) shifted AA metabolism of isolated peritoneal macrophages and d) the GSH synthesis inhibitor BSO, partially reversed the inhibitory effect of urate crystals on PG production. The role of GSH as a modulator of COX was suggested based on *in vitro* findings. This study provide the first *in vivo* evidence that GSH can regulate PG biosynthesis[26].

REFERENCES

1. Terkeltaub, R. A., M. H. Ginsberg. 1988. The inflammatory reaction of crystals. *Rheum. Dis. Clin. North Am.* 14: 353-364
2. Ortiz-Bravo, E., H. R. Schumacher. 1993. Components generated locally as well as serum alter the phlogistic effect of monosodium urate crystals *in-vivo. J. Rheumatol.* 20: 1162-1166
3. Terkeltaub, R. A. 1993. Gout and mechanisms of crystal-induced inflammation. *Current Opin. Rheumatol.* 5: 510-516
4. Abramson, S. B. 1992. Treatment of gout and crystal arthropathies and uses and mechanisms of action of nonsteroidal anti-inflammatory drugs. *Current Opin. Rheumatol.* 4: 295-300
5. Margalit, A., K. L. Duffin, A. F. Shaffer, S. A. Gregory, P. C. Isakson. 1997. Altered arachidonic acid metabolism in urate crystal induced inflammation. *Inflammation.* 21: 205-222
6. Mukherjee, A. B., L. Miele, N. Pattabirman. 1994. Phospholipase A$_2$ enzymes: regulation and physiological role. *Biochem. Pharmacol.* 48: 1-10
7. Piomelli, D. 1993. Arachidonic acid in cell signaling. *Curr. Opin. Cell. Biol.* 5:274-280
8. Herschman, H. R. 1994. Regulation of prostaglandin synthase-1 and prostaglandin synthase-2. *Cancer Metas. Rev.* 13: 241-256

9. Lands, W. E. M., P. J. Marshall, R. J. Kulmacz. 1985. Hydroperoxide availability in the regulation of the arachidonate cascade. In *Adv. Prostagl. Thromb. Leukotr. Res.*, eds. O. Hayaishi, S. Yamamoto, vol. 15, 233-235. New-York: Raven Press

10. Marshall, P. J., R. J. Kulmacz, W. E. M. Lands. 1987. Constraints on prostaglandin biosynthesis in tissues. *J. Biol. Chem.* 262: 3510-3517

11. Kulmacz, R. J., R. B. Pendleton, W. E. M. Lands. 1994. Interaction between peroxidase and cyclooxygenase activities in prostaglandin-endoperoxidase synthase. *J. Biol. Chem.* 269: 5527-5536

12. Landino, L. M., B. C. Crews, M. D. Timmons, J. D. Morrow, L. J. Marnett. Peroxynitrite, the coupling product of nitric oxide and superoxide, activates taglandin biosynthesis. *Proc. Natl. Acad. Sci. USA.* 93: 15069-15074

13. Kulmacz, R. J., L.-H. Wang. 1995. Comparison of hydroperoxide initiator requirements for the cyclooxygenase activities of prostaglandin H synthase-1 and -2 *J. Biol. Chem.* 270: 24019-24023

14. Capdevila, J. H., J. D. Morrow, Y. Y. Belosludtsev, D. R. Beauchamp, R. N. DuBois, J. R. Falck. 1995. The catalytic outcome of the constitutive and the mitogen inducible isoforms of prostaglandin H_2 synthase are markedly affected by glutathione and glutathione peroxidase(s). *Biochemistry.* 34: 3325-3337

15. Tripp, C. S., E. R. Unanue, P. Needleman. 1986. Monocyte migration explains the change in macrophage arachidonate metabolism during the immune response. *Proc. Natl. Acad. Sci. USA.* 83: 9655-9659

16. Masferrer, J. L., B. S. Zweifel, P. T. Manning, S. D. Hauser, K. M. Leahy, W. G Smith, P. C. Isakson, K. Seibert. 1994. Selective inhibition of inducible cyclooxygenase 2 *in vivo* is antiinflammatory and nonulcerogenic. *Proc. Natl. Acad. Sci. USA.* 91: 3228-3232

17. Seibert, K., Y. Zhang, K. M. Leahy, S. D. Hauser, J. L. Masferrer, W. Perkins, L. Lee, P. C. Isakson. 1994. Pharmacological and biochemical demonstration of the role of cyclooxygenase 2 in inflammation and pain. *Proc. Natl. Acad. Sci. USA.* 91:12013-12017

18. Lentner, C. 1984. Pysical Chemistry Composition of Blood Hematology. Somatometric Data. In *Geigy Scientific Tables*, ed. C. Lentner, vol. 3, 205-215.

19. Basel: Ciba-Geigy Meister, A., S. S. Tate. 1976. Glutathione and related g-glutamyl compounds: biosynthesis and utilization. *Ann. Rev. Biochem.* 45: 599-604

20. Morris, P. E., G. R. Bernard. 1994. Significance of glutathione in lung disease 2and implication for therapy. *Am. J. Med. Sci.* 307: 119-127

21. W., A. Meister. 1985. Origin and turnover of mitochondrial glutathione. *Proc. Natl. Acad. Sci. USA.* 82: 4668-4672

22. Meister, A. 1994. Glutathione-ascorbic acid antioxidant system in animals. *J.Biol Chem.* 269: 9397-9400

23. Mårtensson, J., A. Meister. 1992. Glutathione deficiency increases hepatic ascorbic acid synthesis in adult mice. *Proc. Natl. Acad. Sci. U S A.* 89: 11566-11568

24. Shornick, L. P., M. J. Holtzman. 1993. A cryptic, microsomal-type arachidonate 12-lipoxygenase is tonically inactivated by oxidation-reduction conditions in cultured epithelial cells. *J. Biol. Chem.* 268: 371-376

25. Hagmann, W., D. Kagawa, C. Renaud, K. V. Honn. 1993. Activity and protein distribution of 12-lipoxygenase in HEL cells: induction of membrane-association by phorbol ester TPA, modulation of activity by glutathione and 12-HPODE, and Ca^{2+}-dependent translocation to membranes. *Prostaglandins* 46: 471-477

26. Margalit, A., S.D. Hauser, B.S. Zweifel, M.A. Anderson, and P.C. Isakson. 1998 Regulation of prostaglandin biosynthesis *in vivo* by glutathione. *Am. J. Phys.* 274:R294-R302.

26

DIFFERENT EFFECTS OF REACTIVE NITROGEN INTERMEDIATES ON PROSTAGLANDIN E₂ SYNTHESIS IN CULTURED RAT MICROGLIA AND RAW 264.7 CELLS

Cecilia Guastadisegni[1], Luisa Minghetti[2], Alessia Nicolini[2], Elisabetta Polazzi[2], Maria Balduzzi[3] and Giulio Levi[2]

[1]Laboratory of Environmental Hygiene, [2]Neurobiology Section, Laboratory of Pathophysiology, Istituto Superiore di Sanità,Viale Regina Elena 299, 00161 Rome and [3]Toxicology Biomedical Science Section, ENEA, Via Anguillarese 301, 00061 Rome, Italy.

INTRODUCTION

Prostaglandins (PGs) and nitric oxide (NO) play major roles in regulating inflammation, immune functions, blood vessel dilatation and neurotransmission. NO is formed from the terminal guanidino nitrogen atom of L-arginine by the enzyme NO synthase (NOS). The constitutive isoform of NOS is present in endothelial and neuronal cells, whereas the inducible isoform (iNOS) is expressed in macrophages and other cell types after stimulation with cytokines and proinflammatory agents such as lipopolysaccharide (LPS). PGs are generated by metabolic conversion of arachidonic acid (AA) to PGs by the enzyme cyclooxygenase (COX). Also COX has been found in two isoforms, the constitutive enzyme (COX-1) which is expressed in almost every tissue and releases low levels of PGs involved in physiological functions, and the inducible isoform (COX-2), which is the major isoform expressed in inflammatory cells and can synthesize large amounts of PGs in response to inflammatory stimuli. The cross talk between these two pathways is of particular interest at inflammatory sites as the synthesis of prostanoids and NO is often elicited by the same stimuli. Microglial cells, the brain resident macrophages, express both iNOS and COX-2 and produce large amounts of both NO and PGs in response to LPS[1,2]. In a previous study we have shown that NO downregulates microglial COX-2 expression

Eicosanoids and Other Bioactive Lipids in Cancer, Inflammation, and Radiation Injury, 4
Edited by Honn *et al.*, Kluwer Academic / Plenum Publishers, New York, 1999.

169

and PG production[2], yet in other systems, including the widely used monocytic/macrophagic cell line RAW 264.7, NO was reported to stimulate PG synthesis[3].

Here, we attempted to understand the reason for the opposite effects of NO in microglia and RAW 264.7 cells. We found that the interaction between PGs and NO is different in these cell types and is exerted at different levels. In fact, in RAW 264.7 cells NO could affect PG synthesis by enhancing AA availability and downregulating COX-2 expression, while its derivative peroxynitrite upregulated COX-2 expression and PG synthesis. In microglia both NO and peroxynitrite downregulated COX-2 expression.

EFFECTS OF NO ON PGE$_2$ SYNTHESIS, AA RELEASE AND COX-2 EXPRESSION IN MICROGLIAL CELLS

LPS-activated microglial cells release large amounts of prostanoids. We have shown that LPS (10 ng/ml) induces PGE$_2$ release by regulating both AA availability and COX-2 expression. Indeed, LPS significantly increased the release of [^{3}H]AA from prelabelled microglial cells with a time lag of 1 h, and upregulated COX-2 expression and PGE$_2$ production, while the expression of COX-1 was very low both in control and LPS-stimulated cultures[1]. Concomitantly, LPS stimulated the accumulation of nitrite, a stable NO oxidation product[2] (around 100 nmol/mg protein).

The accumulation of nitrite induced by LPS was strongly reduced by the presence of N^G-monomethyl-L-arginine (NMMA) a specific NO-synthase inhibitor . The inhibition of NO production in microglia was accompanied by a two fold increase in PGE$_2$ release. Such increase was similar with two concentrations of NMMA (20 and 200 µM) leading to 70% and 90% inhibition of NO synthesis[2], respectively.

The effect of exogenous NO on PGE$_2$ synthesis in microglial cells was evaluated by using the synthetic donors 3-morpholinosydnonimine (SIN-1) and S-nitroso-N-acetylpenicillamine (SNAP). SIN-1 is a NO generator which also releases O$_2^-$, and these two radicals may combine to produce peroxynitrite (OONO-). On the other hand SNAP is considered to release exclusively NO, as several other S-nitrosothiols[4]. In order to avoid interference with the endogenously produced NO all the experiments were performed in the presence of the NO-synthase inhibitor NMMA (200 µM). Both NO donors were used at a concentration of 100 µM, which produced an accumulation of nitrite in culture media twice as high as that induced by optimal LPS concentrations. In microglial cultures both SIN-1 and SNAP (Fig. 1) significantly inhibited the LPS-induced, but did not affect the basal, PGE$_2$ production.

Moreover, SIN-1 caused a moderate, but statistically significant decrease of both basal (10 ± 3%, mean ± SEM, n=3) and LPS-stimulated (20 ± 5%, mean ± SEM, n=3) [^{3}H]AA release, while SNAP was devoid of effect, indicating that AA release is not a major target of these NO donors. On the other hand, SIN-1 reduced by 47.5% and SNAP almost abolished the LPS-induced expression of COX-2 (Fig. 3A), and neither compound affected the basal (undetectable)

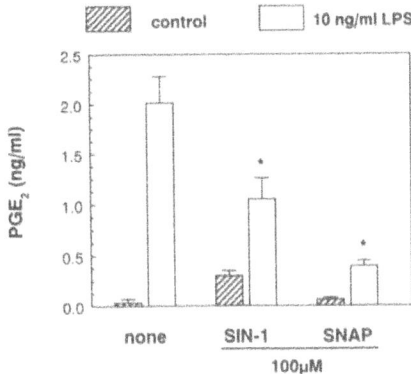

Figure 1. Effect of SIN-1 and SNAP on the basal and LPS-induced PGE₂ production in microglia. Microglial secondary cultures (≥99% pure) were maintained in BME medium, supplemented with 10% fetal calf serum, 2 mM glutamine and 2 µg/ml of gentamicin. Cells were plated in 24-well plates (1.25 x 10⁵ cells/cm²), and incubated for 24 h in the presence of the NOS inhibitor NMMA (200 µM), without or with SIN-1 (100 µM) and SNAP (100 µM). Values are means ± SEM of four independent experiments assayed in triplicate. *p<0.05 as compared to the LPS-stimulated cells.

expression of the enzyme. In conclusion, the concomitant inhibition of both LPS-stimulated COX-2 expression and PGE₂ production by SIN-1 and SNAP, suggests that in microglia both NO and peroxynitrite substantially inhibit COX-2 expression. Moreover, only peroxynitrite moderately depressed AA release.

EFFECTS OF NO ON PGE₂ SYNTHESIS, AA RELEASE AND COX-2 EXPRESSION IN RAW 264.7 CELLS

RAW 264.7 cells spontaneously released detectable levels of both PGE₂ (0.08 ± 0.02 ng/ml, mean ± SEM, n=7) and NO (3.23 ± 1.41 µM, mean ± SEM, n=3, measured as accumulation of nitrite in culture media). Such basal release was remarkably increased to 0.41 ± 0.09 and 0.51 ± 0.13 pg/ml of PGE₂ and 16.22 ± 3.45 and 17.89 ± 1.49 nmol/ml of NO in the presence of 1 and 5 µg/ml of LPS, respectively[5]. The amount of nitrite produced by RAW 264.7 cells stimulated with optimal LPS concentrations (1 µg/ml) was similar to that of microglia, around 100 nmol/mg protein. In RAW 264.7 cells, NMMA (200 and 400 µM) inhibited by 70% NO synthesis without any significant effect on PGE₂ production. This was in contrast to the inhibition of PGE₂ production by NMMA reported by others[6].

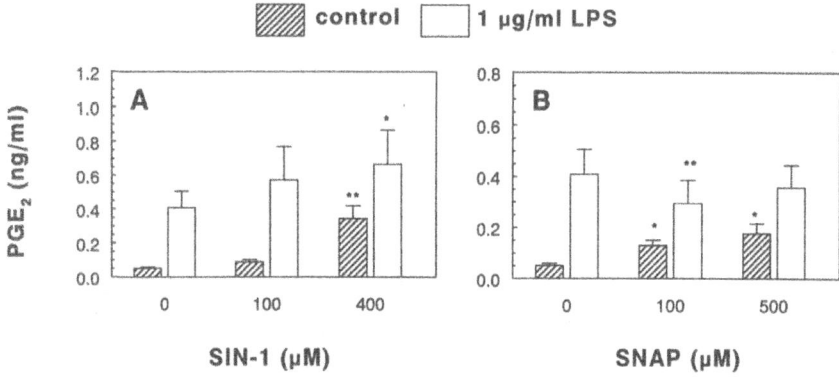

Figure 2. Effect of the NO donors SIN-1 (100 and 400 µM) (A) and SNAP (100 and 500 µM) (B) on the basal and LPS-induced PGE₂ production in RAW 264.7 cells. The mouse monocyte/macrophage cell line was maintained in RPMI 1640 medium supplemented with 100 U/ml penicillin, 100 µg/ml streptomycin, 10% heat-inactivated fetal calf serum, 2 mM of Glutamax I and were plated at a density of 6 x 10⁵ cells/cm² in 24-well plates. Cells were incubated for 24 h in the presence of the NOS inhibitor NMMA (400 µM) without or with 1µg/ml of LPS. Values are means ± SEM of four independent experiments assayed in triplicate. *p<0.05 and **p<0.01 compared with the corresponding control. Modified from ref. 5.

At variance with microglial cultures, in RAW 264.7 cells SIN-1 induced a concentration-dependent increase of basal and LPS-stimulated PGE₂ production (Fig. 2A). The effect of SNAP on basal PGE₂ production was similar to that of SIN-1, while that on LPS-stimulated production was a moderate decrease, with a statistically significant effect only at the lowest concentration (Fig. 2B).

In order to estimate AA release, RAW 264.7 cells were incubated for 20-22 h in complete media supplemented with 0.5 µCi/ml [³H]AA. After exposure (2 h), the supernatants were collected and counted for radioactivity. LPS dose-dependently stimulated [³H]AA release from prelabelled RAW 264.7 cells (50 ± 2% and 90 ± 17% over basal release with 1 and 5 µg/ml of LPS, respectively, mean ± SEM, n=4), in agreement with previous observations in the same cell line[7] and microglial cultures[1]. This effect of LPS was not affected by the presence of NMMA (400 µM).

When we compared the effect of the two NO donors under study on [³H]AA release in RAW 264.7 cells, a different behaviour was observed. In fact, SIN-1 (100 µM) did not affect basal [³H]AA release after 2 or 5 h of incubation and caused a very slight increase in the LPS-induced release. At variance with SIN-1, SNAP (500 µM) markedly increased (>70%) basal [³H]AA release and moderately enhanced the LPS-induced release (25%) suggesting that AA availability could be a regulatory step in the SNAP modulation of PGE₂ synthesis in RAW 264.7 cells. Using yet another NO donor, sodium nitroprusside, both basal and LPS-induced [³H]AA release were reported to be increased in RAW 264.7 cells[7].

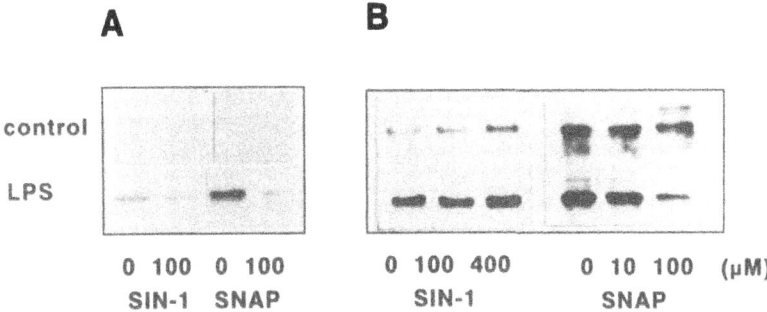

A　　　**B**

control

LPS

　0 100　0 100　　　0 100 400　　 0 10 100　(µM)
　SIN-1　SNAP　　　 SIN-1　　　　　SNAP

Figure 3. Effect of SIN-1 and SNAP on LPS-stimulated COX-2 expression in microglial and RAW 264.7 cells. Microglial (A) and RAW 264.7 cells (B) were incubated for 24 h in complete medium, with or without LPS in the presence of NMMA, SIN-1 or SNAP, at the concentrations indicated. Cell lysates were prepared and equal amounts of protein (25 µg) were analysed by western blot using anti COX-2 specific antibodies (1:500). Protein bands were visualized using horseradish peroxidase-conjugated secondary antibodies and ECL. Determination of COX-2 expression level was performed using the GS-700 Imaging densitometer (Bio-Rad). The intensity of the bands was evaluated semiquantitatively and the variations in percentage were referred to the corresponding control bands (see text below). One experiment representative of four (A) and three (B) is shown.

To study the regulation of COX-2 expression by NO in RAW 264.7 cells, proteins obtained from cell lysates were subjected to western blot analysis (Fig. 3B). At variance with microglia, COX-2 was clearly detectable in unstimulated cells. LPS (1 µg/ml), similar to microglia[2] markedly enhanced (4.8 ± 1.5 fold, mean ± SEM, n=4) COX-2 expression (Fig. 3B). NMMA (400 µM), which enhanced the LPS-induced expression of COX-2 in microglia[2], did not affect the basal nor the LPS-induced COX-2 expression in RAW 264.7 cells (not shown), in agreement with the lack of significant effect of NMMA on LPS-stimulated PGE_2 production. The two NO donors affected in a different way the COX-2 expression in RAW 264.7 cells. The lower concentration of SIN-1 (100 µM) enhanced only slightly the basal expression of COX-2 and did not affect the expression of the enzyme induced by LPS (Fig. 3B), while the higher SIN-1 concentration (400 µM) increased the expression of COX-2, both in basal conditions (44% ± 10) and after LPS stimulation (29% ± 5, mean ± SEM, n=3). After treatment with SNAP, both basal and LPS-stimulated COX-2 expression were depressed in a concentration-dependent way (Fig. 3B).

These observations suggest that in RAW 264.7 cells the stimulatory effect of SIN-1 on both basal and LPS-induced PGE_2 production may be mainly mediated by an increased expression of COX-2 and may be due to the formation of peroxynitrite.

On the other hand, NO released by SNAP has an inhibitory effect on COX-2 expression and the enhancement of basal PGE_2 production is likely to be related to a substantially increased availability of the PGE_2 precursor AA. In cells stimulated with LPS, however, the massive inhibitory effect of SNAP on COX-2 expression may not be counteracted by the small stimulation of AA release observed in this condition, with a resulting depression, albeit small, in the synthesis of PGE_2.

Our observations indicate that the modulation of prostanoid synthesis by NO and NO reactive derivatives can be exerted at different steps in the synthetic pathway, and that the effects of NO may be either direct or mediated by the reactive intermediates formed in an aerobic biological environment[8].

REFERENCES

1. L. Minghetti, and G. Levi, Induction of prostanoid biosynthesis by bacterial lipopolysaccharide and isoproterenol in rat microglial cultures, *J. Neurochem.* 65:2690 (1995).
2. L. Minghetti, E. Polazzi,., A. Nicolini, C. Crèminon, and G. Levi, Interferon-γ and nitric oxide down-regulate lipopolisaccharide-induced prostanoid production in cultured rat microglial cells by inhibiting cyclooxygenase-2 expression, *J. Neurochem.* 66:1963 (1996).
3. M. Di Rosa, A. Ialenti, A. Ianaro, and L. Sautebin, Interaction between nitric oxide and cyclooxygenase pathways, *Prostaglandins Leukot. Essent. Fatty Acids* 54:229 (1996).
4. W.R. Mathews, and S.W. Kerr, Biological activity of s-nitrosothiols: the role of nitric oxide, *J. Pharmacol. Exp. Ther.* 267:1529 (1993).
5. C. Guastadisegni, Minghetti L., Nicolini A., Polazzi E., Ade P., Balduzzi M., Levi G.. Prostaglandin E_2 synthesis is differentially affected by reactive nitrogen intermediates in cultured rat microglia and RAW 264.7 cells, *FEBS Lett.* 413:314 (1997).
6. D. Salvemini, T.P. Misko, J.L. Masferrer, K. Seibert, and M.G. Currie, Nitric oxide activates cyclooxygenase enzymes, *Proc. Natl. Acad. Sci. USA* 90:7240 (1993).
7. R.W. Gross, A.E. Rudolph, J. Wang, C.D. Sommers, and M.J. Wolf, Nitric oxide activates the glucose-dependent mobilization of arachidonic acid in a macrophage-like cell line (RAW 264.7) that is largely mediated by calcium independent phospholipase A_2, *J. Biol. Chem.* 270:14855 (1995).
8. J.S. Stamler, Redox signaling: nitrosylation and related target interactions of nitric oxide, *Cell* 78:931 (1994).

PHOSPHOLIPASES / PAF

27

ACTIVATION OF cPLA$_2$ IN VASCULAR SMOOTH MUSCLE

Edward F. LaBelle[1] and Erzsebet Polyak[2]

[1]Department of Physiology
Allegheny University of the Health Sciences
2900 Queen Lane
Philadelphia, PA 19129
[2]Department of Cell and Developmental Biology
University of Pennsylvania
Philadelphia, PA 19104

INTRODUCTION

Force regulation in vascular smooth muscle is under the control of a variety of lipid-dependent second messengers, such as inositol 1,4,5 trisphosphate and diacylglycerol (Hashimoto et al.,1986; Gu et al., 1991). Our laboratory has demonstrated that norepinephrine can induce force in rat tail artery via the stimulation of IP$_3$ production and the resulting release of calcium ions from intracellular stores (Gu et al.,1991; Somlyo et al,1985). It appears that IP$_3$ production in this tissue is under the control of a G-protein-dependent isoform of phospholipase C known as phospholipase Cb2 (LaBelle and Murray,1990; LaBelle and Polyak,1996). We have also demonstrated that norepinephrine can activate phospholipase D in rat tail artery, with the resulting production of choline and phosphatidic acid from the lipid phosphatidylcholine (Gu et al,1992).

Other investigators have shown that contractile agonists can also stimulate the release of free fatty acids such as arachidonic acid from cultured vascular smooth muscle cells (Lehman et al.,1993; Muthalif et al.,1996). We wanted to determine whether we could repeat this result using intact rat tail artery. We then wanted to determine if this arachidonate release resulted from the

Eicosanoids and Other Bioactive Lipids in Cancer, Inflammation, and Radiation Injury, 4
Edited by Honn *et al.*, Kluwer Academic / Plenum Publishers, New York, 1999.

177

activation of phospholipase C followed by diacylglycerol lipase, or if it resulted from the direct activation of phospholipase A_2 in this tissue. In the event that phospholipase A_2 activation could be detected, we wanted to determine which phospholipase A_2 isoform was activated, $cPLA_2$, $iPLA_2$, or $sPLA_2$.

RESULTS AND DISCUSSION

When segments of rat tail artery were prelabeled with [³H]-arachidonic acid, rinsed extensively to remove extracellular label, and then treated with norepinephrine (10μM), the contractile agonist was shown to stimulate the release of labeled arachidonate from the smooth muscle cells (Figure 1). Arachidonate release from the tissue occured as a linear function of time for 20 min. The identity of the excreted radioactive material as authentic arachidonic acid was confirmed by thin layer chromatography (data not shown). Arachidonic acid release from the tissue was not reduced when the tissue was exposed to nitrogen gas in order to remove the endothelial cells from the preparation (data not shown).

The concentration of norepinephrine that was required to stimulate arachidonate release from the tissue half-maximally was about 200nM.

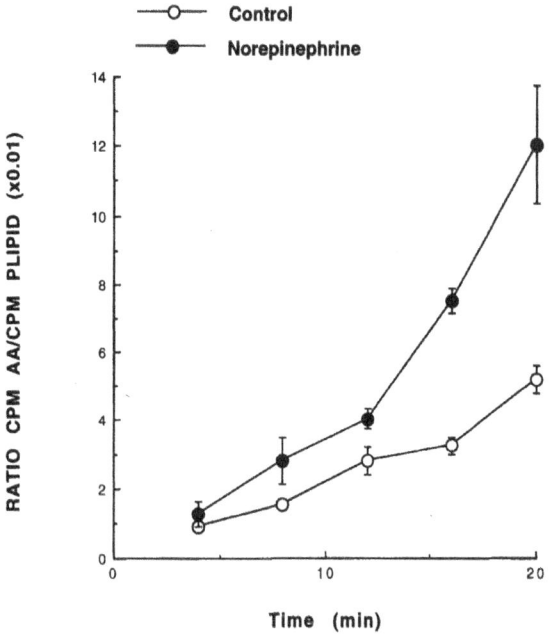

Figure 1 Effect of Norepinephrine on Arachidonate Release. Rat tail artery segments were prelabeled with [³H] arachidonate, rinsed with buffer to remove extracellular label, and then treated for the times indicated with or without NE, and aliquots of the incubation medium measured for radioactivity. The tissue was extracted with chloroform/methanol and the label in cellular lipid determined. The ratio of extracellular arachidonate cpm to total lipid cpm is shown above.

In order to determine if the norepinephrine-sensitive arachidonic acid release from the tissue occured as a direct consequence of PLC activation followed by diacylglycerol lipase activation, we treated rat tail artery with either the PLC inhibitor U73122 or the diacylglycerol lipase inhibitor RHC 80267. Neither U73122 nor RHC 80267 inhibited norepinephrine-induced arachidonate release from rat tail artery (Figure 2). This indicated that arachidonate release from this tissue most likely did not occur as a direct consequence of the activation of PLC and diacylglycerol lipase.

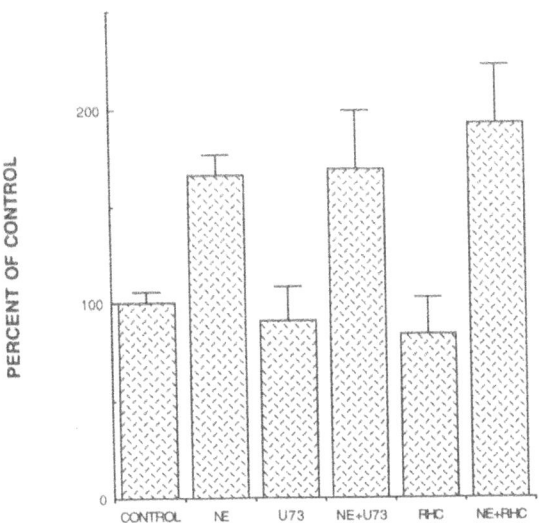

Figure 2   Effect of U73122 and RHC 802676 on Arachidonate Release.   Arachidonate release from tail artery segments was determined as described in the legend of Fig.1 after 10 min treatment with or without NE and 30 min treatment with or without either U73122 or RHC 80267.   Released arachidonate was determined and shown as percent of control released arachidonate.

We sought to determine if norepinephrine could activate an isoform of phospholipase A_2, such as $cPLA_2$, $iPLA_2$, or $sPLA_2$ by treating the tissue with selective inhibitors of these isoforms. The inhibitor HELSS was shown to be incapable of inhibiting norepinephrine-activated arachidonate release (Figure 3), which indicated that norepinephrine failed to activate $iPLA_2$ in rat tail artery. Likewise the inhibitor manoalide failed to block norepinephrine-stimulated arachidonate release, indicating that $sPLA_2$ was not activated by norepinephrine in this system (Figure 3). However the inhibitor $AACOCF_3$ was capable of completely blocking the norepinephrine-stimulated arachidonate release process in rat tail artery, which indicated that norepinephrine activated $cPLA_2$ in this system (Figure 4).

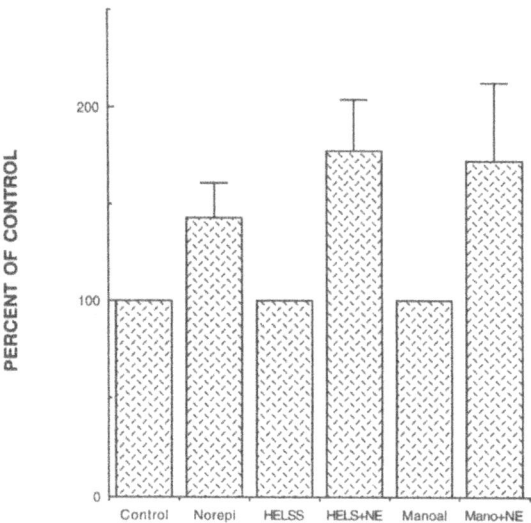

Figure 3 Effect of HELSS and manoalide on arachidonate release. Arachidonate release from rat tail artery was determined as described in the legend of Fig 1 after treatment with or without HELSS or manoalide (30 min) and after 10 min treatment with or without NE.

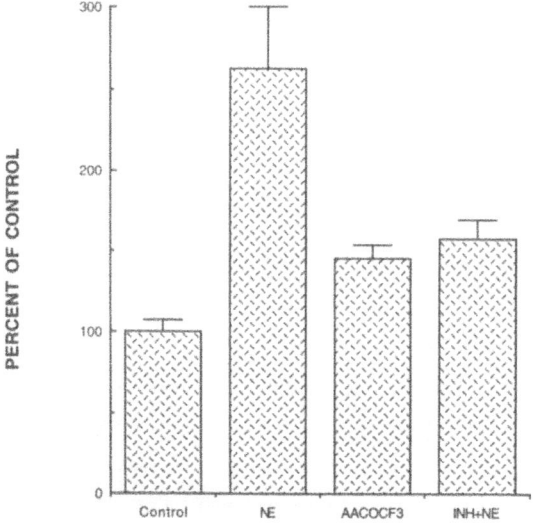

Figure 4. Effect of AACOCF$_3$ on arachidonate release. Arachidonate release from rat tail artery was determined as described in the legend of Fig. 1 after treatment with or without AACOCF$_3$ for 30 min and after 10 min treatment with or without NE.

Other investigators have found evidence that contractile agonists could activate arachidonic acid release from vascular smooth muscle. Gong et al.(1995) have demonstrated that norepinephrine could increase arachidonic acid levels within cells of rabbit femoral artery, but they failed to determine whether or not c-PLA$_2$ was involved. Muthalif et al (1996) have used antisense oligonucleotides directed against cPLA$_2$ to block norepinephrine-activated arachidonic acid release from cultured rabbit vascular smooth muscle cells. Lehman et al (1993) have demonstrated that vasopressin could activate arachidonate release from cultured A-10 smooth muscle cells via a process that seemed to depend on iPLA$_2$. This conclusion differs from our own work as well as the work of Muthalif et al. (1996). This may reflect the difference between vasopressin and norepinephrine, or there may be more complex differences involved.

Arachidonic acid has been shown to serve as a precursor of numerous eicosanoids, such as prostaglandins, thromboxanes, prostacyclin, leukotrienes, HETEs, and EETs (Needleman et al.,1986). These compounds have been shown to influence smooth muscle force and also relaxation by a variety of mechanisms (Gong et al.,1992;1995). Arachidonic acid has also been shown to influence many processes within mammalian cells by itself, without the need for further metabolism. These processes include K$^+$ channel activation (Ordway et al.,1989), calcium channel activation (Wang et al.,1993; van der Zee,1995; Byron et al.,1996), and the inhibition of myosin light chain phosphatase (Gong et al.,1992). It is currently uncertain just which processes are directly sensitive to arachidonate in rat tail artery.

REFERENCES

Byron, K.L., 1996, Vasopressin stimulates Ca^{2+} spiking activity in A7r5 vascular smooth muscle cells via activation of phospholipase A$_2$, *Circ.Res.* 78:813

Gong, M.C., Fuglsang, A. Alessi, D.,Kobayashi, S., Cohen,P.,Somlyo,A.V.and Somlyo A.P., 1992, Arachidonic acid inhibits myosin light chain phosphatase and sensitizes smooth muscle to calcium. *J.Biol.Chem.* 267: 21492

Gong, M.C., Kinter, M.T. Somlyo, A.V., and Somlyo, A.P., 1995, Arachidonic acid and diacylglycerol release associated with inhibition of myosin light chain dephosphorylation in rabbit smooth muscle, *J.Physiol.* 486: 113

Gu,H.,Martin, H., Barsotti, R.J. and LaBelle, E.F., 1991,. Rapid increase in inositol phosphate levels in norepinephrine-stimulated vascular smooth muscle. *Amer.J.Physiol.* 261:C17

Gu,H.,Trajkovic, S. and LaBelle, E.F.,1992, Norepinephrine induced phosphatidylcholine hydrolysis by phospholipases D and C in rat tail artery. *Amer.J.Physiol.* 262: C1376

Hashimoto,T., Hirata, M. Itoh, T., Kanmura,Y. and Kuriyama,H.,1986, Inositol 1,4,5 tris phosphate activates pharmacomechanical coupling in smooth muscle of the rabbit mesenteric artery. *J.Physiol.* 370: 605

LaBelle, E.F. and Murray, B.M., 1990, G protein control of inositol lipids in intact vascular smooth muscle. *FEBS Lett.* 268:91

LaBelle,E.F. and Polyak,E., 1996, Phospholipase Cb2 in vascular smooth muscle *J.Cell Physiol.* 169: 358

Lehman, J.J., Brown, K.A. ,Ramanandham, S., Turk, S.J. and Gross,R.W., 1993,. Arachidonic acid release from aortic smooth muscle cells induced by [Arg8] vasopressin is largely mediated by calcium-independent phospholipase A2. *J.Biol.Chem.* 268:20713

Muthalif, M.M., Benter, I.F., Uddin,M.R. and Malik,K.Y., 1996 Calcium/calmodulin-dependent protein kinase IIa mediates activation of mitogen-activated protein kinase and cytosolic phospholipase A2 in norepinephrine-induced arachidonic acid release in rabbit aortic smooth muscle cells. *J.Biol.Chem.* 271: 30149

Needleman,P., Turk,J., Jakschik,B.A., Morrison,A.R. and Lefkowith, 1986, Arachidonic acid metabolism. *Ann. Rev.Biochem.* 55:69

Ordway, R.W., Walsh,J.V. and Singer, J.T., 1989, Arachidonic acid and other fatty acids directly activate channels in smooth muscle cells. Science 244:1176

Somlyo, A.V., Bond, M.,Somlyo,A.P. and A. Scarpa,A., 1985, Inositol trisphosphate-induced calcium release and contraction in vascular smooth muscle. *Proc.Nat.Acad.Sci.US* 82:5231

van der Zee, L.,Nelemans, A., and den Hertog,A., 1995, Arachidonic acid is functioning as a second messenger in activating the Ca2+ entry process on H1- histaminoceptor stimulation in DDTMF-2 cells.. *Biochem.J.* 305: 859

Wang, X.B., Osugi, T. and Uchida,S., 1993, Muscarinic receptors stimulate calcium influx via phospholipase A2 pathway in ileal smooth muscles. *Biochem.Biophys.Res.Commun.* 93:483

28

ROLE OF TYPE IIA SECRETORY PHOSPHOLIPASE A$_2$ IN ARACHIDONIC ACID METABOLISM

Hiroshi Kuwata, Hisashi Sawada,
Makoto Murakami, and Ichiro Kudo

Department of Health Chemistry
School of Pharmaceutical Sciences
Showa University
1-5-8 Hatanodai, Shinagawa-ku
Tokyo 142, Japan

INTRODUCTION

Two kinetically different pathways that generate prostaglandin (PG) from endogenous arachidonic acid, the immediate and delayed phases, imply the recruitment of different sets of biosynthetic enzymes, the expression and activation of which are tightly regulated by distinct transmembrane signalings. The immediate phase of PG biosynthesis occurs within several minutes of stimulation, is elicited by agonists that mobilize intracellular Ca^{2+} and is characterized by a burst release of arachidonic acid (AA) initiated by Ca^{2+}-dependent phospholipase A$_2$ (PLA$_2$) which is subsequently converted to bioactive PGs by the sequential actions of cyclooxygenase (COX) and terminal PG synthases. The delayed phase of PG biosynthesis is accompanied by the continuous supply of AA and its conversion to PGs, often PGE$_2$, over long culture periods following growth or pro-inflammatory stimuli. Here we report the molecular identification of the PLA$_2$ and COX isozymes involved in the PG biosynthetic pathway in rat fibroblasts and hepatocytes and provide evidence that type IIA secretory PLA$_2$ (sPLA$_2$-IIA) plays an augmentative role in prolonged PG production following proinflammatory stimuli.

PGE$_2$ BIOSYNTHESIS IN RAT FIBROBLASTS

Activation of the rat fibroblastic cell line 3Y1 with interleukin-1β (IL-1β) and tumor necrosis factor α (TNFα) induced delayed PGE$_2$ generation over 6-48 h, which occurred in parallel with the *de novo* induction of sPLA$_2$-IIA and COX-2 (Figure 1, A & B).[1] Studies using an anti-sPLA$_2$-IIA antibody, sPLA$_2$-IIA inhibitors or a sPLA$_2$-IIA-specific antisense oligonucleotide revealed that IL-1β/TNFα-induced delayed PGE$_2$ generation by these cells was largely dependent on inducible sPLA$_2$-IIA, which was functionally linked to inducible COX-2 (Figure 1C). The expression of sPLA$_2$-V and sPLA$_2$-IIC, the two recently identified sPLA$_2$ isozymes, whose genes

Eicosanoids and Other Bioactive Lipids in Cancer, Inflammation, and Radiation Injury, 4
Edited by Honn *et al.*, Kluwer Academic / Plenum Publishers, New York, 1999.

183

map to a chromosomal locus close to the sPLA$_2$-IIA gene,[2] was undetectable in 3Y1 cells before and after cytokine stimulation.

Delayed PGE$_2$ generation was also suppressed markedly by the cPLA$_2$ inhibitor arachidonoyl trifluoromethyl ketone (AACOCF$_3$).[1] Notably, we found that AACOCF$_3$ attenuated cytokine induction of sPLA$_2$-IIA, but not COX-2, expression. AACOCF$_3$ inhibited the initial phase of cytokine-stimulated AA release, and supplementing AACOCF$_3$-treated cells with exogenous AA partially restored sPLA$_2$-IIA expression. In addition, lysophosphatidylcholine (lysoPC), another product of cPLA$_2$ reaction, augmented cytokine-induced sPLA$_2$-IIA expression in synergy with exogenous AA. Furthermore, IL-1β/TNFα-induced sPLA$_2$-IIA expression was greatly reduced by several 12- or 15-lipoxygenase inhibitors, but not by COX or 5-lipoxygenase inhibitors. These results indicate that certain 12- or 15-lipoxygenase products and lysoPC metabolites produced via the cPLA$_2$-dependent pathway in the early phase of cytokine stimulation may be crucial for the subsequent induction of sPLA$_2$-IIA expression and attendant PGE$_2$ generation (Figure 2).

In contrast, Ca^{2+} ionophore-stimulated immediate PGE$_2$ generation, which occurred within 30 min, in 3Y1 cells was regulated predominantly by the constitutive enzymes cPLA$_2$ and COX-1, even when sPLA$_2$-IIA and COX-2 were maximally induced after IL-1β/TNFα treatment (Figure 1D), revealing segregated coupling of discrete PLA$_2$s and COXs in the different phases of PG synthesis.

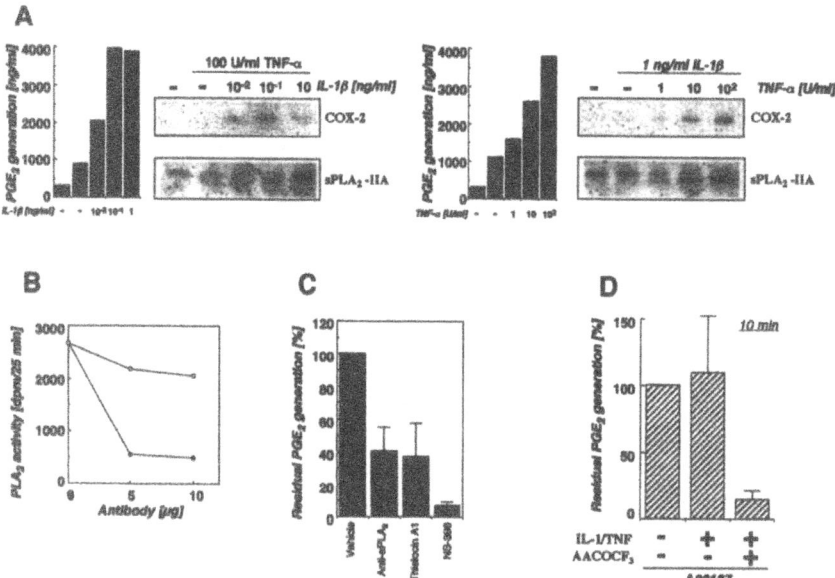

Figure 1. Delayed PGE$_2$ generation by rat fibroblastic 3Y1 cells. (A) Induction of sPLA$_2$-IIA and COX-2 expression and attendant PGE$_2$ generation in 3Y1 cells 24 h after treatment with increasing concentrations of cytokines. (B) Neutralization of induced PLA$_2$ activity with anti-rat sPLA$_2$-IIA antibody. (C) Inhibition of cytokine-induced delayed PGE$_2$ generation by inhibitors or antibodies for sPLA$_2$-IIA and COX-2. (D) A23187-induced immediate PGE$_2$ generation that was not increased by cytokine pretreatment and was suppressed almost completely by a cPLA$_2$ inhibitor.

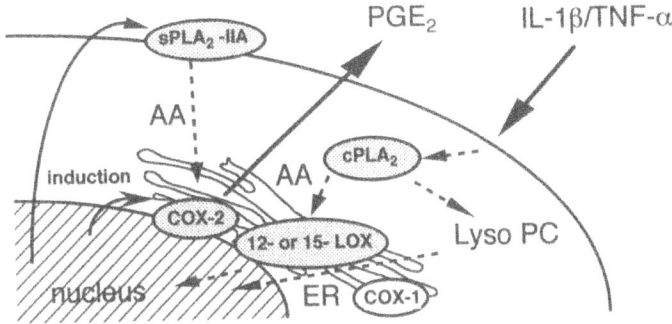

Figure 2. Schematic model for delayed PGE$_2$ generation by rat fibroblastic 3Y1 cells. Stimulation with proinflammatory cytokines activates cPLA$_2$, which produces AA and lysoPC. AA is metabolized to certain 12- or 15-lipoxygenase metabolites and contributes, in concert with lysoPC, to the induction of sPLA$_2$-IIA. sPLA$_2$ is secreted, binds to the cell surface, and provides AA for COX-2, which is also induced by the cytokines, for PGE$_2$ biosynthesis. It is likely that some of the cPLA$_2$-derived AA is also supplied to COX-2. COX-1 is not utilized in the delayed response, but is fully functional when the cells are stimulated with A23187.

cytokine stimulation. Cytokine-induced PGE$_2$ generation was suppressed by inhibitors of sPLA$_2$-IIA, cPLA$_2$ and COX-2, implying functional coupling of these enzymes, as in the case of rat fibroblasts as noted above.

Stimulation of another rat fibroblastic cell line Rat-1 with the combination of IL-1β and TNFα also resulted in *de novo* induction of sPLA$_2$-IIA and COX-2 and a modest increase in the expression of constitutive cPLA$_2$, whereas the expression of other sPLA$_2$ isozymes, including sPLA$_2$s-V and -IIC, was barely detectable (Figure 3A). It is therefore likely that inducible sPLA$_2$-IIA is the dominant sPLA$_2$ isozyme that is functionally linked to inducible COX-2 in cytokine-stimulated rat fibroblastic cell lines.

PGE$_2$ BIOSYNTHESIS IN RAT HEPATOCYTES

We previously reported that TNFα-induced PGE$_2$ generation by rat hepatocytes BRL-3A was dependent upon a heparin-sensitive sPLA$_2$-IIA-like enzyme,[3] yet its molecular identity as well as the downstream COX isozymes involved had remained obscure. RNA blot analysis confirmed that the sPLA$_2$ isozyme induced by IL-1β and TNFα was indeed sPLA$_2$-IIA, and not other sPLA$_2$ isozymes (Figure 3B). Induction of sPLA$_2$-IIA expression occurred 3-24 h after stimulation with these cytokines and was accompanied by the *de novo* induction of COX-2, which peaked at 12 h and declined thereafter. cPLA$_2$ was present in non-stimulated cells and increased gradually after

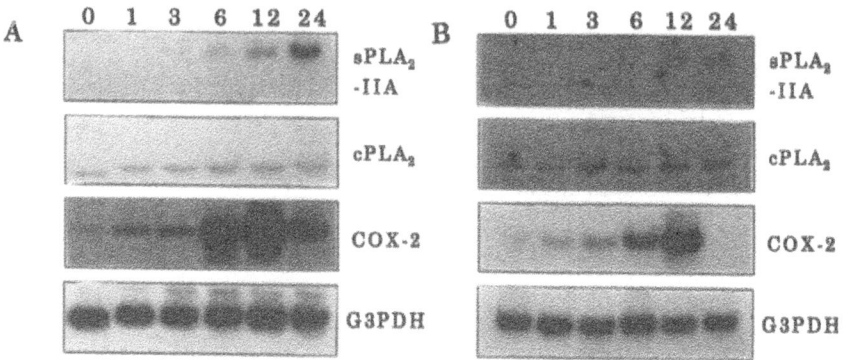

Figure 3. Expression of mRNA for sPLA₂, cPLA₂ and COX-2 enzymes in rat fibroblastic Rat-1 cells (A) and rat liver-derived BRL-3A cells (B) after stimulation with IL-1 and TNF for the indicated periods.

IMPORTANCE OF CELL SURFACE PROTEOGLYCAN-BOUND SPLA₂-IIA

Based upon the observations that solubilization of cell surface-associated sPLA₂-IIA by exogenous heparin led to concomitant reduction of cytokine-stimulated PGE₂ biosynthesis in various cell types such as fibroblasts,[1] hepatocytes[3] and endothelial cells,[4] we have proposed that the association of cytokine-inducible sPLA₂-IIA with cell surfaces via heparan sulfate proteoglycan is important for its effects on cellular functions. To obtain more conclusive evidence for this property of sPLA₂-IIA, we identified the critical domain in sPLA₂-IIA for heparin and cell surface binding and examined its role in cellular PG biosynthesis.[5] Replacement of several conserved Lys residues in the C-terminal region of sPLA₂-IIA by Glu markedly reduced its capacity to bind to heparin and mammalian cell surfaces, without affecting its enzymatic activity toward dispersed phospholipid. CHO cells stably transfected with native sPLA₂-IIA released significantly larger amounts of AA than cells transfected with vector alone during culture for 10 h with IL-1, whereas the ability to enhance arachidonic acid release was impaired in sPLA₂-IIA mutants incapable of binding to cell surfaces. AA released by native sPLA₂-IIA-transfected cells was metabolized to PGE₂ via COX-2 after IL-1 stimulation, confirming a particular functional linkage of sPLA₂-IIA to COX-2. Taken together, these results suggest that sPLA₂-IIA anchored on cell surfaces via its C-terminal heparin-binding domain is involved in the COX-2-dependent delayed PG biosynthesis initiated by proinflammatory cytokines during long-term culture.

sPLA₂-IIA is also capable of augmenting immediate AA metabolism under appropriate conditions. When excess sPLA₂-IIA was added exogenously to cells that had been primed by particular agonists, substantial increase in AA release often occurred.[4,6] Effect of sPLA₂-IIA overexpression on the immediate response is shown in Figure 4. CHO cells transfected with sPLA₂-IIA exhibited increased immediate AA release in response to thrombin compared with that by control cells, whereas replicate cells transfected with sPLA₂-IIA mutant lacking the heparin-binding domain failed to do so.

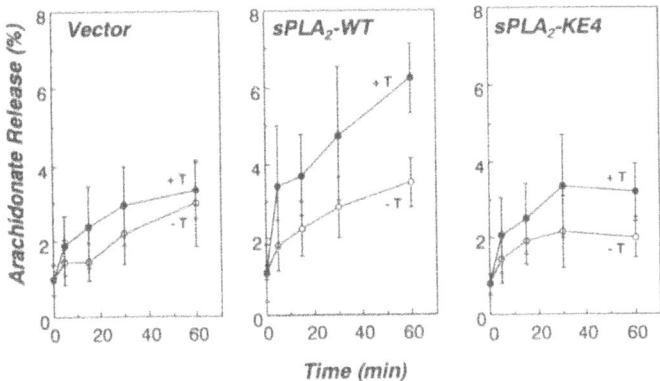

Figure 4. Thrombin-induced immediate AA release from CHO cells transfected with empty vector, native sPLA₂-IIA (WT), or sPLA₂-IIA mutant lacking the heparin-binding domain (KE4). Cells prelabeled with [³H]AA were cultured with (*closed circles*) or without (*open circles*) 1 U/ml thrombin (T) for the indicated times and the radioactivities released into the supernatants were measured. Values are means ± S.D. (n = 3).

SUMMARY

Recent recognition of the rapidly growing sPLA₂ family has led to a suggestion that some of the previously described functions of sPLA₂-IIA need to be reevaluated, since studies based upon enzyme activities and using inhibitors or antibodies against sPLA₂-IIA may not discriminate these sPLA₂s.[2] Our present studies reconfirm the involvement of sPLA₂-IIA in biological responses, demonstrated significant crosstalk between the two Ca²⁺-dependent PLA₂s (cPLA₂ and sPLA₂) where one enzyme is required for the induction of the other, and revealed segregated coupling of discrete PLA₂ and COX enzymes in the different phases of PG biosynthesis. Based upon the analysis of cells derived from sPLA₂-IIA "natural knock-out" mice,[7] it is apparent that sPLA₂-IIA is not essential for the initiation of delayed PGE₂ biosynthesis.[8] However, it is capable of contributing to the delayed response as an enhancer when appropriately induced by proinflammatory stimuli, leading to optimal COX-2-dependent PGE₂ generation. Importantly, in order for sPLA₂-IIA (or related sPLA₂ isozymes) to attack the biological membranes, so-called "membrane rearrangement" should take place in *activated*, but not *resting*, cells.[9] Membrane rearrangement also occurs when cells are undergoing apoptosis, during which acidic phospholipids, the preferred substrates for sPLA₂-IIA, are exposed on the outer leaflet of the plasma membranes.[10] Nonetheless, in view of the dramatically elevated levels of sPLA₂-IIA in inflamed or ischemic sites, it is likely that this extracellular isozyme participates in the expansion of chronic tissue disorders by augmenting generation of proinflammatory eicosanoids or lysophospholipids, depending upon the states of the inflammatory response.

REFERENCES

1. H. Kuwata, Y. Nakatani, M. Murakami, and I. Kudo. Cytosolic phospholipase A_2 is required for cytokine-induced expression of type IIA secretory phospholipase A_2 that mediates optimal cyclooxygenase-2-dependent delayed prostaglandin E_2 generation in rat 3Y1 fibroblasts. *J. Biol. Chem.* 273: 1733 (1998)

2. J.A. Tischfield. A reassessment of the low molecular weight phospholipase A_2 gene family in mammals. *J. Biol. Chem.* 272: 17247 (1997)

3. H. Suga, H., M. Murakami, I. Kudo, and K. Inoue. Participation in cellular prostaglandin synthesis of type II phospholipase A_2 secreted and anchored on cell-surface heparan sulfate proteoglycan. *Eur. J. Biochem.* 218: 807 (1993)

4. M. Murakami, I. Kudo, and K. Inoue. Molecular nature of phospholipase A_2 involved in prostaglandin I_2 synthesis in human umbilical vein endothelial cells: possible participation of cytosolic and extracellular type II phospholipases A_2. *J. Biol. Chem.* 268: 839 (1993)

5. M. Murakami, Y. Nakatani, and I. Kudo. Type II secretory phsopholipase A_2 associated with cell surfaces via C-terminal heparin-binding lysine residues augments stimulus-initiated delayed prostaglandin generation. *J. Biol. Chem.* 271: 30041 (1996)

6. M. Murakami, I. Kudo and K. Inoue. Eicosanoid generation from antigen-primed mast cells by extracellular mammalian group II phospholipase A_2. *FEBS Lett.* 294: 247 (1991)

7. M. MacPhee, K.P. Chepenik, R. Liddell, K.K. Nelson, L.D. Siracusa, and A.M. Buchberg. The secretory phospholipase A_2 gene is a candidate for the *Mom1* locus, a major modifier of Apc^{Min}-induced intestinal neoplasia. *Cell* 81: 957 (1995)

8. M. Murakami, H. Kuwata, H., Y. Amakasu, S. Shimbara, Y. Nakatani, G. Atsumi, and I. Kudo. Prostaglandin E_2 amplifies cytosolic phospholipase A_2 and cyclooxygenase-2-dependent delayed prostaglandin E_2 generation in mouse osteoblastic cells: enhancement by secretory phospholipase A_2. *J. Biol. Chem.* 272: 19891 (1997)

9. M. Murakami, Y. Nakatani, G. Atsumi, K. Inoue, and I. Kudo. Regulatory functions of phospholipase A_2. *Crit. Rev. Immunol.* 17: 225 (1997)

10. G. Atsumi, M. Murakami, M. Tajima, S. Shimbara, N. Hara, and I. Kudo. The perturbed membrane of cells undergoing apoptosis is susceptible to type II secretory phospholipase A_2 to liberate arachidonic acid. *Biochim. Biophys. Acta* 1349: 43 (1997)

29

PHOSPHOLIPASE A₂ IS INVOLVED IN CHEMOTAXIS OF HUMAN LEUKOCYTES

Ulrich Tibes[1], Markus Hinder[1], Werner Scheuer[3], Walter-Gunar Friebe[2], Stephan Schramm[1], Beate Kaiser[3]

[1, 2, 3]Boehringer Mannheim GmbH
[1]Dept. of Preclinical Research
[2]Dept. of Chemical Research
D-68298 Mannheim
[3]Dept. of Molecular Pharmacology
D-82377 Penzberg

INTRODUCTION

Recently, it was shown that phospholipase A_2 (PLA_2) is essential for exposing Mac-1 (CD11b/CD18) on the cell surface of leukocytes[1,2]. Mac-1 is mandatory for leukocyte migration in vitro, and lack of Mac-1 on human leukocytes is associated with failure of in vitro chemotaxis[3]. To further substantiate the contribution of PLA_2 in cell migration we investigated the effects of PLA_2 inhibition on chemotaxis of mixed human leukocytes (MHL) induced by fMLP and the G-protein activator NaF[7-9]. Migration was measured in a modified 48-well micro-chemotaxis Boyden-chamber[6]. As PLA_2 inhibitors manoalide[4], mepacrine[5], and BM 16.2115, a new PLA_2 inhibitor, were employed.

MATERIAL AND METHODS

Test compounds. Mepacrine[5], manoalide[4] and BM 16.2115 were used as PLA_2 inhibitors. The structure of BM 16.2115 (1,6-dioxadispiro[4.1.5.1]trideca-8,11-diene-2,10-dione) is depicted below. Compounds were disolved in DMSO with final concentrations not exceeding 1%.

structure of BM 16.2115

Eicosanoids and Other Bioactive Lipids in Cancer, Inflammation, and Radiation Injury, 4
Edited by Honn *et al.*, Kluwer Academic / Plenum Publishers, New York, 1999.

189

Leukocyte Preparation. Cubital venous blood was drawn from 6 healthy male volunteers (age 20 - 50 years) and anticoagulated by citrate. No drug intake was allowed one week prior to blood sampling. MHL were instantaneously separated by density gradient centrifugation (Polymorphpreb™, 525 x g, 30 min, 5 °C). After separation granulocytes and mononuclear cells were pooled, washed 2 - 3 times in KHS (Krebs-Henseleit Solution) and re-suspended in KHS. Total cell numbers were adjusted to 10^3 cells/ml after last washing. After preparation MHL were immediately used in the chemotaxis assay.

Chemotaxis Assay. Leukocyte migration was assessed in a 48-well micro-chemotaxis Boyden-chamber[6] using polyvinyl-pyrrolidone-free polycarbonate membranes (10 µM) with a pore size of 5 µm. 25 µl of fMLP-supplemented KHS or NaF-stimulated leukocyte suspensions were used as chemoattractants in the lower compartment of the chamber. fMLP concentrations in the lower compartment were adjusted to 1, 2, 4, 8, 10, and 16 nM. In the case of NaF-stimulation, MHL (10^3 cells/ml) were incubated (37 °C, 5 % CO_2 in 95 % O_2) for 15, 30, 45, 60, 75 and 90 minutes in KHS containing 20 mM NaF[7,8,9]. Compensation of osmolality was accomplished by corresponding changes of NaCl and glucose concentrations. 25 µl of this NaF-activated cell suspension were placed in the lower compartment as chemoattractant. In every case, 50 µl (10^3 cells/ml) of MHL were placed in the upper part of the wells on top of the membrane.

After one hour incubation (37 °C, 5 % CO_2 in 95 % O_2) of the cells in the micro-chemotaxis chamber the membranes containing the cells on their surfaces and in the pores were fixed by drying in air and then stained with Hemacolor™. The migrated leukocytes on the lower surface of the membrane were counted according to Zigmond and Hirsch[10] (500 x fold magnification). Leukocytes which had migrated to the lower surface, or nuclei of cells detectable in the filter pores within 5 µm distance from the lower surface, were considered to have passed the membrane and were counted. The numbers of migrated cells were determined in 5 high-power-fields (hpf) of 5 wells, i.e. in 25 hpf. These counts were averaged and defined as single value to be used for statistical evaluations. For treatment with the PLA_2 inhibitors cells were pre-incubated with the compounds or the vehicles for 15 - 20 minutes under the same conditions (37 °C, 5 % CO_2 in 95 % O_2).

Leukotriene B_4 Production. LTB_4 is one of the most potent chemotactic eicosanoids [11,12]. Therefore it was chosen as an indicator of PLA_2 activation by fMLP. MHL were separated as described above, washed twice in sterile NaCl, and after last washing suspended in KHS and adjusted to 2 x 10^6 cells/ml. Cells were stimulated with different fMLP concentrations for 1 h (37 °C, 5 % CO_2 in 95 % O_2). Thereafter cells were separated by centrifugation (400 g, 10 minutes, 0 °C) and LTB_4 concentrations determined in the supernatant by an EIA kit (Advanced Magnetics).

Cell Viability. Cell viability was assessed with the trypan blue exclusion test (Boehringer Mannheim). Leukocytes were incubated in KHS supplemented with the test compound BM 16.2115 or vehicle, and after 20 and 60 min of incubation the trypan blue solution (0.25 % test concentration) was added and vital or dead cells counted (2-3 min).

PLC-ß2 Activity. Effect of BM 16.2115 on PLC-$ß_2$ (phospholipase C-$ß_2$) of human neutrophils was assessed using the Ca^{2+} or ßγ–G protein-activated recombinant human enzyme and analysing IP-3 formation by a RIA[13] (experiments kindly carried out by Prof. Dr. P. Gierschick, University Ulm, Germany).

Inhibition of sPLA$_2$ and cPLA$_2$ Enzymatic Activity by BM 16.2115. Activity of human recombinant sPLA$_2$ (hr-sPLA$_2$, Boehringer Mannheim) was assayed as described elsewere[2,14,15,18] and release of *sn-2* fatty acids was quantified by a test kit (Free Fatty Acids, Boehringer Mannheim). Activity of human cPLA$_2$ isolated from U 937 cell line (kindly provided by Prof. Dr. M. Goppelt-Strübe, University Erlangen, Germany) was evaluated by enzyme assay as previously outlined[2,14,15,18].

Cell assays. Suppression of arachidonic acid (AA) release was evaluated in human thrombocytes, mixed leukocytes, and rat mesangial cells according to literature[2,15,16] (measurements in rat mesangial cells were kindly performed by Prof. Dr. J. Pfeilschifter, University Frankfurt, Germany). Human thrombocytes were labelled by ^{14}C-AA and release of ^{14}C-AA was triggered by human thrombin. Since in A 23187 stimulated rat mesangial cells AA release is mainly accomplished by cPLA$_2$[16], mesangium cells were labelled with ^{14}C-AA and stimulated by A 23187 to release ^{14}C-AA.

In addition to ^{14}C-AA release, rat mesangial cells were activated by IL-1ß for 24 h to induce expression of sPLA$_2$ resulting in PGE$_2$ release[16]. PGE$_2$ was measured by RIA (Amersham).

In mixed human leukocytes ionophore-induced release of LTC$_4$ and PGE$_2$ and its inhibition was analysed as described[2,15,16].

Effects of BM 16.2115 on cyclooxygenase (COX) or 5-lipoxygenase (5-LOX) activities were evaluated in U 937 cells or MHL according to Grossman et al.[17] and as previously described[15]. Briefly, to 10^6 cells/ml 1.25 µM AA (Sigma) was added and 15 min later PGE$_2$ or LTC$_4$ analysed by an EIA (Boehringer Mannheim, Advanced Magnetics, respectively). Cells were incubated with BM 16.2115 for 15 minutes prior to adding AA.

Statistical Methods

Data are expressed as means ± SD, if not otherwise stated. Significant differences were determined using t-test. A value of 2p< 0.05 was considered significant.

RESULTS

Chemotaxis

fMLP-Stimulated Leukocyte Chemotaxis. fMLP stimulated the chemotaxis of MHL along a bell shaped concentration response curve (Figure 1, left part). The peak chemotactic activity was observed at 10 nM fMLP, corresponding to an increase of some 55 - 60 % over control values. Based on these observations all following inhibitor experiments with MHL were carried out at 10 nM fMLP concentration, representing the maximum stimulus.

fMLP-Stimulated Production of Leukotriene B$_4$. LTB$_4$ production, evaluated in the supernatant after one hour incubation of MHL (2 x 10^6/ml) in fMLP-supplemented KHS, increased along a similar bell-shaped concentration response curve (Figure 1, right part) as found for chemotactic migration. The peak of LTB$_4$ production occurred at 10 nM fMLP in aggreement with the fMLP-stimulated chemotaxis. However, the 4 - 5 fold increase of LTB$_4$ was much more pronounced than enhancement of cell migration.

Suppression of fMLP-Stimulated Chemotaxis by PLA$_2$ Inhibitors. 10^3 cells/ml were incubated with vehicle, mepacrine (1, 5, 10, 50, 100 μM), manoalide (1, 5, 10, 50, 100 μM) or BM 16.2115 (0.5, 1, 5, 10, 50 μM) and thereafter transferred into the upper compartment of the chemotaxis chamber. Migration was assessed towards 10 nM fMLP or KHS (control). Compared to KHS 10 nM fMLP increased the cell numbers per hpf for roughly 20 cells, i.e for 60%. All three PLA$_2$ inhibitors completely depressed fMLP-stimulated migration of MHL in a concentration-dependent manner (Figure 2). While mepacrine and manoalide suppressed fMLP-induced component down to the control level obtained in KHS, BM 16.2115 depressed the cell migration even below this control level.

NaF-Stimulated Chemotaxis of Human Leukocytes. Stimulating MHL in KHS for 15 to 90 minutes with 20 mM of the G-protein activator NaF, and using this NaF-activated cell suspension as chemoattractant in the lower compartment of the chamber induced a similar migration of MHL across the membrane like with fMLP. The strength of this chemotactic attraction was time dependent reaching values corresponding to 8 nm fMLP after 90 minutes (Figure 3). Random migration of MHL towards 20 mM NaF in cell free KHS placed as control in the lower compartment was not enhanced. Cell suspensions incubated for up to 90 minutes in KHS only and used as chemoattractant did also not affect migration (data not shown).

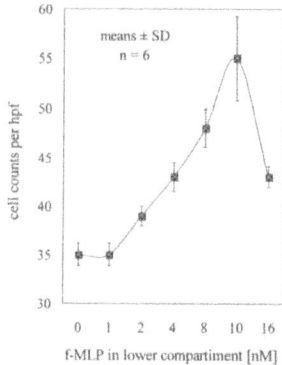 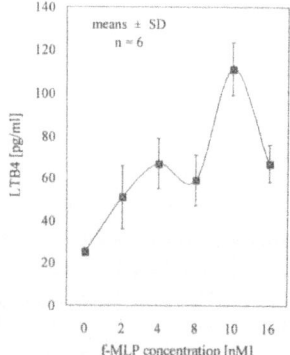

Figure 1. fMLP induced chemotaxis (left) in a modified Boyden chamber and LTB$_4$ production (right) of human mixed leukocytes. With 0 nM fMLP random migration is represented.

Table 1: Suppression by mepacrine of leukocyte migration induced by chemoattractants released from NaF-activated leukocyte suspensions.

| | KHS | cells + NaF | μM mepacrine | | | | |
			1	5	10	50	100
		migrating cells in upper compartment inhibited by mepacrine					
means ± SD	31±1.2	41±1.0	*36±1.5	*35±1.0	*35±2.0	*33±1.2	*33±0.5
% inhibition	100	0	50	60	60	80	80
		chemoattractant releasing cells in lower compartment inhibited by mepacrine					
means ± SD	31±1.2	42±1.0	*37±2.7	*37±2.2	*36±1.7	*35±2.0	*33±1.2
% inhibition	100	0	45	45	55	64	73

Cells in upper or lower compartment of chemotaxis chamber were 20 min pretreated with mepacrine. KHS = cell free control, i.e. KHS + vehicle + 20 mM NaF in lower compartment. Cells + NaF = cells (10^3/ml) in lower compartment incubated with 20 mM NaF and vehicle (90 min). Difference of migration between KHS and NaF defined as 100 %. *significant at least at 2p < 0.025 compared to NaF. Means ± SD of 6 volunteers.

Since the three PLA_2 inhibitors exerted a similar inhibition profile in the chemotaxis assay against fMLP, only mepacrine was used to evaluate inhibition of NaF-effects. Thus, pre-treatment of the migrating MHL placed in the upper compartment with increasing concentrations of mepacrine for 15 - 20 min while stimulating chemotaxis by NaF-activated cell suspension in the lower compartment, depressed migration to a similar extent (Table 1) as observed with fMLP and mepacrine (Figure 2). When cell suspensions pre-treated by mepacrine prior to NaF incubation (90 min) were used as chemoattractant, migration of MHL from the upper compartment was depressed for up to 73 % (Table 1). These data suggest that cell migration itself and NaF-induced release of chemoattractant mediators are both controlled by PLA_2-dependent mechanisms.

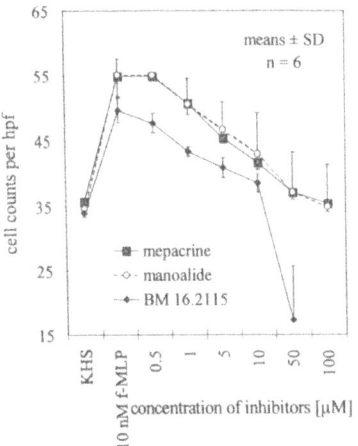 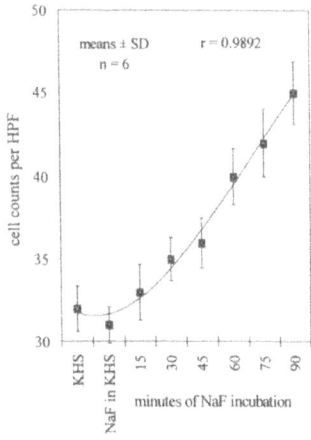

Figure 2 (left). Suppression of f-MLP induced chemotaxis by PLA_2 inhibitors. Peak of migration was at 10 nM fMLP vs control (KHS, = random migration). For all three compounds inhibition of migration was already significant with 1μM compared to peak migration (1μM 2p<0.025; 5μM 2p<0.0013; 10 μM 2p<0.00022; 50 μM 2p<0.00022; 1μM 2p<0.00004). Measurements from 6 subjects per compound.

Figure 3 (right). Increase of chemotactic leukocytes migration by NaF (20 mM) stimuled leukocytes in lower copmpartment. Migration promoting effect of NaF increased with time (significance of correlation 2p<0.001).

BM 16.2115, Evaluation of PLA₂ Inhibitory Potency

Inhibition of sPLA₂, cPLA₂ Enzymatic Activities, Arachidonic Acid, PGE₂, and LTB₄ Release. BM 16.2115 inhibited human recombinant sPLA₂ and human cPLA₂ of U 937 cells with IC_{50}s depicted in Table 2. Mepacrine showed comparable suppression of the enzymes whereas manoalide inhibited sPLA₂ somewhat stronger (Table 2).

Thrombin-stimulated ^{14}C-AA release from labelled human platelets was concentration-dependently inhibited for up to 100 % by BM 16.2115, mepacrine, and manoalide (Table 2). In A 23187-stimulated rat mesangial cells ^{14}C-AA release was attenuated for 60 % (means of two experiments, determinations in duplicate) with 10 μM BM 16.2115, demonstrating significant inhibition of cPLA₂[18].

As a consequence of inhibiting A 23187-induced AA release, LTC₄ and PGE₂ synthesis was suppressed by BM 16.2115, mepacrine and manoalide in concentration dependent manner (IC_{50}s in Table 2).

In IL-1ß activated rat mesangial cells 10 μM of BM 16.2115 depressed PGE₂ release for 97 % demonstrating sPLA₂ inhibition[16].

Thus, in total BM 16.2115 displayed both inhibition of sPLA₂ and cPLA₂ in a similar concentration range.

Table 2. Inhibition of sPLA₂, cPLA₂, arachidonic acid release, PGE₂ and LTC₄ synthesis by BM 16.2115, mepacrine, and manoalide. Table contains IC_{50}s.

rh-sPLA₂	cPLA₂	^{14}C-AA-release	PGE₂ A23187	LTC₄ A23187
		inhibition by BM 16.2115, IC_{50}s, μM		
17	21	9	15	7
		inhibition by mepacrine, IC_{50}s, μM		
25	50	15	5	20
		inhibition by manoalide, IC_{50}s, μM		
1.3	nd	69	3.0	2.5

Means of 2-3 replicates (no SD was calculated) for each parameter with duplicate determinations. ^{14}C-AA-release measured in thrombin stimulated human platelets, PGE₂ and LTC₄ in ionophore activated mixed human leukocytes. cPLA₂ was isolated from U 937 monocytic cell line. nd = not determined.

Failure of BM 16.2115 to Inhibit Cyclooxygenase and 5-Lipoxygenase Activities. In contrast to PLA₂ inhibition BM 16.2115 up to 100 μm failed to suppress COX-dependent PGE₂ synthesis in AA-challenged U 937 cells (Table 3). Similarly, 10 μM of BM 16.2115 did not suppress 5-LOX-dependent LTC₄ synthesis in AA-challenged human leukocytes whereas 100 μM decreased synthesis (Table 3), probably indicating some 5-LOX inhibition at the top concentration.

Failure of BM 16. 2115 to Inhibit PLC-ß2 Activity. BM 16.2115 did not inhibit recombinant PLC-ß₂ of human leukocytes (Table 4). Therefore a significant contribution of PLC-ß2 on fMLP mediated MHL migration and eicosanoid release may be ruled out.

194

Table 3: Failure of BM 16.2115 to inhibit cyclooxygenase (COX) and 5-lipoxygenase (5-LOX) as determined by arachidonic acid-induced PGE_2 and LTC_4 synthesis.

concentration	PGE_2 (pg/ml) U 937 + AA	LTC_4 (pg/ml) U 937 + AA
medium + LPS + vehicle	500	3.0
0 µM BM 16.2115 + AA	2151	329
1 µM BM 16.2115 + AA	2100	nd
10 µM BM 16.2115 + AA	2500	356
100 µM BM 16.2115 + AA	2510	42
100 µM NDGA	2200	2.7
100 µM meclofenamat	428	350

Means from 2 replicates (no SD calculated), determinations in duplicate. nd = not determined.

BM 16.2115 Is not Cytotoxic in MHL. BM 16.2115 did not affect viability of MHL in a concentration range between 1-1000 µM during 20 and 60 min incubation, as was tested by the trypan blue assay. Thus, it can be excluded that the inhibitory activities of BM 16.2115 observed in the various cell systems were related to cytotoxic effects.

CONCLUSION

The present data demonstrate that in fMLP- and NaF mediated chemotactic leukocyte migration PLA_2 is involved in both release of chemoattractant phospholipid mediators, and chemotactic cell migration itself.

Based on results obtained with PLA_2 inhibitors it may be suggested that secretory and cytosolic PLA_2 are engaged in these mechanisms, whereas PLC seems not to be involved.

G-proteins may participate in both chemotactic activation of PLA_2, likely cytosolic PLA_2, and cell migration.

Obviously Mac-1 is important for leukocyte migration and it seems to be controlled by secretory and cytosolic PLA_2[1,2]. Depression of cell migration may contribute to the strong anti-inflammatory activities PLA_2 inhibitors display in vivo[2,15,18].

Table 4. Failure of BM 16.2115 to inhibit Ca^{2+} or ßγ-G-protein activated recombinant human PLC- ß$_2$.

		BM 16.2115 concentration		
control	activated control	1 µM	10 µM	100 µM
		Ca^{2+} activated		
0.9 ± 0.1	7.6 ± 0.8	7.0 ± 0.4	6.5 ± 0.3	6.4 ± 0.3
		G-protein activated		
0.9 ± 0.1	9.4 ± 0.8	7.3 ± 1.9	9.6 ± 2.4	8.4 ± 1.0

Means ± SD, n = 3. Human recombinant PLC-ß$_2$ was either activated by 2 mM Ca^{2+} or by ßγ-G-protein.

ACKNOWLEDGMENT

The authors thank W. Kinle, S. Kippenhan, A. Litters, and R. Vogler for excellent technical assistance, Prof. Dr. P. Gierschick (University Ulm) for evaluation of PLC-ß activities, Prof. Dr. J. Pfeilschifter (University Frankfurt/Main) for investigations in rat mesangial cells, and Prof. Dr. M. Goppelt-Strübe (University Erlangen) for providing human $cPLA_2$.

REFERENCES

[1] Jacobson, P.B., Schrier, D.J., 1993, Regulation of CD11b/CD18 expression in human neutrophils by phospholipase A_2. *Immunol.* 151:5639-5652.

[2] Amandi-Burgermeister, E., Tibes, U., Kaiser, B.M., Friebe, W.G., Scheuer, W., 1997, Suppression of cytokine synthesis, integrin expression and chronic inflammation by inhibitors of cytosolic phospholipase A_2. *Europ. J. Pharmacol.* 326:237-250.

[3] Arnaout, M.A., 1990, Leukocyte adhesion molecules deficiency: Its structural basis, pathophysiology and implications for modulating the inflammatory response. *Immunol. Rev.* 114:145-180.

[4] Glaser, K.B., Mobilio, D., Chang, J.Y., Senko, N., 1993, Phospholipase A_2 enzymes: Regulation and inhibition. *Trends Pharmacol. Sci.* 14:92-98.

[5] Rao, G.N., Lassegue, B., Griendling, K.K., Alexander, R.W., Berk, B.C., 1993, Hydrogen peroxide-induced c-fos expression is mediated by arachidonic acid release. Role of protein kinase C. *Nucleic Acids Res.* 21:1259-1263.

[6] Falk, W., Goodwin, R.H., Leonard, E.J., 1980, A 48-well micro chemotaxis assembly for rapid and accurate measurement of leukocyte migration. *Immunol. Methods* 33:239-247.

[7] Kadiri, C., Cherqui, G., Masliah, J., Rybkine, T., Etienne, J., Béréziac, G., 1990, Mechanism of N-formyl-methionyl-leucyl-phenylalanine and platelet-activating factor-induced arachidonic acid release in guinea pig alveolar macrophages: Involvement of a GTP-binding protein and role of protein kinase A and protein kinase C. *Mol. Pharmacol.* 38:418-425.

[8] Brom, C., Brom, J., König, W., 1991, G-protein activation and mediator release from human neutrophils and platelets after stimulation with sodium fluoride and receptor-mediated stimuli. *Immunology* 73:287-292.

[9] Kozawa, O., Tokuda, H., Miwa, M., Takahashi, Y., Ozaki, N., Oiso, Y., 1992, Mechanism of prostaglandin E_2-induced arachidonic acid release in osteoblast-like cells: Independence from phosphoinositide hydrolysis. *Prostaglandins Leukot. Essent. Fatty Acids* 46:291-295.

[10] Zigmond, S.H., Hirsch, J.G., 1973, Leukocyte locomotion and chemotaxis. New methods for evaluation, and demonstration of a cell-derived chemotactic factor. *Exp. Med.* 137:387-410.

[11] Samuelsson, B., Dahlen, S.E., Lindgren, J.A., Rouzer, C.A., Serhan, C.N., 1987, Leukotrienes and lipoxins: Structures, biosynthesis, and biological effects. *Science* 237:1171-1176.

[12] Ford-Hutchinson, A., Bray, W.M., Doig, M.V., Shipley, M.E., Smith, M.J.H., 1980, Leukotriene B_4 a potent chemokinetic and aggregating substance released from polymorphonuclear leukocytes. *Nature* 286:264-265.

[13] Gierschik, P., Camps, M., 1994, Stimulation of phospholipase C by G-protein $\beta\gamma$ subunits. *Methods in Enzymology* 238:181-195.

[14] Vondran, A., Tibes, U., Borowski, E., Friebe, W.G., Scheuer, W.V., 1997, New type II phospholipase A_2 inhibitors effective in blocking an ongoing enzyme reaction: Inability of manoalide and scalaradial to exert an inhibitory capacity. In: Phospholipase A_2. Basic and Clinical Aspects in Inflammmatory Diseases. Eds Uhl, W., Nevalainen, T.J., Büchler, M.W., *Progress in Surgery* 24: 130-139. Karger, Basel.

[15] Tibes, U., Scheuer, W.V., Thierse, H.J., Burgermeister, E., Schramm, S., Friebe, W.G., Dietz, E., 1997, Role of cytosolic PLA_2, secretory PLA_2 and nitric oxide synthase in inflammation. In: Phospholipase A_2. Basic and Clinical Aspects in Inflammmatory Diseases. Eds Uhl, W., Nevalainen, T.J., Büchler, M.W., *Progress in Surgery* 24:153-167. Karger, Basel.

[16] Pfeilschifter, J., Schalkwijk, C.G., Briner, V.A., van-den-Bosch, H., 1993, Cytokine-stimulated secretion of group II phospholipase A_2 by rat mesangial cells. Its contribution to arachidonic acid release and prostaglandin synthesis by cultured rat glomerular cells. *J. Clin. Invest.* 92:2516-2523.

[17] Grossman, C.J., Wiseman, J., Lucas, F.S., Trevethick, M.A., Birch, P.J., 1995, Inhibition of constitutive and inducible cyclooxygenase activity in human platelets and mononuclear cells by NSAIDs and Cox 2 inhibitors. *Inflamm. Res.* 44:253-257.

[18] Tibes, U., Vondran, A., Rodewald, E., Friebe, W.G., Schäfer, W., Scheuer, W., 1995, Inhibition of allergic and non-allergic inflammation by phospholipase A_2 inhibitors. Int Arch Allergy Immunol 107:432-434.

30

SUPRESSION OF ACUTE EXPERIMENTAL INFLAMMATION BY ANTISENSE OLIGONUCLEOTIDES TARGETING SECRETORY PHOSPHOLIPASE A₂ (sPLA₂) *IN VITRO* AND *IN VIVO* EXPERIMENTS

Ulrich Tibes[1], Sigrid P. Röhr[1], Werner Scheuer[2], Elke Amandi-Burgermeister[2], Anette Litters[1]

[1,2] Boehringer Mannheim GmbH
[1]Dept. of Preclinical Research
D-68305 Mannheim
[2]Dept. of Molecular Pharmacology
D-82377 Penzberg

SUMMARY

In HepG2 cells phosphorothioate modified antisense oligonucleotides against a sequence in the Ca^{2+} binding domain (AS-Ca^{2+}) of type II sPLA₂ mRNA restrained IL-6-induced synthesis of sPLA₂ protein, sPLA₂ mRNA (northern blot), and abolished IL-6 stimulated PGE₂ release. An antisense oligonucleotide corresponding to a sequence in the catalytic domain (AS-Cat) of sPLA₂ was less effective. The antisense oligonucleotides did not affect albumin synthesis in HepG2 cells, additionally demonstrating their specificity. The corresponding AS-Ca^{2+} against a homologous part of the rat sPLA₂ mRNA depressed rat carrageenin oedema for 60-70%. Identical suppression was achieved by specific low molecular weight inhibitors of sPLA₂. Since cyclo- and 5-lipoxygenase inhibitors exerted similar reductions of carrageenin oedema type II sPLA₂ dependent eicosanoid formation seems to be a key cascade in this type of inflammation.

INTRODUCTION

Among the isoforms of PLA₂s described so far[1] the type II secretory 14 kDa sPLA₂ and the type IV cytosolic 82.5 kDa cPLA₂ are believed to be key enzymes during inflammatory processes[1,5,12-14]. PLA₂s are not only involved in the release of nearly all lipidmediators and eicosanoids but also in various other pro-inflammatory mechanisms like O_2^- generation and cytotoxicity[2,5-7],

Eicosanoids and Other Bioactive Lipids in Cancer, Inflammation, and Radiation Injury, 4
Edited by Honn *et al.*, Kluwer Academic / Plenum Publishers, New York, 1999.

199

apoptosis[2,8,9], expression of cytokines[10-16] as well as many other proteins promoting inflammation, cell development, and proliferation [2,5,16,17,18].

Despite substantial efforts the precise roles of Ca^{2+}-dependent sPLA$_2$ and cPLA$_2$ in arachidonic acid release, and their contributions to different kinds of inflammation is still controversial[1-5,14,19,20]. To gain more insight into the functional role of the 14 kDa sPLA$_2$ in acute inflammation we applied the antisense technique to specifically block the translation of sPLA$_2$ protein. Phosphorothioate modified antisense oligonucleotides (PAOs) were designed for hybridization to specific sequences in the type II sPLA$_2$ mRNA. Since PAOs are rather resistant towards various nucleases[21] they are particular suitable for *in vitro* and *in vivo* experiments[21-23]. PAOs against sPLA$_2$ mRNA have mainly been applied for cellular *in vitro* experiments[22] and only few reports of *in vivo* studies are communicated[23]. In the present experiments we evaluated the *in vitro* activity of PAOs in cells and compared it to *in vivo* effects of acute inflammation, using the rat hind paw carrageenin oedema model. Results were comparable to data obtained with low molecular weight inhibitors of sPLA$_2$.

METHODS

In Vitro Experiments

Oligonucleotides. According to the published sequences[2,24] of the human and rat sPLA$_2$ PAOs against sPLA$_2$ mRNA were constructed. One PAO, designated as AS-Ca^{2+}, was complementary to a sequence in the Ca^{2+}-binding domain[22]. A second one was complementary to a sequence in the catalytic domain, and was named AS-Cat. Corresponding sense oligonucleotides, named sense-Ca^{2+} (S-Ca^{2+}) and sense-cat (S-Cat), were used for control. These oligonucleotide sequences are depicted below.

Oligotype	human sPLA$_2$	rat sPLA$_2$
AS-Ca^{2+}	5'-GATCCTCTGCCACCCACGCC-3'	5'-GATCCTCTGCCACCCACACC-3'
S-Ca^{2+}	5'-GGCGTGGGTGGCAGAGGATC-3'	5'-GGTGTGGGTGGCAGAGGATC-3'
AS-Cat	5'-TTGTAGCAACAGTCATGAGT-3'	not evaluated
S-Cat	5'-ACTCATGACTGTTGCTACAA-3'	not evaluated

In the rat oligonucleotides one guanine was exchanged by an adenine nucleotide according to the published sequences of the enzymes[2,24].

Cell Culture. For cell testing of the oligonucleotides the human hepatocarcinoma cell line HepG2 was used (ATCC, Rockville, M.D). HepG2 cultures were kept in Earle's minimal essential medium (with 10 % heat-inactivated fetal calf serum, 1 % non-essential amino acids, 1 % L-glutamine, 1 % sodium pyruvate, 37°C, 5% CO_2, 95% O_2). Cells were passaged every 3 - 4 days by trypsination (0.05 % trypsin, 0.02 % mass/vol EDTA, pH 7.4, Boehringer Mannheim) according to the manufacturer's instruction.

To induce de novo synthesis of sPLA$_2$ protein cells were plated at a density of 10^6 cells/ml, allowed to adhere overnight, and subjected to incubation with IL-6 (1000 U/ml, Boehringer Mannheim) and oligos for up to 96 h the following day, according to Crowl et al.[25]. 60 min after the addition of IL-6 the oligos were added in a concentration of 5 µM. This concentration was tested in pilot experiments to achieve optimal effects on sPLA$_2$ synthesis without being cytotoxic (trypan blue exclusion test, Boehringer Mannheim).

Analysis of sPLA$_2$ Protein. To determine the effects of IL-6 and oligonucleotides in HepG2 cells the resulting changes in sPLA$_2$ protein were measured by a specific ELISA (Boehringer Mannheim) following the manufacturer's instruction.

Prostaglandin E$_2$ as Indicator of sPLA$_2$ Suppression. To estimate reduction of sPLA$_2$ enzyme, PGE$_2$ release (ELISA kit, Boehringer Mannheim) was measured in the supernatant of HepG2 cell cultures.

Analysis of sPLA$_2$ mRNA. To determine the effects of the oligonucleotides on the sPLA$_2$ mRNA northern blot analysis was performed using an sPLA$_2$-specific probe. Total RNA was extracted from the cells by the acid guanidinium thiocyanate method[26]. The total RNA concentration was determined by densitometry after separation on formaldehyd agarose gel. Fractionated RNA was transferred to a nylon membrane and the sPLA$_2$ mRNA was detected by a human sPLA$_2$ antisense transcript (riboprobe). The transcript was labelled by digoxigenin-UTP (DIG-UTP, kit of Boehringer Mannheim) in correspondence with the manufacturer's instruction.

To produce sufficient amounts of this probe an expression plasmid was established. HepG2 RNA was transcribed (with oligo-dt primers) and from this product a 378 bp type II sPLA$_2$-DNA fragment was amplified by PCR (Perkin Elmer kit; Model 392 DNA/RNA synthesizer, Applied Biosystems Inc., Foster City, CA) using 5' and 3' primers designed from the cDNA sequence of human type II PLA$_2$ (*5`-sPLA$_2$ primer*, for SalI sequence: 5'-GATCTGTCGACGATCAAGTTGACGACAGGAAAGG-3'; *3`-sPLA$_2$ primer*, for T7/EcoRI sequences: 5'-GATCTGAATTCTAATACGACTCACTATAGGGCTCCCTCT-GCAGTGTTTATTGG-3'). The type II sPLA$_2$ specific 378 bp-DNA fragment was cloned into the pCMV-sPLA$_2$ expression plasmid modified from a pCMV-A18 system[27] and using standard cloning procedures[28]. Finally, the riboprobe was obtained by digestion and purification (Quiagen kits) of plasmid DNA.

In vivo experiments

Carrageenin-Induced Rat Hind Paw Oedema. Rat (Sprague-Dawley, 300 g, Charles River) hind paw carrageenin oedema was used to monitor effects of oligonucleotides and test compounds on acute inflammation as previously described[14]. Oedema formation of the right hind paws was measured by water plethysmography over 3 – 6 h following carrageenin injection. The differences between paw volume after carrageenin injection and control data were calculated to represent paw swelling and are contained in tables and figures. Oligos (200 µg in 20 µl sterile NaCl) were injected into the right hind paws, i.e. intraplantar like carrageenin, once daily over 4 days. Control group received either vehicle injections (NaCl) or sense oligos. On the 4th day carrageenin oedema was induced at least one hour after last oligonucleotide injection. Low molecular weight inhibitors were given i.p. 30 - 60 min before carrageenin injection.

sPLA$_2$ Inhibitor. For in vivo comparison the low molecular weight sPLA$_2$ inhibitor BM 16.2266 (6,7-dihydroxy-2-oxo-2H-chromen-4-ylmethylsulfanyl)-acetic acid)[12-14,29] was used.

Statistical Methods

Data are expressed as means ± SD or SE. Significant differences were determined using t-test. A value of $p < 0.05$ was considered significant.

RESULTS AND DISCUSSION

In Vitro Experiments

Induction of sPLA₂ Protein Expression in HepG2 Cells by IL-6. Stimulation of HepG2 cells with human IL-6 (1000 U/ml) over 96 h induced secretion of sPLA₂ protein into the culture medium (Table 1). The release continued to increase over 96 h and did probably not reach its maximum within the observation period. Unstimulated HepG2 cells secreted only minute amounts of sPLA₂ protein, i.e. near detection limit.

Co-incubation of HepG2 cells with IL-6 and 5 μM of the PAOs, starting 60 min after IL-6 addition, restrained the enhanced expression of sPLA₂ protein (Figure 1). The AS-Ca²⁺ inhibited sPLA₂ secretion more effectively (75 % after 96 h) than AS-Cat (50 %, 96 h). Therefore the following experiments were mainly continued with the AS-Ca²⁺. Both sense oligos were completely ineffective.

Table 1. Increase of sPLA₂ protein and PGE₂ in IL-6 stimulated HepG2 culture.

	sPLA₂ pg/ml		PGE₂ pg/ml	
		stimulus		
time after starting stimulation	no IL-6	with IL-6, 1000 U/ml	no IL-6	with IL-6, 1000 U/ml
24 h	1	3	210	260
48 h	2	3	260	480
72 h	3	28	265	410
96 h	3	41	265	440

means of n = 2, duplicate determinations.

Antisense Oligonucleotides Inhibited IL-6 Induced PGE₂ Synthesis. To explore the consequences of sPLA₂ suppression on the arachidonic acid cascade, PGE₂ synthesis was measured in the culture medium as an indicator. Table 1 demonstrates that within 48 - 96 h the IL-6 stimulation caused an increase of PGE₂ release. By co-incubation with AS-Ca²⁺ the IL-6-induced PGE₂ release was completely depressed below the control level (Figure 1, right). This AS-Ca²⁺-dependent PGE₂ component was likely related to the type II sPLA₂ activity, whereas the remaining release may be allocated to the activities of some other PLA₂s. Again the control sense oligonucleotide was completely ineffective.

No Effect of Oligonucleotides on Albumin Synthesis. HepG2 cells were stimulated with IL-6 and the albumin secretion was measured (ELISA, Boehringer Mannheim). Table 2 demonstrates that AS-Ca²⁺ and S-Ca²⁺ did not significantly change albumin synthesis.

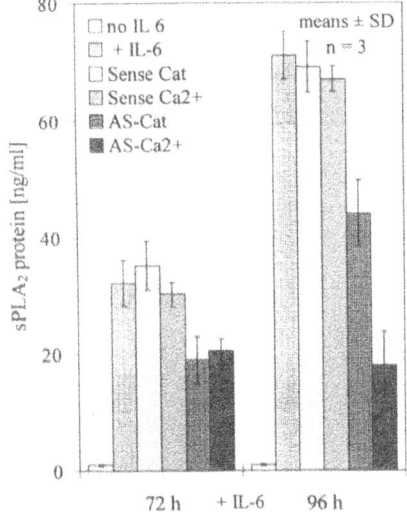 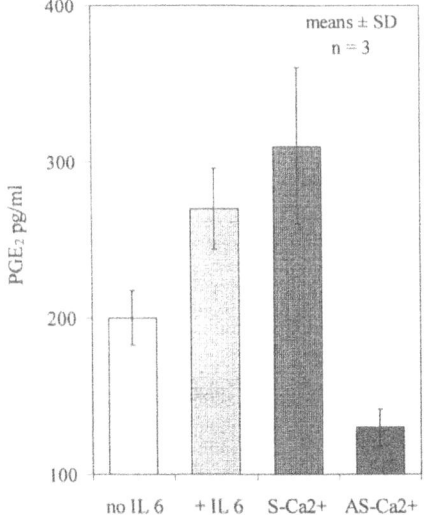

Figure 1. Inhibition of IL-6 stimulated sPLA$_2$ (left) and PGE$_2$ (right, after 48 h) synthesis in HepG2 cells by antisense oligonucleotides versus sPLA$_2$ mRNA. AS-Ca^{2+} = antisense oligo vs Ca^{2+} binding domain, AS-Cat = antisense vs catalytic domain.

Table 2. Effect of oligonucleotides on albumin synthesis in IL-6-stimulated HepG2 cells.

time after stimulation	control	+ IL-6	+ IL-6 + AS-Ca^{2+}	+ IL-6 + S-Ca^{2+}
24 h	5396	4866	4435	5002
48 h	5381	5830	5998	6669
72 h	8351	7925	8286	8238

Antisense (AS-Ca^{2+}) and sense (S-Ca^{2+}) oligos did not change albumin (ng/ml) synthesis. Means of n = 2.

Suppression of sPLA$_2$ mRNA by Antisense Oligonucleotides. To demonstrate the specificity of antisense blockade on mRNA level the northern blot analyses of sPLA$_2$-mRNA are shown in Figures 2 and 3. 3 or 5 μg of total RNA were transferred for separation on the gels. The mRNA of sPLA$_2$ is concentrated at 1.4 kb. Without IL-6 stimulation no sPLA$_2$ mRNA was detectable neither with 3 μg nor 5 μg total RNA (Fig. 2, part A). After IL-6 stimulation intense bands were seen, with 5 μg more pronounced than with 3 μg. To exclude a cross reactivity of the probe with the 18S- and 28S ribosomal RNA, the poly(A)$^+$ mRNA is shown in part B of Figure 2. Only mRNA of sPLA$_2$ were detectable at 1.4 kb with the probe and no 18S- and 28S ribosomal RNAs. IL-6 stimulation caused similar changes of poly(A)$^+$ mRNA like of total RNA.

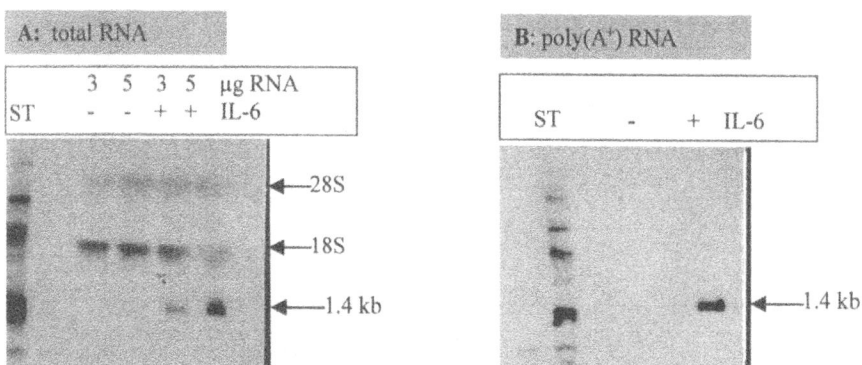

Figure 2. Northern blot of sPLA$_2$. After IL-6 stimulation (+) of HepG2 cells sPLA$_2$ mRNA increased. Without (-) IL-6 no sPLA$_2$ mRNA was detectable. ST = standard RNAs.

Figure 3 demonstrates the effects of PAOs and sense oligos on the appearance of sPLA$_2$ mRNA. Again, without IL-6 stimulation no sPLA$_2$ mRNA was detectable, with IL-6 stimulation the bands were well visible. After co-incubation of IL-6 stimulated cells with AS-Ca^{2+} or AS-Cat the bands of sPLA$_2$ mRNA were completely abolished (part A). In part B failure of the sense oligos to depresss sPLA$_2$ mRNA is shown. Since depression of mRNA by PAOs took place in a period of roughly 48 h, but sPLA$_2$ protein secretion was still present at 76 - 96 h (Figure 1), it may be suggested that secretion of pre-existing vesicular sPLA$_2$ was occurring beyond 48 h.

Antisense Oligonucleotides Did not Inhibit sPLA$_2$ Enzyme In Vitro. In an sPLA$_2$ enzyme assay as previously described[12-14,29,31] the antisense and sense oligonucleotides (5 μM) were no effective inhibitors of the enzyme when activity was monitored over 20 - 40 min. These negative results support the effects of the oligonucleotides on a mRNA level.

Conclusion from Cell Experiments. The data shown so far demonstrate that the used PAOs specifically suppressed type II sPLA$_2$ protein synthesis. However, some time ranging between 48 - 96 h is needed to completely abolish secretion of sPLA$_2$.

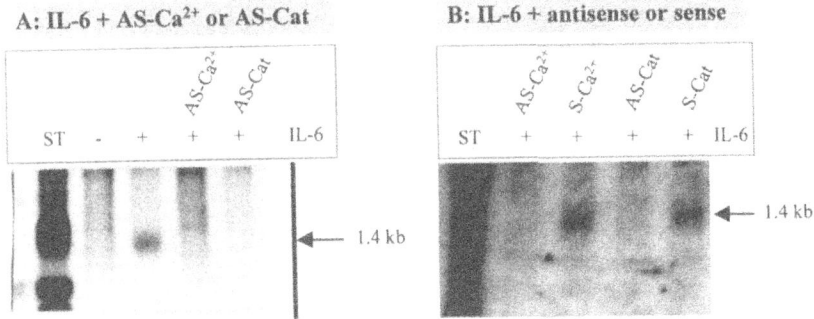

A: IL-6 + AS-Ca²⁺ or AS-Cat

B: IL-6 + antisense or sense

Figure 3: Suppression of sPLA₂ mRNA by antisense oligonucleotides in HepG2 cells (Nothern blot). Cells were stimulated by IL-6 (+) and co-incubated with 5 μM antisense (part A) or sense (part B) oligonucleotides. 48 h later 5 μg of total RNA were analysed on the gels. ST = standard RNAs.

Animal Experiments

4 Days Pretreatment with AS-Ca²⁺ is Needed to Inhibit Carrageenin Oedema. For hybridisation with the sPLA₂ mRNA of rats a PAO complementary to the homologous part of the Ca²⁺ binding domain and a corresponding sense oligonucleotide were synthesized. After application into the right hind paw of 200 μg (o.i.d. over 4 days) of these oligonucleotides dissolved in 20 μl NaCl carrageenin oedema was induced. Results are depicted in Figure 4. After application over 4 days neither antisense nor sense oligonucleotides caused a significant inflammatory reaction in the paws (at time 0 h). Carrageenin injection caused a great acute inflammatory oedema, being nearly the same in NaCl pretreated, un-pretreated (control) or sense treated paws. Pre-treatment with the AS-Ca²⁺ reduced oedema formation significantly for about 60 -7 0 %.

A single application of the AS-Ca²⁺ oligo 24 h before carrageenin injection inhibited oedema formation for only about 22 – 39 % as shown in the right part of Figure 4. Thus, to obtain a full response multiple applications over some days were necessary, in accordance with the cell experiments.

Comparison of Antisense Oligonucleotides with Low Molecular Weight sPLA₂ Inhibitors. The inhibitory activity of the antisense oligonucleotides was compared to that of the low molecular weight sPLA₂ inhibitor BM 16.2266 which had been described previously[12-14]. This specific sPLA₂ inhibitor had no inhibitory activities against cPLA₂, COX or 5-LOX[14]. It depressed carrageenin oedema for about 50-70%, corresponding to the effects of the oligonucleotides, supporting the effectiveness of the antisense oligonucleotide.

Since neutralisation of PGE₂ by a specific antibody inhibited carrageenin oedema to a similar extent like indomethacin[30] it was assumed that mainly the prostaglandin production from the released arachidonic acid was responsible for the acute inflammation. However, when comparing the depression of oedema by indomethacin and ketokonazole, a 5-lipoxygenase inhibitor[31], surprisingly the separate and combined inhibition of prostaglandin and leukotriene cascade induced similar suppression of oedema formation (Table 3). Thus, both cascades contribute in a redundant fashion to this type of acute inflammation. However, since sPLA₂ inhibition surpassed the blockade of cyclo- and 5-lipoxygenase somewhat it may constitute a superior pharmacological principle.

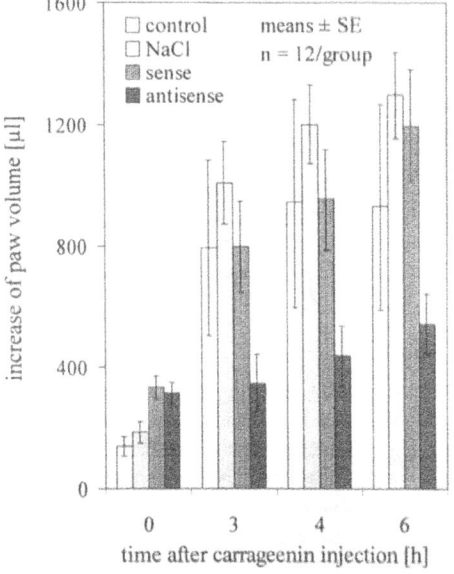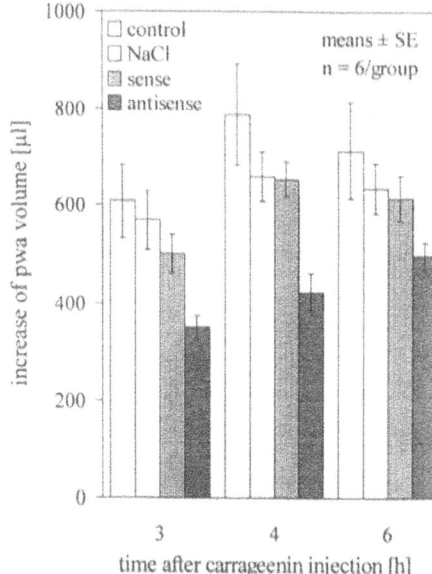

Figure 4. Left part: Inhibition of rat carrageenin oedema by antisense oligonucleotide versus sPLA₂ mRNA (AS-Ca²⁺) after multiple (o.i.d for 4 days) intraplantar application of 200 μg AS-Ca²⁺ prior to carrageenin challenge (3h: 2p<0.002; 4h: 2p<0.0006; 6h: 2p<0.001). Right part: Weaker inhibition of oedema after single application 24 h prior to challenge (3h: 2p<0.006; 4h: 2p<0.002; 6h: 2p<0.04). Oedema formation was measured over 6 h.

Table 3. Inhibition of rat carrageenin oedema by indomethacin, ketokonazole and a combination of both.

treatment of rats	indomethacin 2.5 mg/kg p.o.	ketokonazole 20 mg/kg p.o.	indo 2.5mg p.o.+ keto 20mg p.o.
% medium inhibition over 6h	52	58	38

n = 6 rats per group

ACKNOWLEDGMENT

We thank Dr. H. J. Müller for generously providing the expression plasmid pCMV-A18 and many helpful discussions. We thank Dr. H. Stockinger for the sPLA₂ ELISA, Dr. W.G. Friebe for synthesis of BM 16.2266, and W. Kinle, S. Kippenhan, H. Kern, G. Müller, and I. Schulz for excellent technical assistance.

REFERENCES

1. Dennis, E.A., 1994, Diversity of group types, regulation and function of phospholipase A₂. *J. Biol. Chem.* 269:13057-13060.
2. Murakami, M., Nakatani, Y., Atsumi, G., Inoue, K., and Kudo, I., 1997, Regulatory functions of phospholipase A₂. *Crit. Rev. Immunol.* 17:225-283.
3. Bonventre, J.V., Huang, Z., Taheri, M.R., O'Leary, E., Li, E., Moskowitz, M.A., Sapirstein, A., 1997, Reduced fertility and postischaemic brain injury in mice deficient in cytosolic phospholipase A₂, *Nature* 390:622-625.
4. Uozumi, N., Kume, K., Nagase, T., Nakatani, N., Ishii, S., Tashiro, F., Komagata, Y., Maki, K., Ikuta, K., Ouchi, Y., Miyazaki, J., Shimizu, T., 1997, Role of cytosolic phospholipase A₂ in allergic response and parturition. *Nature* 390:618-622.
5. Tibes, U., and Friebe, W.G., 1997, Phospholipase A₂ inhibitors in development. *Exp. Opin. Invest. Drugs* 6:279-298.
6. Sapirstein, A., Spech, R., Witzgall, R., Bonventre, J.V., 1996, Cytosolic phospholipase A₂ (PLA₂), but not secretory PLA₂, potentiates hydrogen peroxide cytotoxicity in kidney epithelial cells. *J. Biol. Chem.* 271:21505-21513.
7. Hayakawa, M., Ishida, N., Takeuchi, K., Shibamoto, S., Hori, T., Oku, N., Ito, F., Tsujimoto, M., 1993, Arachidonic acid-selective cytosolic phospholipase A₂ is crucial in the cytotoxic action of tumor necrosis factor *J. Biol. Chem.* 268:11290-11295.
8. Wissing, D., Mouritzen, H., Egeblad, M., Poirier, G.G., Jäätelä, M., 1997, Involvement of caspase-dependent activation of cytosolic phospholipase A₂ in tumor necrosis factor-induced apoptosis. *Proc. Nat. Acad. Sci. USA,* 94:5073-5077.
9. Macewan, D.J., 1996, Elevated cPLA₂ levels as a mechanism by which the p70 TNF and p75 NGF receptors enhance apoptosis. *FEBS-Letters* 379:77-81.
10. Rola-Pleszczynski, M., Stankova, J., 1992, Leukotriene B₄ enhances interleukin-6 (IL-6) production and IL-6 messenger RNA accumulation in human monocytes in vitro: transcriptional and posttranscriptional mechanisms. *Blood* 80:1004-1011.
11. Poubelle, P.E., Gingras, D., Demers, C., Dubois, C., Harbour, D., Grassi, J., Rola-Pleszczynski, M., 1991, Platelet-activating factor (PAF-acether) enhances the concomittant production of tumor necrosis factor α and interleukin 1 production by subsets of human monocytes. *Immunol.* 72:181-187.
12. Tibes, U., Vondran, A., Rodewald, E., Friebe, W.G., Schäfer, W., Scheuer, W., 1995, Inhibition of allergic and non-allergic inflammation by phospholipase A₂ inhibitors. *Int. Arch. Allergy. Immunol.* 107:432-434.
13. Amandi-Burgermeister, E., Tibes, U., Kaiser, B.M., Friebe, W.G., Scheuer, W.V., 1997, Suppression of cytokine synthesis, integrin expression and chronic inflammation by inhibitors of cytosolic phospholipase A₂. *Europ. J. Pharmacol.* 326:237-250.
14. Tibes, U., Scheuer, W.V., Thierse, H.J., Burgermeister, E., Schramm, S., Friebe, W.G., Dietz, E., 1997, Role of cytosolic PLA₂, secretory PLA₂ and nitric oxide synthase in inflammation. In: Phospholipase A₂. Basic and Clinical Aspects in Inflammmatory Diseases. Eds Uhl, W., Nevalainen, T.J., Büchler, M.W., *Progress in Surgery* 24: 153-167. Karger, Basel.
15. Baldie, G.D., Kaimakarnis, D., Rotondo, D., 1993, Fatty acid modulation of cytokine release from human monocytic cells. *Biochim. Biophys. Acta.* 1179:125-133.
16. Los, M., Baeuerle, P.A., 1995, IL-2 gene expression and NF-κB activation through CD28 requires reactive oxygen production by 15-lipoxygenase. *EMBO J.* 14:3731-3740.
17. Rao, G.N., Lassegue, B., Griendling, K.K., Alexander, R.W., Berk, B.C., 1993, Hydrogen peroxide-induced c-fos expression is mediated by arachidonic acid release. Role of protein kinase C. *Nucleic Acids Res.* 21:1259-1263.
18. Anderson, K.M., Roshak, A., Winkler, J.D., McCord, M., Marshall, L.A., 1997, Cytosolic 85-kDa phospholipase A₂-mediated release of arachidonic acid is critical for proliferation of vascular smooth muscle cells. *J-Bio. Chem.* 272:30504-30511.
19. Marshall, L.L., Bolognese, B., Winkler, J.D., Roshak A., 1997, Depletion of human monocyte 85-kDa phospholipase A₂ does not alter leukotriene formation. *J. Biol. Chem.* 272:759-765.

31

COMPARISON OF RECOMBINANT TYPES IIA, V AND IIC PHOSPHOLIPASE A2S, THE THREE RELATED MAMMALIAN SECRETORY PHOSPHOLIPASE A2 ISOZYMES

Satoko Shimbara, Makoto Murakami,
Terumi Kambe, and Ichiro Kudo

Department of Health Chemistry
School of Pharmaceutical Sciences
Showa University
1-5-8 Hatanodai, Shinagawa-ku
Tokyo 142, Japan

INTRODUCTION

Secretory phospholipase A2s (sPLA2s) are low molecular weight (~14 kDa) enzymes with a rigid tertiary structure configured by 6-8 disulfide bridges. In mammals, five sPLA2 enzymes have been identified so far.[1] Type I sPLA2 (sPLA2-I) is abundantly expressed in the pancreas, where it functions as a digestive enzyme for dietary phospholipids, and is present in relatively small amounts in several non-digestive organs, where it may act as a regulator of cellular functions via the M-type sPLA2 receptor.[2] Type IIA sPLA2 (sPLA2-IIA; often referred to as sPLA2 in the literature), originally isolated from inflammatory fluids and cells, is induced by proinflammatory stimuli in many if not all cells, and is therefore thought to play a role in inflammatory responses.[3] Type V sPLA2 (sPLA2-V) is expressed in the heart and to a lesser extent in several other tissues, and is the primary sPLA2 expressed in several murine inflammatory cells.[4] Type IIC sPLA2 (sPLA2-IIC) is expressed in rodent testes, but is a non-functional pseudogene in humans.[5] Type X sPLA2 (sPLA2-X), the most recently discovered sPLA2 isozyme, exhibits some features characteristic of both sPLA2s-I and -IIA, and is expressed mainly in immune tissues.[6] A phylogenetic tree derived by aligning the sequences of these sPLA2s reveals that three related isozymes (sPLA2s-IIA, -V and -IIC), the genes for which are tightly linked to human chromosome 1, have emerged from recent gene duplication events, whereas sPLA2s-I and -X, the genes for which map to human chromosomes 12 and 16, respectively, are more distant offshoots.[1,6] Here we

Eicosanoids and Other Bioactive Lipids in Cancer, Inflammation, and Radiation Injury, 4
Edited by Honn *et al.*, Kluwer Academic / Plenum Publishers, New York, 1999.

209

show the enzymatic and functional properties of the three closely related sPLA2s, i.e. sPLA2s-IIA, -V and -IIC, which were overexpressed in human embryonic kidney 293 cells and Sf9 insect cells.

RECOMBINANT EXPRESSION OF SPLA2S

To obtain recombinant sPLA2 isozymes in substantial amounts, the cDNAs of rat, mouse and human sPLA2-IIA, human sPLA2-V, and rat sPLA2-IIC were each inserted into the baculovirus vector pVL1392 or pVL1393 (Pharmingen) in the correct direction. The recombinant plasmids were co-transfected with BaculoGold linearized baculovirus DNA (Pharmingen) into 3 x 10^6 Sf9 cells using calcium phosphate. The cells were cultured at 27°C in Grace's insect medium (Invitrogen) supplemented with lactalbumin hydrolysate, yeastolate and 10% fetal calf serum. The recombinant virus was amplified every 7 days. After the third amplification, the supernatants of the cells transfected with sPLA2-IIA and sPLA2-IIC cDNA contained significant amounts of the respective enzymes, whereas no detectable PLA2 activity was found in cells transfected with sPLA2-V cDNA.

In an attempt to express these sPLA2s in mammalian cells, the cDNAs for mouse sPLA2-IIA, rat sPLA2-V, and rat sPLA2-IIC were subcloned into the mammalian expression vector pCDNA3.1 (Invitrogen), and transfected into human embryonic kidney 293 cells using a lipofection method. To establish stable transfectants, cells transfected with each cDNA were cloned by limiting dilution in 96-well plates in culture medium supplemented with G418. After culture for 2-4 weeks, a number of G418-resistant colonies of sPLA2-IIA or sPLA2-IIC transfectants were formed, whereas most of the cells were dead within 1 week after sPLA2-V transfection. When 293 cells were transfected with rat sPLA2-V cDNA subcloned into pCEP4[4] and selected in the presence of hygromycin, a few colones were grown in 96-well plates. These transfectants were expanded and used in subsequent studies. The low frequency of sPLA2-V transfectants in both insect cells and 293 cells indicates that it may be cytotoxic, but further study is necessay to prove it.

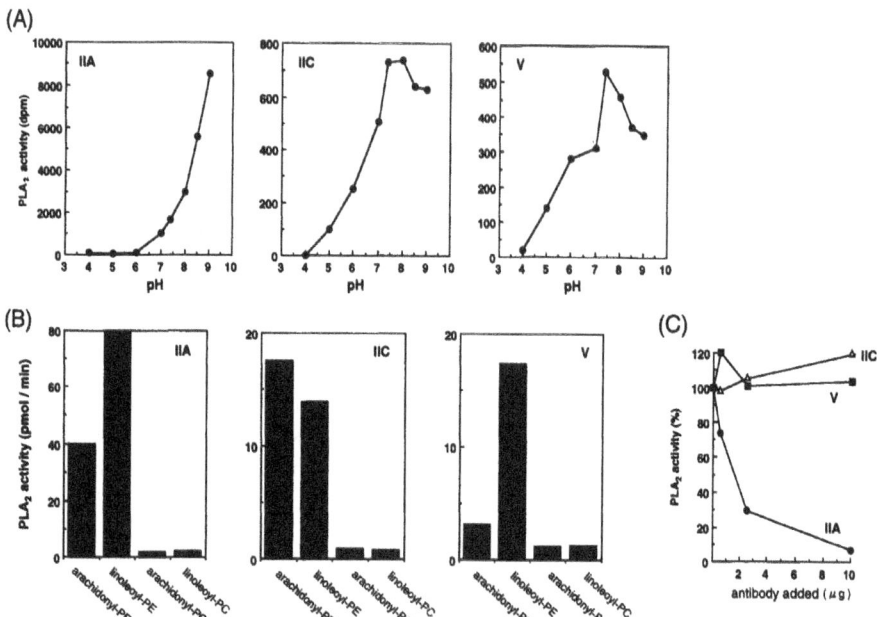

Figure 1. Enzymatic properties of recombinant sPLA2s-IIA, -IIC and -V. (A) pH-dependence using arachidonoyl-PE as the substrate. (B) Substrate specificity assessed at their optimal pH. (C) Effect of the anti-sPLA2-IIA antibody R377 on sPLA2s-V and -IIC activity, using arachidonoyl-PE as the substrate at pH 7.4.

210

ENZYMATIC PROPERTIES

In our PLA_2 assay, which comprizes 100 mM Tris-HCl, 2 mM substrate and 10 mM $CaCl_2$, recombinant $sPLA_2s$-IIA, -V and -IIC all exhibited maximal enzymatic activities in the presence of millimolar Ca^{2+} concentrations, were active at neutral to alkaline pHs (Figure 1A), and hydrolyzed phosphatidylethanolamine (PE) several times more efficiently than phosphatidylcholine (PC) (Figure 1B). $sPLA_2s$-IIA and -IIC showed no apparent fatty acid preference, while $sPLA_2$-V hydrolyzed linoleoty-PE several times more efficiently than arachidonoyl-PE. $sPLA_2$-IIA consistently showed activity several-fold greater than those of $sPLA_2s$-IIC and -V when PE was used as the substrate at pH 7.4. However, these differences in enzymatic characteristics will need more detailed evaluation in a future study using purified enzymes under various assay conditions. Indeed, in the presence of deoxycholate, $sPLA_2$-V hydrolyzes PC more efficiently than PE[4]. The properties of baculovirus-derived recombinant $sPLA_2$-IIA and $sPLA_2$-IIC were essentially identical to those of 293-derived recombinant enzymes.

The $sPLA_2$ inhibitors thielocin A1β[7] and LY311727[8] were fairly selective for $sPLA_2$-IIA, although at higher concentrations they inhibited $sPLA_2$-V as well (Figure 1C). Both p-bromophenacyl bromide, which covalently modifies the catalytic center (His48) of $sPLA_2s$, and dithiothreitol, which disrupts disulfide bridges, strongly inactivated these $sPLA_2s$.

AFFINITY FOR HEPARIN

It is well established that $sPLA_2$-IIA shows high affinity for heparin.[3] Taking advantage of this characteristic, baculovirus-derived recombinant rat and mouse $sPLA_2$-IIAs were each purified to near homogeneity using heparin-Sepharose or sulfate-Cellulofine column, followed by an anti-rat $sPLA_2$-IIA antibody immunoaffinity column (starting from 50 ml of the culture supernatants, approximately 100 mg of recombinant rat and mouse $sPLA_2$-IIAs were obtained). Site-directed mutagenesis revealed that the heparin-binding domain of $sPLA_2$-IIA is located in the C-terminal Lys-rich region[9]. Disruption of several Lys residues in this domain markedly reduced the cell surface association of $sPLA_2$-IIA, indicating that heparan sulfate proteoglycan is the primary acceptor site of $sPLA_2$-IIA on the cell surface. This domain is not necessary for the catalytic activity of $sPLA_2$-IIA toward dispersed phospholipid, but is critical for its ability to enhance stimulus-initiated arachidonic acid release from activated cells, as noted below [9, 10]

When the culture supernatants of the 293 transformants expressing each $sPLA_2$ were applied to the heparin-Sepharose column to compare their heparin-binding capacities, not only $sPLA_2$-IIA but also $sPLA_2$-V showed significant affinity for heparin, being eluted with 0.8 M and 0.4 M NaCl, respectively, whereas most of the $sPLA_2$-IIC was recovered from the flow-through fractions with only a minor (<10%) fraction being eluted with 0.3 M NaCl.[10] Like $sPLA_2$-IIA, $sPLA_2$-V exhibited significant affinity for the sulfate-cellulofine column (Figure 2A) and had the capacity to associate with the cell surface.[10] In contrast, no appreciable binding of $sPLA_2$-IIC to the cell surface was observed, which is consistent with its failure to bind to heparin.

A sequence alignment of the C-termini of the three related $sPLA_2s$ revealed that although none of the Lys residues crucial for $sPLA_2$-IIA binding to heparin are present in $sPLA_2$-V, there is an alternative cluster of cationic residues in the C-terminus of $sPLA_2$-V (Figure 3A), which is located in close proximity to a cationic region of $sPLA_2$-IIA.[9,10] When these residues were replaced by non-charged or negatively charged residues, $sPLA_2$-V became incapable of binding to cell surface proteoglycan (Figure 3B).[10]

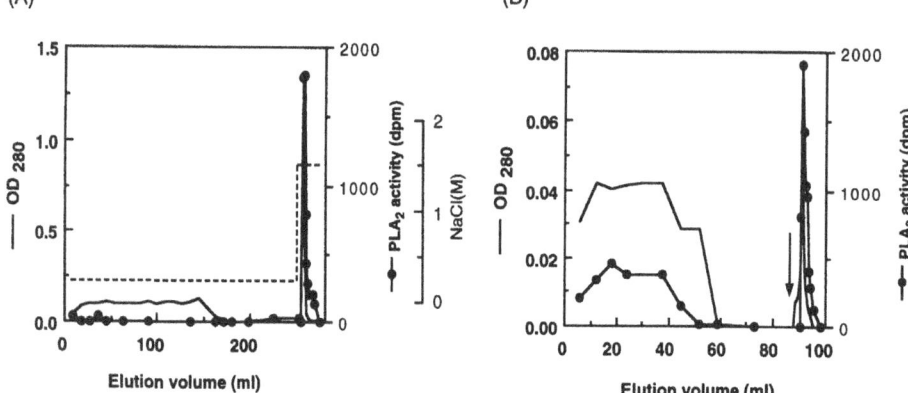

Figure 2. Chromatographic profiles of sPLA2-V. (A) Supernatants of 293 cells overexpressing rat sPLA2-V were applied to a sulfate-cellulofine column. The bound protein was eluted with 10 mM Tris-HCl (pH 7.4) containing 1 M NaCl. (B) Replicate samples were applied to an anti-rat sPLA2-IIA monoclonal antibody-conjugated cellulofine column. The bound protein was eluted with 20 mM glycine-HCl (pH 2.2).

Although the cluster of basic amino acids found in sPLA2-IIA is not present in sPLA2-IIC, two of the four clustered basic amino acids present in sPLA2-V are conserved in it. We therefore asked if the introduction of cationic residues into the corresponding positions would alter the capacity of sPLA2-IIC to associate with the cell surface (Figure 3A). Although one amino acid replacement from Leu95 to Arg or from Glu102 to Lys did not alter the inability of sPLA2-IIC to bind to the cell surface, the simultaneous replacement of these two residues significantly increased its cell binding capability (Figure 3B). These results again imply that the cluster of conserved cationic residues in the C-terminal domain is crucial for the ability of sPLA2s to associate with cell surface proteoglycan. However, this double mutation in sPLA2-IIC resulted in approximately 60% loss of its catalytic function.

IMMUNOCHEMICAL CHARACTERISTICS

It has been reported that some antibodies raised against sPLA2-IIA show significant crossreactivity with sPLA2-V.[11,12] We therefore examined whether several anti-sPLA2-IIA antibodies we had established previously could crossreact with recombinant sPLA2-V and sPLA2-IIC. The rabbit polyclonal anti-rat sPLA2-IIA antibody R377 did not precipitate rat sPLA2-V or sPLA2-IIC appreciably at concentrations sufficient to completely neutralize rat sPLA2-IIA (Figure 1C).[13] This strict specificity to the IIA isozyme reconfirmed that prostanoid biosynthesis by several cell types, which was significantly suppressed by this antibody, is indeed mediated by sPLA2-IIA.[13,14] On the other hand, the monoclonal antibody raised agaist rat sPLA2-IIA, MD7.1,[15] showed significant crossreactivity with sPLA2-V. As shown in Figure 2B, sPLA2-V expressed by 293 transfectants was absorbed to the MD7.1-immunoaffinity column and eluted with an acidic buffer. Another rabbit polyclonal antibody raised against mouse sPLA2-IIA also appeared to crossreact with sPLA2-V; a sPLA2-like activity in C57BL/6J mouse-derived cultured mast cells, in which sPLA2-V, but not sPLA2s-IIA and -IIC, expression was detectable by RT-PCR, was absorbed to an immunoaffinity column conjugated with this antibody.[16]

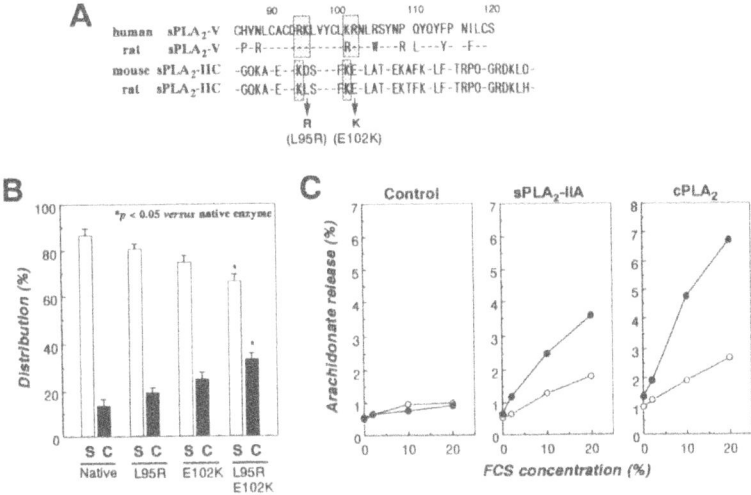

Figure 3. (A) The C-terminal heparin binding domain of sPLA₂-V and its alignment with sPLA₂-IIC. The conserved cationic residues (*boxed*) are crucial for the cell surface association of sPLA₂-V. The amino acid number shown is based on the comparison with the sequence of sPLA₂-I. (B) The percentage distribution of native and mutated sPLA₂-IIC in the culture supernatant (S) and cell membrane-associated (M) fractions of transfected 293 cells, assessed by their enzymatic activities. (C) Release of arachidonic acid from 293 cells. [³H]Arachidonic acid-labeled cells transfected with sPLA₂-IIA, cPLA₂ or control vector were culture for 4 h with (*closed circles*) or without (*open circles*) 1 ng/ml human IL-1β in the presence of the indicated concentrations of FCS.

FUNCTIONS IN ARACHIDONIC ACID METABOLISM IN CELLS

Despite many efforts to prove the involvement of sPLA₂ isozymes, especially sPLA₂-IIA, in arachidonic acid metabolism, confirmation has not been possible until recently because of conflicting results from different experimental systems. Nevertheless, our recent studies have provided convincing evidence that sPLA₂-IIA can promote arachidonic acid metabolism under certain conditions, especially when "membrane rearrangement" is induced by particular stimuli such as the proinflammatory cytokines.[13,14] In addition, sPLA₂-V has recently been shown to act as an effector of arachidonic acid metabolism in mouse macrophages and mast cells.[11,12]

To confirm the biological roles of these sPLA₂s, we assessed arachidonic acid release by 293 cells transfected with each sPLA₂ isozyme. Treating sPLA₂-IIA transfectants, but not control cells, with IL-1 led to significant enhancement of arachidonic acid release in the presence of incremental concentrations of fetal calf serum (FCS) (Figure 3C).[9,10] sPLA₂-V exhibited a similar effect to sPLA₂-IIA, whereas sPLA₂-IIC, which was unable to bind to the cell surface, had no effect.[10] Moreover, various sPLA₂-IIA or sPLA₂-V mutants, in which residues crucial for catalytic activity or for proteoglycan- binding capacity were disrupted, failed to augment IL-1- and FCI-induced arachidonic acid release. These results imply that sPLA₂-IIA and sPLA₂-V, but not sPLA₂-IIC are the 'signaling sPLA₂s', which play an augmentative role in proinflammatory stimulus-initiated, but not basal, release of arachidonic acid, and that both the catalytic and cell surface proteoglycan-binding domains are essential for them to act properly on live cells.

Acknowledgments

We thank Drs. J.A. Tischfield, T. Kamimura, T. Yoshida and R.M. Kramer for providing rat and human sPLA$_2$-V cDNAs, thielocine A1β and LY311727, respectively.

REFERENCES

1. J.A. Tischfield. A reassessment of the low molecular weight phopholipase A$_2$ gene family in mammals. *J. Biol. Chem.* 272: 17247 (1997)
2. K. Hanasaki, and H. Arita. Characterization of a high affinity binding site for pancreatic-type phospholipase A$_2$ in the rat: its cellular and tissue distribution. *J. Biol. Chem.* 267: 6414 (1992)
3. M. Murakami, Y. Nakatani, G. Atsumi, K. Inoue, and I. Kudo. Regulatory functions of phospholipase A$_2$. *Crit. Rev. Immunol.* 17:225 (1997)
4. J. Chen, S.J. Engle, J.J. Seilhamer, and J.A. Tischfield. Cloning and recombinant expression of a novel human low molecular weight Ca^{2+}-dependent phospholipase A$_2$. *J. Biol. Chem.* 269: 2365 (1994)
5. J. Chen, S.J. Engle, J.J. Seilhamer, and J.A. Tischfield. Cloning and characterization of a novel rat and mouse low molecular weight Ca^{2+}-dependent phospholipase A$_2$ containing 16 cysteines. *J. Biol. Chem.* 269: 23018 (1994)
6. L. Cupillard, K. Koumanov, M.-G. Mattei, M. Lazdunski, and G. Lambeau. Cloning, chromosomal mapping, and express ion of a novel human secretory phospholipase A$_2$. *J. Biol. Chem.* 272: 15745 (1997)
7. K. Tanaka, T. Kato, K. Matsumoto, and T. Yoshia. Antiinflammatory action of theilocin A1β, a group II phospholipase A$_2$ specific inhibitor, in rat carrageenan-induced pleurisy. *Inflammation* 17:107 (1993)
8. R.W. Schevitz, N.J. Bach, D.G. Carlson, N.Y. Chirgadze, D.K. Clawson, R.D. Dillard, S.E. Draheim, L.W. Hartley, N.D. Jones, E.D. Mihelich, J.L. Olkowski, D.W. Snyder, C. Sommers, and J.-P. Wery. Structure-based design of the first potent and selective inhibitor of human non-pancreatic secretory phopholipase A$_2$. *Nature Stru. Biol.* 2:458 (1995)
9. M. Murakami, Y. Nakatani, and I. Kudo. Type II secretory phospholipase A$_2$ associated with cell surfaces via C-terminal heparin-binding lysine residues augments stimulus-initiated delayed prostaglandin generation. *J. Biol. Chem.* 271: 30041 (1996)
10. M. Murakami, S. Shimbara, T. Kambe, H. Kuwata, M.V. Winstead, J.A. Tischfield, and I. Kudo. The functions of five distinct mammalian phospholipase A$_2$s in regulating arachidonic acid release: type IIA and type V secretory phospholipase A$_2$s are functionally redundant and act in concert with cytosolic phospholipase A$_2$. *J. Biol. Chem.* in press (1998)
11. M. Balboa, J. Salsinde, M.V. Winstead, J.A. Tischfield, and E.A. Dennis. Novel group V phospholipase A$_2$ involved in arachidonic acid mobilization in murine P388D$_1$ macrophages. *J. Biol. Chem.* 271: 32381 (1996)
12. S.T. Reddy, M.V. Winstead, J.A. Tischfield, and H.R. Herschman. Analysis of the secretory phospholipase A$_2$ that mediates prostaglandin production in mast cells. *J. Biol. Chem.* 272: 13591 (1997)
13. H. Naraba, M. Murakami, H. Matsumoto, S. Shimbara, A. Euno, I. Kudo and S. Oh-ishi. Segregated coupling of phospholipases A$_2$, cyclooxygenases, and terminal prostanoid synthases in different phases of prostanoid biosynthesis in rat peritoneal macrophages. *J. Immunol.* 160:2974 (1998)
14. H. Kuwata, Y. Nakatani, M. Murakami, and I. Kudo. Cytosolic phospholipase A$_2$ is required for cytokine-induced expression of type IIA secretory phospholipase A$_2$ that mediated optimal cyclooxygenase-2-dependent delayed prostaglandin E$_2$ generation in rat 3Y1 fibroblasts. *J. Biol. Chem.* 273: 1733 (1998)
15. M. Murakami, T. Kobayashi, M. Umeda, I. Kudo, I., and K. Inoue. Monoclonal antibodies aganist rat platelet phospholipase A$_2$. *J. Biochem. (Tokyo)* 104: 884 (1988)
16. M. Murakami, K. Tada, S. Shimbara, T. Kambe, H. Sawada, and I. Kudo. Detection of secretory phospholipase A$_2$s related but not identical to type IIA isozyme in cultured mast cells. *FEBS Lett.* 413: 249 (1997)

32

RESPECTIVE ROLES OF THE 14 KDA AND 85 KDA PHOSPHOLIPASE A2 ENZYMES IN HUMAN MONOCYTE EICOSANOID FORMATION

Lisa A. Marshall[1,2], Brian Bolognese[1] and Amy Roshak[1]

[1]Department of Immunopharmacology
SmithKline Beecham Pharmaceuticals
709 Swedeland Road
King of Prussia, PA 19406
[2]Corresponding Author
Tel. (610) 270-6746
Fax. (610) 270-5381
Email: Lisa_A_Marshall@SBPHRD.com

SUMMARY

Human monocytes possess both the cytosolic 85 kDa phospholipase (PLA) A_2 and a 14 kDa PLA_2 and are capable of simultaneously producing prostanoids (PG), leukotrienes (LT) and platelet activating factor (PAF). As the exact roles of the two enzymes in monocyte lipid mediator formation was unclear, both selective PLA_2 inhibitors and antisense were used to elucidate their respective roles. Reduction in 85 kDa PLA_2 cellular protein levels by initiation site-directed antisense (SK 7111) or exposure to the 85 kDa PLA_2 inhibitor, arachidonyl trifluormethyl ketone (AACOCF3), prevented A23187 or zymosan-stimulated monocytes prostanoid formation but not LTC_4 or PAF production. This confirmed the important role of the 85 kDa PLA_2 in prostanoid formation but indicated a less significant role in LT or PAF biosynthesis. Alternatively, treatment of monocytes with the selective, active-site-directed 14 kDa PLA_2 inhibitor, SB 203347, totally inhibited LT and PAF formation, while prostanoid formation was not altered. Addition of 20 uM exogenous arachidonic acid (AA) to monocytes exposed to SB 203347 did not alter A23187-induced LTC_4 generation, indicating that SB 203347 had no effect on downstream AA metabolizing enzymes in this setting. Taken together, these results provide evidence that the 14 kDa PLA_2 provides substrate for monocyte LT and PAF formation, while the 85 kDa PLA_2 plays a more significant role in the formation of PG.

INTRODUCTION

Much work has been directed toward understanding the liberation of arachidonic acid (AA) from human monocyte phospholipids (PL) and its subsequent metabolism to a number of cyclooxygenase (COX) and 5-lipoxygenase (5-LO) products. The first rate-limiting enzyme in eicosanoid formation is phospholipase A_2 (PLA_2, EC 3.1.1.4) which liberates

Eicosanoids and Other Bioactive Lipids in Cancer, Inflammation, and Radiation Injury, 4
Edited by Honn *et al.*, Kluwer Academic / Plenum Publishers, New York, 1999.

215

arachidonic acid (AA) from the *sn*-2 position of cellular PL[1]. The two most studied mammalian forms are the group IIa 14 kDa PLA$_2$, known to exist as both an extracellular and cell-associated form, and the group IV cytosolic 85 kDa-PLA$_2$. Although both enzymes have been extensively studied, the relative contribution of the two enzymes in stimulated-eicosanoid production in a single cell system is poorly understood.

The individual participation of the two distinct, cell-associated sn-2 acylhydrolases in a single cell system has not been fully appreciated because previous reports utilize cell systems which either do not contain both enzymes forms, only generate specific eicosanoid classes and/or readily secrete the 14 kDa PLA$_2$ upon activation. The monocyte/macrophage possess several acylhydrolase activities including the 14 and 85 kDa-PLA$_2$ enzymes[2]. They offer an optimal system for studying the respective roles of the two enzymes in eicosanoid synthesis since they simultaneously produce both LT, PG and PAF upon stimulation with soluble or receptor-mediated stimuli. We have previously reported that monocytes do not secrete the 14 kDa PLA$_2$, even with endotoxin treatment[3]. In addition, monocyte eicosanoid formation is not altered by exposure to the neutralizing mAb (anti-group IIa 14 kDa PLA$_2$ mAb) 3F10[3]. The 14 kDa PLA$_2$ therefore appears to exist predominantly as an intracellular enzyme. This is particularly important as the presence of extracellular secreted enzyme may have stimulatory activity itself, complicating the interpretation of stimuli-induced eicosanoid formation. Indeed, to further assess the localization of the 14 kDa PLA$_2$ enzyme, FACs analysis utilizing the anti-14 kDa PLA$_2$ mAb, 3F10, and a goat anti-mouse IgG-FITC label (1:500, Biosource International, Camarillo, CA) indicated the lack of extracellular surface enzyme.

EVALUATION OF HUMAN MONOCYTE AA RELEASE, EICOSANOID GENERATION AND PAF FORMATION.

To assess the respective roles of the 2 enzymes in the human monocyte system, selective inhibitors were used. This includes the selective 14 kDa inhibitor, SB 203347, which exhibits a 20-40 fold selectivity for the rh 14 kDa PLA$_2$ (IC$_{50}$, 0.5 uM) over the rh 85 kDa PLA$_2$ (IC$_{50}$, 20 uM) and AACOCF3 which selectively inhibits rh 85 kDa PLA$_2$ (IC$_{50}$, 0.1 uM) with a 300-fold greater potency than rh type II 14 kDa PLA$_2$ (IC$_{50}$, 31 uM). Both inhibitors exhibited similar fold selectivity when assessed on the respective monocyte enzymes[4]. The AACOCF3 possesses cyclooxygenase inhibitory activity which interferes with its use as a tool to evaluate the effect of 85 kDa PLA$_2$ inhibition on prostanoid generation. Therefore, initiation site-directed 85 kDa PLA$_2$ antisense (SK 7111) was also employed. We have previously shown that 24hr exposure to SK7111 but not the corresponding sense oligonucleotide (SK 9030) depletes the monocyte of up to 75-90% of 85 kDa PLA$_2$ protein. This was accompanied by a similar reduction in cellular enzymatic activity[3].

Exposure of monocytes to either SB 203347 or AACOCF3 prior to stimulation with A23187 induced concentration-dependent inhibition of AA release[5]. Interestingly, the inhibition pattern of both compounds plateaued at the higher concentrations, and full inhibition could not be achieved, suggesting the contribution of more than one enzyme in the liberation of monocyte AA. Indeed, simultaneous addition of suboptimal concentrations of both compounds resulted in near total reduction of AA released. Given the notable inhibition in cellular AA liberation, both compounds were evaluated for their effects on monocyte eicosanoid generation.

SB 203347 inhibited A23187-induced LT [4] (Figure 1) or PAF formation[5] but not PGE$_2$ production[4], while the AACOCF3 inhibited PGE$_2$ but had no effect on LT or PAF production. This demonstrates the important contribution of the 14 kDa PLA$_2$ in providing substrate for LT and PAF formation, and suggested that the 85 kDa PLA$_2$ and not the 14 kDa PLA$_2$ played the more predominant role in AA liberation for metabolism into prostanoids. Addition of exogenous AA resulted in elevated eicosanoid levels and abrogation of the SB 203347 inhibitory activity, indicating a lack of its action on down stream 5-

lipoxygenase pathway components[4]. As mentioned above, this was not the case for the AACOCF3 which also inhibits COX activity. PGE_2 was still inhibited by the compound when cells were incubated in the presence of the exogenous AA.

The specific role of the 85 kDa PLA_2 was demonstrated using antisense to deplete cellular 85 kDa PLA_2 levels. Antisense SK7111 treatment concentration-dependently reduced both A23187- and zymosan-stimulated PGE_2 formation when compared to stimulated lipofectin controls (Figure 2). Up to 3 uM SK7111 had no significant effect on LTC_4 formation in either system, clearly demonstrating the lack of a prominent role for the 85 kDa PLA_2 in the release of AA for LT formation. Addition of AA (20 uM) to SK7111-treated cells during stimulation with A23187 enhanced PGE_2 formation, and completely abrogated antisense-induced inhibition of PGE_2 (lipofectin, 4.1 ± 1.3 vs. SK7111, 2.1 ± 0.3 (49% inhibition) and AA + lipofectin, 13.8 ± 0.9 vs. AA + SK7111, 16.9 ± 1.6 ng $PGE_2/5 \times 10^6$; mean $\pm$ S.D.; n=3), demonstrating the lack of effect of SK7111 on downstream AA metabolism and the specific action of the antisense[4].

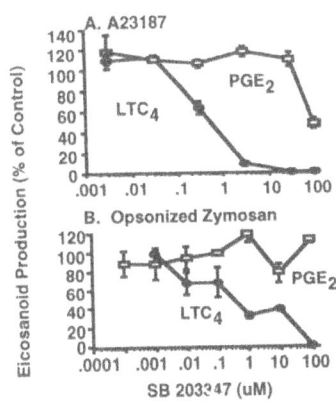

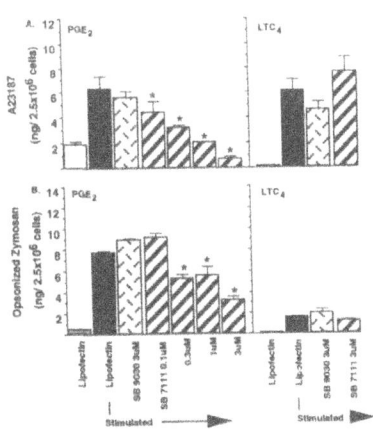

Figure 1. **The effect of SB 203347 on stimulated human monocyte eicosanoid formation.** Human monocytes (5×10^6/ml) were treated with SB 203347 (0.03-100 uM) prior to stimulation with A23187 (1 uM, 15 min) or opsonized zymosan (5mg/mt, 2 hr). The PGE_2 and LTC_4 were measured in cell-free media. Data are expressed as % of stimulated control (panel A; 5.5 ng/ml PGE_2, 55 ng/ml LTC_4; panel B; 1.9 ng/ml PGE_2, 0.9 ng/ml LTC_4). Data represent mean $\pm$ SD, n=3 of one representative of two experiments.

Figure 2. **85 kDa PLA_2 antisense inhibits stimulated monocyte PGE_2 but not LTC_4 formation in a concentration-dependent manner.** Monocytes were treated with 0.1, 0.3, 1, or 3 uM antisense SK7111, SK9030 (3 uM), or lipofectin (5 ug/ml) in serum-free conditions for 18 h. Cells were then exposed to A23187 (1 uM, 7 min, panel A) or opsonized zymosan (5 mg/ml, 2 hr, panel B). PGE_2 and LTC_4 data are expressed as mean $\pm$ SD (n=3). *Indicates significantly different from control at p<0.05 using ANOVA and Duncan's multiple range analysis.

DISCUSSION

We have provided further support for the co-existence of both the 85 kDa PLA_2 and a non-pancreatic 14 kDa PLA_2 in human monocyte as cell-associated enzymes. Monocytes can be induced to co-produce a number of eicosanoid classes by a variety of stimuli and therefore offer an ideal system for the simultaneous study of the two enzymes. Early studies indicated that cell-associated 14 kDa PLA_2 participated in stimulated-AA release and subsequent eicosanoid formation. With the discovery of 85 kDa PLA_2, many re-focused their attention on this enzyme as it exhibited characteristics that one would expect for an enzyme responsible for cellular AA metabolism, i.e., regulation by intracellular (nM) Ca^{2+} levels, phosphorylation, upregulation by growth factors or inflammatory cytokines and a selectivity for AA in the sn-2 position of substrate phospholipid[6]. However, the 14 kDa PLA_2 responds to nM levels of intracellular calcium[7] and readily hydrolyzes AA from the sn-2 position of substrate phospholipid, such as the AA-rich phosphatidylethanolamine, despite the lack of fatty acid specificity noted in vitro[8], indicating that it too could contribute to liberation of cellular AA.

This data provides additional evidence that the 85 kDa PLA_2 primarily supports monocyte prostanoid formation. The data indicate that this is the case in both acute stimuli systems as well as the ligand-activated cell systems previously reported. Alternatively, neither 75-90 % reduction in 85 kDa PLA_2 by antisense nor specific inhibition of its activity with AACOCF3 altered LT or PAF formation. Inhibition of cell-associated 14 kDa PLA_2 with SB 203347 produced the reverse stimulated-eicosanoid profile, i.e., inhibition of LT and PAF while prostanoid biosynthesis was spared. These data support the concept that two distinct enzymes might hydrolyze AA from different pools and/or supply distinct AA metabolizing systems in a single-cell system (Figure 3).

Hypothesis

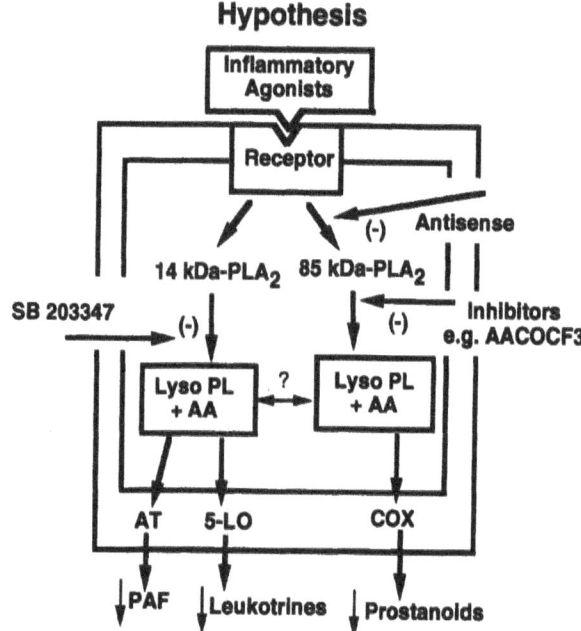

Figure 3. Proposed role of the human monocyte 85 kDa PLA_2 and 14 kDa PLA_2 enzymes in eicosanoid biosynthesis.

References

1. E.A. Dennis, The growing phospholipase A2 superfamily of signal transduction enzymes, *Trends Biochem.* 22:1 (1997).

2. L.A. Marshall and D.W. Morgan, Modulation of arachidonic acid, focus on phospholipase A_2, *Drug News Persp.* (in press) (1998).

3. A. Roshak, G. Sathe, and L.A. Marshall, Suppression of monocyte 85 kDa phospholipase A2 by antisense and effects on endotoxin-induced prostaglandin biosynthesis, *J. Biol. Chem.* 42:25999 (1994).

4. L.A. Marshall, B. Bolognese, J.D. Winkler, and A. Roshak, Depletion of human monocyte 85 kDa phospholipase A_2 does not alter leukotriene formation, *J. Biol. Chem.* 272:759(1997).

5. J.D. Winkler, B.J. Bolognese, A.K. Roshak, C-M. Sung, and L.A. Marshall, Evidence that 85 kDa phospholipase A2 is not linked to CoA-independent transacylase-mediated production of platelet-activating factor in human monocytes, *Biochim. Biophys. Acta. Lipids* 1346:173 (1997).

6. J.D. Clark, A.R. Schievella, E.A. Nalefski, and L.L. Lin, Cytosolic phospholipase A2, *J. Lipid Med. & Cell. Signal.* 12:83 (1995).

7. L.A. Marshall and A. McCarte-Roshak, Demonstration of similar calcium dependencies by mammalian high and low molecular mass phospholipase A_2, *Biochem. Pharm.* 44:1849 (1992).

8. E. Diez, F.H. Chilton, G. Stroup, R.J. Mayer, J.D. Winkler, and A.N. Fonteh, Fatty acid and phospholipid selectivity of different phospholipase A2 enzymes studied by using a mammalian membrane as substrate, *Biochem. J.* 301:721 (1994).

33

MODULATION OF LONG-TERM POTENTIATION IN THE CA1 AREA OF RAT HIPPOCAMPUS BY PLATELET-ACTIVATING FACTOR

Kunio Kato

Department of Physiology
Exploratory Research for Advanced Technology (ERATO)
Japan Science and Technology Corporation (JST)
2-9-3, Shimo-Meguro, Meguro-ku, Tokyo 152, Japan

We have previously shown that platelet-activating factor (PAF: 1-O-alkyl-2-sn-3-glycerophosphocholine) selectively augments EPSCs in cultured hippocampal neurons by a presynaptic mechanism, suggesting that this agent could be involved in hippocampal long-term potentiation (LTP). We have also examined the possible involvement of PAF in LTP in the CA1 region of rat hippocampal slices using the hydrolysis resistant analog, methyl-carbamyl-PAF (C-PAF), and PAF receptor antagonists. When 2 μM BN--52021, a synaptosomal PAF receptor antagonist, was applied, LTP was inhibited and 20 min application of 1 μM MC-PAF in the absence of tetanic stimulation produced a slowly developing potentiation of extracellular recorded EPSP's which persisted for more than 2h. Extra- or intracellular administration of C-PAF also appears to slow the washout of LTP-generating ability which occurs during whole-cell recording, making it possible to induce LTP for as long as 60 min after cell penetration. A decrease in extracellular magnesium (0.1 mM) for 15 min induced LTP which was not inhibited by a PAF antagonist, indicating the existence of PAF dependent and independent LTP. These data suggest that PAF and PAF activated process are important in LTP and modulate the threshold for LTP induction.

INTRODUCTION

We have previously reported that PAF could be involved in the induction of LTP (Cerro et al., 1990, Clark et al., 1992, Kato et al., 1994). PAF has been believed to have two types of receptors, microsomal and synaptosomal. Application of C-PAF, a long-lasting PAF analogue, to the CA1 area of rat hippocampus induces a robust increase in transmitter release, which was verified by a miniature EPSC study in primary culture of hippocampal pyramidal neurons as well as hippocampal slices. Application of the potent antagonist for synaptosomal PAF receptors, 2 μM BN-52021, which is ginkgaloide, blocked LTP induction, however BN-57030, a microsomal PAF receptor antagonist, had no effect whatsoever LTP induction. These observations indicate that endogenous PAF is involved in LTP induction via modulation of the presynaptic site. Since an initial step in LTP induction is triggered at postsynaptically

Eicosanoids and Other Bioactive Lipids in Cancer, Inflammation, and Radiation Injury, 4
Edited by Honn *et al.*, Kluwer Academic / Plenum Publishers, New York, 1999.

221

and the expression of LTP is assumed to occur at a presynaptic site, the existence of a retrograde messenger which is produced at the postsynaptic site and travels to the presynaptic site has been suggested. We have shown that application of C-PAF with relatively high frequency electrical stimulation (theta bursts for 1 sec) to the CA1 area induces NMDA receptor activity-independent LTP. Since PAF is produced downstream from NMDA receptor activation which is generally required for LTP induction, we could assume PAF to be the retrograde messenger for LTP induction.

Questions must be raised as to whether PAF is really produced after tetanic stimulation, the role of theta-burst stimulation during C-PAF-induced LTP, whether application of C-PAF alone induces potentiation, whether PAF acts only at the presynaptic site, and so on. To address these questions, we further investigated the role of PAF by applying C-PAF under various protocols for LTP induction. We found that PAF receptor activation is required for tetanus-induced LTP in which high frequency electrical stimulation is applied, but not for low Mg solution-induced LTP which does not require specific stimulation. Furthermore, we tested the role of PAF at the postsynaptic site by applying C-PAF through a patch pipette and investigated its effect on LTP induction.

MATERIALS AND METHODS

1.   Slice Preparation.
      Hippocampal slices were prepared according to standard procedures. Male albino rats (20-25 days old) were anaesthetized with halothane and decapitated. Hippocampi were rapidly dissected and placed in gassed (95% O2-5% CO2) external solution containing (in mM): 124 NaCL, 3 KCl, 2CaCl, 2 MgSO, 1.25 NaHCO and 10 glucose at 10°C. Transverse slices (500 μM thick) were cut with a rotary tissue slicer, then maintained in an incubation chamber for at least 2h at 30°C. At the time of each experiment, individual slices were transferred to a submersion recording chamber, where they were continuously perfused with extracellular solution (2ml min) at 30°C.

2.   Field and Whole cell recordings.

      Field recordings were obtained from apical dendritic region of CA1 using a 3-8 MΩ electrode filled with 2M NaCl. Whole-cell recordings were obtained from the cell bodies of pyramidal neurons in the CA1 area by the 'blind patch' method using an Axopatch 200A amplifier. Patch electrodes were pulled from 1.2-mm outside diameter borosilicate glass and had a resistance of < 6MΩ after fire-polishing. Pipettes were routinely filled with a solution containing (in mM): 130 cesium methanesulphonate, 10 tetraethylammonium chloride, 5 NaCl, 1 MgCl$_2$, 0.25 1,2-bis (2-aminophenoxy)ethane-N,N,N,N-tetraacetic acid (BAPTA), 10 N-[1-hydroxyethyl]-piperazine-N-[2-ethanesulphonic acid] (HEPES), 2 Mg-ATP and 0.3 Na-GTP, with pH adjusted to 7.25 using CsOH. For whole-cell studies, the neuronal membrane potential was voltage clamped at - 80 mV. During an experiment, the Schaffer collateral-commissural fibers were stimulated every 20 s in the stratum radiatum using a concentric bipolar electrode (24 μM tip, Rhodes Medical Instruments) and 0.3 ms constant-current pulses at an intensity sufficient to evoke a 50% maximal response. LTP was produced by an electrical tetanus administered for 1 s at 100 Hz using the same stimulus intensity for a field recording and paired stimulation (1 Hz electrical stimulation with depolarization of membrane potential from -80 to 0 mV for 2 min) was applied for whole cell recording. Data were analyzed using IBM-based system.

3.   Materials.
      BN-52021 and BN-57030 were generously supplied by N.G. Bazan. PAF and methylcarbamyl PAF (C-PAF) were obtained from Cayman. BN compounds were dissolved

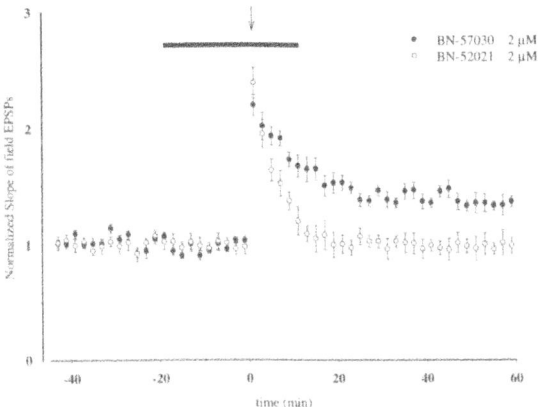

Fig. 1 The effect of PAF receptor blockers on LTP induction

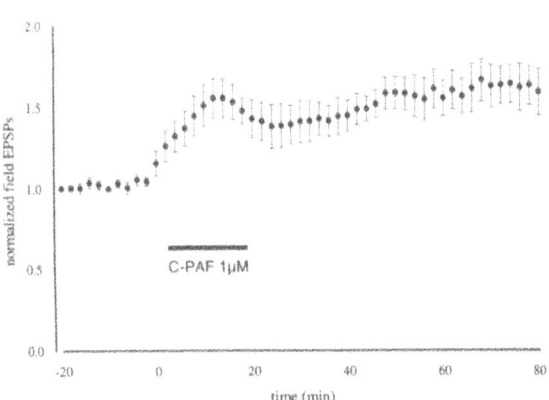

Fig. 2 Application of C-PAF induces LTP.

in DMSO as a stock solution and kept under -30°C. PAF, C-PAF and lyso-PAF were prepared daily in the appropriate solution and sonicated for 15 min immediately before use.

RESULTS AND CONCLUSIONS

Two μM BN-52021, which is a synaptosomal PAF receptor antagonist, blocked LTP induction in CA1 pyramidal neurons, but BN-57030, a microsomal PAF receptor antagonist did not (fig. 1). Application of 0.5-1 μM C-PAF, a long-lasting PAF analogue, for 10 min in the presence of a NMDA receptor antagonist, 50 μM AP5, with theta-burst stimulation induced a slowly developing long-lasting protentiation, suggesting that synaptosomal PAF receptor activity at the synapse may be important in hippocampal LTP induction (Kato et al., 1994). To investigate the role of exogenously applied PAF, 0.2-1 μM C-PAF was circulated in the bath solution for 20 min, resulting in the slowly developing potentiation without theta-burst electrical stimulation which lasted for more than two hours (fig. 2). A selective NMDA receptor antagonist, 50 μM AP-5, blocked this potentiation, suggesting that this LTP is NMDA receptor activity dependent (fig. 3). This further suggests that the robust increase in transmitter release resulting from the application of a high dose of C-PAF may enhance Ca^{2+} influx through NMDA receptors, thereby yielding LTP induction. this potentiation in C-PAF was observed exceptionally in young rats; i.e. younger than 23 years old, but not in rats older than 30 days. PAF has shown to activate tyrosine kinase, PI turnover, protein kinase C, and so on. Some of these mechanisms which declines with aging, such as PI turnover, could be involved in the LTP induced by C-PAF. Coapplication of C-PAF with theta-burst stimulation induced NMDA receptor activity independent LTP induction. The discrepancy between these two distinct types of protocols for inducing LTP, i.e., NMDA dependent and independent LTP, could be explained by the additional effect of theta-burst stimulation rather than by increased transmitter release, presumably via enhancement of Ca^{2+} influx at the presynaptic site, which facilitated the biological processes used in LTP induction.

A different type of protocol is used for LTP induction, such as tetanus (100 Hz for 1 second) and theta-burst (groups of 100Hz, 5 pulses). This variability influences transmitter release during stimulation. LTP is also induced without a specific paradigm of electrical stimulation when NMDA channels are persistently opened by an artificial procedure; decreasing extracellular magnesium which normally closes NMDA channels at the resting membrane potential. When the concentration of magnesium was decreased from 2.0 to 0.1 mM for 20 min with an increase in calcium (from 2.0 to 4.0 mM), sustained potentiation persisting for more than five hours was observed. The low-Mg solution-induced LTP is NMDA receptor activity dependent, since it is completely blocked by 50 μM AP-5. In contrast to tetanus-induced LTP, 5 μM BN-52021, an even high dose than was used in tetanus-induced LTP, had not inhibitory effect whatsoever on low-Mg solution-induced LTP (fig. 4). Since low-Mg solution-induced LTP does not require especially high frequency presynaptic stimulation, the significant role of PAF in tetanus-induced LTP could be explained simply by enhancement of transmitter release during tetanus.

To test the effect of PAF at the postsynaptic site, we diffused C-PAF into the cell body through a patch pipette. After penetrating the cell membrane, the ability to generate LTP is usually lost within 20 min even when abundant energy sources like ATP and GTP are included in the patch pipette (Kato et al., 1993). When 2-5 μM C-PAF is included in the recording pipette, LTP is successfully induced even when tetanus is applied 60 min after membrane penetration with depolarization at a 0 mV holding membrane potential (fig. 5).

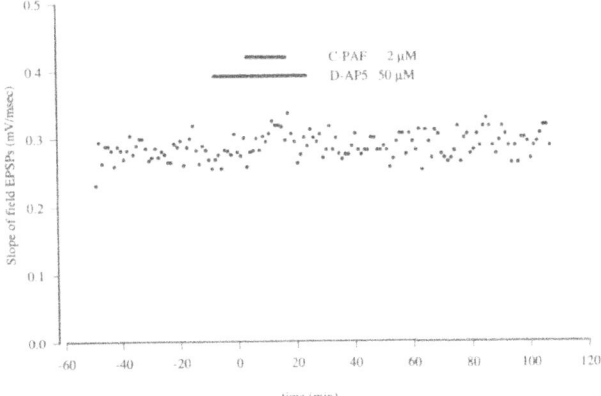

Fig. 3   C-PAF-induced LTP was eliminated by NMDA antagonist, 50 μM D-AP5.

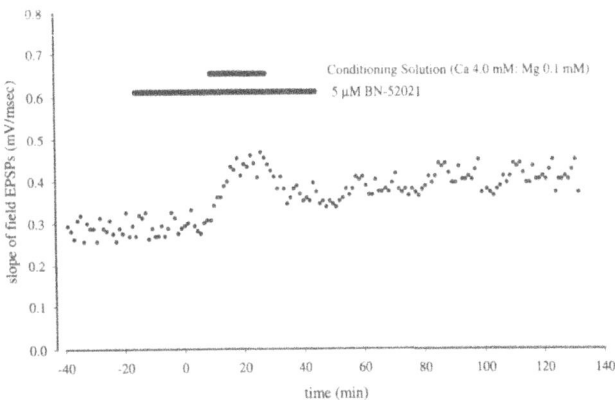

Fig. 4   5 μM BN-52021 did not block low Mg^{2+} solution-induced LTP in pyramidal neurons.

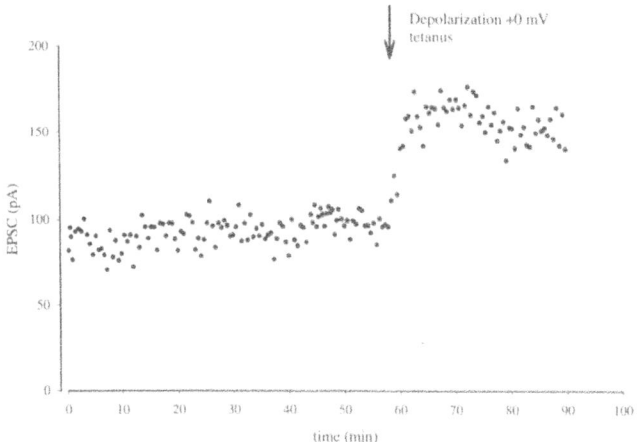

Fig. 5 Intracellular C-PAF reverse washout of the ability to generate LTP

DISCUSSIONS

We have previously discussed the possibility of PAF being a retrograde messenger in LTP induction, since PAF is produced downstream from NMDA receptor activation and enhances transmitter release by via the presynaptic site. The requirement for a retrograde messenger is assumed from the electrophysiological observation that a persistent increase in transmitter release occurred during LTP induction, which was verified by quantum analysis and a miniature EPSC study. However, an alternative interpretation of these statistical analyses has also recently been proposed, i.e., the so-called 'silent synapse' theory. According to this theory, the results of statistical analysis can be explained also by newly-revealed functional receptor which used to be non-functional. If this is the case, the role of PAF might be to generate functional receptors from 'silent synapses'. Since PAF has been shown to change platelet morphology, PAF might also contribute to changes in the structure of synapses and/or the spine such that the functional receptor would be revealed.

PAF receptor activation is required for tetanus-induced LTP in which strong presynaptic facilitation is achieved by high frequency stimulation. However, it is not required for low-Mg solution-induced LTP in which only low frequency test pulses were adequate to elicit LTP. When NMDA channels are opened artificially by low-Mg solution, enhanced calcium influx through NMDA channels triggers LTP induction without a specific paradigm of stimulation at the presynaptic site. Activation of NMDA receptors produces arachidonic acid (AA) and PAF by activating phospholipase A2. These lipid mediators then increase transmitter release and enhance NMDA channel activity. As AA and PAF increase transmitter release and activate NMDA channels, these lipid messengers generate a positive feedback mechanism which is essential for tetanus-induced LTP induction but not necessary for low-Mg solution-induced LTP. From these results, we may conclude that low-Mg solution-induced LTP does not require a retrograde messenger, or that strong activation of NMDA channels can produce other messengers, such as AA or its metabolites that serves as substitutes for PAF.

PAF has been shown to exert several biological effects, such as activation of tyrosine and MAP kinases, and increasing arachidonic acid and PI turnover. These substances are known to be involved in LTP induction. The increment in C-PAF in a patch pipette allows to activation of many biological processes, mentioned above, which could suppress washout of the ability to generate LTP in a whole cell configuration. The role of PAF diffused into the cell on LTP induction could be that it supplies PAF itself exogenously which cannot be produced due to washout of the cell and that it activates several kinds of kinases and increases PI turnover, which may be necessary for LTP induction and could be lost during washout of the cell. Taken the results of all studies into account, the role of PAF could not simply be that of a retrograde messenger. Rather PAF appears to exert many biological actions, including the enhancement and modulation of the threshold of LTP induction.

Production of endogenous PAF occurs in many pathological situations such as immune reactions, ischemia and epilepsy (Birkle, et al., 1988), in which NMDA receptor activation is critical. Physiological phenomena such as LTP are other situation in which PAF is produced. There remain many questions about the role of PAF in LTP, where PAF is produced, the mechanism of the effect of PAF on the transmitter increase, and so on. Further investigation is needed to address these questions.

REFERENCES

Clarke, D.G., Happel, L.T., Zorumski, C.F., and Bazan, N.G. 1992. Enhancement of hippocampal excitatory synaptic transmission by platelet-activating factor. *Neuron* 92:1211-1216.

Cerro, S.del, Arai, A., and Lnch, G. 1990. Inhibition of long-term potentiation by an antagonist of platelet-activating factor. 54:213-217.

Kato, K., Clark, G.D., Bazan, N.G., and Zorumski, C.F. 1994. Platelet-activating factor as a potential retrograde messenger in CA1 hippocampal long-term potentiation. *Nature* 367:175-179.

Birkle, D.L., Kuiran, P., Braquet, P., and Bazan, N.G. 1988. Platelet-activating factor antagonist BN52021 decreases accumulation of free polysaturated fatty acid in mouse brain during ischemia and electroconvulsive shock. *J. Neurochem.* 51(6):1900-1905.

RECEPTORS

34

PPARS: NUCLEAR RECEPTORS FOR FATTY ACIDS, EICOSANOIDS, AND XENOBIOTICS

Pallavi R. Devchand, Annemieke Ijpenberg, Béatrice Devesvergne, and Walter Wahli

Institut de Biologie Animale
Université de Lausanne
Bâtiment de Biologie
CH 1015, Lausanne
Switzerland.

INTRODUCTION

The induction of gene expression in response to hormones and nutrient-derived signaling molecules is important not only for tissue remodeling and cellular differentiation during development, but also for various aspects of homeostasis during adulthood. One mechanism by which cells respond to signaling molecules involves a family of nuclear receptors that activate transcription in a ligand-dependent manner. The peroxisome proliferator-activated receptors (PPARs) are a class of these transcription factors[1 for review]. We have previously identified PPARα as a nuclear receptor for the eicosanoid LTB$_4$ and also for xenobiotics[2,3] such as the lipid-lowering drug Wy 14,643 and the arachidonic acid analogue ETYA. These studies also link, *in vivo*, PPARα function with inflammation control. Here, we describe an assay for evaluating various compounds as ligands for the three PPAR isotypes. These results reveal that PPARs are nuclear receptors for 8S-HETE and polyunsaturated fatty acids, and also highlight the differences in ligand specificity between the different PPAR isotypes.

THE PPAR-RXR HETERODIMER

The three PPAR isotypes (α, β/δ and γ) bind to DNA as heterodimers with the retinoid **X** receptor, RXR (either α, β or γ). The PPAR-RXR heterodimer recognizes a specific DNA sequence termed PPRE for peroxisome proliferator response element. The PPRE comprises a 5' flanking sequence and a core DR-1 element of two hexameric half-sites separated by an adenine spacing nucleotide[5] (Figure 1). As a result of the asymmetric PPRE, the PPAR-RXR heterodimer is oriented with PPAR on the 5' upstream half-site and RXR on the 3' downstream half-site. This is unlike the other RXR-containing heterodimers, where RXR is positioned on the 5' half-site of the DNA response element. Extensive analyses indicate that binding affinities are dependent on the isotype combinations of PPAR and RXR, and can be related to the conservation of the PPRE with respect to the consensus sequence[6]. The PPAR target genes identified to date are associated with energy homeostasis and global regulation of lipids. Taken together with the information on tissue specific expression of the different PPAR isotypes, the above information might aid in evaluating PPAR activity *in vivo*.

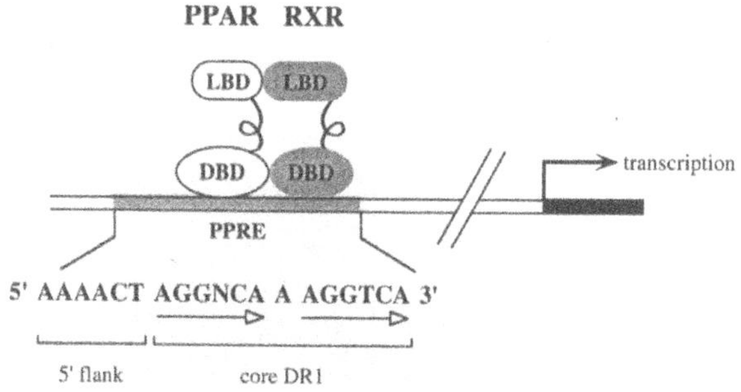

PPAR RXR

5' AAAACT AGGNCA A AGGTCA 3'

5' flank   core DR1

Figure 1. Polarity and sequence specificity of PPAR-RXR binding to DNA. Schematic representation showing the DNA binding domain (DBD), ligand binding domain(LBD) and hinge regions of the two receptors. The complex binds to DNA elements (PPRE, indicated in grey) composed of two half-sites (indicated by the arrows) separated by a single adenine residue and preceded by a conserved region, indicated as the 5' flank. The asymmetric binding site imposes a head-to-tail polarity on the complex, with PPAR binding to the upstream half-site and RXR binding to the downstream half-site. In the presence of the respective ligands of the receptors, transcription of the target gene (indicated in black) is induced.

CARLA IDENTIFIES LIGANDS

The action of nuclear hormone receptors is mediated by cofactors that are thought to bridge the transcription factors to the basal transcriptional machinery. Two classes of cofactors have been identified, coactivators and corepressors. These proteins bind the nuclear hormone receptors (NHRs) in a ligand-dependent manner. For example, the steroid receptor coactivator 1 (SRC-1) binds to classical NHRs, only in the presence of a NHR ligand. Here, we describe the CARLA assay[4]. CARLA uses the association of PPAR with SRC-1 as an indication of whether a test compound is a ligand for PPAR (see Figure 2 for details). This method bypasses the requirement of radiolabeled potential ligands and allows for fast, economical screening of many test compounds.

The CARLA assay was used to evaluate the ligand specificity of the three PPAR isotypes, α, β and γ. Many natural and synthetic compounds were evaluated as PPAR ligands[4]. Figure 3 shows some examples.

Of the fatty acids, PPARs have a preference for dietary polyunsaturated fatty acids (PUFAs, e.g. linoleic and linolenic acid) over monounsaturated fatty acids, but do not show positive interaction with either saturated or dicarboxylic fatty acids. The different PPAR subtypes bind these fatty acids with different affinities.

PPARα and PPARγ have been described as nuclear receptors for leukotriene B$_4$ (LTB$_4$) and the prostaglandin J$_2$ derivative 15d-$\Delta^{12,14}$-PGJ$_2$ respectively[2,7,8]. Interestingly, of the eicosanoids tested the inflammatory mediator 8(S)-HETE binds preferentially to PPARα, at sub-micromolar concentrations. At higher concentrations it also binds to PPARβ and PPARγ. The interaction with 8-HETE is selective for the 8(S) enantiomer. While PPARα binds LTB$_4$ it does not bind the 5(S)-HETE precursor.

PPARs are primarily associated with peroxisome proliferation in rodents, as a response to high concentrations of various xenobiotics. The hypolipidaemic drug Wy 14,643 and the stable arachidonic acid analog ETYA have already been reported as PPARα ligands[2,3], and the antidiabetic thiazolidinedione BRL 49653 as a PPARγ ligand[9]. The CARLA assay has allowed us to identify bezafibrate as the first synthetic ligand selective for Xenopus PPARβ, while it activates the mammalian PPARα.

FROM LIGAND TO FUNCTION

Our CARLA results show that PPARs are nuclear receptors for diverse compounds including some polyunsaturated fatty acids, eicosanoids and xenobiotics. These are exciting findings at many levels. First, it shows that fatty acids can potentially act as hormones that directly interact

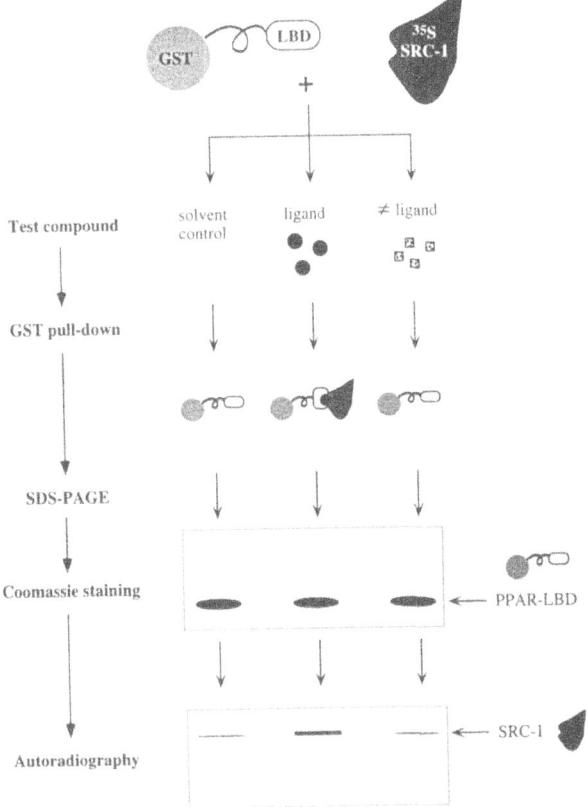

Figure 2. The Co-Activator Dependent Receptor Ligand Assay (CARLA). The CARLA was developed to test putative ligands. It is based on the ability of nuclear hormone receptors to interact with co-activators in the presence of ligand. To this end, the hinge and ligand binding regions (LBD) of the PPAR receptors are bacterially expressed as a fusion protein with GST (Glutathione S-transferase), and partially purified and retained on glutathione beads (Pharmacia). Full length radio-labeled SRC-1 is produced by *in vitro* translation using reticulocyte lysate and ^{35}S-methionine. The SRC-1 and GST-PPAR, are incubated in the presence of the test compounds. In the presence of a ligand, indicated by the black circles, the two proteins associate. In the absence of ligand (i.e. solvent control or a non-ligand substance as indicated by the dotted squares) no association between PPAR and SRC-1 is observed. After incubation, the resulting GST complex is purified by centrifugation of the beads, and analyzed by SDS-PAGE. Coomassie staining of the gel is used to normalize the amount GST-PPAR protein between different reactions. Subsequent autoradiography of the gel reveals if association of GST-PPAR and SRC-1 has occurred, i.e. if the tested compound is a *bona fide* ligand.

with transcription factors in the nucleus. Second, eicosanoids exert their function via both membrane and nuclear receptors. Third, one potential mechanism for the action of clinically relevant xenobiotics such as fibrates and TZDs, is at the transcriptional level via interaction with PPARs.

The identification of ligands has greatly enhanced our understanding of the roles of PPARs in the organism (Figure 4). Studies have concentrated primarily on the α and γ isotypes. In terms of lipid metabolism in the cell, PPARα is associated with lipid catabolism whereas PPARγ is

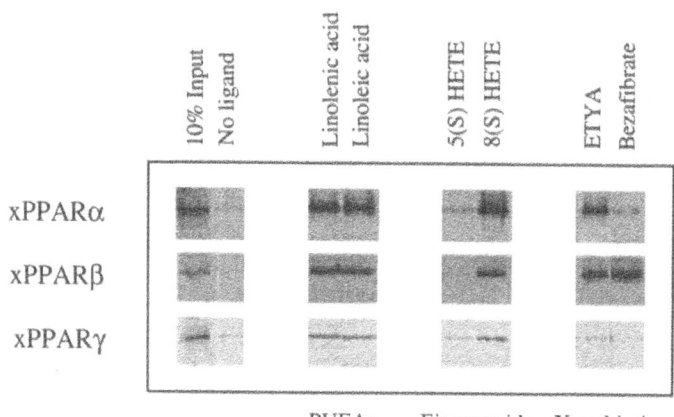

Figure 3. Some polyunsaturated fatty acids, eicosanoids and xenobiotics are PPAR ligands. CARLA results showing that PUFAs, eicosanoids and xenobiotics can behave as PPAR isotype-selective ligands. The specificity of these pull-down reactions is demonstrated by the lack of SRC-1 in the absence of ligand. For reference, the first lane shows the signal obtained with 10% of the SRC-1 input in the reaction.

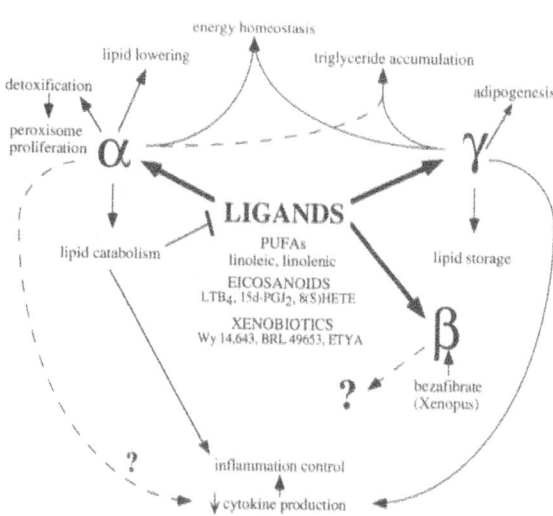

Figure 4. Schematic outline linking identified ligands to known PPAR functions. Direct interactions of ligands (PUFAs, eicosanoids and xenobiotics) with the different PPAR isotypes induce transcription of specific subsets of target genes. The consequences of these interactions are reflected in the biological functions indicated. See text for details.

associated with lipid storage. Very little is known about PPARβ. So far, PPAR functions include: peroxisome proliferation in the liver and kidney, energy homeostasis including lipid lowering and triglyceride accumulation in cells, adipogenesis, detoxification and inflammation control.

Our knowledge of the mechanisms involved in different PPAR actions is increasing rapidly. The first function associated with these receptors is that of detoxification of xenobiotics leading to subsequent peroxisomal proliferation in the liver. We now know that this is PPARα-dependent and the result of induction of the fatty acid ω- and β-oxidation pathways[10]. The identification of the different PUFAs, eicosanoids and xenobiotics by CARLA indicate that all these ligands potentially control their own fate via interaction with the different PPAR isotypes. Interaction with PPARγ can lead to storage. Conversely, interaction with PPARα can lead to degradation via the same negative feedback mechanism described above. This is an important finding as reflected in the case of eicosanoids such as LTB_4 and 8(S)HETE, since by controlling their own degradation they effectively limit the duration of their pro-inflammatory activity. The net result of this PPARα function is anti-inflammatory, as demonstrated *in vivo* with the PPARα knock-out mouse. Recent findings have also tried to link the activation of PPARγ to inflammation, via a decrease in cytokine production when activated macrophages are exposed to PPARγ ligands[11,12]. While these studies are in vitro, if we extrapolate to in vivo conditions one might speculate an anti-inflammatory function for PPARγ. The effect of PPARα activators on cytokine production in different cell types is yet to be determined. It is probable that this is an additional mechanism involved in the prolonged inflammatory response observed in the PPARα knock-out mouse.

Obviously one of the major functions of PPARs is in energy homeostasis. While we have quite a clear picture of some of the different roles of PPARα and PPARγ in this process, we have yet to place PPARβ. The identification of bezafibrate as a Xenopus PPARβ ligand provides an invaluable tool in evaluating the function of this elusive receptor.

SUMMARY

The peroxisome proliferator-activated receptors have enjoyed the spotlight for many reasons. These transcription factors are ligand-inducible nuclear receptors that modulate gene expression in response to a broad spectrum of compounds. The recognition that PPARs are indeed nuclear receptors for polyunsaturated fatty acids, some eicosanoids and also lipid-lowering and antidiabetic drugs, has opened many exciting avenues of research and drug discovery. Recent studies on the PPAR function have extended the role of these transcription factors beyond energy homeostasis to master gene in adipogenesis and also determinants in inflammation control. While rapid advances have been made, it is clear that we are far from a global understanding of the mechanisms and functions of PPARs.

ACKNOWLEDGMENTS

We thank Liliane Michalik for critical reading of the manuscript. This work was funded by Etat de Vaud and grants from the Swiss National Science Foundation to B.D. and W.W.

REFERENCES

1. P.R. Devchand and W. Wahli, PPARα : Tempting fate with fat, in: *Hormones and Signalling* , B.W. O'Malley, ed., Academic Press, San Diego, Vol 1:236-265 (1997).
2. P.R. Devchand, H. Keller, J.M. Peters, M. Vazquez, F.J. Gonzalez, and W, Wahli, The PPARα-leukotriene B4 pathway to inflammation control, *Nature* 384: 39-43 (1996).
3. H. Keller, P.R. Devchand, M. Perroud, and W. Wahli, PPARα SARs derived from species-specific differences in responsiveness to hypolipidemic agents, *Biol. Chem.*: 378(7): 651-655 (1977).
4. G. Krey, O. Braissant, F. L'Horset, E. Kalkhoven, M. Perroud, M.G. Parker, and W. Wahli, Fatty acids, eicosanoids and hypolipidemic agents identified as ligands of PPARs by CARLA, *Mol. Endocrinol* . 11: 779-791, (1997).
5. A. IJpenberg, E. Jeannin, W. Wahli, and B. Desvergne, Polarity and specific sequence requirements of PPAR-RXR heterodimer binding to DNA, *J.Biol. Chem* 272(32)1 20108-20117 (1997).
6. C. Juge-Aubry, A. Pernin, T. Favez, A.G. Burger, W. Wahli, C.A. Meier, and B. Desvergne, DNA-binding properties of PPARs on various natural PPREs, *J Biol Chem.* 272(40): 25252-9 (1997).

7. B.M. Forman, P. Tontonoz, J. Chen, R.P. Brun, B.M. Spiegelman, and R.M. Evans, 15d-Δ−12,14)-PDJ2 is a ligand for the adipocyte determination factor PPAR-gamma, *Cell* 83: 803-12 (1995).
8. S.A. Kliewer, J.M. Lenhard, T.M. Willson, I. Patel, D.C. Morris, and J.M. Lehmann, A prostaglandin J2 binds PPAR gamma and promotes adipocyte differentiation. *Cell* 83: 813-19 (1995).
9. J.M. Lehmann, L.B. Moore, T.A. Smith-Oliver, W.O. Wilkison, T.M. Willson, and S.A. Kliewer, An antidiabetic TZD is a high affinity ligand for PPAR gamma *J. Biol. Chem*. 270, 12953-6 (1995).
10. S.S. Lee, T. Pineau, J. Drago, E.J. Lee, J.W. Owens, D.L. Kroetz, P.M. Fernandez-Salguero, H. Westphal, and F.J. Gonzalez, Targeted disruption of the alpha isoform of the PPAR gene in mice results in abolishment of the pleiotropic effects of PPs, *Mol Cell Biol* 15: 3012-22 (1995).
11. M. Ricote, A.C. Li, T.M. Willson, C.J. Kelly, and C.K. Glass, PPAR-gamma is a negative regulator of macrophage activation, *Nature* 391(6662): 79-82 (1998).
12. C. Jiang, A.T. Ting, and B. Seed, PPAR-gamma agonists inhibit production of monocyte inflammatory cytokines, *Nature* 391(6662): 82-86 (1998).

35

MOLECULAR CLONING AND CHARACTERIZATION OF LEUKOTRIENE B4 RECEPTOR

Takashi Izumi, Takehiko Yokomizo, Toshio Igarashi,
and Takao Shimizu

The Department of Biochemistry and Molecular Biology,
Faculty of Medicine, The University of Tokyo, Hongo 7-3-1,
Bunkyo-ku, Tokyo 113, Japan

INTRODUCTION

Leukotriene B_4 (LTB4) is a potent proinflammatory and chemotactic factor biosynthesized in various tissues, which is involved in inflammation, immune responses and host defense against infection.[1] LTB4 activates inflammatory cells by binding to its cell-surface receptor (BLT). Specific binding for [^{3}H]LTB4 was reported in the membrane fractions of granulocytes, T lymphocytes, alveolar macrophages, spleen, brain, THP-1 cells, and differentiated HL-60 cells.[2-5] By addition of GTPγS, low affinity sites of [^{3}H]LTB4 binding appeared in spleen membranes.[5] LTB4 stimulated a dose-dependent increase in GTP hydrolysis in HL-60 cells.[6] LTB4 induced increase in intracellular calcium, D-*myo*-inositol-1,4,5-triphosphate (InsP3) accumulation, O_2^- production, and activation of GTPase in a pertussis toxin (PTX)-sensitive manner.[7,8] On the other hand, LTB4-induced hyperadhesiveness of vascular endothelial cells was reported to be PTX-insensitive.[9] These data indicated the presence of BLT coupled with G-protein(s). However, BLT had not been purified to homogeneity or cDNA-cloned.

We cloned a cDNA for BLT from HL-60 cells differentiated with retinoic acid using a subtraction strategy.[10] The open reading frame (ORF) of the cDNA encodes a protein of 352 amino acids and is predicted to contain seven membrane-spanning domains. Membrane fractions of Cos-7 cells transiently transfected with an expression construct containing the ORF of the BLT showed specific binding of LTB4 with a K*d* of

Eicosanoids and Other Bioactive Lipids in Cancer, Inflammation, and Radiation Injury, 4
Edited by Honn *et al.*, Kluwer Academic / Plenum Publishers, New York, 1999.

237

0.154 nM. In CHO cells stably expressing BLT, LTB4 elicited many signal transductions such as increase in intracellular calcium, InsP3 accumulation, and inhibition of adenylyl cyclase. The CHO cells revealed a marked chemotactic response toward nM order of LTB4. Calcium increase and InsP3 accumulation induced by LTB4 were partially sensitive to pertussis toxin   (PTX),

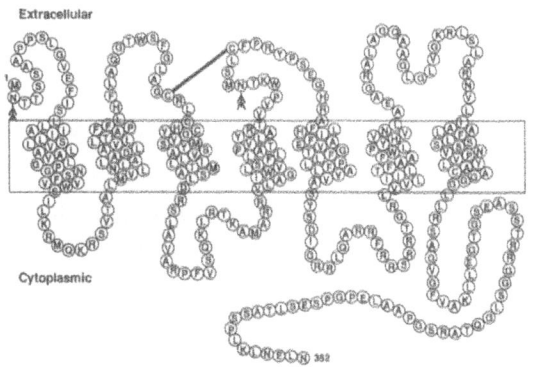

Figure 1. A putative transmembrane structure of the BLT is illustrated.

while chemotaxis and inhibition of adenylate cyclase were completely sensitive to PTX, indicating that BLT transduces LTB4 signals through both PTX-sensitive and -insensitive G proteins.

MOLECULAR CLONING OF LTB4 RECEPTOR (BLT)

When HL-60 human leukemia cells were treated with 1 μM retinoic acid, they were differentiated into granulocyte-like cells with an increase of [3H]LTB4 binding. The binding was maximal after 4 days of differentiation. Two cDNA pools were synthesized from poly(A)+ RNAs from undifferentiated and retinoic acid-differentiated HL-60 cells. A subtracted cDNA was obtained using a polymerase chain reaction (PCR)-Select cDNA subtraction kit (Clontech), subcloned into a T-vector (Promega), and sequenced. Among 66 clones, one clone was found to encode the 3'-UTR of a seven-span orphan receptor previously reported.[11] Two cDNAs (named HL-1 and HL-5) containing the corresponding ORF of this orphan receptor were isolated from a day-3 differentiated HL-60 cDNA library. HL-1 and HL-5 had different 3'-UTR but the same ORF.

Primary Structure of BLT

The primary structure of the amino acid predicted from the ORF of the cloned cDNA is shown in Figure 1. The ORF encodes 352 amino acids. There are two putative N-glycosylation sites (Asn2 and Asn164) at the N-terminal and in the second extracellular loop. Two cysteine residues (Cys90 and Cys168) that might form a disulfide bond to give an appropriate configuration as in most of seven-span receptors exist in the first and second extracellular loops. The Asp64 which is also conserved in most of seven-span receptors is in the second putative transmembrane domain. It lacks an arginine in the seventh transmembrane domain which is conserved in most of prostanoid receptors and was proposed as the binding site of the carboxyl group of prostaglandins.[12] In the cytoplasmic tail, there is a cluster of serine/threonine residues that might be substrates of various kinases, but no cysteine residue which is palmitorylated in some receptors.

The predicted protein has the highest overall homology to rat somatostatin receptor type 3 (SSTR3) (35.8% identity), followed by human SSTR5 (33.1%), human IL-8 receptor (33.0%), human lipoxin A4 receptor (30.7%), and human formyl-peptide receptor (28.6%).

Tissue Distribution of BLT

Northern blotting of various human tissues revealed that the mRNA is most abundant in leukocytes, followed by spleen and thymus. This data is compatible with previous reports that [^{3}H]LTB4 binding was observed in leukocytes, lymphocytes and macrophages.

PROPERTIES OF BLT EXPRESSED IN MAMMALIAN CELLS

The ORF of the cloned receptor was subcloned into a mammalian expression vector pcDNA3 (Invitrogen), and resulting plasmid designated pLTBR. Then, characterization of [^{3}H]LTB4 binding was analyzed using membranes of Cos-7 cells expressing pLTBR, and LTB4-induced signal transduction was studied in CHO cells expressing pLTBR.

Binding properties of Cos-7 membranes expressing pLTBR

The expression vetor pLTBR was introduced into Cos-7 cells, and the binding properties of the cell membranes were analyzed. The Kd value of [^{3}H]LTB4 binding in the Cos-7 membranes was 0.154 nM almost same as that (0.144 nM) of differentiated HL-60 cell membranes. Next, inhibition of 0.25 nM [^{3}H]LTB4 binding to the Cos-7 membranes by

Table 1. Inhibition of 0.25 nM [^{3}H]LTB4 binding to the membrane of pLTBR-transfected Cos-7 cells by various eicosanoids.

Eicosanoids	IC$_{50}$ (nM)
LTB4	0.38
20-OH-LTB4	7.6
12-oxo-LTB4	7.6
12(R)-HETE	30
20-COOH-LTB4	410
6-*trans*-LTB4	500
12(S)-HETE	830
5(S),12(S)-diHETE	1200

The IC$_{50}$ values of PGD$_2$, PGE$_2$, PGF$_{2\alpha}$, 5(S)-HETE, 5(R)-HETE, 15(R)-HETE, 15(S)-HETE, LTD$_4$, and 5-oxo-ETE were over 5 μM.

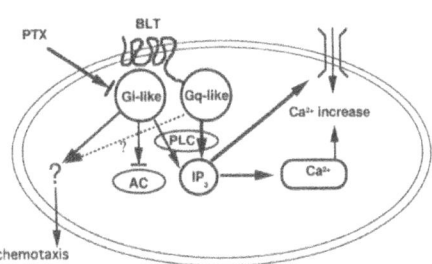

Figure 2. Intracellular signaling mechanisms induced by BLT.

various eicosanoids was carried out. As shown in Table 1, LTB4 inhibited most efficiently (Ki = 0.38 nM), followed by 20-OH-LTB4 and 12-oxo-LTB4, both of that are first-step metabolites in inactivation pathways of LTB4 in leukocytes[13] and tissues[14], respectively. 6-*trans*-LTB4 and 5(*S*),12(*S*)-diHETE, both LTB4 isomers, were much less active than LTB4, and 12(*R*)-HETE was much stronger than 12(*S*)-HETE in the inhibition of [^{3}H]LTB4 binding. These data indicate that 6-*cis*-configurations, the 12(*R*)-hydroxy moiety and methyl group at C20 position of LTB4 are important for binding to BLT.

LTB4-induced Signal Transduction in CHO cells stably expressing pLTBR

One CHO clone named CHO-LTBR was chosen as a representative clone from eight clones stably expressing BLT that responded to 100 nM LTB4 with calcium mobilization. In CHO-LTBR cells, LTB4 induced many intracellular signals as shown Figure 2. LTB4 inhibited forskolin-induced adenylyl cyclase activity, increased InsP3 production and intracellular calcium concentration in dose-dependent manners. The most marked activity of LTB4 in leukocytes is its chemotactic activity. CHO-LTBR cells also showed a potent chemotactic activity towards LTB4. The dose dependency of LTB4 had a bell-shaped pattern, with an optimum concentration of 1-10 nM.

Effects of Pertussis toxin on LTB4-induced signal transductions

Pretreatment of CHO-LTBR with 100 ng/ml pertussis toxin for 12 h completely blocked LTB4-induced inhibition of adenylyl cyclase activity and chemotaxis, but only partially affected LTB4-induced increase of intracellular calcium concentration as shown in Table 2. These data indicate that BLT expressed in the CHO cells couples with both PTX-sensitive (Gi-like) and insensitive (Gq-like) G proteins. In leukocytes and HL-60 cells, LTB4-induced increase of intracellular calcium concentration was PTX-sensitive. This discrepancy may be explained by different expression patterns of G proteins in CHO cells and leukocytes, or by overexpression of BLT in the CHO-LTBR cells. Such difference of PTX-sensitivity in signals was reported in a case of platelet-acivating factor (PAF) receptor. When PAF receptor was expressed in CHO cells, PTX-sensitivity of some signals decreased compared with that in most inflammatory cells.[15] Not all signals induced by LTB4 had been reported to be PTX-sensitive. LTB4-induced hyperadhesiveness of vascular endothelial cell for leukocytes was PTX-insensitive.[9] When partially purified BLT of porcine leukocytes was reconstituted with many types of G proteins (Gi and Go) *in vitro*, BLT was reconstituted with G proteins resulted in recovered [^{3}H]LTB4 binding activities (T. Igarashi unpublished data). BLT may couple to different types of G proteins depending on cell types, and induce signals both in PTX-sensitive and -insensitive manners.

Table 2. Effects of pertussis toxin on 1 nM LTB4-induced signals in CHO-LTBR cells.

Signals	PTX (-)	PTX (+)[a]
Inhibition of forskolin-induced adenylyl cyclase activity	100[b]	- 75[c]
Increase of intracellular calcium concentration	100	67
Chemotaxis	100	0

a: CHO-LTBR cells were pretreated with 100 ng/ml pertussis toxin (PTX) for 12 h.
b: Signals induced by 1 nM LTB4 are presented as 100%.
c: In the PTX-pretreated cells, LTB4 increased forskolin-induced cAMP formation.

CONCLUSIONS

The cDNA encoding a cell-surface receptor for LTB4 (BLT) was isolated. The primary structure reveals that the LTB4 receptor also belongs to the G protein-coupled receptor superfamily. In CHO cells expressing BLT, LTB4 induced many intracellular signals including a potent chemotactic response. Thus, CHO cells have full equipment for cell locomotion, and in the presence of receptors for chemoattractants they can be useful for analyses of chemotaxis. LTB4 can also bind PPARα, an intranuclear transcription factor, and activate it to increase transcription of genes that terminate inflammatory processes.[16] LTB4 is a unique lipid mediator that activates both cell-surface receptor (BLT) and nuclear receptor (PPARα). Our data provide a clue to the signal-transduction system following LTB4 receptor activation and may help in the rational design of therapeutic antagonists for many disorders in which leukocytes and other inflammatory cells play major roles.

ACKNOWLEDGMENTS

This work was supported in part by grants-in-aid from the Ministry of Education, Science, Sports and Culture and the Ministry of Health and Welfare of Japan, and by grants from Sankyo Foundation of Life Science, the Yamanouchi Foundation for Metabolic Disorders, and Human Science Foundation.

REFERENCES

1. B. Samuelsson, S.E. Dahlen, J.A. Lindgren, C.A. Rouzer and C.N. Serhan, Leukotrienes and lipoxins: structures, biosynthesis, and biological effects, *Science*, 237: 1171, (1987)
2. J.P. Cristol, B. Provencal and P. Sirois, Leukotriene receptors, *J Recept Res*, 9: 341, (1989)
3. A.J. de Brum-Fernandes, G. Guillemette, P. Borgeat and P. Sirois, Specific binding sites for leukotriene B4 in guinea pig brain membranes, *J Pharmacol Exp Ther*, 259: 1035, (1991)
4. Y. Harada, Effect of leukotriene B4 on enhancement of superoxide production evoked by formyl-methionyl-leucyl-phenylalanine in myeloid differentiated HL-60 cells: possible involvement of intracellular calcium influx and high affinity receptor for leukotriene B4, *Hiroshima J Med Sci*, 39: 89, (1990)
5. I. Miki, T. Watanabe, M. Nakamura, Y. Seyama, M. Ui, F. Sato and T. Shimizu, Solubilization and characterization of leukotriene B4 receptor-GTP binding protein complex from porcine spleen, *Biochem Biophys Res Commun*, 166: 342, (1990)
6. K.R. McLeish, P. Gierschik, T. Schepers, D. Sidiropoulos and K.H. Jakobs, Evidence that activation of a common G-protein by receptors for leukotriene B4 and N-formylmethionyl-leucyl-phenylalanine in HL-60 cells occurs by different mechanisms, *Biochem J*, 260: 427, (1989)
7. T. Andersson, W. Schlegel, A. Monod, K.H. Krause, O. Stendahl and D.P. Lew, Leukotriene B4 stimulation of phagocytes results in the formation of inositol 1,4,5-trisphosphate. A second messenger for Ca^{2+} mobilization, *Biochem J*, 240: 333, (1986)
8. R.I. Fonteriz, A. Sanchez, F. Mollinedo, E.D. Collado and S.J. Garcia, The role of intracellular acidification in calcium mobilization in human neutrophils, *Biochim Biophys Acta*, 1093: 1, (1991)
9. J. Palmblad, R. Lerner and S.H. Larsson, Signal transduction mechanisms for leukotriene B4 induced hyperadhesiveness of endothelial cells for neutrophils, *J Immunol*, 152: 262, (1994)
10 T. Yokomizo, T. Izumi, K. Chang, Y. Takuwa and T. Shimizu, A G-protein coupled receptor for leukotriene B$_4$ that mediates chemotaxis, *Nature*, 387: 620, (1997)
11. C. Owman, C. Nilsson and S.J. Lolait, Cloning of cDNA encoding a putative chemoattractant receptor, *Genomics*, 37: 187, (1996)
12. S. Narumiya, Structures, properties and distributions of prostanoid receptors. , *Adv Prostaglandin Thromboxane Leukot Res*, 23: 17, (1995)
13. H. Sumimoto and S. Minakami, Oxidation of 20-hydroxyleukotriene B4 to 20-carboxyleukotriene B4 by human neutrophil microsomes. Role of aldehyde dehydrogenase and leukotriene B4 omega-hydroxylase (cytochrome P-450LTB omega) in leukotriene B4 omega-oxidation, *J Biol Chem*, 265: 4348, (1990)
14. T. Yokomizo, T. Izumi, T. Takahashi, T. Kasama, Y. Kobayashi, F. Sato, Y. Taketani and T. Shimizu, Enzymatic inactivation of leukotriene B$_4$ by a novel enzyme found in the porcine kidney. Purification and properties of leukotriene B$_4$ 12-hydroxydehydrogenase, *J Biol Chem*, 268: 18128, (1993)

15. Z. Honda, T. Takano, Y. Gotoh, E. Nishida, K. Ito and T. Shimizu, Transfected platelet-activating factor receptor activates mitogen-activated protein (MAP) kinase and MAP kinase kinase in Chinese hamster ovary cells, *J Biol Chem*, 269: 2307, (1994)
16. P.R. Devchand, H. Keller, J.M. Peters, M. Vazquez, F.J. Gonzalez and W. Wahli, The PPARalpha-leukotriene B4 pathway to inflammation control, *Nature*, 384: 39, (1996)

36

DETERMINANTS OF RECEPTOR SUBTYPE SPECIFICITY IN THE LPA-LIKE LIPID MEDIATOR FAMILY

G. Tigyi[1], D.J. Fisher[1], K. Lilion[1,2], Z. Guo[1], T. Virag[1], G. Sun[3], D.D. Miller[3], K. Murakami-Murofushi[4], s. Kobayashi[5], and J.R. Erickson[6]

[1]Dept. of Physiology and Biophysic and [3]Dept. of Pharmaceutical Sciences, The University of Tennessee, 894 Union Avenue, Memphis, TN 38163, USA
[2]Institute of Enzymology, Biological Research Center of the Hungarian Academy of Sciences, Hungary
[4]Dept. of Biology, Faculty of Science, Ochanomizu University, Japan
[5]Dept. of Pharmaceutical Sciences, Science University of Tokyo, Japan
[6]LXR Biotechnology Inc., Richmond, CA, USA

INTRODUCTION

In addition to polypeptide growth factors, an emerging group of naturally occurring phospholipid growth factors has been discovered[1]. Within this group, lysophosphatidic acid (LPA, 1-acyl-2-lyso-*sn*-glycero-3-phosphate), nature's simplest phospholipid, was found to elicit growth factory-like effects in almost every cell type spanning the phylogenetic tree, from *Dictyostelium* to humans. These biological effects include: 1) the mitogenic[2] or antimitogenic regulation of the cell cycle[3], 2) regulation of Ca^{2+} homeostasis[4]. 3) regulation of cell shape[5], migration[6], and tumor cell invasiveness[7], and 4) the prevention of apoptosis[8]. LPA elicits these effects via multiple G protein-coupled receptors. Here we report that naturally occurring analogs of LPA elicit distinct cellular responses via the selective activation of receptor subligand specificities towards LPA, cyclic-phosphatidic acid (cyclic-PA), and plasmalogenglycerophosphate (alkenyl-GP).

MULTIPLE PHOSPHOLIPIDS WITH LPA-LIKE EFFECTS

LPA elicits oscillatory Cl⁻ currents in *Xenopus* oocytes via the receptor-mediated activation of the inositol trisphosphate-Ca^{2+} second messenger system[4]. In addition to LPA, several other structurally similar phospholipids elicit oscillatory Cl⁻ currents in oocyte (Fig. 1). Glycerophosphates with ether- as well as ester-linked hydrocarbon chains are active in the nanomolar-to-micromolar concentration range. LPA heterologously desensitizes the response to each active compound shown in Fig. 1, suggesting that they act on overlapping receptors.

Eicosanoids and Other Bioactive Lipids in Cancer, Inflammation, and Radiation Injury, 4
Edited by Honn *et al.*, Kluwer Academic / Plenum Publishers, New York, 1999.

245

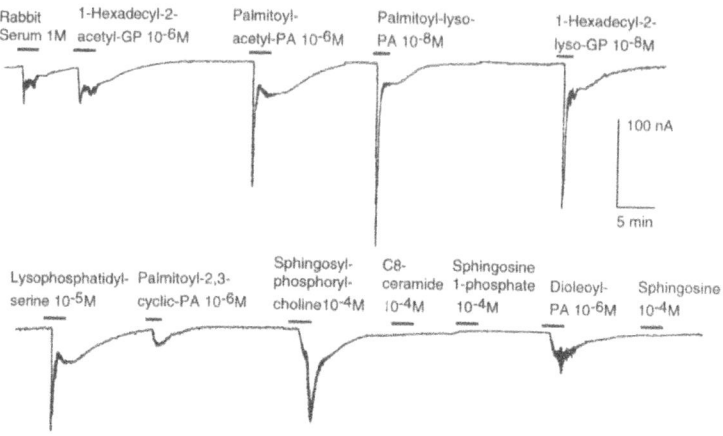

Fig. 1. Multiple lipid mediators elicit oscillatory cl⁻ currents in *Xenopus* oocytes.

Among the sphingolipids, sphingosylphosphorycholine (SPC) was active at high micromolar concentrations. It was found[6] that certain commercial SPC, as well as SPP, preparations (e.g., Sigma, Toronto /Biochemicals), contain a 1-O-*cis*-alk-1'-enyl-2-lyso-*sn*-glycero-3-phosphate (alkenyl-GP) impurity, which is responsible for the LPA-like effects.

LPA, cyclic-PA, and alkenyl-GP are all naturally occurring and chemically stable phospholipids that have different cellular effects. While LPA and alkenyl-GP stimulate proliferation[2,9,10], cyclic-PA inhibits cell growth[11]. Interestingly, LPA can be mitogenic or antimitogenic depending on the cell type[3]. One way to explain such diverse and sometimes opposing effects elicited by LPA, is through multiple receptors with distinct ligand and signal transduction properties that are expressed in a cell-type-specific manner. This idea is in contrast to the current view, which assumes that LPA signals through a single G-protein-coupled receptor[12].

HETEROLOGOUS DESENSITIZATION REVEALS MULTIPLE PHARMA-COLOGICAL SUBTYPES OF RECEPTORS SELECTIVELY ACTIVATED BY NATURALLY OCCURRING LPA ANALOGS

Following homologous desensitization with an agonist, the lack of responsiveness to a heterologous ligand indicates that the two ligands share the same receptor, whereas if the cells are responsive to the heterologous ligand, then the two ligands must activate different receptors. Using heterologous desensitization between LPA, alkenyl-GP, and cyclic-PA in *Xenopus* oocytes, a complex pattern was observed (Fig. 2). Following desensitization with LPA, oocytes were cross-desensitized to alkenyl-GP and cyclic-PA (Fig. 2A). Oocytes desensitized with alkenyl-GP were cross-desensitized to cyclic-PA, but were only partially desensitized to LPA (Fig. 2B). Oocytes desensitized with cyclic-PA remained partially responsive to both alkenyl-GP and LPA (Figure 2C). This pattern of partial heterologous desensitization indicates that oocytes express multiple subtypes of lipid receptors and that these receptors distinguish between LPA, alkenyl-GP, and cyclic-PA.

the pattern of heterologous desensitization in NIH3T3 fibroblasts is similar to that observed in *Xenopus* oocytes. LPA, alkenyl-GP, and cyclic-PA elicit inositol trisphosphate production and a transient elevation of $[Ca^{2+}]_i$. When NIH3T3 cells were desensitized with

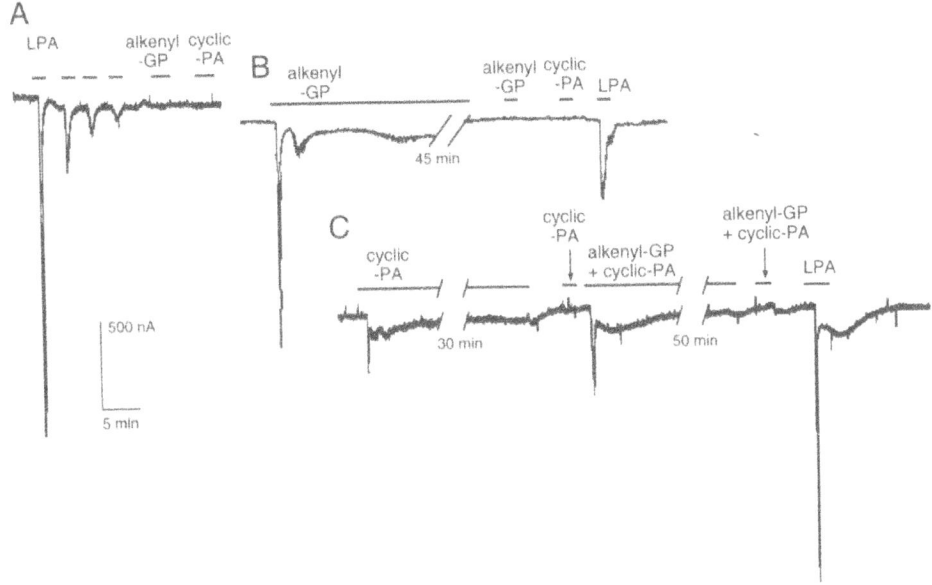

Fig. 2. Heterologous (cross)-desensitization between LPA-like lipid mediators in the *Xenopus* oocyte.

LPA, the cells were no longer responsive to cyclic-PA or alkenyl-GP (Fig. 3A), but the cells remained responsible to the unrelated hormone bradykinin. Cells desensitized with alkenyl-GP were unresponsive to cyclic-PA, but were partially responsive to both alkenyl-GP (Fig. 3B). Cells desensitized with cyclic-PA were partially responsive to both alkenyl-GP and LPA (Fig. 3C). The reduction in the size of Ca^{2+} transients for each ligand was paralleled by a similar reduction in the generation of $InsP_3$ (data not shown).

The complex pattern of heterologous desensitization (Figs. 2 & 3) indicates the expression of multiple receptors in the same cell, which are selectively activated by one or the other ligand. These data also indicate that alkenyl-GP and LPA are promiscuous ligands that activate than a single subset of receptors. Furthermore, this pattern of desensitization indicates the expression of a receptor subtype that is specific for LPA and is not activated by cyclic-PA or alkenyl-GP. To explain this pattern of desensitization, based on the specificity of the three ligands, we propose a model to distinguish among the LPA receptors coupled to the $InsP_3$-Ca^{2+} second messenger system. We propose to designate Type I receptors that are specific for LPA and are not activated by cyclic-PA and alkenyl-GP. Type II receptors are activated by LPA and alkenyl-GP, but not activated by cyclic-PA. Type III receptors are selectively activated by cyclic-PA and also nonselectively activated by alkenyl-GP and LPA.

SIGNALING EVENTS ELICITED BY THE DIFFERENT LPA ANALOGS

NIH3T3 fibroblasts were used to examine the effects of LPA, alkenyl-GP, and cyclic-PA on intracellular cAMP levels. Figure 4A shows the changes in basal cAMP levels elicited by a 10μM concentration of each of the three lipids, or solvent control, in subconfluent cultures of NIH3T3 cells. Following a 5-min treatment with LPA or alkenyl-GP, the basal cAMP level decreased by approximately 25%, whereas cyclic-PA increased cAMP 1.6-fold.

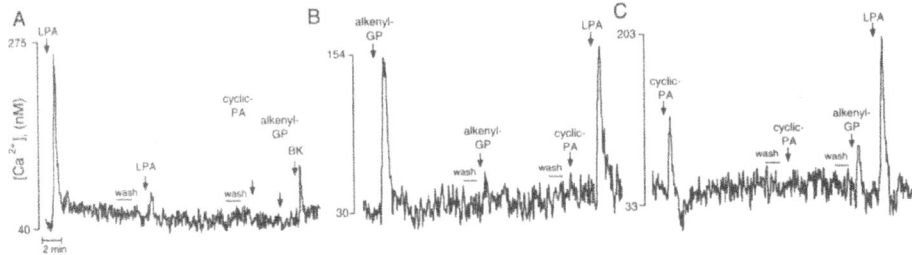

Fig. 3. Heterologous desensitization between LPA-like lipids in NIH3T3 fibroblast cells.

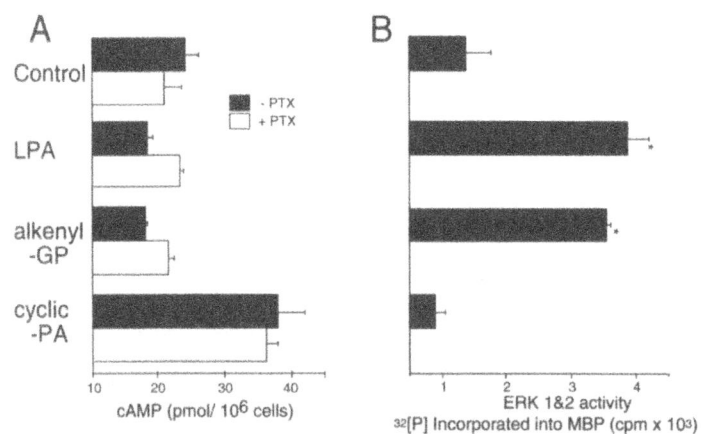

Fig. 4. Differential effects of LPA analogs on cAMP (A) and ERK 1 & 2 activity (B) in NIH3T3 fibroblasts. Lipids were applied at a concentration of 1μM for (A) or 10 μM for (B).

In parallel cultures that were treated for 18h with 100ng/ml of PTX, LPA and alkenyl-GP no longer decreased cAMP, but showed a light increase over control levels. In contrast, the cyclic-PA-stimulated increase in cAMP was unaffected by PTX.

We also examined the activation of the MAP kinases ERK 1 & 2. Following a 10-min treatment, LPA and alkenyl-GP elicited a significant increase in the ERK 1 & 2 activity, with a 2.7- and a 2.5-fold increase over control levels, respectively (Fig. 4B). In contrast, cyclic-PA caused a slight inhibition of the basal ERK 1 & 2 activity.

The three lipids were tested for their effects on the proliferation of NIH3T3 fibroblasts. 10^4 cells/cm^2 were plated in 2% FBS. Twenty-four hours later, the cells were changed over to MCDB-104 medium, containing 25 ng/ml bFGF, 5μg/ml insulin and transferrin, and 10 ng/ml dexamethasone. A 10-μM concentration of each lipid was added, and the cells were counted after 3 d (Fig. 5). LPA caused a 2.3-fold and alkenyl-GP caused a 2-fold increase in cell number, whereas cyclic-PA caused a 27% decrease in cell number. These effects were all statistically significant as compared to the control. The proliferative effects of LPA and alkenyl-GP and the antiproliferative effect of cyclic-PA were all dose-dependent, with apparent EC_{50} values of $1.7\mu M$ for the LPA and $1.5\mu M$ for the alkenyl-GP, and with an apparent IC_{50} of $1.6\ \mu M$ for cyclic-PA.

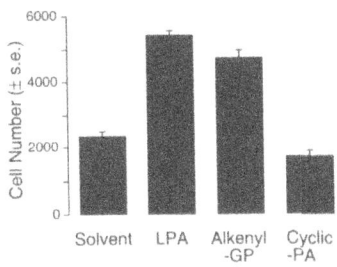

Fig. 5. LPA and alkenyl-GP induce, whereas cyclic-PA inhibits, the proliferation of NIH3T3 cells.

THE PSP24 RECEPTOR IS NOT ACTIVATED BY ALKENYL-GP OR CY-CLIC-PA IN *XENOPUS* OOCYTES

The PSP24 gene has recently been identified as a functional LPA receptor that couples to the phosphoinositide-Ca^{2+} signaling pathway[13]. We have expanded the characterization of this receptor to include alkenyl-GP along with LPA and cyclic-PA (Fig. 6A). Oocytes were either sham-injected with buffer or with cRNA that encodes the PSP24 receptors. Oocytes

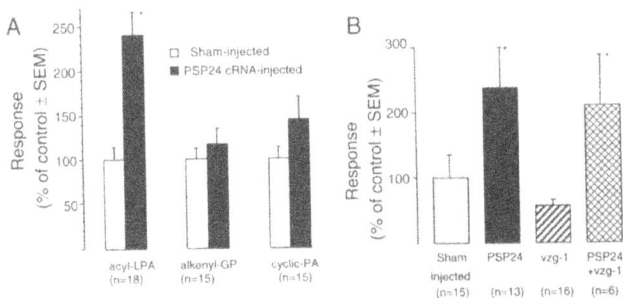

Fig. 6. The PSP24 receptor is selectively activated by LPA but not by alkenyl-GP or cyclic-PA in *Xenopus* oocytes. In contrast, expression of the vzg-1/Edg-2 receptor does not show functional coupling in the oocyte.

over-expressing the PSP24 cRNA showed a 2.5-fold increase over control in response to LPA (10 nM); however, the response to alkenyl-GP and cyclic-PA did not change significantly.

The oscillatory Cl current response to the lipids remained at control level in oocytes injected with cRNA for the vzg-1/Edg-2 LPA receptor[14] (Fig. 6B). When oocytes were co-injected with a mixture of PSP24 and vzg-1/Edg-2 cRNA, the increased response elicited by LPA was similar to that found in oocytes that were injected with the PSP24 cRNA alone. This type of ligand selectivity agrees with that of a predicted type I receptor.

Because the vzg-1/Edg-2 LPA receptor did not functionally couple to C^{2+} mobilization in the oocyte, we developed a heterologous expression system in S. *cerevisiae* cells[15]. In S. *cerevisiae* expressing the Edg-2/vzg-1 receptor, LPA elicits the dose- and time-dependent activation of a FUS1::*lacZ* reporter gene via the yeast mitogen-activated protein kinase signaling pathway. The activities of LPA, alkenyl-GP, and cyclic-PA were compared in the S. *cerevisiae* assay (Fig. 7). All three lipids were effective in activating *lacZ* reporter gene expression, indicating that the vzg-1/Edg-2 receptor shows the ligand selectivity of a predicted type III receptor.

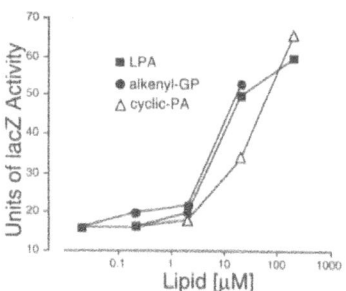

Fig. 7. LPA, alkenyl-GP, and cyclic-PA activate the *lacZ* reporter gene in S. *cerevisiae* expressing the Edg-2/vzg-1 receptor.

Pharmacological and genetic evidence indicates multiple receptors for LPA-like lipid mediators. Although we recognize that the classification of the receptors proposed above is incomplete, the concept of different LPA-like mediators that act on multiple receptors should enhance the current research and offers a more comprehensive explanation of the diverse and cell-type-specific responses activated by LPA. Clearly, more research is necessary to refine this model in two important ways. First, more of the endogenous, as well as synthetic, receptor-specific ligands must be identified; and, second, more receptor genes must be cloned.

ACKNOWLEDGMENTS

This work was supported by grants from NSF (IBN-9722969); LXR Biotechnology Inc.; The Ministry of Culture and Education; the Science and Technology Agency of Japan. GT is an Established Investigator of the American Heart Association. DJF is a Fellow of the AHA Tennessee Affiliate and recipient of an NIH Training Grant (USPHS-HL07746).

REFERENCES

1.   A. Tokumura. A family of phospholipid autacoids: Occurrence, metabolism, and bioactions. *Prog. Lipid. Res.* 34:151 (1995).
2.   E.J. van Corven, A. Groenink, K. Jalink, t. Eichholtz, and W.H. Moolenar. Lyso-phosphatidic acid-induced cell proliferation: Identification and dissection of signaling pathways mediated by G proteins. *Cell* 59:45 (1989).

3.   G. Tigyi, D. Dyer, and R. Miledi. Lysophosphatidic acid possesses dual action in cell proliferation. *Proc. Nat. Acad. Sci. USA* 91:1908 (1994).

4.   G. Tigyi and R. Miledi. Lysophosphatidates bound to serum albumin activate membrane currents in *Xenopus* oocytes and neurite retraction in PC12 pheochromocytoma cells. *J. Biol. Chem.* 267:21360 (1992).

5.   G. Tigyi, D. Fischer, C. Yang, D. Dyer, A. Sebok, and R. Miledi. Lysophosphatidic acid-induced neurite retraction in PC12 cells: Conrol by phosphoinositide-Ca^{2+} signaling and Rho. *J. Neurochem.* 66:537 (1996).

6.   D. Zhou, W. Luini, S. Bernasconi, L. Diomede, M. Salmona, A. Mantovani, and S. Sozzani. Phosphatidic acid and lysophosphatidic acid induce heptotactic migration of human monocytes. *J. Biol. Chem.* 270:25549 (1995).

7.   F. Imamura, t. Horai, M. Mukai, K. Shinkai, M. Sawada, and H. Akedo. Inductin of *in vitro* tumor cell invasion of cellular monolayers by lysophosphatidic acid or phospholipase D. *Biochem. Biophys. Res. Commun.* 193:497 (1993).

8.   S.R. Umansky, J.P. Shapiro, G.M. Cuenco, W.M. Foehr, I.C. Bathurst, and L.D. Tomei. Prevention of rat neonatal cardiomyocyte apoptosis induced by stimulated *in vitro* ischemia and reperfusion. *Cell Ceath Diff.* 4:608 (1997).

9.   K. Liliom, D.J. Fischer, G. Sun, D.D. Miller, J.L. Tseng, D.M. Desiderio, M.C. Seidel, J.R. Erickson, and G. Tigyi. Identificatin of a novel growth factor-like lipid: 1-O-cis-alk-1'-enyl-2lyso-*sn*-glycero-3-phosphage (Alkenyl-GP) that is present in commercial sphingolipid preparations. *J. Biol. Chem.* 273:in press (1998).

10. K. Lilion, Z. guan, J.L. Tseng, D.M. Desiderio, G. Tigyi, and M. Watsky. Growth factor-like phospholipids generated following corneal injury. *Am. J. Physiol. (Cell Physiol.* 43) 274:in press (1998).

11. K. Murakami-Murofushi, K. Kaji, K. Kano, M. Fukuda, M. Shioda, and H. Murofushi. Inhibition of cell proliferation by a unique lysophosphatidic acid, PHYLPA, isolated from *Physarum polycephalum:* signaling events of antiproliferative action by PHYLPA. *Cell Struct. Funct.* 18:363 (1993).

12. W.H. Moolenaar, O. Kranenburg, F.R. Postma, and G. Zondag. Lysophosphatidic acid: G protein signaling and cellular responses. *Curr. Opin. Cell Biol.* 9:168 (1997).

13. Z. Guo, K. Lilion, D.J. Fischer, I.C. Bathurst, D.J. tomei, M.C. Kiefer, and G.Tigyi. Molecular cloning of a h igh affinity receptor for the growth factory-like phospholipid mediator lysophosphatidic acid. *Proc. Natl. Acad. Sci. USA* 93:14367 (1996).

14. J. H. Hecht, J.A. Weiner, S.R. Pots, and J. Chun. Vzg-1 encodes a lysophosphatidic acid receptor expressed in neurogenic regions of the developing cerebral cortex. *J. Cell. Biol.* 135:1071 (1996).

15. J. R. Erickson, J.J. Wu, g. Goddard, K. Kawanishi, G. Tigyi, L.D. Tomei, and M.C. Kiefer. The putative lysophosphatidic acid receptor Edg-2/Vzg-1 functionally couples to the yeast response pathway. *J. Biol. Chem.* 273:1506 (1998).

37

SUBUNITS AND CELLULAR OCCURRENCE OF THE 12(S)-HETE BINDING COMPLEX

Helena Herbertsson, Tobias Kühme and Sven Hammarström

Department of Biomedicine and Surgery
Division of Cell Biology
Linköping University
S-58185 Linköping, Sweden

INTRODUCTION

12(S)-Hydroxyeicosatetraenoic acid is a 12-lipoxygenase product of arachidonic acid[1]. 12(S)-HETE has been shown to be produced in a variety of cells and a major source is platelets[2]. Several other bone-marrow derived cells, such as macrophages, erythrocytes, lymphocytes and neutrophils also produce 12(S)-HETE, usually as a minor metabolite. The same is true for vascular cells, endothelium as well as smooth muscle, cells in the central nervous system and keratinocytes. 12(S)-HETE has diverse effects in these cells: It has been proposed to play a role in platelet activation[3], in the pathogenesis of psoriasis[4, 5], in neurotransmitter peptide-induced hyperpolarization by neuronal cells[6] and in modulating the metastatic potential of melanoma and carcinoma cells[7]. 12-Lipoxygenase has been reported to influence the expression of bcl-2 protein[8], an essential regulator of cell survival and apoptosis.

Lewis lung carcinoma (LLC) cells contain specific high affinity binding sites for 12(S)-HETE[9]. The binding was saturable, reversible and specific for 12(S)-HETE. The Kd was about 0.3 nM and there were approximately 66,000 binding sites per cell. Binding was specific; only at large excess concentrations of 5(S)-HETE, 15(S)-HETE or the linoleic acid metabolite 13(S)-HODE was some displacement of 12(S)-HETE binding observed.

The subcellular distribution of the 12(S)-HETE binding sites was mainly cytosolic (52%); some 30% of 12(S)-HETE binding was observed in mitochondria and *ca* 18% in nuclei. A single binding unit of high molecular weight (apparent $M_r \approx 650$ kDa) was detected when cytosol from prelabeled cells was analyzed by gel permeation chromatography and sucrose density centrifugation[10].

Here we present data considering occurrence of this 650 kDa binding sites in some other cell lines as well as the identification of components of the 12(S)-HETE binding complex. For information on methods, see Herbertsson et al. [11]

Eicosanoids and Other Bioactive Lipids in Cancer, Inflammation, and Radiation Injury, 4
Edited by Honn *et al.*, Kluwer Academic / Plenum Publishers, New York, 1999.

253

OCCURRENCE OF THE 650 kDa 12(S)-HETE BINDING COMPLEX IN VARIOUS CELL LINES

Table 1 shows a comparison of the ability of five cell lines to form a 650 kDa 12(S)-HETE binding complex. The cells were incubated with 0.1 nM 12(S)-[^{3}H]HETE in the presence or absence of unlabeled 12(S)-HETE. After 2 hrs at 4°C cytosol was prepared and analyzed by gel permeation chromatography. The amount of radioactivity in fractions containing the 650 kDa binding complex was compared.

Table 1.

Cell line	650 kDa-bound 12(S)-HETE (fmol/10^6 cells)
LLC (murine Lewis lung carcinoma)	0.22±0.07; n=3
HEL (human erythroleukemia)	0.14±0.04; n=3
TPA-differentiated (0.62 ng/ml, 4 days)	0.18±0.07; n=3
U937 (human promonocytic leukemia)	0.41±0.16; n=4
TPA-differentiated (20 ng/ml, 4-5 days)	0.84±0.31; n=4
3T3-L1 (murine preadipocytes)	0.19±0.03; n=3
Insulin/IBMX/dexamethasone-differentiated	0.03±0.01; n=3
Int407 (human intestinal epithelium)	0.01±0.004; n=3

HEL cells bound 45% 12(S)-HETE in the 650 kDa complex and 55% in a 50 kDa containing fraction. TPA-induction did not affect the binding ability.

U937, promonocytic leukemia cells interestingly bound about twice as much as Lewis lung carcinoma cells. When treated with TPA these cells differentiate to monocyte-like cells; concomitantly the binding capacity was doubled.

3T3-L1 preadipocytes bound approximately the same amount of 12(S)-HETE as Lewis lung carcinoma cells. When these cells had been differentiated to adipocytes the binding capacity was reduced about six times.

In intestine 407 cells there was no detectable 650 kDa complex.

DISSOCIATION OF THE 12(S)-HETE BINDING COMPLEX INTO SUBUNITS

Cytosol, prepared from LLC cells was prelabeled with 12(S)-[^{3}H]HETE and then incubated with apyrase to deplete ATP. An increase (24%, mean value; n=5) of the 650 kDa complex was observed compared to a nonapyrase-treated control (Fig. 1). When ATP was regenerated in apyrase-treated cytosol there was a decrease (23%, mean value; n=5) in the amount of the 650 kDa complex.

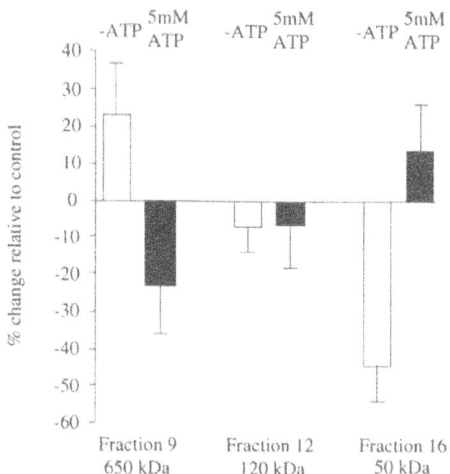

Figure 1. ATP depletion enhances formation of the 650 kDa binding complex. LLC cells were incubated with 0.1 nM 12(S)-[³H]HETE for 2 hrs at 4°C. Cytosol was prepared and divided into three parts; a) control, b) sample in which ATP was depleted and c) sample in which ATP was first depleted and then regenerated. The three samples were analyzed on Superdex™200. Data are presented as mean values ± S.E. from n=5. Reproduced from (11) with permission of the publisher.

When ATP was regenerated, two additional radiolabeled components of apparent molecular weight 120 kDa and 50 kDa, respectively, appeared in addition to the 650 kDa complex (Fig. 2).

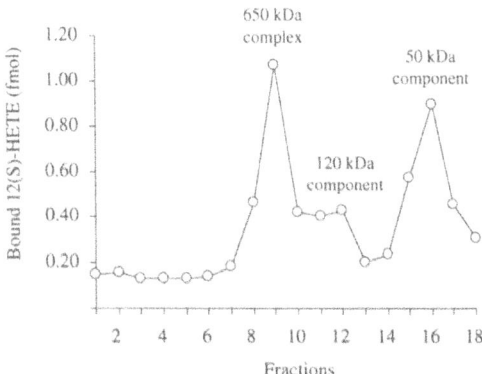

Figure 2. ATP regeneration leads to dissociation of the 650 kDa binding complex to a 50 kDa 12(S)-HETE binding component. Apyrase treated cytosol from cells preincubated with 12(S)-[³H]HETE (0.1 nM, 2 hrs, 4°C) was incubated with an ATP-regenerating system. It was then analyzed on Superdex™200. Reproduced from (11) with permission of the publisher.

The relative amount of the 120 kDa binding component was not significantly influenced by the ATP concentration but the 50 kDa component was increased (14%, mean value; n=5) when ATP was regenerated. A 50 kDa subunit was also observed in the control cytosol that had been incubated for 30 minutes at room temperature without apyrase or ATP regeneration. ATP depletion resulted in a 45% reduction (mean value; n=5) of the 50 kDa component. These results suggest that ATP destabilizes the association of subunits in the 650 kDa 12(S)-HETE binding complex.

IDENTIFICATION OF HSP70 AS A COMPONENT OF THE 12(S)-HETE BINDING COMPLEX

Western blot analyses

Cytosol was prepared from LLC cells which had been preincubated with unlabeled 12(S)-HETE (0.1 nM) or from non-incubated control cells. After fractionation on Superdex™200, the high molecular weight fractions were analyzed on 7.5% SDS-PAGE, electrotransferred to nitrocellulose and incubated with a monoclonal hsp70 antibody followed by HRP-conjugated anti-mouse IgG for ECL-detection. The intensity of the hsp70 band was highest in fraction 9 which also exhibited the highest specific 12(S)-HETE binding (Fig. 3). In control cells a lower concentration of hsp70 was detected compared to cells incubated with 12(S)-HETE. Also in this case the intensity of the hsp70 band was highest in the fraction containing the 650 kDa complex.

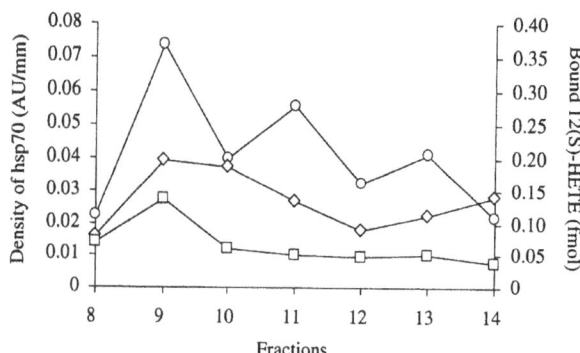

Figure 3. Densitometric measurements of hsp70 immunoreactivity. O LLC cells preincubated with 0.1 nM 12(S)-HETE (2 hrs, 4°C), ◇ nonincubated control cells. □ specific 12(S)-HETE binding is also shown for comparison. Reproduced from (11) with permission of the publisher.

Immunoprecipitation

To immunoprecipitate the 650 kDa complex with hsp70 antibody, cytosol was first depleted of ATP by apyrase to stabilize the 12(S)-HETE binding complex and then incubated with anti-hsp70.

256

After sedimentation of immunocomplexes bound to protein A Sepharose the immunoprecipitate was treated with 0.5 M NaCl to release the 12(S)-HETE binding complex from hsp70. Western blot analyses showed that hsp70 was only present in the protein A Sepharose pellet but not in the remaining cytosol after immunoprecipitation. Gel permeation chromatography after incubation with 12(S)-[³H]HETE showed that 16±4% (mean value ± S.E., n=3) of the 650 kDa complex had coprecipitated with the immunocomplexes while 84% still resided in the hsp70-depleted cytosol. This indicated that hsp70 was associated with the 12(S)-HETE binding complex in substoichiometric amounts at an approximate 1:6 ratio.

CONCLUSIONS

The cytosolic 650 kDa 12(S)-HETE binding complex was detected in all cell lines studied except in intestine 407. Differentiation of promonocytic U937 by TPA doubled the binding capacity which might implicate a role for 12(S)-HETE in the differentiated cell. In contrast, when 3T3-L1 preadipocytes were differentiated to adipocytes, the binding capacity was downregulated. This could indicate that 12(S)-HETE is not needed in differentiated adipocytes.

The data presented suggests that the cytosolic binding complex is multimeric. Part of the 650 kDa complex dissociated into 120 kDa and 50 kDa 12(S)-HETE binding components when the ATP concentration was increased. The 120 kDa 12(S)-HETE binding component which was observed is probably an intermediate in the transformation of the 650 kDa complex to the 50 kDa binding subunit. The 50 kDa polypeptide is presumed to be the actual 12(S)-HETE binding protein.

Western blot and coimmunoprecipitation experiments indicated that hsp70 is part of the 12(S)-HETE binding complex. Although the 12(S)-HETE binding complex has not yet been purified to homogeneity the observation that it coeluted with hsp70 immunoreactivity and that ligand binding increased the hsp70 immunoreactivity of the complex suggested that hsp70 is associated with the 12(S)-HETE binding complex. Hsp70 has intrinsic ATPase activity and ATP binding and hydrolysis results in peptide release. The results of the immunoprecipitation experiments indicated that hsp70 binds substoichiometrically to the 12(S)-HETE binding complex in a ratio similar to what has been reported for the glucocorticoid hormone receptor[12].

Several similarities between the 12(S)-HETE binding complex in carcinoma cells and cytosolic ligand-dependent transcription factors of the steroid/thyroid hormone receptor superfamily seem to exist. The similarities include a cytosolic/nuclear localization, subnanomolar Kd for the ligand and substoichiometric occurrence of hsp70 in a large receptor heterocomplex. Heat-shock proteins are essential parts of the binding complexes for steroid hormones. Hsp70 is involved in transport across intracellular membranes, hsp90 keeps the receptor in a high-affinity state for ligand binding[13] and hsp 56 is probably involved in receptor trafficking to the nucleus, possibly functioning as the nuclear localization signal recognition protein[14].

REFERENCES

1. S. Yamamoto, "Enzymatic" Lipid Peroxidation : Reactions of Mammalian Lipoxygenases. *Free Radic. Biol. Med.* 10: 149 (1991).

2. A.A. Spector, J.A. Gordon, and S.A. Moore. Hydroxyeicosatetraenoic Acids (HETEs). *Prog. Lipid Res.* 27: 271 (1988).
3. F. Sekiya, J. Takagi, T. Usui, K. Kawajiri, Y. Kobayashi, F. Sato, and Y. Saito. 12S-Hydroxyeicosatetraenoic Acid Plays a Central Role in the Regulation of Platelet Activation. *Biochem. Biophys. Res. Commun.* 179: 345 (1991).
4. S. Hammarström, M. Hamberg, B. Samuelsson, E.A. Duell, M. Stawiski , and J.J. Voorhees. Increased Concentrations of Nonesterified Arachidonic Acid, 12L-hydroxi-5,8,10,14-eicosatetraenoic acid, Prostaglandin E_2 and Prostaglandin $F_{2\alpha}$ in Epidermis of Psoriasis. *Proc. Natl. Acad. Sci. USA.* 72: 5130 (1975).
5. T. Ruzicka, The Role of the Epidermal 12-Hydroxyeicosatetraenoic Acid Receptor in the Skin. *Eicosanoids Suppl.* 5: S63 (1992).
6. D. Piomelli, A. Volterra, N. Dale, S. A. Siegelbaum, E. R. Kandel, J.H. Schwartz, and F. Belardetti. Lipoxygenase Metabolites of Arachidonic Acid As Second Messengers for Presynaptic Inhibition of Aplysia Sensory Cells. *Nature* 328: 38 (1987).
7. K.V. Honn, I.M. Grossi, L.A. Fitzgerald, L.A. Umbarger, C.A. Diglio, and J.D. Taylor. Lipoxygenase Products Regulate IRGpIIb/IIIa Receptor Mediated Adhesion of Tumor Cells to Endothelial Cells, Subendothelial Matrix and Fibronectin. *Proc. Soc. Exp. Biol. Med.* 189: 130 (1988).
8. D.G. Tang, Y.Q. Chen, and K.V. Honn. Arachidonate Lipoxygenases as Essential Regulators of Cell Survival and Apoptosis. *Proc. Natl. Acad. Sci. USA.* 93: 5241 (1996).
9. H. Herbertsson, and S. Hammarström. High-Affinity Binding Sites for 12(S)-Hydroxy-5,8,10,14-eicosatetraenoic acid (12(S)-HETE) in Carcinoma Cells. *FEBS Letters.* 298: 249 (1992).
10. H. Herbertsson, and S. Hammarström. Subcellular Localization of 12(S)-Hydroxy-5,8,10,14-eicosatetraenoic Acid Binding Sites in Lewis Lung Carcinoma Cells. *Biochim. Biophys. Acta.* 1244: 191 (1995).
11. H. Herbertsson, T. Kühme, U. Evertsson, J. Wigren, and S. Hammarström. Identification of Subunits of the 650 kDa 12(S)-HETE Binding Complex in Carcinoma Cells. *J. Lipid Res.* 39: 237-244 (1998).
12. E.E. Diehl, and T.J. Schmidt. Heat Shock Protein 70 Is Associated in Substoichiometric Amounts with the Rat Hepatic Glucocorticoid Receptor. *Biochemistry.* 32: 13510 (1993).
13. W.B. Pratt, The Role of Heat Shock Proteins in Regulating the Function, Folding and Trafficking of the Glucocorticoid Receptor. *J. Biol. Chem.* 268: 21455 (1993).
14. M.J. Czar, R.H. Lyons, M.J. Welsh, J.-M. Renoir, and W.B. Pratt. Evidence That the FK506-Binding Immunophilin Heat Shock Protein 56 Is Required for Trafficking of the Glucocorticoid Receptor from the Cytoplasm to the Nucleus. *Mol. Endocrinol.* 9: 1549 (1995).

38

A SUBFAMILY OF G PROTEIN-COUPLED CELLULAR RECEPTORS FOR LYSOPHOSPHOLIPIDS AND LYSOSPHINGOLIPIDS

Edward J. Goetzl and Songzhu An

Departments of Medicine and Microbiology-Immunology
University of California Medical Center
San Francisco, CA 94143-0711

INTRODUCTION: THE BIOCHEMISTRY AND BIOLOGY OF LIPID PHOSPHORIC ACIDS

The lipid phosphoric acids, including lysophosphatidic acid (LPA) and sphingosine 1-phosphate (S1P), are generated from membranes of many different types of stimulated cells. In extracellular fluids, these distinctive phospholipids are potent mediators of growth and many other cellular functions,[1,2] and intracellularly serve as messengers of diverse signals from cell-surface receptors.[3] Both lipid phosphoric acid mediators are characterized by widespread cellular production, often separate synthetic pathways for the intracellular and secreted pools, micromolar maximal concentrations in serum and some other fluids, high levels of binding to albumin and some other serum proteins, and biodegradation by one or more specific enzymatic mechanisms. The multicomponent pathway for biosynthesis of secreted LPA is distinct from those which produce LPA in cellular metabolism, and is one well-described example of the compartmental generation of lipid phosphoric acids. Efficient generation of LPA for secretion requires sequential release of plasma membrane vesicles from stimulated leukocytes, platelets or other cells, exposure of the vesicles to sphingomyelinase, production of phosphatidic acid (PA) by phospholipase (PL) C- and/or D-dependent mechanisms, and conversion of PA to LPA by a secretory PLA2.[4] LPA is enzymatically inactivated by lysophospholipases, acetyltransferases and phosphatidate phosphohydrolases of differing specificities, whereas S1P is cleaved predominantly by a lyase.[2,5] The steady-state tissue and fluid levels of LPA and S1P are determined by the ratio of enzymatic activities of the biosynthetic and degradative systems. LPA has many potent effects on cellular proliferation and differentiation, cellular migration, platelet shape change and aggregation, smooth muscle tone, and activities of membrane ion channels and enzymes.[1] The broad range of effects of S1P is similar to that of LPA, but S1P is a more prominent mediator of endothelial cell adherence and junction formation, and neurite retraction.[2] Early results of studies of lipid phosphoric acids in cell death suggest that S1P and LPA may be

Eicosanoids and Other Bioactive Lipids in Cancer, Inflammation, and Radiation Injury, 4
Edited by Honn *et al.*, Kluwer Academic / Plenum Publishers, New York, 1999.

259

active inhibitors of apoptosis. The capacity to suppress formation of intracellular S1P with selective inhibitors of sphingosine kinase has facilitated investigations of its possible role as an intracellular messenger,[3] whereas any similar functions of LPA and PA are less clear. The different contributions of S1P exhibit cellular specificity. Intracellular signaling functions of S1P are best appreciated in fibroblasts and mast cells, and extracellular mediator activities predominate in platelets, neurons, cardiac myocytes and macrophages.

G PROTEIN-COUPLED CELLULAR RECEPTORS (GPCRs) FOR LPA AND S1P

The existence of GPCRs for lipid phosphoric acid mediators was suggested by ligand structural-dependence of their effects, ligand-induced desensitization of the responses of some cells, and inhibition of the cellular proliferation-promoting and Ca^{++}-mobilizing activities of LPA and S1P by pertussis toxin. The requirement for removal of serum, as a source of the mediators, from cell cultures to elicit optimal responses, an apparent bimodal concentration-dependence of many of their cellular effects with both nanomolar and micromolar maxima, the ability of virtually every type of cell to recognize and respond to LPA and/or S1P, and the very high level of their unspecific binding to cultured cells *in vitro* have complicated the search for lipid phosphoric acid-specific GPCRs. Two very recent developments have permitted the structural definition of one subfamily of highly homologous GPCRs for LPA and S1P. One was the appreciation that expression of at least one neural subset of LPA-dedicated GPCRs is developmentally-regulated in rat brain.[6] Another was the successful application of serum response element-based reporter constructs for expression cloning.[7] This report will review recent discoveries of a series of endothelial differentiation gene (edg)-encoded cellular proteins and the identification of each Egd protein as a distinctive GPCR for LPA or S1P.

Retrovirally-immortalized clones of mitotically-active neurons from the ventricular zone (vz) of rat embryonic cerebral cortex, that lack superficial processes during mitosis and are the source of most cortical neurons in adult brain, selectively expressed mRNA encoding a novel GPCR.[6] The cDNA encoding this GPCR initially was designated vzg-1, and also was referred to as edg-2 based on the highest homology of its amino acid sequence with a human orphan GPCR termed Edg-1 protein, earlier found to be encoded by an immediate-early gene expressed after endothelial cell activation. Rat Edg-2 protein was identified as an LPA receptor based on the finding that overexpression led to increased specific binding of LPA and LPA induction of the rounded shape of mitotic vz neurons. The highest embryonic cerebral expression of edg-2 was in the vz, as assessed by *in situ* hybridization, and the distribution of Edg-2 GPCRs correlated with that of mitotic activity.

LPA stimulation of cellular proliferation is dependent on induction and binding of both serum response factor (SRF) and ternary complex factor (TCF) proteins to the nuclear serum response element (SRE) present in promoter sequences of many growth-related genes.[8] These observations suggested that reporter cells created by transfection with luciferase-containing plasmids bearing multiple tandem copies of SRE would permit expression-cloning of LPA GPCRs. Double transfection of K562 cells with an SRE-luciferase plasmid and pools of cDNA from a human lung library, as well as a large panel of human cDNAs encoding GPCRs homologous to human Edg-1, sheep Edg-2 and many GPCRs of known specificity for eicosanoid and phospholipid mediators led to the identification of human Edg-2 protein as a highly specific GPCR for LPA.[7] Reporter cells showed LPA concentration-dependent responses, but none to other lysophospholipids or a range of sphingolipids. Overexpression of human Edg-2 protein in CHO cells increased specific binding of ^{3}H-LPA.

A concurrent search for other GPCRs specific for LPA or S1P resulted in discoveries of the ligand specificities of Edg-3, -4 and -5 proteins (Table I).[9,10] Human Edg-4 protein is 46% identical and 72% similar in amino acid sequence to human Edg-2 protein. As for Edg-2 protein, Edg-4 is a GPCR for LPA. Jurkat T cell transfectants overexpressing Edg-4 protein show increased specific binding of ^{3}H-LPA and LPA concentration-dependent activation of an SRE-luciferase reporter gene with an EC50 of 10 nM. At 1 uM LPA, the reporter signals transduced by Edg-4 GPCRs had a mean magnitude three-fold higher than those from Edg-2 GPCRs.[10] No signals were elicited in either type of transfectant by S1P or lysophosphatidyl-choline, -ethanolamine or -serine. Reporter signals from both Edg-2 and Edg-4 GPCRs were suppressed significantly by either pertussis toxin or Clostridial C3 exotoxin and a combination of the two together resulted in suppression of 80% to 90%. In contrast to their similar ligand specificity and signaling properties, Edg-2 and Edg-4 are distributed differently in human tissues, as assessed by Northern blots. Edg-2 mRNA is expressed at highest levels in heart and brain tissues, and at moderate levels in gastrointestinal and genitourinary tissues, whereas Edg-4 mRNA is expressed at the highest level in leukocytes and at a moderate level in testicular tissues, but not in brain or heart tissues.

A similar approach identified Edg-3 and Edg-5 (originally designated H218) as GPCRs for S1P.[9] In Jurkat T cells overexpressing Edg-3 protein, 1 uM S1P evoked SRE-luciferase reporter signals more than 2-fold greater than those for sphingosine, dihydro-sphingosine, dihydrosphingosine 1-phosphate and sphingosylphosphorylcholine (SPC), whereas no signal above empty vector background was induced by C6-ceramide, psychosine or LPA. In Edg-5 transfected Jurkat T cells, reporter signals of similar maximal magnitude were elicited by S1P, sphingosine and SPC, that were approximately two-fold higher than those for the corresponding dihydro-sphingolipids, and again none was induced by C6-ceramide, psychosine or LPA. The reporter responses to S1P in both Edg-3 and Edg-5 transfectants were detectable at 1 nM and optimal at 0.1 uM. In contrast, Jurkat T cell transfectants overexpressing Edg-1 GPCRs manifested reporter signals only a mean of 1.7-fold higher than vector alone control levels with 1 uM S1P or SPC. In Xenopus oocytes both Edg-3 and Edg-5 transduced significant increases in calcium efflux evoked by S1P and dihydro-S1P, but not the other lipids. Edg-1 GPCRs in oocytes transduced no detectable calcium efflux. Northern blot analyses showed expression of mRNA encoding Edg-1, Edg-3 and Edg-5 in most human tissues. Edg-3 mRNA was most prominent in heart tissues and also was strongly expressed in blood leukocytes. Edg-5 like Edg-3 was most prominent in heart tissues, but also was highly expressed in CNS, placenta and gonadal tissues.

Table I. Characteristics of the Human Edg GPCRs for LPA and S1P

Tentative Name	Size (amino acids)	Homology Group	Preferred Ligands	Most Prominent Human Tissue Distribution
Edg-1	381	1	S1P (weak)	CV
Edg-2	364	2	LPA	CV, CNS, GON, GI
Edg-3	378	1	S1P	CV, LEUK
Edg-4	382	2	LPA	LEUK, TEST
Edg-5	354	1	S1P	CV, CNS, GON. PL

CV=cardiovascular, CNS=central nervous system, GON=gonadal tissues, GI-gastrointestinal system, LEUK=blood leukocytes, TEST=testes, and PL=placenta.

HOMOLOGY CLUSTERS AND EVOLUTIONARY TREE FOR LPA/S1P GPCRs

The GPCRs for LPA or S1P and other lysosphingolipids, that have been characterized structurally and functionally, may be considered in two major homology clusters based on amino acid sequence identities (Fig. 1). Human Edg-2 and Edg-4 exhibit 46% identity and 72% similarity of amino acid sequences and Edg-1, Edg-3 and Edg-5 exhibit 45-60% identity of amino acid sequences. In contrast, the level of amino acid sequence identity between the two clusters is lower in the range of 31%-34%. Some common structural features, unusual in the superfamily of GPCRs, are shared by all members of the Edg protein subfamily of GPCRs. For example, all Edg proteins lack a cysteine in the first extracellular loop, that is found in most other GPCRs. Each of the Edg proteins also has distinguishing structural elements, that have not yet been related to any aspects of ligand binding or signaling. One example is the substitution of alanine for proline in the seventh transmembrane sequence NPXXY of Edg-4, that is conserved in the other Edg proteins and most other GPCRs. The two clusters of Edg proteins also have partial homology with the cannabinoid receptor family (Fig. 1),[11] suggesting the possibility of a common ancestral gene. The genomic organization of Edg genes has not been fully elucidated, but early data support the relatedness of the two clusters defined by amino acid sequence homologies.

The functional properties of Edg protein GPCRs in each of the two homology clusters so far also support the fundamental biological relevance of this tentative classification (Table I). Edg-2 and Edg-4 protein GPCRs of the first cluster bind LPA specifically and transmit numerous cellular signals in response to LPA, but not lysosphingolipids. Edg-3 and Edg-5 protein GPCRs of the second cluster bind and signal responses to S1P, but not LPA. Edg-1 protein of the second homology cluster transduced weak reporter signals in response to S1P, but not LPA when expressed in Jurkat T cell transfectants. It is expected that more detailed studies of ligand binding and signaling by the Edg-1 protein will verify its role as a GPCR for S1P and possibly other lysosphingolipids. The relationships between Edg protein GPCRs and receptors for cannabinoids have been strengthened by finding that the lipid sn-2 arachidonylglycerol is an endogenous cannabinoid ligand.[12]

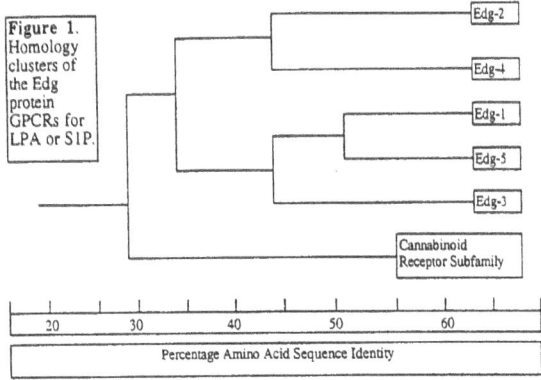

Figure 1. Homology clusters of the Edg protein GPCRs for LPA or S1P.

Figure 1

SUMMARY

The results of molecular cloning and homology searches have identified a minimum of five different proteins of the endothelial differentiation gene (edg) encoded subfamily of GPCRs. Edg protein GPCRs show amino acid sequence identity of 31% to 34% as a subfamily, but contain two homology clusters with greater similarity of structures and functions. One cluster of high amino acid sequence homology includes Edg-2 and Edg-4 proteins, that encode GPCRs for LPA, but not lysosphingolipids. A second homology cluster encompasses Edg-1, Edg-3 and Edg-5. Edg-3 and Edg-5 encode GPCRs specific for S1P, but not LPA. Preliminary data suggest that Edg-1 encodes a GPCR for S1P and one or more other lysosphingolipids, but the signals evoked by S1P alone are far weaker than those transduced by Edg-3 and Edg-5. Similarities of the structures of genes for the respective homology clusters supports this tentative classification of the Edg protein GPCRs. Future research will be directed to completion of the elucidation of genomic organization and signaling pathways, and a greater understanding of the breadth of functional roles of Edg proteins in development and activities of the nervous, cardiovascular, endocrine and immune systems.

ACKNOWLEDGMENTS

The authors would like to thank Bethann Easterly for graphics and manuscript preparation. This work was supported by research grant HL31809 from the National Institutes of Health, National Heart Lung and Blood Institute.

263

REFERENCES

1. W.H. Moolenaar. Lysophosphatidic acid, a multifunctional phospholipid messenger, *J. Biol. Chem.* 270(22):12949 (1995).
2. S. Spiegel and S. Milstein. Sphingolipid metabolites: members of a new class of lipid second messengers, *J. Membrane Biol.* 146:225 (1995).
3. A. Olivera and S. Spiegel. Sphingosine-1-phosphate as second messenger in cell proliferation induced by PDGF and FCS mitogens, *Nature* 365:557 (1993).
4. O. Fourcade, M-F. Simon, C. Viode, N. Rugani, F. Leballe, A. Ragab, B. Fournie, L. Sarda, and H. Chap. Secretory phospholipase A_2 generates the novel lipid mediator lysophosphatidic acid in membrane microvesicles shed from activated cells, *Cell.* 80:919 (1995).
5. A. Wang, R.A. Deems, and E.A. Dennis. Cloning, expression, and catalytic mechanism of murine lysophospholipase I, *J. Biol. Chem.* 272(19):12723 (1997).
6. J. H. Hecht, J.A. Weiner, S.R. Post, and J. Chun. Ventricular zone gene-1 (vzg-1) encodes a lysophosphatidic acid receptor expressed in neurogenic regions of the developing cerebral cortex, *J. Cell Biol.* 135(4):1071 (1996).
7. S. An, M.A. Dickens, T. Bleu, O.G. Hallmark, and E.J. Goetzl. Molecular cloning of the human edg2 protein and its identification as a functional cellular receptor for lysophosphatidic acid, *Biochem. and Biophys. Res. Comm.* 231(3):619 (1997).
8. R. Treisman. Journey to the surface of the cell: Fos regulation and the SRE, *The EMBO Journal.* 14(20):4905 (1995).
9. S. An, T. Bleu, W. Huang, O.G. Hallmark, S.R. Coughlin, E.J. Goetzl. Identification of cDNAs encoding two G protein-coupled receptors for lysosphingolipids, *FEBS Letters* 417:279 (1997).
10. S. An, T. Bleu, O.G. Hallmark, and E.J. Goetzl. Characterization of a novel subtype of human G protein-coupled receptor for lysophosphatidic acid, *J. Biol. Chem.*, in press.
11. M.E. Abood, K.E. Ditto, M.A. Noel, V.M. Showalter, and Q. Tao. Isolation and expression of a mouse CB1 cannabinoid receptor gene. Comparison of binding properties with tose of native CB1 receptors in mouse brain and N18TG2 neuroblastoma cells, *Biochem. Pharmacol.* 53:207 (1997).
12. N. Stella, P. Schweitzer, and D. Piomelli. A second endogenous cannabinoid that modulates long-term potentiation, *Nature* 388:773 (1997).

39

EVIDENCE FOR INVOLVEMENT OF PROSTAGLANDIN I$_2$ AS A
MAJOR NOCICEPTIVE MEDIATOR IN ACETIC ACID-INDUCED
WRITHING REACTION: A STUDY USING IP-RECEPTOR
DISRUPTED MICE

Sachiko Oh-ishi,*[1] Akinori Ueno,[1] Hideki Matsumoto[1], Takahiko
Murata,[2] Fumitaka Ushikubi,[2] and Shuh Narumiya[2]

[1]Department of Pharmacology, School of Pharmaceutical
Sciences, Kitasato University, Minato-ku, Tokyo 108-8641
[2]Department of Pharmacology, Faculty of Medicine, Kyoto
University, Yoshida, Kyoto 606-01

INTRODUCTION

Writhing reaction of mice has been used for the evaluation of analgesics or anti-inflammatory agents (Vinegar et al., 1979).   It has been suggested that there may be a putative involvement of cyclooxygenase metabolites in this model of inflammatory pain. However, it is not yet generally concluded which product of eicosanoids, including PGE$_2$ and PGI$_2$, is a main mediator of this pain model.

Berkenkopf and Weichman (1988) indicated that PGI$_2$ may be a candidate for nociceptive mediator, since 6-keto-PGF$_{1\alpha}$ was found as a major prostanoid in the peritoneal cavity during the writhing reaction of mice induced by acetic acid or phenylquinone.   Therefore, we examined whether PGI$_2$ or PGE$_2$ is a main mediator of writhing reaction of mouse, using prostacyclin (IP)-receptor disrupted mice ((-/-)-mice).

MATERIALS AND METHODS

Experimental animals and Induction of writhing reaction

*Corresponding Author: FAX; +81-3-3442-3875, TEL; +81-3-5791-6252

Eicosanoids and Other Bioactive Lipids in Cancer, Inflammation, and Radiation Injury, 4
Edited by Honn *et al.*, Kluwer Academic / Plenum Publishers, New York, 1999.

265

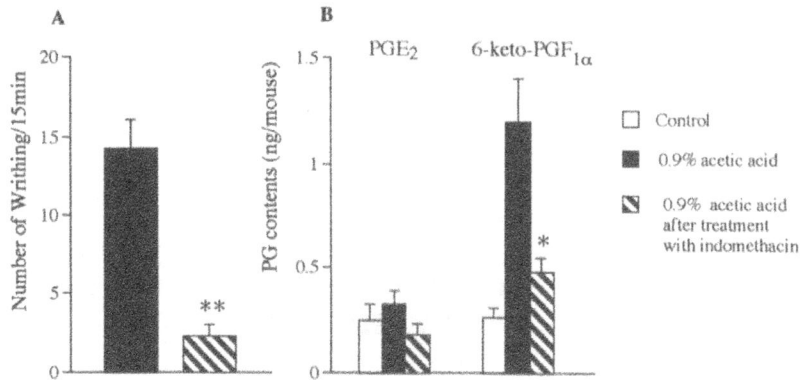

Figure 1. Number of writhing response and PG contents in the peritoneal cavities of
male ICR mice after 0.9% acetic acid (0.05 ml/10 g b.w.) i.p. injection
* and ** indicate significant difference from non-pretreated group at P < 0.05
and P < 0.01, respectively. (n= 5-8)

Writhing reaction was induced by an intraperitoneal injection of 0.9% acetic acid solution
into 5-6 -week-old male ICR mice (SLC Japan, Hammatsu, Japan), or IP-receptor knockout mice
(-/-) and their wild type (+/+) (Murata et al., 1997) at a dose of 5 ml/kg. Indomethacin, 10 mg/kg,
was intaperitoneally injected 30 min before the induction of the writhing .

Determination of prostaglandins in the peritoneal exudates

Prostaglandins were extracted from the peritoneal exudates of mice that had been injected
with 0.9% acetic acid. Mice were exsanguinated at 15 min after the acetic acid injection; and their
peritoneal cavity was washed with Ca^{2+}, Mg^{2+}-free Hanks' balanced salt solution containing 10 μM
indomethacin. The pooled washings were treated and assayed with enzyme immunoassay kits
(Cayman Chemical Co., MI, USA) as described previously (Yamaki and Oh-ishi, 1992;
Matsumoto et al. 1997).

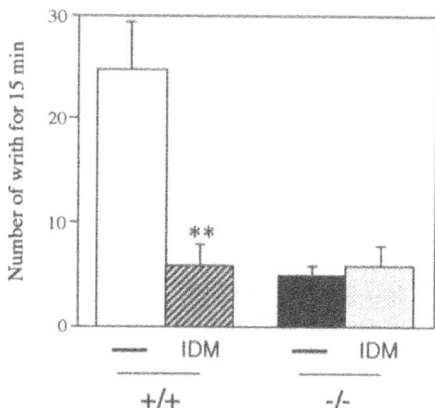

Figure 2. Acetic acid-induced writhing reaction in IP-receptor knockout mice

Acetic acid, 0.9%, was injected into wild type (+/+) and knockout (-/-)mice as described
in the method section. IDM: Indomethacin pretreatment. Figures on each column indicates
number of mice used. ** Significant difference between control group at p < 0.01.
(n=5-8)

Statistical Analysis

Data were expressed as the means with S. E. M. of more than three independent experiments. Statistical analysis was conducted with Student's t-test or one-way analysis of variance followed by Dunnet's t-test.

RESULTS

Writhing reaction induced by 0.9% acetic acid into ICR mice and effect of indomethacin

Numbers of the writhing response was counted during each sequential 5 min periods after the injection of acetic acid and showed a peak at the 5-10 min after the injection of acetic acid. Writhing number accumulated for 15 min was evaluated for the effect of indomethacin as expressed in Fig. 1A. Indomethacin pretreatment significantly attenuated the acetic acid-induced writhing reaction.

Prostaglandins in the peritoneal exudates were measured at 15min after acetic acid injection as shown in Fig.1B. A stable metabolite of PGI_2, 6-keto-$PGF_{1\alpha}$ was detected as the highest one as well as a small amount of PGE_2. Indomethacin also suppressed these levels of prostaglandins.

Withing reaction in IP-receptor disrupted mice

As shown in Fig. 2, number of the writhing reaction in wild type mice (+/+) was almost similar to that of ICR mice, while number of the writh in IP-receptor knockout mice (-/-) was significantly less and the level was almost the same as that of the indomethacin-pretreated wild type mice. Moreover, indomethacin treatment did not further attenuate the level of the writh of the (-/-) mice.

To know the ability of prostaglandin production in the IP-receptor knockout mice we measured prostaglnadins produced in the peritoneal exudates of these mice after acetic acid injection. As shown in Fig. 3, amount of PGE_2 and 6-keto-$PGF_{1\alpha}$ measured at 15 min after the acetic acid injection in the knockout mice (-/-) were almost same as those in wild type (+/+).

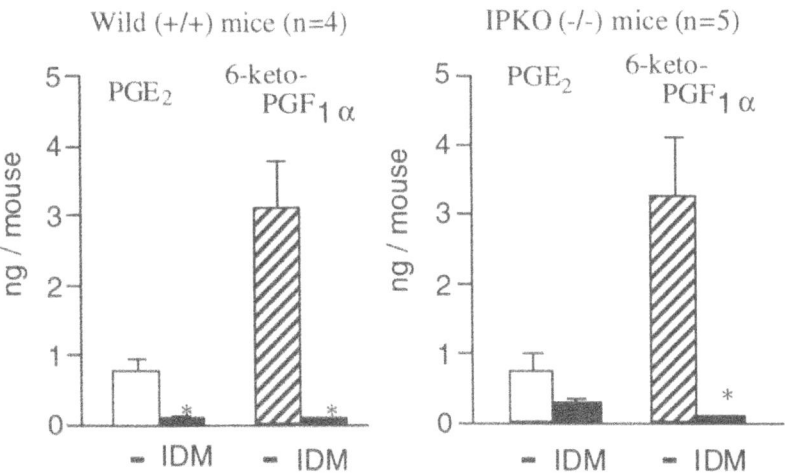

Figure 3. Prostaglandin contents in the peritoneal exudate of IP-receptor knockout mice at 15 min after intraperitoneal injection of 0.9% acetic acid

* indicates statistically significant difference from each non-pretreated group at p < 0.05. IDM: Indomethacin pretreatment, 10 mg/kg

These result indicate that IP-receptor knockout mice express the same ability to produce PGI_2, but its writhing response was reduced.

DISCUSSION

As expected indomethacin-treatment reduced the number of writhing response induced by acetic acid in ICR mice as well as wild type (+/+) mice. And IP-receptor disrupted mice showed less responsiveness to acetic acid and the number of writh is almost the same level as that of indomethacin-treated wild type mice, indicating that cyclooxygenase metabolites responsible to pain response in this model could be mainly PGI_2. Furthermore, (-/-) mice showed almost the same ability to produce prostaglandins as that of wild type mice (Fig. 3). Therefore low response of (-/-) mice to acetic acid is attributable to the deficiency of IP-receptor. This result clearly indicates that PGI_2 is the main nociceptive mediator in the acetic acid-induced writhing reaction. This conclusion well agrees with the report by Berkenkopf and Weichman (1988), who suggested PGI_2 may be a major mediator for writhing response. Moncada et al. (1975) reported PGI_2 and PGE_2 are the most potent agonists for sensitization of peripoheral pain receptors among various cyclooxygenase products. Whether PGI_2 is an only mediator for this writhing response or other prostanoids including PGE_2 could sensitize mouse peritoneal pain receptors remains in future study.

REFERENCES

Berkenkopf,J. and Weichman,B.M.1988. production of prostacyclin in mice following intraperitoneal injection of acetic acid, phenylquinone and zymosan: Its role in the writhing response. *Prostaglandins* 36: 693-709.

Matsumoto, H., Naraba, H., Murakami, M., Kudo, I., Yamaki, K., Ueno, A., Oh-ishi, S., 1997. Concordant induction of prostaglandin E_2 synthase with cyclooxygenase-2 leads to preferred production of prostaglandin E_2 to thromboxane and prostaglandin D2. *Biochem. Biophys. Res. Commun.* 230: 110-114.

Moncada, S., Ferreira, S.H., Vane, J.R., 1975. Inhibition of prostaglandin biosynthesis as the mechanism of algesia of aspirin-like drugs in the dog knee joint. *Eur. J .Pharmacol.* 31: 250-260.

Murata,T., Ushikubi,F., Mmatsuoka,T., Hirata, M., Yamasaki,A., Sugimoto,Y., Ichikawa,A., Aze,Y., Tanaka, T., Yoshida,N., Ueno,A., Oh-ishi,S., Narumiya,S., 1997. Increased thrombotic tendency, decreased inflammatory responses and hypoalgesia in prostacyclin receptor-deficient mice. *Nature* 388: 678-682.

Vinegar, R., Truax, J.F., Selph, J.L., Johnston, P.R., 1979. Antagonism of pain and hyperalgesia, in *Handbook of Experimental Pharmacology Vol. 50/II, Anti-inflammatory drugs* , Vane, J.R., Ferreira, S.H. Eds., Springer-Verlag, Berlin. pp.208-222.

Yamaki, Oh-ishi, S., 1992. Comparison of eicosanoid production between rat polymorphonuclear leukocytes and macrophages: Detection by high-performance liquid chromatography with precolumn fluorescence labeling. *Jpn. J. Pharmacol.* 58: 299-307.

40

LOCAL ANESTHETIC EFFECTS ON TXA$_2$ RECEPTOR MEDIATED PLATELET AGGREGATION USING QUENCHED FLOW AGGREGOMETRY

Christian W. Honemann[1,2], Bernard Lo[2], Jo S. Erera[2], Renate Polanowska-Grabowska[3], Adrean R.L. Gear[3] and Marcel Durieux[2]

[1]Klinik and Poliklinik fur Anasthesiologie and Operative Intensivmedizin Westfalische Wilhelms Universitat Munster, Albert-Schweister-Strasse 33, D-48149 Munster, Germany
[2]Department of Anesthesiology, University of Virginia Health Science Center, P.O. Box 10010, Charlottesville, VA 22906-0010, USA
[3]Department of Biochemistry, University of Virginia Health Sciences Center, Box 400, Charlottesville, VA 22908, USA

INTRODUCTION

Thrombembolic complications are common during and after surgery, and appear to result from hyper-coagulation and hypo-perfusion (the latter often induced by vasoconstriction). After major orthopedic or urologic surgery, deep venous thrombosis is reported to occur in 30 - 80% of patients.[1,2] Similarly, myocardial ischemia[1], induced by microthombosis and/or coronary vasoconstriction, is frequent.[3,4] The mechanisms for this response are poorly understood, but inflammatory mediators - released in response to surgical trauma and inducing thrombosis and vasoconstriction - are likely to be pivotal.

Interestingly, the complications described appear to be minimized if epidural anesthesia, rather than general anesthesia, is used.[5,6] Unfortunately, epidural anesthesia is not always an option (because of patient preference, surgical procedure or coexisting disease), and it is therefore important to know if these beneficial effects can be obtained by other means. Findings that local anesthetics infused intravenously decrease the incidence of thrombotic complications[1] made is hypothesize that some of the beneficial effects of epidural anesthesia can be explained by interactions between inflammatory mediators and local anesthetics present in blood. This is supported by reports that local anesthetics dilate blood vessels[7] and decrease platelet aggregation.[8,9] The site(s) of action of these effects has not been described.

We hypothesized that the thromboxane A$_2$ (TXA$_2$) signaling pathway might be a site of local anesthetic action. TXA$_2$ is a potent platelet aggregator and vasoconstrictor. It is one of the major mediators released during surgery (orthopaedic[10], cardiac[11,12], abdominal[13], vascular[14], and obstetric surgery[15]). has been proposed as a mediator of myocardial ischemia[16,17], as well as coronary vasoconstriction[17] and thrombosis[18]. Thus, TXA$_2$ is likely to play an important role in perioperative thrombosis, and local anesthetics could exert their beneficial actions by interfering with TXA$_2$ signaling.

Eicosanoids and Other Bioactive Lipids in Cancer, Inflammation, and Radiation Injury, 4
Edited by Honn *et al.*, Kluwer Academic / Plenum Publishers, New York, 1999.

269

METHOD

MATERIALS: U-46619 was obtained from Cayman Chemical (Ann Arbor, MI). Lidocaine bupivacaine and all other chemicals were from Sigma Chemicals Inc. (St. Louis, MO). S (+) ropivacaine was a kind gift from ASTRA USA Inc. (Westborough, MA).

Preparation of platelet-rich plasma: After IRB approval, venous blood was obtained by antecubital veni-puncture from volunteers who had not ingested any drugs within 10 days. ACD was used as anticoagulant. Platelet-rich plasma (PRP) was prepared by slow-speed centrifugation (2 x 3 min and 1 x 5 min at 350 g). To inhibit endogenous TXA_2 production indomethacin ($1\mu g/ml$) was added.

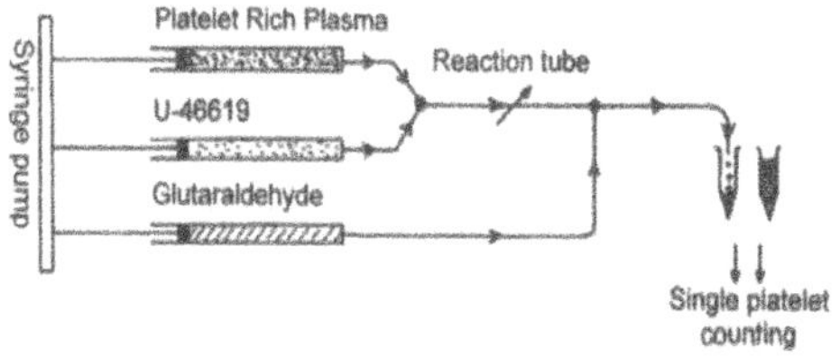

Figure 1 *Principle of quenched flow aggregometry.* A syringe pump was used to force platelets together with an inducing agent through narrowbore tubing to mimic sheer stress forces found in the arterial circulation ($10-50$ dyn/cm^2). Reaction times beginning from 0.1 sec up to 5 sec are typically studied, with quenching at the outlet by glutaraldehyde.

Preparation was done in plastic tubes, which were capped in an atmosphere of 5% CO_2 and 95% air to help preserve pH and platelet function. PRP was stored at room temperature until use (within 4 hours unless indicated otherwise).

Platelet aggregation: Figure 1 shows the basic principle of quenched flow aggregometry[19]. PRP is placed in one plastic syringe and U-46619 in another syringe. A dual drive syringe pump forces these solutions through 0.8 mm ID Teflon tubing before entering a common reaction tube of 0. mm ID, which mimics shear stress forces in the arterial

Table 1: fitting values using the Hill equation for platelet aggregation induced by U-46619 at different reaction times in the quenched flow aggregometry model.

Time (sec)	Parameter	Mean	SEM	r^2
2	Hill	0.28	1.65	
	Ec_{50} (M)	7.5×10^{-6}	8.5×10^{-6}	0.92
	E_{max} (% of same day control)	106.2	16.6	
3	Hill	0.37	0.07	
	Ec_{50} (M)	2.55×10^{-6}	1.08×10^{-6}	0.95
	E_{max} (% of same day control)	103.1	6.0	
4	Hill	0.66	0.08	
	Ec_{50} (M)	1.9×10^{-6}	3.8×10^{-7}	0.97
	E_{max} (% of same day control)	91.2	2.72	
5	Hill	0.64	0.16	
	Ec_{50} (M)	8.7×10^{-7}	3.94×10^{-7}	0.99
	E_{max} (% of same day control)	93.3	6.4	

circulation (10-50 dyn/cm^2). Reaction duration is determined by tubing length and pump rate and the reaction is stopped by quenching agents (e.g. dlutaraldehyde). The reaction mixture is then sampled and the effluent analyzed by a resistive particle counter. Local anesthetics were diluted in Tyrode's solution, added to the PRP in the appropriate concentration and incubated for 10 min. The same amount of Tyrode's solution without anesthetic was added to the control sample. Experiments were performed at 37°C.

RESULTS

TXA$_2$ plays a major role during early (5 sec) platelet activation. Therefore, we chose to study these interactions with quenched flow aggregometry, as it allow exact control of aggregation duration. In addition, physiologic flow and pressure conditions[19].

U-46619 induces platelet aggregation in a concentration dependent manner. Figure 2

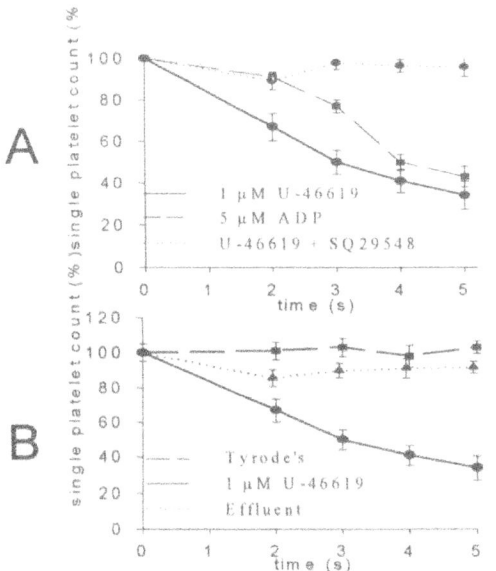

shows the loss of single platelets induced by various concentration of U-46619. We constructed concentration-response curves for U-46619 for different aggregation time points (2,3,4 and 5 sec) to determine the most sensitive duration (table 1). We selected a reaction time of 5 sec and a U-46619 concentration of 1µM (close to the EC$_{50}$: 7.6 x 10^{-7}M, data not shown) for subsequent experiments. As ADP is commonly used for aggregation experiments we compared the aggregation potency of U-46619 with that of ADP[20,21]. Both compounds induced aggregation, with U-46619 being approximately 5-fold more potent than ADP. U-46619 (1µM) reduced single platelet count to 37 ± 8% of control; ADP (5µM) reduced it to

Figure 2 A: Example traces of 1µM U-46619 (dots, solid line) and 5µM AD (squares, dashed line) induced loss of single platelets. The U-46619 induced platelet aggregation could be blocked completely by the specific TXA$_2$ receptor antagonist SQ29548 (dots, dotted line).

B: Dots and solid line represents control platelet aggregation induced by 1µM U-46619. Effluent (prepared as described in the result section induces 15% of loss of single platelets. Tyrode's solution does not activate platelets (dashed line, squares).

41 ± 6% (n=3). SQ29548, a specific TXA$_2$ receptor antagonist, blocked U-46619 (1µM) induced platelet aggregation completely (doted line, figure 2A). Aggregation of TXA$_2$ receptors induces release of intracellular platelet granule contents (ADP, arachidonic acid, platelet activating actor, etc.), which may amplify platelet aggregation[22]. If so, even significant inhibition by local anesthetics of TXA$_2$ signaling might have little effect on aggregation. We therefore investigated how much of the effect of U-46619 is due to secondary mediators. We incubated platelets for 10 min in 1µM U-46619, and then removed platelets by centrifugation. The supernatant (which would contain secreted mediators), mixed with excess TXA$_2$ antagonist to block any effect of

remaining U-46619, was then used as an agonist in the quenched flow system. The effluent induced a 15-20% decrease in single platelet count (figure 2B, dotted line, black triangles). Thus, induced release of secondary mediators does not play an important role in this aggregation model.

Ligation, bupivacaine and ropivacaine inhibit platelet aggregation;: After confirming that Tyrode's solution did not induce platelet aggregation (figure 2B), we tested the effects of lidocaine, bupivacaine and ropivacaine on platelet function. All three anesthetics inhibited TXA$_2$-induced platelet aggregation at 10^{-2} and 10^{-3} M, S(+) ropivacaine (figure 3C) at 10^{-3} and 10^{-4} M. bupivacaine was more potent, inhibiting platelet aggregation at concentrations of 10^{-3}, 10^{-4} and 10^{-5} M (figure 3B).

DISCUSSION

Quenched Flow Aggregometry: Assessment of platelet function is valuable in both scientific and clinical settings. A number of approaches have been used in vitro, fewer in vivo. The turbidometric method first described by Jurgens and Braunsteiner[23] and then developed by O'Brien[24] and Born[25] has assumed an important role for in vitro studies. Other approaches involve changes in filtration pressure (Filteragometer[26]) and the platelet aggregate ratio[27]. In view of the need to study earliest reactions of platelet aggregation (Thromboxane A$_2$ has been proposed to play a major role in early platelet aggregation) we choose the quenched flow model to determine effects of local anesthetics on TXA$_2$ induced platelet aggregation.

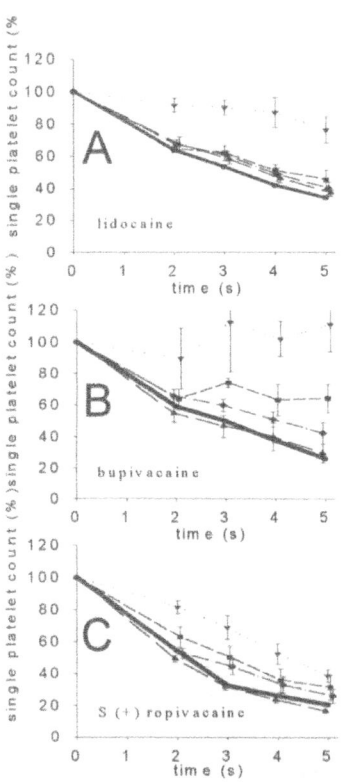

Pharmacology of thromboxane A$_2$ receptors: The amino acid sequence of the human TXA$_2$ receptor has been determined by Hirakata et al[28], cloned from a human placental cDNA library. Nusing et al[29]. recently cloned the gene for the human TXA$_2$ receptor and found no evidence for additional genes. Hence, only one TXA$_2$ receptor type appears to exist. This is unusual, as for most other members of the G protein-coupled receptor superfamily the existence of several receptor subtypes has been demonstrated. For the present study, it has the advantage that our findings are not subtype-dependent.

Physiology of Thromboxane A$_2$ receptors: TXA$_2$ is a potent stimulator of platelet aggregation[30,31], and a constrictor of vascular[30] and respiratory smooth muscle. TXA$_2$ has been suggested to play a role in thrombosis, asthma[32,33], unstable angina, and myocardial infarction[32]. TXA$_2$ receptor antagonists are therefore of considerable therapeutic importance. Preliminary clinical

Figure 3. A: Loss of single platelets induced by 1μM U-46619 in the presence of various concentrations of lidocaine (triangle, dotted line - 100 mM; squares, dashed line - 10 mM; diamonds, dashed - dotted line - 1 mM triangles, long dashed line - 10 μM; solid line, dot - control).

B: Loss of single platelets induced by 1μM U-46619 in the presence of various concentrations of bupivacaine (triangle, dotted line - 1 mM; squares, dashed line - 100 μM; diamonds, dashed - dotted line - 10 μM; triangles, long dashed line - 1μM; solid line, dots - control).

C: Loss of single platelets induced by 1μM U-46619 in the presence of various concentrations of S(+) ropivacaine (triangle, dotted line - 10 mM; squares, dashed line - 100 μM; triangles, long dashed line - 10 μM; solid line, dots - control).

studies suggested that local anesthetics may play a role in preventing postoperative thrombosis. Cooke et al[35]. demonstrated beneficial effects of intravenous lidocaine infusion as a possible means of preventing deep venous thrombosis. This was confirmed in later studies by Modig et al[8, 36-39]. They reported reduced incidence of pulmonary thromoembolism (PTE) due to lumbar epidural anesthesia as well as decreased pulmonary artery pressures. However, the mechanisms, which induce these beneficial effects remained still unclear. thus, acute inhibitory effects of local anesthetics on TXA_2 signaling may be beneficial; additionally, modulation of TXA_2 signaling during epidural anesthesia and analgesia for up to 5 days may result in long term beneficial effects as well[40]. This hypothesis is in agreement with recent data from Beattie et al[41]. Inhibition of TXA_2 synthesis by the nonspecific cyclooxygenase inhibitor ketorolac indeed reduced the risk of myocardial ischemia in their investigation. A molecular basis for this hypothesis is provided by the fact that TXA_2 receptors expressed in Xenopus oocytes or K562 cells are inhibited by local anesthetics in a concentration range found after epidural anesthesia (submitted to American Society of Anesthesiologists, Annual Meeting 1998).

Local anesthetics inhibit platelet aggregation by U-46619: Several studies have shown that local anesthetics (10 - 45 minute incubation) interfere with platelet aggregation[36]. The concentrations used were at least three times higher than clinically measured plasma concentrations. Odoom et al[8,9,42]. found I their studies that the possible thromboprophylatic effect of extradural bupivacaine may result from an inhibitory effect on platelet aggregation which is in addition to the increase in lower limb blood flow. Our study demonstrate that the local anesthetics lidocaine, bupivacaine and ropivacaine inhibit platelet aggregation induced by the TXA_2 agonist U-46619. However, the inhibitory concentrations needed were higher than from those required in previous studies using K562 and Xenopus oocytes. A variety of explanations may exist for this finding. First, local anesthetics are protein-bound to a significant degree[43,43,44], and the free fraction available for interaction with receptors may have been reduced considerably by proteins present in plasma. Second, whereas we studied intracellular Ca release in K562 cells (directly) and oocytes (indirectly), the endpoint of aggregation used in the platelet experiments requires a considerably more complex signaling pathway. We ruled out a major role for secondary extracellular aggregators, released after TXA_2 stimulation. However, it is still possible that the major amplification and feedback steps involved in platelet aggregation may have obscured local anesthetic effects. Alternatively, local anesthetics may modulate some of these other segments of the platelet aggregation pathway. As a first step towards resolving this issue, we will measure intracellular Ca responses to TXA_2 signaling in platelets, in order to reduce signaling complexity to that studied in other models. Third, parameters inherent in our quenched flow aggregometry system may have influenced the results. We choose to study only the early events in platelet aggregation, as TXA_2 is considered most important here[22,45]. However, it is possible that more than 5 sec of aggregation time is required before local anesthetic effects become pronounced. It is also possible that, despite the rheologic advantages of quenched flow aggregometry, platelet behavior is not normal in this system. this could be addressed by replicating out findings in a turbidometric aggregation system. A fourth potential explanation would be a difference in membrane properties between platelets and the other cells studied, which would prevent local anesthetics from reaching their proposed intracellular site of action (see below). At this time, we can not explain the relative insensitivity of our platelet model to local anesthetic action. What can be ruled out is a different receptor subtype: human TXA_2 receptors were used in K562 cells and platelet studies and only a single TXA_2 receptor type is known to exist[46,47,47].

Conclusion: Local anesthetics inhibit platelet aggregation induced by TXA_2 receptor functioning as studied by quenched flow aggregometry. In contrast to studies on local anesthetic effects on TXA_2 signaling in K562 cells and *Xenopus* oocytes the concentrations needed were 3-10 times higher than concentrations measured after epidural anesthesia and analgesia.

REFERENCES

1. Cooke, E.D., Lloyd, M.J., Bowcock, S.A., Pilla, M.A.: Intravenous lignocaine in prevention of deep venous thrombosis after elective hip surgery. Lancet 8042:797-799, 1997.
2. Bradley, J.G., Krugener, G.H., Jager, H.J.: The effectiveness of intermittent plantar venous compression in prevention of deep venous thrombosis after total hip arthroplasty. J. Arthroplasty 8:57-61, 1993.
3. Landesberg, G., Einav, S., Christopherson, R., Beattie, C., Berlatzky, Y., Rosenfeld, B., Eidelman, L.A., Norris, E., Anner, H., Mosseri, M., Cotev, S., and Luria, M.H.: Perioperative ischemia and cardiac complications in major vascular surgery: importance of the preoperative twelve-lead electrocardiogram. j. Vasc Surg 26:570-578, 1997.
4. Spiess, B.D. Ischemia-a coagulation problem. J. Cardiovasc. Pharmacol. 27:S38-41, 1996.
5. Modig, J. The role of lumbar epidural anesthesia as anthithrombotic prophylaxis in total hip replacement. Acta Chir. Scand. 151:589-594, 1984.
6. deLeon-Casasola, O.A., Lema, M.J., Karabella, D., and Harrison, P. Postoperative myocardial ischemia: Epidermal versus intravenous patient-controlled analgesia. A pilot project. Reg. Anesth. 20:105-112, 1995.
7. Borg, T., Modig, J. Potential anti-thrombotic effects of local anesthetics due to their inhibition of platelet aggregation. Acta Anaesth. Scand. 29:739-742, 1985.
8. Odoom, J.A., Dokter, P.W., Sturk, A., Ten, C.W., Sih, I.L. Bovill, J.G. The influence of epidural analgesia on platelet function and correlation with plasma bupivacaine concentrations. Eur. J. Anaesthesiol. 5:305-312, 1988.
9. Henny, C.P., Odoom, J.A., Ten, C.H., Oosterhoff, R.J., Dabhoiwala, N.F., Sih, I.I. Effect of extradural bupivacaine on the haemostatic system. Br. J. Anaesth. 58:301-305, 1986.
10. Vesterqvist, O., Schott, u., Berseus, O., Axelsson, K., and Green, K. In vivo production of thromboxane and prostacyclin in patients following total hip arthroplasty. Scand. J. Clin. Lab. Invest. 48:233-239, 1987.
11. Teoh, K.H., Fremes, S.E., Weisel, R.D., Christakis, G.T., Teasdale, S.J., Madonik, M.M., Ivanov, J., Mee, A.V., and Wong, P.Y. Cardiac release of prostacyclin and thromboxane A2 during coronary revascularization. J. Thorac. Cardiovasc. Surgery 93:120-126, 1987.
12. Faymonville, M.E., Deby-Dupont, G., Larbuisson, R., Deby, C., Bodson, L., Limet, R., and Lamy, M. Prostaglandin E2, prostacyclin, and thromboxane changes during nonpulsatile cardiopulmonary bypass in humans. J. Thorac. Cardiovasc. Surg. 92:858-866, 1986.
13. Schilling, M.K. Gassmann, N., Sigurdsson, G.H., Regli, B., Stoupis, C. Furrer, M., Signer, C., Redaelli, C., Buechler, M.W. Role of thromboxane and leukotriene B4 in patients with acute respiratory distress syndrome after oesophagectomy. Brit. J. /Anaesth. 80:36-40, 1998.
14. Barry, M.C., Kelly, C., Burke, P., Sheehan, S., Redmond, H.P. and Bouchbier-Hayes, D. Immunological and physiological responses to aortic surgery: Effect of reperfusion on neutrophil and monocyte activation and pulmonary function. Br. J. Surgery 84:513-519, 1997.
15. Ylikorkala, O., and Makila, U.M. Prostacyclin and thromboxane in gynecology and obstetrics. Am. J. Obstet. Gynecol. 152:318-329, 1985.
16. Ito, B.R., Roth, D.M. and Engler, R.L. Thromboxane A2 and peptidoleukotrienes contribute to the myocardial ischemia and contractile dysfunction in response to intracoronary infusion of complement C5a in pigs. Circ. Res. 66:596-607, 1990.

17. Montalescot, G. Drobinski, G., Maclouf, J., Lellouche, F., Ankri, A., Moussallem, N. Eugene, L., Thomas, D. and Grosgogeat, Y. Early thromboxane release during pacing-induced myocardial ischemia with angiographically normal coronary arteries. Am. Heart J. 120:1445-1447, 1990.

18. Packham, M.A. Role of platelets in thrombosis and hemostasis. [Review] [56 refs]. Can. J. Physiol. Pharmacol. 72:278-284, 1994.

19. Gear, A.R. Rapid reactions of platelets studied by a quenched-flow approach: Aggregation kinetics. J. Lab. Clin. Med. 100:866-886, 1982.

20. Carty, D.J., Jones, G.D., Freas, D.L., Gear, A.R. Effect of epinephrine on rapid ADP-induced aggregation, protein phosphorylation, and cytoplasmic calcium dynamics of platelets: A quenched-flow study. J. Lab. Clin. Med. 112:603-611, 1988.

21. Jones, G.D., Gear, A.R. Subsecond calcium dynamics in ADP-and thrombin-stimulated platelets: A continuous-flow approach using indo-1. Blood 71:1539-1543, 1988.

22. Packham, M.A., Kinlough-Rathbone, R.L., Mustard, J.F. Thromboxane A2 causes feedback amplification involving extensive thromboxane A2 formation on close contact of human platelets in media with low concentration of ionized calcium. Blood 70:647-651, 1987.

23. Juergens, R., Braunsteiner, H. Zur pathogenese der thrombose. Schweiz Med Wochenschr 80:1388-1395, 1950.

24. O'Brien, J.R. Some results from a new method of study. J. Clin. Pathol. 15:452-455, 1962.

25. Born, G.V.R. Aggregation of blood platelets by adenosine diphosphate and its reversal. Nature 194:927-930, 1962.

26. Hornstra, G., Ten Hoor, F. the filtagrometer. Thromb. Diath. Haemorrh. 34:924-928, 1975.

27. Wu, K.K., Hoak, J.C. A new method for the quantitative detection of platelet aggregates in patients with arterial insufficiency. Lancet 7886:924-929, 1974.

28. Hirakata, M., Hayashi, Y., Ushikubi, F., Yokota, Y., Kageyama, R., Nakanishi, S., Narumiya, S. Cloning and expression of cDNA for a human thromboxane A2 receptor. Nature 349:617-620, 1991.

29. Nuesing, R.M., Hirata, M., Kakizuka, A., Eki, T., Ozawa, K. Narumiya, S. Characterization and chromosomal mapping of the human thromboxane A2 receptor gene. J. Biol. Chem. 268:25253-25259, 1993.

30. Blaise, G.,A., Parent, M., Laurin, S., Omri, A., Reader, T.A., Moutqin, J.M. Platelet-induced vasomotion of isolated canine coronary artery in the presence of halothane or isoflurane. J. Cardiothorac. Vasc. Anesth. 8:175-181, 1994.

31. Yamamoto, Y., Kamiya, K., Terao, S. Modeling of human thromboxane A2 receptor and analysis of the receptor-ligand interaction. J. Med. Chem. 36:820-825, 1993.

32. Keith, J., Spitz, B., Van Asche, F.A. Thromboxane synthetase inhibition as a new therapy for preeclampsia: Animal and human studies minireview. Prostaglandins 45:3-13, 1993.

33. Kurosawa, M. Role of thromboxane A2 synthetase inhibitors in the treatment of patients with bronchial asthma. Clinical Therapeutics 17:2-11, 1995.

34. Schroer, K. The effect of prostaglandins and thromboxane A2 on coronary vessel tone - mechanisms of action and therapeutic implications. European Heart Journal 14:34-41, 1993.

35. Cooke, E.D., Lloyd, M.J., Bowcock, S.A., Pilla, M.A. Intravenous lignocaine in prevention of deep venous thrombosis after elective hip surgery. Lancet 8042:797-799, 1977.

36. Bort, T., Modig, J. Potential anti-thrombotic effects of local anaesthetics due to their inhibition of platelet aggregation. Acta Anaesthesiol. Scanda. 29:739-742, 1985.

37. Modig, J. Influence of regional anesthesia, local anesthetics, and sympathicomimetics on the pathophysiology of deep vein thrombosis. Acta Chir. Scand. Suppl. 550:119-124; discussion 124-127, 1989.

38. Modig, J. The role of lumbar epidural anaesthesia as anithrombotic prophylaxis in total hip replacement. Acta Chir. Scand. 151:589-594, 1985,

39. Henny, C.P., Odoom, J.A., TenCate, J.W., TenCate, R.J.F., Osterhoff, N.F., Dabhoiwala, N.F., and Sih, I.L. Effects of extradural bupivacaine on the hemostatic system. Br. J. Anaesth. 58:301-305, 1986.

40. Masterson, G.R. and Hunter, J.M. Does anaesthesis have long-term consequences? British Journal of Anaesthesia 77(11): 569-571, 1985.

41. Beattie, W.S., Warriner, C.B., Etches, R., Badner, N.H., Parsons, D., Buckley, N., Chan, V., Girard, M. The addition of continuous intravenous infusion of ketorolac to a patient-controlled analgetic morphine regime reduced post-operative myocardial ischemia in patients undergoing elective total hip or knee arthroplasty. Anesth. Analg. 84:715-822, 1997.

42. Odoom, J.A., Sturk, A., Dokter, P.W.C., Bovill, J.G., TenCate, J.W., and Osting, J. The effects of bupivacaine and pipecoloxylidide on platelet function in vitro. Acta Anaesth. Scand. 33:385-388, 1989.

43. Santos, A.C., Arthur, G.R., Wlody, D., De, A.P., Morishima, H.O., Finster, M. Comparative systemic toxicity of ropivacaine and bupivacaine in nonpregnant and pregnant ewes. Anesthesiology 82: 734-740, 1995.

44. Emanuelsson, B.M., Persson, J., Sandin, S., Alm, C., and Gustafsson, L.L. Intraindividual and interindividual variability in the disposition of the local anesthetic ropivacaine in healthy subjects. Ther. Drug. Monit. 19: 126-131, 1997.

45. De, C.F. and Janssen, P.A. Amplification mechanisms in platelet activation and arterial thrombosis. J. Hypertens. Suppl. 8:S87-S93, 1990.

46. Hirakata, M., Hayashi, Y., Ushikubi, F., Yokota, Y., Kageyama, R., Nakanishi, S., and Narumiya, S. Cloning and expression of cDNA for a human thromboxane A2 receptor. Nature 349: 617-620, 1991.

47. Nuesing, R.M., Hirata, M., Kakizuka, A., Eki, T., Ozawa, K., and Narumiya, S. Characterization and chromosomal mapping of the human thromboxane A2 receptor gene. J. Biol. Chem. 268:25253-25259, 1993.

41

VOLATILE AND LOCAL ANESTHETICS INTERFERE WITH THROMBOXANE A₂
RECEPTORS RECOMBINANTLY EXPRESSED IN *XENOPUS* OOCYTES

Christian W. Hönemann[1,2], John A. Arledge[2], Tobias Podranski[1], Hugo Van Aken[1] and Marcel Durieux[2]

[1] Klinik und Poliklinik für Anästhesiologie und operative Intensivmedizin
Westfälische Wilhelms Universität Münster
Albert-Schweitzer-Strasse 33
D-48149 Münster
Germany
[2] Department of Anesthesiology
University of Virginia Health Science Center
P.O. Box 10010
Charlottesville, VA 22906-0010
USA

INTRODUCTION

It has been hypothesized that anesthetics induce malfunctions in some critical membrane proteins. New cell physiology research findings points to direct interactions with proteins. This implies a specific and direct mode of action.

Ligand gated ion channels, like nicotinic acetylcholine receptors (NCh) and N-methyl-D-aspartate (NMDA) receptors[1; 2-5] and γ-aminobutyric acid receptors (GABA)[6, 7-9] as well as [10]voltage dependent Ca^{2+} channels[11; 12] and sodium channels [13-15] have received particular attention. However, interactions with other proteins may also be important, either in modulatory roles or by inducing anesthetic side effects. Investigations of serotonin[16], muscarinic acetylcholine[17 18; 19]and lysophosphatidate (LPA)[20] receptor signaling have demonstrated the existence of interactions between anesthetics and G protein-coupled receptors.

Receptors for lipid mediators are of specific interest in this regard, as a hydrophobic ligand binding domain might be a likely site of anesthetic action. Lipid mediators - such as prostaglandins, leukotrienes and platelet activating factor - are important intracellular (regulation

Eicosanoids and Other Bioactive Lipids in Cancer, Inflammation, and Radiation Injury, 4
Edited by Honn *et al.*, Kluwer Academic / Plenum Publishers, New York, 1999.

277

of gene expression[21]) and intercellular (platelet aggregation[22], smooth muscle contraction[23], pain[24], inflammation[25] and sepsis) messengers. Thromboxane A_2 (TXA₂) is a prominent member of the prostaglandin family and has a variety of actions in the processes described before. As several volatile and local anesthetics exhibit effects opposite to these, we hypothesized that TXA₂ receptor signaling could be a target for anesthetics.

To test our hypothesis we studied the effects of three clinically used volatile and the local anesthetic lidocaine in an isolated receptor expression model, *Xenopus* oocytes, which form a flexible system for the study of recombinantly expressed G protein-coupled receptors and the influence of anesthetics on their functioning[17; 26].

METHODS AND MATERIALS

Animals: The study protocol was approved by the Animal Research Committee at the University of Virginia. Female *Xenopus laevis* toads were obtained from *Xenopus* I (Ann Arbor, MI) and for removal of oocytes anesthetized by immersion in 0.2% 3-amino-benzoic-methyl-ester. Through a 1 cm-long abdominal incision a lobule of ovarian tissue was removed and placed in modified Barth's solution (MBS, containing in mM: NaCl 88, KCl 1, NaHCO₃ 2.4, CaCl₂ 0.41, MgSO₄ 0.82, Ca₂NO₃ 0.3, gentamicin 0.1, and HEPES 15, pH adjusted to 7.6). The wound was closed in two layers and the frog was allowed to recover from anesthesia. Oocytes were defolliculated by gentle shaking in a 1 mg/ml solution of collagenase type Ia (Sigma Chemical, St. Louis, MO) in OR2 solution (containing in mM: NaCl 82.5, KCl 2, MgCl₂ 1, HEPES 5, pH adjusted to 7.5) for 2 hours. After this process the oocytes were returned to MBS. Microscopic observation confirmed the absence of follicle cells.

mRNA synthesis and injection: The rat TXA₂ receptor clone was obtained from Dr. K.R. Lynch (Department of Pharmacology, University of Virginia, Charlottesville, VA) as a cDNA encoding a 343 amino acid protein in the pcDNAI vector (Invitrogen, San Diego, CA). The construct was linearized and transcribed in the presence of capping analog by T7 polymerase, using a commercial RNA preparation kit (mMessage mMachine, Ambion, Austin, TX). Oocytes were injected with 5 ng of mRNA in 30.6 nl sterile water, using an automated microinjector (Nanoject; Drummond Scientific, Broomall, PA). The adequacy of injection was confirmed by noting a slight increase in cell size during injection. The cells were then cultured in MBS for 72 hours before study.

Electrophysiological recording: A single defolliculated oocyte was placed in a recording chamber filled with Tyrode's solution (containing in mM: NaCl 150, KCl 5, CaCl₂ 2, MgSO₄ 1, Dextrose 10, HEPES 10, pH adjusted to 7.4). The cell was voltage clamped (Holding potential −70mV) using a two microelectrode oocyte voltage clamp amplifier (OC725A; Warner Corporation, New Haven, CT), connected to an IBM- compatible computer for data acquisition (DAS-8A/D conversion board: Keithley-Metrabyte, Traunton, MA,) and analysis (OoClamp software[27]).

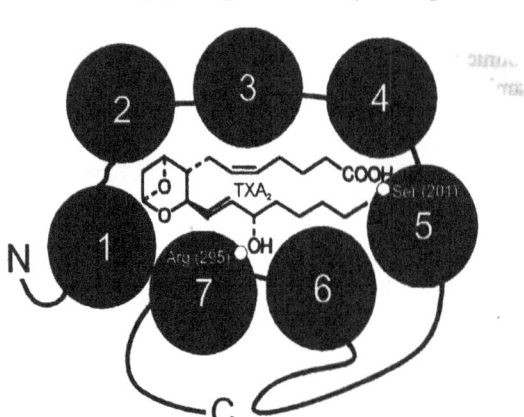

Membrane current was sampled at 125 Hz and recorded for 5 s before and 55 s after administration of the agonist. Agonist (U-46619, Cayman Chemical, Ann Arbor, MI) was delivered as a 30 µl aliquot over a period of 2 seconds. Agonist binding activates intracellular signaling cascades which increase intracellular Ca^{2+} concentration. In the *Xenopus* oocyte, intracellular Ca^{2+} opens endogenous membrane Cl^- channels, resulting in a Cl^- - flux ($I_{Cl(Ca)}$) measured conveniently by voltage clamp. Responses were quantified by integrating the current trace and are reported in microcoulombs. All experiments were performed at room temperature.

Activation of intracellular signaling pathway by intracellular IP3 and GTPγS injection: To study IP3- and GTPγS-induced $I_{Cl(Ca)}$ a third micropipette was inserted into the voltage-clamped oocyte and 50 nl of 2 mM IP_3 (estimated final concentration 100 µM) or 100 mM GTPγS (estimated final concentration 5 mM) was microinjected, thereby activating intracellular signaling pathways. Induced $I_{Cl(Ca)}$ were recorded and analyzed as described above.

Anesthetic administration: Volatile anesthetics were bubbled through a reservoir filled with 40 ml Tyrode's solution for at least 10 min. After equilibration the solution superfused the oocyte at a flow rate of approximately 5 ml/min; measurements were obtained after 10 bath volumes had been exchanged. Anesthetic concentrations in the recording chamber were quantified by gas chromatography (Aerograph 940; Varian Analytical instruments, Walnut Creek, CA). Each oocyte was exposed to a single concentration of anesthetic only. Lidocaine was dissolved in Tyrode's solution to the appropriate concentrations.

Data analysis: Responses are reported as mean ± SEM. Differences among treatment groups were analyzed using Student's t-test. If multiple comparisons were made, data were analyzed using analysis of variance (ANOVA) followed by t-test corrected for multiple comparisons (Bonferroni). $P<0.05$ was considered significant. Concentration-response curves were fit to the Hill equation to calculate EC_{50} for U46619 and IC_{50} for anesthetics.

Materials: The TXA2 receptor agonist U-46619 (Cayman Chemical - Ann Arbor, MI, 5-heptenoic acid, 7-[6(3-hydroxy-1-octenyl)-2-oxabicyclo [2.2.1] hept-5-yl]) and the TXA2 receptor antagonist Bay U 3405 (Bayer AG - Wuppertal, Germany (3R)-3-(4-fluoro-phenylsulfonamido)-1,2,3,4 tetrahydro [4(,4b-3H] - carbazolepropanoic acid) were diluted in 0.1 % fatty acid free bovine serum albumin (BSA; ICN Pharmaceuticals, Inc., Costa Mesa, CA) in Tyrode's solution. Halothane was from Halocarbon Laboratories (River Edge, NJ), isoflurane and sevoflurane were from Abbott International Ltd. (Abbott Park, IL). All other chemicals were from Sigma (St. Louis, MO).

RESULTS AND INTERPRETATION

We have previously shown that oocytes injected with TXA_2 receptor mRNA respond to U-46619 application with transient chloride inward currents[28]. The response is concentration dependent and curve fitting using the Hill equation revealed a half-maximal effect concentration (EC_{50}) of 3.2×10^{-7} M. The recombinantly expressed receptors signal mainly through IP_3 release, as we have been able to show that the responses could be inhibited by intracellular injection of heparin (MW 3000)[28], which specifically inhibits IP_3 receptors in oocytes[29]. The response was inhibited by the specific TXA_2 antagonist Bay U-3405[28]. Therefore the $I_{Cl(Ca)}$ are induced by activation of TXA_2 receptors and result in increased intracellular Ca^{2+} concentration.

Volatile anesthetics interfered with TXA_2 signaling through different mechanisms[28]. In agreement with our hypothesis that the lipophilic binding pocket may be the potential target of anesthetics, the less lipophilic compound sevoflurane was without effects on TXA_2 receptor signaling.

Halothane inhibited TXA_2 signaling in a concentration dependent manner (IC_{50}: 0.47 mM, 0.83%). The inhibitory effect was reversible[28]. The presence of halothane at the IC_{50} concentration shifted the concentration response curve parallel to the right (EC_{50} changed from 3.2×10^{-7} M to 1×10^{-5} M), indicating that the effect is competitive. Therefore, the binding pocket for the agonist (fig. 1), built by the seven lipophilic trans-membrane regions is most likely affected by anesthetic action. This is supported by the fact that we were able to overcome the inhibitory effect of halothane by increasing the agonist concentration and measure the same $I_{Cl(Ca)}$ with and without anesthetic (fig. 2). In agreement with our data on interactions between halothane and TXA_2 signaling, Hirakata et al. demonstrated in human platelets[30] that halothane (3.3 mM) suppressed TXA_2 binding. Halothane markedly increased K_d without significantly altering B_{max}.

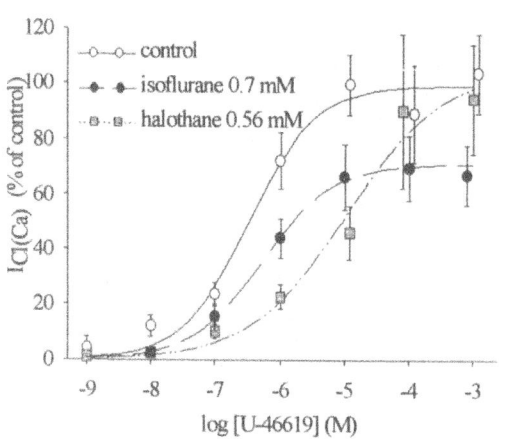

Figure 2: Halothane inhibits TXA_2 signaling in a competitive, isoflurane inhibits in a noncompetitive manner. Halothane shifts the agonist concentration response curve parallel to the right. In contrast isoflurane did not change EC_{50} but decreased the maximum response to 70% of control (with permission from Lippincott-Raven Inc.)[28].

If volatile anesthetics mainly act at the ligand-binding pocket, then we would expect that the less lipophilic compound isoflurane would have a lower inhibitory potency on TXA_2 signaling and shift the concentration response curve parallel to the right. In agreement with our hypothesis isoflurane had a lower inhibitory potency than halothane and inhibited TXA_2 signaling in a concentration dependent manner (IC_{50}:0.69 mM, 1.7%). However, in contrast to halothane the concentration response curve was not shifted parallel to the right in the presence of isoflurane. Isoflurane acted, therefore, in a non-competitive manner. It did not affect the U-46619 EC_{50}

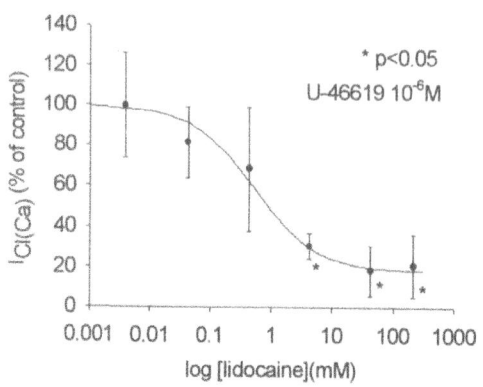

$(5.3 \times 10^{-7}$ M) but decreased maximal signaling to 70% of control. The inhibitory effect could not be abolished by higher agonist concentrations (fig. 2). To determine if our hypothesis can be generalized to other lipophilic compounds, we tested lidocaine and its effects on TXA_2 signaling. Lidocaine inhibits TXA_2 signaling in a concentration dependent manner (IC_{50}: 0.52 mM) (fig.3).

Other compounds that are generally thought to modify biological membrane action may affect thromboxane A_2 receptors as well. Previously, it has been shown that local anesthetics have inhibitory effects on platelet aggregation induced by platelet-activating factor, arachidonic acid (AA) and ADP. Platelet cyclo-oxygenase and lipo-oxygenase converts AA to prostaglandin endoperoxides, thromboxane A_2 (TXA_2) and other prostanoids, resulting in platelet aggregation. To determine if the effect of volatile anesthetics on TXA_2 signaling as described above is transferable to other lipophilic drugs, we used the local anesthetic lidocaine (octanol – buffer partition coefficient 366) and determined its effect on TXA_2 receptor functioning. We found indeed, that lidocaine inhibited TXA_2 receptors recombinantly expressed in *Xenopus* oocytes at a concentration, similar to that of the volatile anesthetics tested (IC_{50}:0.46 mM, 0.01%). This supports our theory that lipophilicity is of major importance for inhibitory action on TXA_2 signaling.

Halothane inhibited TXA_2 receptor function. We analyzed the concentration response curve in the presence of IC_{50} of halothane. It appears that the membrane receptor is most likely the target (competitive interaction). However, to determine if the effect of halothane, isoflurane and lidocaine were specific and not due to nonspecific membrane interactions or inhibition of the chloride channel, we expressed angiotensin (AT_{1A}) and muscarinic m_1 receptors, which both induce $I_{Cl(Ca)}$ in *Xenopus* oocytes through a mechanism that, apart from the receptor and possibly G proteins, is considered to be the same as for TXA_2 receptor signaling. We previously tested isoflurane effects on muscarinic m1 receptors. Isoflurane did not inhibit m1 muscarinic signaling[31]. We published that AT_{1A} angiotensin receptors, sharing the same intracellular signaling cascade as the TXA_2 receptor, are not affected by halothane[20] or lidocaine[26]. Therefore, halothane, isoflurane and lidocaine do not appear to affect the oocyte membrane, intracellular pathways, or the Ca^{2+}-activated Cl⁻channel.

To confirm this data, we tested anesthetic effects (halothane, isoflurane) on intracellular Ca^{2+} release or the Ca^{2+} activated Cl⁻ channel and determined lidocaine effects on $I_{Cl(Ca)}$ induced by microinjection of IP_3 and GTPγS into oocytes. We showed, before, that heparin, an IP_3 receptor antagonist, inhibits TXA_2 induced $I_{Cl(Ca)}$ completely, and thus that TXA_2 induced Ca^{2+} release is mediated mainly by IP_3. Our results indicated that in contrast to currents induced by TXA_2 application, size and kinetics of IP_3 and GTPγS induced responses were not affected by the presence of the anesthetics tested[20; 28; 31]. TXA_2 signaling inhibition by halothane, isoflurane and lidocaine occurs, therefore, early in the signaling pathway, most likely at the receptor or the G protein.

CONCLUSION

Lipid mediators are of major interest for inflammation and related syndromes in the perioperative setting as well. Therefore, it is very important to understand interactions with these commonly used compounds.
Our data indicate that commonly used volatile (halothane, isoflurane) and lidocaine can interfere significantly with thromboxane A_2 receptor signaling, a prototype lipid mediator receptor. The point of action for these compounds is most likely the membrane receptor, because intracellular signaling mechanisms are not effected.

REFERENCES

1. Franks NP, Lieb WR: Stereospecific effects of inhalational general anesthetic optical isomers on nerve ion channels. Science 254:427-430, 1991
2. Davies SN, Alford ST, Coan EJ, Lester RA, Collingridge GL: Ketamine blocks an NMDA receptor-mediated component of synaptic transmission in rat hippocampus in a voltage-dependent manner. Neurosci Lett 92:213-217, 1988
3. Sumikawa K, Matsumoto T, Amenomori Y, Hirano H, Amakata Y: Selective actions of intravenous anesthetics on nicotinic- and muscarinic-receptor-mediated responses of the dog adrenal medulla. Anesthesiology 59:412-416, 1983
4. Hentall ID, Abate KL, Wojcik RS, Andresen MJ: Nicotinic activity in the interpeduncular nucleus of the midbrain prolongs recovery from halothane anesthesia. Neuropharmacology 31:1299-1304, 1992
5. Sumikawa K, Matsumoto T, Ishizaka N, Nagai H, Amenomori Y, Amakata Y: Mechanism of the differential effects of halothane on nicotinic- and muscarinic-receptor-mediated responses of the dog adrenal medulla. Anesthesiology 57:444-450, 1982
6. Mihic SJ, Ye Q, Wick MJ, Koltchine VV, Krasowski MD, Finn SE, Mascia MP, Valenzuela CF, Hanson KK, Greenblatt EP, Harris RA, Harrison NL: - Sites of alcohol and volatile anaesthetic action on GABA(A) and glycine receptors. Nature 389(6649):385-9, 1997
7. Hall AC, Lieb WR, Franks NP: Stereoselective and non-stereoselective actions of isoflurane on the GABA$_A$ receptor. British Journal of Pharmacology 112:906-910, 1994
8. Quinlan JJ, Firestone S, Firestone LL: Isoflurane's enhancement of chloride flux through rat brain gamma-aminobutyric acid type A receptors is stereoselective. Anesthesiology 83:611-615, 1995
9. Saunders PA, Ho IK: Barbiturates and the GABAA receptor complex. Progress in Drug Research 34:261-286, 1990
10. Lin LH, Chen LL, Harris RA: Enflurane inhibits NMDA, AMPA and kainate-induced currents in *Xenopus* oocytes expressing mouse and human brain mRNA. FASEB J. 7:479-485, 1993
11. Pajewski TN, Miao N, Lynch C, Johns RA: Volatile anesthetics affect calcium mobilization in bovine endothelial cells. Anesthesiology 85:1147-56, 1996
12. Pancrazio JJ, Lynch C: Differential anesthetic-induced opening of calcium-dependent large conductance channels in isolated ventricular myocytes. Pflugers Arch 429:134-6, 1994
13. Bennett PB, Valenzuela C, Chen LQ, Kallen RG: On the molecular nature of the lidocaine receptor of cardiac Na+ channels. Modification of block by alterations in the alpha-subunit III-IV interdomain. Circ Res 77:584-592, 1995
14. Nishiguchi E, Hamada N, Shindo J: Lidocaine action and conformational changes in cytoskeletal protein network in human red blood cells. Eur J Pharmacol 286:1-8, 1995
15. O'Leary ME, Chen LQ, Kallen RG, Horn R: A molecular link between activation and inactivation of sodium channels. J Gen Physiol 106:641-658, 1995
16. Lin LH, Leonard S, Harris A: Enflurane inhibits the function of mouse and human brain phosphatidylinositol-linked acetylcholine and serotonin receptors expressed in *Xenopus* oocytes. Molec.Pharmacol 43:941-948, 1993

17. Durieux ME: Halothane inhibits signaling through m1 muscarinic receptors expressed in *Xenopus* oocytes. Anesthesiology 82:174-182, 1995
18. Durieux ME: Inhibition by ketamine of muscarinic acetylcholine receptor function. Anesth. Analg. 81:57-62, 1995
19. Durieux ME: Muscarinic signaling in the central nervous system: Recent developments and anesthetic implications. Anesthesiology 84:173-189, 1996
20. Chan, C. K. and Durieux, M. E. Effects of halothane and isoflurane on lysophosphatidate signaling. Anesthesiology 86;660-669, 1997
21. Kliewer SA, Sundseth SS, Jones SA, Brown PJ, Wisely GB, Koble CS, Devchand P, Wahli W, Willson TM, Lenhard JM, Lehmann JM: Fatty acids and eicosanoids regulate gene expression through direct interactions with peroxisome proliferator-activated receptors alpha and gamma. Proc Natl Acad Sci U S A 94:4318-23, 1997
22. Blaise GA, Parent M, Laurin S, Omri A, Reader TA, Moutquin JM: Platelet-induced vasomotion of isolated canine coronary artery in the presence of halothane or isoflurane. J Cardiothorac Vasc Anesth 8:175-181, 1994
23. Schroer K.: The effect of prostaglandins and thromboxane A_2 on coronary vessel tone - mechanisms of action and therapeutic implications. European Heart Journal 14:34-41, 1993
24. Halushka P.V., Allan C.J., Davis-Bruno K.L.: Thromboxane A_2 receptor. Journal of Lipid Mediators & Cell Signaling 12:361-378, 1995
25. Horrobin DF, Manku MS: Roles of prostaglandins suggested by the prostaglandin agonist/antagonist actions of local anaesthetic, anti-arrhythmic, anti-malarial, tricyclic anti-depressant and methyl xanthine compounds. Effects on membranes and on nucleic acid function. Med Hypotheses 3:71-86, 1977
26. Nietgen G.W., Chan C.K., and Durieux M.E. Inhibition of Lysophosphatidate signaling by Lidocaine and bupivacaine. Anesthesiology 86;1112-1119, 1997
27. Durieux ME: OoClamp: An IBM-compatible software system for the study of receptors expressed in *Xenopus* oocytes. Comput Methods Programs Biomed 41:101-105, 1993
28. Hoenemann C.W., Nietgen G.W., Podranski T., Chan C.K., Durieux M.E.: Influence of Volatile Anesthetics on Thromboxane A_2 Signaling. Anesthesiology 88:440-451, 1998
29. Parys JB, Sernett SW, DeLisle S, Snyder PM, Welsh MJ, Campbell KP: Isolation, characterization, and localization of the inositol 1,4,5-trisphosphate receptor protein in *Xenopus laevis* oocytes. J.Biol.Chem. 267:18776-18782, 1992
30. Hirakata H., Ushikubi F., Toda H., Nakamura K., Sai S., Urabe N., Hatano Y., Narumiya S., Mori K.: Sevoflurane inhibits human platelet aggregation and thromboxane A_2 formation, possibly by suppression of Cyclooxygenase activity. Anesthesiology 85:1447-1453, 1996
31. Nietgen G.W., Hoenemann C.W., Chan C.K., Kamatchi G, Durieux M.E.: Volatile Anaesthetics Have Differential Effects on Recombinant m1 and m3 Muscarinic Acetylcholine Receptor Functioning. Br J Anaesth (submitted)

LEUKOTRIENES AND LIPOXINS

42

ASPIRIN-TRIGGERED 15-EPI-LIPOXIN A4 AND NOVEL LIPOXIN B4 STABLE ANALOGS INHIBIT NEUTROPHIL-MEDIATED CHANGES IN VASCULAR PERMEABILITY

Charles N. Serhan,[1,*] Tomoko Takano,[1] Clary B. Clish,[1]
Karsten Gronert,[1] and Nicos Petasis[2]

[1]Center for Experimental Therapeutics and Reperfusion Injury,
Department of Anesthesia, Brigham and Women's Hospital,
Harvard Medical School, Boston Massachusetts, 02115
[2]Department of Chemistry, University of Southern California,
Los Angeles, California 90089

INTRODUCTION

Leukocyte-dependent endothelial cell injury leads to changes in vascular permeability, edema, and further release of chemoattractants.[1] The hallmark of reperfusion injury is considered to be diminished barrier function of vascular endothelium. Leukotriene B_4 $(LTB_4)^{\ddagger}$ is among the most potent neutrophil stimuli and thus participates in tissue injury via recruiting PMN in pathophysiologic scenarios.[2] Lipoxins are trihydroxytetraene-containing eicosanoids that are, among other in vivo sites, also generated within vascular lumen primarily by platelet-leukocyte interactions by pathways that are activated during multicellular responses such as inflammation, atherosclerosis, and thrombosis (as reviewed in Serhan).[3] Recently, aspirin was shown to trigger the biosynthesis of a new group of compounds termed 15-epi-LX or aspirin-triggered lipoxins that may contribute to some of the beneficial actions ascribed to aspirin.[3] Thus, these two LX branches involving cell-cell interactions within the eicosanoid cascade appear to produce "endogenous stop

*Address correspondence to C.N. Serhan, Ph.D., Director, Center for Experimental Therapeutics and Reperfusion Injury, Brigham and Women's Hospital, 75 Francis Street, Boston, MA 02115.

‡*Abbreviations used in this paper:* ATL, aspirin-triggered lipoxin (15-epi-lipoxin); LO, lipoxygenase; LTB_4, leukotriene B_4; MPO, myeloperoxidase; lipoxin A_4 (LXA_4), 5(S),6(R),15(S)-trihydroxy-7,9,13-trans-11-cis-eicosatetraenoic acid; lipoxin B_4 (LXB_4), 5(S),14(R),15(S)-trihydroxy-6,8,12-trans-10-cis-eicosatetraenoic acid; LX, lipoxin; 15-epi-LXA_4, 5(S),6(R),15(R)-trihydroxy-7,9,13-trans-11-cis-eicosatetraenoic acid; 15(R/S)-methyl-LXA_4, 5(S),6(R),15(R/S)-trihydroxy-15-methyl-7,9,13-trans-11-cis-eicosatetraenoic acid methyl ester; 16-phenoxy-LXA_4, 15(S)-16-phenoxy-17,18,19,20-tetranol-LXA_4 methyl ester; 5(S)-methyl-LXB_4, 5(S),14(R),15(S)-trihydroxy-5(R)-methyl-6,8,12-trans-10-cis-eicosatetraenoic acid; 5(R)-methyl-LXB_4, 5(R),14(R),15(S)-trihydroxy-5(S)-methyl-6,8,12-trans-10-cis-eicosatetraenoic acid; PMN, polymorphonuclear neutrophil.

Eicosanoids and Other Bioactive Lipids in Cancer, Inflammation, and Radiation Injury, 4
Edited by Honn *et al.*, Kluwer Academic / Plenum Publishers, New York, 1999.

287

signals," whereas the 5-lipoxygenase pathway generates leukotrienes, which are proinflammatory mediators.

Lipoxins inhibit human PMN responses including: i) FMLP and LTB$_4$-induced chemotaxis,[4] ii) adhesion and transmigration with endothelial cells,[5] and iii) FMLP-induced transmigration across epithelium.[6] These actions of LXA$_4$ were also recently demonstrated using an acute inflammation model, where PMN infiltration was dramatically inhibited by stable analogs of both LXA$_4$ and aspirin-triggered 15-epi-LXA$_4$. These actions were likely mediated by specific LXA$_4$ receptors on mouse PMNs, since the bioactive LXA$_4$ analogs compete with [^{3}H]-LXA$_4$ binding to LXA$_4$ receptors.[7]

Lipoxin B$_4$ is a positional isomer of LXA$_4$, carrying alcohol groups at carbon 5S, 14R, and 15S positions, instead of the C-5S, 6R, and 15S positions in LXA$_4$. Although LXA$_4$ and LXB$_4$ show similar bioactivities in some systems,[5] in many others they each show distinct actions[8] (see Serhan[3] for recent review). Here, we report that aspirin-triggered LXA$_4$ and novel fluorinated LXA$_4$ as well as LXB$_4$ stable analogs inhibit PMN-directed actions following topical application.

METHODS

Mouse Ear Inflammation

Balb/c mice (6-8 weeks old) were used. The inner side of the right ear was treated with acetone (i.e., control), and the inner side of the left ear was topically treated with the compounds to be tested prepared in acetone. After 5 to 7 minutes, LTB$_4$ (1 μg in acetone) was applied to both ears. At 24 h, punch biopsy samples (6 mm diameter; Acu-Punch®, Acuderm, Inc., Ft. Lauderdale, FL) were obtained, and myeloperoxidase (MPO) activity and PMN infiltration were quantified as in Takano et al.[7] Percent inhibition of PMN infiltration was calculated after background levels of MPO activity present in mouse ear skin were subtracted. For the purpose of quantifying vascular permeability, 0.2 ml of Evans blue (0.5% in PBS^{2-}) was injected intravenously immediately after the topical application of test compounds. After 24 h, punch biopsies (4 mm diameter; Acu-Punch®) were obtained and Evans blue was extracted in formamide (55 °C for 1 hour) and quantified by measuring absorbances at 610 nm with subtraction of reference absorbance at 450 nm. The mean absorbance was 0.13 ± 0.01 (n = 40) for ears exposed to LTB$_4$ and 0.06 ± 0.01 (n=4) for ears exposed to vehicle alone. The difference between these values represented complete vascular permeability, and percent inhibition was calculated from the difference between the right (control) and left (experimental) ears. Statistical analysis was performed using Student's t-test. See Takano et al.[9] for additional details.

RESULTS

LXA$_4$ Stable Analogs Inhibit both PMN Infiltration and Vascular Permeability Changes

5-LO and cyclooxygenase inhibitors are both useful anti-inflammatory agents but each displays some side effects. To explore other approaches, we evaluated LX stable analogs that were designed as mimetics of the endogenous anti-inflammatory actions noted for LXA$_4$[10] and recently for LXB$_4$[11] in vitro. Here,we examined two LXA$_4$ stable analogs and tested their ability to inhibit both PMN infiltration and changes in vascular permeability in vivo. 15(R/S)-methyl-LXA$_4$ that has a methyl group at C-15 position (racemate 15R/S) is an analog of both the aspirin-triggered 15-epi-LXA$_4$ and native LXA$_4$; and 16-phenoxy-LXA$_4$, which has a phenoxy group at C-16 position, is an analog of native LXA$_4$ that prevents

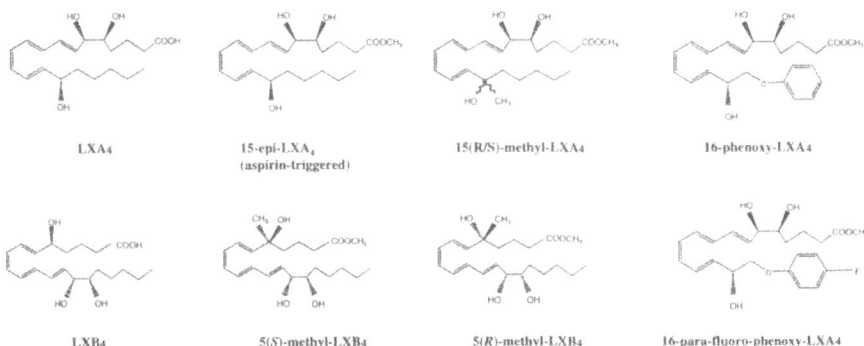

| LXA4 | 15-epi-LXA$_4$ (aspirin-triggered) | 15(R/S)-methyl-LXA4 | 16-phenoxy-LXA4 |

| LXB4 | 5(S)-methyl-LXB4 | 5(R)-methyl-LXB4 | 16-para-fluoro-phenoxy-LXA4 |

Figure 1. Structures of lipoxin stable analogs. In 15(R/S)-methyl-LXA$_4$, the hydrogen at C-15 position was replaced by a methyl group as a racemate, and thus carries properties of both 15-epi-LXA$_4$ and native LXA$_4$. 16-phenoxy-LXA$_4$ has a phenoxy group at C-16, and 16-para-fluoro-phenoxy-LXA$_4$ was fluorinated in the para position. LXB$_4$ is a positional isomer of LXA$_4$ and, in 5(S)- and 5(R)-methyl-LXB$_4$, hydrogen at C-5 position was replaced by a methyl group at the S and R configuration, respectively. Note that these analogs were used as their methyl esters.

enzymatic inactivation with recombinant enzyme in vitro[10] (see structures in Fig. 1). Both analogs act at LXA$_4$ receptors.[7]

When topically applied to mouse ears, these two stable analogs inhibited both PMN infiltration and vascular permeability changes in a concentration-dependent fashion.[9] At 130 nmol per ear, the degree of inhibition of PMN infiltration was more than 90% for both analogs with apparent IC$_{50}$s noted at ~13-26 nmol per ear range for each analog. In the same concentration range, these two LXA$_4$ stable analogs also inhibited the vascular permeability, namely, extravasation of Evans blue. At 130 nmol per ear, the inhibition of vascular permeability change was > 98% for 15(R/S)-methyl-LXA$_4$, and ~ 87% for 16-phenoxy-LXA$_4$, respectively. Moreover, their impact was striking and visible.[9] The inhibition of vascular permeability changes paralleled inhibition of PMN infiltration with both analogs, further implicating PMN involvement in these vascular permeability events and tissue damage.[1,2]

Comparison of LX Stable Analogs with LTB₄ Receptor Antagonist

We next compared three LXA₄ analogs to native LXA₄ and the actions of the LTB₄ receptor antagonist U-75302. In addition, we evaluated the impact of recently designed novel LXB₄ analogs that resist enzymatic inactivation in vitro[11] to determine whether they also possess anti-inflammatory actions (for structures, see Fig. 1). When applied topically at 26 nmol per ear, the stable analogs were three to four times more potent than native LXA₄. Among the five lipoxin stable analogs tested, 15(R/S)-methyl-LXA₄ was the most potent (> 70% inhibition), and its inhibitory actions on PMN infiltration and vascular permeability changes were significantly greater than topically-applied native LXA₄ ($p < 0.05$). A 16-para-fluoro derivative of 16-phenoxy-LXA₄ was prepared for these experiments to assess whether fluorination of the phenoxy ring could enhance potency. Results in Fig. 2A,B indicate that 16-para-fluoro-phenoxy-LXA₄ was also potent and retained the activity at levels comparable to 16-phenoxy-LXA₄.

Both of the two new LXB₄ analogs inhibited PMN infiltration and vascular permeability. The S enantiomer, 5(S)-methyl-LXB₄, was significantly more potent than 5(R)-methyl-LXB₄, indicating a preferred stereoselectivity for inhibition. The rank order of

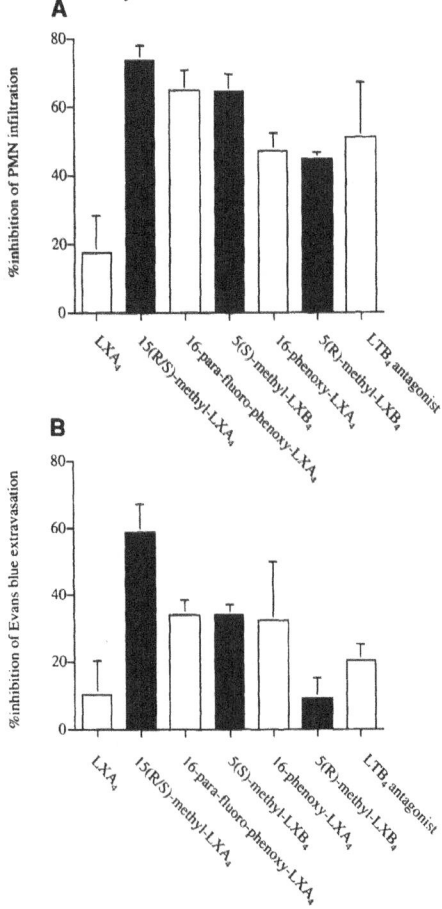

Figure 2. Lipoxin A₄ and lipoxin B₄ stable analogs inhibit both PMN infiltration and vascular permeability change: the rank order of potency. Experiments were performed as in Takano et al.[9] using 26 nmol of compounds per ear, and compared to native LXA₄ and LTB₄ receptor antagonist U-75302. (A) % inhibition of PMN infiltration, (B) % inhibition of vascular permeability change (Evans blue extravasation). Results are mean ± SEM of n=3.

inhibitory potency was 15(R/S)-methyl-LXA$_4$ > 16-para-fluoro-phenoxy-LXA$_4$ ~ 5(*S*)-methyl-LXB$_4$ > 16-phenoxy-LXA$_4$ > 5(*R*)-methyl-LXA$_4$ for both PMN infiltration and vascular permeability changes. Together these results provide further evidence that these LX analogs inhibit vascular permeability changes via blocking of PMN infiltration. Of interest, each of these LX analogs proved to be as potent as, or more potent than, topical application of equivalent amounts of the LTB$_4$ receptor antagonist U-75302 (Fig. 2).

DISCUSSION

The results presented here are the first to demonstrate inhibition of PMN-induced vascular permeability changes by LX analogs and the first evaluation and direct comparison of both novel LXB$_4$ stable analogs and a fluorinated LX analog derivative (Fig. 1). The dose response for two bioactive LXA$_4$ analogs and the rank order of potency for five analogs were essentially parallel in their ability to inhibit both PMN infiltration and vascular permeability changes (Fig. 2). These results suggest that these LX analogs exert anti-inflammatory actions by preventing PMN infiltration and subsequent events that lead to tissues injury. It is of particular interest that the most potent of the analogs, 15(R/S)-methyl-LXA$_4$, had no direct effect on LTB$_4$-triggered proton efflux, whereas the LTB$_4$ antagonist U75302, also used for comparison in topical application experiments, was clearly inhibitory (Fig. 2). The present findings, taken together with previous findings indicating that LXA$_4$ does not compete for specific binding of ^{3}H-LTB$_4$ (4 °C) to human neutrophils expressing LTB$_4$ receptors,[12] and that LXA$_4$ interacts with its own specific site of action,[3,4] it appears that anti-PMN actions of LX analogs reported here are mediated by events independent of LTB$_4$ receptor level competition, which are likely to involve LX specific receptors. Thus, the present findings are consistent with our recent results with LXA$_4$ analogs[7] and the notion that activation of LX specific receptors may represent novel therapeutic targets.

Of particular interest, other well-known proinflammatory mediators such as FMLP, IL-8, PAF, LTD$_4$, and C5a were not typically active in this model and thus were unable to promote either vascular leakage or PMN influx and essentially no visible alterations were noted.[9] Although not topically active in this model, LXA$_4$ inhibits PAF, LTD$_4$, LTB$_4$, and FMLP-directed cellular actions *in vitro*[4,6,10] (reviewed in Serhan).[3] Of particular interest, PGE$_2$ was topically active when applied together with LTB$_4$, and in this setting visibly enhanced this inflammatory response. On its own, PGE$_2$ gave only a marginal response,[9] a finding consistent with results from the hamster cheek pouch microcirculation, where PGE$_2$ potentiates local LTB$_4$-induced extravasation of plasma and leukocyte emigration.[13] In the present experiments, the fluorinated 15-epi-LXA$_4$ analog also proved to be a potent inhibitor of the responses induced by the combined actions of both prostaglandin and leukotriene[9] (Fig. 2), thus providing further evidence for the unique actions of LX, ATL and their analogs. PMA induced both PMN influx and vascular leak but only the PMN influx component was inhibitable by the LXA$_4$ analog. Dexamethasone (20 µg/ear) inhibited both PMA-induced responses (not shown; n=4). Together, these findings indicate that analogs of aspirin-triggered LXA$_4$ are potent inhibitors of PMN-induced vascular leakage and suggest that LX and dexamethasone act via distinct anti-inflammatory mechanisms.

LXB$_4$ is as potent as LXA$_4$ in inhibiting PMN transmigration across monolayers of endothelial cells and inhibiting LTB$_4$-induced adhesion to endothelial cells in vitro.[5] Results in Figure 2 clearly establish that the novel LXB$_4$ analogs also display anti-inflammatory actions in vivo. In this regard, 5(S)-methyl-LXB$_4$, which carries the stereochemistry of native LXB$_4$, was significantly more potent than 5(R)-methyl-containing LXB$_4$ analog for both inhibition of PMN infiltration and vascular permeability change, indicating that stereochemistry of the carbon-5 position is critical for topically-delivered bioactivity. These results also provide further evidence for selective sites of action for LXB$_4$. Its receptor and/or site of action remains to be identified.

The stable LX analogs tested were designed to resist rapid inactivation and proved to resist conversion by cells and isolated recombinant enzymes in vitro.[10,11] It was therefore of interest to determine whether the LX analogs remained within the ear tissues following topical application or if they were effectively cleared. Our results indicate that at 24 h after topical application greater than ~95% of the LX analogs were not present in an extractable or recoverable forms from the biopsies, which suggests that they were either cleared from the biopsied areas or were present within the tissue in a form that was not extractable using the current methodology (see Takano et al.).[9] We also found no evidence using LC/MS/MS work station-based analyses of anticipated local metabolites of these compounds persisting within the 24 h biopsies (data not shown). Along these lines, it is noteworthy that, in addition to local clearance, an alternate explanation may lie with the possibility that LX analogs could have been subject to local metabolism and/or covalent modification that results in their binding to tissue matrix components during the 24 h in vivo interval. Whether such matrix forms of LX analogs exist and whether the LX analogs are in an inactive or bioactive configuration is of interest. Nevertheless, it is clear from the present results that LX analogs are not retained in their native form within the local microenvironment (i.e., ear biopsies), and this may be another useful property of these LX analogs.

Taken together, results of the present study indicate that stable analogs of aspirin-triggered 15-epi-LXA$_4$, LXA$_4$ and LXB$_4$ serve as potent, topically active agents that inhibit PMN recruitment and PMN-mediated changes in vascular permeability. Moreover, they provide additional new tools for investigating the actions of LX and aspirin-triggered 15-epi-LX.

Acknowledgments

These studies were supported in part by National Institutes of Health grant no. GM38765 and DK50305 (C.N.S.) and a grant from Schering AG (C.N.S.; N.P.). T.T. was also supported in part by a fellowship from the Ministry of Education, Science, and Culture of Japan. K.G. is the recipient of a postdoctoral fellowship from the National Arthritis Foundation. We thank Mary Halm Small for expert assistance in the preparation of this manuscript.

REFERENCES

1. J. Varani and P.A. Ward, Mechanism of neutrophil-dependent and neutrophil-independent endothelial cell injury, *Biol. Signals* 3:1 (1994).
2. P. Borgeat and P.H. Naccache, Biosynthesis and biological activity of leukotriene B$_4$, *Clin. Biochem.* 23:459 (1990).
3. C.N. Serhan, Lipoxins and novel aspirin-triggered 15-epi-lipoxins (ATL): a jungle of cell-cell interactions or a therapeutic opportunity?, *Prostaglandins* 53:107 (1997).
4. T.H. Lee, C.E. Horton, U. Kyan-Aung, D. Haskard, A.E. Crea, and B.W. Spur, Lipoxin A$_4$ and lipoxin B$_4$ inhibit chemotactic responses of human neutrophils stimulated by leukotriene B$_4$ and N-formyl-L-methionyl-L-leucyl-L-phenylalanine, *Clin. Sci.* 77:195 (1989).

5. A. Papayianni, C.N. Serhan, and H.R. Brady, Lipoxin A_4 and B_4 inhibit leukotriene-stimulated interactions of human neutrophils and endothelial cells, *J. Immunol.* 156:2264 (1996).

6. S.P. Colgan, C.N. Serhan, C.A. Parkos, C. Delp-Archer, and J.L. Madara, Lipoxin A_4 modulates transmigration of human neutrophils across intestinal epithelial monolayers, *J. Clin. Invest.* 92:75 (1993).

7. T. Takano, S. Fiore, J.F. Maddox, H.R. Brady, N.A. Petasis, and C.N. Serhan, Aspirin-triggered 15-epi-lipoxin A_4 and LXA_4 stable analogs are potent inhibitors of acute inflammation: Evidence for anti-inflammatory receptors, *J. Exp. Med.* 185:1693 (1997).

8. J. Tamaoki, E. Tagaya, I. Yamawaki, and K. Konno, Lipoxin A_4 inhibits cholinergic neurotransmission through nitric oxide generation in the rabbit trachea, *Eur. J. Pharmacol.* 287:233 (1995).

9. T. Takano, C.B. Clish, K. Gronert, N. Petasis, and C.N. Serhan, Neutrophil-mediated changes in vascular permeability are inhibited by topical application of aspirin-triggered 15-epi-lipoxin A_4 and novel lipoxin B_4 stable analogues, *J. Clin. Invest.* 101 (in press, 1998).

10. C.N. Serhan, J.F. Maddox, N.A. Petasis, I. Akritopoulou-Zanze, A. Papayianni, H.R. Brady, S.P. Colgan, and J.L. Madara, Design of lipoxin A_4 stable analogs that block transmigration and adhesion of human neutrophils, *Biochemistry* 34:14609 (1995).

11. J.F. Maddox, S.P. Colgan, C.B. Clish, N.A. Petasis, V.V. Fokin, and C.N. Serhan, Lipoxin B_4 regulates human monocyte/neutrophil adherence and motility: design of stable lipoxin B_4 analogs with increased biologic activity, *FASEB J.* 12 (in press, 1998).

12. S. Fiore, S.W. Ryeom, P.F. Weller, and C.N. Serhan, Lipoxin recognition sites. Specific binding of labeled lipoxin A_4 with human neutrophils, *J. Biol. Chem.* 267:16168 (1992).

13. P. Hedqvist, L. Lindbom, U. Palmertz, and J. Raud, Microvascular mechanisms in inflammation, *Adv. Prostaglandin Thromboxane Leukot. Res.* 22:91 (1994).

43

CLEAVAGE OF LEUKOTRIENE D$_4$ IN MICE WITH TARGETED DISRUPTION OF A MEMBRANE-BOUND DIPEPTIDASE GENE

Geetha M. Habib[1] and Michael W. Lieberman[1,2]

Departments of Pathology[1] and Cell Biology[1,2]
Baylor College of Medicine
Houston, Texas 77030

INTRODUCTION

Leukotriene C$_4$ (LTC$_4$), leukotriene D$_4$ (LTD$_4$), and leukotriene E$_4$ (LTE$_4$), collectively known as cysteinyl leukotrienes, are members of the eicosanoid group of lipid mediators[1]. They have been implicated in a wide variety of acute and chronic inflammatory conditions including asthma, tissue injury, cardiac and liver diseases, and shock[2,3]. Production of LTC$_4$ is initiated by the conjugation of the epoxide intermediate LTA$_4$ with glutathione (GSH). LTC$_4$ is further metabolized to the cysteinyl glycine conjugate of LTA$_4$ known as LTD$_4$ by the actions of two plasma-membrane-bound ectoenzymes, γ-glutamyl transpeptidase (GGT) and γ-glutamyl leukotrienase (GGL)[4,5]. LTD$_4$ is believed to be converted to the less active cysteinyl glycine conjugate of LTA$_4$ called LTE$_4$ by a dipeptidase[6]. The rank order of molar potencies of cysteinyl leukotrienes is LTD$_4$>LTC$_4$>LTE$_4$. LTD$_4$ is considered to be at least 10 to 100-fold more potent than LTE$_4$[7]. Consequently, conversion of LTD$_4$ to LTE$_4$ is a critical step in the cysteinyl leukotriene metabolism. LTC$_4$, LTD$_4$, and LTE$_4$ are eliminated from the blood circulation with initial half-lives of 30-40 s. They are mainly taken up by kidney and liver and excreted into the urine and bile, respectively[8]. Thus, the liver seems to be the major site of their metabolic inactivation.

MICE DEFICIENT IN MEMBRANE-BOUND DIPEPTIDASE

Mice deficient in GGT have been developed and characterized in detail with respect to their ability to metabolize LTC$_4$ to LTD$_4$ and GSH to cysteinyl glycine[5,9]. There is little information available regarding the metabolism of LTD$_4$ and the enzymes involved in its conversion. Although membrane-bound dipeptidase (MBD) has been believed to mediate the metabolism of LTD$_4$, it is by no means certain that MBD is the only dipeptidase involved in the process[6]. Although both GGT and MBD act sequentially to metabolize a number of compounds, their expression patterns are concordant in some tissues and discordant in others[10]. To address this issue in more detail, we have developed mice deficient in MBD and have begun to characterize them. MBD-deficient mice grow and develop normally and are fertile. Since MBD is known to cleave β-lactam antibiotics such as imipenem and carbapenem derivatives, we tested the ability of these mice to hydrolyze glycyl dehydrophenylalanine, a commonly used β-lactam substrate[11]. The MBD-deficient mice lack the ability to cleave β-lactam substrates, indicating that testing for this ability is a specific assay for MBD in mice[12].

ANALYSIS OF CYSTINYL-*BIS*-GLYCINE (CYS-*BIS*-GLY) AND LTD$_4$ CLEAVAGE IN MBD-DEFICIENT MICE

Because MBD has long been implicated as the enzyme responsible for the inactivation of LTD$_4$, we assayed this reaction in MBD-deficient mice. We found that the LTD$_4$-to-LTE$_4$ converting ability of kidney extracts from MBD-deficient mice was only ~ 10-20% of that of kidney extracts from wild-type mice. We also evaluated LTD$_4$-to-LTE$_4$ converting activity in a variety of tissues. The mutant mice demonstrated a differential ability to cleave LTD$_4$, ranging from 10% to 90% depending on the tissue examined. We have previously shown that GGT-deficient mice die prematurely because of an excessive loss of cysteine as uncleaved GSH in their urine[9]. Since GSH is believed to be cleaved by the consecutive actions of GGT and MBD, we hypothesized that MBD-deficient mice may share some of the same characteristics with the GGT-deficient mice. Unexpectedly, MBD-deficient mice showed only a moderate (3 to 5-fold) increase of cys-*bis*-gly in the urine even though most of the cysteinyl glycine found in the renal tubules is in the oxidized form of cys-*bis*-gly (7.9 μM in wild-type versus 33.8 μM in MBD-deficient mice). Homogenates from kidney also retained ~ 40% of their ability to metabolize cys-*bis*-gly into its constituent amino acids, indicating that MBD is not the only enzyme capable of this cleavage (manuscript submitted). These data led us to propose that there are at least two independent, alternative pathways operating in the metabolism of LTD$_4$ and cys-*bis*-gly.

FURTHER CHARACTERIZATION OF LTD$_4$ DIPEPTIDASE IN MBD-DEFICIENT MICE

We investigated the cleavage of LTD$_4$ in MBD-deficient mice in more detail. When we examined MBD expression in different wild-type mouse tissues by northern blotting analysis with a mouse MBD cDNA probe, we found that MBD message levels were undetectable in spleen[10]. We also assayed spleen extracts from wild-type and MBD-deficient mice for β-lactam hydrolysis and failed to detect appreciable activity (data not shown). When LTD$_4$-to-LTE$_4$ converting activity was compared between wild-type and MBD-deficient spleen extracts, we found that MBD-deficient spleen was capable of converting LTD$_4$ to LTE$_4$ to the same extent as wild-type spleen. In fact, the rate of conversion of LTD_4 to LTE_4 in wild-type spleen (specific activity 107.33 nmol LTE_4 formed/h/mg protein) closely mirrored that in MBD-deficient spleen (95.2 nmol LTE_4 formed/h/mg protein), indicating that almost all the activity in the spleen may be accounted for by other LTD_4 dipeptidase (Fig. 1).

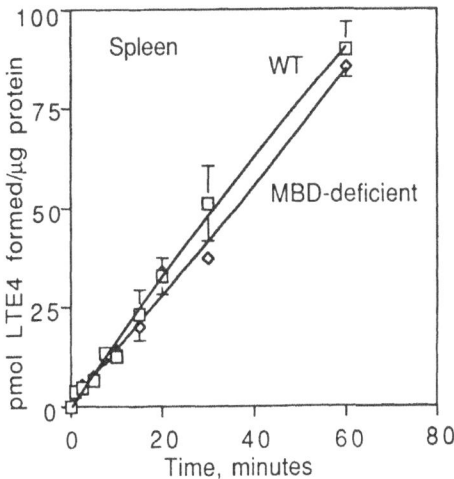

Figure 1. Time-dependent conversion of LTD_4 to LTE_4 by wild-type and MBD-deficient spleen extracts. The assays were carried out in a total volume of 200 μl containing 0.1 M Tris-HCl, pH 8.0, 5 μM LTD_4, and 100 μg spleen homogenate. Reactions were carried out in duplicate and at various time points were terminated by the addition of methanol. The supernatants were passed through Sep-Pak C-18 cartridges before they were analyzed by reverse-phase HPLC as previously described[7].

We then carried out a preliminary characterization of LTD_4 dipeptidases in MBD-deficient mice. Spleen extracts incubated with LTD_4 metabolized ~70% of the LTD_4 to LTE_4 (Fig. 2, top left panel). This LTD_4 catabolism was suppressed to ~16% when 1 mM D-penicillamine, a noncompetitive inhibitor of LTD_4 dipeptidase[13], was included in the incubation medium (Fig. 2, top right panel). This LTD_4-to-LTE_4 conversion was not inhibited by high concentrations of bestatin, an effective inhibitor of aminopeptidase M[12] (Fig. 2, bottom left panel). When captopril, another well-known inhibitor of aminopeptidases that has a structural similarity to D-penicillamine[14], was included in the assay, it had no detectable effect on LTD_4 catabolism (Fig. 2, bottom right panel). Similar results were obtained with wild-type spleen extracts (data not shown).

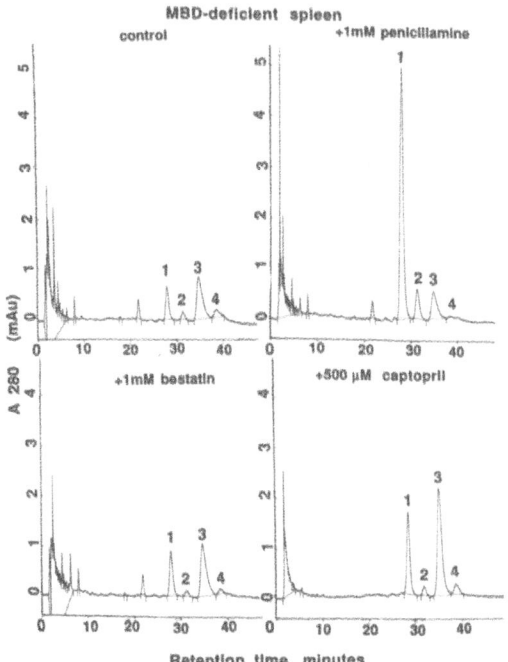

Figure 2. HPLC profiles of LTD$_4$ metabolism by the MBD-deficient homogenates of spleen in the presence and absence of various inhibitors. The reaction conditions were as described in the legend to Fig. 1 except that the homogenates were preincubated in assay medium containing appropriate concentrations of the inhibitor (1 mM D-penicillamine, 1 mM bestatin, or 500 μM captopril) for 30 min at 37°C in the absence of the substrate, the reactions were initiated by the addition of 5 μM LTD$_4$, and incubation continued at 37°C. For all the reactions, the incubation time was 1 h and the reactions contained 100 μg of protein. Peaks 1, 2, 3, and 4 correspond to LTD$_4$, the 11-*trans* isomer of LTD$_4$, LTE$_4$, and an unknown metabolite, respectively.

We also investigated LTD$_4$ cleavage in brain tissue. Earlier reports suggested that normal brain capillaries are rich in GGT, which acts like an enzymatic barrier to the vasoactive effects of LTC$_4$ by cleaving it, whereas the same capillaries do not have any metabolic activity against LTD$_4$[15]. Therefore, we compared LTD$_4$ conversion in wild-type and MBD-deficient mouse brain (Fig. 3, top panel). Brain had very low activity, and both homogenates converted ~2% of LTD$_4$ to LTE$_4$ in 1 h, indicating that it is the other LTD$_4$ dipeptidase that is responsible for this activity in the brain. We used similar assays to compare LTD$_4$ conversions in wild-type and MBD-deficient testis. We found that wild-type testis and MBD-deficient testis converted 54% and 45% of LTD$_4$ to LTE$_4$, respectively (Fig. 3, middle panel), indicating that about half of the LTD$_4$ cleavage activity comes from the other LTD$_4$ dipeptidases. Seminal vesicles have a very high GGT activity but negligible MBD expression[10] and our LTD$_4$ cleavage assays indicated that the other enzyme may significantly contribute to the LTD$_4$ clearance in them (Fig. 3, bottom panel).

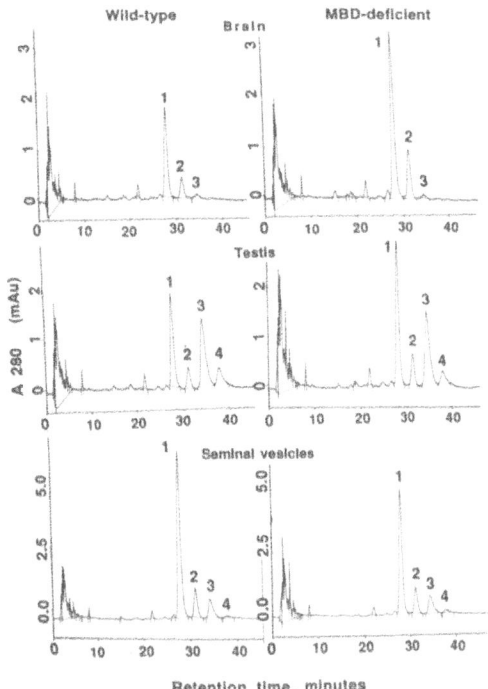

Figure 3. HPLC analyses of LTD₄ metabolism by wild-type and MBD-deficient mice homogenates from brain, testis, and seminal vesicles. The reaction conditions were as described in the legend to Fig. 1. For peak designations, see the legend to Fig. 2.

SIGNIFICANCE OF MULTIPLE LTD₄ DIPEPTIDASES

In light of our data on LTD₄ catabolism in MBD-deficient mice[12] (and these studies) and our earlier observations on LTC₄ metabolism in GGT-deficient mice[5], leukotriene degradation is more complex than originally thought. In most organs that we have investigated, both MBD and the other LTD₄ dipeptidases seem to coexist, the significance of which remains to be elucidated. Future studies on the molecular cloning and identification of new members of the LTD₄ dipeptidase family will provide us with new tools for studying LTD₄ metabolism.

ACKNOWLEDGMENT

This work was supported in part by NIH grant number ES-08668.

REFERENCES

1. W.R. Henderson, Jr., The role of leukotrienes in inflammation, *Ann. Intern. Med.* 121:684 (1994).
2. M. Huber and D. Keppler, Leukotrienes and the mercapturate pathway, in: *Glutathione Conjugation,* H. Sils and B. Ketterer, eds., Academic Press, New York (1988).
3. J. M. Drazen, J.P. Arm, and K.F. Austen, Sorting out the cytokines in asthma, *J. Exp. Med.* 183:1 (1996).
4. D. Keppler, Leukotrienes: biosynthesis, transport, inactivation, and analysis, *Rev. Physiol. Biochem. Pharmacol.* 121:2 (1992).
5. B.Z. Carter, A.L. Wiseman, R. Orkiszewski, K.D. Ballard, C-N. Ou, and M.W. Lieberman, Metabolism of leukotriene C_4 in γglutamyl transpeptidase-deficient mice, *J. Biol. Chem.* 272:12305 (1997).
6. E.M. Kozak and S.S. Tate, Glutathione-degrading enzymes of microvilli membranes. *J. Biol. Chem.* 257:6322 (1982).
7. P.M. O'Byrne, Eicosanoids and asthma. *Ann. NY. Acad. Sci.* 744:251 (1994).
8. D. Keppler, Leukotrienes and other eicosanoids in liver pathophysiology, in: *The Liver Biology and Pathobiology,* 3rd ed., I.M. Arias, J.L. Boyer, N. Fausto, W.B. Jakoby, D.A. Schachter, and D.A. Shafritz, eds., Raven Press, New York (1994).
9. M.W. Lieberman, A.L. Wiseman, Z.Z. Shi, B.Z. Carter, R. Barrios, C-N. Ou, P. Chevez-Barrios, Y. Wang, G.M. Habib, J.C. Goodman, S.L. Huang, R.M. Lebovitz, and M.M. Matzuk, Growth retardation and cysteine deficiency in g-glutamyl transpeptidase-deficient mice, *Proc. Natl. Acad. Sci. USA* 93:7923 (1996).
10. G.M. Habib, R. Barrios, Z.Z. Shi, and M.W. Lieberman, Four distinct membrane-bound dipeptidase RNAs are differentially expressed in the mouse, *J. Biol. Chem.* 271:16273 (1996).
11. H.S. Kim, and B.J. Campbell. b-lactamase activity of renal dipeptidase against N-formidoyl thienamycin, *Biochem. Biophys. Res. Commun.* 108:1638 (1982).
12. G.M. Habib, Z.Z. Shi, A.A. Cuevas, Q. Guo, M.M. Matzuk and M.W. Lieberman, Leukotriene D_4 and cystinyl-*bis*-glycine metabolism in membrane-bound dipeptidase deficient mice, *Proc. Natl. Acad. Sci. USA* In press (1998).
13. M. Huber, and D. Keppler, Inhibition of leukotriene D_4 catabolism by D-penicillamine, *Eur. J. Biochem.* 167:73 (1987).
14. K. Shindo, J.R. Baker, D.A. Munafo, and T.D. Bigby, Captopril inhibits neutrophil synthesis of leukotriene D_4 *in vitro* and *in vivo, J. Immunol.* 153:5750 (1994).
15. T. Inamura, W.M. Pardridge, Y. Kumagai, and K.L. Black, Differential tissue expression of immunoreactive deydropeptidase I, a peptidyl leukotriene metabolizing enzyme, *Prostaglandins leukot. Essent.Fatty acids* 50:8592 (1994).

44

γ-GLUTAMYL LEUKOTRIENASE CLEAVAGE OF LEUKOTRIENE C4

Michael W. Lieberman[1,2], Jefry E. Shields[3], Yvonne Will[3], Donald J. Reed[3], and Bing Z. Carter[1]

Departments of [1]Pathology and [2]Cell Biology
Baylor College of Medicine
Houston, Texas 77030

[3]Department of Biochemistry and Biophysics and
Environmental Health Sciences Center
Oregon State University
Corvallis, Oregon 97331

INTRODUCTION

The cysteinyl leukotrienes are powerful mediators of vaso- and bronchoconstriction, edema formation, and mucus secretion[1]. They appear to play a key role in asthma and may be involved in cardiac and renal disease as well[1-8]. The parent compound, leukotriene C_4 (LTC_4), is formed by the conjugation of leukotriene A_4 with glutathione (GSH). Thus it may be expected that the metabolism of LTC_4 resembles GSH conjugates formed with carcinogens, toxins, and xenobiotics[9]. Until recently, the only enzyme known to metabolize this class of compounds as well as GSH itself was γ-glutamyl transpeptidase (GGT)[10-12]. GSH conjugates including LTC_4 are metabolized to their cysteinylglycine derivatives. In the case of LTC_4 the resulting leukotriene is LTD_4. This compound is the most potent of the cysteinyl leukotrienes and has been found to be a consistently more effective agonist than LTC_4 and 10 to 100 times more effective than LTE_4, a metabolite of LTD_4[13-16]. LTC_4 is synthesized in the liver and is secreted into the bile[15]. It is also produced in peripheral tissues[1, 2, 16]. Conversion of LTC_4 to LTD_4 is not well understood although it is believed to occur extracellularly because GGT is an ectoenzyme[10-17]. We became interested in cysteinyl

Eicosanoids and Other Bioactive Lipids in Cancer, Inflammation, and Radiation Injury, 4
Edited by Honn *et al.*, Kluwer Academic / Plenum Publishers, New York, 1999.

301

leukotriene metabolism when we developed mice deficient in GGT[18]. Our subsequent studies unexpectedly showed that GGT-deficient mice are competent to metabolize LTC$_4$.

MICE DEFICIENT IN γ-GLUTAMYL TRANSPEPTIDASE

Several years ago we developed mice deficient in GGT[18]. These mice are small and grow slowly, achieving a weight of about 8-9 g. They fail to mature sexually, develop cataracts, and begin to die at ~12 weeks of age. As expected, these mice are unable to cleave GSH, and large amounts of the tripeptide are passed into the urine. Normal mouse urine GSH levels are ~6 μM, whereas GGT-deficient mouse levels are ~15,000 μM. Although GGT is the only enzyme known to initiate the breakdown of GSH, GGT-deficient mice paradoxically showed reduced GSH levels in many tissues. In liver, pancreas, and eye, for example, GSH values for GGT-deficient mice were 24%, 51%, and 39% of wild-type values, respectively. We realized early on that one reason for low GSH levels might be that the body was being depleted of cysteine by the large excretion of GSH by the kidneys. When we administered N-acetyl cysteine to GGT-deficient mice in their drinking water starting at weaning age, we were able to reverse the physical changes completely in these mice, and this allowed us to maintain these mice for long periods of time.

CLEAVAGE OF GSH AND LTC$_4$ IN GGT-DEFICIENT MICE

Although the appearance of large amounts of GSH in the urine suggested that GGT-deficient mice were unable to cleave GSH, we assayed homogenates of tissues from wild-type and GGT-deficient mice using [glycine-2-^{3}H]GSH and high performance liquid chromatography (HPLC)[12]. We were unable to detect the formation of [^{3}H]cysteinyl glycine in the GGT-deficient mice. In other studies we found that homogenates from GGT-deficient mice were unable to cleave GSSG (oxidized GSH) and were unable to metabolize γ-glutamyl-p-nitroanilide and γ-glutamyl-4-methoxy-2-naphthylamide. These agents are commonly used to assay GGT biochemically and histochemically, respectively[12, 18].

The common wisdom is that GGT is the enzyme responsible for converting LTC$_4$ to LTD$_4$, and thus one might expect that GGT-deficient mice should be unable to catalyze this reaction. However, we recently demonstrated that GGT and membrane-bound dipeptidase, the two enzymes believed to convert GSH to its constituent amino acids, are not always expressed at similar levels in different tissues[19]. This left open the possibility that GSH metabolism is more complex than generally assumed. In addition, another group of investigators had previously cloned a human gene, termed GGT-rel, which appeared to direct the cleavage of LTC$_4$[11]. Using northern blotting, this group was unable to identify the expression of GGT-rel in the mouse. Homogenates of various organs were examined for cleavage of LTC$_4$ and appearance of LTD$_4$ by HPLC. Homogenates from GGT-deficient mice readily converted LTC$_4$ to LTD$_4$. Figure 1 presents HPLC data showing this conversion in homogenates of spleen, liver, and lung from GGT-deficient and wild-type mice. Interestingly, an appreciable percentage of all conversion is carried out by this new activity, and in these homogenates much of the LTD$_4$ was converted to LTE$_4$ through the action of dipeptidase(s)[12,19]. The authenticity of the LTC$_4$, LTD$_4$, and LTE$_4$ peaks was verified by FAB-MS/MS analysis[12]. To distinguish the newly identified enzyme from GGT and to recognize its action in cleaving a γ-glutamyl group from LTC$_4$, we termed the enzyme γ-glutamyl leukotrienase (GGL).

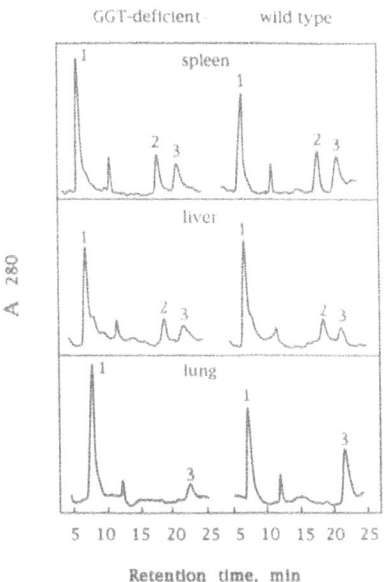

Figure 1. HPLC analysis of LTC$_4$ metabolism by homogenates of spleen, liver, and lung of GGT-deficient (left) and wild-type (right) mice. The reaction conditions are described in reference 12. Peaks 1, 2, and 3 had retention times corresponding to LTC$_4$, LTD$_4$, and LTE$_4$, respectively. For all the reactions, the incubation time was 3 h. For spleen, the reactions contained 50 μg protein. For liver and lung, the reactions contained 200 μg protein.

We examined the cleavage of LTC$_4$ in more detail in different organs of wild-type and GGT-deficient mice (Table 1). In kidney and small intestine, in which GSH cleavage is significant, wild-type mouse homogenates cleaved much more LTC$_4$ than those from GGT-deficient mice. These findings indicate that GGT is the predominant LTC$_4$-cleaving enzyme in these organs. On the other hand, in spleen and liver, ~90% of the cleavage resulted from GGL activity, while in lung only 30% of the cleavage was due to GGL activity. GGL activity was highest in spleen, followed by small intestine, kidney, and pancreas.

We have performed a preliminary characterization of the enzyme prepared from GGT-deficient mice. Like GGT, GGL is membrane bound and may be released by the action of papain. Using papain digestion, ammonium sulfate precipitation, and Sephadex G-150 chromatography, we have partially purified GGL from the small intestine of GGT-deficient mice. The papain-digested fragment has a molecular mass of ~65-70 kD as determined by sucrose velocity sedimentation and is inhibited by Acivicin, a known inhibitor of GGT. GSH does not inhibit GGL activity even when added in 80-fold excess. We measured the K_m of GGL and compared it with that of GGT using LTC$_4$ as a substrate and found that GGL had a K_m of 2.2 X 10^{-6} M, about 10-fold lower than that of GGT.

Table 1. LTC$_4$-converting activities in GGT-deficient and wild type mice*

Organs	GGT-deficient mice	wild-type mice
Spleen	3.24	3.81
Small intestine	1.30	48.47
Kidney	1.25	502.40
Pancreas	0.90	248.50
Liver	0.62	0.66
Lung	0.28	0.78

* LTC$_4$-converting activities are expressed as nmol/h/mg protein.

We also investigated the ability of GGL to cleave a variety of S-substituted GSH compounds (Table 2). We found that GGL will cleave S-decyl GSH, a known competitor for LTC$_4$ in the brain[20]. The K_m for S-decyl GSH cleavage by GGL is approximately 10-fold higher than that for LTC$_4$ (see above). We also investigated the cleavage of a series of S-alkyl GSH compounds ranging from S-methyl GSH to S-nonyl GSH (Table 2). Compounds with chains of six or more alkyls were cleaved, while those with zero (GSH) to five did not appear to be cleaved or were cleaved at a rate too slow for us to detect. We also examined the cleavage of one aromatic GSH derivative, S-(2,4 dinitrophenyl) GSH, but were unable to detect cleavage. It is known that GGT will cleave S-nitrosyl GSH[21], and a standard assay showed similar conversion by GGL (Table 2).

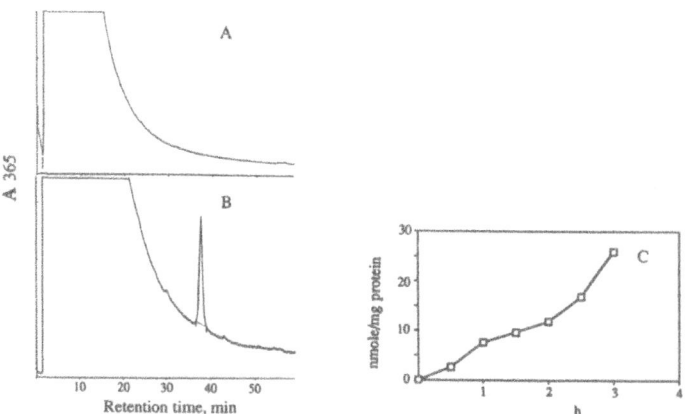

Figure 2. HPLC analysis of S-decyl GSH cleavage. The reaction conditions are described in reference 12. A. S-decyl GSH only. Note that substrate peak is obscured by the solvent front; B. S-decyl GSH incubated with spleen homogenate from GGT-deficient mice (350 µg protein, 3h); C. S-decyl cysteinylglycine formation as a function of time. For reaction conditions see Fig. 2B.

Table 2. Comparison of substrate specificity of GGL and GGT

Substrates	GGL	GGT
LTC_4	+	+
S-decyl GSH	+	+
S-nonyl GSH	+	+
S-octyl GSH	+	+
S-heptyl GSH	+	+
S-hexyl GSH	+	+
S-pentyl GSH	-	+
S-nitrosyl GSH	+	+
GSH	-	+
GSSG	-	+
S-(2,4 dinitrophenyl) GSH	-	+
γ-glutamyl-p-nitroanilide	-	+
γ-glutamyl-4-methoxy-2-naphthylamide	-	+

THE FUNCTION OF γ-GLUTAMYL LEUKOTRIENASE

The function of GGL is not well understood, but a few considerations are worth mentioning. First, the K_m of GGL for LTC_4 is about 10 times lower than that of GGT. This finding suggests that GGL readily converts LTC_4 to LTD_4 and is the more efficient of the two enzymes. Because it appears that LTD_4 is the most active of the cysteinyl leukotrienes, GGL may be important in generating the acute response in diseases like asthma. However, at present we do not know the localization of GGL within individual organs. Such information is essential if we are to understand the role of GGL in LTD_4 generation. In the lung, only about 30% of LTC_4-converting activity is the result of GGL (Figure 1 and Table 1), but if it is expressed at sites related to bronchial smooth muscle, its action might be crucial. In the kidney, GGT is largely confined to the proximal tubules. Yet the actions of LTD_4 in renal disease are unlikely to be initiated or confined to these cells. Knowledge of the intraorgan localization of GGL will be crucial in understanding its role in leukotriene action. Finally, GGL will catalyze the conversion of other substituted GSH compounds to their cysteinylglycine derivatives. It is possible that other biologically important molecules also rely on GGL for activation or detoxification. The challenge of understanding this pathway will be substantial.

ACKNOWLEDGMENT

This work was supported in part by NIH grant ES-07827.

REFERENCES

1. W.R. Henderson, Jr., The role of leukotrienes in inflammation, *Annul. Intern. Med.* 121:684 (1994).
2. J.M. Drazen, J.P. Arm, and K.F. Austen, Sorting out the cytokines of asthma, *J. Exp. Med.* 183:1 (1996).
3. F. Michelassi, L. Landa, R.D. Hill, E. Lowenstein, W.D. Watking, A.J. Petkau, et al., Leukotriene D4: a potent coronary artery vasoconstrictor associated with impaired ventricular contraction, *Science* 217:841 (1982).
4. K.F. Badr, C. Baylis, J.M. Pfeffer, M.A. Pfeffer, R.J. Soberman, R.A. Lewis, et al., Renal and systemic hemodynamic responses to intravenous infusion of leukotriene C4 in the rat, *Circ. Res.* 54:492 (1984).

5. C.C. Lee, R.F. Appleyard, J.G. Byrne, and L.H. Cohn, Leukotrienes D_4 and E_4 produced in myocardium impair coronary flow and ventricular function after two hours of global ischaemia in rat heart, *Cardiovasc. Res.* 27:770 (1993).

6. T. Katoh, E.A. Lianos, M. Fukunaga, K. Takahashi, and K.F. Badr, Leukotriene D_4 is a mediator of proteinuria an glomerular hemogynamic abnormalities in passive heymann nephritis, *J. Clin. Investig.* 91:1507 (1993).

7. R. Petric and A.W. Ford-Hutchinson, Elevated cysteinyl leukotriene excretion in experimental glomerulonephritis, *Kidney International* 46:1322 (1994).

8. W.R. Henderson, Jr., Role of leukotrienes in asthma, *Ann. of Allergy* 72:272 (1994).

9. A. Parkinson, Biotransformation of xenobiotics, in: *Casarett and Doull's Toxicology, The Basic Science of Poison* C.D. Klaassen, ed., McGraw-Hill , New York (1996).

10. A. Meister, and A. Larsson, Glutathione synthetase deficiency and other disorders of the γ-glutamyl cycle, in: *The Metabolic Basis of Inherited Disease*, 6th ed., C.R. Scriver, A.L. Beaudet, W.S. Sly, and D. Valle, eds., McGraw-Hill, New York (1989).

11. N. Heisterkamp, E. Rajpert-De Meyts, L. Uribe, H.J. Forman, and J. Groffen, Identification of a human γ-glutamyl cleaving enzyme related to, but distinct from γ-glutamyl transpeptidase, *Proc. Natl. Acad. Sci. USA* 88:6303 (1991).

12. B.Z. Carter, A.L. Wiseman, R. Orkiszewski, K.D. Ballard, C-N. Ou, and M.W. Lieberman, Metabolism of leukotriene C_4 in γ-glutamyl transpeptidase-deficient mice, *J. Biol. Chem.* 272:12305 (1997).

13. J.P. Arm, and H.L. Tak, Sulphidopeptide leukotrienes in asthma, *Clin. Sci.* 84:501 (1993).

14. M.N. Samhoun, R.M. Conroy, and P.J. Piper, Pharmacological profile of leukotriene E_4, N-acetyl E_4 and four of th$_$ novel ω and β oxidative metabolites in airways of guinea pig and man in vitro, *Br. J. Pharmacol.* 98:1406 (198

15. D. Keppler, Leukotrienes: biosynthesis, transport, inactivation, and analysis, *Rev. Physiol. Biochem. Pharmacol.* 121:2 (1992).

16. W.R. Henderson, Jr., D.B. Lewis, R.K. Albert, Y. Zhang, W.J.E. Lamm, G.K.S. Chiang, F. Jones, P. Eriksen, Y T. Tien, M. Jonas, and E.Y. Chi, The importance of leukotrienes in airway inflammation in a mouse model of asthma, *J. Exp. Med.* 184:1483 (1996).

17. M.W. Lieberman, R. Barrios, B.Z. Carter, G. Habib, R.M. Lebovitz, S. Rajagopalan, A. Sepulveda, Z.Z. Shi, and D.F. Wan, γGlutamyl transpeptidase: What does the organization and expression of a multipromoter gene tell u about its functions?, *Am. J. Pathol.* 147:1175 (1995).

18. M.W. Lieberman, A.L. Wiseman, Z.Z. Shi, B.Z. Carter, R. Barrios, C.N. Ou, P. Chévez-Barrios, Y. Wang, G.M Habib, J.C. Goodman, S.L. Huang, R.M. Lebovitz, M.M. Matzuk, Growth retardation and cysteine deficiency i γ-glutamyl transpeptidase-deficient mice, *Proc. Natl. Acad. Sci. USA* 93:7923 (1996).

19. G.M. Habib, R. Barrios, Z.Z. Shi, and M.W. Lieberman, Four distinct membrane bound dipeptidase RNAs are differentially expressed in the mouse, *J. Biol. Chem.* 271:16273 (1996).

20. A.M. Goffinet, and A. Nugyen, Brain leukotriene C_4 binding sites are S-alkylglutathione binding sites, *Eur. J. Pharmacol.* 161:99 (1989).

45

LTE$_4$ BLOOD LEVELS IN INFANTS WITH CONGENITAL HEART LESIONS

Jean Pierre Gascard and Charles Brink

CNRS ERS 566
Centre Chirurgical Marie Lannelongue
133 av. de la RÈsistance
92350 Le Plessis Robinson
France

INTRODUCTION

The cysteinyl-leukotrienes are metabolites of arachidonic acid via the 5-lipoxygenase pathway.[1] These potent smooth muscle contractile agents are released from the human lung in vitro.[2,3] Since the cysteinyl-leukotrienes are potent vasoconstrictor agents in human pulmonary vessels,[4,5] these mediators may be responsible for pulmonary hypertension.   Previous data have shown that infants with congenital heart lesions frequently have elevated pulmonary arterial pressure. In these patients, there are also acute transitory episodes of increased pulmonary arterial pressure following open heart surgery for correction of congenital heart lesions. Unfortunately, the circulating levels of leukotrienes in infants with pulmonary hypertension is not known. The aim of this study was to measure the LTE$_4$ levels in the circulation of infants and to correlate these levels with the elevated pulmonary arterial pressure.

METHODS

This study involved 18 patients with congenital heart defects. Six infants were diagnosed to have regular form of Tetralogy of Fallot (Fallot, N=6) and were 12 6 months of age at surgical intervention. Twelve patients were diagnosed to have elevated levels of pulmonary arterial pressure (PH). This group included seven subjects with complete atrioventricular septal defect, three had truncus arteriosus and two had ventricular septal defect. All patients with PH (N=12) presented with heart failure despite medical therapy support. However, none of them received medication that could interfere with the arachidonic acid cascade. The age of the PH infants at the surgical intervention was 9 ± 3 months.

Hemodynamic measurements (pulmonary, systemic and atrial pressues) were continuously monitored for each patient using a Hewlett Packard 78354A detector. However, pulmonary arterial pressures (PAP, mmHg) were not measured in Fallot patients.

Surgery

Open heart repair was performed with aid of hypothermic cardiopulmonary bypass. Aortic cross clamping with injection of blood cardioplegia was used during the intra cardiac repair. The mean duration of cardiopulmonary bypass (ECC) in Fallot patients was 69 ± 4 min and 114 ± 9 min in PH subjects. After sternal and pericardial opening, blood samples were drawn from the main pulmonary artery and the left atrium.

Leukotriene E_4 measurements

In order to validate the measurements of LTE_4 levels in whole blood, a preliminary series of experiments were performed on peripheral blood samples obtained from nine normal infants (age, < 2 yrs) and 4 normal adult subjects (age, > 50 yrs). An LTE_4 standard (final concentration 400 ng/ml) was added to each sample and the blood samples were treated as indicated below. However, due to the complexity of the surgical intervention, internal LTE_4 standards were not performed in patients.

All blood samples (0.5 ml) were collected directly in tubes containing methanol at the beginning (Start) of extracorporeal circulation (ECC), 5 min after lung reperfusion and at the end of ECC. Therefore, 3 blood samples were obtained from the pulmonary artery and 3 from the left atrium for each patient. A total of 35 measurements were made in Fallot patients (one sample was not obtained from one patient during ECC) and 67 samples were analyzed in PH subjects (several samples were not obtained from one patient during ECC). The samples were centrifuged and supernatants were added to tubes containing methanol at 10% and this mixture was passed through a Sep-Pak C_{18} cartridge. The extracts containing the lipids were collected on 3 ml of methanol prior to evaporation under argon. A separation of LTE_4 was then performed by HPLC and quantification was performed by immunoenzymatic assay (EIA).

All results are expressed as means ± SEM and no statistical analysis was performed due to the limited number of patients studied.

RESULTS

The data presented in Figure 1 indicate that LTE_4 was detected in peripheral whole blood samples derived from normal infants. However, this metabolite was not detected in blood samples obtained from normal adult subjects.

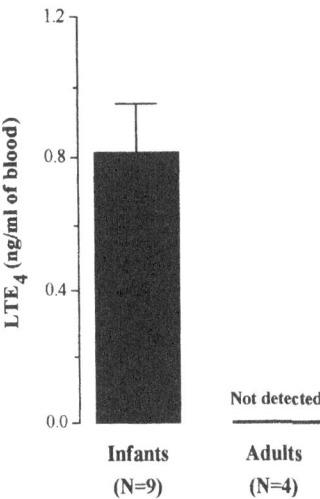

Figure 1. LTE$_4$ levels detected in peripheral blood samples obtained from normal infants and healthy adult subjects. Values are means SEM and the number in parenthesis indicates the number of subjects studied.

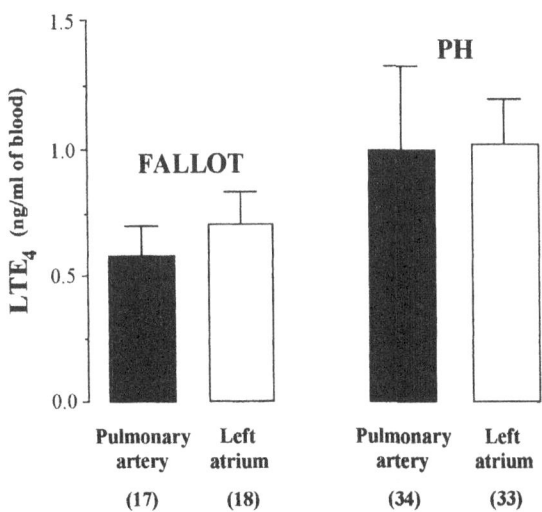

Figure 2. Detection of LTE$_4$ in human blood samples obtained from infants during ECC. Values are means SEM and the number in parenthesis indicates the number of blood samples analyzed in Fallot patients (N=6) and in PH subjects (N=12) during extracorporeal circulation (ECC) in open heart surgery for correction of congenital heart lesions.

The LTE$_4$ levels detected in the peripheral blood samples from Fallot patients was 0.44 0.12 ng/ml blood (range: 0 to 0.76 ng/ml blood, N=6) and was similar to those values obtained from peripheral blood samples derived from PH subjects (1.37 0.67 ng/ml blood, N=12). The range in the LTE$_4$ detected was: 0 to 7.2 ng/ml blood in peripheral blood samples derived from PH subjects. In only one Fallot patient were the LTE$_4$ levels undetectable (value=0) whereas in PH patients 4 patients exhibited undetectable LTE$_4$ levels.

In PH patients there was no correlation between the elevated preoperative pulmonary arterial pressure (PAP, mmHg) and the levels of circulating LTE$_4$ detected in the peripheral blood samples (r= 0.38). The mean pulmonary arterial pressure in the PH patients was 56 4 mmHg.

The total quantities of LTE$_4$ detected in the blood samples obtained from the catheter in the pulmonary artery and in the left atrium during ECC are presented in Figure 2. These results obtained in patients without (Fallot) and with pulmonary hypertension (PH) were similar.

DISCUSSION

In peripheral blood samples obtained from normal healthy infants, infants with Tetralogy of Fallot and infants with PH, the LTE$_4$ levels were similar. These data suggest that neither the surgical intervention nor the pulmonary hypertension were associated with an alteration in the LTE$_4$ blood levels.

While there was a considerable variation in the LTE$_4$ levels detected in infants, the range was similar to that reported by other investigators when this metabolite was measured in other biological fluids obtained from adult patients[6-8]. The observation that LTE$_4$ levels were detectable in peripheral blood samples from infants was in contrast to results derived from similar experiments performed on blood samples obtained from healthy adult subjects. Previous reports have shown that the levels of cysteinyl-leukotrienes extracted from biological fluids of healthy adult subjects were frequently undetectable.[9] The data obtained in peripheral whole blood samples (present report) confirm this observation. In contrast, the methanol extraction procedure for LTE$_4$ measurements in whole blood samples obtained form infants provides some evidence for a significant difference which may exisit between infants and adults. Although this was not the aim of the present investigation, the data provide preliminary information which suggest an alteration of cysteinyl-leukotrienes which may occur during development.

The considerable variation in the levels of LTE$_4$ between patients as well as between infants and adults may be related to the way LTE$_4$ is bound in the human circulation. Unfortunately, little is known about the fixation of cysteinyl-leukotrienes in the human blood. Few investigations have been published dealing with the quantities of leukotrienes in human blood samples and analysis has been principally based on results derived from human plasma[10].

There was no correlation between the elevated pulmonary arterial pressure and the circulating levels of LTE$_4$ detected. Whether or not lung tissue from pulmonary hypertensive patients exhibit elevated levels of LTE$_4$ has not been reported.

These data from a limited number of patients suggest that there may be elevated levels of LTE$_4$ in the circulation of infants, results which are different from similar extractions performed on blood samples from adults where the levels of LTE$_4$ are difficult to detect.

REFERENCES

1. R.C. Murphy, S. Hammarstr^m and B. Samuelsson, Leukotriene C: a slow reacting substance from murine mastocytoma cells, *Proc Natl Acad Sci USA.* 76:4275 (1979).
2. S-E. Dahlèn, G. Hansson, P. Hedqvist, T. Bjorck, E. Granstrom and B. Dahlèn, Allergen challenge of lung tissue from asthmatics elicits bronchial contraction that correlates with the release of leukotriene C_4, D_4 and E_4, *Proc Natl Acad Sci USA.* 80:1712 (1983).
3. B.J. Undem, W.C. Pickett, L.M. Lichtenstein and G.K. Adams, The effect of Indomethacin on immunologic release of histamine and sulfidopeptide leukotrienes from human bronchus and lung parenchyma, *Am Rev Respir Dis.* 138:1183 (1987).
4. C. Labat, J.L. Ortiz, X. Norel, I. Gorenne, J. Verley, T.S. Abram, N.J. Cuthbert, S.R. Tudhope, P. Norman, P.J. Gardiner and C. Brink, A second cysteinyl-leukotriene receptor in human lung, *J. Pharmacol Exp Ther.* 263:800 (1992).
5. J.L. Ortiz, I. Gorenne, J. Cortijo, A. Seller, C. Labat, B. Sarria, T.S. Abram, P.J. Gardiner and C. Brink, Leukotrien receptors on human pulmonary vascular endothelium, *Br J Pharmacol.* 115:1382 (1995).
6. P.J. Manning, J. Rokach, H. Malo, D. Ethier, A. Cartier, Y. Girard, S. Charmeson and P. O'Byrne, Urinary leukotriene E_4 levels during early and late asthmatic responses, *J Allergy Clin Immunol.* 86:211 (1990).
7. T.H. Lee, A.E.G. Crea, V. Gant, B.W. Spur, B.E. Marron, K.C. Nicolaou, E. Reardon, M. Brezinski and C.N. Serhan, Identification of lipoxin A_4, and its relation to the sulfidopeptide leukotrienes C_4, D_4 and E_4 in the bronchoalveolarlavage fluids obtained from patients with selected pulmonary diseases, *Am Rev Respir Dis.* 141:1453 (1990).
8. G.W. Taylor, I. Taylor, P. Black, N.H. Maltby, N. Turner, R.W. Fuller and C.T. Dollery, Urinary leukotriene E_4 after antigen challenge and in acute asthma and allergic rhinitis, *Lancet.* 18:584 (1989).
9. S. Lam, H. Chan, J.C. Leriche, M. Chan-Yeung and H. Salari, Release of leukotrienes in patients with bronchial asthma, *J Allergy Clin Immunol.* 81:711 (1988).
10. A.P. Sampson, C.P. Green, D.A. Spenser, P.J. Piper and J.F. Price, Leukotrienes in the blood and urine of children with acute asthma, Ann NY Acad Sci. 629:437 (1991).

46

EVALUATION OF THE PHARMACOLOGICAL ACTIVITY OF THE PURE CYSTEINYL-LEUKOTRIENE RECEPTOR ANTAGONISTS CGP 45715A (IRALUKAST) AND CGP 57698 IN HUMAN AIRWAYS

Valérie Capra[1], Saula Ravasi[1], Manlio Bolla[2], Serena Viappiani[2], Silvia Pagliardini[1], P.Angelo Belloni[3], Maurizio Mezzetti[4], G.Carlo Folco[2], Simonetta Nicosia[1], G.Enrico Rovati[1]

[1]Laboratory of Molecular Pharmacology
[2]Center for Cardiopulmonary Pharmacology
Institute of Pharmacological Sciences, University of Milan, Via Balzaretti 9, 20133 Milan, Italy;
[3]Niguarda Hospital, Pad. de Gasperis Thoracic Surgery, Milan;
[4]S.Paolo Hospital, Clinical Surgery, Milan.

INTRODUCTION

Cysteinyl-containing leukotrienes (cysteinyl-LTs) produce bronchoconstriction, mucus hypersecretion, and pro-inflammatory effects (Dahlén et al., 1980; Dahlén et al., 1981; Foster & Chan, 1991) that may contribute to airway hyper-responsiveness, mucus plug formation, epithelial cell damage and other changes in airway morphology. The pivotal role of cysteinyl-LTs in the pathogenesis of asthma has stimulated many pharmaceutical companies to develop programs aimed at the discovery of cysteinyl-LT antagonists, either based on LTD_4 structure or unrelated to it, to provide therapeutic agents for use in asthma. Iralukast and CGP 57698 are two recently developed cysteinyl-LT antagonists that have been shown to be potently active in various animal models, both *in vitro* and *in vivo* (Bray et al., 1990; von Sprecher et al., 1991; von Sprecher et al., 1996).

METHODS

Human lung parenchyma membranes (HLPM) and isolated human bronchi (HB) were from macroscopically normal specimens removed at thoracotomy for pulmonary carcinoma. HLPM were prepared as previously described (Rovati et al., 1992). Briefly, specimens were minced and homogenized in Hepes buffer and membrane fraction recovered at 27,000xg. Aliquots were stored at -80°C until use. HB (Bolla et al., 1997) were dissected free of surrounding parenchyma and blood vessels, helically cut and prepared as strips (2-3x10 mm).

[^{3}H]-LTD$_4$ equilibrium binding experiment were performed in HLPM at 25° C for 30 min after preincubating HLPM with the specific antagonist for 15 min (Capra et al., 1998a). Bound ligand was separated by rapid vacuum filtration and radioactivity was measured by liquid scintillation counting. Binding is expressed as the ratio of bound over total ligand concentration (B/T) vs. the log of total concentration.

HB isometric contractions were measured at a tension of 1 g, 37 °C, 95% O$_2$ - 5% CO$_2$ and pH 7.4, after a 60 min equilibration period and administration of cumulative concentrations (10-300 mM) of Ach to check tissue sensitivity. HLPM and HB strips were treated with L-Cysteine to inhibit LTD$_4$ metabolism. The contractile response was expressed as % of the maximal response to 300 mM ACh.

Analysis of binding data and concentration-response curves were performed by using the computer programs LIGAND (Munson and Rodbard, 1980) and ALLFIT (De Lean et al., 1978), respectively. Selection of the best fitting model and evaluation of the statistical significance of the parameter difference was based on the F-test of the extra sum of square principle. A statistical level of significance of $p < 0.05$ was accepted. All the curves shown were computer generated.

RESULTS AND DISCUSSION

Binding studies

LTD$_4$ receptor has been identified in animal and human tissues as a member of the "superfamily" of the seven transmembrane domain receptors. Its transduction mechanism involves at least two G proteins (Crooke et al., 1989; Mong et al., 1987; Sjolander et al., 1990) and, as expected for a G-protein coupled receptor, two different classes of sites (Table 1) have been demonstrated in binding studies (Capra et al., 1998a,b). This result, in disagreement with the result presented in a previous paper (Lewis *et al*, 1985), has been achieved in HL parenchyma (Capra et al, 1988b) by sophisticated binding protocols and extensive computer modeling of binding data (Rovati et al., 1990; Rovati et al., 1991). The two sites represent two affinity states of the same receptor, interconverted by GTP and its analogs, similarly to what demonstrated in guinea pig lung (Aharony et al., 1987; Watanabe et al., 1990; O'Sullivan & Mong, 1989).

Before performing the real binding experiments on HLPM, we followed a theoretical approach (Capra et al., 1998a) to simulate (MacSIMUL, G.E. Rovati and P.J. Munson) the interaction of an unselective receptor antagonist with the two states of a G-protein coupled receptor (De Lean et al., 1980; Lefkowitz et al., 1993) labeled by a selective agonist. The results obtained from real multiligand (Rovati et al., 1990) binding experiments (Figure 1) confirmed the theoretical assumptions, indicating that each antagonist was able to compete for both sites labeled by [^{3}H]-LTD$_4$, without discriminating between them ($K_{i1} = K_{i2}$, Table 1). This demonstrate that iralukast and CGP 57698 are pure receptor antagonist (De Lean et al., 1980; Lefkowitz et al., 1993). Moreover, both K_i's are in the nanomolar range, thus confirming that these compounds are high affinity ligands for the *CysLT* receptor on human lung.

Functional studies on human isolated bronchial strips

The high affinity exhibited by iralukast and CGP 57698 in binding studies on human lung parenchyma seems to correlate well with the potency observed in functional studies performed in human bronchi, although caution should be taken in comparing different tissues. The addition of cumulative concentrations of exogenous LTD$_4$ induced a concentration-dependent contraction of isolated human bronchi with an EC$_{50}$ of 5.9 nM ± 7.5 %CV and with a maximum effect at

Table 1. Binding affinities for the high and low affinity binding sites of the receptor labeled by ^{3}H-LTD4 in HLPM.

Agonist/antagonist	High affinity site	Low affinity site
LTD4	$K_{d1} = 0.054$ nM ± 90 %CV	$K_{d2} = 2.43$ nM ± 83 %CV
Iralukast	$K_{i1} = K_{i2} = 16.6$ nM ± 36 %CV	
CGP57698	$K_{i1} = K_{i2} = 5.7$ nM ± 19 %CV	

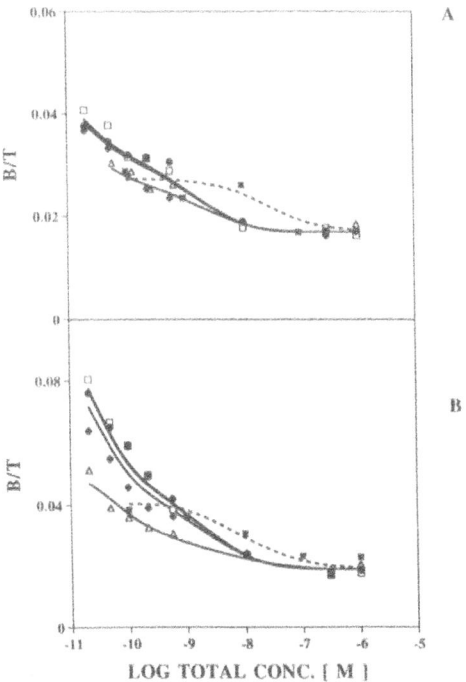

Figure 1. Multiligand curves of iralukast (Panel A) and GCP 57698 (Panel B) vs. [^{3}H]-LTD4. Each panel consists of a control curve (l), a family of three multiligand [^{3}H]-LTD4 curves, each in the presence of a fixed concentration, 0.1(q),1(u), 10(s) nM of the antagonist, and the classical competition curve (k).

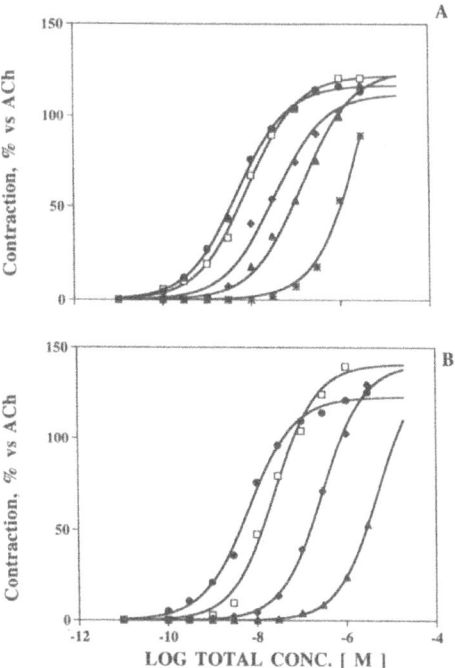

Figure 2. Effect of iralukast (Panel A) and GCP 57698 (Panel B) on LTD$_4$-induced contraction of HB strips. Each panel consists of a control curve of LTD$_4$ (l) and a family of three curves, each in the presence of a fixed concentration, 0.01(q), 0.1(u), 1(s) and 3(k) μM of the antagonist.

approximately 1 mM (Figure 2). The maximal efficacy was 119.7 ± 1.3 %CV of that elicited by 300 μM ACh. Preincubation of the tissue with different concentration of the receptor antagonists iralukast and CGP 57698 caused a parallel rightward shift of the concentration-response curves (Figure 2).

Although from the these curves CGP 57698 seems to be more potent than iralukast (Figure 2 Panel B and A, respectively), their calculated pA$_2$ (8.51 ± 1.6 %CV and 7.77 ± 4.3 %CV, respectively) are not significantly different. The data plotted according to Schild show regression lines with slopes of 1.08 and 0.99 for CGP 57698 and iralukast, respectively, not significantly different from unit. Thus, statistical evaluation would suggest pure competitive antagonism.

It is well known that immunological challenge of human bronchial airways is correlated with the release of histamine and cysteinyl-LTs (Dahlén et al., 1983). A polyclonal antibody against human IgE was used to trigger bronchoconstriction in the normal bronchial tissue, dependent on the mobilization of the endogenous precursor, i. e. arachidonic acid, and on cysteinyl-LTs (Viganò et al., 1990). The addition of cumulative concentrations of anti-IgE (0.01-3.0 mg/ml) caused a progressive increase in bronchial tone, slow in onset and long-lasting, reaching a plateau at 1.0-3.0 μg/ml. Both iralukast and CGP 57698 were able to inhibit the allergen-induced contractions of isolated human bronchi, by significantly reducing the maximal response in a concentration-dependent manner (Capra et al., 1998a). Furthermore, in these experiment CGP 57698 appeared to be 10-30 fold more potent than iralukast in preventing the increase in bronchial tone (data not shown).

CONCLUSIONS

In conclusion, the results of the present *in vitro* investigation indicate that iralukast and CGP 57698 are pure and potent *CysLT* receptor antagonists in HLPM in receptor binding studies. Moreover, both compounds, with comparable potencies, are specific inhibitors of LTD_4- and antigen-induced cysteinyl-LT dependent contraction of isolated human bronchi, thus confirming that they might be useful tools in the therapy of asthma and other allergic diseases.

Acknowledgments

The authors wish to acknowledge Dr. M. A. Bray and Dr. A. von Sprecher (Research Department, Pharmaceutical Division, CIBA-GEIGY Ltd., CH-4002 Basel, Switzerland) for providing LTD_4 and the antagonists and for helpful discussion.

REFERENCES

Aharony, D., Falcone, R.C. and Krell, R.D., 1987, Inhibition of ^{3}H-leukotriene D_4 binding to guinea pig lung receptors by the novel leukotriene antagonist ICI 198,615, *J. Pharmacol. Exp. Ther.* 243:921.

Bolla, M., Caruso, P., Giossi, M., Folco, G.C., Civelli, M. and Sala, A., 1997, Comparative analysis of isolated human bronchi contraction and biosynthesis of cysteinyl leukotrienes using a direct 5-lipoxygenase inhibitor, *Biochem. Pharmacol.* 54:437.

Bray, M.A., Anderson, W.H., Subramanian, N., Niederhauser, U., Kuhn, M., Erard, M. and von Sprecher, A., 1990, CGP 45715A: a leukotriene D4 analogue with potent peptido-LT antagonist activity, *Adv. Prostaglandin Thromboxane Leukot. Res.* 21:503.

Capra, V., Bolla, M., Belloni, P.A., Mezzetti, M., Folco, G.C., Nicosia, S. and Rovati, G.E., 1998a, Pharmacological characterization of the cysteinyl-leukotriene antagonists CGP 45715A, (iralukast) and CGP 57698 in human airways *in vitro*, *Br. J. Pharmacol.* 123:590.

Capra, V., Nicosia, S., Ragnini, D., Mezzetti, M., Keppler, D. and Rovati, G.E., 1998b, Identification and characterization of two cysteinyl-leukotriene high affinity binding sites with receptor characteristics in human lung parenchyma, *Mol. Pharmacol.* in press.

Crooke, S.T., Mattern, M., Sarau, H.M., Winkler, J.D., Balcarek, J., Wong, A. and Bennett, C.F., 1989, The signal transduction system of the leukotriene D_4 receptor, *Trends Pharmacol. Sci.* 10:103.

Dahlén, S.E., Biörk, J., Hedqvist, P., Arfors, K.E., Hammarström, S., Lindgren, J.A. and Samuelsson, B., 1981, Leukotrienes promote plasma leakage and leukocyte adhesion in postcapillary venules: in vivo effects with relevance to the acute inflammatory response, *Proc. Natl. Acad. Sci. USA* 78:3887.

Dahlén, S.E., Hansson, G., Hedqvist, P., Björck, T., Granström, E. and Dahlén, B., 1983, Allergen challenge of lung tissue from asthmatics elicits bronchial contraction that correlates with the release of leukotrienes C_4, D_4 and E_4, *Proc. Natl. Acad. Sci. USA* 80:1712.

Dahlén, S.E., Hedqvist, P., Hammarström, S. and Samuelsson, B., 1980, Leukotrienes are potent costrictors of human bronchi, *Nature* 288:484.

De Lean, A., Munson, P.J. and Rodbard, D., 1978, Simultaneous analysis of families of sigmoidal curves: application to bioassay, radioligand assay, and physiological dose-response curves, *Am. J. Physiol.* 235:E97.

De Lean, A., Stadel , J.M. and Lefkowitz, R.J., 1980, A ternary complex model explains the agonist-specific binding properties of the adenylate cyclase-coupled b-adrenergic receptor, *J. Biol. Chem.* 255:7108.

Foster, A. and Chan, C., 1991, Peptide leukotriene involvement in pulmonary eosinophil migration upon antigen challenge in the actively sensitized guinea pig, *Int. Arch. Allergy Immunol.* 96:279.

Lefkowitz, R.J., Cotecchia, S., Samama, P. and Costa, T., 1993, Constitutive activity of receptors coupled to guanine nucleotide regulatory proteins, *Trends Pharmacol. Sci.* 14:303.

Lewis, M.A., Mong, S., Vessella, R.L. and Crooke, S.T., 1985, Identification and characterization of leukotriene D_4 receptors in adult and fetal human lung, *Biochem. Pharmacol.* 34:4311.

Mong, S., Hoffman, K., Wu, H.L. and Crooke, S.T., 1987, Leukotriene-induced hudrolysis of inositol in guinea pig lung: mechanism of signal transduction for leukotriene-D_4 receptors, *Mol. Pharmacol.* 31:35.

Munson, P.J. and Rodbard, D., 1980, LIGAND: A Versatile Computerized Approach for Characterization of Ligand-Binding Systems, *Anal. Biochem.* 107:220.

O'Sullivan, B.P. and Mong, S., 1989, Binding of radiolabeled high affinity antagonist to leukotriene D4 receptor in guinea pig lung membranes: interconvertion of agonist-receptor binding affinity states, *Mol. Pharmacol.* 35:795.

Rovati, G.E., Giovanazzi, S., Mezzetti, M. and Nicosia, S., 1992, Heterogeneity of binding sites for ^{3}H-ICI 198,615 in human lung parenchyma, *Biochem. Pharmacol.* 44:1411.

Rovati, G.E., Rabin, D. and Munson, P.J., 1991, Analysis, design and optimization of ligand binding experiments, in *Horizon in Endocrinology, Vol II,* M. Maggi and E.V. Geenen, ed., Serono Symposia Publication from Raven Press, New York.

Rovati, G.E., Rodbard, D. and Munson, P.J., 1990, DESIGN: computerized optimization of experimental design for estimating K_d and B_{max} in ligand binding experiments. (II. Simultaneous analysis of homologous and heterologous competition curves, blocking and "multiligand" dose-response surfaces), *Anal. Biochem.* 184:172.

Sjolander, A., Gronroos, E., Hammarström, S. and Andersson, T., 1990, Leukotriene D_4 and E_4 induce transmembrane signaling in human epithelial cells, *J. Biol. Chem.* 265:20976.

Viganò, T., Crivellari, M.T., Mezzetti, M. and Folco, G.C., 1990, Preparation of human and animal lung tissue for eicosanoid research, in *Methods in Enzymology.* ed. R.C. Murphy and F.A. Fitzpatrick, ed., Academic Press, Inc., San Diego.

von Sprecher, A., Beck, A., Sallmann, A., Breitenstein, W., Wiestner, H., Brokatzky-Geiger, J., Eisler, K., Gamboni, R., Rosslein, L., Schlingloff, G., Cohen, N.C., Bach, L.-H., Anderson, G.P., Subramanian, N. and Bray, M.A., 1991, CGP 45715A, a potent peptidoleukotriene antagonist based on the structure of LTD_4, *Med. Chem. Res.* 1:195.

von Sprecher, A., Gerspacher, M., Beck, A., Kimmel, S., Wiestner, H.R., Anderson, G.P., Niederhauser, U., Subramanian, N. and Bray, M.A., 1996, CGP 57698: a simple, highly potent peptidoleukotriene, pLT) antagonist of the quinoline type, *Prostaglandins Leukot. Essent. Fatty Acids* 55:P27.

Watanabe, T., Shimizu, T., Miki, I., Sakanaka, C., Honda, Z., Seyama, Y., Teramoto, T., Matsushima, T., Ui, N. and K., K., 1990, Characterization of the guinea pig lung membrane leukotriene D_4 receptor solubilized in an active form, *J. Biol. Chem.* 265:21237.

47

EVIDENCE FOR A CARBOCATION INTERMEDIATE IN THE ENZYMATIC TRANSFORMATION OF LEUKOTRIENE A4 INTO LEUKOTRIENE B4

Martina Andberg, Mats Hamberg, and Jesper Z. Haeggström

Department of Medical Biochemistry and Biophysics
Karolinska Institutet, S-171 77 Stockholm, SWEDEN.

INTRODUCTION

Leukotriene (LT) A_4 hydrolase is a soluble enzyme that catalyzes the production of $5S,12R$-dihydroxy-6,14-cis-8,10-$trans$-eicosatetraenoic acid (LTB$_4$) from the transient allylic epoxide $5S$-$trans$-5,6-oxido-7,9-$trans$-11,14-cis-eicosatetraenoic acid (LTA$_4$). LTB$_4$ elicits chemotaxis and adherence of leukocytes in only nM concentrations and is regarded as an important chemical mediator in a variety of inflammatory disorders.

Work over the past years have demonstrated that LTA$_4$ hydrolase contains a catalytic zinc and exhibits a peptide cleaving activity[1]. Furthermore, we have used a combination of computer-assisted sequence comparisons and site-directed mutagenesis to identify and characterize functionally important amino acids of LTA$_4$ hydrolase[1]. With this methodology, Glu-296 was found to be critical for the enzyme's aminopeptidase activity, in which it probably acts as a general base. Likewise, a conserved proton donor motif, centered around Tyr-383 was identified in LTA$_4$ hydrolase[2]. Mutagenetic analysis, in which Tyr-383 was exchanged for a Phe, His, or Gln, revealed that this residue is indeed important for the peptidase activity and may act as a proton donor[3]. In the present report, we describe a more detailed characterization of the mutated proteins [Y383F], [Y383H] and [Y383Q]LTA$_4$ hydrolase, which revealed an additional enzyme activity towards LTA$_4$ leading to the formation of $5S,6S$-DHETE. From the stereochemistry of this product, we can draw important mechanistic conclusions regarding the enzyme catalyzed epoxide hydrolysis.

Eicosanoids and Other Bioactive Lipids in Cancer, Inflammation, and Radiation Injury, 4
Edited by Honn *et al.*, Kluwer Academic / Plenum Publishers, New York, 1999.

EXPERIMENTAL PROCEDURES

The mutants [Y383F/H/Q]LTA₄ hydrolase were constructed by a PCR method on mouse cDNA[3]. Recombinant proteins were expressed in *E. coli* and purified on FPLC[3].

For assay of the epoxide hydrolase activity, aliquots of 10 μg purified mutated enzymes in 100 μl, 50 mM Tris-Cl, pH 8.0 were incubated with the substrate LTA₄ (5-100 μM) at RT. After 15 s the reaction was stopped by addition of 2 vol. MeOH and a known amount of the internal standard prostaglandin B₁. After solid-phase extraction the product formation was analyzed by RP-HPLC at 270 nm on a column (Novapak C₁₈; Waters) eluted with acetonitrile/methanol/water/acetic acid (30:35:35:0.01, v/v; 1.2 ml/min). Quantifications were made as described[4]. Extinction coefficients of 40 000 M⁻¹ x cm⁻¹ and 56 000 M⁻¹ x cm⁻¹ were used for 5S,6S-DHETE and LTB₄, respectively.

For the stereochemical analysis, standards of methyl *erythro-* and *threo-*5,6-dihydroxy-eicosanoates were produced by *cis-* and *trans-* hydroxylation, respectively, of methyl (5Z) eicosenoate (50 mg)[4]. The *threo-* and *erythro-*5,6-dihydroxyeicosanoates were analyzed as methyl esters/trimethylsilyl ethers on GC/MS and had retention times of 15.9 and 16.2 min, respectively, and exhibited similar mass spectra with prominent ions at m/z 471 (M⁺-OCH₃), 299 (Me₃SiO⁺=CH-(CH₂)₁₃-CH₃) and 203 (Me₃SiO⁺=CH-(CH₂)₃-COOCH₃). Further experimental details are described elsewhere[4].

RESULTS

Formation of a novel metabolite of LTA₄

Reverse-phase HPLC analysis of products formed when [Y383F], [Y383H], and [Y383Q] mouse LTA₄ hydrolase, were incubated with LTA₄, revealed large amounts of a novel peak, denoted D (Fig. 1). Wild type enzyme and a control mutant [G386A]LTA₄ hydrolase only produced minute amounts of this material. The retention time of the material corresponding to this peak (compound D) suggested that it could be a 5,6-DHETE[5].

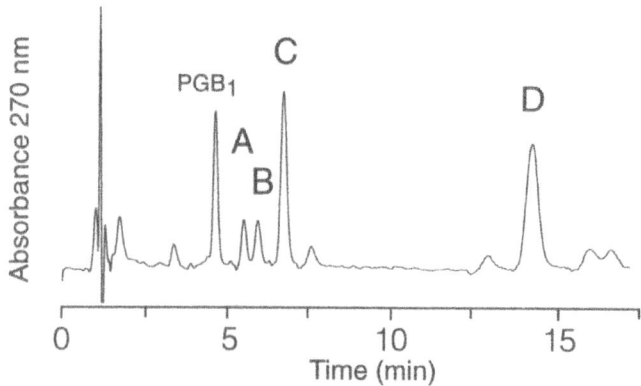

Fig 1. RP-HPLC chromatogram of products formed by [Y383Q]LTA₄ hydrolase incubated with LTA₄. Peaks A-D represent, 6-*trans*-LTB₄, 12-epi-6-*trans*-LTB₄, and LTB₄, respectively.

Structure of compound D

UV analysis of material under peak D revealed a conjugated triene spectrum (in MeOH) typical of leukotrienes with λ_{max} = 274 nm and shoulders at 263 nm and 285 nm. These values differ by 1-2 nm from previously published data for 5,6-DHETE[6] which most likely reflects differences in the experimental conditions and the instrumentation.

The mass spectrum of the methyl ester/trimethylsilyl ether derivative of compound D had prominent ions at m/z 494 (M[+]), 479 (M-15), 463 (M-31), 404 (M-90), 393 (M-101; loss of $\cdot(CH_2)_3$-COOCH$_3$), 291 (M-203; loss of $\cdot$CH(OSiMe$_3$)-(CH$_2$)$_3$-COOCH$_3$), 225, 203 (base peak; Me$_3$SiO[+]=CH-(CH$_2$)$_3$-COOCH$_3$), 171 (203-32), 147 (Me$_3$SiO[+]=SiMe$_2$), 129 (Me$_3$SiO[+]=CH-CH=CH$_2$), and 113 (203-90), compatible with the presence of a tetraunsaturated C$_{20}$ fatty acid with hydroxyl groups at C5 and C6. Incubations of [Y383Q]LTA$_4$ hydrolase (6.9 μM) with LTA$_4$ (69 μM) were carried out in 50 mM Tris-Cl, pH 8, containing 84% H$_2$[18]O. The mass spectrum of compound D revealed that the ion at m/z 291 (M-203), which carries the C6 hydroxyl group, was greatly reduced in intensity and was accompanied by an abundant ion at m/z 293. In contrast, the ion at m/z 203, containing the C5 hydroxyl group, appeared at the same ratio versus the ion at m/z 205, as in the spectrum of compound D generated in H$_2$[16]O.

Upon stereochemical analysis, a hydrogenated sample of compound D (methyl ester-trimethyl silyl ether) exhibited almost identical GC retention time and mass spectrum as those of authentic methyl *threo*-5,6-dihydroxyeicosanoate. Since H$_2$[18]O was incorporated at C6, the hydroxyl group at C5 will retain its *S* configuration from LTA$_4$ and the hydroxyl group at C6 must be in the *S* configuration. Hence, the stereochemistry of the *vic*-diol in compound D is 5*S*,6*S*.

A sample of compound D (1.6 μg) was incubated with soybean lipoxygenase (2.2 μg in 500 μl 0.1 M sodium borate buffer, pH 9 at RT). When the reaction was followed by UV, the conjugated triene spectrum of compound D (λ_{max}=275 nm in buffer) was completely shifted within 20 min to a conjugated tetraene spectrum (λ_{max}=303 nm). Since soybean lipoxygenase requires a *cis,cis*-1,4-pentadiene structure for activity, we conclude that the Δ^{11} and Δ^{14} double bonds are both in *cis*-configuration.

Finally, HPLC and GC/MS analysis of synthetic 5*S*,6*S*-dihydroxy-7,9-*trans*-11,14-*cis*-eicosatetraenoic acid revealed practically identical results as for compound D.

Based on these analytical data, compound D is assigned the structure 5*S*,6*S*-dihydroxy-7,9-*trans*-11,14-*cis*-eicosatetraenoic acid (5*S*,6*S*-DHETE).

Catalytic properties of mutated enzymes

The relative formation of 5*S*,6*S*-DHETE and LTB$_4$ varied among the different mutants. Thus, at 26 μM LTA$_4$, [Y383F], [Y383H], and [Y383Q]LTA$_4$ hydrolase produced 66, 175, and 150 % compound D, respectively.

Since [Y383Q]LTA$_4$ hydrolase had the highest epoxide hydrolase activity, it was selected for further kinetic analysis. Apparent kinetic constants were determined for the formation of 5*S*,6*S*-DHETE and LTB$_4$ by [Y383Q]LTA$_4$ hydrolase as well as the formation of LTB$_4$ by wild type enzyme (Table 2). The K$_m$ for LTA$_4$ was determined to 31 and 61 μM for the formation of LTB$_4$ and 5*S*,6*S*-DHETE, respectively, by [Y383Q]LTA$_4$ hydrolase. Furthermore, the maximal initial reaction velocity (V$_{max}$) was higher for the formation of 5*S*,6*S*-DHETE as compared to LTB$_4$ (932 and 649 nmol mg^{-1} min^{-1}). The specificity constant (k_{cat}/K$_m$) was also higher (34 x 10^3 s^{-1} M^{-1}) for the conversion of LTA$_4$ into 5*S*,6*S*-DHETE, suggesting that the active site of [Y383Q]LTA$_4$ hydrolase is better suited for this reaction than for the hydrolysis of LTA$_4$ into LTB$_4$ (k_{cat}/K$_m$ = 12 x 10^3 s^{-1} M^{-1}). The values of K$_m$, V$_{max}$, k_{cat}, and k_{cat}/K$_m$ for wild type

Table 1. *The formation of 5S,6S-DHETE and LTB₄ by the mutants of Tyr-383.* [Y383F/H/Q]LTA₄ hydrolase (10 μg) or wild type and [G386A]LTA₄ hydrolase (2.5 μg) were incubated with 26 μM LTA₄ (100 μl 50 mM Tris-Cl, pH 8; 15 s; RT).

Enzyme	Mutation	5S,6S-DHETE/LTB₄
Wild type	None	3 %
[Y383F]	Tyr → Phe	66 %
[Y383H]	Tyr → His	175 %
[Y383Q]	Tyr → Gln	150 %
[G386A]	Gly → Ala	4 %

enzyme were determined to 5 μM, 1030 nmol x mg^{-1} min^{-1}, 1.18, and 236 x 10^3 s^{-1} M^{-1}, respectively. Thus, wild type enzyme exhibited a higher affinity for LTA₄ and the active site seemed better adapted for hydrolysis of this substrate. On the other hand, if both products formed by [Y383Q]LTA₄ hydrolase are considered, *i.e.*, 5S,6S-DHETE and LTB₄, the mutated enzyme turns over LTA₄ more efficiently (k_{cat} = 1.8) than does wild type enzyme (k_{cat} = 1.2). When LTA₄ methyl ester (26 μM) was tested as substrate for [Y383Q]LTA₄ hydrolase, no conversion into the esterified derivatives of LTB₄ or 5S,6S-DHETE could be observed.

DISCUSSION

Recently, computer assisted sequence comparisons and mutational analysis have been used to identify Tyr-383 as a potential proton donor in a general base mechanism for the peptidase activity of LTA₄ hydrolase[2, 3]. Since three mutants in pos 383 converted LTA₄ into LTB₄ we concluded that Tyr-383 was not critical for the epoxide hydrolase activity of LTA₄ hydrolase. However, a

Table 2 *Kinetic constants for the formation of 5S,6S-DHETE and LTB₄ by [Y383Q]LTA₄ hydrolase and wild type LTA₄ hydrolase. Wild type (2.5 μg) or [Y383Q]LTA₄ hydrolase (10 μg) were incubated with LTA₄ (5-100 μM) in 100 μl 50 mM Tris-Cl, pH 8 (15 s; RT).*

	[Y383Q]LTA₄ hydrolase		Wild type LTA₄ hydrolase
	LTB₄	5S,6S-DHETE	LTB₄
K_m (μM)	61	31	5
V_{max} (nmol mg^{-1} min^{-1})	649	932	1030
k_{cat} (s^{-1})	0.75	1.07	1.18
k_{cat}/K_m (s^{-1} M^{-1})	12 x 10^3	34 x 10^3	236 x 10^3

more detailed analysis of the catalytic consequences of the amino acid changes led to the unexpected finding that all mutated enzymes, and in particular [Y383Q]LTA$_4$ hydrolase, generated a second enzymatic product in addition to the expected LTB$_4$ (Fig. 1). Several different approaches were used to solve the structure of the novel enzymatic metabolite, termed compound D. Thus, from analysis with RP-HPLC, UV spectrometry, GC/MS of material generated in the presence or absence of H$_2^{18}$O, sterechemical analysis, soybean lipoxygenase digestion, and comparison with a synthetic standard, compound D was assigned the tentative structure 5S,6S-dihydroxy-7,9-*trans*-11,14-*cis*-eicosatetraenoic acid (5S,6S-DHETE; Fig. 2).

5S,6S-DHETE is a major enzymatic product generated by mutants of Tyr-383

All mutants converted LTA$_4$ into 5S,6S-DHETE. Considering the relative formation of 5S,6S-DHETE (versus LTB$_4$) they had the following order of efficiency [Y383H] > [Y383Q] > [Y383F]LTA$_4$ hydrolase. For [Y383Q]LTA$_4$ hydrolase, the relative formation of 5S,6S-DHETE versus LTB$_4$ was 150 % and thus 5S,6S-DHETE was the dominating product. This is noticeable since 5,12-dihydroxy acids are by far the most abundant metabolites obtained via non-enzymatic hydrolysis as well as hydrolysis catalyzed by wild type LTA$_4$ hydrolase, suggesting that the *vic*-diols are thermodynamically unfavored products.

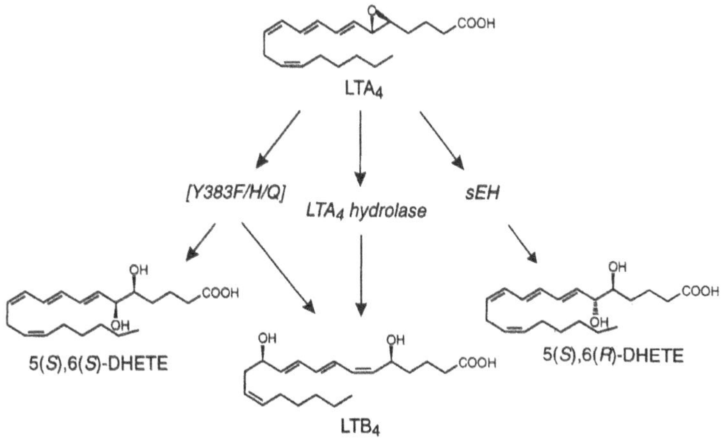

Fig 2. The structure and formation of compound D. The mutated enzymes [Y383F/H/Q]LTA$_4$ hydrolase converts LTA$_4$ into both LTB$_4$ and 5S,6S-DHETE, whereas wild type enzyme and sEH produces LTB$_4$ and 5S,6R-DHETE, respectively.

[Y383Q]LTA$_4$ hydrolase turns over LTA$_4$ more efficiently than wild type enzyme

At saturating concentrations of LTA$_4$, the specific epoxide hydrolase activity of [Y383Q]LTA$_4$ hydrolase (only counting LTB$_4$) was estimated to approx. 60% of the control[3]. However, if the formation of 5S,6S-DHETE is taken into account, the turnover of LTA$_4$ is in fact higher for [Y383Q]LTA$_4$ hydrolase, *i.e.*, k_{cat} = 1.8 s^{-1} versus 1.2 s^{-1} for wild type enzyme (Table 2). When the kinetic data for the formation of LTB$_4$ and 5S,6S-DHETE were plotted separately, no significant differences were found, except for a somewhat higher V_{max} value for the *vic*-diol formation, as expected.

Evidence for a carbocation intermediate in the enzymatic hydrolysis of LTA$_4$

From experiments with H$_2$^{18}O, it was inferred that the nucleophilic attack of water during formation of 5S,6S-DHETE, was directed towards C6 according to an S$_N$1 or S$_N$2 reaction. Considering the S-configuration of the hydroxyl group at C6, an S$_N$2 reaction, with concomitant chiral inversion, would not be possible since the epoxide oxygen of the substrate is already in the 6S configuration. Thus, enzymatic hydrolysis of LTA$_4$ into 5S,6S-DHETE must occur via an S$_N$1 reaction involving a carbocation intermediate. This intermediate would be planar at C6 allowing an enzyme-directed nucleophilic attack from either side of the molecule. Since [Y383F], [Y383H], and [Y383Q]LTA$_4$ hydrolase could produce not only 5S,6S-DHETE but also LTB$_4$ it seems very likely that hydrolysis of LTA$_4$ into LTB$_4$ proceeds according to the same mechanism. This conclusion was further corroborated by the fact that [Y383Q]LTA$_4$ hydrolase makes both products with indistinguishable reaction kinetics. Nevertheless, we can not rule out the possibility that LTB$_4$ could be formed via an S$_N$2, or rather S$_N$2' reaction, an interpretation which however seems unlikely since it would imply that the mutants can operate simultaneously at C6 and C12 of LTA$_4$ via two distinct enzymatic mechanisms. Hence, the formation of 5S,6S-DHETE by the mutated enzymes, represents the first experimental evidence that formation of LTB$_4$ by wild type LTA$_4$ hydrolase follows an S$_N$1 mechanism. In addition, this reaction mechanism would be in agreement with the mechanism for non-enzymatic hydrolysis of LTA$_4$[7] and conforms to the general rule that enzymes reduce the activation energy for chemical reactions which also occur spontaneously.

Potential relationship to soluble xenobiotic epoxide hydrolase

Recent work including, *e.g.*, X-ray crystallographic analysis of structurally related enzymes, as well as biochemical and mutational analyses, have identified soluble and microsomal epoxide hydrolases (sEH and mEH) as members of the α/β-fold family of hydrolases[8]. For LTA$_4$ hydrolase, there is not enough data available to conclusively determine whether or not it belongs to the same class of enzymes. However, LTA$_4$ has been shown to be an excellent substrate for sEH (but not for mEH) which converts the allylic epoxide into 5S,6R-DHETE, *i.e.*, an epimer at C6 of the vicinal diol produced by [Y383Q]LTA$_4$ hydrolase[6]. Hence, the subtle structural changes at the active site of [Y383Q]LTA$_4$ hydrolase, shift the positional specificity of the stereospecific hydrolysis such that the mutated enzyme begins to mimic the action of sEH (Fig 2). One may speculate that this functional resemblance, caused by a single amino acid change, is a sign of structural similarity between the active sites of sEH and LTA$_4$ hydrolase.

Acknowledgements. The authors thank Anders Wetterholm for discussions and advice and to Eva Ohlson for excellent technical assistance. This study was supported by funds from the Swedish Medical Research Council (03X-10350), The European Union (BMH4-CT960229), and Konung Gustav V:s 80-årsfond.

REFERENCES

1. J.Z. Haeggström, A. Wetterholm, J.F. Medina, and B. Samuelsson, Leukotriene A_4 hydrolase: Structural and functional properties of the active center, *Journal of Lipid Mediators* 6:1 (1993).
2. M. Minami, H. Bito, N. Ohishi, H. Tsuge, M. Miyano, M. Mori, H. Wada, H. Mutoh, S. Shimada, T. Izumi, K. Abe, and T. Shimizu, Leukotriene A_4 hydrolase, a bifunctional enzyme: Distinction of leukotriene A_4 hydrolase and aminopeptidase activities by site-directed mutagenesis at Glu-297, *FEBS Lett.* 309:353 (1992).
3. M. Blomster, A. Wetterholm, M.J. Mueller, and J.Z. Haeggström, Evidence for a catalytic role of tyrosine 383 in the peptidase reaction of leukotriene A_4 hydrolase, *Eur. J. Biochem.* 231:528 (1995).
4. M. Blomster Andberg, M. Hamberg, and J.Z. Haeggström, Mutation of tyrosine 383 in leukotriene A_4 hydrolase allows formation of 5S,6S-dihydroxy-7,9-*trans*-11,14-*cis*-eicosatetraenoic acid: Implications for the epoxide hydrolase mechanism, *J. Biol. Chem.* 272:23057 (1997).
5. P. Borgeat, and B. Samuelsson, Metabolism of arachidonic acid in polymorphonuclear leukocytes. Structural analysis of novel hydroxylated compounds, *J. Biol. Chem.* 254:7865 (1979).
6. J. Haeggström, J. Meijer, and O. Rådmark, Leukotriene A_4 : Enzymatic conversion into 5,6-dihydroxy-7,9,11,14-eicosatetraenoic acid by mouse liver cytosolic epoxide hydrolase, *J. Biol. Chem.* 261:6332 (1986).
7. P. Borgeat, and B. Samuelsson, Arachidonic acid metabolism in polymorphonuclear leukocytes: unstable intermediate in formation of dihydroxy acids, *Proc. Natl. Acad. Sci. USA* 76:3213 (1979).
8. M. Arand, H. Wagner, and F. Oesch, Asp333, Asp495, and His523 form the catalytic triad of rat soluble epoxide hydrolase, *J. Biol. Chem.* 271:4223 (1996).

48

A RANDOM RAPID EQUILIBRIUM MECHANISM FOR LEUKOTRIENE C$_4$ SYNTHASE

Namrata Gupta,[1] Michael J. Greeser, and Anthony W. Ford-Hutchinson

Merck Frosst Centre for Therapeutic Research
P.O. Box 1005, Pointe Claire - Dorval
Quebec, Canada H9R 4P8

[1]Department of Pharmacology and Therapeutics
McGill University
3655 Drummond, Montreal
Quebec, Canada H3G 1Y6

INTRODUCTION

Glutathione S-transferases (GSTs) comprise a family of ubiquitous enzymes mainly responsible for xenobiotic metabolism, drug biotransformation, and protection against peroxidative damage via the catalysis of reduced glutathione (GSH) with hydrophobic electrophiles[1]. Predominantly cytosolic[2], four known membrane-bound forms of GSTs have been documented, now part of a novel supergene family designated MAPEG (Membrane Associated Proteins in Eicosanoid and glutathione Metabolism)[3]. A member of this class of enzymes is leukotriene (LT) C$_4$ synthase and unlike all other GSTs, has minimal activity for conventional GST substrates, and is committed to the biosynthesis of cysteinyl leukotrienes (LTC$_4$, LTD$_4$, LTE$_4$) via the conjugation of GSH to LTA$_4$[4].

Cysteinyl leukotrienes comprise the slow reacting substance of anaphylaxis and elicit their biological responses through G-protein coupled receptors[5]. These lipid-derived mediators are implicated in the pathway of inflammatory diseases, in particular, human bronchial asthma as a consequence of their ability to elicit pulmonary smooth muscle contraction, bronchial hyperresponsiveness, mucus hypersecretion, vasoconstriction, vascular permeability and infiltration of myeloid cell types such as esinophils[6,7,8]. As LTC$_4$ synthase catalyzes the first committed step in the formation of these biologically active mediators, this enzyme becomes an attractive therapeutic target for the treatment of asthma.

As a potential therapeutic target, numerous studies have been conducted at Merck Frosst to elucidate the effects of blocking the activity of LTC$_4$ synthase and other proteins involved in the leukotriene biosynthetic pathway, which have further substantiated their involvement in disease states mentioned. As a result of these studies, an inhibitor, L-699,333 was described to be potent

Eicosanoids and Other Bioactive Lipids in Cancer, Inflammation, and Radiation Injury, 4
Edited by Honn *et al.*, Kluwer Academic / Plenum Publishers, New York, 1999.

327

against five-lipoxygenase, an enzyme upstream from LTC$_4$ synthase, and a weak LTC$_4$ synthase inhibitor[9,10]. We conducted further elaborate studies including structure activity relationship studies which revealed cross-reactivity between various inhibitors and proteins in the leukotriene biosynthetic pathway. These studies also showed that the most potent synthetic inhibitor of LTC$_4$ synthase was in face L-699,333 in low μmolar range[11].

Through the use of this inhibitor, we have characterized and documented the kinetic mechanism of GSH conjugation to LTA$_4$ by LTC$_4$ synthase[12]. Kinetic mechanisms of GST-catalyzed reaction are quite complex and isoenzyme dependent. Therefore the kinetic mechanism may provide a further criterion to distinguish among GST isoenzyme classes. In addition, mechanistic characterization and knowledge of substrate interaction with one another or with an inhibitor and the enzyme may shed light on the mechanism of drug actions in use and those in development. Among the possible prevailing two substrate mechanisms area a ping-pong double-displacement mechanism which entails addition of one substrate followed by an intermediate product formation before the entry of the second substrate; an ordered compulsory which dictates a fixed sequence of events; and a random mechanism where the two substrates in question interact with the enzyme in either order, thereby having two possible pathways. The nature of the interaction of the inhibitor with the substrates and the enzyme may also aid in the determination of the mechanism of the catalysis. The conjugation reaction mechanism for many GST's, mainly cystolic, have previously been documented[13-18]. The kinetic mechanism for rat liver microsomal GST, the closest counterpart studied to human LTC$_4$ synthase has been reported to obey a random sequential mechanism[19]. Here we summarize our results from our previously documented characterization of the mechanism of GSH conjugation to LTA$_4$ by LTC$_4$ synthase with the use of the inhibitor, L-699,333[12].

RESULTS AND DISCUSSION

The mechanism of GSH conjugation to LTA$_4$ by LTC$_4$ synthase was investigated by calculation of enzymatic rates of reactions from initial velocities which were derived from experiments where the concentrations of the two substrates were varied simultaneously, ranging from 5 to 20 μm LTA$_4$ and from 0.1 to 8 mM GSH. The enzymatic activity was monitored over a period of 0 to 15 minutes with 6 varying incubation time points. The rate data which was an 8 x 8 matrix for a total of 64 data points was then fit to pertinent two substrate kinetic equations using the program Grafit which essentially generates kinetic constants and provides the best fit of the model to the data, and yields the standard deviation of each value on the basis of given estimates of the affinities of the substrates towards the enzyme. The two substrate kinetic equations were developed based on proofs and derivations o accommodate the nature of this interaction. Aside from fits of the data to the equations, secondary plots of the rate data as a function of substrate concentrations can shed light onto the mechanism since the trends of the plot are unique from one mechanism to another. For example, a reciprocal plots of the rate data as a function of substrate concentration will demonstrate a parallel pattern of lines for a ping-pong mechanism as opposed to an intersecting pattern for a random mechanism.

The effect of substrate concentrations on the velocities is reproduced in Figure 1 which depicts a double-reciprocal plot of the rate data versus GSH concentration. At lower LTA$_4$ concentrations (5-50 μM), the reciprocal lines appear to be parallel, suggestive of a ping-pong type of mechanism. These lines, however, at higher concentrations begin to represent the data as an intersecting pattern of lines which is more typical of a random mechanistic model.

In order to distinguish between these mechanisms and other such as the compulsory ordered mechanism, the whole data set was fit to various equations which dictate a particular model. These attempts failed to generate satisfactory curve fits whereas the rapid equilibrium mechanism proved to provide the best fit, with kinetic constants in the expected range to accommodate the affinities of the substrates towards the enzyme's active site. Figure 2 illustrates the best fit of the data generated by the model of a random mechanism represented by Equation 1.

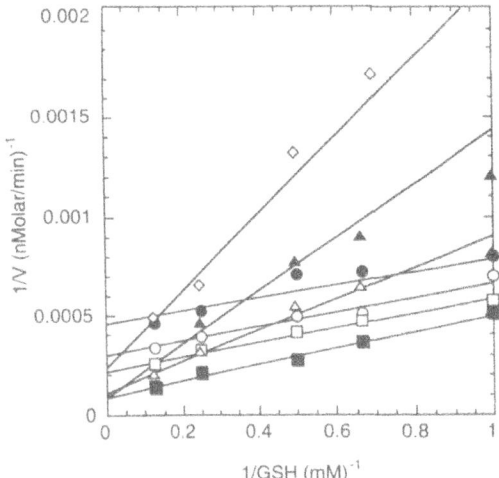

Figure 1. *Double-reciprocal plot of the initial rate data at a range of substrate concentrations for human LTC₄ synthase.* Enzyme activity was measured by the formation of LTC_4 in 150 µl incubation mixtures using reduced GSH and LTA_4 (free acid form) stabilized with 5 mg/ml BSA, as substrates and enzyme purified from THP-1 cells at a concenration of 0.03 µg/ml. The mixtures were incubated at 6 incubation times from 0 to 15 min at 25° C in 0.1 M potassium phosphate pH 7.4 buffer in the presence of 50 mM serine-borate complex. the reactions were then terminated by an equal volume (150 µl) of cold (4° C) acetonitrile:methanol:acetic acid (at 50:50:1, v/v/v). The mixtures were then allowed to stand for a minimum of either 30 nmin or overnight at 4° C. Precipitated proteins were removed by centrifugation for 20 min at 16,000 xg. The resulting supernatant, containing the leukotriene products (250 µl), was then analyzed by RP-HPLC and peaks of interest were identified by their retention time compared with synthetic standards. With GSH as the varied substrate, LTA_4 concentrations were ranged from (•)5, (O) 15, (□) 25, (■) 50, (Δ) 75, (▲) 100, to (◊) 150 µM.

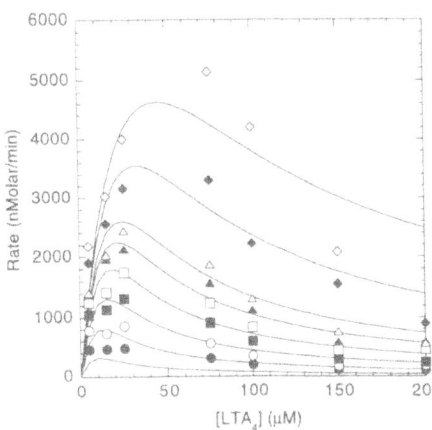

Figure 2. *Initial rate data at the range of substrate concentrations for human LTC₄ synthase.* With LTA_4 as the varied substrate, the range of GSH concentrations were (•) 0.10, (O) 0.30, (■) 0.60, (□) 1.0, (▲) 1.5, (Δ) 2.0, (◆) 4.0, and (◊) 8.0mM. The conditions of the assay were essentially as described in Figure 1. The calculated rates were curve-fitted using the program Grafit, and the curves drawn through the data points represent those best predicted by fitting the data to a random rapid equilibrium mechanism influenced by substrate inhibition. *(Reprinted from biochimica et biophysica Acta, Vol. 1391, No. 2, N. Gupta, M.J. Gresser and A.W. Ford-Hutchinson, "Kinetic Mechanism of Gluthathione Conjugation of Leukotriene A4 by Leukotriene C4 Synthase", pp. 157-168, 1998, (ref. 12) with kind permission from Elzevier Science NL, Sara Burgerhartstraat 26, 1055 KV Amsterdam, the Netherlands).*

$$V= K_{cat}[E_t]/(K_{GSH}K_{LTA4}/[GSH][LTA_4] + K_{GSH}/[GSH] + K_{LTA4}/[LTA_4] + 1 + K_{GSH} [LTA_4]/[GSH]K_{iLTA4}) \qquad (1)$$

329

As can be seen, the activity of the enzyme declines beyond an LTA$_4$ concentration of 50 μM, indicative of substrate inhibition. To ascertain this phenomenon, we determined the critical micellar concentration of LTA$_4$ to be well above the highest concentration of LTA$_4$ used in this study (200μM), confirming a lack of a non-specific detergent effect which may have caused the observed inhibition.

In order to elucidate the mode of the interaction of the reversible inhibitor, L- 699,333 with the enzyme and the substrates, we measured the enzymatic activity over a range of inhibitor concentrations while varying the concentration of one substrate and maintaining a fixed concentration of the second substrate. A non-linear least-squares analysis computer program was used to compute kinetic constants by direct fits of the inhibition data to the appropriate Michaelis-Menton equations which represent four inhibition models, namely, competitive, non-competitive, mixed, and uncompetitive. This analysis revealed the inhibitor to be in direct competition with the GSH binding site (G-site) and in a non-competitive manner with the LTA$_4$ site (hydrophobic site), which are two major, non-overlapping substrate binding sites at the active centre of GSTs[20].

In view of all the pieces of data given thus far, we devised a model which represents a random, rapid equilibrium mechanism influenced by inhibition by substrate LTA$_4$ and the synthetic inhibitor, L-699,333 as depicted in Scheme 1. Equation 2 represents the overall model taking into account the two inhibitory influences.

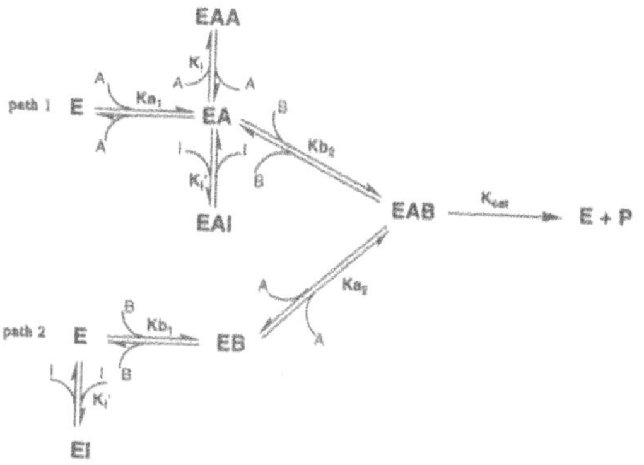

Scheme 1. *A schematic of the random rapid equilibrium mechanism of Gluthathione conjugation to LTA$_4$ catalyzed by LTA$_4$synthase with substrate inhibition imparted by LTA$_4$ and reversible inhibition due to L-699,333.* The letters denote the following entities: A=LTA$_4$, B=GSH, E= LTC$_4$ synthase, I=L-699,333 and the kinetic constants represent the rate of catalysis of each indicated reaction. Binding of I to E is not shown in the upper pathway, indicating that binding of I to GSH site has no influence on binding of LTA$_4$ to the LTA$_4$ catalytic site.
(Reprinted from Biochimica et Biophysica Acta, Vol. 1391, No. 2, N. Gupta, M.J. Gresser and A.W. Ford-Hutchinson, "Kinetic Mechanism of Gluthathione Conjugation of Leukotriene A4 by Leukotriene C4 Synthase", pp. 157-168, 1998, (ref. 12) with kind permission from Elzevier Science-NL, Sara Burgerhartstraat 25, 1055 KV Amsterdam, The Netherlands).

$$V= K_{cat}[E_T]/(((Kb_2Ka_1/[A][B]) + (Kb_2/[B])) (1 + [I]/K_i') + Ka_2/[A] + 1 + (Kb_2[A]/[B]K_i)) \qquad (2)$$

The mechanism dictates two pathways for substrate binding. In path 1, LTA$_4$ is shown to bind first and is subject to substrate inhibition to yield the EAA complex. The inhibitor is shown to be in direct competition with the substrate inhibition. Path 2 assumes GSH binding first with the inhibitor in direct competition with GSH for the respective binding site. Either pathway converges to form the EAB complex which in turn yields he product LTC$_4$ and the enzyme free for another catalytic turnover. Assuming that all binding reactions are in rapid equilibrium, and also that Ka$_1$ =

Ka$_2$ and Kb$_1$ = Kb$_2$, the overall data was fit to Equation 2 which gave rise to the following kinetic constants; V$_{max}$ = 9.8 $\pm$ 0.08 Molar/min, K$_a$ = 55 μM, K$_b$' = 0.4 $\pm$ 0.1 mM, K$_i$ = 5.8 $\pm$ 2.9 μM and K$_i$'of 0.08 $\pm$ 0.04 μM.

SUMMARY

Kinetic studies performed on the conjugation reaction catalyzed by LTC$_4$ synthase proved to conform to a random rapid equilibrium mechanism which was further substantiated by competition patterns ruling out other possible mechanisms. Most cytosolic Gst's investigated to date appear to follow a random kinetic mechanism although are mainly responsible for detoxification purposes. Conversely, LTC$_4$ synthase possesses a very different biological role yet still follows a similar mechanism. Therefore, it can be concluded that most GSTs function in a consistent manner regardless of their biological function. Of interest are the mechanisms of the other members of the MAPEG family which in some respects are closer to conventional GSTs than to LTC$_4$ synthase, yet they remain to be deciphered.

REFERENCES

1.　　V. Daniel, Glutathione S-transferases: gene structure and regulation of expression.*Crit Rev Biochem Mol Biol.* 28(3):173 (1993).
2.　　S. Tsuchida and K. Sato, Glutathione transferases and cancer, *CRC Crit Rev Biochem Mol biol* 27:337 (1992).
3.　　A.W. Ford-Hutchinson and P.J. Jakobsson, Enzymes involved in the production of leukotrienes and related molecules, in: *SRS-A to Leukotrienes: New Concept and Targets for Terapy*, I. Rodger, J. Blotting and S.E. Dahlen, eds., Kluwer Academic Publishers (1998).
4.　　D.W. Nicholson, A. Ali, M.W. Klemba, N.A. Munday, R.J. Zamboni, and A.W. Ford-Hutchinson, Human leukotriene C4 synthase expression in dimethyl sulfoxide-differentiated U937 cells, *J Biol Chem* 267:17849 (1992).
5.　　K.M.Metters, Leukotriene receptors, *J Lipid Mediat Cell Signal* 12:413 (1995).
6.　　B. Samuelsson, Leukotrienes: mediators of immediate hypersensitivity reactions and inflammation, *Science* 220:568 (1992).
7.　　P.J. Piper, Formation and actions of leukotrienes, *Physiol Rev* 64:744 (1984).
8.　　A.W. Ford-Hutchinson, Leukotriene B4 in inflammation, *Crit Rev Immunol* 10:1 (1990).
9.　　J.H. Hutchinson, D. Riendeau, c. Brideau, C. Chan, J.P. Falgueyret, J. Guay, T.R.Jones, C. Lepine, D. Macdonald, C.S. McFarlane, H. Piechuta, J. Scheigetz, P. Tagari, M., Therien, and Y. Girard, Thiopyrano [2,3,4-cd]indoles as 5-lipoxygense inhibitors: synthesis, biological profile, and resolution of 2-[2-[1-(4-chlorobenzyl)-4-methyl-6[(5-phenylpryridin-02-yl)methoxy]-4, 5-dihydro-1H-thiopyrano [2,3,4-cd]indol-2-yl]ethoxy]butanoic acid, *J Med Chem* 37(8):1153 (1994).
10.　　J.H. Hutchinson, S. Charleson, J.F. Evans, J.P. Falgueyret, K. Hoogsteen, T.R. Jones, C. Lepine, D. Macdonald, C.S. McFarlane, D.W. Nicholson, H. Piechuta, D. Riendeau, J. Scheigetz, M. Therien, and Y. Girard, Thiopyranol[2,3,4-c,d]indoles as inhibitors of 5-lipoxygenase, 5-l lipoxygenase-activating protein, and leukotriene C4 synthase. *J Med Chem* 38(22):4538 (1995).
11.　　N. Gupta, D.W. Nicholson, and A.W. Ford-Hutchinson, Pharmacological cross-reactivity between 5-lipoxygenase-activating protein, 5-lipoxygenase, and leukotriene C4 synthase, *Can J Physiol Pharmacol* 75:1212 (1997).

12. N. Gupta, M.J. Gresser, and A.W. Ford-Hutchinson, Kinetic mechamism of gluthathione conjugation to leukotriene A4 by leukotriene C4 synthase, *Biochim biophys Acta* 1391(2):157 (1998).

13. S.S. Tang, and G.G. chang, Steady-state kinetics and chemical mechanism of octopus hepatopancreatic glutahione transferase, *Biochem J* 309:347 (1995).

14. A.E.A. Thumser, and K.M. Ivanetich, Kinetic mechanism of human erythrocyte acidic isoenzyme rho, *Biochem Biophys Acta* 1203:115(1993).

15. W.R. Vorachek, W.R. Pearson, and G.S. Rule, Cloning, expression and characterization of class-mu glutathione transferase from human muscle, the product of GST4 locus, *Proc Nat Acad Sci USA* 88:4443 (1991).

16. K.M. Ivanetich, and R.D. Goold, A rapid equilibrium random sequential bi-bi mechanism for human placental glutathione S-transferase, *Biochim Biophys Acta* 998:7 (1989).

17. P.R. Young, and A.V. Briedis, Purification and kinetic mechanism of the major glutathione S-transferase from bovine brain, *Biochem J* 257:541 (1989).

18. I. Jakobson, P. Askelof, M. Warholm, and b. Mannervik, A steady-state-kinetic mechanism for glutathione S-transferase A from rat liver. A model involving kinetically significant enzyme-product complexes in the forward reaction, *Eur J Biochem* 77:253 (1977).

19. C. Andersson, F. Piemonte, E. Mosialou, R. Weinander, T.H. Sun, G. Lundqvist, A.E.P. Adang, and R. Morgenstern, Kinetic studies on rat liver microsomal glutathione transferase: consequences of activation, *Biochim Biophys Acta* 1247:277 (1995).

20. T.H. Rushmore, and C.B. Pickett, Glutathione S-transferase, structure, regulation, and therapeutic implications, *J Biolog Chem* 268(16):11475 (1993).

ISOPROSTANES

49

FORMATION OF REACTIVE PRODUCTS OF THE ISOPROSTANE PATHWAY:
ISOLEVUGLANDINS AND CYCLOPENTENONE ISOPROSTANES

L. Jackson Roberts, II,[1] Cynthia J. Brame,[1] Yan Chen,[1] Jason D. Morrow[1],
and Robert G. Salomon,[2]

[1]Departments of Pharmacology and Medicine
Vanderbilt University
Nashville, TN 37232

[2]Department of Chemistry
Case Western Reserve University
Cleveland, OH 44106

INTRODUCTION

In 1990 we reported the discovery of the formation of prostaglandin (PG) F_2-like compounds *in vivo* by a free radical mechanism independent of the cyclooxygenase enzyme.[1] These compounds have been termed F_2-isoprostanes (F_2-IsoP's). The formation of these compounds proceeds via PGH_2-like bicyclic endoperoxide intermediates which are reduced to form F-ring IsoP's. Four F_2-isoP regioisomers are formed, each of which is comprised of eight racemic diastereomers. It is well known that PGH_2 endoperoxide is an unstable molecule which undergoes rearrangement in aqueous media to form PGE_2 and PGD_2. We recently reported that glutathione is a key molecule responsible for the reduction of the IsoP endoperoxides.[2] However, the reduction of the IsoP endoperoxides is not completely efficient. In this regard, we have found that E-ring and D-ring IsoP's are produced *in vivo* as rearrangement products of the IsoP endoperoxides.[3]

PGE_2 and PGD_2 are also relatively unstable molecules which can undergo dehydration to form cyclopentenone derivatives, PGA_2 and PGJ_2, respectively. The dehydration of these PG's is induced by alkaline conditions but also can be catalyzed by proteins such as albumin.[4,5] PGA_2 and PGJ_2 have been shown to exert interesting biological properties including inhibition of NfκB, induction of heat shock protein transcription, and modulation of cellular proliferation.[6] In addition, these are reactive molecules which can undergo Michael addition reactions with thiols and form protein adducts.[7-9]

More recently, Salomon and colleagues also reported the formation of γ-ketoaldehydes, termed

Eicosanoids and Other Bioactive Lipids in Cancer, Inflammation, and Radiation Injury, 4
Edited by Honn *et al.*, Kluwer Academic / Plenum Publishers, New York, 1999.

335

levuglandin (LG) E_2 and LGD_2, as rearrangement products of PGH_2.[10] LG's have also been shown to be highly reactive compounds which rapidly adduct to protein and form protein-protein and protein-DNA crosslinks.[11,12]

Because the IsoP endoperoxides have been found to undergo rearrangement *in vivo* to form E_2/J_2-IsoP's, we explored whether E_2/D_2-IsoP's also undergo dehydration *in vivo* to form A_2/J_2-IsoP's (Figure 1A). In addition, we explored whether the IsoP endoperoxides also undergo rearrangement to form E_2/D_2-IsoLG's (Figure 1B). Our interest in the possibility that cyclopentenone IsoP's and IsoLG's may be produced as products of the IsoP pathway derives from the reactive nature of these compounds, which if formed, may mediate some of the adverse sequella of oxidant injury.

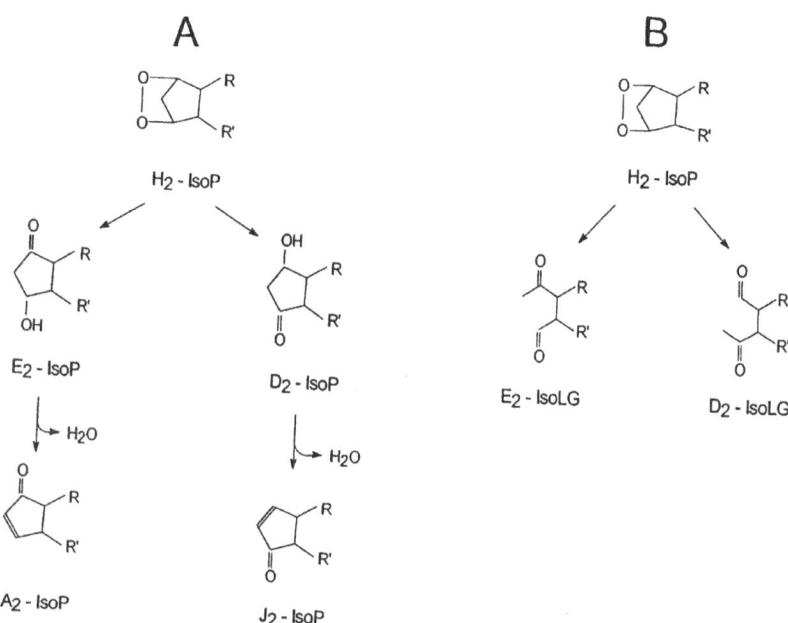

Figure 1. (A) Formation of cyclopentenone IsoP's, A_2-IsoP's and J_2-IsoP's, by dehydration of E_2-IsoP's and D_2-IsoP's, respectively, which are formed by rearrangement of H_2-IsoP endoperoxides. (B) Formation of E_2-IsoLG's and D_2-IsoLG's by rearrangement of H_2-IsoP endoperoxides. Although not shown for simplicity, four regioisomers of each IsoP and IsoLG are formed. Each E_2/D_2-IsoP regioisomer is comprised of 8 racemic diastereomers and each E_2/D_2-IsoLG regioisomer is comprised of 4 racemic diastereomers.

ANALYSIS OF ISOLG'S AND CYCLOPENTENONE ISOP'S

Methods for the analysis and identification of IsoLG's and cyclopentenone IsoP's utilized extraction by C_{18} SepPak cartridge and modifications of TLC purification and mass spectrometric procedures previous used for the analysis and identification of IsoP's.[13] LG's are unstable molecules and therefore are initially stabilized by conversion to a *bis* O-methyloxime derivative prior to extraction and purification. Generation of compounds *in vitro* for analysis was accomplished by oxidation of arachidonic acid with Fe/ADP/ascorbate.[3] Experiments in which the formation of cyclopentenone IsoP's was assessed *in vivo*, an oxidant injury to the liver was induced in rats by adminstration of CCl_4.[1,14] The amount of cyclopentenone IsoP's produced esterified in lipids in the liver was assessed by quantifying free compounds in a bee venom PLA_2 hydrolysate of a Folch lipid extract of liver.[3] In experiments assessing the formation of IsoLG adducts to ApoB-100 protein in LDL, LDL was isolated, delipidated, and digested to individual amino acids by sequential treatment with pronase and leucine aminopeptidase as described.[15,16] Oxidation of LDL was accomplished by incubation of LDL with 0.1 mM 2,2'-azobis(2-amidinopropane)HCL (AAPH) for 4 hr.

RESULTS

Formation Cyclopentenone IsoP's

Analysis for cyclopentenone IsoP's following oxidation of arachidonic acid *in vitro* by negative ion chemical ionization mass spectrometry as a pentafluorobenzyl ester, O-methyloxime, trimethlsilyl ether derivative resulted in a series of $M-CH_2C_6F_5$ peaks at m/z 434 with a similar GC retention time as the internal standard, $[^2H_4]$ PGA_2. Evidence was obtained that these compounds represented cyclopentenone IsoP's from analyses as a deuterated trimethylsilyl ether dervative, deuterated O-methyloxime derivative, and following catalytic hydrogenation which indicated that all these compounds had one carbonyl group, one hydroxyl group, and three double bonds. Further, a significant proportion of the compounds formed a piperidyl enol trimethylsilyl ether derivative when treated with BSTFA and piperidine, a derivative that is specific for A-ring cyclopentenone prostanoids[17]. In addition, electron impact ionization mass spectra were also obtained which were consistent with these compounds being cyclopentenone IsoP's.

We explored whether these compounds were formed *in vivo* by analyzing for their presence esterified in lipids in the liver of control rats and rats treated with CCl_4 to induce an oxidant injury to the liver. Interestingly, these compounds were detectable in normal rat liver at levels of 5.1 ± 2.3 ng/g liver and increased dramatically by a mean of 24-fold in the livers of rats that had been treated with CCl_4. It was important to be certain that the cyclopentenone IsoP's detected *in vivo* did not arise from dehydration of E_2/D_2-IsoP's *ex vivo* during sample processing. Evidence that these compounds were not formed *ex vivo* was obtained by the finding that there was no appreciable generation of PGA_2 or PGJ_2 when PGE_2 and PGD_2 were added to a liver homogenate and processed in an identical fashion as samples in which cyclopentenone IsoP's were measured. We then compared the relative abundance of cyclopentenone IsoP's formed *in vivo* with the amounts of E_2/D_2-IsoP's formed . The amount of cyclopentenone IsoP's present esterified in liver was a mean of 18 % and 20 % of the amount of E_2/D_2-IsoP's measured in normal and CCl_4 treated rat liver, respectively.

Formation of Isolevuglandins

Analysis for the formation of IsoLG's following oxidation of arachidonic acid *in vitro* as a pentafluorobenzyl ester, *bis*-O-methyloxime, trimethylsilyl ether derivative resulted in a series of M-·CH$_2$C$_6$F$_5$ peaks at m/z 481 with a similar GC retention time as the internal standard, *bis*-[^{2}H$_3$]-O-methyloxime LGE$_2$. Analyses as a deuterated O-methyloxime derivative, deuterated trimethylsilyl ether derivative, and following catlytic hydrogenation indicated that all these compounds contained two carbonyl groups, one hydroxyl group, and two double bonds. Analysis by electron impact ionization mass spectrometry revealed multiple mass spectra consistent with these compounds being IsoLG's. Interestingly, the amounts of IsoLG's formed during arachidonic acid *in vitro* (1241 ± 388 ng/mg arachidonate) was greater than the amounts of F$_2$-IsoP's formed (785 ± 289 ng/mg arachidonate) and only somewhat less than the amount of E$_2$/D$_2$-IsoP's generated (1826 ± 413 ng/mg arachidonae).

We then explored whether these compounds are also formed *in vivo*. Exhaustive attempts to detect the formation of IsoLG's *in vivo* were not successful. In addition, we were also unsuccessful in detecting the formation of these compounds *in vitro* in any simple biological system, *e.g.* oxidized liver microsomes and oxidized LDL. We speculated that this might be attributed to the remarkable proclivity of LG's to adduct with protein. To evaluate this hypothesis, we assessed the rate of adduction of LGE$_2$ to albumin by quantifying the disappearance of free LGE$_2$ during an incubation of a ~1:1 molar mixture of LGE$_2$ and albumin at 37°C. LGE$_2$ was found to adduct to albumin with extreme rapidity, being essentially complete by 5 min. This rate of adduction of LGE$_2$ to albumin exceeds the rate of adduction of 4-hydroxynonenal to albumin by several orders of magnitude.[18] This was a remarkable finding in that 4-hydroxynonenal is thought to be one of the most reactive aldehydes that are generated as products of lipid peroxidation. Nonetheless, this finding suggested strongly that our failure to detect the formation of free IsoLG's in biological systems could be attributed to the extreme rapidity with which these molecules adduct to proteins.

Therefore, we undertook the development of an LC electrospray ionization mass spectrometric method to detect these compounds in the form of a protein adduct. Initially, studies were carried out to identify the nature of the adducts formed with lysine. Previously, Salomon and colleagues had provided evidence that LG's form a pyrrole adduct with the ε-amino group of lysine residues.[19] Analysis for the presence of a LGE$_2$-lysine pyrrole adduct following incubation of LGE$_2$ with lysine did reveal an m/z 463 peak consistent with the MH$^+$ ion of the pyrrole adduct, but this was an inconsistent finding. However, we did consistently detect the presence of compounds at m/z 479 and m/z 495, which are16 and 32 Da higher than the MH$^+$ ion for the LG-lysine pyrrole adduct. Evidence confirming that these were lysine adducts was obtained by noting a shift of m/z 479 to m/z 485 and m/z 495 to m/z 501 following incubation of LGE$_2$ with [^{13}C$_6$] lysine. Pyrroles are known to be easily oxidized.[20] Therefore, these compounds would be consistent with oxidation of the LG/pyrrole adduct to lactam and hydroxylactam adducts (Figure 2).

We then carried out analyses for these adducts following oxidation of arachidonic acid in the presence of lysine *in vitro*. This resulted in multiple lactam and hydroxylactam peaks, as would be expected from the formation of multiple IsoLG compounds during oxidation of arachidonic acid.

We then developed an approach to detect IsoLG protein adducts. For these analyses, protein was incubated sequentially with pronase and leucine aminopeptidase, which digests the protein into single amino acids. Following digestion of protein, the IsoLG lysine adducts are analyzed following their extraction using a C$_{18}$-SepPak cartridge to remove the unadducted free amino acids, which are not retained. Proof of concept that the formation IsoLG protein adducts could be detected using this approach was obtained from the analysis of native LDL and oxidized LDL for IsoLG adducts. No adducts could be detected in native (unoxidized) LDL but intense m/z 479 and m.z 495 signals were detected at a LC retention volume indicative of the formation of IsoLG lactam and hydroxylactam adducts of the ApoB-100 protein isolated from oxidized LDL. This finding was also of interest in that

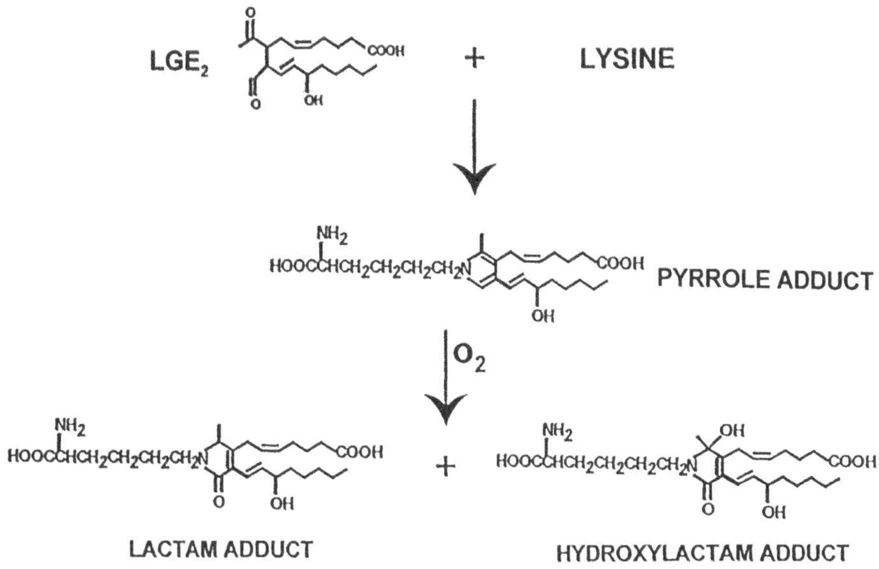

Figure 2. Formation of a LGE$_2$-lysine pyrrole adduct which readily undergoes oxidation to lactam and hydroxylactam adducts.

previous attempts to detect the formation of free IsoLG's in oxidized LDL were unsuccessful, consistent with the highly reactive nature of IsoLG's.

DISCUSSION

These studies have elucidated the formation of two novel series of reactive compounds that are produced as products of the IsoP pathway, namely cyclopentenone IsoP's (A$_2$/J$_2$-IsoP's) and E$_2$/D$_2$-IsoLG's. Free radical induced lipid peroxidation is known to generate reactive aldehydes, such as malondialdehyde and 4-hydroxynonenal. These reactive aldehydes are thought to be important mediators of some of the adverse sequella of oxidant injury. Adduction of reactive compounds to key biomolecules may disrupt normal cellular function and adduction to DNA may be mutagenic. Further, adduction of aldehyde products of lipid peroxidation to the ApoB-100 protein on LDL is thought to be the key process which converts LDL to an atherogenic form that is recognized by scavenger receptors on macrophages leading to uptake of LDL and foam cell formation. Thus, these newly identified reactive products of the IsoP pathway may participate as mediators of oxidant injury and the conversion of LDL to an atherogenic form. The latter possibility is of relevance since it has been shown that incubation of LDL with LGE$_2$ converts LDL to an atherogenic form and does so with an efficiency that exceeds that of 4-hydroxynonenal and malondialdehyde by several orders of magnitude.[21] In this regard, it is of interest that we were able to identify the presence of IsoLG ApoB-100 adducts in oxidized LDL.

The development of an LC electrospray mass spectrometric method to detect the formation of IsoLG protein adducts provides the necessary tool to begin to assess the formation of IsoLG's *in vivo* in settings of oxidatnt injury. In addition, studies aimed at further delineating the biological actions of cyclopentenone IsoP's and IsoLG's should contribute importantly to our understanding of the role of these reactive compounds as potential mediators of oxidant injury.

ACKNOWLEDGMENTS

This work was supported by grants GM 42056 and DK 48831 from the National Institutes of Health.

REFERENCES

1. J.D. Morrow, K.E. Hill, R.F, Burk, T.M. Nammour, K.F. Badr, and L.J. Roberts, II, A series of prostaglandin F_2-like compounds produced *in vivo* in humans by a non-cyclooxygenase free radical catalyzed mechanism, *Proc. Natl. Acad. Sci. USA* 87:9383 (1990).

2. J.D. Morrow, L.J. Roberts, II, V.C. Daniel, V.C. Mirotchnechenko, L. Swift, and R.F. Burk, Comparison of the formation of D_2/E_2-isoprostanes to F_2-isoprostanes *in vitro* and *in vivo*: effect of oxygen tension and glutathione, *Arch. Biochem. Biophys.* in press.

3. J.D. Morrow, T.A. Minton, C.R. Mukundan, M.D. Campbell, W.E. Zackert, V.C. Daniel, K.F. Badr, I.A. Blair, and Roberts, L.J., II, Free radical-induced generation of isoprostanes *in vivo*: evidence for the formation of D-ring and E-ring isoprostanes, *J. Biol. Chem.* 269:4317 (1994).

4. F.A. Fitzpatrick, and M.A. Wynalda, Albumin-catalyzed metabolism of prostaglandin D_2, *J. Biol. Chem.* 258:11713 (1983)

5. Fizpatrick F.A., W.F. Liggett, and M.A. Wynalda, Albumin-eicosanoid interactions: a model system to determine their attributes and inhibition, *J. Biol. Chem.* 259:2722 (1984).

6. M. Fukushima, Biological activities and mechanisms of action of PGJ_2 and related compounds: an update, *Prostaglandins Leukotrienes Essent. Fatty Acids*, 47:1 (1992).

7. J. Atsmon, B.J. Sweetman, S.W. Baertschi, T.M. Harris, and L.J. Roberts, II, Formation of thiol conjugates of 9-deoxy-^{9},^{12}(E)-prostaglandin D_2 and ^{12}(E)-prostaglandin D_2, *Biochemistry* 29:3760 (1990).

8. J. Atsmon, M.L. Freeman, M.J.Meridith, B.J. Sweetman, and L.J. Roberts, II, conjugation of 9-deoxy-^{9},^{12}(E)-prostaglandin D_2 with intracellular goutathione and enhancement of its antiprolifeative activity by glutathione depletion, *Canc. Res.* 50:1879 (1990).

9. K.V. Honn, and L.J. Marnett, Requirement of a reactive alpha, beta-unsaturated carbonyl for inhibition of tumor growth and induction of differentiation by "A" series prostaglandins, *Biochem. Biophys. Res. Commun.* 129:34 (1985)

10. R.G. Salomon, and D.B. Miller, Levuglandins: isolation, characterization, and total synthesis of new secoprostanoid products from prostaglandin endoperoxides, *Adv. Prostaglandin Thromboxane Leukotriene Res.* 15:323 (1985).

11. R.S. Iyer, S. Ghosh, and R.G. Salomon, Levuglandin E_2 crosslinks proteins, *Prostaglandins* 37:471 (1989).

12. K.K. Murthi, L.R. Friedman, N.L. Olenick, and R.G. Salomon, Formation of DNA-protein crosslinks in mammalian cells by levuglandin E_2, *Biochemistry* 32:4090 (1993).

13. J.D. Morrow, and L.J. Roberts, II, Mass spectrometric method for quantification of prostanoids (F_2-isoprostanes) produced by a non-cyclooxygenase free radical catalyzed mechanism, *Methods Enzymol.* 233:163 (1994).

14. J.D. Morrow, J.A. Awad, T. Kato, K. Takahashi, K.F. Badr, L.J. Roberts, II, and R.F. Burk, Formation of novel non-cyclooxygenase derived prostanoids (F_2-isoprostanes) in carbon tetrachloride hepatotoxicity, an animal model of lipid peroxidation, *J. Clin. Invest.* 90:2502 (1992).

15. G.L. Miles, P.A. Lane, and P.K. Weech. *A Guidebook to Lipoprotein Techniques*, Elsevier Science Publishing (1984).

16. A.P. Decaprio, E.J. Alajos, and P. Weber, Covalent binding of a neurotoxic n-hexane metabolite: conversion of primary amines to substituted pyrrole adducts by 2,5-hexanedione, *Tox. Appl. Pharmacol.* 65:440 (1982).

17. J. Rosello, and E. Gelpi, Gas chromatographic and mass spectrometric identification of new specific derivatives of prostaglandins A and E:application to prostaglandin profiling in general, *J. Chromatogr. Sci.* 16:166 (1978).

18. H. Esterbauer, R.J. Schaur, and H. Zollner, Chemistry and biochemistry of 4-hydroxynonenal, malondialdehyde, and related aldehydes, *Free Rad. Biol. Med.* 11:81 (1991).

19. R.S. Iyer, M.E. Kobieski, and R.G. Salomon, Generation of pyrroles in the reaction of levuglandin E_2 with proteins, *J. Org. Chem.* 59:6038 (1994).

20. E. Hoft, A.R. Katritzky, and M.R. Nesbit, The autoxidation of alkylpyrroles, *J. Org. Chem.* 32:3330 (1968).

21. G. Hoppe, G. Subbanagounder, J. O'Neil, R.G. Salomon, and H.F. Hoff, Macrophage recognition of LDL modified by levuglandin E_2, an oxidation product of arachidonic acid, *Biochem Biophys. Acta* 1344:1 (1997).

50

FORMATION OF NOVEL ISOPROSTANE-LIKE COMPOUNDS FROM DOCOSAHEXAENOIC ACID

Jason D. Morrow[1], Andrew R. Tapper[1], William E. Zackert[1], James Yang[1], Stephanie C. Sanchez[1], Thomas J. Montine[2], and L. Jackson Roberts[1]

Departments of Medicine[1] and Pathology[2]
Vanderbilt University School of Medicine
Nashville, TN 37232-6602 USA

INTRODUCTION

Isoprostanes (IsoPs) are prostaglandin (PG)-like compounds that are formed non-enzymatically *in vivo* by free radical-induced peroxidation of arachidonic acid.[1] Formation of these compounds proceeds through bicyclic endoperoxide intermediates resembling PGH_2. The endoperoxide intermediates are then reduced to PGF_2-like compounds termed F_2-IsoPs or undergo rearrangement to form D-ring and E-ring IsoPs (D_2/E_2-IsoPs)[2] and thromboxane-like compounds (isothromboxanes).[3] Unlike cyclooxygenase derived PGs in which arachidonic acid must first be released from phospholipids prior to oxygenation, IsoPs are formed completely *in situ* esterified in lipids and are subsequently released by a phospholipase(s) A_2.[4] Since the identification of these compounds, we have accumulated a large body of evidence showing that quantification of IsoPs provides an accurate measure of lipid peroxidation *in vitro* and *in vivo*.[5,6] Further, at least three of these compounds possess biological activity[1,2,7] and thus may mediate some of the adverse effects associated with oxidant stress.

Docosahexaenoic acid ($C22:6\omega3$, DHA) is a polyunsaturated fatty acid that is highly enriched in the brain, particularly in grey matter, where it comprises approximately 30% of the total fatty acids in aminophospholipids.[8,9] Although DHA is concentrated in neurons, these cells are incapable of synthesizing it. Rather, DHA is generated by astrocytes and subsequently secreted and taken up by neurons.[10] Deficiencies of DHA, particulary early in life, are associated with abnormalities in brain function.[11]

Eicosanoids and Other Bioactive Lipids in Cancer, Inflammation, and Radiation Injury, 4
Edited by Honn *et al.*, Kluwer Academic / Plenum Publishers, New York, 1999.

343

IsoP-like compounds can theoretically be formed by the free radical-induced peroxidation of any fatty acid containing at least three interrupted methylene groups. Thus, we initially sought to determine whether F_2-IsoP-like compounds are generated during oxidation of DHA *in vitro*. Because DHA is enriched in the central nervous system, we propose to name these compounds neuroprostanes (NPs). We then sought to determine whether NPs could be detected esterified in lipids of normal rat brain and whether levels increased following exposure of rat brain to an oxidant stress *in vitro*. Finally, we undertook studies to determine whether NPs could be quantified in cerebrospinal fluid from normal humans and whether levels are increased in patients with Alzheimer's disease, a chronic neurodegenerative disorder associated with increased oxidant stress in the central nervous system.[12]

METHODS

DHA was oxidized *in vitro* using Fe/ADP/ascorbate as described.[3] DHA-derived compounds (F_4-NPs) were purified by C18 SepPak solid phase extraction and thin layer chromatography using modifications of the method described for the purification of F_2-IsoPs.[1] Briefly, instead of scraping 1cm below to 1 cm above where $PGF_{2\alpha}$ migrates on TLC for analysis of F_2-IsoPs, the area scraped was extended to 3 cm above where $PGF_{2\alpha}$ migrates. Compounds were then quantified by gas chromatography/stable isotope dilution mass spectrometry using $[^2H_4]PGF_{2\alpha}$ as an internal standard. F_4-NPs were analyzed as pentfluorobenzyl ester, trimethylsilyl ether derivatives with selected ion monitoring of the $M-CH_2C_6F_5$ ions m/z 593. The corresponding m/z 573 ion was monitored for the $[^2H_4]PGF_{2\alpha}$ internal standard.

Brains were excised from normal rats and extracted lipids subjected to oxidation *in vitro* using the water soluble oxidant AAPH (2,2'-azobis(amidinopropane), a source of peroxyl radicals.[13]

Cerebrospinal fluid was obtained from subjects *post mortem* diagnosed with probable Alzheimer's disease. Control subjects were age matched without evidence of neurological disease. Statistical analysis was performed using the unpaired t-test.

RESULTS

Oxidation of DHA *in vitro* resulted in the formation of multiple m/z 593 peaks that eluted from the GC column over approximately a 90 sec period beginning about 30 sec after the elution of the $[^2H_4]PGF_{2\alpha}$ internal standard (Figure 1). F_4-NPs would be expected to have a longer GC retention time than $PGF_{2\alpha}$ since the former contain two more carbon atoms. The total amont of F_4-NPs formed was approximately 5 μg/g DHA. Analysis of these compounds as $[^2H_9]$ trimethylsilyl ether derivatives indicated the presence of three hydroxyls as would be expected. Further, catalytic hydrogenation revealed these compounds contained four double bonds; hence the term "F_4"-NPs, in keeping with PG nomenclature. Analysis of the compounds by electron ionization mass spectrometry as methyl ester trimethylsilyl ether derivatives yielded multiple mass spectra with an expected molecular ion at m/z 608 and ions resulting from common losses, *e.g.* of trimethylsilanol.

Anticipated prominent regioisomer-specific ions were also detected resulting from α-cleavage of the trimethylsiloxy substituents on the side chains of different regioisomers. Collectively these data proved convincing evidence that F_2-IsoP-like compounds, F_4-NPs, are generated as products of free radical-induced peroxidation of DHA *in vitro*.

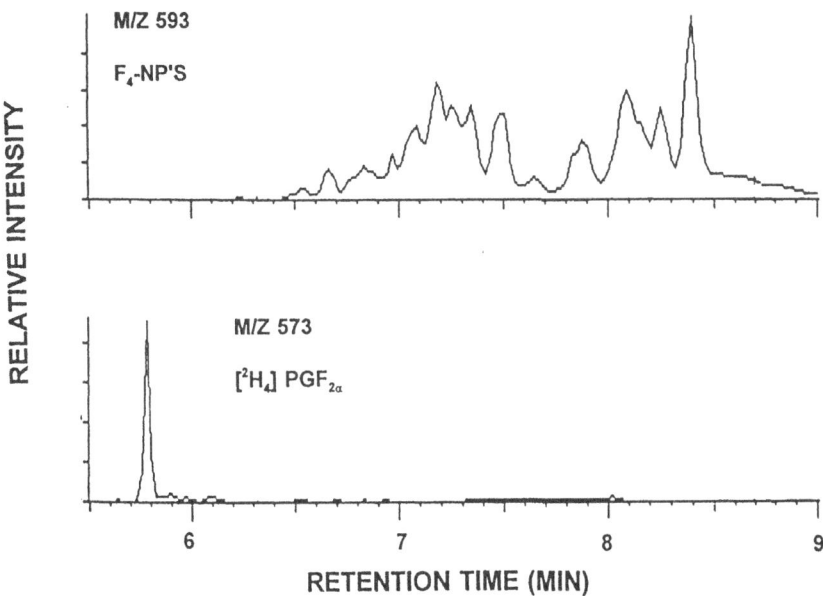

Figure 1. Selected ion current chromatograms obtained from the analysis of F_4-NPs generated during Fe/ADP/ascorbate-induced oxidation of DHA *in vitro*. The series of peaks in the m/z 593 ion current chromatogram represents putative F_4-NPs and the single peak in the m/z 573 chromatogram represents the $[^2H_4]PGF_{2\alpha}$ internal standard.

Compounds esterified in lipids extracted from brain tissues of normal rats were then analyzed for F_4-NPs *in vivo*. Rat brain lipids were also incubated with AAPH for 60 min *in vitro* to assess whether oxidation of brain tissue induced formation of these compounds. Interestingly, levels of F_4-NPs were readily detectable esterified in normal rat brain at levels of 7.0 ± 1.4 ng/g liver and increased up to approximately 20-fold following oxidation with AAPH.

Although F_4-NPs can be readily detected esterified in brain lipids, the utility of such measures to assess oxidative injury in the central nervous system would be restricted to animal models of neurological disorders or brain samples obtained post-mortem from humans. We therefore examined whether F_4-NPs could be detected in cerebrospinal fluid obtained from four patients with Alzheimer's disease and three age-matched control subjects. F_4-NPs were detected in 1 ml of cerebrospinal fluid from control subjects at a level of 64 ± 8 pg/ml. Interestingly, concentrations of F_4-NPs measured in patients with Alzheimer's disease were nearly 2-fold higher (110 ± 12 pg/ml), $p<0.05$.

DISCUSSION

These studies have provided evidence that free radical-induced peroxidation of DHA *in vitro* results in the formation of F₂-IsoP-like compounds termed F₄-NPs. F₄-NPs can also be detected esterified in normal rat brain and levels increase strikingly following oxidation of brain lipids *in vitro*. In addition, these compounds can be measured in cerebrospinal fluid from normal humans and quantities increase significantly in patients with Alzheimer's disease, a disorder associated with oxidant stress in the central nervous system.

The potential relevance of these findings encompasses several areas. Several IsoPs studied thus far have been found to exert potent biological activity as vasoactive agents and evidence suggests that the might interact with a unique receptor.[1,2,7] Thus, the possibility exists that NPs derived from DHA might also exert biological activity. Further, there are potential structural ramifications associated with the formation of these compounds esterified in membrane lipids. IsoP-containing phospholipids have been shown to be remarkably distorted molecules. Similarly, the analogous compounds derived from DHA would also be very distorted. Therefore, the formation of NPs in neuronal membranes would be expected to have significant effects on membrane fluidity, intergrity and function, which are well known sequelae of oxidant injury.

Considerable interest also relates to the possibility that quantification of F₄-NP in brain tissue and cerebrospinal fluid might provide a unique and accurate marker to assess oxidative injury in the central nervous system. The fact that levels of F₄-NPs are readily detectable in rat brains and human cerebrospinal fluid, and that levels are increased in patients with Alzheimer's disease, suggests that quantification of these compounds could be utilized to explore the role of free radicals in other neurological disorders in which enhanced oxidant stress has been implicated.

ACKNOWLEDGEMENTS

Supported by NIH grants DK48831, GM42056, and GM15431.

REFERENCES

1. J.D. Morrow, K.E. Hill, R.F. Burk, T.M. Nammour, K.F. Badr, and L.J. Roberts, A series of prostaglandin F₂-like compounds are produced *in vivo* by a non-cyclooxygenase free radical catalyzed mechanism, *Proc. Natl. Acad. Sci. U.S.A.* 87:9383 (1990).
2. J.D. Morrow, T.A. Minton, C.R. Mukundan, M.D. Campbell, W.E. Zackert, V.C. Daniel, K.F. Badr, I.A. Blair, and L.J. Roberts, Free radical induced generation of isoprostanes *in vivo*: Evidence for the formation of D-ring and E-ring isoprostanes, *J. Biol.Chem.* 269:4317 (1994).
3. J.D. Morrow, J.A. Awad, W.E. Zackert, V.C. Daniel, and L.J. Roberts, Free radical induced generation of thromboxane-like compounds (isothromboxanes) *in vivo*, *J. Biol. Chem.* 271:23185 (1996).

4.   J.D. Morrow, J.A. Awad, H.J. Boss, I.A. Blair, L.J. Roberts, Non-cyclooxygenase derived prostanoids (F_2-isoprostanes) are formed *in situ* on phospholipids, *Proc. Natl. Acad. Sci. U.S.A.* 89:10721 (1992).

5.   J.D. Morrow, J.A. Awad, T. Kato, K. Takahashi, K.F. Badr, L.J. Roberts, and R.F. Burk, Formation of novel non-cyclooxygenase-derived prostanoids (F_2-isoprostanes) in carbon tetrachloride hepatotoxicity, *J. Clin. Invest.* 90:2502 (1992).

6.   J.D. Morrow and L.J. Roberts, The isoprostanes: unique bioactive products of lipid peroxidation, *Prog. Lipid Res.* 36:1 (1997).

7.   P. Kunapuli, J.A. Lawson, J.A. Rokach, and G.A. FitzGerald, Functional characterization of the ocular $PGF_{2\alpha}$ receptor: activation by the isoprostane 12-epi-$PGF_{2\alpha}$, *J. Biol. Chem.* 272:27147 (1997).

8.   N. Salem, H.-Y. Kim, and J.A.Yergey, Health Effects of Polyunsaturated Fatty Acids in Foods, Academic Press, New York (1986).

9.   E.R. Skinner, C. Watt, J.A.O. Besson, and P.V. Best, Differences in fatty acid composition of grey and white matter of different regions of the brains of   patients with Alzheimer's disease and control subjects, *Brain.* 116:717 (1993).

10.   S.A Moore, E. Yoder, S. Murphy, G.R. Dutton, and A.A. Spector, Astrocytes, not neurons, produce docosahexaenoic acid and arachidonic acid, *J. Neurochem.* 55:518 (1991).

11.   W.E. Connor, M. Neuringer, S. Reisbick, Essential fatty acids: the importance n-3 fatty acids in the retina and brain, *Nutr. Rev.* 50:21 (1992).

12.   W.R. Markesbery, Oxidative stess hypothesis in Alzheimer's disease, *Free Radic. Biol. Med. 23: 134 (1997).*

13.   S.M. Lynch, J.D. Morrow, L.J. Roberts, and B. Frei, Formation of non-cyclooxygenase derived prostanoids (F_2-isoprostanes) in plasma and low density lipoprotein exposed to oxidative stress *in vitro, J. Clin. Invest.* 93:998 (1994).

LYSOPHOSPHATIDIC ACID (LPA) AND LPA RECEPTORS

51

cDNA CLONING, EXPRESSION AND CHROMOSOMAL LOCALIZATION OF TWO HUMAN LYSOPHOSPHATIDIC ACID ACYLTRANSFERASES

Christine Eberhardt, Patrick W. Gray, and Larry W. Tjoelker

ICOS Corporation
Bothell, WA 98021

INTRODUCTION

Lysophosphatidic acid (1-acyl-*sn*-glycero-3-phosphate, LPA) is both a biochemical intermediate in phospholipid metabolism and a potent bioactive lipid with diverse activities that range from the physiologic to the pathophysiologic. As an intermediate, LPA can be formed by acylation of *sn*-glycerol-3-phosphate or by acylation of dihydroxyacetone phosphate (DHAP) followed by reduction of the acyl-DHAP[1]. The LPA can be acylated further by the action of LPA acyltransferase (LPAAT) to form phosphatidic acid (PA), which preceeds diacylglycerol and CDP-diacylglycerol formation. LPA has also gained notoriety as a bioactive mediator that modulates mitogenesis, cell differentiation, platelet aggregation, actin cytoskeleton remodeling, monocyte chemotaxis, smooth muscle contraction, and neurite retraction[2]. *In vitro* experiments suggest that LPA can also impact immune cell functions such as proliferation and IL-2 production[3]. The phospholipid may also participate in the pathophysiology of neurodegenerative processes by causing vasoconstriction as well as impairment of glutamate and glucose uptake by astrocytes[4,5]. In addition, LPA is a potent promoter of tumor cell growth and invasion[6,7].

By converting LPA to PA, LPAAT plays a central role in LPA biology. Genetic and molecular biological approaches have facilitated cloning of genes encoding LPAAT activity from a number of species representing plants, prokaryotic organisms, yeast, and nematodes. This prompted us to explore the expressed sequence tag (EST) database of GenBank for uncharacterized mammalian homologues of the plant and microbial LPAATs. In this report we

Eicosanoids and Other Bioactive Lipids in Cancer, Inflammation, and Radiation Injury, 4
Edited by Honn *et al.*, Kluwer Academic / Plenum Publishers, New York, 1999.

351

describe the EST-based cloning of two distinct human LPAAT cDNAs and characterize their tissue expression patterns. We also describe the enzymatic activity and subcellular localization of the expressed proteins as well as the chromosomal addresses of the cognate genes. Several other reports describing these cDNAs have also recently appeared. The first, by West et al. (1997)[8], established the LPAAT-α/β nomenclature. This was followed by our characterization of the β isozyme[9]. Others have since described a mouse LPAAT-α homolog[10] and further characterization of the human LPAAT-α[11,12].

IDENTIFICATION OF LPAAT cDNAs.

The coconut LPAAT cDNA sequence[13] was used to conduct a TBLASTN search of the EST database for mammalian homologues. A number of ESTs derived from human tissue cDNA libraries displayed significant sequence homology with the plant sequence. Alignment of the ESTs revealed that two distinct genes were represented. To obtain full-length cDNA clones, we used the EST sequences to design PCR primers for screening a human monocyte-derived macrophage cDNA plasmid library[14]. Confirmatory clones were subsequently obtained from a human cardiac muscle cDNA library. The nucleotide sequences of these clones confirmed the existence of two distinct but closely related cDNAs that are predicted to encode a pair of LPAAT isozymes. The open reading frame of one clone (called LPAAT-α) is predicted to encode a polypeptide of 283 amino acids while the other (LPAAT-β) is predicted to encode a 278 amino acid protein. The two isoforms display 47% amino acid sequence identity with each other and variable levels of homology with other known or predicted acyltransferases (ranging from 33% with a *C. elegans* LPAAT to 23% with the coconut enzyme).

ENZYMATIC ACTIVITY

The homology shared with LPAATs of very diverse species suggested an acyltransferase function for the new human cDNAs. To test this, we modified the coding region of the human cDNAs by attaching sequence encoding the FLAG tag (Kodak) to the 3' end and subcloning the resulting fragment into the eukaryotic expression vector pcDNA3 (Invitrogen). COS7 cells transfected with these constructs produced protein that was readily detectable when a FLAG-specific antibody was used to probe Western blots of cell lysates. To assay for acyltransferase activity, we incubated lysates of the transfected cells with various acyl donors such as myristoyl coenzyme A, palmitoyl coenzyme A, stearoyl coenzyme A, and arachidonyl coenzyme A (Sigma). Acyl acceptor glycerolipids that were tested included 1-palmitoyl-2-hydroxy-*sn*-glycero-3-phosphate, 1-palmitoyl-2-hydroxy-*sn*-glycero-3-phosphocholine, and 1-alkyl-2-hydroxy-*sn*-glycero-3-phosphocholine (Avanti). As Table I indicates, both enzymes readily transfer acyl groups from the CoA donor to LPA. Under the conditions of the assay, lysophosphatidylcholine cannot serve as an acyl acceptor. The enzymes, particularly the α isozyme, also exhibit an acyl donor preference. Under these assay conditions, arachidonyl CoA is preferred although the saturated shorter chain donors can also be utilized.

Table 1. Acyltransferase activity of human LPAAT isozymes.

Acyl Donor[1]	Acyl Acceptor[2]	Relative Acyltransferase Activity[3]	
		LPAAT-α	LPAAT-β
C20:4	C16-LPA	5.8	20.6
C20:4	C16-LPC	1.0	1.0
C18:0	C16-LPA	1.3	8.7
C16:0	C16-LPA	2.7	9.3
C14:0	C16-LPA	2.4	10.6

[1]1.3 μM 14C-labeled and 40 μM nonlabeled acyl CoA per reaction
[2]20 μM lysophosphoglyceride per reaction
[3]Fold increased activity over endogenous COS7 activity

STRUCTURE AND SUBCELLULAR LOCALIZATION

The sequence homology between the acyltransferases of the various species is largely confined to highly conserved blocks of amino acids that are located between residues 165-205 of the human enzymes. Nevertheless, hydropathy profiles of the human isozymes bear striking resemblance to those of LPAATs from other species. All of these enzymes share three to four hydrophobic regions that are predicted to serve as transmembrane domains. Prior studies of LPAAT activity in eukaryotic cells have indicated that the activity is associated with the endoplasmic reticulum (ER)[1]. The presence of the hydrophobic and possibly transmembrane domains would be consistent with an integral association of both isozymes with the ER membrane. To test this, we transfected COS7 cells with the expression constructs described above and found that all LPAAT activity remained associated with the cells; none could be detected in the culture medium. To determine the cellular location of the recombinant enzymes, we fixed and permeabilized the transfected cells and incubated them with a FITC-labeled FLAG-specific antibody (Kodak). The resulting signal displayed a pattern characteristic of ER morphology. An identical staining pattern with a carbocyanine dye (DiOC$_6$(3)) suppports the conclusion that the recombinant proteins localize to the ER. Together, these data suggest that both LPAAT isozymes are integral ER membrane proteins.

mRNA TISSUE EXPRESSION PATTERNS

Northern blot analyses demonstrated that mRNAs encoding both human LPAAT isozymes are widely but differentially expressed. Message for both isozymes could be detected in each of 16 tissues examined. The LPAAT-α message was particularly abundant in testes, skeletal muscle, and kidney but was barely detected in thymus, lung, or liver. In contrast, highest expression of LPAAT-β was seen in the liver. LPAAT-β mRNA was also abundant in the heart, small intestine, lung, skeletal muscle, and pancreas but was expressed at very low levels in the thymus and placenta. Northern blot analysis of LPAAT-β expression in the human myeloid-like cell lines THP-1, HL-60, and U937 suggested that the gene is constitutively expressed in these cells. Treatment of any of these cell lines with PMA or treatment of HL-60 cells with DMSO had no effect on message levels. These expression patterns are consistent with the hypothesis that LPAAT activity is essential for tissue viability.

GENOMIC STRUCTURE AND CHROMOSOMAL LOCALIZATION

A BLASTX search of GenBank using the human LPAAT-α cDNA as the query revealed that its genomic sequence has been determined[15]. Interestingly, the gene is comprised of six exons embedded within the class III MHC locus on chromosome 6, position p21.3. To consider the evolutionary history of the two human isozymes, we isolated a 20 kb genomic clone that hybridized to the LPAAT-β cDNA. Exon-specific sequencing revealed that the β isozyme is also encoded by six exons, the boundaries of which are exactly homologous with those of the LPAAT-α isozyme. Fluorescent *in situ* hybridization analysis of human chromosomes using the LPAAT-β genomic clone as a probe pinpointed the address of the gene to chromosome 9, position q34.3[9]. The conserved structure of the two genes suggests that one arose from the other as a result of a partial chromosomal duplication event. This is supported by a comparison of genes surrounding the LPAAT genes. On chromosome six, LPAAT-α is flanked by genes encoding NOTCH 4, the homeobox-containing protein PBX2, and collagen XI, α2. On chromosome nine, LPAAT-β is flanked by a similar array of genes: NOTCH homolog 1, a homeobox-containing protein LHX3, and collagen V, α1.

SUMMARY

In this report we describe a pair of human LPAAT isozymes. These isozymes are encoded by distinct genes located on different chromosomes, but share sequence homology, substrate specificity, and intracellular location. The biological value of maintaining the two closely related LPAAT genes in the human genome is not clear. We find that both isozymes are widely expressed, although expression levels do diverge significantly in tissues such as the liver, placenta, testes, and pancreas. We also find that, at least in the artificial system of over-expression in COS7 cells, both isozymes localize to the ER membrane. Thus, distinct tissue-specific or subcellular compartment-specific roles for the two isozymes are not supported by the current experimental evidence. It does remain possible that induction of expression or subcellular translocation of one or the other isozyme may distinguish their functions. A survey of a limited number of acyl CoA substrates indicates that the two isozymes display similar substrate specificities, although slight differences are suggested by the data. However, extensive analysis of both isozymes with multiple substrates in the same assay system will be required to detect physiologically relevant differences in substrate specificity.

LPA and PA are central intermediates in phospholipid biogenesis. Furthermore, they have the capacity to mediate signaling both between and within cells. The importance of these mediators is reflected in the growing body of literature dedicated to unraveling the mechanistic basis for their actions. Until recently, the field has been hampered by a dearth of reagents appropriate for the molecular dissection of the LPA and PA metabolic and signaling pathways in eukaryotes. However, the recent cloning of possible LPA receptors[16, 17, 18] will promote further understanding of LPA signaling. Similarly, the recent appearance of LPAAT homologs in the EST database has prompted a flurry of reports describing their characterization[8-12]. These clones will afford opportunity for defining the function of LPAAT in eukaryotic phospholipid metabolism.

REFERENCES

1. W.R. Bishop and R.M. Bell, Assembly of phospholipids into cellular membranes: biosynthesis, transmembrane movement and intracellular translocation, *Annu. Rev. Cell Biol.* 4:579 (1988).
2. W.H. Moolenaar, Lysophosphatidic acid, a multifunctional phospholipid messenger, *J. Biol. Chem.* 270:12949 (1995).
3. Y. Xu, G. Casey, and G.B. Mills, Effect of lysophospholipids on signaling in the human Jurkat T cell line, *J. Cell. Physiol.* 163:441 (1995).
4. G. Tigyi, L. Hong, M. Yakubu, H. Parfenova, M. Shibata, and C.W. Leffler, Lysophosphatidic acid alters cerebrovascular reactivity in piglets, *Am. J. Physiol.* 268:H2048 (1995).
5. J.N. Keller, M.R. Steiner, M.P. Mattson, and S.M. Steiner, Lysophosphatidic acid decreases glutamate and glucose uptake by astrocytes, *J. Neurochem.* 67:2300 (1996).
6. F. Imamura, T. Horai, M. Mukai, K. Shinkai, M. Sawada, and H. Akedo, Induction of *in vitro* tumor cell invasion of cellular monolayers by lysophosphatidic acid or phospholipase D., *Biochem. Biophys. Res. Comm.* 193:497 (1993).
7. Y. Xu, X.-J. Fang, G. Casey, and G.B. Mills, Lysophospholipids activate ovarian and breast cancer cells, *Biochem. J.* 309:933 (1995).
8. J. West, C.K. Tompkins, N. Balantac, E. Nudelman, B. Meengs, T. White, S. Bursten, J. Coleman, A. Kumar, J.W. Singer, and D.W. Leung, Cloning and expression of two human lysophosphatidic acid acyltransferase cDNAs that enhance cytokine-induced signaling responses in cells, *DNA Cell Biol.* 16:691 (1997).
9. C. Eberhardt, P.W. Gray, and L.W. Tjoelker, Human lysophosphatidic acid acyltransferase: cNDA cloning, expression, and localization to chromosome 9q34.3, *J. Biol. Chem.* 272:20299 (1997).
10. K. Kume and T. Shimizu, cDNA cloning and expression of murine 1-acyl-*sn*-glycerol-3-phosphate acyltransferase, *Biochem. Biophys. Res. Comm.* 237:663 (1997).
11. A.C. Stamps, M.A. Elmore, M.E. Hill, K. Kelly, A.A. Makda, and M.J. Finnen, A human cDNA with homology to non-mammalian lysophosphatidic acid acyltransferases, *Biochem. J.* 326:455 (1997).
12. B. Aguado and R. D. Campbell, Characterization of a human lysophosphatidic acid acyltransferase that is encoded by a gene located in the class III region of the human major histocompatibility complex, *J. Biol. Chem.* 273:4096 (1998).

13. D.S. Knutzon, K.D. Lardizabal, J.S. Nelsen, J.L. Bleibaum, H.M. Davies, and J.G. Metz, Cloning of a coconut endosperm cDNA encoding a 1-acyl-*sn*-glycerol-3-phosphate acyltransferase that accepts medium-chain-length substrates. *Plant Physiol.* 109:999 (1995).

14. L.W. Tjoelker, C. Wilder, C. Eberhardt, D.M. Stafforini, G.Dietsch, B.Schimpf, S. Hooper, H. Le Trong, L.S. Cousens, G.A. Zimmerman, Y. Yamada, T.M. McIntyre, S.M. Prescott, and P.W. Gray. Anti-inflammatory properties of a platelet-activating factor acetylhydrolase, *Nature* 374:549 (1995).

15. L. Rowen, C. Dankers, D. Baskin, J. Faust, C. Loretz, M.E. Ahearn, A. Banta, T. Spies, and L Hood, Sequence determination of 300 kilobases of the human class III MHC locus, GenBank Accession U89336 (1997).

16. J.H. Hecht, J.A. Weiner, S.R. Post, and J. Chun, Ventricular zone gene-1 (vzg-1) encodes a lysophosphatidic acid receptor expressed in neurogenic regions of the developing cerebral cortex, *J. Cell Biol.* 135:1071 (1996).

17. Z. Guo, K. Liliom, D.J. Fischer, I.C. Bathurst, L.D. Tomei, M.C. Kiefer, and G. Tigyi, Molecular cloning of a high-affinity receptor for the growth factor-like lipid mediator lysophosphatidic acid from *Xenopus* oocytes, *Proc. Natl. Acad. Sci. USA* 93:14367 (1996).

18. S. An, M.A. Dickens, T. Bleu, O.G. Hallmark, and E.J. Goetzl, Molecular cloning of the human Edg2 protein and its identification as a functional cellular receptor for lysophosphatidic acid, *Biochem Biophys Res Commun.* 231:619 (1997).

52

THE FIRST CLONED AND IDENTIFIED LYSOPHOSPHOLIPID (LP) RECEPTOR GENE, *VZG-1*: IMPLICATIONS FOR RELATED RECEPTORS AND THE NERVOUS SYSTEM

Jerold Chun

Department of Pharmacology
Neurosciences and Biomedical Sciences Program
School of Medicine, University of California, San Diego
9500 Gilman Drive
La Jolla, CA   92093-0636

INTRODUCTION

In this piece I present some of the biological issues and technical approaches that led to the cloning and functional identification of the first lysophosphlipid receptor gene, named "ventricular zone gene-1" or "*vzg*-1" that serves as a receptor for lysophosphatidic acid (LPA). Based on its prominent expression in the mammalian central nervous system (CNS), it is likely that this and related receptor-ligand interactions represent a novel signaling system for brain development and function.   Moreover, this identity has provided a rationale for examining the same and/or structurally similar ligands on homologous orphan receptors either published or in the databases, and indicates that lysophosphlipid receptors form a distinct subfamily of the G-protein coupled receptor (GPCR) superfamily.

A key feature during the embryonic development of the brain is the generation of neurons from discrete proliferative regions of the neural tube. The largest such region gives rise to neurons of the cerebral cortex, and because of its location over the cerebral ventricles (the space filled with cerebral spinal fluid), this region has been named the "ventricular zone" or "VZ"(Boulder Committee, 1970). One regulatory mechanism that affects the production of young neurons from the VZ is peptide growth factor stimulation, acting through receptor tyrosine kinases (Kazlauskas, 1994; Temple and Qian, 1995). Studies of the developing cortex, however, indicated that other soluble factors might also exist that could affect the output of cells from these regions of neuroproliferation (Kilpatrick and Bartlett, 1993), including factors associated with cell membranes (Temple and Davis, 1994). In contrast to receptor tyrosine kinases, the 7-transmembrane domain receptors for G-protein coupled receptors (GPCRs) had unknown existence in the embryonic cortex, which led to an examination of such genes by using degenerate polymerase chain reaction (PCR) techniques for GPCRs on specific cell lines that had been derived from the embryonic cortex, and that expressed genes consistent with an early neuroblast identity (Chun and Jaenisch, 1996).

Eicosanoids and Other Bioactive Lipids in Cancer, Inflammation, and Radiation Injury, 4
Edited by Honn *et al.*, Kluwer Academic / Plenum Publishers, New York, 1999.

357

Cloning strategy identifies *vzg-1*, a novel GPCR gene

The strategy to identify GPCR genes of relevance to the VZ was based on the use of novel cell lines derived from this embryonic region of neuroproliferation. This step was critical since 1) the embryonic cortex is not readily accessible (it must be accessed via intrauterine manipulations), 2) the amount of tissue from a single animal is minuscule, 3) the cells within the VZ are not uniform but consist of several lineages at varying stages of differentiation, and 4) biologically relevant cells provide an analytical system that could allow the functional study of isolated genes/gene products. These issues led to the development of a way to produce cells of stable phenotype that approximated that of VZ neuroblasts (Chun and Jaenisch, 1996). These clonal, murine cell lines were produced by a sequential "2-hit" approach using different, retrovirally-transduced oncogene combinations. The cell lines expressed neuronal but not glial markers, as well as a cerebral cortex (telencephalic) gene called BF-1 (Xuan *et al.*, 1995). Other commonly used cell lines could not be rationally used since there are extremely few "neuronal" cell lines that are well characterized, and those that are (e.g. neuroblastoma (Augusti-Tocco and Sato, 1969; Klee and Nirenberg, 1974) or PC12 (Greene, 1978; Greene and Tischler, 1976) cells) were produced or derived from neural crest, an embryonic lineage that gives rise to the *peripheral* nervous system (along with other, non-CNS tissues). Using a variety of subtractive or targeted approaches with the cortical cell lines, new genes/gene products that are expressed in the CNS have been identified (Hecht *et al.*, 1996; Weiner and Chun, 1997a; Weiner and Chun, 1997b).

One gene family of especial interest was that encoding GPCRs. Degenerate PCR was utilized at different stringencies and cycle numbers to amplify possible GPCR gene fragments from the cortical neuroblast cell line cDNA; this strategy has been used successfully in other systems to identify novel GPCRs (Buck and Axel, 1991; Hecht *et al.*, 1996; Libert *et al.*, 1989). Several hundred PCR fragments were isolated from the initial screens. These fragments were isolated by agarose gel electrophoresis and as a primary screen, were examined for expression using *in situ* hybridization within the embryo. This approach identified several candidate receptor gene species that demonstrated expression within distinct, anatomical regions of the embryonic cortex.

Fragments demonstrating an interesting hybridization signal were sequenced to ascertain that they were of the GPCR family. A sequence of particular interest was isolated using several different but related PCR primer combinations and stringency conditions. The *in situ* hybridization pattern was remarkable for its localization with in the VZ, and this led to its naming as *vzg-1* (J. Chun, University of California-Mexico (MEXUS) Symposium on Neurobiology, Mexico City 1993; Hecht et al., Soc. Neurosci. Abstr. 21: 1289, 1995, (Hecht *et al.*, 1996)). The gene fragments were used to identify full-length coding cDNAs, that, at the time of its sequencing, represented a novel, putative GPCR gene on the basis of containing 7 transmembrane domains, and on its homology to other 7-transmembrane-domain receptors. Comparison of its sequence to those within BLAST-accessible databases revealed highest homology to the CB1 cannabanoid receptor, that interacts with anandamide and a human orphan receptor identified from studies on endothelial cells called "endothelial differentiation gene-1 (*edg-1*)"(Hla and Maciag, 1990).

Functional assay of *vzg-1* by overexpression in cortical cell lines

The *in situ* expression pattern of *vzg-1* indicated that a new GPCR-signal could have important influences on the fate of cells generated within the VZ. It was therefore of extreme importance - and difficulty, as reflected by the many cloned orphan receptors that include the many olfactory receptors (Buck and Axel, 1991) - to identify the ligand for this receptor. To determine the ligand, a strategy that was based on the use of VZ cell lines (Chun and Jaenisch, 1996) from which *vzg-1* was first cloned was employed. Mammalian expression vectors constructed in the sense, antisense orientations for *vzg-1* along with empty vector controls were used. Following transfection, the cells were assayed for detectable changes, the most obvious of which was morphology. Based on the biological phenomenon of neurite retraction and cell rounding observed in the VZ for many decades (Sauer, 1935), a change in cell shape from elongated to round would be considered relevant.

From these studies, "cell rounding" was indeed apparent. Moreover, it required the presence of both the sense strand constructs, as well as the presence of serum. The bioactive agent was further observed to be heat-stable. Additional clues came from the literature on neuroblasts in culture that indicated the growth and survival of putative VZ neuroblasts depended on peptidergic growth factors like bFGF, yet additionally required still unknown factors present in serum or as components of cell membrane fractions (Ghosh and Greenberg, 1995; Temple and Davis, 1994; Temple and Qian, 1995). These properties of the *vzg-1* ligand, coincident with the morphological changes observed in VZ neuroblasts, also correlated with the operation of cell proliferation, differentiation and/or death. Based on these data, lysophospholipids became a leading candidate, that then led to identification of LPA as a high affinity ligand for the receptor encoded by *vzg-1*.

Lysophosphlipids

Application of lysophospholipids through extracellular exposure has been shown to produce cellular responses of varying types, including cytoskeletal rearrangements of actin. Two of the most prominent agents, lysophosphatidic acid (LPA which is 1-acyl-sn-glycerol-3-phosphate) or sphingosine-1-phosphate (S1P, which is 1-phosphate-2-amino-4-cis-octadecene-1,3-diol) produce overlapping types of responses, and studies on the responses and signaling paths have been reviewed (Durieux and Lynch, 1993; Gaits *et al.*, 1997; Moolenaar, 1995a; Moolenaar, 1995b; Moolenaar *et al.*, 1992; Moolenaar *et al.*, 1997). As a major example of lysophospholipid (LP) effectors, LPA studies had been hampered by a missing element in understanding the mechanisms of extracellular responses: the absence of a cloned receptor. It should be noted that some data supported non-receptor mechanisms for LPA action (e.g. the lack of LPA steropecificity (Simon *et al.*, 1982; Sugiura *et al.*, 1994), and the structural similarity of LPA to detergents furthered this notion, as LPA could have been disturbing the cell membrane. The lack of a proven receptor thus left open the issue of a specific LPA receptor located on the cell surface. The initial properties observed using overexpression of *vzg-1*, however, strongly supported the hypothesis that it encoded a receptor for LPA, with responses that could not be accounted for by no-receptor mechanisms (Hecht *et al.*, 1996).

Characteristics of the LPA receptor VZG-1

Following the identification of *vzg-1* as a receptor for LPA, numerous approaches were used to disprove or prove this assignment. Using overexpression in the cell lines from which it was cloned (Hecht *et al.*, 1996), combined with heterologous expression studies in cell lines from distinct lineages (Fukushima *et al.*, 1998), proof for the initial identity has been achieved (summarized in Fig. 1). Further support for this identity has been reported recently using the human *vzg-1* homologue overexpressed in CHO cells (An *et al.*, 1997), as well as using heterologous expression analyses in yeast (Erickson *et al.*, 1998).

Other homologous receptors

Several orphan receptors with homology percentages ranging from approximately the upper 30s to over 60 have been cloned by ourselves and others. Based on both the amino acid predicted sequence (J. Chun, J.J.A. Contos and D. Munroe, *Cell Biochem. Biophys.* in press) and the genomic structure (J.J.A. Contos and J. Chun, *Genomics*, in press), these genes fall into 2

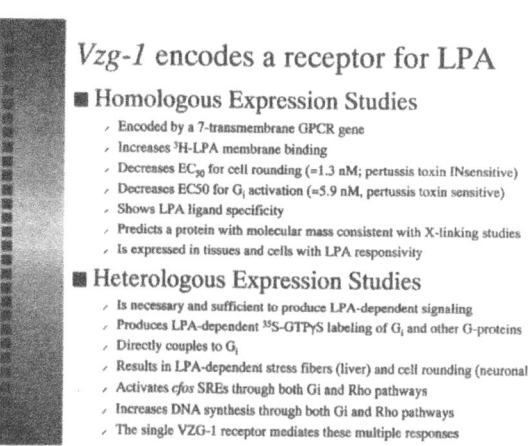

Vzg-1 encodes a receptor for LPA

■ Homologous Expression Studies

- Encoded by a 7-transmembrane GPCR gene
- Increases ^{3}H-LPA membrane binding
- Decreases EC_{50} for cell rounding ($\approx$1.3 nM; pertussis toxin INsensitive)
- Decreases EC50 for G_i activation ($\approx$5.9 nM, pertussis toxin sensitive)
- Shows LPA ligand specificity
- Predicts a protein with molecular mass consistent with X-linking studies
- Is expressed in tissues and cells with LPA responsivity

■ Heterologous Expression Studies

- Is necessary and sufficient to produce LPA-dependent signaling
- Produces LPA-dependent ^{35}S-GTPγS labeling of G_i and other G-proteins
- Directly couples to G_i
- Results in LPA-dependent stress fibers (liver) and cell rounding (neuronal)
- Activates *cfos* SREs through both Gi and Rho pathways
- Increases DNA synthesis through both Gi and Rho pathways
- The single VZG-1 receptor mediates these multiple responses

Figure 1. Characteristics of the LPA receptor VZG-1/LP$_{A1}$ from homologous and heterologous expression studies.

groups, one related to *vzg-1* and (renamed *lp*$_{A1}$ or "lysophospholipid A1" receptor) and a second group epitomized by endothelial differentiation gene-1, *edg-1* (renamed *lp*$_{B1}$) that has hallmarks of a receptor of sphingosine-1-phosphate (Tim Hla, personal communication). This existence of several LP receptors that have different tissue distributions and related yet distinct ligands/ligand affinities and/or cellular responses may in part account for the range of LP-related phenomena in the literature. It is likely that other related receptors will be identified

Vzg-1 is expressed within the nervous system

Northern blots revealed a single *vzg-1* mRNA band of 3.8 kb in the embryonic and postnatal CNS (Hecht *et al.*, 1996). Expression is predominantly within the cerebral cortex, and during embryonic life, is present predominantly within the VZ. A second pattern of hybridization (Weiner, Hecht and Chun, Soc. Neurosci. Abstr. 23: 1689 and submitted) is observed during the initial week after birth where it is most prominent in developing fiber tracts. Double-labeling techniques have demonstrated that expression is within oligodendrocytes. Expression in the peripheral nervous system has also been observed (J.A. Weiner and J. Chun, submitted). The clear presence of *vzg-1* in the nervous system suggested that physiological stimulation by LPA could take place, and studies of neural cells from these different locations using electrophysiological and cell biological techniques demonstrate that cells expressing *vzg-1* do indeed respond to LPA (Dubin and Chun, in preparation). The actual biological roles of this signaling path remain to be determined. In this light it is of especial note that many homologous orphan receptors are also expressed in the nervous system (G. Zhang, J.J.A. Contos, J.A.

Weiner, N. Fukushima, Y. Kimura and J. Chun, submitted); this suggests that this LP family of receptors and related ligands likely has hitherto unrecognized roles in CNS development and function and pathology. In support of this view, it is notable that the cannabanoid receptor CB1 has significant homology to members of the LP family. The direct relevance of this homology to the LP family has been underscored by the discovery of a second endogenous cannabanoid ligand (Stella *et al.*, 1997), sn-2 arachidonylglycerol, that has similar structural elements compared to lysophospholipids. The known CNS effects of cannabanoid receptor stimulation raise new venues for the biology of receptor-mediated LP signaling.

Acknowledgments

I thank Drs. Erickson and Hla for sharing unpublished data, Dr. J. A. Weiner for reading the manuscript, and Ms. Carol G. Akita for expert technical help. This research was supported by the NIMH.

REFERENCES

An, S., Dickens, M. A., Bleu, T., Hallmark, O. G., and Goetzl, E. J., 1997, Molecular cloning of the human Edg2 protein and its identification as a functional cellular receptor for lysophosphatidic acid, *Biochem Biophys Res Commun* 231:619.

Augusti-Tocco, G., and Sato, G., 1969, Establishment of functional clonal lines of neurons from mouse neuroblastoma, *Proc. Natl. Acad. Sci. USA* 64:312.

Boulder Committee, 1970, Embryonic vertebrate central nervous system: revised terminology., *Anat. Rec.* 166:257.

Buck, L., and Axel, R., 1991, A novel multigene family may encode odorant receptors: a molecular basis for odor recognition, *Cell* 65:175.

Chun, J., and Jaenisch, R., 1996, Clonal cell lines produced by infection of neocortical neuroblasts using multiple oncogenes transduced by retroviruses, *Mol. Cell. Neurosci.* 7:304.

Durieux, M. E., and Lynch, K. R., 1993, Signalling properties of lysophosphatidic acid, *Trends Pharmacol Sci* 14:249.

Erickson, J. R., Wu, J. J., Goddard, J. G., Tigyi, G., Kawanishi, K., Tomei, L. D., and Kiefer, M. C., 1998, Edg-2/Vzg-1 couples to the yeast pheromone response pathway selectively in response to lysophosphatidic acid, *J. Biol. Chem.* 273:1506.

Fukushima, N., Kimura, Y., and Chun, J., 1998, A single receptor encoded by *vzg-1 /lpA1/edg-2* couples to G-proteins and mediates multiple cellular responses to lysophosphatidic acid (LPA)., *Proc. Natl. Acad. Sci. USA* in press.

Gaits, F., Fourcade, O., Le Balle, F., Gueguen, G., Gaig, B., Gassama-Diagne, A., Fauvel, J., Salles, J. P., Mauco, G., Simon, M. F., and Chap, H., 1997, Lysophosphatidic acid as a phospholipid mediator: pathways of synthesis, *FEBS Lett* 410:54.

Ghosh, A., and Greenberg, M. E., 1995, Distinct roles for bFGF and NT-3 in the regulation of cortical neurogenesis, *Neuron* 15:89.

Greene, L. A., 1978, Nerve growth factor prevents the death and stimulates the neuronal differentiation of clonal PC12 pheochromocytoma cells in serum-free medium, *J. Cell Biol.* 78:747.

Greene, L. A., and Tischler, A. S., 1976, Establishment of a noradrenergic clonal line of rat adrenal pheochromocytoma cells which respond to nerve growth factor, *Proc. Natl. Acad. Sci. U S A* 73:2424.

Hecht, J. H., Weiner, J. A., Post, S. R., and Chun, J., 1996, Ventricular zone gene-1 (vzg-1) encodes a lysophosphatidic acid receptor expressed in neurogenic regions of the developing cerebral cortex, *J. Cell Biol.* 135:1071.

Hla, T., and Maciag, T., 1990, An abundant transcript induced in differentiating human endothelial cells encodes a polypeptide with structural similarities to G-protein-coupled receptors., *J. Biol. Chem.* 265:9308.

Kazlauskas, A., 1994, Receptor tyrosine kinases and their targets, *Curr. Opin. Genet. Dev.* 4:5.

Kilpatrick, T. J., and Bartlett, P. F., 1993, Cloning and growth of multipotential neural precursors: requirements for proliferation and differentiation, *Neuron* 10:255.

Klee, W. A., and Nirenberg, M., 1974, A neuroblastoma times glioma hybrid cell line with morphine receptors, *Proc. Natl. Acad. Sci. USA* 71:3473.

Libert, F., Parmentier, M., Lefort, A., Dinsart, C., Van Sande, J., Maenhaut, C., Simons, M. J., Dumont, J. E., and Vassart, G., 1989, Selective amplification and cloning of four new members of the G protein-coupled receptor family, *Science* 244:569.

Moolenaar, W. H., 1995a, Lysophosphatidic acid signalling, *Curr. Opin. Cell Biol.* **7**:203.

Moolenaar, W. H., 1995b, Lysophosphatidic acid, a multifunctional phospholipid messenger, *J. Biol. Chem.* 270:12949.

Moolenaar, W. H., Jalink, K., and van Corven, E. J., 1992, Lysophosphatidic acid: a bioactive phospholipid with growth factor-like properties, *Rev. Physiol. Biochem Pharmacol* 119:47.

Moolenaar, W. H., Kranenburg, O., Postma, F. R., and Zondag, G. C., 1997, Lysophosphatidic acid: G-protein signaling and cellular responses, *Curr. Opin . Cell Bio.* 9:168.

Sauer, F. C., 1935, Mitosis in the neural tube, *J. Comp. Neurol.* 62:377.

Simon, M. F., Chap, H., and Douste-Blazy, L., 1982, Human platelet aggregation induced by 1-alkyl-lysophosphatidic acid and its analogs: a new group of phospholipid mediators?, *Biochem. Biophys. Res. Commun.* 108:1743.

Stella, N., Schweitzer, P., and Piomelli, D., 1997, A second endogenous cannabinoid that modulates long-term potentiation, *Nature* 388:773.

Sugiura, T., Tokumura, A., Gregory, L., Nouchi, T., Weintraub, S. T., and Hanahan, D. J., 1994, Biochemical characterization of the interaction of lipid phosphoric acids with human platelets: comparison with platelet activating factor, *Arch. Biochem. Biophys.* 311:358.

Temple, S., and Davis, A. A., 1994, Isolated rat cortical progenitor cells are maintained in division in vitro by membrane-associated factors, *Development* 120:999.

Temple, S., and Qian, X., 1995, bFGF, neurotrophins, and the control or cortical neurogenesis, *Neuron* 15:249.

Weiner, J. A., and Chun, J., 1997a, Maternally derived immunoglobulin light chain is present in the fetal mammalian CNS, *J. Neursci.* 17:3148.

Weiner, J. A., and Chun, J., 1997b, *Png-1*, a nervous system-specific zinc finger gene, identifies regions containing postmitotic neurons during mammalian embryonic development, *J. Comp. Neurol.* 381:130.

Xuan, S., Baptista, C. A., Balas, G., Tao, W., Soares, V. C., and Lai, E., 1995, Winged helix transcription factor BF-1 is essential for the development of the cerebral hemispheres, *Neuron* 14:1141.

SIGNAL TRANSDUCTION

53

INVOLVEMENT OF PROTEIN KINASE C, P38 MAP KINASE AND ERK IN ARACHIDONIC ACID-STIMULATED SUPEROXIDE PRODUCTION IN HUMAN NEUTROPHILS

Charles S.T. Hii[*], Zhi H. Huang[*], Andrea Bilney[#], Kathryn Stacey[*], Andrew W. Murray[#], Deborah A. Rathjen[*], and Antonio Ferrante[*]

Department of Immunopathology[*], Women's and Children's Hospital, North Adelaide, South Australia 5006 and School of Biological Sciences[#], Flinders University of South Australia, Bedford Park, SA 5042, Australia.

INTRODUCTION

Arachidonic acid (AA) is a second messenger which is released by the action of phospholipase A_2 in activated cells[1]. When added exogenously, AA is biologically active in a wide variety of cells. For example, AA inhibits gap junctional permeability between adherent cells[2], stimulates superoxide production, degranulation and upregulates CD11b/CD18 expression in human neutrophils[3], stimulates insulin secretion[4], modulates the permeability of ion channels[5-7] and stimulates gene transcription[8]. However, the molecular mechanisms via which the actions of AA are mediated are poorly understood.

In in vitro assays, AA and other polyunsaturated fatty acids have been demonstrated to stimulate the activity of protein kinase C (PKC)[9,10]. We have previously demonstrated that AA and other polyunsaturated fatty acids also stimulate the activity of the extracellular signal-regulated protein kinase (ERK) in rat liver epithelial WB cells[11]. ERK and the closely related p38 and jun N-terminal kinase (JNK) are members of the mitogen-activated protein (MAP) kinase family of kinases[12]. These kinases are activated when cells are exposed to growth factors, cytokines and/or various forms of stress[12,13]. Activation of ERK, JNK and p38 MAP kinases are achieved through the dual phosphorylation of threonine and tyrosine residues in the TXY motif by upstream MAP kinase kinases. MAP kinases have been proposed to regulate a diverse range of biological functions, including cytokine production, cell growth, differentiation and death[12,13]. We now report that p38, ERK and PKC are involved in regulating AA-stimulated superoxide production in human neutrophils.

Eicosanoids and Other Bioactive Lipids in Cancer, Inflammation, and Radiation Injury, 4
Edited by Honn *et al.*, Kluwer Academic / Plenum Publishers, New York, 1999.

365

EXPERIMENTAL PROCEDURES

Cell preparation and kinase assays

Human neutrophils were isolated from healthy volunteers by the single-step method of Ferrante and Thong[14]. To assay for kinase activation/phosphorylation, neutrophils were incubated in the presence of AA for the times indicated. Incubations were terminated by removing the medium and washing the cells once with HBSS (4°C). Cells were lysed[15] and dual phosphorylation of p38 was determined by Western blotting using the anti-ACTIVE™ p38 antibody (Promega). For ERK assays, pelleted cells were sonicated, centrifuged (100,000g x 30 min) and ERK in the supernatant was partially purified[11]. Kinase activity was determined by the incorporation of ^{32}Pi into myelin basic protein[11]. The supernatants were also Western blotted using the anti-ACTIVE™ ERK antibody (Promega). To study PKC translocation, neutrophils were sonicated, centrifuged (30 min x 100,000g) and particulate fraction-associated PKC was extracted with 2% Triton X-100[11]. After centrifugation, the supernatants were Western blotted with isozyme-specific anti-PKC antibodies[11].

Superoxide production

Superoxide production was measured by monitoring the chemiluminescence resulting from the oxidation of lucigenin (9,9',-bis-N-methyl-acridinium nitrate). Briefly, neutrophils in HBSS were preincubated with DMSO (0.1% v/v), PD98059 and/or SB203580 for 45 min, and/or GF109203X for 10 min at 37°C. After preincubation, lucigenin in HBSS (250 µM) was added together with AA (20 µM) or ethanol (0.1% v/v). The cells were immediately placed in a water-jacketed luminometer chamber (Model 1250 with MultiUse software, Bio-Orbit Oy, Turku, Finland) and the resulting chemiluminescence (mV) recorded immediately after the addition of the lucigenin. Results are expressed as the maximum rate of superoxide production achieved (mV) during a 1 min period.

Statistical analysis

Where appropriate, differences were analysed by ANOVA or unpaired Student's T test and were considered significant when $p<0.05$.

RESULTS AND DISCUSSION

AA stimulated the dual phosphorylation of p38 MAP kinase in human neutrophils

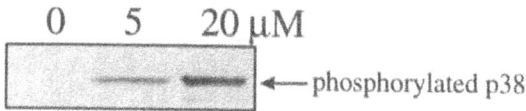

Fig 1. Stimulation of p38 dual phosphorylation by 20:4n-6
Neutrophils (3x10^7 cells in 30 ml HBSS) were incubated with AA for 4 min at 37°C, lysed and fractions were Western blotted with anti-ACTIVE™ p38 antibody.

(Fig 1). The effect of AA on p38 dual phosphorylation was transient, with maximal phosphorylation being observed at 5 min after the addition of AA (Fig 2). Stimulation of p38 dual phosphorylation by AA was not diminished by either nordihydroguaiaretic acid or indo-

Fig 2. Transient stimulation of p38 dual phosphorylation by AA
Neutrophils were incubated with AA (20 μM) for up to 30 min and soluble fractions were prepared and Western blotted using the anti-ACTIVE™ p38 antibody.

methacin (data not shown), suggesting that 20:4 ω6 *per se* exerted these effects.

AA also stimulated the dual phosphorylation of ERK. Thus, two bands with Mr of approx. 43-and 44-kDa were detected predominantly in samples from AA-stimulated cells

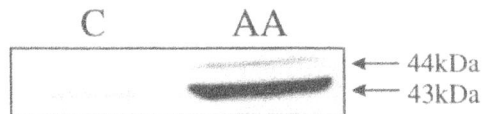

Fig 3. Stimulation of ERK phosphorylation by AA in human neutrophils
Neutrophils were stimulated with AA (20 μM, 10 min) and cytosolic fractions were prepared and Western blotted with the anti-ERK antibody.

(Fig 3). Kinase activity assays revealed that the effect of AA on ERK activity was more persistent that its effect on p38 phosphorylation (Fig 4).

Since polyunsaturated fatty acids have been demonstrated to activate PKC *in vitro*[9,10] and to stimulate the translocation of PKC in WB cells[11], we investigated whether PKC was involved in the activation of the p38 and ERK cascades by AA in neutrophils. Neutrophils contain PKC α, βI, βII, δ and ζ[16]. In unstimulated neutrophils, a substantial amount of PKC βII was detected in a particulate fraction and this was increased after incubation with AA (Fig 5). AA also caused a small amount of PKC α and βI to associate with the particulate fraction (Fig 5). Neither PKC δ nor ζ was detected in the particulate fraction (data not shown). An involvement of PKC in the activation of MAP kinases by

367

AA was confirmed by the observation that GF109203X, an inhibitor of classical PKC isozymes[17], attenuated the stimulatory effect of AA on ERK activity by >85% (data not shown). GF109203X also

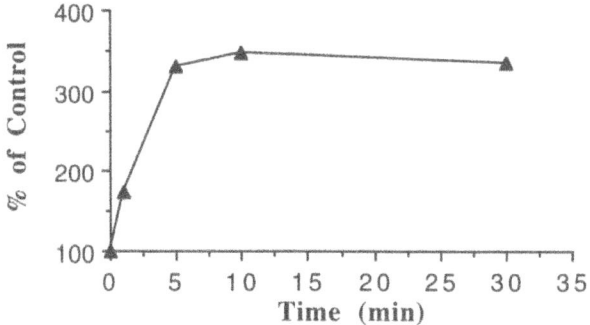

Fig 4. Kinetics of AA-stimulated ERK activity in human neutrophils Neutrophils were incubated with AA for the times indicated. ERK in cytosolic fractions was partially purified and the activity assayed. Results are means of duplicate assays.

caused a modest reduction in the ability of AA to stimulate the appearance of dual-phosphorylated p38 (data not shown).

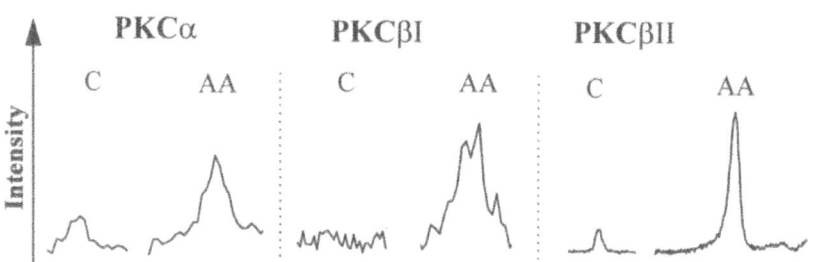

Fig 5. AA stimulated the translocation of PKCα, βI and βII Neutrophils were stimulated with AA (20 μM) for 3 min and fractions were prepared and Western blotted using anti PKC isozyme-specific antibodies. Immunoreactive bands were subjected to densitometry scanning.

The above data suggest that p38, ERK and PKC may play a role in mediating the effect of AA on superoxide production. To confirm this, cells were preincubated with SB203580, PD98059 and GF109203X either alone or in combination, to inhibit the activity of p38, MEK (MAP kinase/ERK kinase) and PKC, respectively, before being challenged with AA. These inhibitors have been demonstrated to be relatively specific in

their actions at the concentrations being tested in this study[18,19]. Each inhibitor partially inhibited AA-stimulated superoxide production in a concentration-dependent manner (Fig 6). None of the inhibitors were able to totally inhibit AA-stimulated superoxide production. At low inhibitor concentrations, these inhibitors, when added simultaneously, inhibited AA-stimulated superoxide production in an additive manner (Fig 7). The maximum inhibition achievable was approximately 66% (Fig 7), even when the concentration of each of the inhibitors was
increased by 2 fold (data not shown)

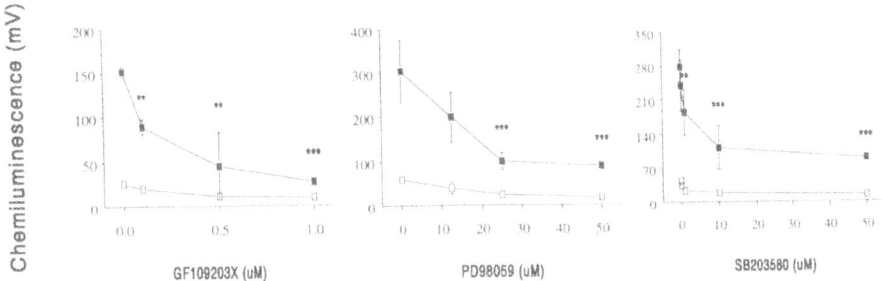

Fig 6. Inhibition of AA-stimulated superoxide production by kinase inhibitors Neutrophils were preincubated with PD98059, SB203580 or GF109203X and superoxide production in response to AA was determined. Results are means ± 3 determinations. Significance of difference between the presence and absence of inhibitor: ** p<0.01; *** p<0.001

These data demonstrate that AA stimulated the phosphorylation and/or activity of p38, ERK and PKC in human neutrophils. Activation of p38 and ERK was dependent, to some degree

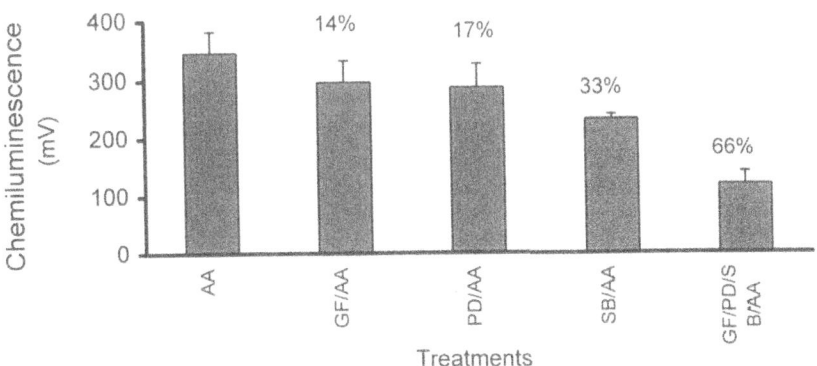

Fig 7. Inhibition of superoxide production by a combination of kinase inhibitors Neutrophils were preincubated with PD98059 (12.5 μM), SB203580 (5 μM) and GF109203X (0.2 μM) and superoxide production in response to AA was determined. Results are means ± 9 determinations. Numbers on top of the bars represent % inhibition with respect to the response observed in the absence of inhibitors

on PKC activation. Our data strongly suggest that p38, ERK and PKC mediate, at least in part, the actions of AA on neutrophil superoxide production. The inability of SB203580, PD98059 and GF109203X to totally inhibit AA-stimulated superoxide production suggests that other signalling pathways are also involved in regulating superoxide production.

REFERENCES

1. E.A. Dennis, The growing phospholipase A_2 superfamily of signal transduction enzymes, *TIBS*. 22: 1 (1997).
2. C.S.T. Hii, A. Ferrante, S. Schmidt, D.A. Rathjen, B.S. Robinson, A. Poulos, and A.W. Murray, Inhibition of gap junctional communication by polyunsaturated fatty acids in WB cells: evidence that connexin 43 is not hyperphosphorylated, *Carcinogenesis*. 16: 1505 (1995).
3. A. Ferrante, C.S.T. Hii, Z.H. Huamg and D.A. Rathjen, Regulation of neutrophil function by fatty acids, in: The Neutrophils: New Outlook for the Old Cell, D. Gabrilovich, ed., Imperial College Press, In Press (1998).
4. A.M. Band, P.J. Jones, and S.L. Howell, Arachidonic acid-induced insulin secretion from rat islets of Langerhans, *J. Mol. Endocrinol*. 8: 95 (1992).
5. S.V. Smirnov and P.I. Aaronson, Modulatory effects of arachidonic acid on the delayed rectifier K^+ current in rat pulmonary arterial myocytes. Structural aspects and involvement of protein kinase C, *Circ. Res*. 79: 20 (1996).
6. S.J. Wieland, Q. Gong, H. Poblete, J.E. Fletcher, L.Q. Chen and R.G. Kallen, Modulation of human muscle sodium channels by intracellular fatty acids is dependent on the channel isoform, *J. Biol. Chem*. 271: 19037 (1996).
7. T.E. DeCoursey and V.V. Cherny, Potential, pH, and arachidonic acid gate H^+ ion currents in human neutrophils, *Biophys. J*. 65: 1590 (1993).
8. S.D. Clarke. and D.B. Jump, Dietary polyunsaturated fatty acids regulation of gene transcription, *Prog. Lipid Res*. 32: 139 (1993).
9. K. Murakami, K. and A. Routtenberg, Direct activation of purified protein kinase C by unsaturated fatty acids (oleate and arachidonate) in the absence of phospholipids and Ca^{2+}, *FEBS Lett*. 192: 189 (1985).
10. L.C. McPhail, C.C. Clayton and R. Snyderman, A potential second messenger role for unsaturated fatty acids: activation of Ca^{2+} dependent protein kianse, *Science*. 224: 622 (1984).
11. C.S.T. Hii, A. Ferrante, Y. Edwards, Z.H. Huang, P.J. Hartfield, D.A. Rathjen, A. Poulos and A.W. Murray, Activation of mitogen-activated protein kinase by arachidonic acid in rat liver epithelial WB cells by a protein kinase-dependent mechanism, *J. Biol. Chem*. 270: 4201 (1995).
12. J.M. Kyriakis and J. Avruch, Sounding the alarm: protein kinase cascades activated by stress and inflammation, *J. Biol. Chem*. 271: 24313 (1996).
13. R. Seger and E.G. Krebs, The MAP kinase cascade, *FASEB J*. 9: 726 (1995).
14. A. Ferrante and Y.H. Thong, Separation of mononuclear and polymorphonuclear leukocytes from human blood by the one-step Hypaque-Ficoll method is dependent on blood column height, *J. Immunol. Methods* 48: 81 (1982).
15. N. Sakata, H.R. Patel, N. Terada, A. Aruffo, G.L. Johnson and E.W. Gelfand, Selective activation of c-jun kinase mitogen-activated protein kinase by CD40 in human B cells, *J. Biol. Chem*. 270: 30823 (1995).
16. J.I. Smallwood and S.E. Malawista, Protein kinase C isotypes in human neutrophil cytoplasts, *J. Leukoc. Biol*. 51: 84 (1992).
17. D. Toullec, D. Pianetti, H. Coste, P. Bellevergue, Grand, T. Perret, M. Ajakane, V. Baudet, P. Boissin, F. Lorielle et al, The bisindolylmaliemide GF109302X is a potent and selective inhibitor of protein kinase C,*J. Biol. Chem*. 266: 15771 (1991).
18. D.R. Alessi, A. Cuenda, P. Cohen, D.T. Dudley and A.R. Saltiel, PD98059 is a specific inhibitor of the activation of miyogen-activated protein kinase kinase *in vitro* and *in vivo*, *J. Biol. Chem*. 270: 27489 (1995).
19. A Cuenda, J. Rouse, Y.N. Doza, R. Meier, P. Cohen, T.F. Gallagher, P.R. Young and J.C. Lee, SB203580 is a specific inhibitor of a MAP kinase homologue which is stimulated by cellular stresses ansd interleukin 1, FEBS Lett. 346: 229 (1995).

54

13(S)-HPODE AUGMENTS EPIDERMAL GROWTH FACTOR SIGNAL
TRANSDUCTION BY ATTENTUATING EGF RECEPTOR DEPHOSPHORYLATION

Wayne C. Glasgow[*], Rutai Hui, Shiranthi Jayawickreme, Julie
Angerman-Stewart, Bing-Bing Han, and Thomas E. Eling

[*]Division of Basic Medical Sciences, Mercer University School
of Medicine, Macon, Georgia 31207 and Laboratory of Molecular
Carcinogenesis, NIEHS/NIH, Research Triangle Park, North
Carolina 27709

INTRODUCTION

The binding of epidermal growth factor (EGF) to its specific cell surface receptor (EGFR) activates the intrinsic receptor tyrosine kinase activity. The EGFR phosphorylates other key signaling proteins initiating a cascade of phosphorylation events that culminate in activation of transcription factors involved in mitogenesis. In Syrian hamster embryo (SHE) fibroblasts, the addition of EGF to serum-starved quiescent cells initiates cell proliferation and stimulates the metabolism of linoleic acid to 13(S)-hydroperoxyoctadecadienoic acid (HpODE)[1]. The biosynthesis of 13(S)-HpODE by an apparent 15-lipoxygenase is regulated by the EGFR tyrosine kinase, suggesting that formation of this lipid mediator is a component of the EGF mitogenic signaling pathway in these cells[2]. Blocking 13(S)-HpODE formation with lipoxygenase inhibitors results in a dramatic attenuation of EGF-dependent DNA synthesis[1]. Other SHE cell mitogenic factors, including fibroblast growth factor, insulin, and platelet-derived growth factor, do not activate linoleic acid metabolism suggesting that simulation of this pathway is unique to EGF[3].

Addition of exogenous 13(S)-HpODE to quiescent cells enhances EGF-dependent DNA synthesis in a SHE cell phenotype which expresses tumor suppressor genes (supB[+]), but has no effect in a neoplastic variant that lacks tumor suppressor genes (supB[-])[1]. Further structural characterization of the lipid-dependent enhancement of mitogenesis indicated that 13(S)-

Eicosanoids and Other Bioactive Lipids in Cancer, Inflammation, and Radiation Injury, 4
Edited by Honn *et al.*, Kluwer Academic / Plenum Publishers, New York, 1999.

371

HpODE/HODE has exclusive activity[4]. For example, the 13(R)-HODE enantiomer and the analogous 15-lipoxygenase metabolites of arachidonic acid, 15(S)-HpETE/HETE, were not active. This finding suggests a specific interaction of 13(S)-HpODE with either a specific receptor or with a intracellular signal transduction protein of the EGF signaling pathway. Potentially, this site of interaction has been deleted or mutated in the supB⁻ cell line.

In studies designed to define the molecular basis for these observations, we examined the relationship between 13(S)-HpODE enhancement of EGF-dependent mitogenesis and 13(S)-HpODE modulation of critical tyrosine phosphorylation events involved in EGFR signal transduction. In this report, we found 13(S)-HpODE to specifically increase the magnitude and duration of EGF-dependent phosphorylation of EGFR, GTPase activating protein (GAP), and mitogen activated protein (MAP) kinase in supB⁺ cells, but not in supB⁻ cells. The differential response to 13(S)-HpODE between supB⁺ and supB⁻ cells is not due to differences in EGFR number, direct modulation of EGFR kinase activity, or changes in the rate of EGFR internalization and degradation. However, in studies of EGFR dephosphorylation in intact cells, 13(S)-HpODE stimulated a concentration dependent attenuation of the rate of EGFR dephosphorylation in supB⁺ cells, with no effect on EGFR dephosphorylation in supB⁻ cells.

EXPERIMENTAL PROCEDURES

Experiments were performed with SHE cell lines 10WsupB⁺8 and 10WsupB⁻1, as described previously[1]. Cells were maintained at 37°C in a humidified 5% CO_2, 95% air atmosphere. The cells were cultured in IBR containing 10% fetal bovine serum and gentamicin (10 μg/ml). Trypsin (0.05%) was used to subculture the cells. Both variants were grown under identical conditions and passaged at 70-80% confluence. Passage 10-17 was used in these experiments.

After reaching 70-80% confluence on 100-mm dishes, cells were serum-deprived for 20 h to synchronize cell cycles in G_0. All experiments were done in serum-free media at 37°C, unless stated otherwise. Cells were treated with 13(S)-HpODE (10 nM to 10 μM) for 1 h prior to the addition of EGF (10 ng/ml). After stimulation by EGF, cells were harvested at various stated time points.

Western blot analyses using anti-phosphotyrosine, anti-EGFR, anti-GAP, and anti-MAP kinase antibodies were performed as described[5]. Immunoprecipitation experiments and the MAP kinase activity assay were performed as described previously[5]. The method of Griswold-Prenner et al. was used for the EGFR kinase activity assay[6]. EGFR down-regulation was measured using the procedure described by Sorkin et al.[7]

To measure EGFR dephosphorylation in intact cells, cells were made quiescent by serum deprivation and then incubated with 13(S)-HpODE for 1 h at 37°C. EGF (10 ng/ml) was added and the cells were incubated for 1 min of treatment. The EGFR kinase activity was blocked by the addition of 50 μM methyl 2,5-dihydroxycinnamate, a protein tyrosine kinase inhibitor, and the cells were harvested at the times indicated. To harvest, the cells were placed on ice, media was removed, and the cells were washed with ice-cold PBS containing sodium orthovanadate (2 mM). The media was removed and the cells lysed immediately with boiling sample buffer and boiled for 5 min. Sample buffer consisted of 5 mM sodium phosphate, pH 6.8, 2% sodium dodecyl sulfate, 0.1 M dithiothreitol, 10% glycerol, 5% β-mercaptoethanol, 0.2% bromphenol blue, 1 mM sodium orthovanadate, and 1 mM sodium fluoride. The lysate was dispersed with a 25-guage needle and centrifuged. Supernatants were analyzed by Western blots with the anti-phosphotyrosine monoclonal antibody. Gels were loaded with lysate from 4×10^5 cells.

RESULTS AND DISCUSSION

The phosphorylation and dephosphorylation of the EGFR was examined in both supB$^+$ and supB$^-$ cell lines ± 13(S)-HpODE. The EGFR was rapidly tyrosine-phosphorylated in response to EGF with maximum phosphorylation observed at 1-2 min after the addition of EGF with a rapid decline in the signal. The duration of EGF-dependent tyrosine phosphorylation appears to be greater in the supB$^-$ variant when compared to the supB$^+$ cells. In the supB$^+$ cells, 13(S)-HpODE (1 μM) clearly stimulated an increase in autophosphorylation of EGFR at the 1-10 min time points. In contrast, 13(S)-HpODE did not alter the phosphotyrosine state of the EGFR in supB$^-$ cells. In addition to modulating EGFR tyrosine phosphorylation, the treatment of supB$^+$ cells with 13(S)-HpODE enhanced the phosphotyrosine content of a distinct band at 120 kDa (immunoreactive with anti-GAP antibodies). This biological effect of the linoleate metabolite on EGFR phosphorylation can be demonstrated even more dramatically in experiments conducted at room temperature rather than 37°C. Shifting to lower temperatures slows down the rate of EGFR dephosphorylation and ligand-induced receptor internalization and degradation. In this analysis, 13(S)-HpODE increased the magnitude and duration of the tyrosine phosphorylation of both EGFR and GAP. Western immunoblots using antibodies specific for GAP indicated that the level of this protein did not change during the course of the experiment. Thus, the 13(S)-HpODE-stimulated changes noted with the anti-phosphotyrosine immunoblots appear to be due to an effect on the tyrosine phosphorylation state of GAP and not due to alterations in protein levels.

Initiation of the EGF signal transduction pathway leads to the downstream activation of MAP kinases, which are likely to mediate many of the nuclear activation events associated with EGF-dependent mitogenesis. We thus measured tyrosine phosphorylation of p42 and p44 MAP kinase stimulated by EGF in the presence and absence of 13(S)-HpODE (1 μM). A biphasic response to EGF was observed, with initial tyrosine phosphorylation noted at early time points (2-30 min) followed by a down-regulation of the signal. A second phase of MAP kinase tyrosine phosphorylation occurs at 90-100 min. The addition of 13(S)-HpODE enhanced the magnitude and prolonged the tyrosine phosphorylation of MAP kinase at all time points examined in the supB$^+$ cell line. 13(S)-HpODE did not alter phosphorylation of either p42 or p44 in the supB$^-$ variant. To determine if the observed changes in the phosphotyrosine state of MAP kinase correlates with changes in MAP kinase activity, the activity was measured in immunoprecipitates from supB$^+$ SHE cells. Quiescent cells were treated with EGF for 2 or 90 min ± 13(S)-HpODE (pre-treatment for 1 h prior to EGF). The MAP kinase activity of the immunoprecipitates was determined by measuring the incorporation of $[\chi\text{-}^{33}P]ATP$ into myelin basic protein. At the 2 min time point, the combination of EGF and 13(S)-HpODE increased

MAP kinase activity 1.5-fold compared with corresponding cells incubated with EGF alone. After 90 min of treatment, a greater than 2-fold increase in MAP kinase activity was found in the 13(S)-HpODE treated cells.

One possible explanation for the up-regulation of the EGF signaling pathway in the supB[+] cells by 13(S)-HpODE is a selective increase in the EGFR kinase activity. EGFR kinase activity in SHE cell lysates was measured by estimating the incorporation of $[\chi\text{-}^{33}P]ATP$ into angiotensin II. Little or no difference was observed in the EGFR kinase activity of the two SHE cell lines. We also did not observe any direct effects *in vitro* of 13(S)-HpODE on modulation of EGFR kinase activity.

The linoleic metabolite could also act by altering the rate of EGFR internalization. Using the method described previously[7], we found the rate and extent of receptor internalization to be very similar in supB[+] and supB[-] cells. Moreover, pre-treatment of cells with 13(S)-HpODE did not change the rate and extent of EGFR internalization and degradation.

Since the 13(S)-HpODE effect to augment the EGFR signal was not due to direct increases in kinase activity or to changes in receptor degradation pathways, we focused on examining modulation of EGFR dephosphorylation. The rate of EGFR dephosphorylation was estimated using a modification of the assay described by Bohmer *et al.*[8] EGFR dephosphorylation occurred very rapidly in the supB[+] cell line reaching completion after 2 min. In contrast, the dephosphorylation of the EGFR was significantly slower in the supB[-] variant reaching completion 6-8 min after EGF binding. The effect of 13(S)-HpODE on EGFR dephosphorylation was examined following preincubation of the cells for 1 h with the linoleate metabolite. As shown in Figure 1, 13(S)-HpODE significantly altered the rate of receptor dephosphorylation in the supB[+] cells. In the presence of 13(S)-HpODE, EGFR dephosphorylation was not complete until 7 min after EGF addition, while in the absence of 13(S)-HpODE, dephosphorylation was complete by 3 min. Treatment with increasing concentrations of 13(S)-HpODE (10 nM to 10 µM) resulted in a dose-dependent increase in the intensity of EGFR tyrosine phosphorylation. No detectable effect on EGFR dephosphorylation was observed in the supB[-] cell line throughout the entire 13(S)-HpODE dose range.

13(S)-HpODE appears to up-regulate the EGF pathway in supB[+] cells by inhibiting dephosphorylation of the EGFR, resulting in an increased activation of the EGFR tyrosine kinase

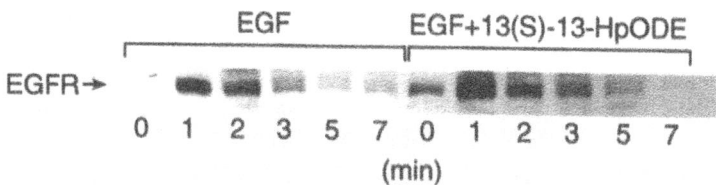

Figure 1. Effects of 13(S)-HpODE on dephosphorylation of the EGFR in supB[+] cells. After growth arrest by serum deprivation, cells were treated for 1h at 37°C with 1 µM 13(S)-HpODE. Cells were then stimulated with EGF (10 ng/ml) for 1 min and then incubated with 50 µM methyl 2,5-dihydroxycinnamate to stop further tyrosine kinase activity. At the indicated times the cells were harvested and tyrosine phosphorylation of EGFR was estimated by an anti-phosphotyrosine Western immunoblot (as described in Exprerimental Procedures section). The results are representative of three different experiments.

and downstream kinase signaling. The lack of a response to 13(S)-HpODE in the supB⁻ variant and the differences in the EGF-dependent tyrosine phosphorylation in the two SHE cell lines suggest that during neoplastic progression of the SHE cells, gene deletions or mutations of EGF signaling proteins have occurred with the loss of tumor suppressor phenotype. One of these proteins appears to be the site or "receptor" responsible for the modulation of the EGF signaling pathway by 13(S)-HpODE. The mechanism for the attenuation of EGFR dephosphorylation by 13(S)-HpODE is not known and is made more complex by a poor general understanding of EGFR dephosphorylation. Negative regulation of growth factor receptor by receptor-directed protein tyrosine phosphatases has been suggested. Differences in functional activity of tyrosine phosphatases specific for the EGF pathway could provide an explanation for the phenotypic difference between the two SHE cell lines and may be a target for the enhancement of EGFR signaling by 13(S)-HpODE. Our current hypothesis is that 13(S)-HpODE modulates the interaction of a tyrosine phosphatase with the EGFR. On-going studies are designed to test this hypothesis.

REFERENCES

1. Glasgow, W.C., Afshari, C.A., Barrett, J.C., and Eling, T.E. (1992) *J. Biol. Chem.* **267**: 10771-10779
2. Glasgow, W.C., Hill, E.M., McGown, S.R., Tomer, K.B., and Eling, T.E. (1996) *Mol. Pharmacol.* **49**: 1042-1048
3. Angerman-Stewart, J., Eling, T.E., and Glasgow, W.C. (1995) *Arch. Biochem. Biophys.* **318**: 378-386
4. Glasgow, W.C. and Eling, T.E. (1994) *Arch. Biochem. Biophys.* **311**: 286-292
5. Glasgow, W.C., Hui, R., Everhart, A.L., Jayawickreme, S., Angerman-Stewart, J., Han, B.-B., and Eling, T.E. (1997) *J. Biol. Chem.* **272**: 19269-19276
6. Griswold-Prenner, I., Carlin, C.R., and Rosner, M.R. (1993) *J. Biol. Chem.* **268**: 13050-13054
7. Sorkin, A., Helin, K., Waters, C.M., Carpenter, G., and Beguinot, L. (1992) *J. Biol. Chem.* **267**: 8672-8678
8. Bohmer, F.D., Bohmer, A., Obermeier, A., and Ullrich, A. (1995) *Anal. Biochem.* **228**: 267-273

55

ROLE OF LIPOXYGENASE IN THE REGULATION OF GLUCOSE TRANSPORT IN AORTIC VASCULAR CELLS

Shlomo Sasson, Ana Davarashvili, and Reuven Reich.

Department of Pharmacology, School of Pharmacy, Faculty of Medicine, Hebrew University, Jerusalem, IL-91120, Israel

INTRODUCTION

Hyperglycemia is closely involved in early and late metabolic, morphologic and pathologic changes leading to the atherosclerosis in diabetes. High glucose levels inhibit replication of vascular endothelial cells (VEC)[1-2], but stimulate the proliferation of vascular smooth muscle cells (VSMC)[1-5]. Hyperglycemia has also been shown to modify extracellular matrix components in blood vessels and to attenuate their production by the vascular cells[6-10]. Some of these effects result from non-enzymatic glycation of extracellular matrix components and cell-surface proteins by extracellular glucose[11-12]. Further, intracellular glucose, or its metabolites, affect cell function by modifying cellular metabolism and/or glycation of intracellular proteins[13-15] Hyperglycemia also increases oxygen free radical formation in cells and plasma of diabetic patients. These can cause lipid oxidation and damage proteins through cross-linking and fragmentation and be involved in the initiation and progression of the atherosclerotic process [16].

Glucose transport in vascular cells

Glucose entry into animal cells is mediated by members of the facilitated diffusion glucose transporters that are ubiquitously expressed in mammalian cells[17]. Molecular cloning techniques have identifies more than 6 members of glucose transporters with distinct tissue distribution, biochemical properties and modes of regulation[17-19]. We have shown earlier that VSMC and

Eicosanoids and Other Bioactive Lipids in Cancer, Inflammation, and Radiation Injury, 4
Edited by Honn *et al.*, Kluwer Academic / Plenum Publishers, New York, 1999.

377

VEC express predominantly glucose transporter-1 (GLUT-1)[20]. This protein, quantified by Western blot analysis, is more abundant in VSMC than in VEC. (GLUT-1) is the most widely distributed and is expressed to some extent in most tissues[21]. Among its regulators are: oncogenes, tumor promoters, growth factors, ATP, glucose and insulin[17].

We have shown previously a differential autoregulation of glucose transport and transporters in aortic VEC and VSMC[20]. Glucose transport activity in VSMC is inversely and reversibly regulated by glucose. Exposure of bovine and human VSMC to high glucose concentration (20 mM) decreased the $Vmax$ for hexose uptake, whereas the Km remained unchanged. It was decreased by nearly 50% in VSMC when medium glucose was elevated from 5.5 to 22.2 mM glucose. The hexose transport system in VEC exhibited lower hexose transport capacity compared with VSMC and showed no adaptation to changes in ambient glucose upon acute exposure (24 hr) to high glucose containing medium. Recently, we have examined the effect of chronic exposure of vascular cells to hyperglycemic conditions. The cells were exposed to 20 mM glucose for 3-5 generations in culture. VSMC exhibited the same down-regulatory process of glucose transport as observed under acute exposure to high glucose levels. Unlike VEC exposed acutely to hyperglycemic conditions, when maintained chronically exposed to high glucose levels these cells developed the autoregulatory mechanism and reduced the rate of glucose transport like VSMC. The alterations in total content and plasma membrane-associated GLUT-1 in both cell type, correlate with the changes observed in transport activity. We suggest that the ability of VSMC and VEC to downregulate the rate of glucose transport in response to hyperglycemia serves as a cellular protective mechanism against the adverse affects of increased intracellular glucose.

Arachidonic acid metabolism and diabetes

Vascular disease in diabetes has been linked to enhanced production of eicosanoids [22]. Several reports indicate that arachidonic acid metabolites of the lipoxygenase (LOX) pathway play a role in diabetes-related complications[23]. Antonipillai et al.[24] showed that the urinary excretion rate of 12-HETE in Type II diabetics was higher than in nondiabetic patients. Similarly, the over-all production of 12-HETE was also found to be increased in streptozotocin-diabetic rats[25]. Wang and Powell found increased levels of HETEs in aortae of atherosclerotic rabbits [26]. 12-LOX products activate oncogenes, including *c-fos* and *c-jun* and specific isoforms of PKC in VSMC[27-28]. Since HETEs are mitogenic, proinflamatory and vasoconstrictive the increased production of 12-HETE early in diabetes mellitus may play a role in the etiology of diabetes vascular disease [29,30].

There are three mammalian lipoxygenases that can oxygenate carbon 5,12 or 15 in arachidonic acid, thus, termed as 5-, 12- and 15-LOX, generating 5-, 12- and 15-HPETE, respectively. All LOXs contain a single non-heme iron per molecule of the enzyme. It is assumed that the Fe(III) enzyme must be oxidized first to Fe(II) form to become activate [31]. This property of LOXs makes them susceptible to inactivation by reactive radicals such as nitric oxide, hydroxyl radicals or superoxide (O_2^-) by reducing the Fe(III) to the Fe(II) form [32,33]. On the other hand, hydrogen peroxide (H_2O_2) stimulates LOXs activity by oxidizing the Fe(II) form to Fe(III)[33,34]. In addition, some small molecular weight iron chelators inhibit LOXs by interacting with their iron center[35]. Two forms of 12-LOX were identified in mammalian tissues: the platelet type 12-

LOX and the leukocyte type 12-LOX[36,37]. Both enzymes share 65% amino acid homology. The human 15-LOX shares 86% homology to the porcine type 12-LOX[38].

Several investigators have found a direct association between hyperglycemia-induced alterations in VSMC function and 12-LOX and its product, 12-HETE. Natarajan et al.[39] and Kim et al[40] showed that human and porcine VSMC express the leukocyte type 12-LOX, but not the platelet type 12-LOX. No 15-LOX mRNA has been detected in human or porcine SMC and human VEC [40-41]. The basal expression of 12-LOX mRNA is very low in porcine aortic SMC grown and maintained at 5.5 mM glucose but is clearly enhanced in cells exposed to high glucose levels [39,42]. Also, the content of cell-associated 12-HETE is significantly higher in SMC under hyperglycemic conditions in comparison to those under normoglycemic conditions [39, 42]. Further, VEC cultured under hyperglycemic conditions produce increased amounts of HETEs [43].

Despite these findings on the effects of arachidonic acid metabolites and their role in diabetes-related complication in vascular cells[44], little is known about their role in the autoregulation of glucose transport in vascular cells. Since this autoregulatory process seems to be a natural defense mechanism we decided to study the role of arachidonic acid metabolites on the glucose transport system in vascular cells under normal and hyperglycemic conditions.

RESULTS

We studied the effects of cyclooxygenase and lipoxygenase inhibition on hexose transport in VEC and VSMC[45]. Indomethacin, an inhibitor of cyclooxygenase I, and nimesulide, an inhibitor of the inducible cyclooxygenase II [46], had no effect on the rate of hexose transport in both types of cells (data not shown). However, esculetin, an inhibitor of the lipoxygenase pathway increased the rate of hexose transport in both VEC and VSMC in a dose- and time-dependent manner. Table 1 shows the glucose-induced down-regulation of glucose transport in VSMC and VEC exposed chronically to 20 mM glucose, where the rate of hexose transport was reduced by 54 and 73%, respectively, in comparison to cells under 5 mM glucose. Inhibition of LOXs with esculetin increased the rate of hexose transport ~2-fold in both types of cells maintained at 20 mM glucose, while its effect on cells exposed to normal glucose levels was minimal. Half maximal and maximal effects of esculetin were observed at 50 and 100μM, within 8 and 12 hrs, respectively (data not shown).

Among the three lipoxygenase enzymes, 5-LOX was not involved in this action of esculetin, since a specific inhibitor of this enzyme, SC-41661, and an inhibitor of its intracellular translocation, MK-886, produced no noticeable effects on the rate of hexose transport (Fig. 1). Therefore, 5-HETE and subsequent leukotrienes synthesis seem not to be involved, but inhibition of 12-HETE and/or 15-HETE production by esculetin appears to increase the rate of glucose transport in these cells. In parallel to its effects on the transport capacity, esculetin increased the total content of GLUT-1 protein and its abundance in the plasma membrane of both types of cells (Fig. 2).

Table 1: Effects of glucose and esculetin on hexose transport in VSMC and VEC

| | dGlc uptake | | |
	CONTROL	+ESCULETIN	% INCREASE
VSMC			
5 mM glucose	2.16±0.09	2.80±0.09	129
20 mM glucose	0.99±0.07	1.85±0.05	187
VEC			
5 mM glucose	0.41±0.04	0.45±0.02	111
20 mM glucose	0.11±0.01	0.25±0.03	230

Bovine aortic VSMC and VEC were maintained in culture for 5 generations under 5 or 20 mM glucose. The cells were incubated for 16 hrs in the absence or the presence of 50μM esculetin and the rate of [^{3}H]2-deoxyglucose uptake was determined, as described[20]. Uptake values are given in nmol/10^6 cells/min units, Mean±SEM (n=3). In these experiments control cells were also incubated with 5 mM D-glucose+15 mM L-glucose, to rule out the possibility that the effects observed at 20 mM D-glucose resulted from increased osmotic pressure in the cell cultures. The rate of 2-deoxyglucose uptake measured in these cells were similar to that observed in cells exposed to 5 mM D-glucose only (data not shown).

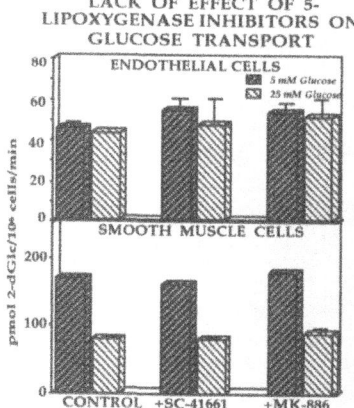

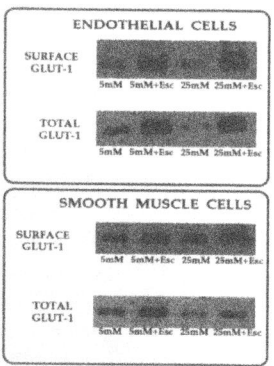

Figure 1. Bovine aortic SMC and EC were incubated at 5 and 25 mM D-glucose for 24 hrs. The incubation medium was then changed to fresh medium without or with 50 µM SC-41661 (a 5-LOX inhibitor) or with 50 µM MK-886 (a 5-LOX translocation inhibitor). The rates of 2-deoxyglucose uptake were determined after 24 hr incubation. Mean±SEM, (n=3).

Figure 2: Bovine aortic SMC and EC were incubated at 5 and 25 mM D-glucose ± 100 µM esculetin for 24 hrs. Total GLUT-1 content was determined by Western blotting in total cell membrane pellets. of Cell surface GLUT-1 was determined by cell surface biotinylation[47] followed by PAGE and Western blotting with rabbit anti GLUT-1 antiserum.

CONCLUSIONS

Our results indicate that HETE metabolites of 12- and/or 15-LOX are involved in the autoregulation of the glucose transport system in vascular cells. Inhibition of these enzymes results in increased cellular content and plasma membrane abundance of GLUT-1.
The fact that HETEs production is increased in tissues of type II diabetes patients in general, and in vascular cells in particular, establishes the hypothesis that glucose-dependent regulation of the expression and/or activity 12- and 15-LOX is an important factor in a cellular protective mechanism that limits cell damage in face of ambient hyperglycemia.

Acknowledgments

S. Sasson and R. Reich are members of the David R. Bloom Center for Pharmacy at the Hebrew University of Jerusalem. This work was supported by a grant from the David R. Bloom Center for Pharmacy at the Hebrew University of Jerusalem.

REFERENCES

1. Stout RW, Glucose inhibits replication of cultured human endothelial cells, *Diabetologia* ,23:436-439,(1982).
2. Lorenzi M, Cagliero E, Toledo S, Glucose toxicity for human endothelial cells in culture: delayed replication, disturbed cell cycle, and accelerated death, *Diabetes* ,34:102-110, (1985).
3. Kaiser N, Tur-Sinai A, Hasin M, Cerasi E, Binding, degradation and biological activity of insulin in vascular smooth muscle cells, *Am. J. Physiol.* , 249:E292-E299, (1985).
4. Pfeifle B, Ditschuneit HH, Ditschuneit H, Insulin as a cellular growth regulator of rat arterial smooth muscle cells in vitro, *Horm. Metab. Res.* ,12:381-385 , (1980).
5. Natarajan R, Gonsales N, Xu L, Nadler JL, Vascular smooth muscle cells exhibit increased growth in response to elevated glucose, *Biochem. Biophys. Res. Commun.* , 187:552-560, (1992).
6. Sank A, Wei D, Reid J, et al, Human endothelial cells are defective in diabetic vascular disease, *J. Surg. Res.* , 57:647-653, (1994).
7. Cagliero E, Maiello M, Boeri D, Roy S, Lorenzi M, Increased expression of basement components in human endothelial cells cultured in high glucose, *J. Clin. Invest.* 82:735-738, (1988).
8. Osterby R. . In: Rifkin H, Porte D, eds. Diabetes Mellitus: Theory and Practice. Amsterdam: Elsevier, 1990:220-233.
9. Kreisberg JL, Hyperglycemia and microangiopathy. Direct regulation by glucose of microvascular cells, *Lab. Invest.* , 67:416-426, (1992).
10. Bucala R, Vlassara H, Advanced glycosylation end products in diabetic renal and vascular disease, *J. Kidney Dis.* , 26:875-888, (1995).
11. Danne T, Spiro MJ, Spiro RG, Effect of high glucose on type IV collagen production by cultured glomerular epithelial, endothelial, and mesangial cells, *Diabetes* 42:170-177, (1993).
12. Ono H, Umeda F, Inoguchi T, Ibayashi H, Glucose inhibits prostacyclin production by cultured aortic endothelial cells, *Thromb. Haemost.* , 60:174-177, (1991).
13. Baumgartner-Parzer SM, Wagner L, Pettermann M, Grillari J, Gessl A, Waldhausl W, High glucose triggered apoptosis in cultured endothelial cells, *Diabetes* 44:1323-1327, (1995).
14. Brownlee M, Non-enzymatic glycosylation of macromolecules: Prospects of pharmacological modulation, *Diabetes* , 41:57-60, (1992).
15. Brownlee M, Glycation products and the pathogenesis of diabetic complications, *Diabetes Care* 15:1835-1843, (1992).

16. Brownlee M, Cerami A, Vlassara H, Advanced glycosylation end products in tissue and biochemical basis of diabetic complication, *N. Eng. J. Med.* , 318:1315-1322, (1988).

17. Mueckler M, facilitative glucose transporters, *Eur. J. Biochem.* 219:713-725, (1994).

18. Gould GW, Bell GI, Facilitative glucose transporters: an expanding family, *Trends Biochem. Sci.* , 15:18-22, (1990).

19. Davis A, Meeran K, Cairns MT, Baldwin S A, Peptide-specific antibodies as probes of the orientation of glucose transporter in human erythrocyte membrane, *J. Biol. Chem.* , 262:9347-9352, (1987).

20. Kaiser N, Sasson, S, Feener E, Boukobza-Vardi N, Higashi S, Moller DE, Davidheiser S, Przyblyski RJ, King GL. Differential regulation of glucose transport and transporters by glucose in vascular endothelial and smooth muscle cells. *Diabetes* 42:80-89 (1993).

21. Thorens B, Facilitated glucose transporters in epithelial cells, *Annu. Rev. Physiol.* 55:591-608, (1993).

22. De Rubertis F, Craven P, Activation of protein kinase C in glomerular cells in diabetes. Mechanism and potential links to the pathogenesis of diabetic glomerulopathy, *Diabetes* , 43:1-8, (1994).

23. Dai FX, Deiderich A, Skopec J, Deiderich DJ, Diabetes-induced endothelial dysfunction in streptozotocin-treated rats: role of prostaglandin endoperoxides and free radicals, *Am. Soc. Nephrol.* , 4:1327-1336, (1993).

24. Antonipillai I, Nadler J, Vu EJ, Bughi S, Natarajan R, Horton R, A 12-lipoxygenase product, 12-hydroxyeicosatetraenoic acid is increased in diabetes with incipient and early renal disease, *J. Clin. Endocrinol. Metab.* , 81:1940-1945, (1996).

25. Antonipillai I, Jost-Vu E, Natarajan R, Nadler J, Horton R, Renin response to 12-hydroxyeicosatetraenoic acid is increased in diabetic rats, *Diabetes* , 44:321-325, (1995).

26. Wang T, Powell W, Increased levels of monohydroxy metabolites of arachidonic and linoleic acids in LDL and aorta from atherosclerotic rabbits, *Biochem. Biophys. Acta* , 1084:129-138, (1991).

27. Rao GN, Lasseque B, Griendling KK, Alexander R W, Hydrogen peroxide stimulates transcription of c-jun in vascular smooth muscle cells: role of arachidonic acid, *Mol. Cell. Biol.* , 10:6683-6689, (1993).

28. Haliday E, Ramesha C, Ringold G, TNF induces c-fos via novel pathway requiring the conversion of arachidonic acid to a lipoxygenase metabolite, *EMBO J.* , 10:109-115, (1991).

29. Setty BNY, Graeber JE, Stuart MJ, The mitogenic effects of 15- and 12-hyroxyeicosatetraenoic acids on endothelial cells may be mediated via diacylglycerol kinase inhibition, *J. Biol. Chem.* , 262:17613-17622, (1987).

30. Natarajan R, Gonzales N, Hornsby PJ, Nadler J, Mechanism of angiotensin II-induced proliferation in bovine adrenal cortical cells, *Endocrinology* , 131:1174-1180, (1992).

31. Nelson MJ, Seitz SP, The structure and function of lipoxygenase, *Curr. Opin. Struct. Biol.* , 4:878-884, (1994).

32. Muller K, Gawlik I, Effects of reactive oxygen species on the biosynthesis of 12(S)hydroxyeicosatetraenoic acid in mouse epidermal homogenate, *Free Radical Biol. Med.* , 23:321-330, (1997).

33. Maccarrone M, Corasaniti MT, Guerrieri P, Nistico G, Agro AF, Nitric oxide-donor compounds inhibit lipoxygenase activity, *Biochem. Biophys. Res. Commun.* , 219:128-133, (1996).

34. Kulkarni AP, Mitra A, Chaudhuri J, Byczkowski JZ, Richards I, Hydrogen peroxide; A potent activator of dioxygenase activity of soybean lipoxygenase, *Biochem. Biophys. Res. Commun.* , 166:417-423, (1990).

35. Abeysinghe R, Roberts PJ, Ciiper CE, MacLean KH, Hider RC, Porter JB, The environment of the lipoxygenase iron binding site explored with novel hydroxypyridione iron chelators, *J. Biol. Chem.* , 271:7965-7972, (1966).

36. Yoshimoto T, Suzuki H, Yamamoto S, Takai T, Yokoyama C, Tanabe T, Cloning and sequence analysis of the cDNA of arachidonate lipoxygenase of porcine leukocytes, *Proc. Natl. Acad. Sci. USA* , 87:2142-2146, (1990).

37. Funk CD, Furci L, FitzGerlad GA, Molecular cloning, primary structure and expression of the proliferation in bovine adrenocortical cells, *Endocrinology* , 131:1174-1180, (1992).

38. Yoshimoto T, Arakawa T, Suzuki H, et al, Molecular cloning and expression of human arachidonate 12-lipoxygenase, *Biochem. Biophys. Res. Comm.* , 172:1230-1235, (1990).

39. Natarajan R, Gu JL, Rossi J, et al, Elevated glucose and angiotensin II increase 12-lipoxygenase activity and expression in porcine aortic smooth muscle cells, *Proc. Natl. Acad. Sci. USA* , 90:4947-4951, (1993).

40. Kim JA, Gu JL, Natarajan R, Esteban J, Berliner JA, Nadler JA, A leukocyte-type 12-lipoxygenase is expressed in human vascular and mononuclear cells: Evidence for upregulation by angiotensin II, *Arterioscl. Thromb. Vasc. Biol.* , 15:942-948, (1995).

41. Lopez S, Vila L, Breviaro F, Castellarnau C, Interleukin-1 increased 15-hydroxytetraenoic acid formation in cultured human endothelial cells, *Biochim. Biophys. Res. Comm.* , 1170:17-24, (1993).

42. Natarajan R, Bai W, Rangarajan V, et al, Platelet-derived growth factor BB mediated regulation of 12-lipoxygenase in porcine aortic smooth muscle cells, *J. Cell. Physiol.* , 169:391-400, (1996).

43. Brown ML, Jakubowski JA, Leventis L, Deykin D, Sigal E, Elevated glucose alters eicosanoids release from porcine aortic entothelial cells, *J. Clin. Invest.* , 82:2136-2141, (1992).

44. Kadish J, Endothelium, fibrinolysis, cardiac risk factors, and prostaglandins: a unified model of atherogenesis, *J. Med. Hypotheses* , 45:205-213, (1995).

45. Sasson S, Reich R, Kaiser N, Davarashvili A, Cerasi E, Role of eicosanoids in the regulation of glucose transport in aortic vascular cells, *Diabetologia* , 39 (Supp. 1):A174, (1996).

46. Vane JR, NSAIDs, COX-2 inhibitors, and the gut, *Lancet* , 346:1105-1106, (1995).

47. Sasson S, Kaiser N, Dan-Goor M, Oron R, Koren S, Wertheimer E, Unluhizarchi K, Cerasi E. Substrate autoregulation of glucose transport: hexose-6-phosphate mediates the cellular distribution of glucose transporters. *Diabetologia* 40:30-38 (1997).

56

FUNCTIONS OF THE p21-ACTIVATED PROTEIN KINASES (PAKS) IN NEUTROPHILS AND THEIR REGULATION BY COMPLEX LIPIDS

Dwight Robinson[1], RiYun Huang[1], Jian P. Lian[2], Alex Toker[2] and John A. Badwey[2,3]

[1]Arthritis Unit, Massachusetts General Hospital,
[2]Boston Biomedical Research Institute, and the
[3]Department of Biological Chemistry and
Molecular Pharmacology, Harvard Medical School
Boston, MA 02114

INTRODUCTION

Neutrophils stimulated with the chemoattractant fMet-Leu-Phe (fMLP) exhibit rapid activation of four renaturable protein kinases with molecular masses of $\underline{ca}$. 69, 63, 49 and 40 kDa.[1,2,3] The 69 and 63 kDa kinases were subsequently identified as p21-activated protein kinases (Paks)[4,5]. These enzymes can be conveniently assayed by their ability to undergo renaturation after SDS/PAGE and catalyzing the phosphorylation of a peptide substrate fixed within a gel[2,3] (e.g., Fig. 1). The peptide substrate utilized corresponds to amino acid residues 297-331 of the 47 kDa subunit of the NADPH-oxidase complex (p47-*phox*).[2,3] The preferred recognition/consensus sequence for Pak is - (K/R)RXS - where X can be an acidic, basic or neutral amino acid.[6] There are three sites in the p47-*phox* peptide which contain this sequence (ser-304, -320 & -328).

Paks undergo autophosphorylation/activation upon interacting with the active (GTP-bound) forms of the small GTPases (p21) Rac or Cdc42.[7] Association/targeting of Pak to the membrane can also result in a GTPase-independent autophosphorylation/activation of this kinase.[8,9] The N-terminal region of Pak 1 contains a Cdc42/Rac interactive binding motif (CRIB region) (i.e., ISXPXXXXFXHXXHVG) that mediates the binding of Rac/Cdc42 to Pak[10] and a proline rich sequence (PPAPP; residues 12-16) that mediates a specific interaction between Pak and the small adaptor protein Nck.[11] In addition, the C-terminal region of Pak contains a specific binding site (SSLXPLXXXA) for the β-subunit of heterotrimeric G-proteins.[12] Nck and/or the β-subunit of complex G proteins may mediate the association of Pak with the membrane.[8,11,12] Activation of Pak is thought to involve the release of an inhibitory region from the Mg·ATP substrate site followed by autophosphorylation/displacement of a pseudosubstrate sequence from the active site.[13] Numerous autophosphorylation events in the N-terminal region of the kinase may facilitate disassociation of the phosphorylated pseudosubstrate sequence from the peptide/substrate binding grove for low affinity substrates.[13] The critical

Eicosanoids and Other Bioactive Lipids in Cancer, Inflammation, and Radiation Injury, 4
Edited by Honn *et al.*, Kluwer Academic / Plenum Publishers, New York, 1999.

385

autophosphorylation site/putative pseudosubstrate sequence (-KRST-) (residues 419-422)[13,14] conforms to the consensus

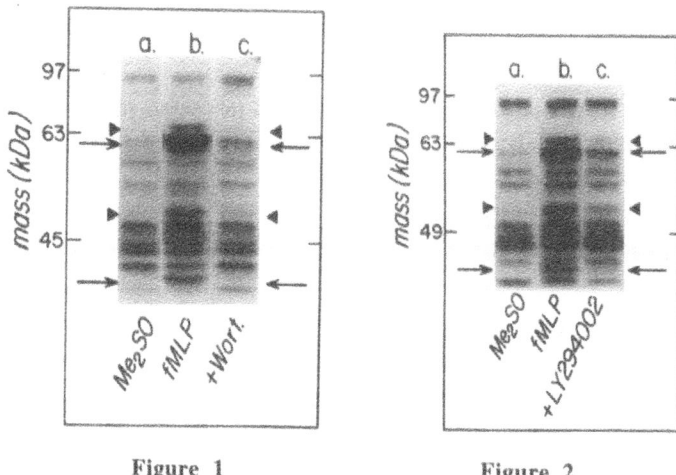

Figure 1   Figure 2

Figure 1. Effects of wortmannin on the activation of the 69, 63 and 49 and 40 kDa renaturable protein kinases in neutrophils. Cells were incubated with 100 nM wortmannin for 5 min before stimulation with 1.0 μM fMLP for 30 sec. The autoradiograms shown were from: (a) unstimulated cells; (b) stimulated cells and (c) cells treated with wortmannin prior to stimulation. The 63 and 40 kDa renaturable protein kinases are designated by the arrows, whereas the 69 and 49 kDa kinases are designated by arrowheads.

Figure 2. Effects of LY294002 on the activation of the 63 and 69 kDa Paks in neutrophils. Cells were incubated with 50 μM LY294002 for 5 min prior to stimulation with 1.0 μM fMLP for 30 sec. The autoradiograms shown were from: (a) unstimulated cells, (b) stimulated cells and (c) cells treated with LY294002 prior to stimulation. Symbols are defined in Figure 1.

sequence described above and lies within the "activation segment" found in many protein kinases that are boarded by the sequences DFG (residues 406-408) and APE (residues 431-433).[15] A constitutively active Pak can be generated by substituting threonine 422 with a glutamic acid residue.[14]

What are some of the functions of Pak in neutrophils and other cell-types? Paks can catalyze the phosphorylation/activation of p47-phox[4] and myosin-1.[16] Phosphorylation of p47-phox is critical for the production of superoxide and hydrogen peroxide during phagocytosis,[2] whereas the existence and possible functions of myosin-1 in phagocytic leukocytes remain unknown. Studies in *Dictyostelium* suggest that myosin-1 is involved in the extension and/or retraction of actin rich membranous projections (e.g., pseudopods).[17] Microinjection of activated Pak1 into Swiss 3T3 cells induces filopodia and membrane-ruffles.[18] In contrast, overexpression of constitutively activated Pak mutants into HeLa cells causes disassembly of focal adhesion complexes but not the formation of filopodia or ruffles.[14] The reasons for these differences are not known, but may reflect, at least in part, the different mutants of Pak employed in these investigations. Pak1 can interact with an essential component of the Ras signalling pathway that is distinct from Rac or Cdc42, and a catalytically inactive form of Pak inhibits transformation of rat

fibroblasts by the Ras oncogene.[19] Perhaps most important, Paks are involved in the activation/potentiation of several distinct MAP kinase cascades. Transfection of constitutively activated Pak can partially activate p38 MAP kinase and potentiates the ability of wild type Raf-1 and certain growth factors to stimulate ERKs and MEKs in various cell types.[14,20] Selective antagonists of p38 MAP kinase
(SB203580) and MEK (PD50980) (which blocks activation of ERK1 and ERK2) have recently been shown to inhibit arachidonate release, chemotaxis, phagocytosis, degranulation and superoxide production by neutrophils.[21,22] Thus, Paks are likely to participate in a variety of neutrophil responses.

REGULATION OF PAKS BY PHOSPHATIDYLINOSITOL 3-KINASE

Neutrophils stimulated with fMLP exhibit a rapid but transient activation of the 63 and 69 kDa Paks along with the renaturable 49 and 40 kDa kinases.[2,3] While the identities of the 49 and 49 kDa enzymes are not known, they may be kinases analogous to Pak that are selectively activated by Cdc42.[5] The 63 and 69 kDa Paks exhibit maximum activation within 15 sec of cell stimulation followed by significant inactivation at 3 to 5 min.[2] In contrast, maximum activation of Pak in vitro with either Rac/Cdc42-GTP or phospholipid requires several minutes to 1 hr .[9, 23] Considerable effort has been invested in exploring the signal transduction mechanism(s) that trigger the rapid activation of Paks in neutrophils. Activation of all four of the renaturable kinases appears to be under the control of a single stimulatory pathway that is sensitive to antagonists of heterotrimeric G-proteins (pertussis toxin), type 1 and/or 2A protein phosphatases (okadaic acid, calyculin A), tyrosine kinases (erbstatin, genistein) and inhibitors of phosphoinositide 3-kinase (PI 3-K).[1, 24, 25] The latter antagonists are perhaps the most informative because they can unify much of the pharmacological data on the activation of Paks. The microbial alkaloid wortmannin and the chromone LY294002 block activation of the 63 and 69 kDa Paks (along with the 49 and 40 kDa kinases) in fMLP-stimulated neutrophils (Figures 1 & 2). Wortmannin alkylates and irreversibly inactivates PI 3-K.[25] In contrast, LY294002 reversibly inhibits PI 3-K by competing with ATP for its substrate binding site.[25] Thus, two structurally and mechanistically distinct antagonists of PI 3-K block activation of Paks.

Neutrophils stimulated with fMLP exhibit a rapid and transient increase in phosphatidylinositol-3,4,5-trisphosphate (PIP_3)[26] that is temporally consistent with PIP_3 serving as a critical second messenger in the activation of Paks. An isoform of PI 3-K has been described in leukocytes that is activated synergistically by tyrosine-phosphorylated peptides and G$\beta\gamma$.[27] Localization of this lipid kinase upstream of Pak could account for the sensitivity of the Paks in neutrophils to pertussis toxin, antagonists of tyrosine kinases and inhibitors of PI 3-K. Antagonists of tyrosine kinases block PIP_3 formation in neutrophils by ca. 70%.[28] Interestingly, expression of membrane targeted, constitutively activated PI 3-K in a monoblastic cell line results in continuous phosphorylation of p47-*phox*.[29] Inhibitors of PI 3-K (wortmannin, LY294002) reduce phosphorylation of p47-*phox* and block superoxide production by fMLP-stimulated neutrophils.[25] How might PIP_3 function in the activation of Pak? While several cellular targets for PIP_3 have recently been described,[30] two may be particularly relevant in relation to the above discussion. PIP_3 may promote the activation of Rac and/or Cdc42 by stimulating the GTP/GDP exchange activity of the relevant guanine nucleotide exchange factor(s),[30,31] and/or by promoting the transloclation/activation of 3-phosphoinositide-dependent protein kinase 1
(PDK1),[32] protein kinase B (PKB/Akt)[30,31,32] or some other ser/thr kinase to the plasmalemma where it may catalyze the phosphorylation and rapid activation of Pak. PDK1 catalyzes the phosphorylation of thr-308 in the activation loop of PKB.[32] PKB undergoes rapid activation in fMLP-stimulated neutrophils with kinetics very similar to those observed for the 63 and 69 kDs Paks.[33] The consensus sequence for thr-308 in PKB [i.e.,-T(P)FCGTPEYLAPE][34] is very similar to the C-terminal sequence of thr-422 in Pak [i.e., T(P)MVGTPYWMAPE].[7]

EFFECTS OF PHOSPHOLIPIDS AND SPHINGOLIPIDS ON PAK

Pak and Pak-related kinases (e.g., myosin-1 heavy chain kinase) can also undergo autophosphorylation/activation after binding certain acidic phospholipids (e.g., phosphatidate)[9,35,36]. Phospholipids obviate the need for activated Rac/Cdc42[9] and are likely to be responsible for the Rac/Cdc42-independent activation of Pak that occurs when this kinase is targeted to the plasmalemma.[8,9] This phospholipid-dependent mechanism of Pak activation may be physiologically relevant since fMLP-stimulated neutrophils exhibit large increases in phosphatidate as a result of the activation of phospholipase D.[37] Neutrophils stimulated with fMLP also exhibit increases in ceramides and sphingoid bases,[38,39] both of which can block activation of the 63 and 69 kDa Paks in these cells.[40] C_2-ceramide prevents translocation of Cdc42 to the membrane of fMLP-stimulated granulocytes.[37] Sphingosine may interact with products of PI 3-K in the membrane (or other acidic lipids) and block the formation of signalling complexes involving these lipids.[41] Thus, Paks can integrate signals from a variety of biologically active lipids.[41] We are currently exploring these possibilities.

ACKNOWLEDGMENTS

Studies described in this chapter from the authors laboratories were supported by NIH grants AR43518, DK50015 and AI23323.

REFERENCES

1. J. Ding and J.A. Badwey, Neutrophils stimulated with a chemotactic peptide or a phorbol ester exhibit different alterations in the activities of a battery of protein kinases, *J. Biol. Chem.* 268:5234 (1993).
2. J. Ding and J.A. Badwey, Stimulation of neutrophils with a chemoattractant activates several novel protein kinases that can catalyze the phosphorylation of peptides derived from p47-*phox* and MARKS, *J. Biol. Chem.* 268:17326 (1993).
3. S. Grinstein, W. Furuya, J.R. Butler, and J. Tseng, Receptor mediated activation of multiple serine/threonine kinases in human neutrophils, *J. Biol. Chem.* 268:20233 (1993).
4. U.G. Knaus, S. Morris, H.-J. Dong, J. Chernoff, and G.M. Bokoch, Regulation of human leukocyte p21-activated kinases through G-protein coupled receptors, *Science* 269: 221 (1995).
5. J. Ding, U.G. Knaus, J.P. Lian, G.M. Bokoch, and Badwey, J.A. The renaturable 69- and 63-kDa protein kinases that undergo rapid activation in chemoattractant-stimulated guinea pig neutrophils are p21-activated kinases, *J. Biol. Chem.* 271:24869 (1996).
6. P.T. Tuazon, W.C. Spanos, E.L. Gump, C.A. Monnig, and J.A. Traugh, Determinants for substrate phosphorylation by 21-activated protein kinase (γ-Pak), *Biochemistry* 36:16059 (1997).
7. E. Manser, T. Leung, H. Salihuddin, Z.-S. Zhao, and L. Lim A brain serine/threonine protein kinase activated by Cdc42 and Rac1, *Nature* 367:40 (1994).
8. W. Lu, S. Katz, R. Gupta, and B.J. Mayer, Activation of Pak by membrane localization mediated by Nck SH3 domain, *Current Biol.* 7:85 (1997).
9. G.M. Bokoch, A.M. Reilly, R.H. Daniels, C.C. King, A. Olivera, S. Spiegel, and U.G. Knaus, A GTPase-independent mechanism of Pak activation: Regulation by sphingosine and other biologically active lipids, *J. Biol. Chem.* - In press (1998).
10. P.D. Burbelo, D. Drechsel, and A. Hall, A conserved binding motif defines numerous candidate target proteins for both Cdc42 and Rac GTPases, *J. Biol. Chem.* 270:29071 (1995).

11. M.L. Galisteo, J. Chernoff, Y.-C. Su, E.Y. Skolnick, and J. Schlessinger, The adaptor protein Nck links receptor tyrosine kinases with the serine-threonine kinase Pak1, *J. Biol. Chem.* 271:20997 (1996).

12. T. Leeuw, C. Wu, J.D. Schrag, M. Whiteway, D.Y. Thomas, and E. Leberer, Interaction of a G-protein β-subunit with a conserved sequence in Ste20/Pak family protein kinases, *Nature* 391:191 (1998).

13. G.E. Benner, P.B. Dennis, and R.A. Masaracchia, Activation of an S6/H4 kinase (Pak65) from human placenta by intramolecular and intermolecular autophosphorylation, *J. Biol. Chem.* 270:21121 (1995).

14. L. Lim, E. Manser, T. Leung, and C. Hall, Regulation of phosphorylation pathways by p21 GTPases. The p21 Ras-related Rho subfamily and its role in phosphorylation signalling pathways, *Eur. J. Biochem.* 242:171(1996).

15. L.N. Johnson, M.E.M. Noble, and D.J. Owen, Active and inactive protein kinases: structural basis for regulation, *Cell* 85:149 (1996).

16. C. Wu, V. Lytvyn, D.Y. Thomas, and E. Leberer, The phosphorylation site for Ste20-like protein kinases is essential for the function of myosin-1 in yeast, *J. Biol. Chem.* 272:30623 (1997).

17. K.D. Novak, and M.A.Titus, Myosin I overexpression impairs cell migration, *J. Cell Biol.* 136:633 (1997).

18. M.A. Sells, U.G. Knaus, S. Bagrodia, D.M. Ambrose, G.M. Bokoch, and J. Chernoff, Human p21-activated kinase (Pak1) regulates actin organization in mammalian cells, *Current Biol.* 7:202(1997).

19. Y. Tang, Z. Chen, D. Ambrose, J. Liu, J.B. Gibbs, J. Chernoff, and J. Field, Kinase deficient Pak1 mutants inhibit Ras transformation of rat-1 fibroblasts. *Mol. Cell Biol.* 17:4454 (1997).

20. J.A. Frost, H. Steen, P. Shapiro, T. Lewis, N. Ahn, P.E. Shaw, and M.A. Cobb, Cross-cascade activation of ERKs and ternary complex factors by Rho family proteins, *EMBO J.* 16:6426 (1997).

21. I. Hazan, J. Dana, Y. Granot, and R. Levy, Cytosolic phospholipase A_2 and its mode of activation in human neutrophils by opsonized zymosan. Correlation between 42/44 kDa mitogen-activated protein kinase, cytosolic phosholipase A_2 and NADPH oxidase, Biochem. J. 326:867(1997).

22. Y.-L. Zu, J. Qi, A. Gilchrist, G.A. Fernandez, D. Vazquez-Abad, D.L. Kreutzer, C.-K. Huang, R.I. Sha'afi, p38 Mitogen-activated protein kinase activation is required for human neutrophil function triggered by TNF-α or fMLP stimulation, *J. Immunol.* 160:1982 (1998).

23. G.A. Martin, G. Bollag, F. McCormick, and A. Abo, A novel serine kinase activated by Rac1/Cdc42 Hs-dependent autophosphorylation is related to Pak65 and Ste20, *EMBO J.* 14:1970 (1995).

24. J.H. Brumell, and S. Grinstein, Serine/threonine kinase activation in human neutrophils: relationship to tyrosine phosphorylation, *Am. J. Physiol.* 267:C1574 (1994).

25. J. Ding, C.J. Vlahos, R. Liu, R.F. Brown, and J.A. Badwey, Antagonists of phosphatidylinositol 3-kinase block activation of several novel protein kinases in neutrophils, *J. Biol. Chem.* 270:11684 (1995).

26. A.E. Traynor-Kaplan, B.L. Thompson, A.L. Harris, P. Taylor, G.M. Omann, and L.A. Sklar, Transient increase in phosphatidylinositol 3, 4-bisphosphate and phosphatidylinositol trisphosphate during activation of human neutrophils, *J. Biol. Chem.* 264:15668 (1989).

27. X. Tang and C.P. Downes, Purification and characterization of G$\beta\gamma$-responsive phosphoinositide 3-kinases from pig platelet cytosol, *J. Biol. Chem.* 272:14193 (1997).

28. A. Ptasznick, E.R. Prossnitz, D. Yoshikawa, A. Smrcka, A.E. Traynor-Kaplan, and G.M. Bokoch, A tyrosine kinase signalling pathway accounts for the majority of phosphatidylinositol 3, 4, 5-trisphosphate formation in chemoattractant-stimulated human neutrophils, *J. Biol. Chem.* 271:25204 (1996).

29. S.A. Didichenko, B. Tilton, B.A. Hemmings, K. Ballmer-Hofer, and M. Thelen, Constitutive activation of protein kinase B and phosphorylation of p47-*phox* by a membrane-targeted phosphoinositide 3-kinase, *Current Biol.* 6:1271 (1996).

30. A. Toker and L.C. Cantley, Signalling through the lipid products of phosphoinositide-3-OH kinase, *Nature* 387:673 (1997).

31. J. Han, K. Luby-Phelps, B. Das, X. Shu, Y. Xia, R.D. Mosteller, U.M. Krishna, J.R. Falck, M.A. White, and D. Broek, Role of substrates and products of PI 3-kinase in regulating activation of rac-related guanosine triphosphatases by Vav. *Science,* 279:558 (1998).

32. J. Downward, Lipid-regulated kinases: some common themes at last, *Science,* 279:673 (1998).

33. B. Tilton, M. Andjelkovic, S.A. Didichenko, B.A. Hemmings, M. Thelen, G-Protein-coupled receptors and Fcγ-receptors mediate activation of Akt/protein kinase B in human phagocytes, *J. Biol. Chem.* 272:28096 (1997) .

34. N. Pullen, P.B. Dennis, M. Andjelkovic, A. Dufner, S.C. Kozma, B.A. Hemmings, and G.A. Thomas, Phosphorylation and activation of p70^{s6k} by PDK1, *Nature* 279:707 (1998).

35. H. Brzeska, T.J. Lynch, and E.D. Korn, *Acanthamoeba* myosin 1 heavy chain kinase is activated by phosphatidylserine-enhanced phosphorylation, *J. Biol. Chem.* 265:3591 (1990).

36. H. Brzeska, J. Szczepanowska, J. Hoey, and E.D. Korn, The catalytic domain of *Acanthamoeba* myosin 1 heavy chain kinase. II. Expression of active catalytic domain and sequence homology to p21-activated kinase (PAK), *J. Biol. Chem.* 271:27056 (1996).

37. A. Abousalham, C. Liossis, L. O'Brien, and D.N. Brindley, Cell-permeable ceramides prevent the activation of phospholipase D by ADP-ribosylation factor and RhoA, *J. Biol. Chem.* 272:1609 (1997).

38. T. Nakamura, A. Abe, K.J., Balazovich, D., Wu, S.J., Suchard, L.A. Boxer, and J.A. Shayman,Ceramide regulates oxidant release in adherent human neutrophils, *J. Biol. Chem.* 269:18384 (1994).

39. E. Wilson, E. Wang, R.E. Mullins, D.J. Uhlinger, D.C. Liotta, J.D. Lambeth, and A.H. Merrill, Jr., Modulation of free sphingosine levels in human neutrophils by phorbol esters and other factors, *J. Biol. Chem.* 263:9304 (1988).

40. J.P. Lian, R.-Y. Huang, D.R. Robinson, and J.A. Badwey, Regulation of the p21-activated protein kinases by sphingolipids, *Molec. Biol. of the Cell,* 8-S:453a (abstract number 2637) (1997).

41. A.H. Merrill, Jr., Cell regulation by sphingosine and more complex sphingolipids, *J. Bioenerg. Biomembr.* 23:83 (1991).

57

INHIBITION OF NEUROFIBROMIN AND p120 GTPASE ACTIVATING PROTEIN (GAP) BY DIETARY FATTY ACIDS

Joung H. Lee[1], Jyoti A. Harwalkar[1], Sophia S. Bryant[2], Vidyodhaya Sundaram[3], Richard Jove[2], and Mladen Golubic[1]

[1]Department of Neurosurgery, and [3]Molecular Biology, Cleveland Clinic Foundation, Cleveland, Ohio
[2]Moffitt Cancer Center, 12902 Magnolia Drive, Tampa, Florida

ABSTRACT

Neurofibromin and p120 GTPase activating protein (p120 GAP) down-regulate the activity of cellular Ras proteins. How the activity of these two proteins is controlled is not yet clear. In this study, we analyzed the effects of eight nutritionally relevant fatty acids on GTPase stimulatory activity of full-length neurofibromin and p120 GAP. The fatty acids tested were: saturated stearic acid, monounsaturated oleic acid, three ω-6 and three ω-3 polyunsaturated fatty acids. The analysis was performed by Ras immunoprecipitation GTPase assay. The full-length p120 GAP expressed in insect Sf9 cells and the immunoaffinity purified full-length neurofibromin were used. Neurofibromin was readily inhibited by stearic and oleic acid, but p120 GAP was not inhibited even at high concentrations (>80 μM). Neurofibromin was also inhibited by low concentrations of all the polyunsaturated fatty acids tested (IC_{50} of 6 to 16 μM). p120 GAP was 2-3 fold less sensitive to inhibition by these fatty acids. The GTPase stimulatory activity of neurofibromin was also inhibited by arachidonic and oleic acid in the presence of a lipid mixture representing the major lipid components of the cell membrane. Chimeric proteins of neurofibromin and p120 GAP were used to determine that differential sensitivity to fatty acid inhibition maps to the catalytic domain of the proteins. These results indicated that fatty acids can modulate the GTPase function of the c-Ha-Ras protein by inhibiting the GTPase stimulatory activity of two Ras regulators, full-length neurofibromin and p120 GAP, at physiologically relevant concentrations *in vitro*.

INTRODUCTION

Many of the signals that direct nucleated cells to proliferate are channeled through an evolutionarily conserved, universal signaling pathway with the Ras family of proteins at its center[1]. Ras proteins, membrane-associated molecular switches, cycle between an inactive GDP-bound state and an active GTP-bound state. Stimulation by extracellular polypeptide growth factors results in enhanced exchange of bound GDP for free GTP, resulting in accumulation of biologically active Ras-GTP complexes. Once in a GTP-bound state, Ras proteins can interact with several cellular proteins known as downstream target effectors[2]. The signals generated by these Ras-GTP-specific interactions lead to Ras-dependent biological cellular functions, such as proliferation, differentiation and cell migration[1]. Ras-GTP converts to its inactive, Ras-GDP form by GTPase Ras activity. Normally the intrinsic GTPase activity of purified Ras is low, but it can be dramatically enhanced by Ras-specific GTPase activating proteins (GAP) such as p120 GAP and neurofibromin, the protein product of the tumor suppressor gene mutated in Neurofibromatosis type 1 (NF1)[3].

Aberrant functions of the proteins that transduce proliferative signaling through Ras, including Ras proteins themselves, are common in human cancers. Many natural tumors contain either activating point mutations in members of the ras gene family, or they overexpress the wild type allele. Nearly all activating point mutations found in ras genes in numerous types of human tumors decrease intrinsic GTPase activity of Ras and render it insensitive to stimulation by GTPase activating proteins[4,5]. The transforming activity of mutant Ras is, therefore, considered a consequence of constitutively activated Ras in its GTP-bound state. Thus, the functional loss of GTPase activating proteins to regulate negatively the activity of Ras proteins might be important in the tumorigenesis process. Supporting this idea, mutations of the NF1 and p120 GAP genes and reduced protein levels and catalytic activity were described in NF1-associated and sporadic tumors[6,7]. Furthermore, lipids produced soon after mitogenic stimulus (phosphatidic acid, arachidonic acid, and phosphatidylinositols) were shown to inhibit the ability of neurofibromin present in rat brain cell extracts[8,9], and the bacterially expressed catalytic domains of p120 GAP, neurofibromin, and homologous yeast protein IRA2[10,11] to stimulate the GTPase activity of Ras. Whether fatty acids, other than arachidonic acid, effect the GTPase stimulatory activity of full-length neurofibromin and p120 GAP and whether such an inhibition occurs in the presence of membrane lipids has not been established.

MATERIAL AND METHODS

The purification of full-length neurofibromin was performed as described earlier[12]. The construction of baculovirus recombinants (full-length p120 GAP, truncated neurofibromin containing amino acids 1-1614, and p120 GAP-neurofibromin, GAP-NF1 chimera), and the preparation of cell lysates used in this study was described earlier[13].

The Ras GTPase stimulatory activity of neurofibromin and p120 GAP in the presence or absence of dietary fatty acids was determined by Ras immunoprecipitation assay, as described earlier[10,14]. The fatty acids tested were: saturated stearic acid (SA, 18:0), monounsaturated oleic acid (OA, 18:1), three ω-6 fatty acids, linoleic acid (LA, 18:2), γ-linolenic acid (GLA, 18:3) and arachidonic acid (AA, 20:4), and three members of ω-3 fatty acids, α-linolenic acid (LNA, 18:3), eicosapentaenoic acid (EPA, 20:5), and docosahexaenoic acid (DHA, 22:6) (Sigma Chemicals and Matreya). Phosphatidylcholine (PC), phosphatidylserine (PS), and phosphatidylethanolamine (PE) derived from bovine brain were purchased from Sigma Chemicals. Fatty acids and lipids were dissolved in chloroform at 5 mg/ml and stored at -20⁰C. The necessary amount of fatty acid solution was dried under a stream of nitrogen in a glass tube. The thin layer of fatty acid coating the glass was resuspended in 500 μl of 0.1 M Tris-HCl, pH=7.5 and sonicated at 6 W output for 10 seconds in a water/ice bath by inserting a titanium microtip under a stream of nitrogen (Fisher, Model 60 Sonic Dismembrator). Fatty acids were prepared fresh for each assay and immediately used. Stock solution of stearic acid was prepared in 100% ethanol, appropriately diluted in 100% ethanol, and added directly to the reaction (1.5 μl/60 μl reaction). Ethanol itself did not affect either the intrinsic or GTPase activating protein-stimulated GTPase activity of c-Ha-Ras.

Inhibition of GTPase stimulatory activity of neurofibromin and p120 GAP by added fatty acids was calculated according to the formula: % inhibition = [(GAP+FA)-(GAP)/(Ras-GAP)] X 100. The percentage of Ras-GTP observed in reactions with both added GTPase activating protein and fatty acid (GAP+FA) was first subtracted from the value observed in reactions with added GTPase activating proteins only (GAP). The resulting value was divided by the percentage of GTP bound to Ras in the absence of all other added materials minus the percentage observed in the presence of added GTPase activating protein only [Ras-GAP]. None of the lipids tested altered the intrinsic GTPase activity of c-Ha-Ras.

RESULTS AND DISCUSSION

The effects of dietary fatty acids on the GTPase stimulatory activity of Ras regulators, neurofibromin and p120 GAP, are particularly interesting because a high intake of dietary fats has been linked with an increased risk of prostate, colon, breast, lung, and pancreatic cancer[15], and signal transduction through Ras proteins is important in these fat-related malignancies[4,5]. Our group and others have shown that arachidonic acid and certain lipids inhibit the GTPase enhancing activity of the catalytic fragments of p120 GAP and neurofibromin[8,10]. However, catalytic fragments of neurofibromin and p120 GAP represent only about 10% and 30% of the full-length molecules, respectively. Since the remaining portion could alter the effects of fatty acids, and the catalytic fragments are not normally expressed in cells, we analyzed full-length p120 GAP and neurofibromin.

Because neurofibromin and p120 GAP have different catalytic properties, the amount of GTPase stimulatory activity for neurofibromin was normalized to that of p120 GAP and the ability of fatty acids to inhibit each protein was determined. In contrast to readily inhibited neurofibromin, stearic and oleic acid did not inhibit p120 GAP activity, even at high concentrations (Table 1, Figure 1). Neurofibromin was inhibited by low concentrations of all polyunsaturated fatty acids tested. Although p120 GAP was 2-3-fold less sensitive to inhibition by these fatty acids than neurofibromin, its IC$_{50}$ values were still in micromolar range (17 to 33 μM). The local concentrations of polyunsaturated fatty acids at the plasma membrane, where stimulation of GTPase activity of Ras occurs, are likely to be higher than the intracellular concentrations of 25-100 μM described for arachidonic acid[16]. The growth-inhibitory prostaglandin A_2[17] had little effect on GTPase stimulatory activity of both proteins (Table 1).

Table 1. IC_{50} of GTPase stimulatory activity of GAPs on c-Ha-Ras[1,2,3].

	SA	OA	LA	GLA	AA	LNA	EPA	DHA	PGA_2
NF1	16	6	8	10	6	11	16	6	>160
p120 GAP	>350	>160	27	33	17	32	32	19	>160
NF1(1-1614)	ND	6	10	ND	4	ND	15	7	ND
GAP-NF1	ND	7	13	ND	7	ND	17	9	ND

[1]Values are expressed in μM.
[2]Abbreviations are as follows: IC_{50}, fatty acid concentration at which 50% inhibition was observed; SA, stearic acid; LA, linoleic acid; GLA, γ-linolenic acid; AA, arachidonic acid; LNA, α-linolenic acid; EPA, eicosapentaenoic acid; DHA, docosahexaenoic acid; ND, not determined; GTPase, guanosine triphosphatase.
[3]Reprinted by permission from reference 14.

To examine whether sequences outside the catalytic domain of neurofibromin and p120 GAP are responsible for their differential sensitivity to fatty acids, we analyzed two additional molecules expressed in insect Sf9 cells. The human neurofibromin truncated at the amino acid position 1614 contains the complete amino terminus of the protein, including the catalytic domain, while p120 GAP-neurofibromin chimera protein, (GAP-NF1), contains amino acids 1-691 of human p120 GAP fused to the catalytic domain of human neurofibromin[13]. When the GTPase stimulatory activities of truncated neurofibromin and GAP-NF1 chimera were examined in the presence of fatty acids, both proteins were inhibited as efficiently as neurofibromin (see Figure 1 for oleic and eicosapentaenoic acid and Table 1 for other fatty acids). These experiments therefore, indicated that the differences observed in the inhibitory sensitivity by dietary fatty acids between full-length p120 GAP and neurofibromin are mostly caused by differences in the catalytic region of the two proteins.

Because neurofibromin was inhibited by all fatty acids analyzed, we wanted to determine whether such inhibition also occurs in the presence of lipids that constitute the major lipid components of the cell membrane. The obtained results show that baculovirus neurofibromin (amino acids 1-1614) was poorly inhibited by the mixture of bovine brain-derived phosphatidylcholine (PC), phosphatidylethanolamine (PE), and phosphatidylserine (PS) prepared in a ratio of 9:5:1. These lipids are present in such ratio in neurons, cells which express high levels of neurofibromin[18,19]. The ability of 10 μM arachidonic and oleic acid to inhibit neurofibromin in the presence of 34 μM membrane phospholipid mixture was only slightly reduced compared to the inhibition obtained when arachidonic and oleic acid were incubated alone with neurofibromin (compare bars 4 and 5 vs. 6 and 7 of the Figure 2). We conclude that these fatty acids can further inhibit the GTPase stimulatory activity of neurofibromin when tested in the presence of the major lipid components of the cell membrane.

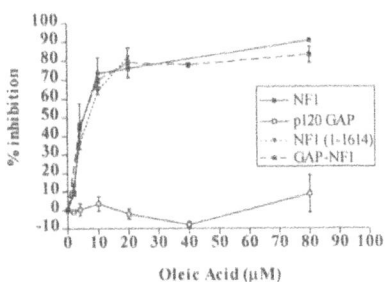

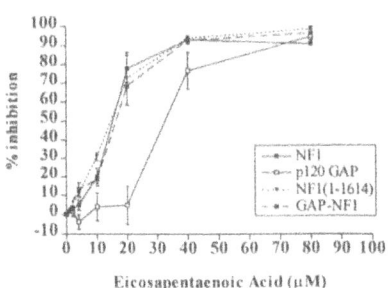

Figure 1. Inhibition of the GTPase stimulatory activity of full-length neurofibromin (NF1, closed squares), p120 GAP (open squares), truncated neurofibromin [NF1(1-1614), triangles], and chimera protein GAP-NF1 (stars) by oleic acid (left panel), and eicosapentaenoic acid (right panel). For oleic acid, values are means +/- SD of four experiments in the case of neurofibromin and p120 GAP, and two experiments in the case of truncated neurofibromin and GAP-NF1 molecules. Data for four molecules tested in the presence of eicosapentaenoic acid represent means +/- SD of four experiments. The differences between p120 GAP and neurofibromin inhibition were significant for eicosapentaenoic acid at concentrations of 4 μM, 10 μM, and 20 μM (p<0.01, paired Student's t-test) and for oleic acid at all concentrations analyzed (p<0.01). This figure is reprinted by permission from reference 14.

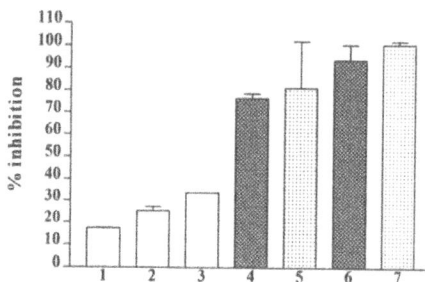

Figure 2. GTPase stimulatory activity of neurofibromin is inhibited by oleic and arachidonic acid in the presence of membrane lipid mixture containing phosphatidylcholine (PC), phosphatidylethanolamine (PE), and phosphatidylserine (PS) in a ratio of 9:5:1. Lanes are as follows: 1, 34 μM (PC+PE+PS); 2, 44 μM (PC+PE+PS); 3, 68 μM (PC+PE+PS); 4, 34 μM (PC+PE+PS) + 10 μM AA; 5, 34 μM (PC+PE+PS) + 10 μM OA; 6, 10 μM AA; 7, 10 μM OA.

The recent elucidation of the three-dimensional structure of the complex between human Ha-Ras and catalytic domain of p120 GAP suggests that the hydrophobic interactions between residues of the phosphate-binding loop and switch regions I and II of Ras and conserved residues in GAPs are likely to be a target of inhibitory fatty acids[20]. The results of this study, provide further support for the idea that fatty acids might be physiological regulators of Ras-specific GTPase activating proteins. Not only did inhibition of GTPase activating proteins occur at physiologically relevant, micromolar, concentrations, but their GTPase stimulatory activity was inhibited by dietary fatty acids in the presence of a lipid mixture representing the major lipid components of the cell membrane. At present, it is not known whether differences in sensitivity to fatty acid inhibition between full-length neurofibromin and p120 GAP are the result of altered affinity for Ras or if in the presence of fatty acids these proteins associate normally with Ras, but fail to stimulate the hydrolysis of GTP.

Because neurofibromin and p120 GAP can negatively regulate activity of normal but not oncogenically activated Ras proteins[1], diets rich in polyunsaturated fatty acids are predicted to promote selectively cells with normal Ras. For example, polyunsaturated fatty acids enhanced proliferation of normal mouse mammary epithelial cells stimulated by epidermal growth factors[21] In the rat mammary cancer model Lu et al., showed a preferential promotional effect produced by the consumption of high fat diets containing either ω-6 or ω-3 polyunsaturated fatty acids on carcinogen-initiated cells which contain the wild type, normal c-Ha-Ras[22]. It is, therefore, important to note that an elevated expression of normal cellular Ras proteins is common in human fat-related cancers of breast and prostate[5,23,24]. In such an intracellular environment, the inhibition of Ras-specific negative regulators by dietary fatty acids could be an important mechanism by which fatty acids effect tumorigenesis.

Acknowledgments

This work has been supported by an American Institute for Cancer Research grant (#94B63) awarded to M.G.

REFERENCES

1. D.R. Lowy, and M.B. Willumsen, Function and regulation of Ras, *Ann Rev Biochem* 62:851-891 (1993).
2. M.E. Katz, and F. McCormick, Signal transduction from multiple Ras effectors, *Curr Opinion Genet Develop* 7:75-79 (1997).
3. M.S. Boguski, and F. McCormick, Proteins regulating Ras and its relatives, *Nature* 366:643-653 (1993).
4. J.L. Bos, Ras oncogenes in human cancer: A review, *Cancer Res* 49:4682-4689 (1989).
5. J.K. Field, and D.A. Spandidos, The role of *ras* and *myc* oncogenes in human solid tumors and their relevance in diagnosis and prognosis (review), *Anticancer Res* 10:1-22 (1990).
6. A. von Deimling, W. Krone, and G.A. Menon, Neurofibromatosis type 1: pathology, clinical features and molecular genetics, *Brain Pathol* 5:153-162 (1995).
7. E. Friedman, P.V. Gejman, G.A. Martin, and F. McCormick, Nonsense mutations in the C-terminal SH2 region of the GTPase activation protein (GAP) gene in human tumours, *Nature Genet* 5:242-247 (1993).
8. G. Bollag, and F. McCormick, Differential regulation of rasGAP and neurofibromatosis gene product activities, *Nature* 351:576-579 (1991).
9. M. Golubic, M. Roudebush, S. Dobrowolski, A. Wolfman, and D.W. Stacey, Catalytic properties, tissue and intracellular distribution of neurofibromin, *Oncogene* 7:2151-2159 (1992).
10. M. Golubic, K. Tanaka, S. Dobrowolski, D. Wood, M-H. Tsai, M. Marshall, F. Tamanoi, and D.W. Stacey, The GTPase stimulatory activities of the neurofibromatosis type 1 and the yeast IRA2 proteins are inhibited by arachidonic acid. *EMBO J* 10:2897-2903 (1991).
11. B.A. Sermon, J.F. Eccleston, R.H. Skinner, and P.N. Lowe, Mechanism of inhibition by arachidonic acid of the catalytic activity of Ras GTPase-activating protein. *J Biol Chem* 271:1566-1572 (1996).
12. J. Yoder-Hill, M. Golubic, and D.W. Stacey, A conserved region of c-Ha-Ras is required for efficient GTPase stimulation by GTPase activating protein but not neurofibromin, *J Biol Chem* 270:27615-27621 (1995).
13. S.S. Bryant, A.L. Mitchell, F. Collins, W. Miao, M. Marshall, and R. Jove, N-terminal sequences contained in the src homology 2 and 3 domains of p120 GTPase-activating protein are required for full catalytic activity toward Ras. *J Biol Chem* 271:5195-5199 (1996).
14. M. Golubic, J.A. Harwalkar, S.S. Bryant, R. Jove, and J.H. Lee, Differential regulation of neurofibromin and p120 GTPase activating protein (GAP) by nutritionally relevant fatty acids, *Nutr. Cancer*, 30:97-107 (1998).
15. World Cancer Research Fund and American Institute for Cancer Research. Fat and cholesterol, in: *Food, Nutrition and the Prevention of Cancer: a Global Perspective*, American Institute for Cancer Research, Washington, D.C. (1997).
16. B.A. Wolf, J. Turk, W.R. Sherman, and M.L. McDaniel, Intracellular Ca^{++} mobilization by arachidonic acid, *J Biol Chem* 261:3501-3511 (1986).
17. M. Hitomi, J. Shu, D. Strom, S.W. Hiebert, M.L. Harter, and D.W. Stacey, Prostaglandin A$_2$ blocks the activation of G1 phase cyclin-dependent kinase without altering mitogen-activated protein kinase stimulation, *J Biol Chem* 271:9376-9383 (1996).
18. R.O. Calderon, B. Attema, and G.H. DeVries, Lipid composition of neuronal cell bodies and neurites from cultured dorsal root ganglia, *J. Neurochem.* 64:424-429 (1995).
19. M.M. Daston, H. Scrable, M. Norlund, A.K. Sturbaum, L.M. Nissen, and N. Ratner, The protein product of the neurofibromatosis type 1 gene is expressed at highest abundance in neurons, Schwann cells and oligodendrocytes, *Neuron*, 8:415-428 (1992).
20. K. Scheffzek, M.R. Ahmadian, W. Kabash, L. Weismueller, A. Lautwein, F. Schmitz, and A. Wittinghofer, The Ras-RasGAP complex: Structural basis for the GTPase activation and its loss in oncogenic Ras mutants, *Science* 277:333-338 (1997).
21. G.K. Bandyopadhyay, W. Imagawa, and S. Nandi, Role of GTP-binding proteins in the polyunsaturated fatty

acid stimulated proliferation of mouse mammary epithelial cells, *Prostaglandins, Leukotrienes and Essential Fatty Acids* **52**:151-158 (1995).

22. J. Lu, C. Jiang, S. Fontaine, and H.J. Thompson, Ras may mediate mammary cancer promotion by high fat, *Nutr Cancer* 23:283-290 (1995).

23. D.M.A. Watson, R.A. Elton, W.L. Jack, J.M. Dixon, U. Chetty, and W.R. Miller, The H-ras oncogene product p21 and prognosis in human breast cancer, *Breast Cancer Res Treatment* 17:161-169 (1990).

24. P.H. Gumerlock, U.R. Poonamallee, F.J. Meyers, and R.W. White, Activated *ras* alleles in human carcinoma of prostate are rare, *Cancer Res* 51:1632-1637 (1991).

BIOACTIVE LIPIDS IN INFLAMMATION

58

SPINAL SYNTHESIS AND RELEASE OF PROSTANOIDS AFTER PERIPHERAL INJURY AND INFLAMMATION

David M. Dirig,[1] and Tony L. Yaksh[1,2]

Departments of Anesthesiology[2] and Pharmacology[1]
University of California, San Diego
La Jolla, CA 92093

INTRODUCTION

After peripheral injury and inflammation, the close relationship between stimulus intensity and response to acute injurious stimuli is altered. Peripheral injury and inflammation increase a subject's responsiveness to noxious stimuli (i.e. pain reports, paw withdrawal latency, or electrophysiological recordings) relative to the un-injured state (Raja et al., 1988) . This hypersensitivity can be expressed as a decreased response latency to a noxious stimulus (hyperalgesia) or a nocifensive response (i.e. pain report or escape attempts) to an innocuous stimulus (allodynia). These altered stimulus-response relationships may result from peripheral as well as central mechanisms; this review will focus on spinal prostanoid synthesis and hyperalgesia after peripheral injury and inflammation.

Consistent with the behavioral effects, recording from small primary afferents revealed that injury or inflammation evokes spontaneous activity in otherwise silent sensory afferent axons (Schaible et al., 1987) as well as a left shift in the relationship between afferent discharge rate and stimulus intensity (Kocher et al., 1987) . This peripheral hypersensitivity of the altered primary afferent can be explained, in part, by a local release of pro-inflammatory substances, such as bradykinin, cytokines, and prostaglandins which can powerfully stimulate/sensitize the peripheral terminal (Baccaglini and Hogan, 1983).

Increased activity in peripheral sensory afferents after injury and nociceptor sensitization can also change spinal processing of sensory input, leading to spinally-mediated hyperalgesia. Repetitive activation of small caliber primary afferents (C-fibers) increases the excitability of spinal neurons (Woolf, 1983) . Peripheral injury generates a state of spinal facilitation in which a moderate stimulus evokes a profound discharge in dorsal horn nociceptive neurons (Dickenson and Sullivan, 1987) . This increased input-to-output ratio or spinal facilitation of peripheral sensory input increases ascending neuronal activity (relative to the un-injured state). This increase in spinal responsiveness after injury is consistent with neuronal activity evoked by a more intense stimulus in the un-injured state. Thus, a moderate stimulus in an un-injured subject is interpreted as a severe noxious stimulus after tissue injury due to spinal facilitation of this peripheral input (increased input-to-output ratio).

Eicosanoids and Other Bioactive Lipids in Cancer, Inflammation, and Radiation Injury, 4
Edited by Honn *et al.*, Kluwer Academic / Plenum Publishers, New York, 1999.

401

Several animal models have been developed in which peripheral injury or inflammation increases sensory afferent activity and evokes spinally-mediated hyperalgesia (Yaksh, 1997). Use of these models, in conjunction with intrathecal (IT) administration of receptor ligands and enzyme inhibitors, has lead to a working hypothesis that spinally-mediated hyperalgesia is due to i) release of excitatory amino acids (EAA's) and Substance P (SP) from the primary afferent central terminal, ii) activation of several classes of EAA and SP receptors on higher-order neuronal terminals within the dorsal spinal cord, and iii) the subsequent activation of enzymatic cascades downstream of receptor activation leading to the synthesis of prostanoids.

Since the seminal studies of Vane (1971), Smith and Willis (1971), Ferreira (1971), and Moncada (1973), it has been accepted that i) prostanoid synthesis by cyclooxygenase (COX) in the periphery is essential for the development of inflammation and tissue injury-associated hyperalgesia, and ii) non-steroidal anti-inflammatory drugs (NSAIDs) are analgesic and anti-pyretic due to their inhibition of COX activity. Vane (1971) and Smith and Willis (1971) revealed that NSAIDs such as acetylsalicylic acid and indomethacin, which were known to alter the hyperalgesia that occurred secondary to inflammation (e.g. tissue injury), inhibited COX. This observation provided a unifying link in explaining the ability of NSAIDs to normalize the otherwise lowered thresholds (i.e. hyperalgesia) observed in the face of local tissue injury without altering normal nociceptive thresholds (Ferreira et al., 1971; Moncada et al., 1973). Despite this explanation, there were a number of elements that were inconsistent with the initial hypothesis. Notably, agents such as acetaminophen were poor anti-inflammatory drugs, but were considered active anti-hyperalgesic agents. Additionally, it was clear that the anti-pyretic actions of these agents could stem from a central site of thermoregulatory action. Later, it became apparent from studies utilizing supraspinal (Ferreira et al., 1978) and spinal administration of NSAIDs (Yaksh, 1982) that there were central sites of NSAID action that implicated central prostaglandin synthesis in nociception and hyperalgesia. There is a growing literature supporting the hypothesis that spinally-derived prostanoids may be essential for a component of hyperalgesia associated with tissue injury and inflammation. This is suggested by studies demonstrating i) reversal of tissue injury-associated hyperalgesia by spinal administration of COX inhibitors; and ii) prostanoid release from the spinal cord after peripheral injury.

SPINALLY-DELIVERED COX INHIBITORS

If spinal synthesis and release of prostanoids is necessary for the development or maintenance of spinal-mediated hyperalgesia, then inhibition of spinal prostaglandin synthesis should abrogate such states. This is indeed the case; inhibition of spinal COX reduces or eliminates behavioral correlates of hyperalgesia. Yaksh (1982) demonstrated that IT administration of NSAIDs reduced acetic acid-evoked writhing in mice. In support of these initial studies, intrathecal pre-treatment with the NSAID, S(+)-ibuprofen, blocked thermal hyperalgesia induced in rats by IT SP or NMDA (Malmberg and Yaksh, 1992b). Additionally, a series of NSAIDs (given intrathecally before tissue injury) reversed hyperalgesia with relative potency ratios that mirrored their structure-activity relationships for COX inhibition (Malmberg and Yaksh, 1992a), but at doses 100-1000 fold lower than the effective systemic doses. While these findings argue for involvement of spinal prostanoid synthesis in spinal facilitation and hyperalgesia, the fact that NSAIDs reverse hyperalgesia, but do not increase nociceptive thresholds in control animals suggests that this facilitated state of processing involves spinal mechanisms distinct from those of acute nociception.

Multiple COX Isozymes

Understanding of prostanoid synthetic cascades has undergone a revolution in the past decade following the discovery of multiple COX isozymes (COX-1 and COX-2). Originally, it was believed that COX-1 was constitutively expressed while COX-2 was upregulated as an immediate-early gene (Kujubu et al., 1991) in response to cellular stimuli, such as hippocampal N-methyl-d-aspartate (NMDA) receptor activation (Yamagata et al., 1993) or effects of carrageenan-mediated inflammation on macrophages (Tomlinson et al., 1994).

The relevance and limitations of studies described above employing spinal administration of NSAIDs, however, becomes clear when it is appreciated that NSAIDs inhibit both COX-1 and COX-2 with varying degrees of discrimination (Meade et al., 1993), whereas recently developed compounds that are specific for COX-2 without inhibition of COX-1 (Seibert et al., 1994; Gierse et al., 1996). The regulation and tissue specificity of COX expression is an area of intense interest (For details, see Smith and DeWitt, 1996) and has direct implications for the understanding of spinally-mediated hyperalgesia.

While investigations into the complex regulation of cellular prostaglandin synthetic machinery have shown an overall pattern of constitutive COX-1 and inducible COX-2 in the periphery (however, see Harris et al., 1994), central expression of the isozymes has been less clear. As in the periphery, COX-1 is expressed constitutively in the brain (Breder et al., 1992), but there is also evidence of constitutive COX-2 expression within specific brain nuclei (Breder et al., 1995) as well as the spinal cord (Beiche et al., 1996, Dirig et al., unpublished observations). Importantly, Beiche and colleagues reported that both COX-1 and COX-2 mRNA were constitutively expressed in rat spinal cord, and COX-2 levels were elevated after carrageenan-mediated paw inflammation. One potential role of COX-2 and central prostanoid synthesis in hyper-responsive states is suggested by studies showing upregulation of COX-2 neuronal content after NMDA-dependent hippocampal synaptic activity (Yamagata et al., 1993). Given the role of spinal NMDA receptor activation in the generation of hyper-responsive states (for review, see Yaksh, 1993), an increase in COX-2 expression after NMDA receptor activation suggests a secondary source of prostanoids within the same time course as the increased neuronal responsiveness after injury. Thus, prostanoids synthesized by spinal COX-2 (either constitutive or induced) may be involved in spinal NMDA receptor-dependent hyperalgesic states.

While it is generally accepted that NSAIDs act via inhibition of peripheral COX activity, there are also central components to the mechanisms of action of NSAIDs (McCormack, 1994; Yaksh et al., 1998). These observations illustrate that further investigation remains to discern not only which isozyme(s) of the enzyme are the source of prostanoids, but whether this source is peripheral, spinal, supra-spinal, or (most likely) a combination of these. Spinal NSAIDs block the development of hyperalgesia, but this blockade may be due to an inhibition of COX-1, COX-2, or both isozymes. If COX-2 is involved in spinal facilitation, then recently developed COX-2 inhibitors given spinally should be as effective as NSAIDs.

Oral administration of COX-2 inhibitors reduces carrageenan-induced paw edema as well as c-fos expression within the dorsal horn (Buritova et al., 1996), but oral administration does not allow determination of peripheral versus spinal prostanoid sources. Intrathecal administration of the non-specific COX inhibitor, S(+)-ibuprofen, or the specific COX-2 inhibitor, SC58125, before paw carrageenan-induced inflammation blocked the development of thermal hyperalgesia, as assessed by decreases in rat hindpaw withdrawal latency (Dirig and Yaksh, 1997b). Similar results have recently been reported by Yamamoto and Nozaki-Taguchi (1997), using intrathecal administration of the COX-2 inhibitor, NS398. These data suggest constitutive spinal COX-2 expression may be necessary for the development or maintenance of thermal hyperalgesia, but cannot exclude a role for spinal COX-1. Future studies will require the use of specific COX-1 inhibitors. Thus, beyond the question of peripheral versus central prostanoid actions, recent developments refine the question to a determination of which COX isozymes are constitutive and/or inducible in the periphery as well as the central nervous system. Both COX isozymes may contribute to peripheral, spinal, and supraspinal prostanoid synthesis which occurs after tissue injury or inflammation and is implicated in the development and maintenance of central facilitation and hyperalgesia.

SPINAL PROSTAGLANDIN RELEASE

Inhibitor studies suggest the involvement of spinal prostanoids in hyperalgesia but cannot address which prostanoids are produced. Microdialysis and *in vitro* spinal tissue superfusion studies have addressed this question and further substantiate the working hypothesis that spinal prostanoids are involved in hyperalgesia after peripheral injury. Microdialysis studies are performed in unanesthetized, unrestrained rats that have been chronically instrumented with intrathecal lumbar injection (Yaksh and Rudy, 1976) and microdialysis catheters (Marsala et al., 1995). By combining such microdialysis methods with the hyperalgesic models described above, relative changes in spinal cerebrospinal fluid levels and behavioral measures of hyperalgesia can be correlated. Using such methods, Malmberg and Yaksh (1995) demonstrated that spinal PGE2 microdialysate levels increase after paw injection with formalin, and the elevated PGE2 levels were reversed by intrathecal pre-treatment with NSAIDs. Accordingly, Yang et al. (1996) reported similar findings 24 hours after induction of knee joint inflammation in rats, suggesting that spinal prostanoid synthesis and release are associated with hyperalgesia after tissue injury.

Thus, a number of studies have demonstrated that during a protracted peripheral injury associated with hyperalgesia and central facilitation, there is a concomitant increase in spinal, extracellular, extravascular PGE2 levels. Such results have also been observed using an *in vitro* spinal cord superfusion model. Figure 1 depicts changes in spinal tissue perfusate concentration of PGE2 after peripheral injury. Resting PGE2 levels are elevated 5-72 hours after induction of inflammation (intra-articular kaolin/carrageenan). Not only are resting PGE2 levels elevated, capsaicin-evoked PGE2 increases are greater from tissue of injured animals than from naive animals. Interestingly, resting prostacyclin (PGI2) levels were not changed by peripheral injury, but capsaicin still increased PGI2 levels (data not shown).

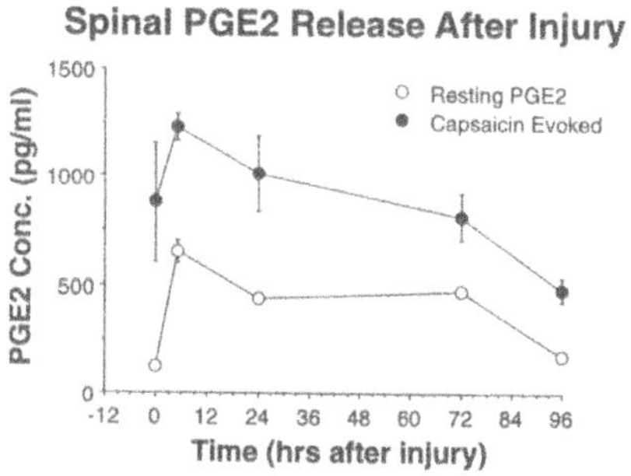

Spinal PGE2 Release After Injury

Figure 1: Time course of changes in resting and evoked PGE2 levels from an *in vitro* spinal tissue superfusion preparation. Data are presented as perfusate concentration vs. time after induction of knee joint inflammation (intra-articular kaolin-carrageenan). Open circles depict the dramatic increase in resting PGE2 levels 5-96 hours. after knee joint injection. Closed circles demonstrate 10μM Capsaicin-evoked PGE2 release over the same time course. Data from naive animals presented at T=0. See Dirig et al. (1997a, 1997b) for methodology.

Beyond the data described above, *in vitro* studies, using either cell culture or spinal tissue superfusion, suggest several possible neuronal effects of PG receptor activation on central neuronal processing and synaptic efficacy. PGE2 increases calcium influx in cultured spinal oligodendrocytes (Soliven et al., 1993) and avian sensory neurons (Nicol et al., 1992), and PGE2 increases tetrodotoxin-resistant sodium influx in rat sensory neurons (Gold et al., 1996). This increased sodium influx is sensitive to the μ-opiate receptor agonist, DAMGO (Gold and

Levine, 1996) . Given the synergy of opiates and NSAIDs in decreasing hyperalgesia after formalin injection (Malmberg and Yaksh, 1993) , prostaglandins may decrease the threshold for activation of opiate receptor-expressing, primary afferent terminals.

Also suggestive of an interplay between sensory system transmitters and prostanoids, PGE2 or PGI2 exposure potentiates capsaicin-evoked release of SP from cultured rat sensory neurons (Hingtgen and Vasko, 1994) and rat spinal tissue slices (Vasko, 1995) . Interestingly, Substance P itself can evoke release of PGE2 from spinal cord tissue, and both the basal and evoked release from spinal tissue of rats after 24 hours of knee joint inflammation are elevated compared to PGE2 release from the spinal cords of naive rats (Dirig and Yaksh, 1997a). A similar phenomenon is shown in Figure 1 with increases in basal and capsaicin-evoked PGE2 release from spinal tissue of injured rats relative to naive. A positive feedback loop is suggested from these studies, where increased primary afferent activity with peripheral inflammation sensitizes spinal neurons and increases synaptic glutamate and neuropeptide release from primary afferent terminals. These transmitters act on post-synaptic cells to evoke synthesis and release of prostanoids. Given that prostaglandins can potentiate the evoked release of neuropeptides, prostanoids may feed back on the primary afferent terminal to increase evoked transmitter release from pre-synaptic terminals. In this fashion, any subsequent action potential that reaches the sensory afferent terminal would result in an increased neuropeptide release, a greater excitation of post-synaptic neurons, and increased ascending neuronal activity relative to the peripheral stimulus (i.e. an increased input/output ratio), and thus spinal prostanoid-mediated central facilitation of peripheral sensory input (See Figure 2).

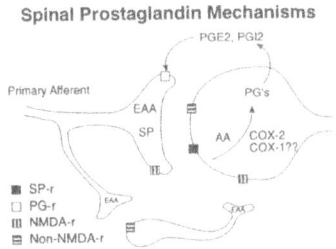

Figure 2. Cartoon depicts possible receptors (squares) associated with prostanoid-mediated spinal facilitation of peripheral input. After tissue injury, increased primary afferent activity leads to protracted terminal release of Substance P (SP) and Excitatory Amino Acids (EAA), activation of higher-order neurons, and liberation of arachidonate (AA) and conversion to prostanoids (PG's) by COX-1 and/or COX-2. Prostanoids can feedback to potentiate depolarization-evoked transmitter release, increasing receptor activation and ascending spinal neuronal activity.

CONCLUSION

Clinical pain syndromes are rarely examples of states mediated by acute afferent input. Treatment of cancer pain, tissue injury-associated pain, or post-surgical trauma cases are common concerns for the clinician, and these syndromes have more in common with the protracted pain models described above than with studies of acute nociception, in that the protracted afferent barrage associated with tissue injury and inflammation can produce tonic changes in sensory processing at spinal and supra-spinal levels. One essential aspect of such protracted nociceptive states is the development of central facilitation or centrally-mediated hyperalgesia. The pre-clinical data clearly emphasizes that repetitive small afferent input after tissue injury will generate a state of facilitated processing which has a unique pharmacology involving a neuropeptide and excitatory amino acid receptors. A component of this centrally-mediated hyperalgesia is dependent upon the spinal synthesis and release of prostaglandins. Regardless of the possible clinical applicability of the spinal NSAIDs, the convergent insights described above from a large number of laboratories emphasize the probable role played by spinal cyclooxygenase products in modulating spinal nociceptive processing. These insights suggest the potential benefits of i) the development of centrally acting agents which alter the receptor mediated actions of prostanoids, and ii) the investigation of enzyme inhibitor effects which can isozyme-specifically target prostaglandin synthetic cascades.

Of particular importance, the recent demonstration in the periphery, brain, and spinal cord of multiple cyclooxygenase isozymes, as well as the subsequent development of selective COX-2 inhibitors may provide yet another approach for alteration of the facilitatory role played by cyclooxygenase products in the human post-injury pain state. However, questions still remain regarding the specific prostanoid species and receptor subtypes involved. Additionally, the role of multiple COX isozymes (both constitutive as well as inducible) still remains a future avenue of investigation with respect to spinal prostaglandin synthesis and release associated with tissue-injury and hyperalgesia.

Acknowledgments Research supported in part by Public Health Service grants NIDA02110 (TLY), and National Research Service Award NIDA05726 (DMD).

REFERENCES

Baccaglini, P.I. and Hogan, P.G., 1983, Some rat sensory neurons in culture express characteristics of differentiated pain sensory cells, *Proc Natl Acad Sci U S A*. 80:594.

Beiche, F., Scheuerer, S., Brune, K., Geisslinger, G., and Goppelt-Struebe, M., 1996, Up-regulation of cyclooxygenase-2 mRNA in the rat spinal cord following peripheral inflammation, *Febs Lett*. 390:165.

Breder, C.D., Dewitt, D., and Kraig, R.P., 1995, Characterization of inducible cyclooxygenase in rat brain, *J Comp Neurol*. 355:296.

Breder, C.D., Smith, W.L., Raz, A., Masferrer, J., Seibert, K., Needleman, P., and Saper, C.B., 1992, Distribution and characterization of cyclooxygenase immunoreactivity in the ovine brain, *J Comp Neurol*. 322:409.

Buritova, J., Chapman, V., Honore, P., and Besson, J.M., 1996, Selective cyclooxygenase-2 inhibition reduces carrageenan oedema and associated spinal c-Fos expression in the rat, *Brain Res*. 715:217.

Dickenson, A.H. and Sullivan, A.F., 1987, Subcutaneous formalin-induced activity of dorsal horn neurones in the rat: differential response to an intrathecal opiate administered pre or post formalin, *Pain*. 30:349.

Dirig, D.M., Hua, X.Y., and Yaksh, T.L., 1997a, Temperature dependency of basal and evoked release of amino acids and calcitonin gene-related peptide from rat dorsal spinal cord, *J Neurosci*. 17:4406.

Dirig, D.M., Konin, G.P., Isakson, P.C., and Yaksh, T.L., 1997b, Effect of spinal cyclooxygenase inhibitors in rat using the formalin test and in vitro prostaglandin E2 release, *Eur J Pharmacol*. 331:155.

Dirig, D.M. and Yaksh, T.L., 1997a, Basal and evoked in vitro PGE2 release from rat spinal cord are elevated after chronic knee joint inflammation (Abstract), *Soc. for Neurosci*. 27:71.

Dirig, D.M. and Yaksh, T.L., 1997b, Spinal and systemic cyclooxygenase (COX) inhibitors suppress paw carrageenan-evoked thermal hyperalgesia in rats (Abstract), *Anesthesiology*. 87:A721.

Ferreira, S.H., Lorenzetti, B.B., and Correa, F.M., 1978, Central and peripheral antialgesic action of aspirin-like drugs, *Eur J Pharmacol*. 53:39.

Ferreira, S.H., Moncada, S., and Vane, J.R., 1971, Indomethacin and aspirin abolish prostaglandin release from the spleen, *Nature New Biol*. 231:237.

Gierse, J.K., McDonald, J.J., Hauser, S.D., Rangwala, S.H., Koboldt, C.M., and Seibert, K., 1996, A single amino acid difference between cyclooxygenase-1 (COX-1) and -2 (COX-2) reverses the selectivity of COX-2 specific inhibitors, *J Biol Chem*. 271:15810.

Gold, M.S. and Levine, J.D., 1996, DAMGO inhibits prostaglandin E2-induced potentiation of a TTX-resistant Na+ current in rat sensory neurons in vitro, *Neurosci Lett*. 212:83.

Gold, M.S., Reichling, D.B., Shuster, M.J., and Levine, J.D., 1996, Hyperalgesic agents increase a tetrodotoxin-resistant Na+ current in nociceptors, *Proc Natl Acad Sci U S A*. 93:1108.

Harris, R.C., McKanna, J.A., Akai, Y., Jacobson, H.R., Dubois, R.N., and Breyer, M.D., 1994, Cyclooxygenase-2 is associated with the macula densa of rat kidney and increases with salt restriction, *J Clin Invest*. 94:2504.

Hingtgen, C.M. and Vasko, M.R., 1994, Prostacyclin enhances the evoked-release of substance P and calcitonin gene-related peptide from rat sensory neurons, *Brain Res*. 655:51.

Kocher, L., Anton, F., Reeh, P.W., and Handwerker, H.O., 1987, The effect of carrageenan-induced inflammation on the sensitivity of unmyelinated skin nociceptors in the rat, *Pain*. 29:363.

Kujubu, D.A., Fletcher, B.S., Varnum, B.C., Lim, R.W., and Herschman, H.R., 1991, TIS10, a phorbol ester tumor promoter-inducible mRNA from Swiss 3T3 cells, encodes a novel prostaglandin synthase/cyclooxygenase homologue, *J Biol Chem*. 266:12866.

Malmberg, A.B. and Yaksh, T.L., 1992a, Antinociceptive actions of spinal nonsteroidal anti-inflammatory agents on the formalin test in the rat, *J Pharmacol Exp Ther*. 263:136.

Malmberg, A.B. and Yaksh, T.L., 1992b, Hyperalgesia mediated by spinal glutamate or substance P receptor blocked by spinal cyclooxygenase inhibition, *Science*. 257:1276.

Malmberg, A.B. and Yaksh, T.L., 1993, Pharmacology of the spinal action of ketorolac, morphine, ST-91, U50488H, and L-PIA on the formalin test and an isobolographic analysis of the NSAID interaction, *Anesthesiology*. 79:270.

Malmberg, A.B. and Yaksh, T.L., 1995, Cyclooxygenase inhibition and the spinal release of prostaglandin E2 and amino acids evoked by paw formalin injection: a microdialysis study in unanesthetized rats, *J Neurosci*. 15:2768.

Marsala, M., Malmberg, A.B., and Yaksh, T.L., 1995, A chronic spinal dialysis catheter for use in the unanesthetized rat: Methodology and application, *J. Neurosci. Meth*. 62:43.

McCormack, K., 1994, The spinal actions of nonsteroidal anti-inflammatory drugs and the dissociation between their anti-inflammatory and analgesic effects, *Drugs*. 5:28.

Meade, E.A., Smith, W.L., and DeWitt, D.L., 1993, Differential inhibition of prostaglandin endoperoxide synthase (cyclooxygenase) isozymes by aspirin and other non-steroidal anti-inflammatory drugs, *J Biol Chem*. 268:6610.

Moncada, S., Ferreira, S.H., and Vane, J.R., 1973, Prostaglandins, aspirin-like drugs and the oedema of inflammation, *Nature*. 246:217.

Nicol, G.D., Klingberg, D.K., and Vasko, M.R., 1992, Prostaglandin E2 increases calcium conductance and stimulates release of substance P in avian sensory neurons, *J Neurosci*. 12:1917.

Raja, S.N., Meyer, R.A., and Campbell, J.N., 1988, Peripheral mechanisms of somatic pain, *Anesthesiology*. 168:571.

Schaible, H.G., Schmidt, R.F., and Willis, W.D., 1987, Convergent inputs from articular, cutaneous and muscle receptors onto ascending tract cells in the cat spinal cord, *Exp Brain Res*. 66:479.

Seibert, K., Zhang, Y., Leahy, K., Hauser, S., Masferrer, J., Perkins, W., Lee, L., and Isakson, P., 1994, Pharmacological and biochemical demonstration of the role of cyclooxygenase 2 in inflammation and pain, *Proc Natl Acad Sci U S A*. 91:12013.

Smith, J.B. and Willis, A.L., 1971, Aspirin selectively inhibits prostaglandin production in human platelets, *Nature New Biol*. 231:235.

Smith, W.L. and DeWitt, D.L., 1996, Prostaglandin Endoperoxide H Synthases- 1 and -2, *Advances in Immunology*. 62:167.

Soliven, B., Takeda, M., Shandy, T., and Nelson, D.J., 1993, Arachidonic acid and its metabolites increase Cai in cultured rat oligodendrocytes, *Am J Physiol*. 264:C632.

Tomlinson, A., Appleton, I., Moore, A.R., Gilroy, D.W., Willis, D., Mitchell, J.A., and Willoughby, D.A., 1994, Cyclo-oxygenase and nitric oxide synthase isoforms in rat carrageenin-induced pleurisy, *Br J Pharmacol*. 113:693.

Vane, J.R., 1971, Inhibition of prostaglandin synthesis as a mechanism of action for aspirin-like drugs, *Nature New Biol*. 231:232.

Vasko, M.R., 1995, Prostaglandin-induced neuropeptide release from spinal cord, *Prog Brain Res*. 104:367.

Woolf, C.J., 1983, Evidence for a central component of post-injury pain hypersensitivity, *Nature*. 306:686.

Yaksh, T.L., 1982, Central and peripheral mechanism for the antialgesic action of acetylsalicylic acid, in: *Acetylsalicylic Acid: New Uses for an Old Drug*, J.M. Barnet, J. Hirsh, and J.F. Mustard, ed., Raven Press, New York.

Yaksh, T.L., 1993, The spinal pharmacology of facilitation of afferent processing evoked by high-threshold afferent input of the postinjury pain state, *Curr Opin Neurol Neurosurg*. 6:250.

Yaksh, T.L., 1997, Pre-clinical models of nociception, in: *Anesthesia: Biologic Foundations*, T.L. Yaksh, C. Lynch, W. Zapol, M. Maze, J.F. Biebuyck, and L.J. Saidman, ed., Lippincott-Raven, Philadelphia.

Yaksh, T.L., Dirig, D.M., and Malmberg, A.B., 1998, Mechanism of action of nonsteroidal anti-inflammatory drugs, *Canc Inv*. 16:In Press.

Yaksh, T.L. and Rudy, T.A., 1976, Chronic catheterization of the spinal sub-arachnoid space, *Physiol. Behav*. 17:1031.

Yamagata, K., Andreasson, K.I., Kaufmann, W.E., Barnes, C.A., and Worley, P.F., 1993, Expression of a mitogen-inducible cyclooxygenase in brain neurons: regulation by synaptic activity and glucocorticoids, *Neuron*. 11:371.

Yamamoto, T. and Nozaki-Taguchi, N., 1997, Role of spinal cyclooxygenase (COX)-2 on thermal hyperalgesia evoked by carageenan injection in the rat, *Neuroreport*. 8:2179.

Yang, L., Marsala, M., and Yaksh, T.L., 1996, Characterization of time course of spinal amino acids, citrulline, and PGE2 release after carrageenan / kaolin-induced knee joint inflammation: a chronic microdialysis study, *Pain*. 67:345.

408

59

PROSTAGLANDIN E2 AS A MODULATOR OF LYMPHOCYTE MEDIATED INFLAMMATORY AND HUMORAL RESPONSES

Kuljeet Kaur[1], Sarah G. Harris[1,2], Josue Padilla[1,2,3], Beth A. Graf[1], and Richard P. Phipps[1,2,]

[1]University of Rochester Cancer Center,
[2]Department of Microbiology and Immunology,
[3]Eastman Dental Center,
University of Rochester School of Medicine and Dentistry,
601 Elmwood Avenue, Rochester, New York, 14642, USA

INTRODUCTION

Prostaglandins are a family of structurally related small lipid molecules that can regulate cellular growth, differentiation, and homeostasis. Prostaglandins are primarily derived from arachidonic acid, which is released from the membrane by phospholipases in response to a variety of extrinsic stimuli. Constitutive and inducible forms of cyclooxygenase in concert with isomerses convert arachidonic acid into different prostaglandins including prostaglandin E_2 (PGE_2). PGE_2 was previously known as an immunosuppressive prostaglandin as it inhibits T cell production of IL-2 and B lymphocyte production of IgM. Recently a new concept emerged that PGE_2 is a critical modulator of B and T cell function and is not necessarily immunosuppressive. The focus of this paper is to present new developments that indicate PGE_2 functions by promoting antibody driven responses at the expense of inflammatory responses.

IMMUNOREGULATORY EFFECT OF PGE_2 ON B LYMPHOCYTES

PGE_2 is synthesized by macrophages, certain dendritic cells and fibroblasts and is present in the microenvironment of B and other lymphocytes. PGE_2 weakly inhibits the proliferation of B cells. In contrast PGE_2 strongly enhances cytokine-mediated immunoglobulin class switching in B lymphocytes to IgE, IgG1 and IgG2a. IL-4 synergizes with PGE_2 to enhance the production of

Eicosanoids and Other Bioactive Lipids in Cancer, Inflammation, and Radiation Injury, 4
Edited by Honn *et al.*, Kluwer Academic / Plenum Publishers, New York, 1999.

409

the immunoglobulins IgE and IgG1. Combinations of IFNγ and PGE₂ promote the production of IgG2a. These effects were mediated by increases in cAMP as shown by the use of cAMP elevating and cAMP inhibiting agents (1).

PGE₂, like most prostaglandins, exerts its effects through interactions with specific G protein-coupled receptors (2). PGE₂ receptors are categorized into four subtypes- EP1, EP2, EP3 and EP4. PGE receptors were known to be present in many tissues including kidney, intestine and brain. We have demonstrated that lymphoid cells also possess functional EP receptors (3). The ability of PGE₂ to stimulate immunoglobulin class switching is mediated primarily by EP2 and EP4 subtype receptors as confirmed by the use of agonists for these receptors. EP2 and EP4 receptors function by increasing cAMP. cAMP inhibits some early events in B cell activation, while promoting differentiation to antibody-secreting plasma cells (3-5).

PGE₂ MODULATES THE PRODUCTION OF Th1 AND Th2 CYTOKINES

CD4⁺ T lymphocytes can be divided into at least two subsets based on cytokine secretion profiles and functions. Th1 cells produce the cytokines IFNγ and IL-2 and promote the induction of inflammatory and cell mediated responses. Th2 cells, however, produce the cytokines IL-4, IL-5 and IL-10 and promote mainly humoral type responses, including the production of IgE. The development of a Th1 or Th2 phenotype in response to a pathogen is extremely important to the outcome of the immune challenge. For example, in *Leishmania major* infection, a Th1 response is associated with parasite destruction, whereas a Th2 response results in death of the host (6, 7). However, in many helminth infections, a Th1 response exacerbates the infection while a Th2 response results in helminth eradication (8). The types of cytokine or immune modulators present during T cell activation can dictate whether a Th1 or Th2 response to a pathogen ensues (9).

PGE₂ plays a critical role in the development of a particular Th response. PGE₂ inhibits the production of Th1 but not Th2 cytokines from mature mouse T cell clones (10). PGE₂ inhibits the production IL-2 and IFNγ from mature Th1 cells and it enhances the production of cytokines IL-5 and IL-10 from Th2 cells (11). IL-12 is considered the key mediator cytokine for driving naive Th cells to a Th1 phenotype. PGE₂ inhibits IL-12 production in whole blood cultures (12). Thus PGE₂ promotes the development of a Th2 response by inhibiting the production of Th1 cytokines and ultimately favoring the production of Th2 cytokines (Fig.1).

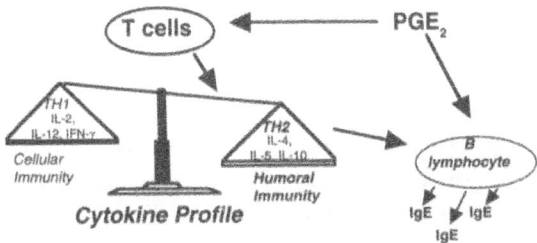

Figure 1. PGE2 can tip the immune balance toward Th2.

IMPORTANCE OF THE BALANCE OF Th1, Th2 FOR HUMAN DISEASE

Elevated levels of PGE_2 *in vivo* play an important role in various diseases. Patients who develop atopy tend to mount Th2 and IgE responses to the allergen and have elevated levels of PGE_2 compared to non-allergic individuals. Patients with Hodgkin's disease, hyper-IgE syndrome, Hyperprostaglandin E syndrome (Bartter's Syndrome), AIDS and trauma patients also have a higher level of PGE_2 and Th2 cytokines compared to healthy individuals (13-15). Therefore, *in vivo* evidence correlates with the assertion that PGE_2 can shift the balance of the immune response to one favoring Th2.

CONCLUSIONS

PGE_2 is a potent lipid molecule that can act on the immune system to promote the development of humoral, or Th2 responses. PGE_2 exerts its effect by elevating the second messenger cAMP. PGE_2 acts on antigen presenting cells, mature differentiated Th cells and B lymphocytes to induce the development of Th2 responses at the expenses of Th1 responses. The ability to use PGE_2 to modulate the immune response would be beneficial for the development of therapeutics aimed at skewing an unfavorable Th response to a more favorable one.

Acknowledgments

This work was supported by USPHS grants DE11390, CA11198, HL007216, DE07061.

REFERENCES

1. Roper, R.L., J.W. Ludlow, and R.P. Phipps. 1994. Prostaglandin E2 inhibits B lymphocyte activation by a cAMP-dependent mechanism: PGE-inducible regulatory proteins. *Cellular Immunology* 154, no. 1:296-308.
2. Negishi, M., Y. Sugimoto, and A. Ichikawa. 1993. Prostanoid receptors and their biological actions. *Progress in Lipid Research* 32, no. 4:417-34.
3. Fedyk, E.R., J.M. Ripper, D.M. Brown, and R.P. Phipps. 1996. A molecular analysis of PGE receptor (EP) expression on normal and transformed B lymphocytes: coexpression of EP1, EP2, EP3beta and EP4. *Molecular Immunology* 33, no. 1:33-45.
4. Fedyk, E.R., and R.P. Phipps. 1996. Prostaglandin E2 receptors of the EP2 and EP4 subtypes regulate activation and differentiation of mouse B lymphocytes to IgE-secreting cells. *Proceedings of the National Academy of Sciences of the United States of America* 93, no. 20:10978-83.
5. Roper, R.L., D.M. Brown, and R.P. Phipps. 1995. Prostaglandin E2 promotes B lymphocyte Ig isotype switching to IgE. *Journal of Immunology* 154, no. 1:162-70.

6. Swihart, K., U. Fruth, N. Messmer, K. Hug, R. Behin, S. Huang, G. Del Giudice, M. Aguet, and J.A. Louis. 1995. Mice from a genetically resistant background lacking the interferon gamma receptor are susceptible to infection with Leishmania major but mount a polarized T helper cell 1-type CD4+ T cell response. *Journal of Experimental Medicine* 181, no. 3:961-71.

7. Launois, P., T. Ohteki, K. Swihart, H.R. MacDonald, and J.A. Louis. 1995. In susceptible mice, Leishmania major induce very rapid interleukin-4 production by CD4+ T cells which are NK1.1-. *European Journal of Immunology* 25, no. 12:3298-307.

8. Bancroft, A.J., K.J. Else, and R.K. Grencis. 1994. Low-level infection with Trichuris muris significantly affects the polarization of the CD4 response. *European Journal of Immunology* 24, no. 12:3113-8.

9. O'Garra, A., and K. Murphy. 1996. Role of cytokines in development of Th1 and Th2 cells. *Chemical Immunology* 63:1-13.

10. Betz, M., and B.S. Fox. 1991. Prostaglandin E2 inhibits production of Th1 lymphokines but not of Th2 lymphokines. *Journal of Immunology* 146, no. 1:108-13.

11. Snijdewint, F.G., P. Kalinski, E.A. Wierenga, J.D. Bos, and M.L. Kapsenberg. 1993. Prostaglandin E2 differentially modulates cytokine secretion profiles of human T helper lymphocytes. *Journal of Immunology* 150, no. 12:5321-9.

12. van der Pouw Kraan, T.C., L.C. Boeije, R.J. Smeenk, J. Wijdenes, and L.A. Aarden. 1995. Prostaglandin-E2 is a potent inhibitor of human interleukin 12 production. *Journal of Experimental Medicine* 181, no. 2:775-9.

13. Sherry, R.M., J.I. Cue, J.K. Goddard, J.B. Parramore, and J.T. DiPiro. 1996. Interleukin-10 is associated with the development of sepsis in trauma patients. *Journal of Trauma* 40, no. 4:613-6; discussion 616-7.

14. DiPiro, J.T., T.R. Howdieshell, J.K. Goddard, D.B. Callaway, R.G. Hamilton, and A.R. Mansberger, Jr. 1995. Association of interleukin-4 plasma levels with traumatic injury and clinical course. *Archives of Surgery* 130, no. 11:1159-62; discussion 1162-3.

15. DiPiro, J.T., T.R. Howdieshell, R.G. Hamilton, N.F. Adkinson, Jr., and A.R. Mansberger, Jr. 1996. Increased plasma IgE levels in patients with sepsis after traumatic injury. *Journal of Allergy & Clinical Immunology* 97, no. 1 Pt 1:135-6.

60

LIPID MEDIATORS STIMULATE REACTIVE OXYGEN SPECIES FORMATION IN IMMORTALIZED HUMAN KERATINOCYTES

Rachel Goldman, Sandra Moshonov, and Uriel Zor

Department of Biological Chemistry
The Weizmann Institute of Science
Rehovot 76100, Israel

INTRODUCTION

Cell signaling is accompanied by rapid remodeling of membrane lipids by activated phospholipases (PLA$_2$, PLC, PLD and sphingomyelinase) with the generation of bioactive lipids that can serve as intra- and/or extracellular mediators (Serhan, et al., 1996). These lipid mediators include eicosanoids, platelet activating factor (PAF), diacyl glycerides, lyso-phosphatidylcholine (LPC), phosphatidic acid (PA), lyso-PA, ceramide and other newly-discovered autacoids. They are important in a wide range of cell-cell communication processes, such as host defense, inflammation, ischemia reperfusion and homeostasis.

One of many cell types which produces PAF, one of the most potent mediators in the inflammatory system, is the human keratinocyte. These cells, and the immortalized human keratinocyte cell line, HaCaT, also express functional PAF receptors: 10 pM to 1 μM PAF increased intracellular Ca^{+2} in both cell types (Travers, et al., 1995). LPC likewise elevated intracellular free calcium concentration ([Ca^{2+}]i), and evidence was presented in macrophages that it was effected via PAF receptors (Ogita, et al., 1997). LPC, a major component of atherogenic lipids, stimulates vascular smooth muscle cell proliferation, whereas ceramide, which is formed upon activation of sphingomyelinase inhibits cell division and triggers apoptosis.

Reactive oxygen species (ROS) have many physiological and pathological roles. They are implicated in aging (Ames, et al., 1995), in a wide range of neurological diseases such as Alzheimer's and Parkinson's, and in the initiation of tumors (Dreher and Junod, 1996). All cells generate ROS during the course of normal metabolism and recent evidence indicates that many cytokines, hormones and growth factors augment their immediate, transient or long term expression (Bae, et al, 1997; Ohba, et al, 1994; Moore and Brightman, 1991; Krieger and Kather, 1995; and Goldman, et al, 1998). A growing body of evidence points to ROS as being a second messenger in the activation of enzymes, transcription factors, growth, differentiation and apoptosis (Goldman, et al, 1998; Sen and Packer, 1996); Stevenson, et al, 1994; and Suzuki, et al, 1997).

Non-phagocytes produce low amounts of ROS compared to phagocytes, which are endowed with a cell specific, tightly regulated NADPH oxidase system (Henderson a Chappel, 1996). Several sites of ROS production have been described in non-phagocytic cells (plasma membranes, mitochondria, endoplasmic reticulum, peroxisomes) each site being associated with its own set of oxidative enzymes.

We have shown (Goldman, et al, 1998) that HaCaT cells generate ROS in response to epidermal growth factor (EGF), bradykinin, thapsigargin and the Ca^{2+}-ionophore A23187, agonists that interact with different primary cell targets. In all cases the presence of extracellular calcium was essential for ROS production, which invariably was also associated with a rise in $[Ca^{2+}]_i$. The enzyme(s) involved in ROS formation were inhibited by diphenylene iodonium (DPI) indicating dependence on flavin adenine dinucleotide (FAD).

In the present study we show that like the growth factors and hormones, lipid mediators stimulate ROS production in HaCaT cells and this too is dependent on a rise in $[Ca^{2+}]_i$. Thus, clearly, ROS could act as a second messenger in pathways regulated by lipid mediators.

MATERIALS AND METHODS

HaCaT cells (Boukamp, et al, 1988; Fusenig, DKFZ) were grown in 15 cm diameter plates in Dulbecco's modified Eagle's medium (DMEM; Gibco) supplemented with 10 % fetal calf serum and antibiotics and passaged twice a week. When confluent (35×10^6 cells), the cells were harvested using trypsin-EGTA solutions, washed, counted and resuspended (2×10^6 cells/ml) in Hank's balanced salt solution (HBSS) containing bovine serum albumin (BSA, 0.2 %). To detect the chemiluminescence (CL) elicited by ROS, isoluminol (6-amino-2,3-dihydro-1,4-phthalazinedione, 100 μM) and horseradish peroxidase (HRPO, 1U/ml) were added. The cells (0.5 ml) were then preincubated at 37°C for 20 minutes before adding the agonist, and the CL response was monitored at half- and one-minute intervals in a Lumac/3M biocounter, model 2080.

RESULTS AND DISCUSSION

Lipid Mediator-Induced ROS Generation

Each of the three lipid mediators we examined in depth, PAF, LPC and C6-ceramide, evoked a time- and dose-dependent CL response in HaCaT cells (Figure 1). The peak response was attained within 30 to 45 sec and decayed within 10 min. The maximum ranged between 20 and 50×10^3 cpm. The ED50 of PAF was about 60 pM and that of LPC was in the μM range. As this assay system requires addition of peroxidase, which does not penetrate the cell (no CL was observed in its absence), the CL is dependent upon the formation and release into the medium of peroxides such as hydrogen peroxide.

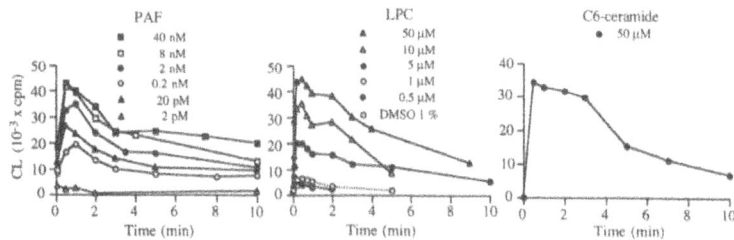

Figure 1. Kinetics of the chemiluminescence (CL) response to PAF, LPC and C6-ceramide.

Isoluminol is oxidized by H_2O_2 or O_2^-, forming an unstable endoperoxide which releases photons on its decomposition to a more electronically stable state (Faulkner and Fridovich, 1993). O_2^- can be rapidly converted intracellularly to H_2O_2 by superoxide dismutase so the actual species generated in the cells in response to specified agonists have not been defined. There was a low background CL from resting HaCaT cells and a small response to 1% DMSO.

In addition to those already mentioned, we have assessed the potential to generate ROS of other lipid mediators known to activate metabolic cascades (Figure 2). Arachidonic acid (AA) had a small activating effect, while its esterified metabolite anantamide (AE) had none. The dialkylglycerols, 1-oleyl-2-acetyl-sn-glycerol (OAG) and 1,2-dioctanoyl-sn-glycerol (diC8), which activate PKC, had a relatively small effect whereas L-α-phosphatidic acid (PA) and lyso-PA (LPA) evoked a large response. C2-ceramide showed comparable activity to C6-ceramide, while the dihydro-C6-ceramide derivative, inactive in other systems, was also inactive with respect to ROS formation. Likewise, the inactive metabolite of PAF, lyso-PAF, did not stimulate ROS formation.

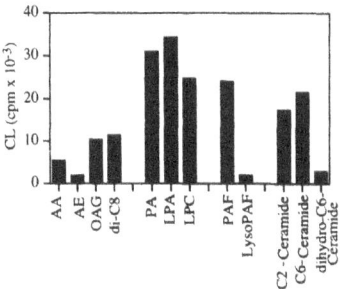

Figure 2. Chemiluminescence (CL) in response to lipid mediators. Peak values were reached 30 to 45 sec after agonist administration. All mediators were added at a concentration of 50μM except for PAF (2 nM) and lyso-PAF (0.2 μM). Abbreviations are: AA, arachidonic acid; AE, anantamide or arachidonyl ethanolamide; OAG, 1-oleyl-2-acetyl-sn-glycerol; di-C8, 1,2-dioctanoyl-sn-glycerol; PA, L-α-phosphatidic acid; LPA, lyso-PA; LPC, lysophsphatidylcholine; PAF, platelet activating factor; C2-ceramide, N-acetylsphingosine; C6-ceramide, N-hexanoylsphingosine.

[Ca^{2+}]i Dependence of ROS Formation

We next determined the dependence of the CL response on intra- and extracellular Ca^{2+} concentration. Ten minutes' pretreatment with BAPTA/AM (40μM) which penetrates into the cell and is hydrolyzed to the intracellular Ca^{2+}-chelator, BAPTA, fully suppressed the CL response to PAF (Figure 3). When HaCaT cells were resuspended in HBSS/BSA lacking Ca^{2+}, there was still a small response to PAF. However, on removal of the residual Ca^{2+} by adding 20 μM EGTA to the medium, the CL response disappeared entirely. Addition of 1 mM Ca^{2+} to cells incubated for at least 120 min in medium lacking Ca^{2+} resulted in a CL response to PAF comparable to that obtained under control conditions (not shown). Thus deprivation of extracellular Ca^{2+} did not undermine the ability of the cells to generate ROS upon Ca^{2+} replenishment. The same pattern was observed using LPC as the agonist (not shown). The H_2O_2-luminol-HRPO system itself is not Ca^{2+}-dependent. Incubation of luminol-HRPO in medium containing 3 mM EGTA for 20 min did not affect the CL response to H_2O_2.

415

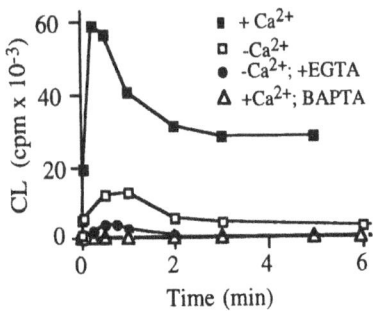

Figure 3. Ca^{2+}-dependence of PAF-induced chemiluminescence (CL). The response to 2 nM PAF was monitored in HBSS (devoid of Ca^{2+}) containing 1mM $MgCl_2$ and 0.2% BSA ($-Ca^{2+}$) with either of the following additives: 20μM EGTA (+EGTA), 40μM BAPTA/AM added 10 min before PAF (BAPTA), 1.5 mM Ca^{2+} ($+Ca^{2+}$).

Agonist-Receptor Specificity

The receptors for PAF and ceramide are not recognized by their derivatives, lyso-PAF and dihydro-C6-ceramide. Unlike the parent molecule, they were inactive in signal transduction pathways and were unable to evoke a CL response (Figure 2). LPC has been reported to effect its response via PAF receptors, but whereas the PAF receptor antagonists, BN 52021 and CV 6209, blocked the response to PAF, they did not affect the response to LPC. An LPC-induced rise in CL did not interfere with a subsequent response to PAF; similarly, cells which had reacted to PAF still responded normally to LPC (Table 1). In this series of experiments too, the antagonists affected only the response to PAF, whether it was given before or after the LPC. PAF antagonists did not affect the CL response to thapsigargin or calcium ionophore A23187.

Table 1. Specificity of PAF-induced chemiluminescence

Exp.	Antagonist[1]	Agonist[2]	cpm[4] (×10⁻³)	Agonist[3]	cpm[4] (×10⁻³)
1	-	PAF	16.7	LPC	9.6
2	BN52021	PAF	0.3	LPC	7.2
3	CV 6209	PAF	0.3	LPC	11.1
4	-	LPC	18.3	PAF	21.9
5	BN52021	LPC	14.5	PAF	1.9
6	CV 6209	LPC	20.5	PAF	2.9

[1]Antagonists were added 10 min before the first agonist[2]. The second agonist[3] was added 10 min after the first.

[4]Peak values of chemiluminescence induced by each agonist at 30 to 45 sec. The concentrations of antagonists and agonists were: BN52021, 30 μM; CV 6209, 2 μM; PAF, 2 nM; LPC, 10 μM.

Source of ROS

The potential sources of peroxides (H_2O_2) and superoxide anion range from normal mitochondrial energy metabolism, NAD(P)H oxidases, lipoxygenases, cyclooxygenases to flavin oxidases and microsomal cytochrome systems. In order to identify or negate any of these sources we tested inhibitors associated with the various systems on lipid-stimulated CL. We found that the ROS generating enzyme(s) is probably FAD-dependent since diphenylene iodonium (DPI) inhibited the CL response. Administered to cells 10 min before the agonist, DPI (6μM) inhibited by 60 to 100 % the activation by PAF and LPC. As DPI inhibits NO synthase (Stuehr, et al, 1991) and the peroxynitrite anion (·ONOO⁻) which results from the reaction between ·NO and ·O2⁻ induces luminol-dependent CL (Radi, et al, 1993), we tested two NO synthase inhibitors. Neither N-methyl-D-arginine nor aminoguanidine (100 μM, 20 min preincubation) affected the CL response to PAF. 1-Aminobenzotriazole (50 μM), a "suicide" substrate for cytochrome P450, was also without effect on CL, as were inhibitors of lipoxygenases, MK886 (20 μg/ml), and of cyclooxygenases, ETYA (100 μM). However, a different lipoxygenase inhibitor, AA861 (20 μM) did inhibit PAF-stimulated CL by 55% and 72% at 10 μM and 50 μM inhibitor, respectively. MAFP (20μM), an inhibitor of cPLA$_2$, did not block the CL response. Of the mitochondrial electron transport inhibitors, rotenone (20 μM), which blocks the NADH-Q reductase complex at the beginning of the electron transport chain, was without effect. However, preincubation for 20 min with 1 mM cyanide, which reduces the activity of mitochondrial respiratory components (including cytochrome oxidase) led to 90 to 100 % inhibition of PAF induced CL. All the above mentioned inhibitors were preincubated with the cells for 10 to 20 min before adding PAF, and had no inhibitory effect on the H_2O_2-luminol-HRPO system itself. Thus, NO synthase, lipoxygenases, cyclooxygenases, cPLA$_2$ and mitochondrial NADH-Q reductase are unlikely to be the lipid-stimulated ROS-producing enzymes. NAD(P)H oxidase and/or mitochondrial cytochrome oxidases, however, are distinctly possible sources of ROS in this system.

Mitochondria are affected by local changes of Ca^{2+} and sustained increases in its concentration change respiratory function and are thought to play a major role in apoptosis. Recent evidence suggests that ceramide has a direct effect on the mitochondrial electron transport chain leading to the generation of ROS (Garcia, et al, 1997). It has also been shown in intact cells that the ceramide-induced activation of transcription factors and apoptosis is mediated by mitochondrial derived ROS (Quillet, et al, 1997). Together, these studies emphasize the potential importance of this site in the determination of the redox state of cells stimulated by lipid, or other, mediators.

Acknowledgment

This research was supported by a grant from the Israeli National Council for Research and Development (NCRD) and the Deutsches Krabsforschungszentrum (DKFZ).

REFERENCES

1. Serhan, C. N., J. Z. Haeggstrom, and C. C. Leslie. 1996. Lipid mediator networks in cell signaling: update and impact of cytokines. *FASEB-J 10:1147-58.*
2. Travers, J. B., J. C. Huff, M. Rola-Pleszczynski, E. W. Gelfand, J. G. Morelli, and R. C. Murphy. 1995. Identification of functional platelet-activating factor receptors on human keratinocytes. *J Invest Dermatol 105:816-23.*
3. Ogita, T., Y. Tanaka, T. Nakaoka, R. Matsuoka, Y. Kira, M. Nakamura, T. Shimizu, and T. Fujita. 1997. Lysophosphatidylcholine transduces Ca2+ signaling via the platelet- activating factor receptor in macrophages. *Am J Physiol 272:H17-24.*
4. Dreher, D., and A. F. Junod. 1996. Role of oxygen free radicals in cancer development. *Eur J Cancer 32a:30-8.*

5. Ames, B. N., M. K. Shigenaga, and T. M. Hagen. 1995. Mitochondrial decay in aging. *Biochim-Biophys-Acta 1271:165-70.*
6. Bae, Y. S., S. W. Kang, M. S. Seo, I. C. Baines, E. Tekle, P. B. Chock, and S. G. Rhee. 1997. Epidermal growth factor (EGF)-induced generation of hydrogen peroxide Role in EGF receptor-mediated tyrosine phosphorylation. *J Biol Chem 272:217-221.*
7. Ohba, M., M. Shibanuma, T. Kuroki, and K. Nose. 1994. Production of hydrogen peroxide by transforming growth factor-beta 1 and its involvement in induction of egr-1 in mouse osteoblastic cells. *J-Cell-Biol 126:1079-88*
8. Morre, D. J., and A. O. Brightman. 1991. NADH oxidase of plasma membranes. *J Bioenerg Biomembr 23:469-89.*
9. Krieger Brauer, H. I., and H. Kather. 1995. The stimulus-sensitive H2O2-generating system present in human fat-cell plasma membranes is multireceptor-linked and under antagonistic control by hormones and cytokines. *Biochem-J 307:543-8 i*
10. Goldman, R., S. Moshonov, and U. Zor. 1998. Stimulation of reactive oxygen species formation in a human keratinocyte cell line : role of calcium. *Arch. Biochem. Biophys. 350:10-8.*
11. Sen, C. K., and L. Packer. 1996. Antioxidant and redox regulation of gene transcription. *FASEB-J 10:709-20.*
12. Stevenson, M. A., S. S. Pollock, C. N. Coleman, and S. K. Calderwood. 1994. X-irradiation, phorbol esters, and H2O2 stimulate mitogen-activated protein kinase activity in NIH-3T3 cells through the formation of reactive oxygen intermediates. *Cancer Res 54:12-5.*
13. Suzuki, Y. J., H. J. Forman, and A. Sevanian. 1997. Oxidants as stimulators of signal transduction. *Free Radical Biology & Medicine 22:269-285.*
14. Henderson, L. M., and J. B. Chappel. 1996. NADPH oxidase of neutrophils. *Biochim Biophys Acta 1273:87-107.*
15. Boukamp, P., R. T. Petrussevska, D. Breitkreutz, J. Hornung, A. Markham, and N. E. Fusenig. 1988. Normal keratinization in a spontaneously immortalized aneuploid human keratinocyte cell line. *J Cell Biol 106:761-71.*
16. Faulkner, K., and I. Fridovich. 1993. Luminol and lucigenin as detectors for O2.-. *Free-Radic-Biol-Med 15:447-51 i*
17. Stuehr, D. J., O. A. Fasehun, N. S. Kwon, S. S. Gross, J. A. Gonzalez, R. Levi, and C. F. Nathan. 1991. Inhibition of macrophage and endothelial cell nitric oxide synthase by diphenyleneiodonium and its analogs. *Faseb J 5:98-103.*
18. Radi, R., T. P. Cosgrove, J. S. Beckman, and B. A. Freeman. 1993. Peroxynitrite-induced luminol chemiluminescence. *Biochem-J 290:51-7.*
19. Garcia Ruiz, C., A. Colell, M. Mari, A. Morales, and J. C. Fernandez Checa. 1997. Direct effect of ceramide on the mitochondrial electron transport chain leads to generation of reactive oxygen species. Role of mitochondrial glutathione. *J-Biol-Chem 272:11369-77.*
20. Quillet Mary, A., J. P. Jaffrezou, V. Mansat, C. Bordier, J. Naval, and G. Laurent. 1997. Implication of mitochondrial hydrogen peroxide generation in ceramide-induced apoptosis. *J-Biol-Chem 272:21388-95.*

418

61

NITRIC OXIDE SYNTHESIS IS INCREASED IN PERIODONTITIS

Michael Matejka[1], Christian Ulm[1], Lukas Partyka[2], Martin Ulm[3], and Helmut Sinzinger[2]

[1] Department of Oral Surgery, School of Dentistry,
[2] Clinic for Nuclear Medicine
[3] Department of Gynaecology,
   University of Vienna

INTRODUCTION

A number of potential regulatory mechanisms have been described that may be involved in the control of cell function in the inflammatory periodontium. Among others, an involvement of growth factors and cytokines in these processes is of particular importance. There is also evidence for a wide range of effects of mechanical stimulation and the central role of prostaglandins (PG)[1,2].

PG production may be stimulated in the gingiva[3-6]. While constitutive cyclooxygenase (COX-1) seems to regulate normal PG synthesis, the inducible isoenzyme (COX-2) appears to be involved in the synthesis induced by cytokines (e.g. interleukin 1β, TNFα), endotoxin, and growth factors.

PG and their metabolites were found in periodontal tissue. In periodontitis, an elevated PG (E_2, I_2) production by gingival tissue resulting in an increase in crevicular sulcus fluid has been observed[2]. The crevicular fluid PGE_2 may serve as a predictor of periodontal attachment loss[2]. PG exert other effects that contribute to the pathology of periodontitis, e.g. inhibition of human gingival fibroblast growth and synthesis of collagen and noncollagenous protein[7]. They have also been shown to contribute to osteolytic processes caused by gingival inflammation[8].

However, the activating mechanism of COX and the sequence of events, which finally results in osteolysis, still need to be elucidated.

Prostacyclin (PGI$_2$) and nitric oxide (NO)[9], are two potent endogenous vasodilators that co-regulate the vascular tone and permeability via cAMP and cGMP. On the other hand, they play an important role in inflammatory process induction and prolongation[9-11].

NO is synthesized in the presence of O$_2$ from L-arginine by conversion to L-citrulline[9]. The enzyme catalyzing this reaction is either the Ca^{2+}-dependent constitutive NO synthase (cNOS) or inducible NOS (iNOS) produced under the influence of lipopolysaccharide and/or cytokines.

Recent studies suggest the involvement of NO in the regulation of mineralized tissue function and the potential role of NO in maintenance of the ligament space[1].

While an increased PG (E$_2$, I$_2$) production in periodontitis has been well documented, no data are available yet on NO synthesis in intact and inflammatory gingival tissues *in vivo*. Therefore, the hypothesis of this study was that inflammatory gingival tissue produces not only more PGI$_2$ but also more NO.

MATERIAL AND METHODS

Patients: Gingival tissue was obtained from 21 patients with moderate periodontitis (AAP Class III, 7 females, 14 males, 27-55 yr.), which underwent periodontal flap surgery. Tissue samples from 16 healthy controls (7 females, 9 males, 24-50 yr.) were obtained during removal of excessive periodontal mucosal tissue during routine surgical treatment. Diabetics and smokers were excluded. None of the patients and controls received any drug known to influence the PG- and NO system for a period of at least 2 weeks.

The tissue samples were excised and immediately placed in cold (0°C) phosphate buffered saline (pH 7.4). They were then frozen and stored at -70°C until analyzed for the amino-acids L-arginine, L-citrulline, L-cysteine, and PGI$_2$. All assays were done within 6 months from sampling. The following analyses were performed.

Assay of amino-acids after extraction of samples

Gingival tissue samples were homogenized and mixed with borate buffer and 9-fluorenylmethyl chloroformate for precolumn derivatization. After about 40 seconds, the mixture was extracted with pentane. The extraction was repeated twice. The aqueous solution with the amino-acid derivatives was then ready for high-performance liquid chromatography (HPLC). The levels of the three investigated amino-acids (L-arginine, L-citrulline, and L-cysteine) were analyzed using reversed phase HPLC with a fluorescence detector[12]. The intraassay and interassay coefficient of variation was 1.2±0.2 % and 2.5±0.6 %, respectively.

Bioassay of PGI$_2$

Venous blood was withdrawn from healthy volunteers (6 males, aged 25-46 yr., with no medication for at least 2 weeks) and anticoagulated (1:10 v/v) with 3.8% sodium citrate. Platelet-rich plasma (PRP) was prepared by centrifugation at 150g and 22°C for 5 minutes.

Platelet aggregation was induced in 600 µl PRP samples by addition of 100µl adenosine diphosphate (ADP, Boehringer Mannheim, Germany) to obtain the final concentration of 1 µM. A Born aggregometer was used.

The incubate of gingival tissue was prepared by adding tissue samples with a wet weight of 16 - 64 mg into the 300 µl Tris-HCl buffer at 22°C for 3 minutes. One minute before the instillation of ADP into PRP, either 100 µl of Tris-HCl buffer (pH 7.4) or 100 µl of the gingival tissue incubate were added. The inhibitory effect of the incubation buffer on ADP-induced platelet aggregation was compared with that of the buffer control. The extent of antiaggregatory action (maximum amplitude of the aggregation response curve) of the biological activity (PGI$_2$) was compared with that of a synthetic PGI$_2$ standard (The Upjohn Company, Kalamazoo, MI, USA), quantified, and expressed in pg PGI$_2$ / mg gingival tissue / min.

Intraassay coefficients of variation were 3.7 $\pm$ 1.0 %, whereas interassay coefficients of variation were 5.9 $\pm$ 1.8 %. The biologically active compound was characterized as PGI$_2$ by inactivation with boiling, inhibition of cyclooxygenase with acetylsalicylic acid, 15-hydroxyperoxy-arachidonic acid and by using a specific anti-serum, as previously described[6].

Statistical analysis

The mean value and standard deviation (SD) for each assay were calculated, and a Kolmogoroff-Smirnov test was carried out to examine whether the differences between the mean values were significant (nominal level of significance: $p < 0.05$). The zero hypothesis was proved with the Mann-Whitney-U-test. Moreover, the Spearman correlation co-efficient and the Kendall correlation were calculated to detect possible monotonous associations between the different parameters.

RESULTS

Inflamed human gingival tissue contained significantly more L-arginine ($p < 0.05$) and L-citrulline ($p < 0.05$) than did the control samples. The results of the analysis are listed in table 1. Furthermore, the synthesis of PGI$_2$ was significantly increased in the "inflamed gingiva" group ($p < 0.05$). However, L-cysteine concentrations in patients with inflamed gingival tissue did not differ from those in controls ($p = 0.101$).

The "inflamed gingiva" group showed a significant correlation between L-arginine and L-cysteine ($p < 0.05$) and between L-arginine and PGI$_2$ ($p < 0.05$).

DISCUSSION

NO prevents platelet aggregation, modifies the response of vascular smooth muscle, and maintains the integrity of the vascular compartment. Similarly to other vascular beds, NO has been shown to increase vascular conductance in oral structures - gingiva, tongue, and submandibular glands[13]. This "physiologically" secreted NO is produced by cNOS, mainly in the endothelial cells.

Table 1. Amino-acids concentration and PGI_2 production in human gingival samples. (* p<.05 - differences statistically significant vs. controls - Kolmogoroff-Smirnov test).

pg/mg/min	Arginine	Citrulline	Cysteine	PGI_2
Noninflamed gingiva				
MEAN	1.734	0.501	0.257	4.017
SD	0.334	0.122	0.112	0.628
Inflamed gingiva				
MEAN	*2.723	*1.259	0.325	*7.206
SD	0.546	0.432	0.134	1.403

However, in periodontitis, NO is probably produced by iNOS, secreted from a variety of cells and its actions are directed on maintaining the inflammation. NO is produced by NOS from L-arginin and O_2. Stechiometrical by-product of this reaction is L-citrulin. NO is highly reactive and therefore difficult to assess directly. In the present study L-citrulline, which correlates with NO formation, was determined to assess NO production[14]. Method used in the present study provides sensitive and non-radioactive measure for NOS activity assay in small tissue samples and in tissues with low to moderate levels of enzyme activity[15, 16].

Other studies have shown that interleukin-1β, tumor necrosis factor-α, and Escherichia coli lipopolysaccharide cause a dose- and time-dependent increase in nitrite (NO_2^-), the stable metabolite of nitric oxide (NO), in conditioned media of osteoblast-like cell cultures over 48 hours[17]. This means that not only monocytes/macrophages or leukocytes, but also gingival tissue cells may be induced to produce NO by proinflammatory cytokines and bacterial endotoxin.

The present study confirms the findings of Matejka[6], who reported that inflamed gingival tissue has an increased capacity to generate prostacyclin. It is still unclear whether PGI_2 in periodontitis has a mainly osteolytic effect or helps to sustain the inflammatory process because of its vasoactive action. We found that inflamed gingival tissue showed not only an increased PGI_2 synthesis but also a higher content of L-arginine and L-citrulline. These findings suggest that changes in NO concentrations occur in patients with inflamed gingival tissue. This may be important pathophysiologically, because both NO and PGI_2 are potent vasodilators and play a crucial role in the development and persistence of inflammatory processes. However, as has been observed in other tissues, NO, aside from other substances (leukotriene, interleukin, bradykinin, etc.) activates cyclooxygenase[11, 18, 19], thus increasing the release of both PGI_2 and

PGE$_2$. The mechanisms that result in COX activation by NO are still unknown. It is also unclear whether the stimulation of PG synthesis relevant to periodontitis takes place through interleukins 1α and 1β and via the activation of inducible COX, as has been described elsewhere[20].

The regulation of COX activation by NO might be an important mechanism to modulate the initial inflammatory response. Further studies will be needed, which address the interactions between all of these mechanisms in order to determine the key factors that may control periodontal cell function. For the future, an understanding of these interactions might to lead to important clinical developments in periodontal therapy.

REFERNCES

1. F.J. Hughes, Cytokines and cell signaling in the periodontium. *Oral. Dis.* 1:259 (1995).
2. S. Offenbacher, B.M. Odle and T.E. van Dyke, The use of crevicular fluid prostaglandin E$_2$ levels as a predictor of periodontal attachment loss. *J. Periodont. Res.* 21:101 (1986).
3. D. Richards and R.B. Rutherford, The effects of interleukin 1 on collagenolytic activity and prostaglandin-E secretion by human periodontal - ligament and gingival fibroblast. *Archs. Oral Biol.* 33:237 (1988).
4. C.A. Dinarello, Interleukin 1 and its Biologically Related Cytokines. *Adv. Immunol.* 14:153 (1989).
5. P. Stashenko, P. Fujiyoshi, M.S. Obernesser, L. Prostak, A.D. Haffajee and S. Socransky, Levels of Interleukin 1-β in tissue from sites of active periodontal disease. *J. Clin. Periodontol.* 18:548 (1991).
6. M. Matejka, C. Ulm, A. Nell, P. Solar, B. Riegler and H. Sinzinger, Stimulation der Prostaglandin I$_2$ - Synthese durch Interleukine im periodontalen Gewebe. *Z. Stomatol.* 91:205 (1994).
7. H. Arai, Y. Nomura and M. Kinoshita et al., Response of human gingival fibroblasts to prostaglandins. *J. Periodontal. Res.* 30:303 (1995).
8. P. Stashenko, J.J. Jandinski, P. Fujioshi, J. Tyna and S.S. Socransky, Tissue levels of bone resorptive cytokines in periodontal disease. *J. Periodontol.* 62: 504 (1991).
9. S. Moncada, R.M.J. Palmer, E.A. Higgs, Nitric oxide: physiology, pathophysiology and pharmacology. *Pharmacol. Rev.* 43:109 (1991).
10. R. Katzenschlager, K. Weiss. W. Rogatti, M. Stelzeneder and H. Sinzinger, Interaction between Prostaglandin E1 and Nitric Oxide (NO). *Thrombosis Res.* 62:299 (1991).
11. D. Salvemini, T.P. Misko, J.L. Masferrer, K. Seibert, M.G. Currie and P. Needleman, Nitric oxide activates cyclooxygenase enzymes. *Proc. Natl. Acad. Sci. USA* 90:7240 (1993).
12. S. Einarsson, B. Josefsson and S. Lagerkvist, Determination of aminoacids with 9-fluorenyl-methylchloroformate and reversed-phase high-performance liquid chromatography. *J. Chromatogr.* 282:609 (1983).

13. A. Fazekas, J.L. Matheny, G.I. Roth amd D.R. Richardson, Effect of nitric oxide inhibition on capsaicin-elicited vasodilation in the rat oral circulation. *Res. Exp. Med. Berl.* 194:357 (1994).

14. M. Carlberg, Assay of neuronal nitric oxide synthase by HPLC determination of citrulline. *J. Neurosci. Methods* 52:165 (1994).

15. K.U. Johansson and M. Carlberg, NADPH-diaphorase histochemistry and nitric oxide synthase activity in deutocerebrum of the crayfish, <u>Pacifastacus leniusculus</u> (Crustacea, Decapoda). *Brain Res.* 649:36 (1994).

16. R. Elofsson, M. Carlberg, L. Moroz, L. Nezlin and D. Sakharov, Is nitric oxide (NO) produced by invertebrate neurones? *Neuroreport* 4:279 (1993).

17. P.D. Damoulis and P.V. Hauschka, Cytokines induce nitric oxide production in mouse osteoblasts. *Biochem. Biophys. Res. Commun.* 201:924 (1994).

18. J.P. Warren, M.L. Coughlan and T. Williams, Endotoxin-induced vasodilatation in anaesthetized rat skin involves nitric oxide and prostaglandin synthesis. *Br. J. Pharmacol.* 106:953 (1992).

19. V. Rettori, M. Gimeno, K. Lyson and S.M. Mc Cann, Nitric oxide mediates norepinephrine induced prostaglandin E2 release from the hypothalamus. *Proc. Natl. Acad. Sci. USA* 89:11543 (1992).

20. J.A. Corbett, M.A. Sweetland, J.L. Wang, J.R. Lancaster Jr. and M.L. McDaniel, Nitric oxide mediates cytokine-induced inhibition of insulin secretion by human islets of Langerhans. *Proc. Natl. Acad. Sci.* USA 90:1731 (1993).

62

AUGMENTATION EFFECTS OF HIGH GLUCOSE ON ENDOTOXIN-INDUCED NITRIC OXIDE PRODUCTION IN MURINE MACROPHAGES

Soo H. Lee, Hyun Goo Woo, Ji Y. Kim and Chang-Hyun Moon

Department of Physiology, School of Medicine
Ajou University
Suwon 442-749, Korea

INTRODUCTION

Nitric oxide (NO) is a highly effective molecule with a multitude of biological effects ranging from vasodilation to cytotoxicity. Expression of the inducible nitric oxide synthase (iNOS) in macrophage following exposure to cytokines or endotoxins is frequently associated with a generalized or localized inflammatory response to resulting from infection or tissue injury. The production of NO and the expression of iNOS are known to be modulated by a variety of factors. Recent study showed that endotoxin induced iNOS expression is enhanced in murine macrophages and mesangial cells by high glucose. Although the activation of protein kinase C (PKC) is suggested as a possible mechanism for the augmenting effect of high glucose (Sharma,et al, 1995), little is known about underlying intracellular mechanisms.

It is quite conceivable that high glucose mediates its adverse effects via multiple mechanisms, since glucose and its metabolites are utilized by numerous pathways. Glucose metabolism via sorbitol pathway is known to increase the $NADH/NAD^+$ ratio in the cytosol, which increases de novo synthesis of diacylglycerol, and in turn, activates PKC. Thus, we designed the experiments to determine whether these metabolic schemes contribute to the glucose enhancing effect on endotoxin induced nitrite production

MATERIALS AND METHODS

Materials

E.coli lipopolysaccharide (0111:B4), H-7, staurosporine, phorbol-12-myristic acetate (PMA), sodium pyruvate and the other reagents were purchased from Sigma. Zopolrestat was kindly gifted from Pfizer Inc. Polyclonal antibody to iNOS was from UBI.

Eicosanoids and Other Bioactive Lipids in Cancer, Inflammation, and Radiation Injury, 4
Edited by Honn *et al.*, Kluwer Academic / Plenum Publishers, New York, 1999.

425

Cell Culture

Macrophage like cell line RAW 264.7 was grown in Dulbecco's Modified Eagles Medium (Gibco-BRL) supplemented with 10 % fetal bovine serum (Gibco-BRL), and antibiotics (100 U/ml of penicillin and 100 μg/ml of streptomycin). They were plated in 96 well plate (2~3 × 10^5 cells per well) and stimulated with various stimuli as indicated.

Assay for Nitrite Production

Nitrite was measured spectrophotometrically with Griess reagent. At the end of the incubation, 100μl of the culture medium was mixed with an equal volume of Griess reagent (1 part of 0.1% naphtylethylenediamine hydrochloride and 1 part of sulfanilamide in 10% phosphoric acid). The reaction was completed after 10 min of incubation, the absorbance at 540nm was measured and the nitrite concentration was determined using a curve calibrated on sodium nitrite standards.

Assay for iNOS Protein

iNOS protein was analyzed by immunoblotting with the anti-iNOS antibody. The cells were incubated in 6 well plates and stimulated with various stimuli for 24 hours and lysed with 200 μl of buffer containing 150 mM NaCl, 10 mM Na_2HPO_4 (pH 7.4), 1% sodium deoxycholate, 1 % NP-40, 0.1 % SDS, 2 mM EDTA, 20 μM leupeptin, 1μM pepstatin A and 2mM phenylmethylsulfonyl fluoride (PMSF). The lysates were then incubated on ice for 10 min and centrifuged at 12,000g for 10 min at 4. And the supernatants were subjected to 7.5 % SDS-PAGE using buffer system of Laemmli. The separated proteins were electrophroretically transferred to polyvinylidene difluoride (PVDF) membranes. The blots were soaked for 1 hour in 5 % non-fat dried milk and then incubated with the anti-iNOS antibody (1:1,500). The iNOS proteins were visualized by incubation with color development reagent (Bio-Rad) containing bromochloroindolyl phosphate and nitro blue tetrazolium.

Data Analysis

Statistical differences ($p < 0.05$) between mean values were determined by one–way analysis of the variance followed by Student's t test.

RESULTS AND DISCUSSION

It is well accepted that the activation of PKC via sorbitol pathway is associated with many of diabetic complications. The incidence of septic shock is much greater in diabetes than in non-diabetes patients, which is suggested to be due to the increased production of NO in diabetes. Recently, it was reported that instillation of high concentrations of glucose into macrophage like RAW 264.7 cells enhanced the NO expression induced by endotoxin, and also suggested that it may be mediated by PKC activation (Sharma et al., 1996). In this study, we hypothesized that the activation of PKC via sorbitol pathway could explain glucose-enhancing effect on endotoxin induced NO production.

The addition of the PKC inhibitors, H-7 or staurosporin, with LPS for 48 hours, diminished

LPS induced NO production regardless of glucose concentrations (Fig. 1). But PMA, a potent PKC activator, did not show any significant effect on enhanced production of NO by high glucose when simultaneously added with LPS for 48 hours (Fig. 2). And also, preincubation with PMA for 30 min before LPS treatment did not affect the NO production (Table 1). Prolonged preincubation of cells with PMA for 24 hours, which is known to downregulate PKC activity, reduced the NO production to control level in both normal and high glucose treated groups (Table 2). From these results, we couldn't get concrete evidence for the possibility that the activation of PKC is specifically involved in glucose enhancing effect on endotoxin induced NO production.

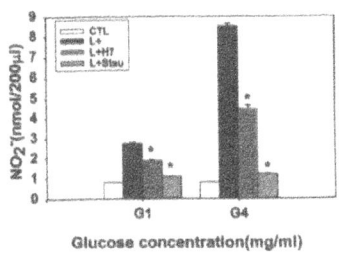

Figure 1. Effects of PKC inhibitors on nitrite production in RAW 264.7 stimulated with LPS. Confluent RAW 264.7 cells were exposed to LPS (10 ug/ml) in the presense or absence of H-7 (50 uM) and staurosporin (100 nM). After an additional 48 hrs of stimulation, the conditioned medium was assayed for nitrite as described under Materials and Methods. Data are means ± S.E of three separate experiments. Stau:Staurosporine(100nM) *p < 0.01 vs LPS

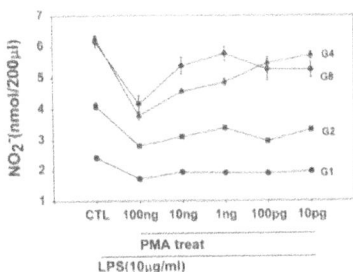

Figure 2. Effects of PMA on nitrite production in RAW 264.7 cell s stimulated with LPS. Confluent RAW 264.7 cells were exposed to LPS in the presence or absence of PMA for various concentrations. After additional 48 hrs of stimulation, the conditioned medium was assayed for nitrites as described under Materials and Methods. Data are means ± S.E of three separate experiments. *p < 0.01 vs LPS

Table 1. Effects of pretreatment with PMA on LPS induced nitrite accumulation

Glucose (mg/ml)	Nitrite Accumulation (nmol/200ml)			
	Control	PMA	LPS	LPS+PMA
G1	1.38±0.18	1.31±0.13	1.78±0.13	1.77±0.23
G4	1.42±0.17	1.47±0.20	7.29±0.35	6.91±0.18

Confluent RAW 264.7 cells were exposed to PMA (100ng/ml) 30 min before LPS (10ug/ml) treatment in different glucose concentrations (G1: 1mg/ml , G4: 4mg/ml) for 48 hrs Results are means ± S.E. of three experiments.

Table 2. Effects of prolonged incubation with PMA for 24 hours

Glucose (mg/ml)	Nitrite Accumulation (nmol/200ml)					
	Group I			Group II		
	Control	PMA	LPS	Control	PMA	LPS
G1	0.67±0.07	0.90±0.03	1.62±0.03	0.66±0.01	0.68±0.01	0.82±0.02
G4	0.64±0.00	0.98±0.02	4.99±0.18	0.64±0.01	0.68±0.00	0.78±0.05

In group I, confluent RAW 264.7 cells were exposed to PMA (100 ng/ml) or LPS (10ug/ml) for 36 hrs. And in group II, cells were pre-incubated with PMA (100 ng/ml) for 24 hrs and further incubated as group I for 36 hours in different glucose concentrations (G1: 1mg/ml , G4: 4mg/ml). Results are means ± S.E. of three experiments.

In order to address the possible contribution of glucose metabolic schemes to the augmentation of NO production, cells were treated with an aldose reductase inhibitor, zopolrestat, or sodium pyruvate, which is expected to decrease the altered NADH/NAD$^+$ ratio in the cells exposed to high glucose. Neither zopolrestat nor sodium pyruvate normalize NO production suggesting that, at least in RAW 264.7 cells, increase in glucose metabolism via sorbitol pathway does not contribute to the augmentation of LPS induced NO production by high glucose (Fig 3,4). These results are supported by western blot analysis. iNOS expression was greatly increased in high glucose group, which was not affected by the treatment of zopolrestat (Fig 5).

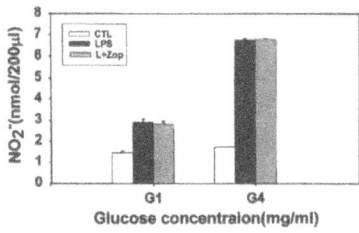

Figure 3. Effects of zoplrestat. Confluent RAW 264.7 cells were exposed to LPS in the presence or absence of zopolrestat in different glucose concentrations (1 mg/ml, 4 mg/ml). After 48 hrs of incubation, the conditioned medium was assayed for nitrites as described under Materials and Methods. Data are means ± S.E of three separate experiments. Zop: zopolrestat

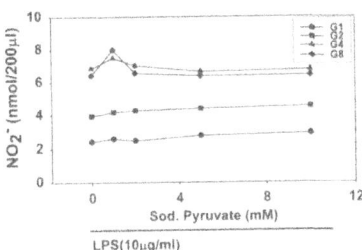

Figure 4. Effects of sodium pyruvate. Confluent RAW 264.7 cells were exposed to LPS in the presence or absence of various concentrations of sodium pyruvate as shown in the figure. After additional 48 hrs of stimulation, the conditioned medium was assayed for nitrites as described under Materials and Methods. Data are means ± S.E of three separate experiments. *p < 0.01 vs LPS

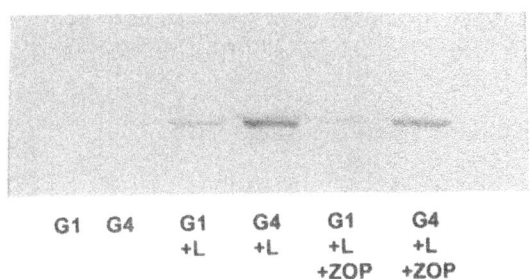

Figure 5. Effects of zopolrestat on the iNOS expression. Confluent RAW 264.7 cells were exposed to LPS in the presence or absence of zopoplrestat (50 nM) in normal and high glucose medium. After incubation for 24 hrs, total cell extracts were resolved by SDS-PAGE (8 % gel), and probed with iNOS antibody as described under Materials and Methods. The iNOS protein band is about 125 KDa. Zop:zopolrestat

It has been reported that there may be differences in tissue and species specificity in relation to PKC activation and iNOS expression. The activation of PKC by phorbol esters promotes the induction of iNOS in rat peritoneal macrophages and hepatocytes. On the other hand, the activation of PKC was reported to inhibit IL-1β induction of NO production in mesangial cells (Muhl and Pfeilschifter, 1994; Geng et al., 1994). These reports imply the possibility that different cells, expressing different subtypes of PKC family, employ this enzyme in different regulatory loop. And also, it is well known that individual PKC isoenzymes are differentially regulated even in the same cell type. In this regard, it is conceivable that glucose enhancing effect might be due to multiple mechanisms, including differential regulation of PKC isotypes. In fact, with our experimental protocol, it was unable to discriminate the possible roles of PKC subtypes in iNOS expression in response to LPS and high glucose. Thus, although the role of sorbitol pathway was excluded, there still remains the possibility that the activation of PKC via another pathway is involved in glucose enhancing effect on endotoxin induced NO production.

REFERENCES

Geng, Y.J., Wu, Q. and Hansson, G.K., 1994, Protein kinase C activation inhibits cytokine-induced nitric oxide synthesis in vascular smooth muscle cells, *Biochim. Biophys. Acta*, 1223:125

Graier, W.F., Simecek, S., Kukovetz, W.R., Kostner, G.M., 1996, High D-glucose-induced changes in endothelial Ca^{++}/EDRF signaling are due to generation of superoxide anions. *Diabetes* 45:1386.

Muhl, H., Pfeilshifter, J., 1994, Possible roles of protein kinase C-ε isozyme in inhibition of interleukin 1β induction of nitric oxide synthase in rat renal mesangial cells, *Biochem. J.*, 303:607.

Sharma, K., Danoff, T.M., DePierro, A., Ziyadeh., 1995, Enhanced expression of inducible nitric oxide in murine macrophages and glomerular mesangial cells by elevated glucose levels : possible mediation via protein kinase C, *Biochem. Biophys. Res. Comm.* 207 (1): 80.

63

1,2-DIACYLGLYCEROL HYDROPEROXIDE INDUCES THE GENERATION AND RELEASE OF SUPEROXIDE ANION FROM HUMAN POLYMORPHONUCLEAR LEUKOCYTES

Yorihiro Yamamoto,[1] Yasuhiro Kambayashi,[2] Takashi Ito,[2] Keiichi Watanabe,[3] and Minoru Nakano[2]

[1] Research Center for Advanced Science and Technology, University of Tokyo, Komaba, Meguro-ku, Tokyo 153

[2] Department of Photon and Free Radical Research, Japan Immunoresearch Laboratories, Takasaki 370

[3] Department of Pathology, Tokai University School of Medicine, Isehara 259-11, Japan

INTRODUCTION

Protein kinase C (PKC) plays a crucial role in receptor-mediated signal transduction affecting a diverse range of cellular responses such as cell proliferation, differentiation, and tumor promotion. PKC is known to be activated by calcium ion and lipids such as phosphatidylserine and 1,2-diacylglycerol, and directly by phorbol 12-myristate 13-acetate (PMA). We have recently reported that 1,2-dilinoleoylglycerol hydroperoxide (1,2-LLG-OOH) and hydroxide (1,2-LLG-OH) activate PKC isolated from rat brain almost as efficiently as does PMA.[1] 1,2-LLG-OOH and 1,2-LLG-OH can activate PKC even in the absence of calcium ion and phosphatidylserine.[1] It is therefore suggested that 1,2-diacylglycerol hydroperoxide (1,2-DAG-OOH) and hydroxide (1,2-DAG-OH) may activate the PKC-dependent signal transduction system if these components are released from oxidized biomembranes by the action of phospholipase C.[1] Recently, we have demonstrated that oxidized 1,2-diacylglycerol can act as a biochemical messenger in cellular systems by showing that 1,2-DAG-OOH activates human polymorphonuclear leukocytes (PMNs).[2] The activation of PMNs was assessed by measuring the superoxide anion released from PMNs. The production of superoxide anion was measured by a chemiluminescence method using a sea-firefly *cypridina* luciferin analogue, 2-methyl-6-(p-methoxyphenyl)-3,7-dihydroimidazo [1, 2-a] pyrazin-3-one (MCLA), as a luminescence reagent.3

EXPERIMENTAL

1-Palmitoyl-2-linoleoylphosphatidylcholine (PLPC), 1,3-dilinoleoylglycerol (1,3-LLG), bacillus cereus phospholipase C (type XI), soybean lipoxidase (type 1B), PMA, chelerythrine chloride, tert-butyl hydroperoxide, and superoxide dismutase (SOD) were purchased from Sigma Chemical Co. Staurosporine and 1-(5-isoquinolinesulfonyl)-2-methylpiperazine dihydrochloride (H-7) were purchased from Funakoshi (Tokyo). Linoleic acid hydroperoxide was prepared as previously described.4 MCLA was purchased from Tokyo Kasei Kogyo (Tokyo). Solvents and other reagents were of the highest analytical grade available.

PLPC hydroperoxide (PLPC-OOH) and its hydroxide (PLPC-OH) were prepared as described previously.[5] PLPC, PLPC-OOH, and PLPC-OH were individually hydrolyzed with Bacillus cereus phospholipase C in 50 mM aqueous Tris-HCl (pH 7.4)/methanol (60/40, v/v) containing 5 mM calcium chloride at 37°C for 60 min. The resulting 1-palmitoyl-2-linoleoylglycerol (PLG), PLG hydroperoxide (PLG-OOH), and PLG hydroxide (PLG-OH) were extracted with chloroform/methanol (2/1, v/v) and the lower layer was concentrated using a rotary evaporator. PLG, PLG-OOH, and PLG-OH were purified by HPLC (Superiorex ODS column; 20 x 250 mm, 5 μm, Shiseido, Japan) using methanol/2-propanol (85/15, v/v) as the mobile phase at a flow rate of 8 ml/min; their elution times were 33.7, 16.7, and 16.7 min, respectively. 1,3-LLG hydroperoxide (1,3-LLG-OOH) was prepared by the aerobic autoxidation of 1,3-LLG for 3 days, and was purified from unreacted 1,3-LLG by HPLC as described above.

Peripheral blood from a healthy human volunteer was drawn into a syringe containing heparin (20 units/ml). Leukocytes were isolated by sedimentation in the presence of 2.0% dextran followed by brief hypotonic lysis of residual erythrocytes.[6] The resultant leukocytes (1.0 x 10[7] cells/ml) containing >90% PMN were suspended in Hanks' balanced salt solution (HBSS) containing 1.3 mM calcium ion and 0.9 mM magnesium ion, and were kept at 0°C for no longer than 3 hr prior to use.

A glass cuvette containing PMNs (1.5 or 2.5 x 104 cells/ml) and MCLA (1 μM) in 2.0 ml of HBSS was placed in a luminescence reader (Type BLR-301, Aloka, Tokyo) and the emission of light was recorded.[7] After keeping the cuvette at 37°C under aerobic conditions for 3 min,10 μl of a 1.6 mM PLG-OOH in methanol (final concentration: 8 μM) was added to the cuvette by a micro syringe. PLG-OH, PLG, 1,3-LLG-OOH, and 1,3-LLG were added similarly instead of PLG-OOH. 20 μl of a 3.2 μM PMA in dimethyl sulfoxide/HBSS (1/1, v/v) was also added in a control experiment. After the addition of stimulants, SOD (final concentration: 0.5 μM) was added to assess the role of superoxide anion in MCLA-dependent light emission from PMNs. PKC inhibitors were added 5 min before the addition of PLG-OOH or PMA.

RESULTS AND DISCUSSION

Figure 1A shows that the addition of PLG-OOH (8 μM) immediately induced a strong light emission from PMNs in the presence of MCLA at 37°C under aerobic conditions. This chemiluminescence is dependent on MCLA, since no detectable photon emission was observed in the absence of MCLA (data not shown). The chemiluminescence intensity sharply increases to reaching a maximum and then the signal decays with time. The stimulant-induced maximal luminescence intensity (MLI) as calculated by subtracting the baseline luminescence intensity from the maximal intensity as shown in Figure 1A. The addition of a catalytic amount of SOD (0.5 µM) suppressed the chemiluminescence intensity to a level which was smaller than the baseline luminescence intensity, indicating the involvement of superoxide anion in the MCLA-dependent chemiluminescence from human PMNs.

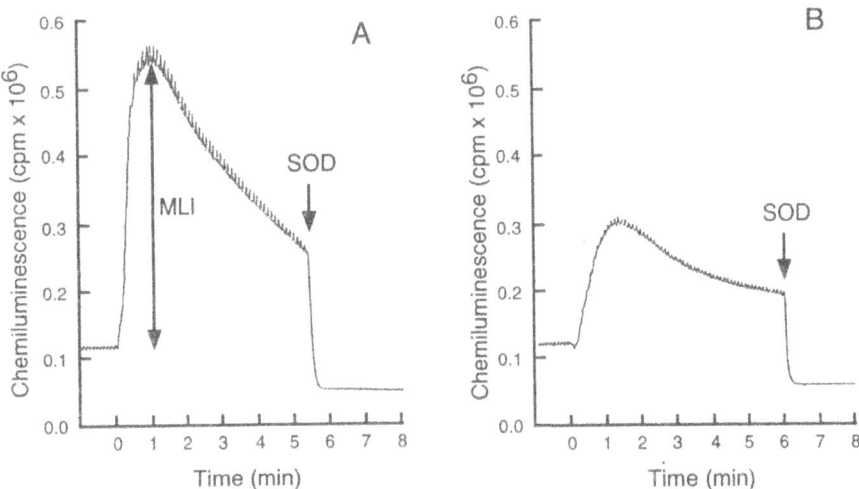

Figure 1. Light emission from human PMNs (1.5 x 104 cells/ml) stimulated by (A) 8 µM PLG-OOH and (B) 32 nM PMA in the presence of 1 µM MCLA in HBSS at 37°C under aerobic conditions. The stimulant was added at time zero and the addition of SOD (0.5 µM) is indicated by the arrows. (From reference 2 with permission).

PMA also stimulated the MCLA-dependent light emission from PMNs and this was suppressed by the addition of SOD (Figure 1B), as observed previously.[3] Superoxide anion can reduce the oxidized form of cytochrome c[8] and the initial rate (maximal rate) of the reduction can be determined by measuring the increase in absorption of the reduced form of cytochrome c at 550 nm (e_{max} = 2.11 x 10^4 M^{-1} cm^{-1}).[8] It was found that PMA-stimulated MLI at 10^5 counts/min (cpm) corresponded to the release of 0.095 nmol superoxide anion per minute. Thus, it was calculated that 8 µM PLG-OOH induced the release of superoxide anion from PMNs (3 x 10^4 cells) at a maximal rate of 0.36 nmol/min. It is noteworthy that cytochrome c cannot be used for the measurement of superoxide anion production from PMNs stimulated by PLG-OOH since cytochrome c rapidly decomposes PLG-OOH (unpublished data).

Next, we examined the effect of PLG-OOH concentration on MCLA-dependent light emission from PMNs. MLI from PMNs in the presence of MCLA increased with increasing PLG-OOH concentration to a maximum level at about 4 µM.[2] Figure 2 shows that 8 µM PLG-OOH induced the MCLA-dependent light emission from PMNs more strongly than the same concentration of 1,3-LLG-OOH. Moreover, PLPC-OOH, linoleic acid hydroperoxide, tert-butyl hydroperoxide, and hydrogen peroxide were much less efficient in stimulating human PMNs than PLG-OOH (data not shown), suggesting that the chemical structure of 1,2-diacylglycerol hydroperoxide is an essential feature for the stimulation of PMNs. Figure 2 also shows that 1,3-LLG, PLG, and PLG-OH did not stimulate the release of superoxide anion from PMNs as efficiently as PLG-OOH. There was a significant difference between PLG-OOH and PLG-OH with respect to stimulation of PMNs.

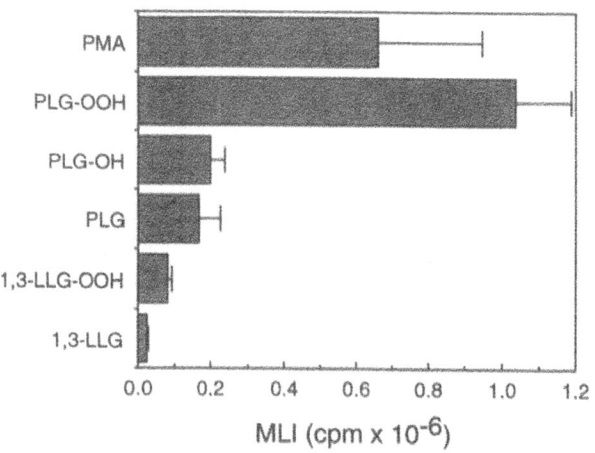

Figure 2. MLI from human PMNs (2.5 x 104 cells/ml) stimulated by 8 µM lipid derivatives or 32 nM PMA in the presence of 1 µM MCLA in HBSS at 37°C under aerobic conditions. Horizontal bars indicate standard deviation (n = 3). (From reference 2 with permission).

In order to verify the involvement of PKC in the stimulation of PMNs by PLG-OOH, the effects of PKC inhibitors, chelerythrine chloride (20 µM), staurosporine (0.5 µM), and 1-(5-isoquinolinylsulfonyl)-2-methylpiperazine dihydrochloride (H-7, 200 µM), were examined. These inhibitors did not affect the viability of PMNs and were added to PMNs 5 min before the addition of stimulants, PLG-OOH and PMA. Chelerythrine chloride and staurosporine nearly suppressed all PMA-induced MLI from human PMNs while H-7 was effective by about 70%. These PKC inhibitors also suppressed PLG-OOH-stimulated MLI from human PMNs but to a lesser extent than stimulated by PMA.[2] These observation suggest that PLG-OOH stimulates PMNs primarily by activating PKC, but also that other activation system may be involved. It is also important to identify which PKC isozyme is responsible for PLG-OOH-induced PMN activation; further investigation is being considered.

Phosphatidylcholine-specific phospholipase C (PC-PLC), which catalyzes the hydrolysis of phosphatidylcholine to 1,2-diacylglycerols, has been found in PMNs, lymphocytes, and hepatic cells.[9] Bacillus cereus PC-PLC can hydrolyze oxidized and unoxidized phosphatidylcholine.[10] This was also confirmed in this study since PLG, PLG-OH, and PLG-OOH were prepared by the hydrolysis of PLPC, PLPC-OH, and PLPC-OOH, respectively, using Bacillus cereus PC-PLC. Although it is necessary to demonstrate that the mammalian PC-PLC hydrolyzes the oxidized phosphatidylcholine, 1,2-DAG-OOH are expected to be formed as a result of oxidative stress. In fact, the oxidation of hepatocyte phospholipids and plasma lipoproteins gave PC-OOH as the major product.[11,12]

The role of oxygen radicals in the aetiology of carcinogenesis has not been fully investigated. However, it would be reasonable to expect that oxygen radicals would likely persist longer under conditions of oxidative stress presenting an increased risk of oxidative damage to DNA. If these oxidative modifications are unrepaired, they can cause errors in genetic replication having the potential to initiate carcinogenesis. Oxygen radicals have also been demonstrated to convert carcinogenic precursors to the ultimate active carcinogen[13] and, further to these roles in the initiation of cancer, oxygen radicals may also be involved in tumor promotion. We have demonstrated that oxidized 1,2-diacylglycerol which is formed from oxidized biomembranes by phospholipase C activates protein kinase C having the ability to act as an endogenous tumor promoter.[1,2] These data are thus consistent with chronic oxidative stress (redox imbalance) being an important cause of carcinogenesis. We have demonstrated that oxidative stress is evident after contracting hepatitis in humans[14] and in LEC rats[15] which is a condition predisposed to the onset of liver cancer. In fact, chronic inflammation is recognized as one of the major contributing causes of cancer.[16]

Acknowledgements

This work was supported by Grants-in-Aid for Scientific Research from the Ministries of Education, Science and Culture, and Health and Welfare of Japan.

REFERENCES

1. S. Takekoshi, Y. Kambayashi, H. Nagata, T. Takagi, Y. Yamamoto, and K. Watanabe, Activation of protein kinase C by oxidized diacylglycerols, Biochem. Biophys. Res. Commun. 217:654 (1995).
2. Y. Yamamoto, Y. Kambayashi, T. Ito, K. Watanabe, and M. Nakano, 1,2-Diacylglycerol hydroperoxides induce the generation and release of superoxide anion from human polymorphonuclear leukocytes, FEBS Lett. 412:461 (1997).
3. M. Nakano, Determination of superoxide radical and singlet oxygen based on chemiluminescence of luciferin analogs, Methods Enzymol. 186:585 (1990).
4. Y. Yamamoto, M. H. Brodsky, J. C. Baker, and B. N.Ames, Detection and characterization of lipid hydroperoxides at picomole levels by high-performance liquid chromatography, Anal. Biochem.160:7 (1987).

5. Y. Nagata, Y. Yamamoto, and E. Niki, Reaction of phosphatidylcholine hydroperoxide in human plasma: The role of peroxidase and lecithin:cholesterol acyltransferase, Arch.Biochem. Biophys. 329:24 (1996).

6. H. Rosen and S. J. Klebanoff, Chemiluminescence and superoxide production by myeloperoxidase-deficient leukocytes, J. Clin. Invest. 58:50 (1976).

7. M. Nakano, K. Sugioka, Y. Ushijima, and T. Goto, Chemiluminescence probe with cypridina luciferin analog, 2-methyl-6-phenyl-3,7-dihydroimidazo[1,2-a]pyrazin-3-one, for estimating the ability of human granulocytes to generate O2-, Anal. Biochem. 159:363 (1986).

8. J.M.McCord and I. Fridovich, Superoxide dismutase, J. Biol. Chem. 244:6049 (1969).

9. M. Lucas, V. Sanchez-Margalet, C. Pedrerea, and M.L. Bellido, A chemiluminescence method to analyze phosphatidylcholine-phospholipase activity in plasma membrane preparations and in intact cells, Anal. Biochem. 231:277 (1995).

10. C.R. Wagner and N. A. Porter, A phospholipase C protocol for phospholipid peroxidation analysis,Chem. Res. Toxicol. 1:274 (1988).

11. Y. Kambayashi, S. Yamashita, E. Niki, and Y.Yamamoto, Oxidation of rat Liver phospholipids: Comparison of pathways in homogeneous solution, in liposomal suspension and in whole tissue homogenates, J. Biochem. 121:425 (1997).

12. B Frei, R. Stocker, and B. N. Ames, Antioxidant defenses and lipid peroxidation in human blood plasma, Proc. Natl. Acad. Sci. USA 85:9748 (1988).

13. C. Ji and L. J. Marnett, Oxygen radical-dependent epoxidation of (7S,8S)-dihydroxy-7,8-dihydrobenzo[a]pyrene in mouse skin in vivo, J. Biol. Chem. 267:17842 (1992).

14. Y. Yamamoto and S. Yamashita, Plasma ratio of ubiquinol and ubiquinone as a marker of oxidative stress, Molec. Aspects Med. 18:s79 (1997).

15. Y Yamamoto, H. Sone, S. Yamashita, Y. Nagata, H. Niikawa, K. Hara, and M. Nagao, Oxidative stress in LEC rats evaluated by plasma antioxidants and free fatty acids, J. Trace Elem. Exp. Med. 10:129 (1997).

16. .N Ames, L. S. Gold, and W. C. Willett, The causes and prevention of cancer, Proc. Natl. Acad. Sci. USA 92:5258 (1995).

64

KINETIC EVALUATION OF ENDOGENOUS LEUKOTRIENE B4 ANDE4 ACUTE ACTIVATION OF INFLAMMATORY CELLS IN THE RABBIT

Antonio Celardo, Giuseppe Dell'Elba, Stefano Manarini, Virgilio Evangelista, Giovanni de Gaetano, Chiara Cerletti

Istituto di Ricerche Farmacologiche "Mario Negri", Department of Vascular Medicine and Pharmacology , "G. Bizzozero" Laboratory of Platelet and Leukocyte Pharmacology, Consorzio Mario Negri Sud, 66030 S. Maria Imbaro, Italy

INTRODUCTION

Leukotrienes (LTs) are lipid mediators synthesised by the 5-lipoxygenase (5-LO) pathway in leukocytes[1], which are implicated both in the pathogenesis of inflammatory processes[2-4] and in ischemic heart diseases[5-8].

Both in animals[9-14] and humans[14-17] the administration of exogenous LTs showed a rapid elimination from circulation but almost no information are available on the rate of LTs biosynthesis and their kinetics when endogenously produced during pathophysiological events such as acute inflammatory reactions or thrombosis. Intravenous bolus administration of the inflammatory peptide FMLP in the rabbit resulted in an immediate leukopenia and thrombocytopenia, due to the recruitment of circulating cells in the lungs and in the production of thromboxane A2 and LTs[18-20].

This model mimics an acute inflammatory situation. We determined, according to Gibaldi[21], the kinetic parameters of endogenously produced LTB4 and LTE4 in comparison with the kinetics of exogenous administration of either compound. We also measured the urinary excretion of LTE4, N-acetyl-LTE4 (N-ac-LTE4) and 11-*trans*-LTE4 metabolites. The knowledge of the kinetic parameters of LTB4 and LTE4 in an *in vivo* acute model of inflammation may be useful in understanding the pathophysiological role of LTs, as well as in testing drug designed to modulate their biosynthesis and biological effects.

METHODS

Animals (New Zealand male rabbits) were anaesthetised by intramuscular injection of chlorpromazine (25 mg/kg) and ketamine chloridrate (50 mg/kg).

They were cannulated in the right jugular vein and in the right commune carotid artery for i.v. administration of FMLP (30 nmol/kg), LTB4 (20 nmol/kg), LTC4 (10 nmol/kg) and LTE4 (10 nmol/kg) standards, and blood sampling, respectively. Twenty-four hours before and after FMLP or LTs administration, urine were collected from rabbit housed in metabolic cages. This work was performed according to the "Guidelines for use of animals in biomedical research"[22].

Eicosanoids and Other Bioactive Lipids in Cancer, Inflammation, and Radiation Injury, 4
Edited by Honn *et al.*, Kluwer Academic / Plenum Publishers, New York, 1999.
437

Extraction and determination of LTs

Blood samples were collected on an inhibitory mixture to prevent *ex vivo* metabolism, and incubated with γ-glutamyl-transpeptidase (γ-GTP) and leucine-aminopeptidase (LAP) to allow LTC4 transformation in LTE4 via LTD4, and extracted from blood. Urine were collected in a cylinder stored at -20 °C. Blood and urine samples were analysed by h.p.l.c. as reported[23]. LTs' concentrations under the h.p.l.c. detection limit were measured by specific enzyme immunoassay (Cayman Chemicals, Ann Arbor, MI, USA).

Cell count and TxB2 determination

Peripheral white blood cells were immediately counted in a Burker chamber by optical microscopy after staining with Turk's solution. This allows to distinguish polymorphonuclear leukocytes (PMNL) from mononuclear cells. Platelet number was evaluated after hypotonic lysis of erythrocytes. Aliquots of blood were centrifuged at 14,000 x g for 3 min and plasma stored at -20 °C until assayed for TxB2, which was measured by specific radio immunoassay.

RESULTS AND DISCUSSION

The kinetic data for LTB4 and LTE4 are reported in table 1.

Table 1. Kinetic parameters of LTB4 and LTE4 after i.v. administration of FMLP (30 nmol/kg) or LTB4 (20 nmol/kg) and LTE4 (10 nmol/kg) in rabbit.

Parameter	Endogenous		Exogenous	
	LTB4	LTE4	LTB4	LTE4
$t_{1/2}$ (min)	24.4±6.7	36.9±13.0	2.5	2.3
k (1/min)	0.030±0.006	0.21±0.070	0.277	0.301
Cmax (pmol/ml)	84.2±60.0	162.2±51.4		
tmax(min)	2±1	5±4		
C_0 (pmol/ml)			118.9	148.3
AUC (pmol min /ml)	2178±1591	7627±3052	470	610
Cls (ml/min/kg)			44.3	20.8
Urinary excretion (0-24h)				
LTE4 (pmol/kg)		2308±508		2400
N-ac-LTE4 (pmol/kg)		172±153		nd
11-*trans*- LTE4 (pmol/kg)		768±413		19.1

Figures are means±SD (n=5) for endogenous production of LTs and individual data for exogenous. k, elimination rate constant; t1/2, half life; tmax, time to reach Cmax; Cmax, maximum experimental concentration; tmax, time to reach the maximum concentration; C_0, extrapolated concentration to time zero; AUC, area under the blood concentration-time curve; Cls, systemic clearance; nd, not detectable.

The time course of LTs concentrations, white blood cells, platelets and TxB2 are reported in Figure 1.

Intravenous administration of LTB4 and LTE4 to individual animals showed blood half-lives of 2.5, 2.3. min, respectively (Table 1). After i.v. administration, to single animal, of LTC4 the half life of LTE4 was 4.7 min. Within 1 min from FMLP administration, circulating PMNL and mononuclear cells fall to zero and 45 % of the basal value (Figure 1A). Platelets are reduced to 40 % of the basal value and TxB2 rises meanly from 2.33 to 66.3 pmol/ml 3 min after FMLP administration (Figure 1B). Urinary excretion of LTE4, N-ac-LTE4, and 11-t*rans*-LTE4 after

438

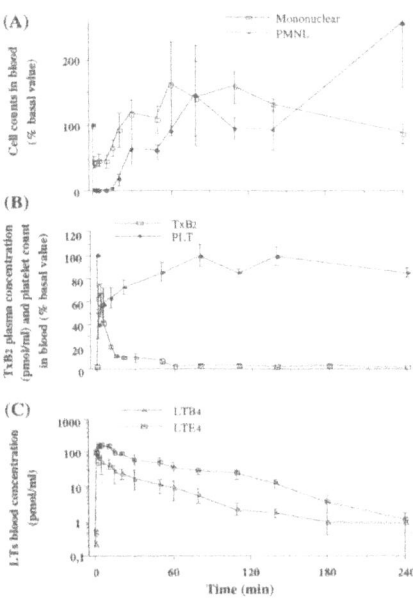

Figure 1. Changes in the counts of PMNL and mononuclear cells (A), TxB2 and platelets (PLT) (B), LTB4 and LTE4 (C) in rabbit blood following i.v. administration of FMLP (30nmol/kg).

FMLP or LTE4 i.v. administration are reported in Table 1. The data, analysed for two 12-h intervals after FMLP administration show that 90% of LTE4, virtually 100 % of N-ac-LTE4 and 15 % of 11-*trans*-LTE4 were excreted in the first time interval. Urinary excretion of LTE4 0-24 h before treatment was 133±45 pmol/kg (mean±SD; n=8).

After i.v. administration of LTE4 (10 nmol/kg) 14.4 % of injected dose was recovered in 0-24 h interval. The urinary excretion of LTE4 after LTC4 administration was 0.5 % of the injected dose. Many studies describe the kinetics of elimination of exogenous administration of LTB4 in rabbit[9], LTC4 or LTE4 in, rat[10], dog[13] or monkey [11,12,14], as in humans[14-17] and show that the clearance of LTs from vascular compartment was very rapid (between 3 and 5 min), suggesting an efficient mechanism for LTE4 removal from blood.

The transformation of LTC4 into LTE4, N-ac-LTE4, and their omega-and beta-oxidation metabolites represents the main metabolic pathway of peptido leukotrienes (pLTs). The induction of an acute systemic inflammation in the rabbit by FMLP administration produced a rapid fall of circulating cells (white cells and platelets), LTB4 and LTE4 release into bloodstream, and pLT metabolites excretion in the urine. In this model we evaluated the kinetic parameters of endogenous LTs. The results show that the elimination half-lives of endogenous leukotrienes are significantly higher than those calculated after direct injection of LTs. LTB4 and LTE4 are rapidly produced, reaching the maximal blood concentrations at 2 and 5 min, after FMLP,respectively.

The rapid synthesis of LTB4 induced by FMLP in rabbit whole blood *in vitro* (not shown) suggests that also *in vivo* activation of 5-LO is rapid and complete within few minutes. On the other hand, the evidence that creatinine clearance, measured by h.p.l.c. in the rabbit was 41.0±21.9 and 44.6±17.5 mg/kg/24 h (mean±SD before and after FMLP treatment, respectively, indicates a normal kidney function after FMLP administration and allows to exclude a reduced renal excretion of endogenously produced LTs. Although it cannot be excluded that an inflammatory situation generated by FMLP could influence LTs catabolic pathway other than renal excretion, the prolonged elimination of LTs may be due to the rate of entry into the bloodstream from the tissue(s) where they are synthesised. Concomitantly with LTB4 and LTE4 production, after FMLP administration, there was a marked granulocytopenia, thrombocytopenia and increased TxB2 levels in plasma within 10 min. At the same time there was a pronounced bronchoconstriction, fall in arterial blood pressure, and electrical heart dysfunction (not shown). These symptoms appear few minutes after FMLP administration concomitantly with the burst of LTs and TxB2 production, suggesting that these metabolites may contribute to the generation of symptoms.

CONCLUSIONS

Elucidation of the kinetics of entry in the bloodstream and clearance of LTs in animal models of diseases is important in view of deeper knowledge of their role in pathophysiological processes. The high clearance of LTs from circulating blood suggests that these mediators can act in the district of the body where they are produced. However in some situations these mediators may reach pharmacological blood concentrations able to produce deleterious effect in organs far from the site of their production.

ACKNOWLEDGMENTS

This work was supported by the Italian National Research Council (Convenzione CNR - Consorzio Mario Negri Sud).
The authors thank Ms. Maria Nigro for help in preparing the manuscript.

REFERENCES

1. M. Köller, W. Schönfeld, J. Knöller, K.D. Bremm, W. König, B. Spur, A. Crea, and W. Peters, The metabolism of leukotrienes in blood plasma studied by high-performance liquid chromatography, *Biochim Biophys Acta.* 833:128 (1985).
2. R.A. Lewis, K.F. Austen, and R.J. Soberman, Leukotrienes and other products of the 5-lipoxygenase pathway, *N Engl J Med.* 323:645 (1990).
3. X.S. Chen, J.R. Sheller, E.N. Johnson, and C.D. Funk, Role of leukotrienes revealed by targeted disruption of the 5-lipoxygenase gene, *Nature.* 372:179 (1994).
4. J.L. Goulet, J.N. Snouwaert, A.M. Latour, T.M. Coffman, and B.H. Koller, Altered inflammatory responses in leukotriene-deficient mice, *Proc.Natl.Acad.Sci. USA.* 91:12852 (1994).
5. S. Barst and K. Mullane, The release of a leukotriene D4-like substance following myocardial infarction in rabbits, *Eur J Pharmacol.* 114:383 (1985).
6. M. Tada, T. Kuzuya, S. Hoshida, and M. Nishida, Arachidonate metabolism in myocardial ischemia and reperfusion, *J Mol Cell Cardiol.* 20 (Suppl. II):135 (1988).
7. M. Carry, V. Korley, J.T. Willerson, L. Weigelt, A.W.Ford-Hutchinson, and P. Tagari, Increased urinary leukotriene excretion in patients with cardiac ischemia, *Circulation.* 85:230 (1992).
8. G. Rossoni, A. Sala, F. Berti, T. Testa, C. Buccellati, C. Molta, R. Muller-Peddinghaus, J. Maclouf, and G.C. Folco, Myocardial protection by the leukotriene synthesis inhibitor BAY X1005: importance of transcellular biosynthesis of cysteinyl-leukotrienes, *J Pharmacol Exp Ther.* 276:335 (1996).
9. S. Marleau, N. Dallaire, P. E. Poubelle, and P. Borgeat, Metabolic disposition of leukotriene B4 (LTB4) and oxidation-resistant analogues of LTB4 in conscious rabbits, *Br J Pharmacol.* 112:654 (1994).
10. C. Denzlinger, S. Rapp, W. Hagmann, and D. Keppler, Leukotrienes as mediators in tissue trauma, *Science.* 230:330 (1985).
11. C. Denzlinger, A. Guhlmann, P.H. Scheuber, D. Wilker, D.K. Hammer, and D. Keppler, Metabolism and analysis of cysteinyl leukotrienes in the monkey, *J Biol Chem.* 261:15601 (1986).
12. P. Tagari, A. Foster, D. Delorme, Y. Girard, and J. Rokach, Metabolism and excretion of exogenous [3H]-LTC4 in primates, *Prostaglandins.* 37:629 (1989).
13. P. Tagari, A. Becker, C. Brideau, R. Frenette, V. Sadl, E. Thomas, P. Vickers, and A. Ford-Hutchinson, Leukotriene generation and metabolism in dogs: inhibition of biosynthesis by MK-0591, *J Pharmacol Exp Ther.* 265:416 (1993).
14. M. Huber, J. Müller, I. Leier, G. Jedlitschky, H.A. Ball, K.P. Moore, G.W. Taylor, R. Williams, and D. Keppler, Metabolism of cysteinyl leukotrienes in monkey and man, *Eur J Biochem.* 194:309 (1990).
15. N.H. Maltby, G.W. Taylor, J.M. Ritter, K. Moore, R.W. Fuller, and C.T. Dollery, Leukotriene C4 elimination and metabolism in man, *J Allergy Clin Immunol.* 85:3 (1990).
16. A. Sala, N. Voelkel, J. Maclouf, and R.C. Murphy, Leukotriene E4 elimination and metabolism in normal human subjects, *J Biol Chem.* 265:21771 (1990).

17. J. Maclouf, C. Antoine, R. De Caterina, R. Sicari, R.C. Murphy, P. Patrignani, S. Loizzo, and C. Patrono, Entry rate and metabolism of leukotriene C4 into vascular compartment in healthy subjects, *Am J Physiol*. 263:H244 (1992).

18. M.A. Boukili, M. Bureau, V. Lagente, J. Lefort, A. Lellouch-Tubiana, E. Malanchere, and B.B. Vargaftig, Pharmacological modulation of the effects of N-formyl-L-methionyl-L-leucyl-L-phenylalanine in guinea-pig: involvement of the arachidonic acid cascade, *Br J Pharmacol*. 89:349 (1986).

19. N. Berend, C.L. Armour, and J.L. Black, Formyl-methionyl-leucyl-phenilalanine causes bronchoconstriction in rabbits, *Agents and Actions*. 17:466 (1985).

20. A.C. Issekutz, M. Ripley, and J.R. Jackson, Role of neutrophils in the deposition of platelets during acute inflammation, *Lab Invest*. 49:716 (1983).

21. M. Gibaldi and D. Perrier, 2nd ed., *Pharmacokinetics*, Marcel Dekker, New York, (1982).

22. A.R. Giles, Guidelines for the use of animals in biochemical research, *Thromb Haemost*. 58:1078 (1987).

23. A. Celardo, G. Dell'Elba, Z.M. Eltantawy, V. Evangelista, and C. Cerletti, Simultaneous determination of leukotriene B4 and E4 in whole blood and of leukotriene E4 in urine of rabbit by reversed-phase high-performance liquid chromatography, *J Chromatogr B*. 658:261 (1994).

65

LIPOPOLYSACCHARIDE- AND LIPOSOME-ENCAPSULTAED MTP-PE- INDUCED FORMATION OF EICOSANOIDS, NITRIC OXIDE AND TUMOR NECROSIS FACTOR-α IN MACROPHAGES

P. Dieter[1], U. Hempel[1], B. Malessa[1]. E. Fitzke[2], T.A. Tran-Thi[2], J. MacLouf[3], C. Créminon[3], Y. Kanaoka[4], and Y. Urade[4]

[1]TU Dresden, Institut für Physiologische Chemie, Dresden, Germany
[2]Universität Freiburg, Institut für Biochemie und Molekularbiologie, Freiburg, Germany
[3]Inserm U. 348, Hôpital Lariboisière, Paris, France
[4]Osaka Bioscience Institute, Department of Molecular Behavioral Biology, Osaka, Japan

INTRODUCTION

Kupffer cells (liver macrophages) possess the ability to secrete a wide array of biologically active compounds including reactive oxygen species, nitric oxide (NO), eicosanoids and cytokines[1-4]. The synthesis and release of these mediators has been shown to be elicited by a number of agents including phagocytotic material (zymosan, viruses, bacteria), tumor promoting agents (phorbol ester) and other biologically active substances[1-5].

Liver macrophages can also be activated by treatment with cytokines, lipopoly-saccharide (LPS) or muramyl peptides as it was shown in *in vitro* and *in vivo* studies[6-8]. In contrast to LPS, liposomal-encorporated N-acety-muramyl-L-alanyl-D-isoglutaminyl-L-alanine-2-[1',2'-dipalmitoyl-sn-glycero-3']-ethylamide (MTP-PE), has been shown to prevent and eradicate experimentally-induced septicemia and other LPS-induced inflammatory diseases[9]. Furthermore, MTP-PE has been used successfully treating relapsed osteosarcoma in humans, preventing HIV replication in macrophages and inhibiting lung metastases[10-12]. Both agents, when given intravenously accumulate almost quantitatively in Kupffer cells. Therefore, Kupffer cells are thought to be the main target for the biological effects of LPS and MTP-PE. However, the cellular and molecular basis of the action of LPS and MTP-PE has not been investigated in detail so far.

In an earlier study we showed that both agents induce a release of tumor necrosis factor (TNF)-α,

NO and eicosanoids but not a generation of superoxide[13]. Furthermore, we demonstrated that both agents are unable to induce changes in intracellular levels of inositol phosphates and calcium, and do not lead to an activation of protein kinase (PK)C isoenzymes. In contrast to LPS, MTP-PE induces an elevation of cellular diacylglycerol, whereas LPS but not MTP-PE elicits a predominant synthesis and release of prostaglandin (PG)E$_2$[13].

In this study, we determine and compare components of the signal transduction pathways used by both agents, leading to the formation of eicosanoids, NO and TNF-α in primary cultures of Kupffer cells. Furthermore we present RT-PCR and Western Blot studies showing the effect of both agents on the levels of mRNA's encoding TNF-α, NO synthases (NOS), cyclooxygenases (COX), cytosolic phospholipase (cPL)A$_2$ and PGD$_2$ synthase and corresponding protein levels.

RESULTS

Signal transduction components

LPS and MTP-PE induce an activation of mitogen-activated protein (map) kinase (Figure 1). The transcription factor AP-1 is activated by LPS within 30 min (Figure 2); MTP-PE has only a minor effect on AP-1 which becomes detectable only after 5 h. The transcription factor NF-κB is also activated by LPS after 30 min (Figure 2). MTP-PE has no effect on NF-κB at this time-point but 5 h and 24 h after the addition of MTP-PE, NF-κB also becomes strongly activated.

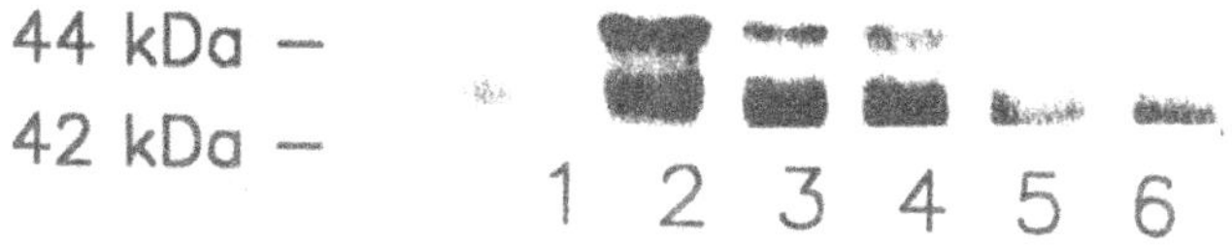

44 kDa —
42 kDa —

1 2 3 4 5 6

Figure 1. Map kinase activity in control cells (lane 1), cells incubated for 10 min with 500 ng/ml LPS (lane 2) and 25 µg/ml MTP-PE for 10 min (lane 3), 30 min (lane 4), 2 h (lane 5) and 5 h (lane 6).

1 2   3 4 5 6 7

Figure 2. AP-1 (upper panel) and NF-κB (lower panel) activites in control cells (lanes 1 and 3) and cells incubated with 500 ng/ml LPS for 30 min (lanes 2) and 25 µg/ml MTP-PE for 30 min (lanes 4), 2 h (lanes 5), 5 h (lanes 6) and 24 h (lanes 7).

RT-PCR experiments

In an earlier paper we showed that LPS and MTP-PE induce a synthesis and release of TNF-α, NO and eicosanoids[13]. Here, we show the effects of both agents on corresponding mRNA levels. Resting Kupffer cells contain detectable levels of mRNA's encoding COX-1, TNF-α, cNOS, COX-2, cPLA₂, PGD₂ synthase and iNOS (Figure 3). After stimulation with LPS, the levels of mRNA's encoding TNF-α, iNOS, COX-2 and cPLA₂ increase in a time-dependent way while no effect is seen for mRNA levels encoding cNOS and COX-1. A small decrease is observed for mRNA encoding PGD₂ synthase (Figure 4). In contrast to LPS, MTP-PE has no effect on mRNA's encoding cPLA₂ and PGD₂ synthase. Similar to LPS, MTP-PE increases the levels of mRNA's encoding TNF-α, iNOS and COX-2, however, in a different magnitude and different kinetic.

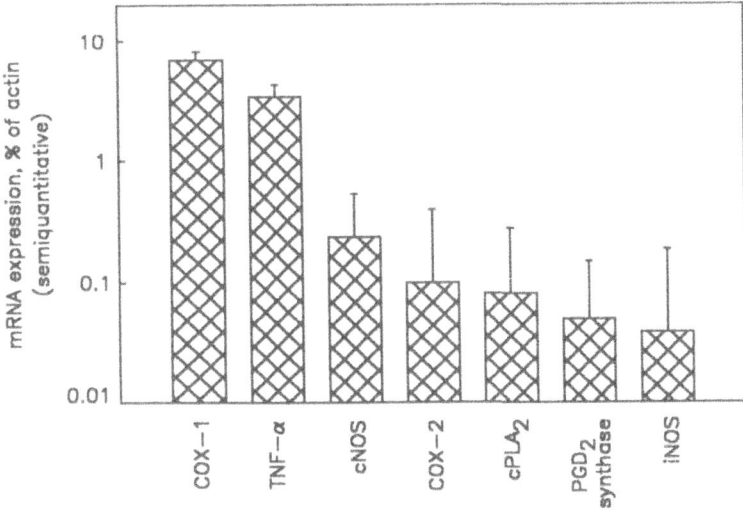

Figure 3. mRNA levels (after RT-PCR amplification) encoding COX-1 and COX-2, TNF-α, cNOS and iNOS, cPLA₂ and PGD₂ synthase in resting Kupffer cells.

Western Blot studies

Consistent with the RT-PCR findings, the level of cPLA₂ protein is enhanced by LPS but not by MTP-PE (Figure 5). The level of COX-1 protein is slightly decreased by LPS while MTP-PE shows an opposite effect. The effects on COX-1 protein levels must take place on the level of translation or protein stability since the two agents show no effects on corresponding mRNA's (Figure 4). Resting cells contain no detectable levels of COX-2 protein (Figure 5). However, LPS induces an expression of COX-2 protein which can be detected after 8 h and reaches a maximum after 24 h. MTP-PE also induces an expression of COX-2 protein after 8 h, but in contrast to LPS, COX-2 protein levels increase up to 48 h to an aboud 5 fold higher value compared to LPS at this time point. Since LPS has a more pronounced effect on the mRNA level encoding COX-2 than MTP-PE (Figure 4), the higher levels of COX-2 protein found after MTP-PE must be due to enhanced translation or enhanced protein stability. The level of PGD$_2$ synthase protein is not changed by LPS or MTP-PE (Figure 5). Resting cells contain no detectable levels of cNOS and iNOS proteins (Fig. 5). LPS does not induce the expression of detectable levels of cNOS protein (data not shown); however, iNOS protein becomes detectable 6 h after addition of LPS and reaches a maximum between 12 h and 24 h (Figure 5).

DISCUSSION

Here we demonstrated that LPS and MTP-PE induce a rapid activation of map kinase (within 10 min). Since both agents do not activate PKC isoenzymes[13], map kinase activation occurs probably via tyrosine kinases. LPS induces a rapid activation of the transcription factors AP-1 and NF-κB (within 30 min). In contrast, activation of both transcrition factors by MTP-PE is observed only after 5 h, and the effect on AP-1 is very weakly expressed. RT-PCR studies showed that resting Kupffer cells express low levels of mRNA's encoding cNOS, COX-2, cPLA$_2$, PGD$_2$ synthase and iNOS while COX-1 and

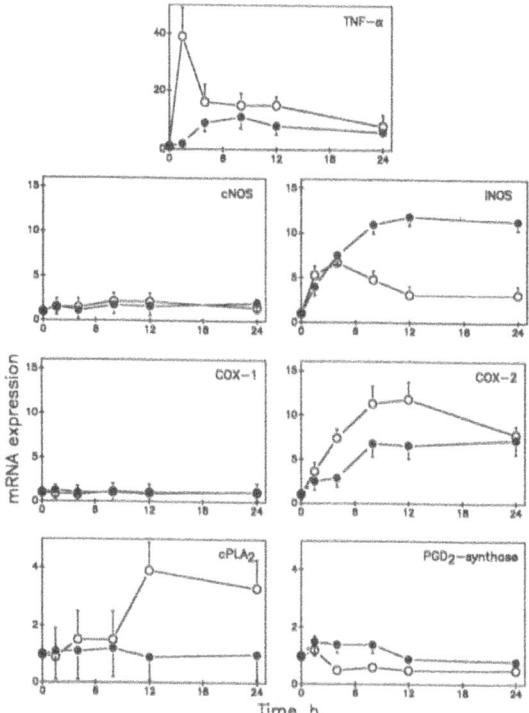

Figure 4. mRNA levels (after RT-PCR amplification) encoding COX-1 and COX-2, TNF-α, cNOS and iNOS, cPLA$_2$ and PGD$_2$ synthase in Kupffer cells stimulated without (time 0), with 500 ng/ml LPS (open circles) and 25 µg/ml MTP-PE (closed circles) for different time periods.

TNF-α mRNA's are expressed at higher levels. Incubation of Kupffer cells with LPS leads to an enhanced expression of mRNA's encoding TNF-α, iNOS, COX-2 and cPLA₂, to an enhanced synthesis of TNF-α-[13], iNOS-, COX-2- and cPLA₂.- proteins and to an enhanced synthesis and release of NO and eicosanoids[13]. LPS has no effect on mRNA's encoding cNOS and COX-1. mRNA encoding PGD₂ synthase is slightly decreased by LPS which might result in an enhanced PGE₂ synthesis[13]. In contrast to LPS, MTP-PE has no effect on mRNA's encoding cPLA₂ and PGD₂ synthase. Additionally, the level of cPLA₂ protein is not changed by MTP-PE. Furthermore, the effect of MTP-PE on mRNA's encoding TNF-α, iNOS and COX-2 is different to LPS with respect to time-dependencies and magnitudes. The latter observation is consistent with differences of synthesis and release of TNF-α protein[13], NO and eicosanoids induced by both agents. In conclusion, it might be that the observed differences reflect differences in the potencies of these two agents as immunomodulators, especially in the pathophysiology of septic shock.

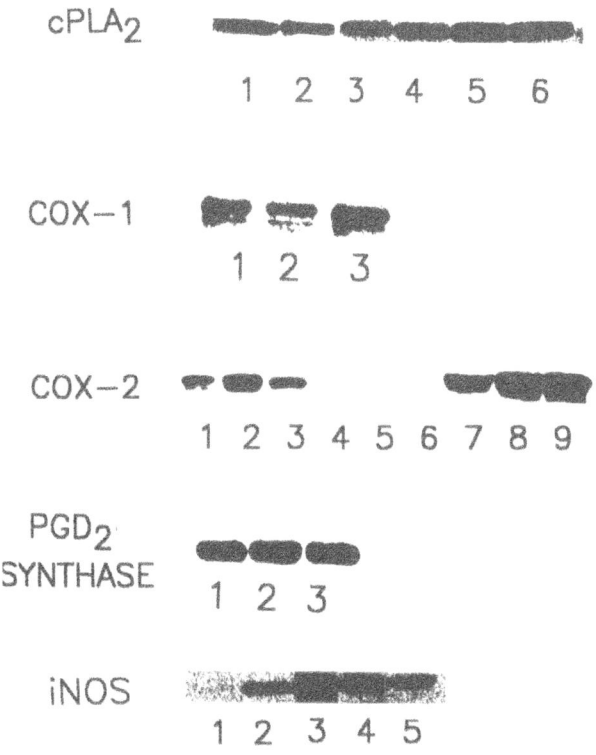

Figure 5. Western Blot analysis of cPLA₂, COX-1, COX-2, iNOS and PGD₂ synthase in Kupffer cells. cPLA₂: 1, control cells; 2, MTP-PE 24h; 3-6, LPS 2h, 4h, 9h, 24h. COX-1: 1, control cells; 2, LPS 24h; 3, MTP-PE 24h. COX-2: 5, control cells. 1-4, LPS 48h, 24h, 8h, 4h; 6-7, MTP-PE 4h, 8h, 24h, 48h. iNOS: 1, control cells; 2-4, LPS, 6h, 12h, 24h, 48h. PGD₂ synthase: 1, control cells; 2, LPS 24h; 3, MTP-PE 24h.

Acknowledgments

We gratefully acknowledge the technical assistance of Mrs. A. Kolada and S. Kamionka, and the assistance of K. Asman in the preparation of the manuscript.

REFERENCES

1. . Decker, Biologically active products of stimulated liver macrophages, *Eur. J. Biochem. 192:245 (1990).*
2. T. K Gaillard, A. Mülsch, H. Klein, and K. Decker. Regulation by PGE$_2$ of cytokine-elicited nitric oxide synthesis in rat liver macrophages, *Biol. Chem. Hoppe-Seyler 373: 897, 1992).*
3. M. Grewe, R. Gausling, K. Gyufko, R. Hoffmann, and K. Decker, Regulation of the mRNA expression for tumor necrosis factor-α in rat liver macrophages, *J. Hepatol. 20:811 (1994).*
4. P. Dieter, and E. Fitzke, Formation of diacylglycerol, inositol phosphates, arachidonic acid and its metabolites in macrophages, *Eur. J. Biochem. 218:753 (1994).*
5. K.J. Busam, A. Schulze-Specking, and K. Decker, Endotoxin-refractory liver macrophages secrete tumor necrosis factor-α upon viral infection, *Biol. Chem. Hoppe-Seyler 372:157 (1990).*
6. C.R. Gardner, A.J. Wassermann, and D.L. Laskin, Differential sensitivity of tumor targets to liver macrophage-mediated cytotoxicity, *Cancer Res. 47:6686 (1987).*
7. T. Decker, M.L. Lohmann-Matthes, U. Karck, T. Peters, and K. Decker, Comparative studies of cytotoxicity, tumor necrosis factor, and prostaglandin release after stimulation of rat Kupffer cells, murine Kupffer cells, and murine inflammatory liver macrophages, *J. Leukoc. Biol. 45:139 (1989).*
8. R.M.J. Hoedemakers, H.W.M. Morselt, G.L. Scherphof, and T. Daemen, Secretion pattern of the rat liver macrophage population following activation with liposomal muramyl dipeptide in vivo and in vitro, *J. Immunother. 15:265 (1994).*
9. H.W.L. Ziegler-Heitbrock, B. Passlick, E. Käfferlein, P.G. Coulie, and J.R. Izbicki, Protection against lethal pneumococcal septicemia in pigs is associated with decreased levels of interleukin-6 in blood, *Infect. Immun.60: 1692 (1992).*
10. W.E. Fogler, and I.J. Fidler, Comparative studies of free and liposome-encapsulated nor-muramyl dipeptide,or muramyl tripeptide phosphatidylethanolamine (^{3}H-labeled) with human blood cells, *Int. J. Immunopharmacol. 9:141 (1987).*
11. S. Sone, T. Utsugi, P. Tandon, and M. Ogawara, A dried preparation of liposomes containing muramyl tripeptide phosphatidylethanolamine as a potent activator of human blood monocytes to the antitumor state, *Cancer Immunol. Immunother. 22:191 (1986).*
12. I.J. Fidler, and E.S. Kleinermann, Clinical application of phospholipid liposomes containing macrophage activators for therapy of cancer metastasis, *Adv. Drug. Del. Rev. 13:325 (1994).*
13. P. Dieter, P. Ambs, E. Fitzke, H. Hidaka, R. Hoffmann, and H. Schwende, Comparative studies of cytotoxicity and the release of TNF-α, nitric oxide, and eicosanoids of liver macrophages treated with lipopolysaccharide and liposome-encapsulated MTP-PE, *J. Immunol. 155:2595 (1995).*

66

LTA4 HYDROLASE EXPRESSION DURING GLOMERULAR INFLAMMATION: CORRELATION OF IMMUNOHISTOCHEMICAL LOCALIZATION WITH CYTOKINE REGULATION

Angel Montero, Susumu Uda, Karen A. Munger and Kamal F. Badr

Center for Glomerulonephritis. Renal Division, Emory University and Atlanta VA Medical Center
1840 Southern Lane
Decatur, GA, 30033
Phone: 404-325-5132
Fax: 404-325-5723

INTRODUCTION

Leukotriene B_4 (LTB_4) is the most potent chemotactic substance described for polymorphonuclear leukocytes (PMN) and it also increases PMN aggregation and adhesion to endothelial cells (Samuelson et al., 1987). Glomerular LTB_4 synthesis is enhanced early in the course of several forms of glomerular injury, particularly those characterized by PMN or macrophage infiltration (for review, see Badr, 1992). Most of the studies performed agree on the transient pattern of LTB_4 production following glomerular injury (Badr, 1992, Katoh et al., 1994). LTB_4 production is enhanced during the first 24 hours of injury, and after this period its levels are undetectable in most proliferative and non-proliferative glomerulopathies.

Leukotriene A_4 hydrolase, which catalizes the final reaction in LTB_4 synthesis, is a 70 KD protein (Radmark and Haeggström, 1990) which shares an "unique" aminoacid sequence common among Zn^{2+} metallohydrolases (Jorgeneel et al., 1989). Besides its hydrolase activity, an aminopeptidase activity has been described for LTA_4 hydrolase (Orning et al., 1994). In the kidney LTA_4 hydrolase mRNA is expressed in macrophages and in glomerular mesangial and endothelial cells (Imai et al., 1987; Makita et al., 1992).

LTB_4 synthesis during glomerulonephritis could be regulated by cytokines released by both lymphocytes and macrophages in several ways. It has been described that IFN-γ enhances the release of 5-LO products from activated macrophages (Meslier et al., 1992) and that IL-2-evoked vascular permeability is mediated through the secondary release of LTB_4 (Klausner et al., 1990). On the other hand, we and others have described that IL-13 and IL-4 (two TH-2-derived cytokines) induce 15-lipoxygenase (15-LO) mRNA expression as well as synthesis of the enzyme in human monocytes/macrophages (Conrad et al., 1990; Levy et al., 1993; Nassar et al. 1994), in association with an increase in 15-(S)-HETE generation (Nassar et al. 1994), which, in turn, suppresses LTB_4 generation by leukocytes and inflamed glomeruli (Fischer et al., 1992). The possibility of the regulation of LTA_4 hydrolase gene expression by the aforementioned cytokines is yet to be studied.

In this study we investigated the regulation of LTA$_4$ hydrolase expression in human monocytes by IFN-γ (a TH-1-derived cytokine) and by two cytokines produced by TH-2 lymphocytes, IL-4 and IL-13, as well as the release of LTB4, the final product of LTA$_4$ hydrolase action. We also determined the glomerular localization of LTA$_4$ hydrolase in rats treated with accelereated nephrotoxic serum nephritis

MATERIALS AND METHODS

Materials. Recombinant human IL-1β, IL-4 and IL-13 were obtained from R&D Systems (Minneapolis, MN). IFN-γ was purchased from Boehringer Manhein (Indianapolis, IN). The concentrations used in our experiments were always approximately 10 times the ED$_{50}$, according to the manufacturers instructions.
Monocyte isolation and incubation. Monocyte isolation was performed as previously described (Nassar et al., 1994). Freshly isolated monocytes were incubated in the presence or absence of cytokines (37^0 C, 5% CO$_2$) for 36 hours.
RNA extraction and analysis. Total RNA from monocytes was purified by the Chomczynsky method (1987) and LTA$_4$ hydrolase expression was analyzed by Northern Blot following methods established in our laboratory (Makita et al., 1992). Results are expressed as relative densities which were determined by dividing the signal density of each RNA sample hybridized to LTA$_4$ hydrolase to that hybridized to GAPDH probe.
Expression of LTA$_4$ hydrolase protein. Determination of LTA$_4$ hydrolase protein was performed by Western blot as previosly described (Nassar et al., 1997). Rabbit anti-human LTA$_4$ hydrolase antibody (Cayman Chemical, Ann Arbor, MI) was used as first antibody.
Measurement of LTB$_4$ release. LTB4 release by cytokine-stimulated monocytes was measured by ELISA as previusly described by our group (Nassar et al., 1997).
In vivo studies. male Spregue-Dawley rats were preimmunized with sheep IgG in CFA for 10 days, then injected with sheep anti-rat GBM serum. Animals were sacrified at 3 hr, 24 hr, 48 hr and 7 days after disease induction. Kidney sections were incubated with rabbit anti-human LTA4 hydrolase antibody, and stained by avidin-biotin immunoperoxydase system.
Glomeruli were isolated by mechanical sieving, proteins were isolated and analized by Western blot as decribed above.

RESULTS

Leukotriene A$_4$ hydrolase monocyte expression and LTB$_4$ production were affected by both TH-1 and TH-2 derived cytokines.
Figure 1 shows the effect of IFN-γ (100 pM) on the expression on LTA$_4$ hydrolase mRNA. This cytokine increases significantly (p<0.05, n=8) LTA$_4$ hydrolase mRNA expression after 36 hours of incubation (nearby 150% of basal values). On the other hand, incubation of human peripheral blood monocytes with IL-1β (100 pM) did not affect significantly the expression of LTA$_4$ hydrolase mRNA in these cells.
The effect of TH-2 derived cytokines in LTA$_4$ hydrolase mRNA expression can be seen in figure 2. The densitometry of the Northern blots (n=8) shows that both, IL-4 (100 pM) and IL-13 (10 nM) reduced significantly (p<0.05) this expression when compared with basal values (to approximately 60% and 40% of basal values respectively at 36 hours).
The incubation with cytokines also alters the cellular levels of immunoreactive LTA$_4$ hydrolase in monocytes, as measured by western blot analysis. The quantitative effect of the cytokines on the protein levels of LTA$_4$ hydrolase was similar to the one described for mRNA. Incubation with IL-4 or IL-13 decreased significantly basal levels of the protein, whereas the incubation of monocytes with γ-IFN for 36 hours increased the levels seen in these cells. No significant changes in cellular levels of immunoreactive LTA$_4$ hydrolase were seen after monocyte incubation with IL-1β.

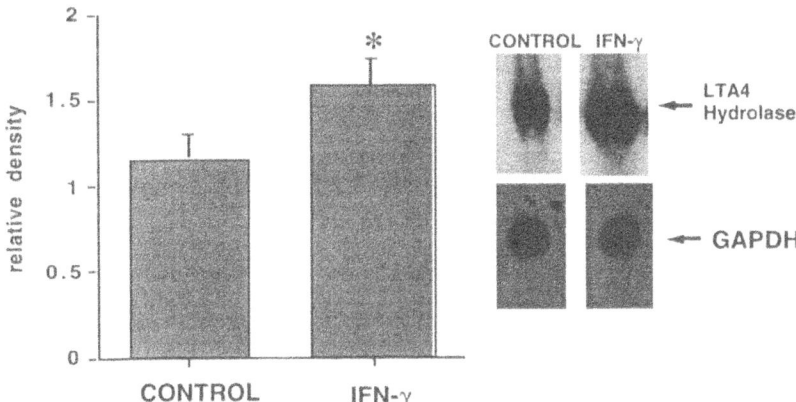

Figure 1.- Expression of LTA₄ hydrolase mRNA in human peripheral blood monocytes incubated with IFN-γ (100 pM). N = 8. * p < 0.05 compared with control values (RPMI supplemented with FCS 10%). Data are mean ± SEM. Insert shows representative Northern blots hybridized with LTA₄ hydrolase and GAPDH radiolabelled probes.

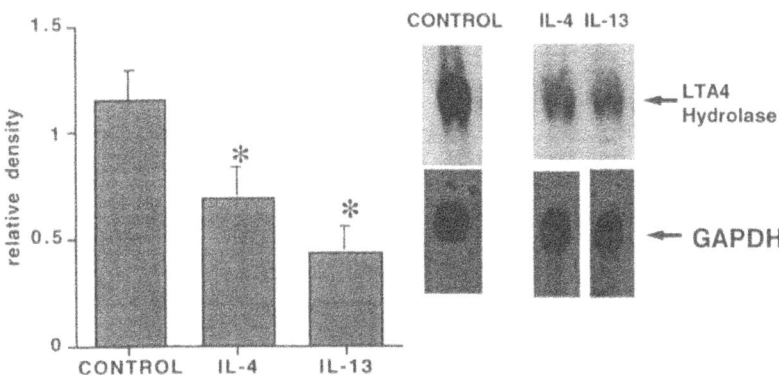

Figure 2.- Expression of LTA₄ hydrolase mRNA in human peripheral blood monocytes incubated with IL-4 (100 pM) and IL-13 (10 nM) for 36 hours. N = 8. * p < 0.05 compared with control values (RPMI supplemented with FCS 10%). Data are mean ± SEM. Insert shows representative Northern blots hybridized with LTA₄ hydrolase and GAPDH radiolabelled probes.

No significant changes in GAPDH expression mRNA were observed after incubating human monocytes with any of the cytokines used. The expression of this housekeeping gene was used to normalize LTA$_4$ hydrolase expression.

The incuabtion of monocytes with cytokines also alters the cellular levels of immunoreactive LTA$_4$ hydrolase, as measured by Western blots analysis. The quantitative effect of the cytokines on the protein levels was similar to the one described for mRNA. Incubation with IL-4 or IL-13 decreased significantly basal levels of the protein, whereas the incubation of monocytes with IFN-γ for 36 hours increased the levels of protein expression. No significant changes in cellular levels of immunoreactive LTA4 hydrolase were seen after monocyte incubation with IL-1-β.

Finally, we analyzed the production of LTB$_4$, the final product of LTA$_4$ hydrolase action. LTB$_4$ leves were also altered by the incubation of monocytes in the presence of pro- and anti-inflammatory cytokines. The basal production of LTB$_4$ was significantly increased by the incubation with either IL-1β or γ-IFN (Figure 3A). On the other hand, "anti-inflammatory" cytokines (IL-4 and IL-13) decreased the basal production of LTB$_4$ after 36 hours of incubation (Figure 3B).

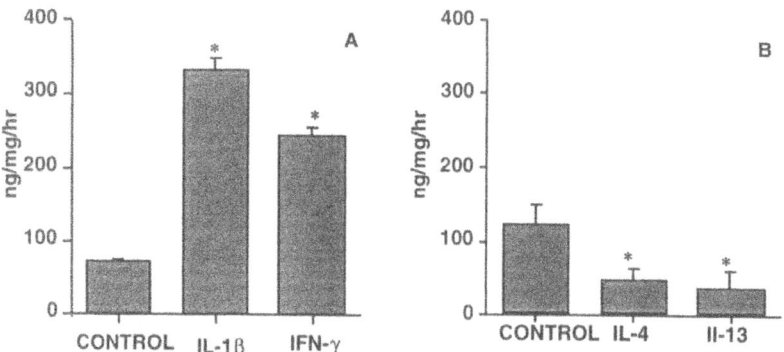

Figure 3.- LTB4 production by activated monocytes after incubation with (A) IL-1β (100 pM) and IFNγ (100 pM) and (B) IL-4 (100 pM) and IL-13 (10 nM) for 36 hours. Results are expressed as ng of LTB$_4$ produced per mg of monocyte cellular protein per hour. N = 10. * p < 0.05 compared with control values (RPMI supplemented with FCS 10%). Data are mean ± SEM.

Glomeruli from rats treated with anti-GBM serum expresed LTA$_4$ hydrolase protein. The expression was detected at 3 hr and decreased after 24 hr. Immunohystology studies also showed an expression of LTA$_4$ hydrolase at 3 hr, but the glomerular expression of the enzyme was totally absent in all parenchimal and infiltrating cells at 24 hours. Unexpectedly, however, intense staining for LTA$_4$ hydrolase was observed in proximal tubules brush border in all nephritic kidneys.

DISCUSSION

Our results show that TH-1 and TH-2 derived cytokines reciprocally regulate LTA_4 hydrolase expression in human monocytes. IFN-γ (a TH-1-derived cytokine) increases LTA_4 hydrolase expression, while IL-4 and IL-13 (TH-2-derived cytokines) decrease this expression. On the other hand, IL-1β, a cytokine produced by macrophages, does not affect significantly the expression of mRNA encoding for LTA_4 hydrolase although it does increase LTB_4 production.

These data agree with previous reports showing opposing roles for TH-1 and TH-2 derived cytokines in other inflammatory processes (Lakkis and Coelho, 1995). The synthesis of LTB_4 is increased during the first 24 hours after the induction of the experimental glomerulonephritis, falling to basal or subbasal levels thereafter (Badr, 1992). No explanation has been given to this surprising result, since, at 24 hours, glomeruli are infiltrated by macrophages/monocytes which, on a per cell basis, constitute a rich source of LTB_4 (Lewis and Austen, 1984).

Our data, showing a differential effect of TH-1 and TH-2 derived cytokines on the expression of LTA_4 hydrolase, could explain partially this temporal pattern. An increase in the glomerular gene expression of IL-4 and IL-13 after the first acute phase of nephrotoxic serum nephritis has been demonstrated previously (Katoh et al., 1994; Lakkis and Cruet, 1993). This increased expression of TH-2 derived cytokines follows the early rise in leukotriene biosynthesis within the glomerulus, which is accompanied by severe structural and functional changes attributable to the vasoactive effects of peptidyl leukotrienes (LTC_4 and LTD_4) and to the leukoattractant effect of LTB_4 (Nassar and Badr, 1995).

During the second phase, leukotriene synthesis is decreased and an increase in the expression of mRNA for rat 12/15-LO lipoxygenase has been reported (Katoh et al., 1994). A product of this dual lipoxygenase, 15-(S)-HETE markedly decreases LTB_4 production by activated leukocytes and inflamed glomeruli (Fisher et al., 1992). IL-4 and IL-13 decrease LTA_4 hydrolase protein levels after incubating monocytes with these cytokines. This inhibition could explain the rapid decrease in LTB_4 production after its initial increase, as the balance between pro and anti-inflammatory cytokines favors a suppression of LTB_4 synthesis, which, in turn, would limit PMN infiltration (Samuelson et al., 1987). Recent *in vivo* data indicates that these changes in the profile of eicosanoid and cytokine expression in acute glomerulonephritis occur in tandem with demonstrable infiltration by T-lymphocytes (Lakkis et al., 1996).

Our in vitro data could explain the observation of an increase in glomerular expression of LTA_4 hydrolase in the first hours after induction of glomerulonephritis in rats. The decreased observed after the first 24 hours may be due to the downregulation exerted by TH-2 cytokines. The unexpected presence of LTA_4 hydrolase in tubule brush border can not be explained on the basis of cytokine regulation; however, this presence may contribute to the development of intertitial inflammation, a known consequence of glomerular injury.

The lack of effect of IL-1β on monocyte LTA_4 hydrolase expression is of interest since this cytokine induces an increase in the synthesis of LTB_4 by these cells. The potential explanation is that IL-1β induces the expression of other enzymes responsible for the synthesis of LTB_4 such as 5-LO or acts through other as yet undefined mechanisms. In a recent study, we have indeed demonstrated the capacity of IL-1b to induce 5-LO gene expression in human monocytes (Nassar et al., 1997).

REFERENCES

Badr, K.F. Five-Lipoxygenase products in glomerular immune injury. *J. Am. Soc. Nephrol.* 3: 907-915; 1992.

Chomczynsky, P., and N. Sacchi. Single-step method of RNA isolation by acid guanidium thiocyanate-phenol-chlorophorm extraction *Anal. Biochem.* 162: 156-159, 1987.

Conrad, D.J., H. Kuhn, M. Mulkins, E. Highland, and E. Sigal. Specific inflammatory cytokines regulate the expression of human monocytes 15-lipoxygenase. *Proc. Nat. Acad. Sci. U.S.A.* 89: 217-221, 1990.

Fischer D.B., J.W. Christman, and K.F. Badr. Fifteen-S-hydroxyeicosatetraenoic acid (15-S-HETE) specifically antagonizes the chemotactic action and glomerular synthesis of leukotriene B4 in the rat. *Kidney Int.* 41: 1155-1160, 1992.

Imai, E., R.L. Hoover, N. Makita, C.D. Funk, and K.F. Badr. Localization and relative abundance of 5-lipoxygenase, 15-lipoxygenase, 12-lipoxygenase and leukotriene A_4-hydrolase gene expression in cultured glomerular cells. *J. Am. Soc. Nephrol.* 1: 443, 1987.

Jorgeneel, C., J. Bouvier, and A. Bairoch. A unique signature identifies a family of zinc-dependent metallopeptidases. *FEBS Lett.* 242: 211-214, 1989.

Katoh, T., F.G. Lakkis, N. Makita and K.F. Badr. Co-regulated expression of glomerular 12/15-lipoxygenase and interleukin-4 mRNAs in rat nephrotoxic nephritis. *Kidney Int.* 46: 341-349, 1994.

Klausner , J.M., G. Goldman, Y. Skornik, R. Valeri, M. Inbar, D. Shepro, and H.B. Hechman. Interleukin-2-induced lung permeability is mediated by leukotriene B4. *Cancer* 66: 2357-2364, 1990.

Lakkis, F.G., F.K. Baddoura, E.N. Cruet, K.R. Parekh, M. Fukunaga, and K.A. Munger. Anti-inflammatory lymphokine mRNA expression in antibody-induced glomerulonephritis. *Kidney Int.* 49: 117-126, 1996.

Lakkis, F.G., and S.N. Coelho. The role of cytokines in inflammatory glomerular injury. *Miner. Electrolyte. Metab.* 21: 250-261, 1995.

Lakkis, F.G., and E.N. Cruet. Cloning of rat interleukin-13 cDNA and analysis of IL-13 gene expression in experimental glomerulonephritis. *Biochem. Biophys. Res. Commun.* 197: 612-618, 1993.

Levy, B.D., M. Romano, H.A. Chapman, J.J. Reilly, J. Draazen, and C.N. Serhan. Human alveolar macrophages have 15-lipoxygenase and generate 15(S)-hydroxy-5,8,11-cis-13-trans-eicosatetraenoic acid and lipoxyns. *J. Clin. Invest.* 92: 1572-1579, 1993.

Lewis, R.A., and K.F. Austen. The biologically active leukotrienes. J. Clin. Invest. 73: 889-897, 1984.

Makita, N., C.D. Funk, E. Imai, R.L. Hoover, and K.F. Badr. Molecular cloning and functional expression of rat leukotriene A_4 hydrolase from glomerular mesangial cells using polymerase chain reaction. *FEBS Lett.* 299: 273-277, 1992.

Meslier, N., A.J. Aldrich, and T.D. Bigby. Effect of interferon-g on the 5-lipoxygenase pathway of rat lung macrophages. *Am. J. Respir. Cell. Mol. Biol.* 6: 93-99, 1992.

Nassar, G.M., and K.F. Badr. Role of leukotrienes and lipoxygenases in glomerular injury. *Miner. Electrolyte. Metab.* 21: 262-270, 1995.

Nassar, G.M., A. Montero, M. Fukunaga, and K.F. Badr. Contrasting effects of proinflammatory and T-helper lymphocyte subset-2 cytokines on the 5-lipoxygenase pathway in monocytes. *Kidney Int.* 51: 1520-1528, 1997.

Nassar, G.M., J.D. Morrow, L. Jackson Roberts, F.G. Lakkis, and K.F. Badr. Induction of 15-lipoxygenase by interleukin 13 in human blood monocytes. *J. Biol. Chem.* 269: 27631-27634, 1994.

Orning, L., J.K. Gierse, and F.A. Fitzpatrick. The bifunctional enzyme Leukotriene-A_4 hydrolase is an arginine aminopeptidase of high efficiency and specificity. *J. Biol. Chem.* 269: 11269-11273, 1994.

Radmark, O, and J. Haeggstrom. Properties of leukotriene A_4-hydrolase. *Adv. Prost. Trombox. Leukot. Res.* 20: 35-4, 1990.

Samuelson, B., S.E. Dahlen, J.A. Lindren, C.A. Rouzer, and C.N. Serhan. Leukotrienes and lipoxins: Structures, biosynthesis and biological effects. *Science* 237: 1171-1176, 1987.

67

12-LIPOXYGENASE PRODUCTS INCREASE MONOCYTE:ENDOTHELIAL INTERACTIONS

Catherine C. Hedrick[1], Mary D. Kim[1], Rama D. Natarajan[2], and Jerry L. Nadler[2]

[1]Division of Cardiology, University of California Los Angeles, CA 90095
[2]Department of Diabetes, Endocrinology, and Metabolism,
City of Hope National Medical Center, Duarte, CA 91010

INTRODUCTION

Diabetes is strongly associated with an accelerated rate of development of atherosclerosis[1]. Evidence suggests that elevated serum glucose levels may contribute to atherogenesis in diabetic patients by promoting oxidation of lipoproteins, propagation of free radicals, and production of advanced glycation end-products (AGE)[2] [3] [4]. However, the exact mechanisms of how hyperglycemia accelerates atherogenesis are not understood.

Atherosclerosis can be described as a disease of chronic inflammation. In the formation of an early atherosclerotic plaque, LDL becomes trapped within the artery wall and undergoes mild oxidation to form minimally-modified LDL (MM-LDL)[5]. MM-LDL is a potent inducer of genes involved in recruitment of monocytes to the intima[6]. One of the key events in the early stages of atherosclerotic lesion formation is the recruitment of monocytes into the vessel wall. This recruitment of monocytes is initiated by MM-LDL[6]. Studies show that MM-LDL may be formed by the transfer of oxidized lipids from artery wall cells to the trapped LDL[5].

Leukocyte-type 12-lipoxygenase (12-LO) incorporates molecular oxygen to arachidonic acid to form both 12(S)- and 15(S)-hydroxyeicosatetraenoic acid (HETE)[7]. Several studies have implicated a role for lipoxygenases in oxidation of LDL[8] [9]. Parthasarathy et al. have shown that endothelial cell-derived LO activity is involved in the oxidation of LDL[8]. From these studies the authors concluded that LDL oxidation occurs either through lipoxygenase oxidation of cellular lipids which are then incorporated into LDL; or via direct LO-dependent oxidation of LDL lipids. Furthermore, studies targeted specifically toward LO expression have shown influence of LO activity in oxidation of LDL[10] [11]. Reaven et al. have recently shown that fibroblasts which overexpress human 15-LO can form MM-LDL in vitro[9]. 12-LO has also been shown to oxidize

Eicosanoids and Other Bioactive Lipids in Cancer, Inflammation, and Radiation Injury, 4
Edited by Honn *et al.*, Kluwer Academic / Plenum Publishers, New York, 1999.

455

LDL in vitro[7]. Studies by Sun *et al.* have shown that macrophages deficient in 12-LO have decreased ability to oxidize LDL in vitro[11]. 12-LO is further implicated as having a key role in LDL oxidation in studies by Sparrow *et al.* which suggested that neither 5-LO nor 15-LO were required for LDL oxidation[12].

Therefore, a potential mechanism in which hyperglycemia may augment the formation of such oxidized lipids is through upregulation of lipoxygenases. In the present study, we sought to investigate whether hyperglycemia may act through the 12-LO pathway in mediating early events of atherogenesis.

Glucose accelerates monocyte:endothelial interactions and enhances 12-(S)-HETE production.

We have recently demonstrated that glucose can significantly increase the binding of monocytes to human aortic endothelial cells (HAEC) in vitro[13]. This induction of binding by glucose was specific for monocytes; neutrophils did not adhere[13].

We also found that human aortic endothelial cells (HAEC) cultured in 25mM glucose for 7-10 days had significantly greater (4-fold) release of 12(S)-HETE into media compared to HAEC cultured in normal levels (5mM) of glucose. This observation suggested that chronic exposure of endothelial cells to elevated glucose conditions caused upregulation of 12-lipoxygenase activity. These data support similar results observed by others which show that both PAEC[14] and PASMC[15] cultured in chronic high glucose have increased production of 15(S)- and 12(S)-HETE, respectively .

12-LO products accelerate monocyte:endothelial interactions.

We next wanted to determine whether 12(S)-HETE and 15(S)-HETE, could influence monocyte adhesion to endothelium, a key early event in atherogenesis. We found that treatment of HAEC with either 12(S)-HETE or 15(S)-HETE for 4h at 37 C significantly induced monocyte adhesion to HAEC by 3-fold over untreated cells (75 ± 5 or 75 ± 3 vs. 26 ± 1 monocytes bound, respectively, P<0.001. This induction of adhesion by 12(S)-HETE and 15(S)-HETE was specific for monocytes, and not neutrophils (similar to that reported above for glucose). This is consistent with the presence of monocytes or monocyte-derived macrophages in aortic lesions and the complete absence of neutrophils[16]. Previous studies have found evidence for a leukocyte-type of 12-LO in HAEC[17].

Lipoxygenase inhibitors block glucose-mediated monocyte adhesion.

To further study whether glucose influences monocyte:endothelial interactions through upregulation of 12-LO, we examined whether inhibition of 12-LO activity directly could block monocyte adhesion to endothelial cells using lipoxygenase inhibitors. Phenidone is a non-selective lipoxygenase inhibitor while baicalein is a more selective 12-LO inhibitor. Both phenidone and baicalein blocked the increase of monocyte adhesion to HAEC grown in elevated glucose levels (60 ± 8 and 45 ± 5 vs. 95 ± 8 monocytes bound, respectively, p<0.001) . These inhibitors at these concentrations did not affect monocyte adhesion in HAEC cultured in normal glucose (5 mM) conditions. These data further suggested that glucose mediates adhesion of monocytes to endothelial cells through upregulation of 12-LO activity.

Glucose and 12-LO products mediate monocyte:endothelial interactions through similar pathways

It is possible that glucose and 12-LO products may mediate monocyte adhesion to the endothelium through similar signaling pathways. One pathway is through activation and upregulation of endothelial cell surface molecules, such as E-selectin, VCAM-1, and ICAM-1. Our previous findings showed that chronic exposure of HAEC to high glucose did not induce endothelial cell surface expression of VCAM-1, E-selectin, or ICAM-1, the major molecules involved in tethering of monocytes to endothelium[15]. To investigate the molecules involved in mediating the binding of monocytes to endothelial cells in the presence of 12(S)- and 15(S)-HETEs, an ELISA was performed for E-selectin, VCAM-1, and ICAM-1. HAEC treated with 12(S)- or 15(S)-HETE for 4 h did not significantly induce VCAM-1, ICAM-1, or E-selectin (data not shown). These data indicate that endothelial cell surface adhesion molecules were not upregulated by products of 12-LO. A second pathway is through activation of integrins on monocytes. Integrin activation has been found to play a major role in adhesion of monocytes to endothelium[18]. Soluble chemoattactants from the local vessel microenvironment or from endothelial cell surface molecules activate monocytes to produce integrins[19 20]. These integrins allow monocytes to stably adhere to endothelial cells. Two major integrins found on monocytes are very late antigen-4 (VLA-4), now known as 4 1, and LFA-1, also known as L 2[21]. The integrin 4 1 has been shown to interact with both VCAM-1 and fibronectin (specifically the CS-1 region of fibronectin) on endothelium[22]. The integrin L 2 interacts with ICAM-1 and ICAM-2 on endothelium[31]. We therefore evaluated the effects of 1 and 2 integrins in both glucose and LO product-induced monocyte binding to endothelial cells. Incubation of monocytes with either CD49d antibodies to block 1 integrins which mediate VCAM-1 binding, or with TS1-18 antibodies to block 2 integrins which mediate ICAM-1 binding, abrogated adhesion of monocytes to HAEC treated with 12(S)-HETE (60±5 and 60±4 vs 85±3 monocytes bound, respectively, p<0.005), and also abrogated adhesion of monocytes to HAEC treated with high glucose (66±4 and 48±4 monocytes bound, respectively vs 90±4 monocytes bound for high glucose, p<0.001). Thus, our data indicate that both glucose and 12-LO products may mediate monocyte:endothelial interactions through a similar pathway of monocyte integrin activation. However, it is possible that other mechanisms may also be involved in glucose and LO product-induced monocyte adhesion. Several studies have shown that exogenous 12(S)-HETE can activate protein kinase C (PKC) in vitro in microvascular endothelial cells[23] and in tumor cells[24 25]. Honn and colleagues have suggested a pathway by which such activation of PKC can, in turn, activate 12-LO to produce endogenous 12(S)-HETE and stimulate activation of integrins[32 33]. Furthermore, Tang et al. have shown that 12(S)-HETE increased v 3 expression in endothelial cells[32], and Trikha et al. have shown increased expression of 3 integrins in tumor cells exposed to 12(S)-HETE[33]. Other factors that affect monocyte rolling such as platelet endothelial cell adhesion molecule-1 (PECAM)[26] and P-selectin[27], or endothelial-derived monocyte chemokines such as MCP-1[6] could certainly also be involved. Additional studies will be needed to evaluate these possibilities.

Lipoxygenases have been implicated as proatherogenic enzymes capable of directly oxidizing polyunsaturated fatty acids, such as arachidonic acid. We hypothesize that hyperglycemia upregulates 12-LO activity, which oxidizes arachidonic acid to form 12(S)- and

15(S)-HETE. These HETEs may be either released from the endothelial cells or may be incorporated into phospholipids on HAEC plasma membranes, and can induce monocyte adhesion to endothelium. It is possible that other products of the LO pathway may even be more potent activators of monocyte adhesion. A very proinflammatory compound, 12-HETre can increase monocyte adhesion at 10^{-10} and 10^{-11} M concentrations (data not shown). Additional studies will be needed to evaluate the most active lipids generated by the 12-LO enzyme.

SUMMARY

In summary, we suggest that hyperglycemia causes upregulation of 12-lipoxygenase activity. The increased production of 12-LO products, 12(S) and 15(S)-HETE, activates monocyte integrins which result in enhanced adhesion of monocytes to endothelium. The binding of monocytes to endothelium is a key early event in development of atherosclerosis. Upregulation of this process by vascular cells exposed to chronic elevations in glucose may be one explanation for the accelerated atherosclerosis observed in patients with Type 2 diabetes.

Acknowledgments

The authors would like to thank Dr. Judith A. Berliner for helpful advice and encouragement. This work was supported by National Institutes of Health grant PO1-HL55798 and by a grant from the Juvenile Diabetes Foundation.

[1]REFERENCES

1.Steiner G. 1985. Atherosclerosis, the major complication of diabetes. *Adv Exp Med Biol.* 189:277-297.

[2]Kawamura M, Heinecke JW, and Cahit A. 1994. Pathophysiological concentrations of glucose promote oxidative modification of LDL by a superoxide-dependent pathway. *J Clin Invest.* 94: 771-778.

[3]Bellomo G, Maggi E, Poli M, Agosta FG, Bollati P, and Finardi G. 1995. Autoantibodies against oxidatively modified LDL in NIDDM. *Diabetes.* 44: 60-66.

[4]Knecht KJ, Dunn JA, McFarland KF, McCance DR, Lyons TJ, Thorpe SR, and Baynes JW. 1991. Effect of diabetes and aging on carboxymethyllysine levels in human urine. *Diabetes.* 40: 190-196.

[5]Berliner JA, Territo MC, Sevanian A, Ramin S, Kim JA, Barnshad B, Esterson M, and Fogelman AM. 1990. Minimally modified LDL stimulates monocyte:endothelial interactions. *J Clin Invest.* 85: 1260-1266.

[6]Cushing S, Berliner JA, Valente AJ, Territo MC, Navab M, Parhami F, Gerrity R, Schwartz CJ, and Fogelman AM. 1990. Minimally-modified LDL induces monocyte chemotactic protein-1 in human endothelial cells and smooth muscle cells. *Proc Natl Acad Sci USA.* 87: 5134-5138.

[7]Roy P, Roy SK, Mitra A, and Kulkarni AP. 1994. Superoxide generation by LO in the presence of NADH and NADPH. *Biochem Biophys Acta.* 1214: 171-179.

[8]Parthasarathy S, Wieland E, and Steinberg D. 1989. A role for endothelial cell lipoxygenase in the oxidative modification of LDL. *Proc Natl Acad Sci USA*. 86: 1046-1050.

[9]Rankin SM, Parthasarathy S, and Steinberg D. 1991. Evidence for a dominant role of lipoxygenases in oxidation of LDL by mouse peritoneal macrophages. *J Lipid Res*. 32: 449-456.

[10]Sigari F, Lee C, Witztum JL, and Reaven PD. 1997. Fibroblasts that overexpress 15-LO generate bioactive and minimally-modified LDL. *Arterioscler. Thromb. Vasc Biol*. 17: 3639-3645.

[11]Sun D, and Funk CD. 1996. Disruption of 12/15 LO expression in peritoneal macrophages. *J Biol Chem*. 271: 24055-24062.

[12]Sparrow CP, and Olszewski J. 1992. Cellular oxidative modification of LDL does not require lipoxygenases. *Proc Natl Acad Sci USA*. 89: 128-131.

[13]Kim JA, Berliner JA, Natarajan RD, and Nadler JL. 1994. Evidence that glucose increases monocyte binding to human aortic endothelial cells. *Diabetes*. 43: 1103-1107.

[14]Brown ML, Jakubowski JA, Leventis LL, and Deykin D. 1988. Elevated glucose alters eicosanoid release from porcine aortic endothelial cells. *J Clin Invest*. 82: 2136-2141.

[15]Natarajan R, Gu J-L, Rossi J, Gonzales N, Lanting L, Xu L, and Nadler JL. 1993. Elevated glucose and angiotensin II increases 12-lipoxygenase activity and expression in porcine aortic smooth muscle cells. *Proc Natl Acad Sci USA*. 90: 4947-4951.

[16]Berliner JA, Navab MN, Fogelman AM, Frank JS, Demer LL, Edwards PA, Watson AD, and Lusis AJ. 1995. Atherosclerosis: basic mechanisms. Oxidation, inflammation and genetics. *Arterioscler. Thromb. Vasc. Biol*. 91: 2488-2496.

[17] Kim JA, Gu J-L, Natarajan R, Berliner JA, and Nadler JL. 1995. A leukocyte type of 12-LO is expressed by human vascular and mononuclear cells: evidence for upregulation by angiotensin II. *Arterioscler. Thromb. Vasc. Biol*. 15: 942-948

[18]Hynes RO. 1992. Integrins: versatility, modulation, and signaling in cell adhesion. *Cell*. 69: 11-25.

[19]Bevilacqua MP. 1993. Endothelial-leukocyte adhesion molecules. *Annu Rev. Immunol*. 11: 767-804.

[20]Butcher EC. 1991. Leukocyte:endothelial cell recognition: three (or more) steps to specificity and diversity. *Cell*. 67: 1033-1036.

[21]McEver RP. 1992. Leukocyte-endothelial cell interactions. *Current Opin Cell Biol*. 4: 840-849.

[22]Springer TA. 1990. Adhesion receptors of the immune system. Nature. 346: 425-434.

[23]Tang DG, Diglio CA, Bazaz R, and Honn KV. 1995. Transcriptional activation of endothelial cell integrin v by protein kinase C activator 12(S)-HETE. *J Cell Sci*. 108: 2629-2644.

[24]Trikha M and Honn KV. 1997. Role of 12-lipoxygenase and protein kinase C in modulating the activation state of the integrin aIIbB3 on human tumor cells. In: *Eicosanoids and Other Bioactive Lipids in Cancer Inflammation and Radiation Injury 3*. KV Honn, editor. Plenum Press, New York., 55-60.

[25]Liu B, Khan WA, Hannun YA, Timar J, Taylor JD, Lundy S, Butovich I, and Honn KV. 1995. 12(S)-HETE and 13(S)-HODE regulation of protein kinase C in melanoma cells: role of receptor-mediated hydrolysis of inositol phospholipids. *Proc Natl Acad Sci USA*. 92: 9323-9327.

[26]Muller WA, Weigl SA, Deng X, Phillips DM. 1993. PECAM-1 is required for transendothelial migration of leukocytes. *J Exp Med*. 178: 449-460.

[27]Frenette PS, Johnson RC, Hynes RO, and Wagner DD. 1995. Platelets roll on stimulated endothelium in vivo: interaction mediated by endothelial p-selectin. *Proc Natl Acad Sci USA*. 92: 7450-7454.

LINOLEIC AND LINOLENIC ACIDS

68

LINOLEIC ACID METABOLITES IN HEALTH AND DISEASE

Michael R. Buchanan

Department of Pathology
McMaster University
Hamilton, Ontario L8N 3Z5

INTRODUCTION

A common factor in the progression of metastasis, atherosclerosis and inflammation is excessive cell growth prompted by both endogenous or exogenous stimuli; e.g. ocogenes, cytokines and thrombin. Excessive cell growth in all of these pathologic responses requires multiple cell cell interactions. There is an abundance of literature suggesting that arachidonic acid metabolites derived from both the cyclooxygenase and lipoxygenase pathways, influence these processes, but it is only recently that our attention has included the importance of linoleic acid metabolism and the relative roles of linoleic and arachidonic acid metabolites in these disease processes.

13-HODE, HETEs AND PGI$_2$ and METASTASIS

Honn et al [1,2] and our laboratory have investigated the role(s) of the lipoxygenase-derived monohydroxides of linoleic acid and arachidonic acid, 13-hydroxyoctadecadienoic acid (13-HODE) and 5-, 12- or 15-hydroxyeicosatetraenoic acid (5-, 12-, 15-HETE) from the perspective of cell cell interactions [3,4]. Initially, Honn el al [5] proposed that the cyclooxygenase-derived arachidonic acid metabolite released from endothelial cells, PGI$_2$, played an important antimetastatic role and inhibited platelet/tumour cell/vessel wall interactions which, if uninhibited, facilitated tumour cell extravasation and secondary metastasis into extravascular tissue. Then, we reported that both tumour cells and vascular wall cells also synthesized linoleic acid via the lipoxygenase pathway to 13-HODE, predominately under unstimulated conditions [6,7]. When these cells were perturbated or injured, 13-HODE synthesis decreased and arachidonic acid derived 12- or 15-HETE synthesis increased. The relative endogenous levels of these two types of monohydroxide in tumour cells correlated with tumour cell adhesivity and metastatic potential [7]. Honn et al [8] pursued this issue further and found that

Eicosanoids and Other Bioactive Lipids in Cancer, Inflammation, and Radiation Injury, 4
Edited by Honn *et al.*, Kluwer Academic / Plenum Publishers, New York, 1999.

463

exogenously-added monohydroxides also influenced tumour cell adhesivity. 13-HODE inhibited tumour cell adhesivity whereas 12-HETE enhanced tumour cell adhesivity. The underlying mechanism for how these metabolites influences cell cell adhesivity, however, remains controversial. Some have suggested that these metabolites act as secondary messengers, influencing the PKC pathway [9], while others have suggested that they are primary messengers and acted directly, altering the conformational structure of cellular adhesive receptors, thereby inhibiting or enhancing the ability of those integrins to recognized their specific ligands[3, 10]. In addition, other investigators have suggested that 13-HODE influences phosphorylation of certain growth factors [11], but how these results would explain altered cell adhesivity is not entirely clear.

13-HODE and HETEs in THROMBOSIS

A number of studies by our laboratory suggest that endogenous 13-HODE downregulates endothelial cell surface adhesivity, thereby rendering the endothelium biocompatable with circulating blood cells [3, 6-7, 10]. Increasing vascular 13-HODE level both experimentally and clinically, decreases vessel wall reactivity to platelets [6, 12-13]. We also demonstrated that when purified 13-HODE is inserted into liposomes containing the endothelial cell-derived vitronectin receptor, the ability of the vitronectin receptor to bind with its specific ligands decreases in a concentration-dependent manner, whereas when 12- or 15-HETE is added, there is increased ligand binding to both the vitronectin receptor and to the platelet derived glycoprotein IIbIIIa fibrin(ogen) receptor [10]. Moreover, these effects are achieved in the absence of any enzymatic pathway, suggesting to us that 13-HODE and the HETEs act directly on the receptors, either by interacting with the active adhesive site(s) on each receptor or by interacting with the transmembrane lipophylic site and inducing a conformational change which alters the exposure of the active site to its ambient surroundings. The latter possibility seems more consistent with both ours and Honn's studies, notwithstanding their observations that there is a concommitant activation of PKC and other pathways.

13-HODE, HETEs and INFLAMMATION

A number of studies have suggested that linoleic acid-derived 13-HODE and PGE_1 decrease inflammation in a variety of dermatological problems associated with defective Δ 6 desaturase activity, excessive epidermal cell proliferation and epidermal barrier function [14-16]. In contrast, elevating HETE levels appears to exacerbate these inflammatory processes. The mechanisms for how these metabolites exert their effects has not been entirely established. We have demonstrated that agents which enhance cyclic AMP, result in increased linoleic acid turnover and subsequent 13-HODE synthesis in vascular tissue [4, 17]. It is possible that prostanoids such as PGE_1 which has a longer half life than PGI_2, enhance cyclic AMP more effectively than PGI_2, thereby indirectly elevating 13-HODE synthesis and facilitating downregulation of nuclear PKC/MAP-kinase and subsequent cell proliferation [18].

THE 13-HODE/HETE PATHWAY in CELL CELL INTERACTIONS

All of the aforementioned studies highlight a number of common steps in different disease processes and the common opposing effects of 13-HODE and the HETEs; namely, excessive cell adhesivity and accelerated cell proliferation result in increased metastatic, thrombogenic and inflammatory responses. Increasing 13-HODE synthesis downregulates the adhesive responses whereas the HETEs upregulate the adhesive responses. Whether 13-HODE and/or the HETEs inhibit or enhance cell proliferation directly or indirectly by altering cell adhesivity, however, still needs to be clarified. Unfortunately, most studies which demonstrate an enhancing effect of 13-HODE on cell proliferation, regardless of the disease model, have not compared the (anti)proliferative effects of 13-HODE with the (anti)proliferative effects of the arachidonate monohydroxides. Nonetheless, these common features in these 3 disease processes suggest that 13-HODE/HETE regulation of cell cell adhesivity is a generalizable pathway of cell cell interaction regulation, irrespective of the disease entity. If so, a better understanding of how that pathway can be modified, may assist us in the development of more effective therapeutic strategies.

13-HODE:HETEs, PRIMARY TUMOUR GROWTH and SECONDARY METASTASIS

Based upon the above studies, we hypothesized that increasing either vascular wall or tumour cell 13-HODE level will decrease both tumour growth and metastasis, whereas increasing vascular wall or tumour cell HETE levels will increase primary tumour growth and secondary metastasis. The basic rationale for the former is that 13-HODE decreases cell cell adhesivity, a step necessary for cell proliferation, migration and intra- and extravasation. The rationale for the latter is that the HETEs have the opposite effect.

To test these possibilities, we fed Wistar rats a diet rich in linoleic acid or in aracidonic acid for 1 month. Half of the rats were injected intraperitoneally with Walker #256 carcino-sarcoma tumour cells (TCs, American Tissue Culture Association). The TCs were passaged 3 times over the month of diet treatment into corresponding diet fed rats (TC donors). The other half of specific diet fed rats acted as TC recipient rats. Thus, half of the TC recipient rats were fed a fatty acid rich diet and injected intramuscularly with TCs obtained from chow diet-fed donor rats, and the other half of the TC recipient rats were fed the chow diet and injected intramuscularly with TCs obtained from the fatty acid rich diet fed TC donors.

Two weeks later, the rats were killed and both hind limbs were skinned and transected at the same anatomical site. We then determined the weight and volume of the tumour present, using the non tumour leg as the reference control. Volume was calculated by measuring the longest and shortest diameter of the tumour and then using the formula - volume $= d^2 \times 1 \times \pi \div 6$ [19]. The aortae were removed, extracted and analysized for vessel wall 13-HODE and fatty acid levels, using HPLC. Similarly TC fatty acid and 13-HODE/HETE levels were determined.

The lungs were harvested and processed histologically to quantify the TC metastasis. Briefly, histological thin sections were prepared and stained with hematoxylin and eosin. Each section was mounted on a glass slide and then projected onto a Microplan II digitized tablet. The areas occupied by the tumour were determined using an image analyser, and expressed as a per cent of the total cross sectional area. Analysis were made on 3 longitudinal sections of each rat lung.

This experimental design allowed us study the effects of these fatty acid rich diets on TC and host tissue fatty acid metabolism, and TC growth and TC metastatic potential in normal chow fed rats injected with TCs obtained from specific fatty acid rich diet fed TC donors, and the effects of the diets on the TC recipient host tissue, independent of a diet effect on the TCs *per se*. The linoleic acid rich diet contained >50 % more linoleic acid that either the regular chow or arachidonic acid rich diets (**Table 1**, next page). This resulted in a preferential uptake of linoleic acid into both the TCs and the vascular tissue as compared to arachidonic acid uptake. More importantly, the ratio of linoleic acid:arachidonic acid increased in both TCs and vascular tissue in linoleic acid rich diet fed animals.

Consequently, the ratios of the basal 13-HODE:HETE level were altered accordingly. The ratio of basal 13-HODE:HETE in the TCs increased from 6:1 in the chow fed rats to 9:1 in the linoleic acid rich diet fed rats (**Table 2**, next page). The incorporation of linoleic acid into the TCs and aortae correlated significantly with the increases in 13-HODE levels in the different diet-fed rats, r = 0.9864, p < 0.001 and r = 0.9947, p < 0.001, respectively.

When TCs obtained from chow diet fed donors were injected into chow fed recipient rats, the leg tumours formed weighed ≈ 18 gms and occupied ≈ 44 cm³ in volume (**Table 3i**). After two weeks, a sufficient number of cells had escaped from the primary leg tumour and metastasized in the lung, occupying ≈ 2 % of its space. Primary tumour size, measured in both weight and volume, was greater in the linoleic acid rich diet fed TC recipients; but secondary metastasis decreased by 50 %, p < 0.025 (**Table 3i**). In contrast, primary tumour growth

Table 1. Diet, Aorta, and TC Linoleic (LA) and Arachidonic (AA) Levels in the Diet Fed Rats.

	CHOW DIET	LA-RICH DIET	AA-RICH DIET
DIET LA Level	21 ± 9	32 ± 1 (52%↑)	18 ± 1
DIET AA Level	11 ± 2	14 ± 2	19 ± 2 (72 %↑)
TC LA Level	18 ± 1	27 ± 1 (50%↑)	15 ± 1
TC AA Level	7 ± 1	7 ± 1	7 ± 1
AORTA LA Level	13 ± 1	25 ± 3 (92%↑)	10 ± 2
AORTA AA Level	1 ± 1	1 ± 0	1 ± 3

*Data are expressed as a percentage of the total fatty acid profile, means ± sem, n = 12 /group. The numbers in ()
indicate the % increase above the chow control.*

Table 2. Basal level of 13-HODE and HETEs in TCs and Aortae Obtained From the Chow, Linoleic Acid-Rich and Arachidonic Acid-Rich Diet Fed Rats.

	CHOW	LA	AA
TC 13-HODE*	7.5 ± 0.8	22.5 ± 3.0	11.9 ± 1.2
TC HETEs*	1.2 ± 0.0	2.4 ± 0.0	2.7 ± 0.0
RATIO 13-HODE:HETEs	6:1	9:1	4:1
AORTA 13-HODE**	33 ± 6	60 ± 14	37 ± 9
AORTA HETES	ND	ND	ND

*Data are expressed as ng/10⁹ TCs, means ± sem, n = 12. **Data are expressed as ng/cm², means ± sem, n = 12. ND
= not detectable.*

Table 3. Primary Tumour Growth and Secondary Metastasis: Relative effects of Diet on Host and Tumour Cell Fatty Acid Metabolism

	CHOW	LA	AA
i) TCs injected into Diet fed rats			
Tumour (wt -gm)	18 ± 2	$32 \pm 1^*$	19 ± 1
Tumour (Vol - cm^3)	44 ± 3	$55 \pm 5^*$	46 ± 1
Metasatasis (% Lung area)	2.0 ± 0.6	$1.3 \pm 0.3^*$	4.8 ± 1.1
ii) TCs Obtain from Diet-Fed Donors			
Tumour (wt -gm)	21 ± 3	$7 \pm 4^{**}$	29 ± 3
Tumour (Vol - cm^3)	44 ± 2	$20 \pm 4^{**}$	40 ± 2
Metasatasis (% Lung area)	2.2 ± 0.6	$1.2 \pm 0.4^{**}$	$3.2 \pm 0.6^*$

Data are expressed as means $\pm$ sem, n = 12. *p< 0.025; **p < 0.005.

proceeded in the arachidonic acid rich diet fed rats to the same extent as in the control chow diet fed animals, but secondary metasatsis increased significantly.

When TCs were obtained from linoleic acid rich diet fed rats and injected into chow diet fed TC recipients, both primary tumour growth and secondary metastasis decreased significantly. In contrast, decreasing the 13-HODE:HETE ratio in TCs obtained from the arachidonic acid rich diet fed rats, increased both primary tumour growth and secondary metastasis (**Table 3ii**).

DISCUSSION

The observations that linoleic acid levels increased in the aortae of linoleic acid rich diet fed rabbits is similar to previous observations in rabbits and humans[12-13]. It is interesting to note a similar preferential uptake in the TCs, and that increasing arachidonic acid levels in the diet had no affect on basal arachidonic acid levels, either in the TCs or in the vascular tisse. However, it should also be noted that fatty acid incorporation into both the aortae and TCs was studied under basal conditions. The possibility that preferential uptake of arachidonic acid under stimulated conditions takes place, cannot be excluded, particularly since arachidonic acid metabolism takes precedence over linoleic acid metabolism in vascular and tumour tissue following stimulation or perturbation[4, 7, 12, 17]. It is also possible that arachidonic acid turnover is increased in the arachidonic acid rich diet fed animals, thereby maintaining steady state tissue arachidonic acid levels but increased HETE synthesis. Consistent with that possibility, the 13-HODE:HETE ratio decreased in those animals fed the arachidonic acid rich diet, and both tumour growth and secondary metastasis increased significantly. These effects were exacerbated more when the 13-HODE:HETE ratio was decreased in the TCs *per se*, consistent with previous studies[7].

The observations that primary tumour growth and secondary metastasis decreased significantly when the TC 13-HODE:HETE ratio was increased in the linoleic acid rich diet fed TC donor rats, is also consistent with previous studies[7]. It is unclear therefore, why primary tumour growth increased in the linoleic acid rich diet fed TC recipient animals. One possibility is that host tissue derived linoleic acid metabolites exacerbates tumour growth. This possibility, however, is not consistent with any of the other finding. An alternate explanation which seems more plausible is, that increased vascular 13-HODE levels render the endothelial cell lining less adhesive (as demonstrated previously [12, 17]) such that the TCs are unable to adhere to them, and, therefore, migration to secondary sites is inhibited. This possibility is supported by other other observations [12, 17], but needs to be confirmed in this animal model.

In summary, there is increasing evidence that the relative levels or ratio of 13-HODE:HETEs in assorted cells influences subsequent cell cell interactions, cell proliferation and progression of the pathophysiologic responses in many disease entities. It is suggested that more research is warranted to better understand the underlying mechanisms by which 13-HODE, HETEs and other linoleic and arachidonic acid metabolites exert their effect(s).

REFERENCES

1. K.V. Honn and L.J. Marnett, Prostaglandin, thromboxane, and leukotriene biosynthesis: Target for antitumor and antimetastatic agents, in: *Novel Approaches to Cancer Therapy*, K.V. Honn and L.J. Marnett, eds., Academic Press, New York (1984).
2. X. Gao, D.J. Grignon, T. Chbihi, A. Zacharek, Y.Q. Chen, W. Sakr, A.T. Porter, J. D. Crissman, J.E. Pontes, I.J. Powell, and K.V. Honn, Elevated 12-lipoxygenase mRNA expression correlates with advanced stage and poor differentiation of human prostate cancer, *Urology* 48: 227 (1995).
3. M.R. Buchanan and E. Bastida, Endothelium and underlying membrane reactivity with platelets, leukocytes and tumor cells: Regulation by the lipoxygenase-derived fatty acid metabolites, 13-HODE and HETEs, *Med Hypothesis* 27: 317 (1988).
4. T.A. Haas, M.-C. Bertomeu, E. Bastida, M.R. Buchanan, Cyclic AMP regulation of endothelial cell triglyceride turnover, 13-hydroxyoctadecadienoic acid (13-HODE) synthesis and endothelial cell thrombogenicity, *Biochim Biophys Acta* 1031: 174 (1990).
5. K.V. Honn, D.G. Menter, J.M. Onoda, B.W. Steinert, C.A. Diglio, J.D. Taylor, and B.F. Sloane, Tumor cell-platelet-endothelial cell interactions and prostaglandin metabolism, *Adv Prost Thromb Leuko Res* 17B: 981(1987).
6. M.R. Buchanan, T.A. Haas, M. Lagarde, and M. Guichardant, 13-Hydroxyoctadecadienoic acid is the vessel wall chemorepellant factor, LOX, *J Biol Chem* 260: 16056 (1985).
7. E. Bastida, M.-C. Bertomeu, T.A. Haas, L. Almirall, D. Lauri, F.W. Orr, and M.R. Buchanan, Regulation of tumor cell adhesion by intracellular 13-HODE:15HETE ratio, *J Lipid Med* 2: 281 (1990).
8. I.M. Grossi, L.A. Fitzgerald, L.A. Umbarger, K.K. Nelson, C.A. Dilio, J.D. Taylor, and K.V. Honn, Interleukin-1α induced vitronectin receptor expression and tumor cell endothelial cell adhesion, *FASEB* 4: A1134 (1990).
9. K.V. Honn, D.G. Tang, X. Gao, I.A. Butovich, B. Lui, J. Timar, and W. Hagmann, 12-lipoxygenases and 12(S)-HETE in cancer metastasis, *Cancer Metast Rev* 13: 365 (1994).
10. M.R. Buchanan, P. Horsewood, and S.J. Brister, Regulation of endothelial cell/ and platelet/ receptor-ligand binding by the 12- and 15-lipoxygenase monohydroxides, 12-, 15-HETE and 13-HODE, *Prost Leuko Essential Fatty Acids*, In Press (1998).
11. W.C. Glasgow, R. Hui, S. Jayawichreme, J. Angerman-Stewart, B.-B. Han, and T.E. Eling, The linoleic acid metabolite (13S)-hydroperoxyoctadecadienoic acid, augments the epidermal growth factor receptor signalling pathway by attenuation of receptor phosphorylation: Differential responses in Syrian hamster embryo tumor suppressor phenotypes, *J Biol Chem* 272:12269 (1997)
12. M.-C. Bertomeu, G.L. Crozier, T.A. Haas, M. Fleith, and M.R. Buchanan, Selective effects of dietary fats on vascular 13-HODE synthesis and platelet/vessel wall interactions, *Thromb Res* 59: 819 (1990).

13. S.J. Brister, T.A. Hass, M.-C. Bertomeu, J. Austin, and M.R. Buchanan, 13-HODE synthesis in internal mammary arteries and saphenous viens: Implications in cardiovascular surgery, *Adv Prost Thromb Leuko Res* 21: 667 (1990).

14. S.J. Taub and S.J. Zakon, Use of unsaturated fatty acids in the treatment of eczema, *JAMA* 105: 1697 (1995).

15. A. Bjorneboe, E. Soyland, G.E. Bjorneboe, G. Rajka, and C.A.Devron, Effect of n-3 fatty acid supplement to patients with atopic dermatitis, *J Intern Med (Suppl)* 225: 233 (1989).

16. C.C. Miller and V.A. Ziboh, Induction of epidermal hyperproliferation by topical n-3 polyunsaturated fatty acids on guinea pig skin linked to decrease levels of 13-hydroxyoctadecadienoic acid (13-HODE), *J Invest Dermatol* 94: 353 (1990).

17. E. Weber, T. A. Haas, T.H. Mueller, W.G. Eisert, J. Hirsh, M. Richardson, and M.R. Buchanan, Relationship between vessel wall 13-HODE synthesis and vessel wall thrombogenecity following injury. Influence of salicylate and dipyridamole treatment, *Thromb Res* 57:383 (1990).

18. V.A. Ziboh, I. Mani, and L. Iversen, Upregulation of nuclear PKC/MAP-kinase in a guinea pig model of epidermal hyperproliferation: Suppression by 13-(S)-hydroxyoctadecadienoic acid, *Proc 5th Intl Conf Eicosanoid & Other Bioactive Lipid in Cancer, Inflammation and Related Diseases:* 14 (Sept 1997).

19. H.I. Robins, W.H. Dennis, J.S. Slattery, T.A. Lange, and M.B. Yatvin, Systematic lidocaine enhancement of hyperthermia-induced tumor regression in transplantable murine tumor models, *Cancer Res* 43:3187 (1983).

69

TOXICITY OF LINOLEIC ACID METABOLITES

Jessica F. Greene & Bruce D. Hammock

Departments of Entomology and Environmental Toxicology
University of California at Davis
Davis, CA 95616

INTRODUCTION

Leukotoxin, *cis*-9,10-epoxyoctadeca-12(Z)-enoic acid (LTX), and isoleukotoxin, *cis*-12,13-epoxyoctadeca-9(Z)-enoic acid (iLTX), are the monoepoxides of linoleic acid (*cis*-9,12-octadecadienoic acid). In the systems so far examined two primary metabolites are LTX diol, *threo*-9,10-dihydroxy-octadeca-12(Z)-enoic acid, and iLTX diol, *threo*-12,13-dihydroxyoctadeca-9(Z)-enoic acid. See figure 1. The first mention of LTX, *per se*, in the literature was in 1986 (1), but the two isomers of epoxyoctadecenoic acid have been discussed since before 1971. LTX and iLTX are found endogenously in both animals and plants. While their defensive role in plants seems fairly well established (2), their role in animals is less clear. What is clear is that LTX and iLTX are formed from linoleic acid, usually when the organism is under less than optimal conditions, and that they have a wide range of effects.

It is somewhat surprising that LTX and iLTX have so many effects. While many compounds containing the epoxide functionality are known to be extremely reactive due to the ease with which the highly strained three-membered oxirane ring can be opened, fatty acid epoxides are relatively stable and are poor alkylating agents (3). Epoxides can be degraded by reaction with glutathione to form conjugates, or water to form *vic*-diols (however, it should be noted that hydrolysis of fatty acid epoxides by water alone is extremely slow). See figure 1. These reactions usually represent detoxification pathways and may be catalyzed by glutathione S-transferases or epoxide hydrolases (4).

In this laboratory, we regularly screen a number of epoxide containing compounds for toxicity in an eukaryotic expression system. We generally find that cells transfected with

Eicosanoids and Other Bioactive Lipids in Cancer, Inflammation, and Radiation Injury, 4
Edited by Honn *et al.*, Kluwer Academic / Plenum Publishers, New York, 1999.

471

epoxide hydrolase are resistant to the cytotoxic and genotoxic effects of most epoxide containing compounds. However, when we screened LTX and iLTX, we found that they weren't cytotoxic unless the cells were expressing soluble epoxide hydrolase (5). This suggests that LTX diol and iLTX diol are the toxic agents, rather than their epoxide precursors. The identification of the final toxic metabolite has provoked a lively debate between our lab and Ishizaki's group in Japan, which has done the majority

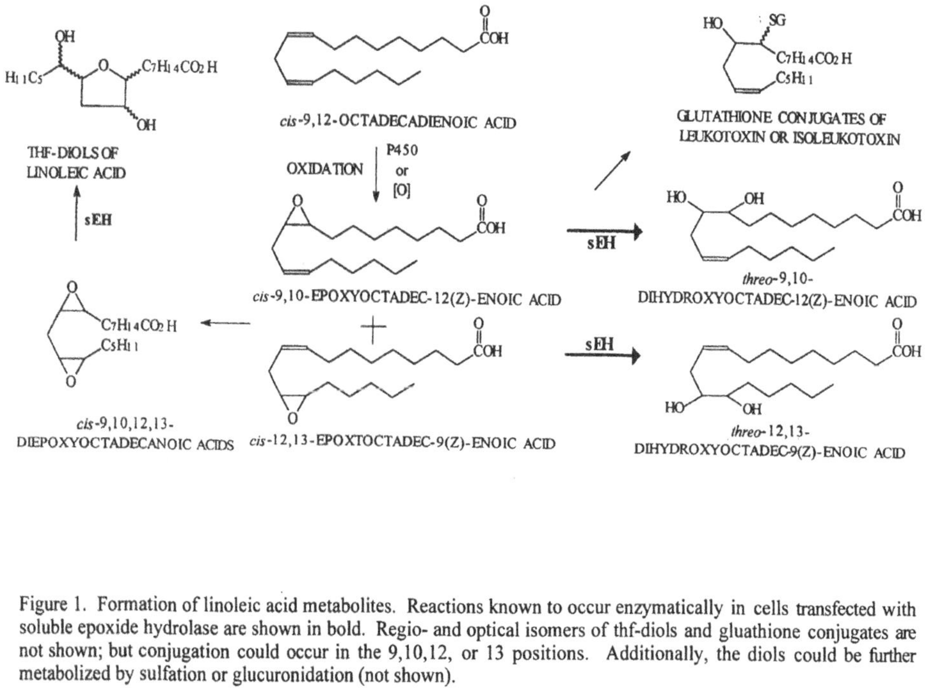

Figure 1. Formation of linoleic acid metabolites. Reactions known to occur enzymatically in cells transfected with soluble epoxide hydrolase are shown in bold. Regio- and optical isomers of thf-diols and gluathione conjugates are not shown; but conjugation could occur in the 9,10,12, or 13 positions. Additionally, the diols could be further metabolized by sulfation or glucuronidation (not shown).

of the work on LTX (6) (7). We hypothesize that LTX diol and iLTX diol account directly for many of the effects observed with LTX.

It is interesting that in the last two generations we have gone from a predominately steric acid based diet to a predominately linoleic acid based diet (8). Since specific P450s and sEH are known to be induced by common industrial and environmental compounds, diet could increase the danger of exposure to LTX, iLTX and their metabolites in part of the population (9) (10). It is also important to remember that despite the massive amount of work that has been done on the arachidonic acid metabolites and their role in the body, linoleic acid constitutes the majority of the unsaturated fatty acids found in the body.

In this paper, we will review what is known about the synthesis, effects, metabolism, and mechanisms of action of LTX and iLTX in animals.

SYNTHESIS OF LEUKOTOXIN & ISOLEUKOTOXIN

Ozawa *et al.* found in 1986 (1) that exposure of rats to pure oxygen for 60 hrs resulted in the formation of LTX and iLTX in the lung lavage of these rats. They also found that when linoleic acid was added to leukocytes isolated from the same lung lavage, LTX and isoleuktoxin were formed. Thus, they named the compounds "leukotoxins." While lung lavage from their control rat didn't appear to contain any LTX or iLTX, it is unclear whether the isolated leukocytes were tested for the presence of LTX and iLTX before addition of linoleic acid. It seems reasonable that if the rats are, in fact, biosynthesizing LTX and iLTX, that some amount of these compounds would still be present after isolation of the leukocytes. They also mention, without showing data, that they have seen non-enzymatic conversion of LTX to iLTX *in vitro*. They suggest that iLTX is formed as a rearrangement product from LTX, but this is a very unlikely chemical reaction.

Also in 1986, Hayakawa *et al.* (11) found that LTX and iLTX could be generated by addition of linoleic acid to various cell types, including human and canine blood neutrophils and guinea pig peritoneal neutrophils. They reported that addition of linoleate to lipoxygenase *in vitro* resulted in hydroperoxy linoleate which was changed to LTX on addition of sonicated neutrophils; however, the data were not shown. They also found that addition of calcium and calcium ionophore significantly increased the amount of LTX biosynthesized by guinea pig neutrophils. They again suggested that iLTX is a by-product of LTX based on the fact that they found higher concentrations of LTX than iLTX in the guinea pig neutrophils as well as higher concentrations of 9-hydroxy-linoleate than 12-hydroxy-linoleate. Unequivocal identification of oxylipins by mass spectrometry is quite difficult. Hayakawa et al. (11) report that 9-hydroxylinoleate or 12-hydroxylinoleate are formed from the corresponding epoxide by epoxide hydrolase. Rearrangements of some epoxides under anhydrous acid can lead to allylic alcohol and reduction can lead to alcohols; however it is very unlikely that either of these products were produced by epoxide hydrolase. Exposure of epoxides to strong acid in their extraction procedure can cause some compounds to undergo complex rearrangements.

Iwase *et al.* (12) tested the hypothesis that LTX and iLTX were formed *in vivo* by reactive oxygen species other than hydroxyl radical. They made a monoepoxide generating system by combining cytochrome c with a hydrogen peroxide generating system. When they added linoleic acid to this system they found that monoepoxide formation increased in direct proportion with the amount of H_2O_2 or hypoxanthine added. Addition of catalase or radical scavenger, *para*-nitrosodimethyl-aniline, reduced the formation of the monoepoxides. However, elimination of superoxide dismutase or addition of mannitol did not reduce monoepoxide formation. They explained these somewhat contradictory results by the possibilities that H_2O_2 was generated even in the absence of superoxide dismutase and that, based on their relative lipophilicities, mannitol is not as effective a hydroxyl radical scavenger as *para*-nitrosodimethylaniline. They concluded that linoleic acid is probably an effective hydroxyl radical scavenger and that in conditions when linoleic acid and cytochrome c are both present, LTX toxicity may occur. These conditions could include ischemia-reperfusion induced cardiac injury and inhibition of mitochondrial respiration.

TOXIC EFFECTS OF LEUKOTOXIN AND ISOLEUKOTOXIN

LTX and iLTX have been reported to have a number of toxic effects, both *in vitro* and *in vivo*. These range from relaxation of smooth muscle to a decrease in mitochondrial respiration to death of the organism.

Ozawa *et al.* (1) and Hayakawa *et al.*(11) both in 1986, found that LTX decreased the rate of state III O_2 consumption in mitochondria at concentrations between 10^{-3} and 10^{-4} M, but consumption at concentrations between 10^{-4} and 10^{-5} M. They also found that iLTX had the same effect but was only half as potent as LTX. Additionally, they saw a dose-dependent relaxation of smooth muscle cells from guinea pig stomach, which stopped quickly after addition of LTX. This suggested to them that the action took place on the membrane surface, rather than within the cell itself.

In 1988, Fukshima *et al.* (13) looked at the effect of LTX and other free fatty acids on the cardiovascular health of dogs. They found that 50 mg/kg i.v. LTX depressed the cardiac function of dogs as measured by aortic flow, dP/dt, systolic and diastolic aortic pressure, and change in heart rate. They also found that free unsaturated fatty acids had similar, but weaker effects.

In 1995, Ishizaki *et al.* (14), (15), (16) published three papers detailing the interaction of LTX and iLTX with nitric oxide synthase. Essentially, they treated isolated perfused rat lungs with various concentrations of LTX and N^G-monomethyl-L-arginine, a putative nitric oxide synthase inhibitor, or superoxide dismutase, or oxyhemoglobin. Pretreatment with any of these compounds decreased the toxic effect due to LTX alone. This suggested to them that LTX was either activating nitric oxide synthase, resulting in pulmonary edema, or that nitric oxide synthase was a marker for lung injury.

In 1988 Jia-ning *et al.* (17) saw a significant increase over controls in various parameters indicating lung cell toxicity after injecting rats with 100 µmol/kg LTX and sacrificing them after ten minutes. They also saw a significant increase in these parameters when they injected the rats with 50 µmol/kg LTX and sacrificed after twelve hours. Using a semi-quantitative measure of histological change, they found that for rats sacrificed at ten minutes, both 500 µmol/kg and 200 µmol/kg LTX resulted in histological changes, including intravascular congestion and coagulation and alveolar exudation, edema, hemorrhage and emphysema. For the rats in the subacute group (100 µmol/kg, sacrifice at six hours) they saw essentially the same changes with the addition of infiltration of neutrophils. These changes are similar to those seen in human patients presenting with adult respiratory distress syndrome.

This is particularly interesting since LTX was associated with toxicity in patients who had symptoms of multiple organ failure, of which adult respiratory distress syndrome is one part. Hayakawa *et al.* (18) and Ozawa *et al.* (19) both published papers in 1990 addressing this. They found that patients with severe burns (>50% body surface area) seemed to recover from the primary shock of the burns, but subsequently became severely ill or died. The patients displayed pulmonary edema, cardiac failure and coagulation abnormalities, all of which are consistent with a diagnosis of multiple organ failure. LTX was present in the plasma of all four patients tested. The concentrations varied between 11.4 nmol and 37 nmol LTX/ml serum. The researchers hypothesized that LTX could be the cause of all these toxic events, having been synthesized by neutrophils in response to the shock of the burn and then spread throughout the body.

In 1994 Kosaka *et al.* (20) correlated high plasma LTX levels in burn patients with death. They saw that patients without extensive burns (<70% body surface area) had a low but fairly steady concentration of plasma LTX over time. The maximum concentration reached was 22.6 ±11.9 nmol/ml at 5 weeks post injury. In contrast, patients with extensive burns (70% body surface area) had two peak plasma LTX levels; one initially and one beginning at week 3. The concentrations varied between 99.6 ±45.1 nmol/ml immediately after injury to 29.8 ±9.6 nmol/ml at day 14 and back up to 98.7 ±24.6 nmol/ml 6 weeks post injury. The control patients had less than 2 nmol/ml plasma LTX. The mortality rate for patients with extensive burns was 66.7%, while it was 22.2% for those without extensive burns. They also found that the mortality rate was significantly higher (61%) for patients with peak plasma LTX concentrations greater than 30 nmol/ml, while the mortality rate was only 8% for patients with peak plasma LTX concentrations less than 30 nmol/ml. They saw a correlation between peak plasma LTX concentrations in the late phase and burn surface area; they also saw a mild correlation between burn surface area and early phase peak plasma LTX levels. They concluded that LTX was either causing the death of these patients or exacerbating their condition.

METABOLITE TOXICITY: LEUKOTOXIN DIOL & ISOLEUKOTOXIN DIOL

In order to examine possible methods for ameliorating LTX toxicity, Moghaddam *et al.*, in 1997, (5) exposed eukaryotic cells expressing recombinant human soluble epoxide hydrolase to LTX and its isomer, iLTX. Surprisingly, they found that LTX and iLTX were only toxic to those cells expressing soluble epoxide hydrolase (LC_{50}=210 µM). Cells expressing control enzymes were not affected by LTX or iLTX. They had expected LTX diol and iLTX diol (*threo*-9,10-dihydroxyoctadec-12-(Z)enoic acid and *threo*-12,13-dihydroxyoctadec-9-(Z)enoic acid, respectively) to be non-toxic metabolites. The results suggested otherwise, so they challenged the cells with LTX diol and iLTX diol. They found that the diol compounds were equally cytotoxic (LC_{50}=180 µM) regardless of which enzyme the cell was expressing. This is a much higher concentration than the 30 nmol/ml plasma LTX seen by Kosaka *et al.* (20) in burn patients. However, it is unclear if the plasma LTX concentrations are of the same magnitude as what might be found in the affected areas. In the same paper, Moghaddam *et al.* reported that 300 μM LTX diol and iLTX diol, but not LTX or iLTX, decreased net transepithelial ion transport and increased paracellular permeability in five hours in rat pulmonary alveolar epithelial cells. They also saw mortality in rats within two hours with 35 mg/kg LTX diol but no mortality with up to 100 mg/kg LTX (injection via cardiac puncture). In mice (injection via the tail vein), 30 % mortality occurred within 4 minutes with 200 mg/kg LTX diol. In contrast, it took 18 to 24 hours for them to see 25 % mortality with 400 mg/kg LTX.

When eukaryotic cells expressing soluble epoxide hydrolase were challenged with 200 μM [^{14}C]-labeled LTX and iLTX, the only metabolites which correlated with toxicity were LTX diol and iLTX diol. The diols accounted for 85% of the radioactivity. Toxicity could be lessened by pre-incubation of soluble epoxide hydrolase expressing cells with a soluble epoxide hydrolase inhibitor. Interestingly, endogenous glutathione levels decrease after incubation with LTX; however, the glutathione conjugate did not appear to be formed in an appreciable concentration. This suggests that the diol may be the final toxic metabolite, but some downstream event may be affecting glutathione levels.

Moran *et al.*, also in 1997 (21), argued that since renal failure often occurs in multiple organ failure, renal cells were a good model for testing the toxicity LTX, iLTX and their respective diols. They found that 1 mM LTX diol and iLTX diol caused 42 % death, as measured by lactate dehydrogenase release, whereas, 1 mM LTX and iLTX did not. They also noted, however, that lower concentrations of the diol failed to result in cell death even after 6 hours. They determined that the diols were not causing oxidative stress to the cells by noting that malondialdehyde levels were not significantly different than those in control cells, and that addition of an iron chelator or an antioxidant failed to protect cell death induced by LTX diol and iLTX diol. They mentioned that LTX diol and iLTX diol reduced basal oxygen consumption and completely depressed ouabain-sensitive oxygen consumption. Ouabain-sensitive oxygen consumption is the result of Na$^+$/K$^+$-ATPase activity and active Na$^+$ transport. They did not see direct inhibition of the Na$^+$/K$^+$-ATPase by the diols however.

CONCLUSIONS

The work originally done on LTX and iLTX found a number of toxic effects, but failed to look for possible metabolites. It seems likely, because much of the original research was done in whole animals, that soluble epoxide hydrolase was present. The work done by Moghaddam and Moran suggests that LTX diol and iLTX diol are probably the toxic agents. There remains the possibility, however, that the toxic agent is a further metabolite. The epoxide moiety suggests possible conjugation with glutathione. Patients with adult respiratory distress syndrome often have a deficiency of glutathione in their lung lavage (22). This could be due to glutathione conjugation of LTX or iLTX or some downstream event of toxicity initiated by the epoxides.

Whatever the toxic agent, be it the epoxide compound, diol compound or some as yet unknown metabolite, the mechanism of toxicity continues to prove elusive. In all cases, cells display common symptoms of toxicity: depression of mitochondrial respiration, influx of ions, decrease of membrane potential, release of lactate dehydrogenase and swelling of cells. These are all symptomatic of necrotic cell death. The work by Ishizaki (14) and Iwase (12) suggest that toxicity could be due to creation of various reactive oxygen species, such as hydroxyl radical or nitric oxide, but Moran's (21) work suggests that oxidative stress is not a cause of toxicity.

It seems clear, however, that whether these compounds or their metabolites are toxic, or simply a marker of a toxic event, their study could add a great deal to the field of bioactive lipids. Most previous work on bioactive lipids has concentrated on the eicosanoic acids. This has proved an extremely rich field. It is important to remember, however, that linoleic acid is one of the most abundant fatty acids in the body. It seems quite possible that oxygenated metabolites of linoleic acid could be chemical mediators in a cascade analogous to the arachidonic acid cascade.

REFERENCES

1. T. Ozawa, *et al.*, *Biochem Biophys Res Com* **134**, 1071-1078 (1986).
2. T. Kato, *et al.*, *Tet Lett* **24**, 4715-4718 (1983).
3. S. M. Mumby, *et al.*, *Pest. Biochem. Phys.* **11**, 275-284 (1979).
4. B. Borhan, *et al.*, *Anal Biochem* (1996).
5. M. F. Moghaddam, *et al.*, *Nat Med* **3**, 562-566 (1997).
6. T. Ishizaki, *et al.*, *Nat. Med.* **3**, 592 (1997).
7. B. D. Hammock, *Nat. Med.* **3**, 592 (1997).
8. A. Stephen & N. Wald, *Am J Clin Nut* **52**, 457-469 (1990).
9. B. D. Hammock, *et al.*, *Tox. Appl. Pharm.* **71**, 254-265 (1983).
10. G. Gibson, *et al.*, *Biochem. Soc. Trans.* **18**, 97-99 (1990).
11. M. Hayakawa, *et al.*, *Biochem Biophys Res Com* **137**, 424-430 (1986).
12. H. Iwase, *et al.*, *Biochem Biophys Res Com* **216**, 483-8 (1995).
13. A. Fukushima, *et al.*, *Cardio Res* **22**, 213-218 (1988).
14. T. Ishizaki, *et al.*, *Biochem Biophys Res Com* **210**, 133-7 (1995).
15. T. Ishizaki, *et al.*, *Am J Physiol* **269**, L65-70 (1995).
16. T. Ishizaki, *et al.*, *Am J Physiol* **268**, L123-8 (1995).
17. H. Jia-Ning, *et al.*, *Lung* **166**, 327-337 (1988).
18. M. Hayakwa, *et al.*, *Biochem Int* **21**, 573-579 (1990).
19. T. Ozawa, *et al.*, *Adv Prost Throm Leuk Res* **21**, 569-572 (1990).
20. K. Kosaka, *et al.*, *Mol Cell Biochem* **139**, 141-8 (1994).
21. J. H. Moran, *et al.*, *Toxicol. Appl. Pharmacol.* **146**, 53-59 (1997).
22. E. R. Pacht, *et al.*, *Chest* **100**, 1397-1403 (1991).

70

DRAMATIC INCREASE OF LINOLELIC ACID PEROXIDATION PRODUCTS BY AGING, ATHEROSCLEROSIS , AND RHEUMATOID ARTHRITIS

Wolfgang Jira and Gerhard Spiteller

Department of Organic Chemistry I, University of Bayreuth
D-95440 Bayreuth, FAX: 0921-552671, e-mail: gerhard.spiteller@uni-bayreuth.de

INTRODUCTION

Oxidation of low density lipoproteins (LDL) was recognized to be involved in atherogenesis [1,2] and rheumatoid arthritis [3]. This process is connected with lipid peroxidation (LPO) of polyunsaturated fatty acids (PUFAs). The process is induced mainly not enzymatically [4,5]. In nonenzymatic LPO processes of PUFAs any activated CH_2-group of a PUFA is attached with about the same probability. The main unsaturated PUFA in LDL is linoleic acid (LA). As a consequence hydroperoxides of LA are expected to be main oxidation products of LDL. Hydroperoxides of PUFAs (LOOH) suffer easily reduction to corresponding hydroxy fatty acids (LOH) in biological surroundings: Thus linoleic acid generates 9-hydroxy-10,12-octadecadienoic acid (9-HODE) and 13-hydroxy-9,11-octadecadienoic acid (13-HODE) (Fig. 1):

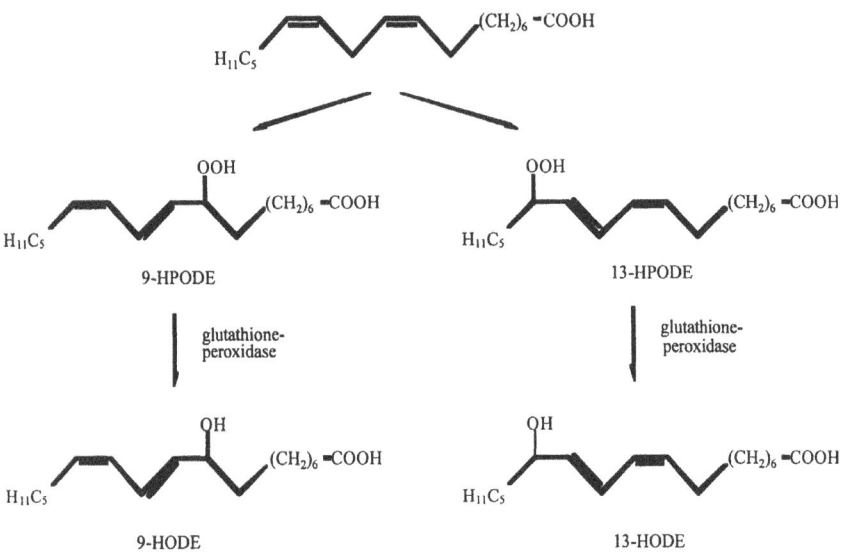

Fig. 1: Formation of 9- and 13-HODE

9- and 13-HODE are rather stable against further oxidation [6], they are accumulated in LPO processes. As a consequence HODEs occur in atherosclerotic plaques [7]. They were also recognized after artificial oxidation of LDL by Cu^{2+} [8]. Therefore we expected that HODEs might be accumulated in LDL of patients suffering from diseases which are known to be in relation to LPO, e.g. in atherosclerosis and rheumatoid arthritis, especially since LA is enriched in LDL for a factor of 7 compared to AA [9].

METHODS

Blood samples were taken after fasting over night in the morning. Lipoproteins (VLDL, LDL and HDL) were precipitated from 10 ml of human blood serum following a method of Leiss et al. [10]. The lipoproteins were lyophilized, weighed and 6-hydroxyheptadecanoic acid was added as internal standard. The lipid extraction followed the method of Bligh and Dyer [11]. The mixture was treated with H_2/PtO_2 to reduce the hydroperoxyl-groups and to hydrogenate double bonds [12]. After seperation of the catalyst the fatty acids were converted to their methylesters with diazomethane. Separation of methylesters of fatty and hydroxy fatty acids was achieved by column-chromatography on silicagel. Hydroxy groups of hydroxy fatty acid methylesters were trimethylsilylated with MSTFA. Derivatized monohydroxy acids were identified and quantified by mass spectrometry ion tracing and measuring the intensity of the α-cleavage fragment ions. Their ion currents were compared with those of the α-cleavage products of the internal standard [13].

RESULTS

The content of 9-HODE in LDL was determined in young volunteers and old atherosclerotic patients. Large differences in 9-HODE up to 100fold amounts were recognized. Occasionally also samples of healthy older individuals were investigated, these also showed an increase in 9-HODE compared to samples of young individuals. Consequently the 9-HODE content was investigated in relation to age. The results of these investigations are outlined in Fig. 2.

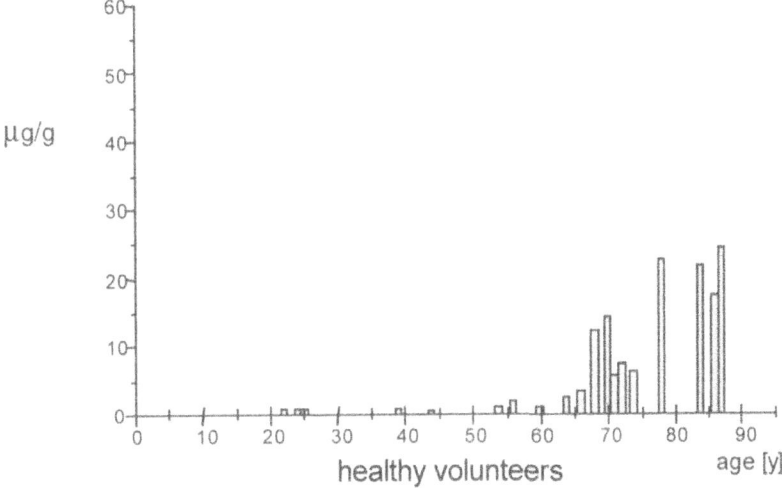

Fig. 2: Content of 9-HODE in LDL of healthy volunteers

According to these results the 9-HODE content increases with age. Therefore only the LDL content of healthy individuals and atherosclerotic patients of the same age group is comparable: Young atherosclerotic patients showed a 10-50 fold higher 9-HODE content compared to age matched healthy persons (Fig. 3). These differences become lower with increasing age. The old „healthy" individuals showed no easily detectable signs of atherosclerosis. Nevertheless an ultrasound investigation of the carotid arteries of the obviously healthy volunteers with high content of 9-HODE in LDL revealed that nearly all of them had either plaques or showed a thickening of the intima-media complex.

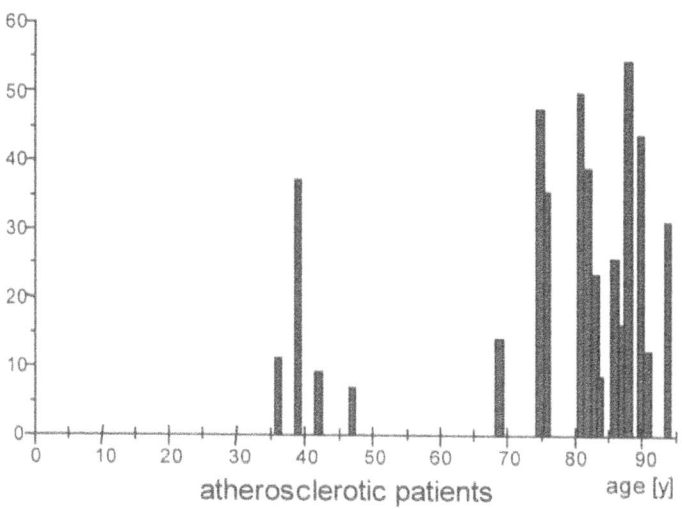

Fig. 3: Content of 9-HODE in LDL of atherosclerotic patients

A similar investigation originating from samples of patients suffering from rheumatoid arthritis indicated also a dramatic increase of 9-HODE compared to healthy age matched controls (Fig. 4).

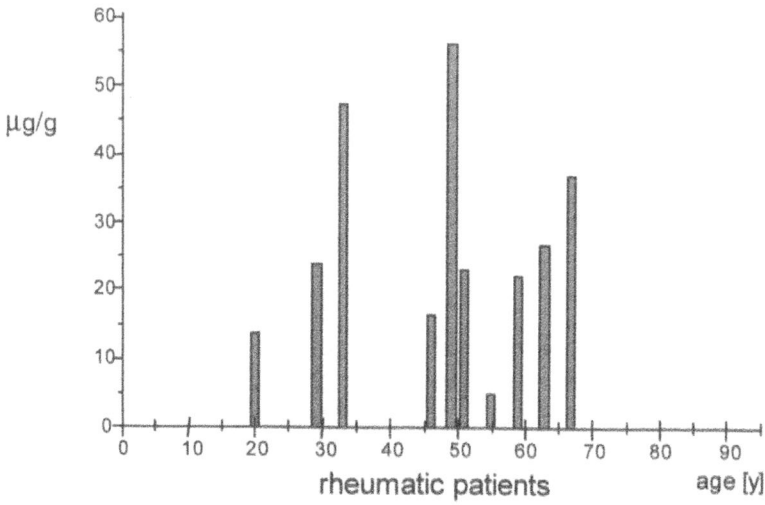

Fig. 4: Content of 9-HODE in LDL of patients suffering from rheumatoid arthritis

CONCLUSION

1) HODEs are the most abundant LPO products. Their amounts exceed other LPO markers e.g. malondialdehyde or isoprostanes for a factor of about 100-1000. Consequently HODEs are excellent suited to recognize LPO processes.

2) The increase of HODEs in LDL is not restricted to atherosclerotic patients but seems to be connected with an increased amount of dying cells caused by inflammatory processes.

3) Lipid peroxides derived from linoleic acid are of much higherl importance for development of diseases than those derived from arachidonic acid.

4) 9- and 13-HODE induce the release of interleukin 1β from macrophages [14]. In addition 9-HODE was shown to be an equal strong proinflammatory mediator as leucotrienes [15]. Therefore the contribution of linoleic acid oxidation products in the development of inflammatory processes ought to be considered in the future.

LITERATURE

1. Steinbrecher, U.P., Parthasarathy, M., Free Radic. Biol. Med. **9**, 155 (1990).
2. Morel, D.W., DiCorleto, P.E., Chisolm, G.M., Atherosclerosis **4**, 357 (1984).
3. Winyard, P. G., Tatzber, F., Esterbauer, H., Kus, M. L., Blake, D. R., Morris, C. J., Ann. Rheum. Dis. 52, 677 (1993).
4. Folcik, V. A., Nivar-Aristy, R. A., Krajewski, L. P., Cathcart, M. K., J. Clin. Invest. 96, 504 (1995).
5. Kühn, H., Belkner, J., Zaiss, S., Fährenklemper, T., Wohlfeil, S., J. Exp. Med. 179, 1903 (1994).
6. Spiteller, P., Spiteller, G., Chem. Phys. Lipids 89, 131 (1997).
7. Brooks, C.J.W., Harland, W.A., Steel, G., Gilbert, J.D., Biochim. Biophys. Acta 202, 563 (1970).
8. Lenz, M. L., Hughes, H., Mitchell, J. R., Via, D. P., Guyton, J. R., Taylor, A. A., Gotto, A. M., Jr. Smith, J. Lipid Res. 31, 1043 (1990).
9. Esterbauer, H., Gebicki, J., Puhl, H., Juergens, G., Free Radic. Biol. Med. 13, 341 (1992).
10. Leiss, O., Murawski, U., Egge, H., J. Clin. Chem. Clin. Biochem. 17, 619 (1979).
11. Bligh, E.C., Dyer, W.J., Can. J. Biochem. Physiol. 37, 911 (1959).
12. Nikkari, T., Malo-Ranta, U., Hiltunen, T., Jaakkola, O., Ylä-Herttuala, S., J. Lipid Res. 36, 200 (1995).
13. Lehmann, W.A., Stephan, M., Fürstenberger, G., Anal. Biochem. 204, 158 (1992).
14. Ku, G., Thomas, C. E., Akeson, A. L., Jackson, R. L., J. Biol. Chem. 267, 14183 (1992).
15. Moch, D., Schewe, T., Kuehn, H., Schmidt, D., Buntrock, P., Biomed. Biochim. Acta 49, 201 (1990).

71

MODULATION OF ATHEROGENESIS BY DIETARY GAMMA-LINOLENIC ACID

Yang-Yi Fan[1], Kenneth S. Ramos[2], and Robert S. Chapkin[1]

[1]Faculty of Nutrition and Molecular and Cell Biology Group,
[2]Department of Veterinary Physiology and Pharmacology,
Texas A&M University, College Station, TX 77843

INTRODUCTION

Atherosclerosis is responsible for 50% of all the mortality in the United States, Japan and Europe (Ross, 1993). Atherosclerotic lesions result from an excessive inflammatory-fibroproliferative response to various forms of insult to the endothelium and smooth muscle of the arterial wall (Badimon, 1993; Ross, 1993, 1995). A large number of growth regulatory factors participate in this pathogenesis, with dietary fat recognized as one of the most prevalent risk factors for disease occurrence. However, future research must be directed toward better understanding the association between dietary fat intake and coronary disease as the current diet-lipid-heart hypothesis may be overly simplistic. Since considerable debate remains regarding the distinct biological activities of individual polyunsaturated fatty acids, the major objective of our research has been to evaluate the anti-atherogenic potential of dietary gamma-linolenic acid (GLA, 18:3n-6).

Although GLA (18:3n-6) administration has been used in the management of chronic inflammatory-hyperproliferative disorders, inhibition of platelet aggregation and reversal of hypertension (Willis, 1989; Horrobin, 1990), its effects on human monocyte/macrophage metabolism and function remain poorly defined. Because blood mononuclear cells are associated with atherosclerotic lesion formation (Gerrity, 1981; Badimon, 1993), we have focused attention on the study of GLA metabolism in the mouse peritoneal macrophage. A plethora of well documented similarites between the mouse macrophage and human monocyte/macrophage systems make the mouse peritoneal macrophage an excellent model system (Nathan, 1985).

Macrophages are present at all stages of atherosclerosis (Ross, 1993). They are considered to be the principal inflammatory mediator in the atheromatous plaque environment. In regard to mitogenesis, macrophages can secrete IL-1, nitric oxide, TNF-α, TGF-β and a mitogenic factor similar to PDGF, macrophage-derived growth factor (MDGF), thus playing a role in the vascular response of smooth muscle cell (SMC) proliferation (Ross, 1993; Zhang, 1993). Macrophages also secrete eicosanoids which have been demonstrated to regulate arterial SMC phenotype and proliferative capacity. Specifically, prostaglandin E_1 (PGE$_1$) and prostacyclin (PGI$_2$) inhibit asynchronous, cycling (growing) SMC proliferation (Owen, 1986;

Fan, 1996a). When PGE_1 or PGI_2 are added to asynchronous cycling SMC (i.e., PGE_1 is added to cells in G_1, S, G_2, and in M) this causes a pronounced elevation in cellular cAMP levels which results in growth arrest (Nilsson, 1984; Loesberg, 1985; Fan, 1996a). Therefore, whether macrophages are beneficial or pathogenic would depend on the balance between the levels of anti- and pro-atherogenic mechanisms achieved by these cells.

In many animal tissues and cells, 18:2n-6 is converted to arachidonic acid (AA, 20:4n-6) by an alternating sequence of $\Delta6$ desaturation, chain elongation and $\Delta5$ desaturation, in which hydrogen atoms are selectively removed to create new double bonds and then two carbon atoms are added to lengthen the fatty acid chain (Willis, 1989). Reports by Chapkin (1988) have demonstrated that peritoneal macrophages lack $\Delta6$ desaturase activity, the enzyme responsible for the conversion of 18:2n-6 into GLA (18:3n-6). In addition, aortic SMCs have little or no ($\Delta5$ or $\Delta6$) desaturase activity (Morisaki, 1985). These observations suggest that the availability of AA (20:4n-6) in macrophages and SMCs cannot be modulated at the level of local synthesis from 18:2n-6, its major dietary essential fatty acid antecedent. We have also shown that peritoneal macrophages *in vitro* can synthesize very low levels of AA (20:4n-6) from dihomo-gamma-linolenic acid (DGLA, 20:3n-6), and therefore possess minimal $\Delta5$ desaturase activity (Chapkin, 1991).

Until recently it was not known if dietary GLA could augment macrophage production of PGE_1, a potent anti-proliferative eicosanoid derived from DGLA. Using combined solid phase extraction, HPLC and RIA methodologies, we determined for the first time that GLA alimentation is capable of significantly ($P<0.05$) increasing PGE_1 synthesis in mouse peritoneal macrophages (Fan, 1992) with no change in AA-drived PGE_2. Since dietary GLA is capable of enhancing mouse macrophage production of PGE_1 (Fan, 1992, 1995ab), it may therefore be capable of down-regulating vascular SMC proliferation during conditions associated with increased macrophage deposition in the vascular wall.

MODULATION OF MACROPHAGE ATHEROGENIC POTENTIAL POTENTIAL BY DIETARY GLA

Effect of Dietary GLA on Vascular SMC Proliferation

SMCs and macrophages are two of the major reactive cell types involved in the progression of atherosclerosis. It is well known that macrophage-derived growth regulatory molecules can influence SMC proliferation (Badimon, 1993), and dietary lipids may influence the profile of these growth factors. Whether SMC proliferation can be directly influenced by dietary lipids irrespective of the modulation of macrophages has not been demonstrated. A series of experiments were therefore undertaken to determine if dietary GLA has a direct effect on SMC cytokinetics (Fan, 1995ab). We demonstrated that aortic SMC (isolated from mice fed GLA) proliferation in the absence of macrophages or in the presence of macrophages isolated from chow-fed mice, was not significantly ($P>0.05$) influenced by dietary GLA. These data indicate that dietary GLA has no direct effect on vascular SMC proliferation (Fan, 1995b), and that modulation of eicosanoid production in SMCs themselves is not responsible for the observed differences in proliferative behavior.

Development of an In Vitro Macrophage-SMC Co-culture Model

In order to study the modulatory effects of macrophages on SMC proliferation, a co-culture model system was developed (Fan, 1995a). In this system, macrophages and SMCs are separated by a semipermeable membrane with a 30-kDa cutoff (Figure 1). Macrophage-derived soluble factors such as eicosanoids are allowed to pass through the membrane and influence SMC behavior while direct cell-to-cell contact is precluded. The use of co-cultures instead of strict conditioned media allows for the examination of stable, as well as labile, biologic response modifiers of the vascular response.

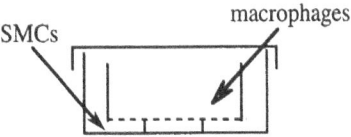

Figure 1. Macrophage-SMC co-culture model. Peritoneal macrophages are seeded onto 25-mm tissue culture inserts (upper chamber) and aortic SMCs are seeded on 35-mm tissue culture dishes (lower chamber). After adherence to the inserts, macrophages are placed inside 35-mm dishes containing SMCs. ^{3}H-thymidine is added to SMC cultures for DNA synthesis measurement. Other endpoints of the proliferative response have also been evaluated using this model.

Dietary GLA Modulates Macrophage-SMC Interactions

Since regulation of the cytokine/growth factor network by eicosanoids may represent an important aspect of the arterial response to injury and the progression of intimal hyperplasia (Ross, 1993), we examined the ability of peritoneal macrophages isolated from GLA-fed mice to influence mouse thoracic aortic SMC proliferation (Fan, 1995a; Fan, 1996b) in vitro. Mice were fed diets containing 10% (w/w) corn oil (CO, containing no GLA) or primrose oil (PO, containing 10% GLA) for 2 wk. Resident peritoneal macrophages were isolated and seeded on a semi-permeable membrane with a 30 kDa cut-off. Macrophages were preincubated with or without 50 μM indomethacin (a cyclooxygenase inhibitor) or 50 μM L655,238 (a 5-lipoxygenase inhibitor) for 30 min, and subsequently co-cultured with naive murine aortic SMCs (isolated from chow-fed mice) grown on culture dishes. DNA synthesis in SMCs and prostaglandin formation in co-culture supernatants were measured at the end of a 39 h incubation period. SMC DNA synthesis was inhibited by GLA treatment relative to the control CO diet containing no GLA (Figure 2). A 4-fold increase in the levels of PGE_1, a potent anti-proliferative eicosanoid derived from GLA, was observed in the PO group relative to control CO group (Figure 3). PGE_2 (derived from AA) synthesis was significantly ($P<0.05$) lower in PO co-cultures relative to CO. Although PGE_1 and PGE_2 share the ability to inhibit vascular SMC proliferation, the biological effects of PGE_1 are approximately 20 times stronger than PGE_2 (Nilsson, 1984). In addition, no difference was observed with regard to levels of 6-keto-$PGF_{1\alpha}$ (prostacyclin, derived from AA). Macrophage inhibition of SMC DNA synthesis in GLA group was abolished by indomethacin, but not L655,238 (Figure 2). Addition of exogenous PGE_1 (100 nM) reversed the effect of indomethacin. These data indicate that macrophages isolated from animals supplemented with dietary GLA reduce SMC DNA synthesis in a cyclooxygenase-dependent manner (Figure 2).

Dietary GLA reduces SMC proliferation, in part via a PGE_1-cAMP-dependent pathway

In order to identify the specific prostaglandin(s) critical for the anti-proliferative effect of dietary GLA, a panel of specific antibodies for macrophage-derived prostaglandins and growth factors were examined. Specifically, anti-PGE_1 antiserum treatment (1:50 or 1:100) blocked the ability of GLA-enriched macrophages to down-regulate SMC proliferation, a response reversed by exogenous PGE_1 treatment (Fan, 1997). These data suggest the important role of PGE_1 on this dietary modulation. The biological effect of PGE_1 on SMC proliferation has been linked to its ability to modulate intracellular cAMP levels (Fan 1996a, Nilsson 1984, Owen 1986). Our

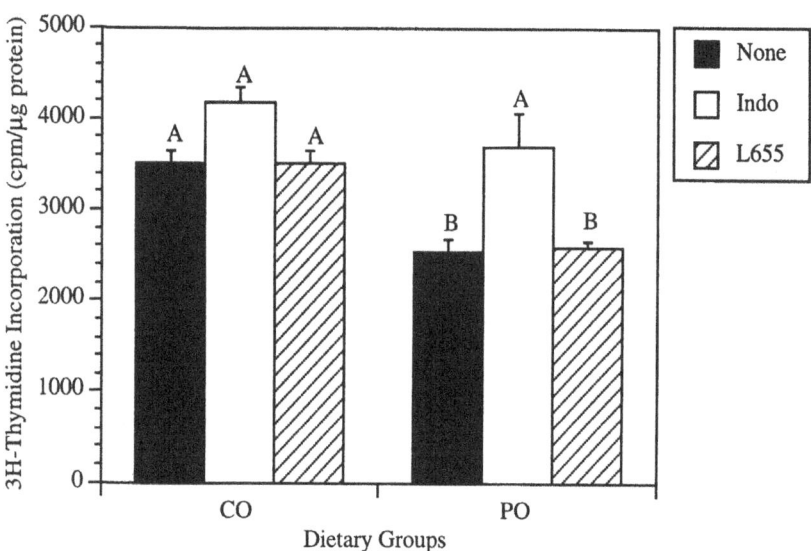

Figure 2. Combined effect of dietary lipids and eicosanoid synthase inhibitors on the ability of macrophages to modulate SMC DNA synthesis. Macrophages were isolated from mice fed corn oil (CO, contains no GLA) or primrose oil (PO, contains GLA) and preincubated with vehicle (None), indomethacin (Indo), or L655,238 (L655) for 30 min. Macrophages were washed and subsequently co-cultured with naive SMCs (isolated from chow-fed mice) and incubated in the presence of ^{3}H-thymidine for 39 h. SMC DNA synthetic rates were measured at the end of the incubation period. Results are expressed as means ± SEM of 3 replicate cultures per group. Values not sharing common superscripts are significantly different (P < 0.05).

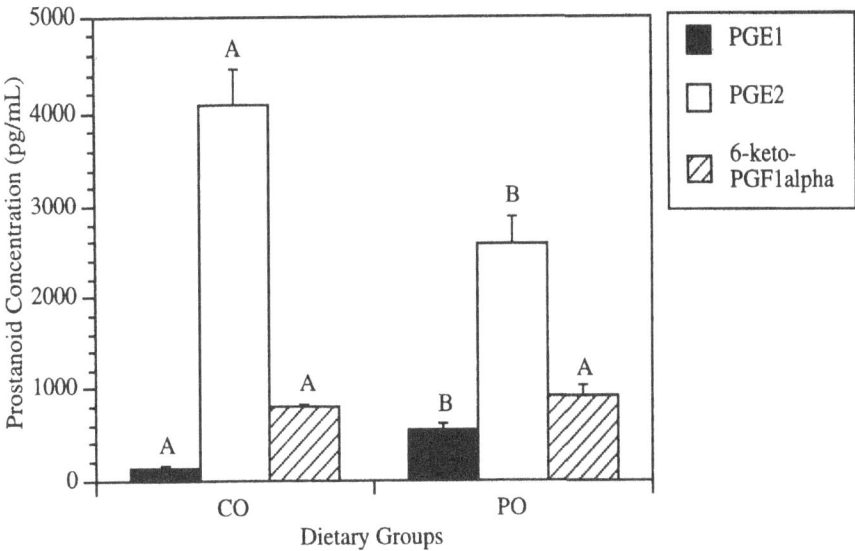

Figure 3. Effect of dietary lipids on macrophage-SMC prostanoid synthesis. Peritoneal macrophages were isolated from mice fed different dietary lipids, co-cultured with naive aortic SMCs (isolated from chow-fed mice) and incubated in the presence of ^{3}H-thymidine for 39 h. Incubation supernatants were collected at the end of incubation period and processed for PGE$_1$, PGE$_2$ and 6-keto-PGF$_{1\alpha}$ analysis. Results are expressed as means ± SEM of 3 replicate cultures per group. Values with different superscripts are significantly different (P<0.05).

data demonstrated that macrophages isolated from mice fed GLA induced SMC intracellular cAMP levels in a biphasic fashion. In addition, exogenous PGE$_1$ (1 nM - 10 µM) exerted a similar biphasic cAMP response in SMCs (Fan, 1997). This is noteworthy in view of the role of cAMP, an important second messenger, in the regulation of SMC proliferation (Sadhu, 1993).

RETARDATION OF ATHEROSCLEROTIC PROGRESS BY DIETARY GLA

Since dietary GLA reduces SMC DNA synthesis and proliferation, we have hypothesized that this lipid source will favorably modulate the atherogenic process. However, it remains unclear whether dietary GLA affords a protective effect on atherosclerotic lesion progression *in vivo*. Over the last decade, the transgenic mouse model has been aggressively utilized for the elucidation of the common and complex polygenic mechanisms of cardiovascular diseases (Lusis, 1993; Rubin, 1994). Therefore, we used the apolipoprotein E (ApoE) knock-out mouse model (Piedrahita, 1992) to evaluate the *in vivo* effect of dietary GLA on atherosclerotic lesion development and progression. These animals develop atherosclerotic lesions similar to humans as a function of dietary fat content, and therefore, have been recognized as an important animal model for studying the effect of dietary lipids on atherosclerosis (Reddick, 1994).

Male Apo E knock-out mice (4 wk old) were fed diets containing 10% (w/w) corn oil (CO) [control diet, 0 mol% GLA] or primrose oil (PO) [10 mol% GLA] ± cholesterol (1.25%, w/w) for 27 wk. No major lesions were found in the thoracic aortas at any of the diets examined when mice were fed cholesterol-free diet for the first 15 wk. Mice fed the control CO diet developed 10-fold larger atherosclerotic lesions in the thoracic aorta, relative to the PO-fed mice at the end of the 27 wk feeding period. The average medial layer thickness of the vessel wall in the CO group was 1.5-fold greater ($P < 0.05$) than PO group. These results indicate that dietary GLA can retard the development of atherosclerosis.

SUMMARY

Data from our in vitro studies indicate that macrophages isolated from mice fed GLA-enriched diets inhibit vascular SMC proliferation via a PGE_1-cAMP dependent mechanism. Since SMC proliferation is one of the main events implicated in the pathogenesis of atherosclerosis (Ross, 1993), this anti-proliferative effect observed by dietary GLA is noteworthy. In vivo studies have established that dietary GLA is capable of retarding the atherosclerotic lesion formation in ApoE knock out mice, an animal model that develops atherosclerosis similar to humans (Reddick, 1994). We propose that dietary GLA has the potential to inhibit SMC proliferation leading to retardation of atherosclerotic lesion formation, and therefore favorable modulation of the atherogenic process.

ACKNOWLEDGMENTS

Supported in part by Scotia Pharmaceuticals Ltd, and NIHES 04849.

REFERENCE

Badimon, J.J., Fuster, V., Chesebro, J.H., and Badimon, L., 1993, Coronary atheroslcerosis, a multifactorial disease, *Circulation* 87[suppl II]:II3.

Chapkin, R.S., Somers, S.D., and Erickson, K.L., 1988, Inability of murine peritoneal macrophages to convert linoleic acid into arachidonic acid, *J. Immunol.* 140:2350.

Chapkin, R.S., and Coble, K.J., 1991, Utilization of gammalinolenic acid by mouse peritoneal macrophages, *Biochim. Biophys. Acta* 1085:365.

Fan, Y.-Y., and Chapkin, R.S., 1992, Mouse peritoneal macrophage prostaglandin E_1 synthesis is altered by dietary gamma-linolenic acid, *J. Nutr.* 122:1600.

Fan, Y.-Y., Chapkin, R.S., and Ramos, K.S., 1995a, A macrophage-smooth muscle cell co-culture model: applications in the study of atherogenesis, *In Vitro Cell. Dev. Biol.* 31:492.

Fan, Y.-Y., Ramos, K.S., and Chapkin, R.S., 1995b, Dietary γ-linolenic acid modulates macrophage-vascular smooth muscle cell interactions. Evidence for a macrophage-derived soluble factor that downregulates DNA synthesis in smooth muscle cells, *Arterioscler. Thromb. Vasc. Biol.* 15:1397.

Fan, Y.-Y., Ramos, K.S., and Chapkin R.S., 1996a, Cell cycle-related inhibition of mouse vascular smooth cell proliferation by prostaglandin E_1: Relationship between prostaglandin E_1 and intracellular cAMP levels, *Prost. Leuk. Essent. Fatty Acids* 54:101.

Fan, Y.-Y., Chapkin, R.S., and Ramos, K.S., 1996b, Dietary lipid source alters murine macrophage/vascular smooth muscle cell interactions in vitro, *J. Nutr.* 126:2083.

Fan, Y.-Y., Ramos, K.S., and Chapkin, R.S., 1997, Dietary γ-linolenic acid enhances mouse macrophage-derived prostaglandin E_1 which inhibits vascular smooth muscle cell proliferation, *J. Nutr.* 127:1765.

Gerrity, R.G., 1981, The role of the monocyte in atherogenesis, *Am. J. Pathol.* 103:181.

Horrobin, D.F., 1990, Gammalinolenic acid, *Rev. Contemp. Pharmacotherapy* 1:1.

Loesberg, C., Van Wijk, R., Zandbergen, J., Van Aken, W.G., Van Mourik, J.A., and De Groot PhG, 1985, Cell cycle-dependent inhibition of human vascular smooth muscle proliferation by prostaglandin E_1, *Exp. Cell Res.* 160:117.

Lusis, A.J., 1993, The mouse model for atherosclerosis, *Trends. Cardiovasc. Med.* 3:135.

Morisaki, N., Kanzaki, T., Fujiyama, Y., Oksawa, I., Shirai, K., Matsuoka, N., Saito, Y., and Yoshida, S., 1985, Metabolism of n-3 polyunsaturated fatty acids and modification of phospholipids in cultured rabbit aortic smooth muscle cells, *J. Lipid Res.* 26:930.

Nathan, C.F., and Cohn, Z.A., 1985, *Textbook of Rheumatology*, W.B. Saunders Company, Philadelphia.

Nilsson, J., and Olsson, A.G., 1984, Prostaglandin E_1 inhibits DNA synthesis in arterial smooth muscle cells stimulated with platelet-derived growth factor, *Atherosclerosis* 53:77.

Owen, N.E., 1986, Effect of prostaglandin E_1 on DNA synthesis in vascular smooth muscle cells, *Am. J. Physiol.* 250:C584.

Piedrahita, J.A., Zhang, S.H., Hagaman, J.R., Oliver, P., and Maeda, N., 1992, Generation of mice carrying a mutant apolipoprotein E gene inactivated by gene targeting in embryonic stem cells, *Proc. Natl. Acad. Sci. (U.S.A)* 89:4471.

Reddick, R.L., Zhang, S.H., and Maeda, N., 1994, Atherosclerosis in mice lacking apo E. Evaulation of lesional development and progression, *Arteroscl. & Thromb.* 14:141.

Ross, R., 1993, The pathogenesis of atherosclerosis: a perspective for the 1990's, *Nature* 362:801.

Ross, R., 1995, Cell biology of atherosclerosis, *Annu. Rev. Physiol.* 57:791.

Sadhu, D.N., and Ramos, K.S., 1993, Cyclic AMP inhibits c-Ha-ras protooncogene expression and DNA synthesis in rat aortic smooth muscle celss, *Experientia* 49:567.

Willis, A.L., and Smith D.L., 1989, *New Protective Roles for Selective Nutrients*, Alan R. Liss, Inc., New York.

Zhang, H., Downs, E.C., Lindsey, J.A., Davis, W.B., Whisler, R.L., and Cornwell, D.G., 1993, Interactions between monocyte/macrophage and vascular smooth muscle cell; stimulation of mitogenesis by a soluble factor and of prostanoid synthesis by cell-cell contact, *Arterioscler. & Thromb.* 13:220.

72

THE SELECTIVE CYTOTOXICITY OF γ-LINOLENIC ACID (GLA) IS ASSOCIATED WITH INCREASED OXIDATIVE STRESS

Sujata Vartak[1], Mike E.C. Robbins[1] and Arthur A. Spector[2]

[1]Radiation Research Laboratory, Department of Radiology,
[2]Department of Biochemistry,
University of Iowa,
Iowa City, IA 52242.

INTRODUCTION

Gamma linolenic acid (GLA, 18:3n6) supplementation has been reported to suppress the growth of tumor cells *in vitro* and *in vivo*[1,2]. GLA supplementation selectively decreases the survival of rat astrocytoma cells and increases their radiosensitivity; normal rat astrocytes are not affected[3]. In order to investigate whether other fatty acids also exert similar cytotoxic effects on tumor cells, we have studied the effect of several polyunsaturated fatty acids (PUFA) on the survival of normal and malignant glial cells.

The cytotoxic action of PUFA is thought to be mediated predominantly *via* lipid peroxidation and free radical generation.[3,4] Studies indicate that eicosanoid and leukotriene synthesis also play an important role in tumor cell proliferation and metastasis.[5] Further, GLA and other PUFA modulate prostanoid synthesis in cells.[6] To investigate the mechanism(s) by which GLA exerts a cytotoxic effect on astrocytoma cells and increases their radiation sensitivity, we studied the role of lipid peroxidation, prostanoids, and leukotrienes on the cytotoxic action of GLA and its enhancement of radiation sensitivity.

MATERIALS AND METHODS

Cell and Culture Conditions

The tumor cell lines tested were the 36B10 malignant rat astrocytoma and the C6 rat glioma. The "normal" rat astrocytes were obtained as primary cell cultures isolated from 1-2 day old

Eicosanoids and Other Bioactive Lipids in Cancer, Inflammation, and Radiation Injury, 4
Edited by Honn *et al.*, Kluwer Academic / Plenum Publishers, New York, 1999.

Sprague Dawley rat pups.[7] All cell types were incubated at 37°C under a humidified atmosphere of 95% air : 5% CO_2. The astrocytoma cells were cultured in Dulbecco's modified Eagle's medium (DMEM) and the C6 cells were grown in 1:1 mixture of DMEM and Ham's F-12 medium. The astrocytes were cultured in Eagle's minimum essential medium. For all cell types the medium was supplemented with 10% fetal bovine serum.

Cell Survival Analyses and Cytotoxicity Studies

The effect of fatty acids on the survival of normal and malignant glial cells was determined using the fluorescent dye Calcein AM. Rat astrocytes (2500 cells/well) or astrocytoma cells (250 cells/well) were seeded onto 96-well plates. After an overnight incubation, the cells were supplemented with 45 μM of the fatty acids. The viability of the cells was assessed after 3 days. Cells were incubated with calcein-AM (live cell fluorescence stain) at a final concentration of 1 μM for 30 min at room temperature. The fluorescence then was measured using a microplate reader; live cell populations were characterized by an intense fluorescence in the 530 nm region.

To investigate whether altered prostanoid or leukotriene synthesis was involved in GLA cytotoxicity and/or the modified radiation response of GLA-supplemented cells, the cyclooxygenase inhibitor ibuprofen (100 μM) and lipoxygenase inhibitor nordihydro-guaretic acid (NDGA, 1-5 μM) were used. Cells were supplemented with GLA and/or the inhibitors for 24 h prior to irradiation with 5 or 10 Gy from a [137]Cs γ-source. The effect of these inhibitors on 36B10 cell survival was determined using the clonogenic cell survival assay.

Determination of 8-Isoprostane Levels

The 8-isoprostane (ISP) assay[8] was used to determine whether increased oxidative stress played a role in GLA-cytotoxicity alone or when combined with radiation. Astrocytoma cells and astrocytes were supplemented with 45 μM GLA and exposed 5 or 10 Gy radiation. The medium was tested for ISP 1 or 48 h after irradiation using an EIA kit.

RESULTS

Effect of Fatty Acids on the Survival of Normal and Malignant Rat Astrocytes

None of the fatty acids used reduced the viability of the "normal" astrocytes (Fig. 1). In contrast, GLA and DGLA supplementation decreased the viability of the 36B10 and C6 cells. GLA supplementation reduced the survival of the 36B10 and C6 cells to 66 ± 9.86% and 67.3 ± 3.56%, respectively, of the unsupplemented controls.

8-Isoprostane Levels of 36B10 Cells Supplemented With GLA

Irradiating the 36B10 astrocytoma cells with 5 Gy did not affect the ISP level in the cells (Fig. 2). However, exposure of the cells to 10 Gy γ rays caused significant increases in their ISP levels. GLA supplementation increased the level of ISP in the astrocytoma cells. This increase in

ISP level following GLA supplementation was further increased when the cells were exposed to 10 Gy radiation.

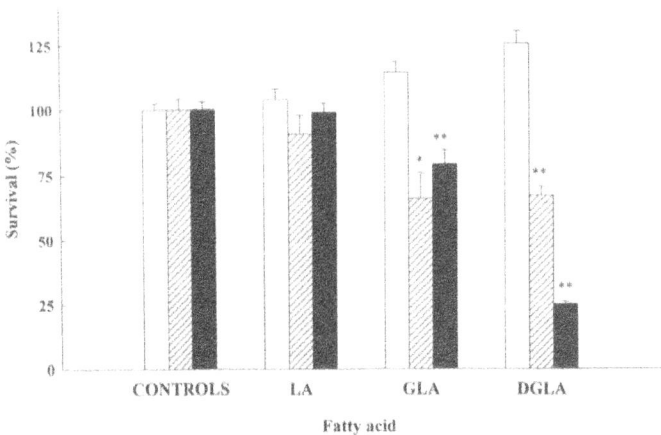

Figure 1. Effect of 45µM fatty acid supplementation on the survival of □ astrocytes, □ 36B10 astrocytoma cells and ■ C6 glioma cells assessed using the calcein assay. Bars indicate mean ± S.E.. from 5 separate observations. * p<0.05 and ** p< 0.01; Significantly different from respective controls.

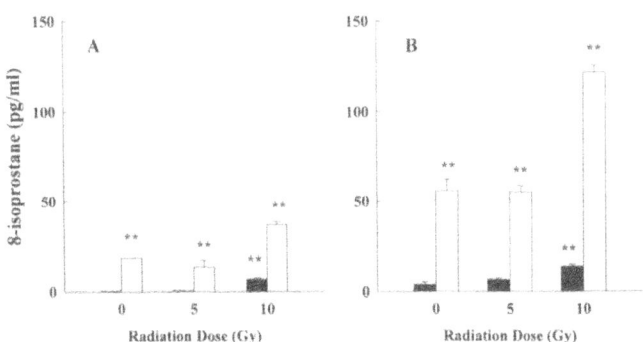

Figure 2. GLA and radiation increase ISP generation in 36B10 cells. Cells were supplemented with 45 µM GLA, □, and exposed to 0-10 Gy γ-radiation. The ISP levels were measured A) 1 hr or B) 48 hr after supplementation and/or irradiation. Control cells, ■, received no GLA supplementation. Values represent mean ± S.E. from 3 separate cultures. ** Significantly different from respective controls; p<0.01.

Effect of Cyclooxygenase and Lipoxygenase Inhibitors on the Radiation Response of GLA Supplemented 36B10 Cells

Ibuprofen alone had no effect on the survival of the astrocytoma cells (data not shown) and addition of ibuprofen plus GLA did not affect GLA cytotoxicity when no radiation was administered (Fig. 3). However, incubating the cells with ibuprofen along with GLA completely blocked the GLA induced enhancement of radiation response of 36B10 cells at doses of either 5 or 10 Gy (Fig. 3). Similar results were obtained with indomethacin. NDGA, at concentrations of 1-5 μM, did not affect astrocytoma cell survival. Also, incubation of the astrocytoma cells with GLA and NDGA did not affect GLA cytotoxicity or the GLA-induced increase in radiation sensitivity (data not shown).

DISCUSSION

The present results show that GLA is selectively cytotoxic to the glioma cells but not to the "normal" astrocytes. The increase in ISP formation following GLA supplementation and irradiation in the 36B10 astrocytoma cells suggests that the cytotoxic action of GLA alone and the GLA-mediated increase in radiation sensitivity is the result of increased lipid peroxidation and/or free radical generation. However, the ability of the cyclooxygenase inhibitor ibuprofen to block the GLA-mediated increase in astrocytoma cell radiosensitivity suggests that increased prostanoid synthesis also may be involved in the radiation effect.

The selective cytotoxic effect of GLA also may reflect the reduced ability of the neoplastic astrocytoma cells to protect themselves against lipid peroxidation and/or free radical generation resulting from their increased membrane PUFA content. Increased ISP levels in the 36B10 cells following GLA supplementation and radiation indicates increased oxidative stress in these cells. Antioxidant enzymes such as superoxide dismutase (SOD), catalase and glutathione peroxidase are the cells primary mechanisms for scavenging free radicals. Studies have shown decreased activities of these enzymes in many tumor cells.[9] Preliminary results in our laboratory show that the manganese (Mn) and copper-zinc SOD levels in 36B10 cells are significantly lower than those of astrocytes, and supplementation of 36B10 cells with GLA failed to alter the SOD levels in these malignant cells (unpublished data). Our findings that "normal" astrocytes were unaffected by GLA suggests that they are able to deal with the increased oxidative stress resulting from PUFA supplementation. The mechanism might involve a PUFA-mediated increase in antioxidant enzyme activity. We have observed that GLA supplementation increases the activity of MnSOD in rat astrocytes.[10]

Oxidative stress in the PUFA-enriched 36B10 cells further increased following irradiation. We have previously observed that providing 36B10 cells with the antioxidant trolox not only blocked the cytotoxic effect of GLA, but also blocked the GLA-mediated increase in glioma cell radiosensitivity.[3] Thus, our results support the hypothesis that the cytotoxic effect of GLA in tumor cells is, at least in part, the result of increased generation of free radicals and lipid peroxidation. In contrast, LA supplementation did not lead to either an increase in ISP levels or reduced cell survival.[3]

There appear to be contradictory views on the role of prostanoid and leukotriene synthesis in PUFA-mediated cytotoxicity. Thus, while studies indicate that eicosanoid and leukotriene synthesis play an important role in tumor cell proliferation,[5] other groups have reported that prostanoid and leukotriene synthesis are not important in PUFA-induced tumor cell cytotoxicity.[11] The inability of the cyclooxygenase inhibitor ibuprofen to block the cytotoxic effect of GLA alone as observed in the present study supports the view that prostanoid and leukotriene synthesis are not important in GLA-cytotoxicity. However, ibuprofen did block the enhanced radiation sensitivity of GLA enriched 36B10 cells. Thus, while the cytotoxic effect of GLA alone is probably due to increased oxidative stress, the augmentation of radiation response of GLA-supplemented astrocytoma cells might also involve an increase in prostanoid synthesis. However, the exact mechanism(s) by which prostanoids affect GLA-mediated increase in radiation sensitivity is not known and needs to be investigated.

In summary, the present results indicate that GLA is selectively cytotoxic to 36B10 and C6 malignant rat glioma cells; "normal" rat astrocytes are not affected. Malignant gliomas are extremely radioresistant and difficult to treat.[12] Moreover, damage to surrounding normal tissues is a dose limiting factor in the treatment of these tumors.[13] The use of GLA as a therapeutic adjunct may increase the efficacy of radiotherapy when it is used in the treatment of some malignant brain tumors.

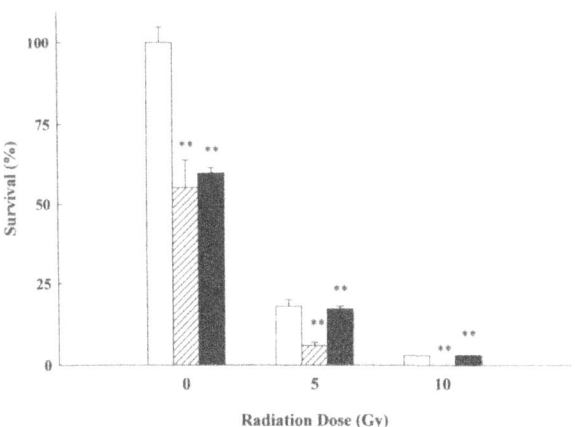

Figure 3. Ibuprofen prevents GLA enhancement of glioma cell radiosensitivity but not GLA-mediated cytotoxicity. 36B10 cells were supplemented with 45 μM GLA ▢, or 100 μM ibuprofen + GLA ■ and then exposed to 5 or 10 Gy γ-radiation. Cells were continued on supplemented media throughout the experiment. Control cultures, ▢, received no supplement. Values represent mean ± S.E. from 6 separate cultures. ** Significantly different from unsupplemented group; p< 0.01.

ACKNOWLEDGMENTS

This work was supported by grants from Scotia Pharmaceuticals Ltd., UK and NIH, CA 66081.

REFERENCES

1. P. Sangeetha and U.N. Das, Cytotoxic action of *cis*-unsaturated fatty acids on human cervical carcinoma (HeLa) cells: relationship to free radicals and lipid peroxidation and its modulation by calmodulin antagonists, *Cancer Lett.* 63: 189 (1992).
2. S.H. Abou El-Ela, K.W. Prasse, R. Carroll, and O.R. Bunce, Effects of dietary primrose oil on mammary tumorigenesis induced by 7, 12- dimethylbenz (a) anthracene, *Lipids.* 22: 1041 (1987).
3. S. Vartak, R. McCaw, C.S. Davis, M.E.C. Robbins, and A.A. Spector, γ-Linolenic acid (GLA) is cytotoxic to 36B10 malignant rat astrocytoma cells but not to "normal" rat astrocytes, *Br. J. Cancer.* In Press. (1998).
4. G.W. Ells, K.A. Chisholm, V.A. Simmons, and D.F. Horrobin, Vitamin E blocks the cytotoxic effect of γ-linolenic acid when administered as late as the time of onset of cell death- insight into the mechanism of fatty acid induced cytotoxicity, *Cancer Lett.* 98: 207 (1996).
5. M. Earashi, M. Noguchi, K. Kinoshita, and M. Tanaka, Effects of eicosanoid synthesis inhibitors on the *in vitro* growth and prostaglandin E and leukotriene B secretion of a human breast cancer cell line, *Oncology.* 52: 150 (1995).
6. O.R. Bunce and S.H. Abou-El-Ela, Eicosanoid synthesis and ornithine decarboxylase activity in mammary tumors of rats fed varying levels and types of N-3 and/or N-6 fatty acids, *Prostagl. Leukotr. and Essential Fatty Acids.* 41: 105 (1990).
7. S. Murphy. Generation of astrocyte cultures from normal and neoplastic central nervous system, in: *Methods in Neuroscience. Vol. 2*, P.M. Conn, ed., Academic Press, New York. p.33 (1990).
8. J.D. Morrow, K.E. Hill, R.F. Burk, T.M. Nammour, K.F. Badr, and L.J. Roberts, A series of prostaglandin F2-like compounds are produced *in vivo* in humans by a non-cyclo-oxygenase, free radical catalyzed mechanism, Proc. Natl. Acad. Sci. 87: 9383 (1990).
9. L.W. Oberley, M.L. Mccormick, E. Sierra-Rivera, and D. Kasemset-St-Clair, Manganese superoxide dismutase in normal and transformed human embryonic lung fibroblasts, *Free Rad. Biol. Med.* 6: 379 (1989).
10. G.D. Gimun, S.A. Moore, L.W. Oberley, and M.E.C. Robbins, Polyunsaturated fatty acids (PUFA) can selectively increase MnSOD levels in normal central nervous system (CNS) derived cells. (Abstract), *44th Annual Meeting of the Radiation Research Society.* Apr. 14-17. (1996).
11. U.N. Das, Tumoricidal action of *cis*-unsaturated fatty acids and their relationship to free radicals and lipid peroxidation, *Cancer Lett.* 56: 235 (1991).
12. S. Phuphanich, S. Ferrall, and H. Greenberg, Long-term survival in malignant glioma. Prognostic factors, *J. Florida Med. Assoc.* 80: 181 (1993).
13. G.A. Sheline, W.M. Wara, and V. Smith, Therapeutic irradiation and brain injury, *Intl. J. Radiat. Oncol. Biol. Phys.* 6: 1215 (1980).

73

γ-LINOLENIC ACID (GLA)-MEDIATED CYTOTOXICITY IN HUMAN PROSTATE CANCER CELLS

Mike Robbins,[1] Kamran Ali,[1] Ryan McCaw,[1] Josalyn Olsen,[1] Sujata Vartak,[1] and David Lubaroff[2]

[1]Radiation Research Laboratory
Department of Radiology and [2]Urology
University of Iowa
Iowa City, IA 52242

INTRODUCTION

The role of dietary fats, particularly the n-6 polyunsaturated fatty acids (PUFAs) derived from dietary consumption of linoleic acid (LA; 18:2n-6) in prostate cancer has been viewed primarily in terms of their positive association.[1] Such a view is oversimplistic. Thus, while studies have demonstrated that supplementation of human prostate cancer cells (HPCCs) *in vitro* or *in vivo* with either LA or arachidonic acid (AA; 20:4n-6) results in stimulation of PCC growth,[2,3] not all n-6 PUFAs have a stimulatory effect on PCC growth. Indeed, γ-linolenic acid (GLA; 18:3n-6) has been shown to reduce the growth of human DU-145 PCCs implanted in nude mice.[4] Although the mechanism for this GLA-mediated effect is only partially understood, free radical generation and excessive lipid peroxidation subsequent to GLA supplementation is thought to play a major role.[5]

Therefore, we hypothesized that supplementation of HPCCs with GLA would result in an increase in the n-6 PUFA content of cellular lipids, with a concomitant reduction in malignant cell growth and survival. Our results indicate that despite an essentially identical increase in n-6 PUFA content within the PCC membranes of 4 different HPCC lines, PC3, LNCaP, GW, and A63, the response of these PCC lines to GLA varied considerably. LNCaP and GW cells showed pronounced reductions in cell survival and proliferation, PC3 cells showed an intermediate response, and A63 cells were unaffected.

Eicosanoids and Other Bioactive Lipids in Cancer, Inflammation, and Radiation Injury, 4
Edited by Honn *et al.*, Kluwer Academic / Plenum Publishers, New York, 1999.

MATERIALS & METHODS

Cell and Culture Conditions

The HPCC lines used included PC3, LNCaP, and two sublines of LNCaP, GW (can secrete prostate serum antigen (PSA); androgen dependent) and A63 (can secrete PSA; androgen independent). The PC3, LNCaP and GW cell lines were grown in RPMI 1640 medium supplemented with 10% fetal bovine serum (FBS) at 37^0C in a humidified atmosphere of 95% air: 5% CO_2. The androgen-independent PCC line A63 was grown in RPMI 1640 medium supplemented with charcoal-dextran-stripped FBS in which all steroids had been removed.

Lipid Extraction and Assays

The fatty acid composition of the HPCCs was determined by gas-liquid chromatography. Incorporation of fatty acids into the cells was determined by adding 30 or 60 μM of GLA to the medium. The monounsaturated fatty acid, oleic acid (OA, 18:1n-9), was used as a fatty acid control. After incubation for 24 hours, lipids were extracted and tranesterified with 12% BF_3 in methanol at 95^0C for 45 min. The resulting fatty acid methyl esters were extracted into n-heptane and the fatty acids separated using gas chromatography.[6]

Cell Survival Analyses

To determine GLA-induced changes in HPCC growth cells (500/well) were seeded onto 96-well plates containing 200 μL medium. After an overnight incubation, cells were supplemented with 45 μM GLA or OA and allowed to grow for 7 days. Viability was assessed using the fluorescent dye calcein-AM; cells were washed using PBS, and incubated with calcein-AM at a final concentration of 1 μM for 45 min at room temperature. Fluorescence then was measured using a microplate reader (FL500, Bio-Tek Instruments Inc.), live cells are characterized by an intense fluorescence at 530 nm. Cell proliferation was also assessed using the CyQUANT® cell proliferation kit (Molecular Probes, Eugene, OR), with sample fluorescence measured using a microplate reader.

RESULTS

Incorporation of Fatty Acids

Incubation of HPCCs with either GLA or OA supplemented media resulted in dose-and fatty acid-dependent changes in cellular fatty acid composition. Table 1 shows the effect of 30-60 μM GLA supplementation on the fatty acid composition of GW cells. Incubating GW cells with GLA led to significant (p< 0.001) dose-dependent increases in GLA and in particular its elongation product dihomo-γ-GLA (DGLA, 20:3n-6). Little change was observed in levels of AA. The total percentage of fatty acids present in the cell lipids as PUFAs also showed a significant increase from control levels of $25.2 \pm 1.6\%$ up to levels of $38.1 \pm 0.6\%$ and $43.5 \pm 1.0\%$ (p< 0.001) after incubation with 30 and 60 μM GLA, respectively. A similar pattern of change was observed after GLA supplementation in the other HPCC lines; the percentage of HPCC lipids present as PUFAs increased

Table 1. Effect of GLA supplementation on the fatty acid composition of GW cells

Fatty acid	GLA concentration (μM)		
	0	30	60
16:0	27.4 ± 1.3	22.3 ± 0.4*	21.9 ± 0.7*
16:1	3.4 ± 0.6	1.5 ± 0.1	1.5 ± 0.1
18:0	12.7 ± 0.5	13.0 ± 0.5	11.1 ± 0.6
18:1n-9	31.3 ± 0.4	25.2 ± 0.3***	22.0 ± 0.3***
18:2n-6	1.8 ± 0.1	1.6 ± 0.1	1.2 ± 0.2
18:3n-6	2.2 ± 0.1	3.5 ± 0.1***	9.2 ± 0.1***
20:3n-6	2.7 ± 0.2	13.3 ± 0.1***	14.8 ± 0.5***
20:4n-6	7.9 ± 0.3	9.5 ± 0.1*	7.5 ± 0.2
22:4n-6	2.4 ± 0.4	3.0 ± 0.4	4.1 ± 0.8
20:5n-3	0.9 ± 0.1	0.6 ± 0.1	0.6 ± 0.1
22:5n-3	3.5 ± 0.3	3.4 ± 0.3	3.3 ± 0.7
22:6n-3	3.9 ± 0.4	3.1 ± 0.4	2.9 ± 0.6
Total SFA	40.1 ± 1.7	35.2 ± 0.3*	33.0 ± 0.9*
Total MUFA	34.7 ± 0.2	26.7 ± 0.3***	23.5 ± 0.2***
Total PUFA	25.2 ± 1.6	38.1 ± 0.6**	43.5 ± 1.0**

Values represent mean ± SE from 3 separate cultures. Cells were grown in 30-60 μM of GLA enriched medium for 24 h prior to lipid extraction. Statistical analysis was performed using Student's t-test, p values *< 0.05, **, < 0.01, ***< 0.001; SFA, saturated fatty acids; MUFA, monounsaturated fatty acids; PUFA, polyunsaturated fatty acid.

approximately two-fold (data not shown). In contrast, incubating HPCCs with OA did not cause an alteration in PUFA composition; increases were observed in the levels of OA alone (data not shown).

Effects of Fatty Acids on Cell Survival

Although incubating the 4 HPCC lines with GLA led to essentially similar changes in overall fatty acid composition, the effect of these changes on their subsequent survival varied considerably between the cell lines. Thus, in the LNCaP and GW cells, survival determined 7 days after the start of supplementation with 45 μM GLA was markedly reduced to levels that were only 60.4 ± 7.0% and 63.6 ± 3.8%, respectively, of those seen in control cells (p< 0.001). A smaller, but statistically significant decrease also was noted in the PC3 cells, where survival was reduced to 81.0 ± 2.5% of controls (p< 0.01). In contrast, no statistically significant reduction in survival was noted in the A63 cells (Figure 1). A similar pattern of results was obtained from the cell proliferation assay (Figure 2). Both the LNCaP and GW cells appeared to be very sensitive to GLA; cell proliferation was reduced to values that were only 31.4 ± 7.6% and 20.8 ± 3.1%, respectively, of those seen in the control cells (p < 0.001). The PC3 cells again showed a smaller decrease; cell proliferation was reduced to 66.3 ± 6.3% of that observed in control cells (p<0.001). In contrast, A63 cell proliferation did not appear to be affected by GLA supplementation. Incubating the HPCC lines with 45 μM OA enriched media did not affect cell survival or proliferation as assessed by the calcein-AM or CyQUANT assay, respectively (Figures 1 and 2).

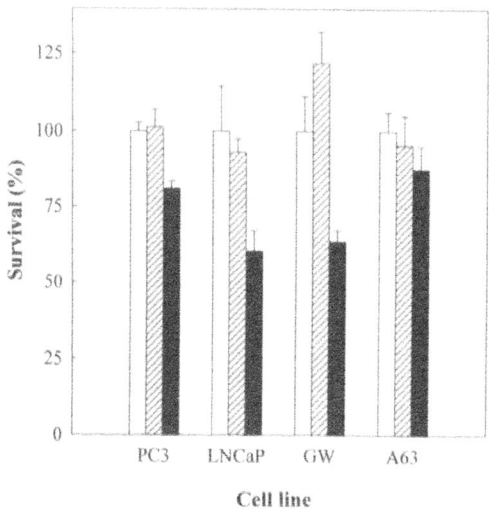

Figure 1. Effect of GLA and OA supplementation on the survival of HPCCs assessed using the calcein-AM assay. Cells were supplemented with 45 μM of GLA (■) or OA (diagonally striped bar) for 7 days. Control cells (□) received no supplemental fatty acid. Bars indicate values obtained from 5 separate cultures (mean ± SE).

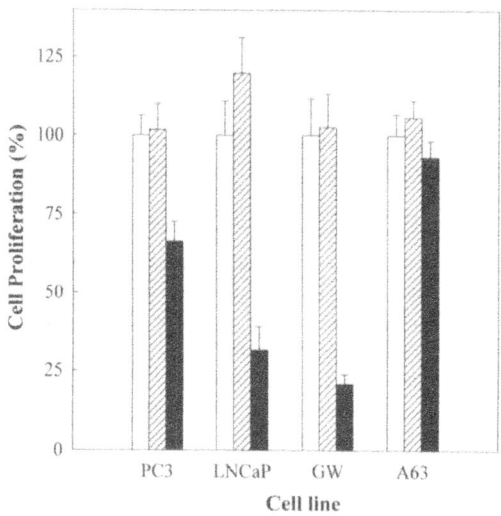

Figure 2. Effect of GLA and OA supplementation on the proliferation of HPCCs assessed using the CyQUANT assay. Cells were supplemented with 45 μM of GLA (■) or OA (diagonally striped bar) for 7 days. Control cells (□) received no supplemental fatty acid. Bars indicate values obtained from 5 separate cultures (mean ± SE).

DISCUSSION

Unlike saturated and monounsaturated fatty acids that can be synthesized *de novo*, PUFAs, made up of n-3 and n-6 fatty acids are derived entirely from the diet. Terrestrial plants synthesize the first member of the n-6 PUFA series, LA; within the body LA is metabolized by a series of alternating oxidative desaturation and elongation steps principally to AA. While all mammalian cells can interconvert the fatty acids within each series by elongation, desaturation and retroconversion, albeit to varying degrees, the two series are not interchangeable.[7] Thus, tissue PUFA content will depend to a large extent on the diet; **we are what we eat.**

The same principle applies to malignant cells, which derive all their PUFAs from the host. Thus, incubating malignant cells in culture with varying levels of different PUFAs leads to specific, PUFA-dependent, changes in malignant cell PUFA content.[8] The present findings confirm a similar modification of cellular PUFA content following incubation of HPCCs with GLA. It is interesting to note that while GLA supplementation resulted in dose-dependent increases both in GLA and its elongation product DGLA, the percentage of n-6 PUFAs present as AA remained essentially unchanged. The increase in the AA elongation product docosatetraenoic acid (DTA, 22:4n-6) suggests that any AA formed from desaturation of DGLA was rapidly converted to DTA. These changes are similar to those seen after administration of GLA to humans. Determination of plasma fatty acid composition revealed a consistent and significant increase in plasma DGLA levels, associated with a less marked increase in AA values.[9] Thus, it appears possible to modulate levels of specific PUFAs by varying the precursor supplied either in the medium or the diet.

GLA supplementation has been shown to lead to profound and selective reductions in tumor cell growth.[10] Although the mechanism for this cytotoxic effect is only partially understood, free radical generation and lipid peroxidation subsequent to PUFA supplementation is believed to play a major role. Supplementation of tumor cells with PUFAs leads to an increase in the oxidizability of cell membranes, increased free radical generation and lipid peroxidation.[11] Moreover, this PUFA-mediated cytotoxicity can be blocked by the addition of antioxidants.[5] However, the current findings indicate that despite similar GLA-induced alterations in HPCC fatty acid composition, there was marked variation in the response of the various cell lines to GLA. Thus, whereas a GLA-induced reduction in tumor cell growth was indeed observed in the PC3, LNCaP and GW HPCC lines, the growth of A63 HPCCs was not affected by GLA supplementation. We have observed similar variation in the response of human breast cancer cells lines to GLA supplementation (unpublished observations). These results suggest that the cytotoxic effect of GLA was not simply due to increased oxidizability of the PCC membranes.

An alternative hypothesis is that GLA-mediated cytotoxicity might reflect alterations in eicosanoid metabolism.[12] Recent data suggest that AA plays an important role in promoting PCC growth. Inhibitors of eicosanoid metabolism inhibit PCC growth in vitro,[1] and increases in AA turnover and prostaglandin E_2 synthesis have been detected in human malignant prostate tissue.[13] GLA supplementation with resultant increase in DGLA relative to AA would result in production of an alternative substrate for cyclooxygenase, leading to a decrease in PGE_2 synthesis and reduction in PCC growth. However, this hypothesis fails to explain the apparent insensitivity of the A63 cells to GLA supplementation, despite a similar increase in the DGLA:AA ratio as that seen in the other PCC lines. Moreover, we have failed to demonstrate an AA-induced promotion of HPCC growth in these 4 cell lines (unpublished observations).

In summary, our findings suggest that the role of n-6 PUFAs in prostate cancer should not be viewed merely in terms of their positive association. The effect of particular PUFAs is variable;

GLA is cytotoxic to some, but not all HPCCs. This variation in the response of HPCCs to GLA cannot be explained in terms of variations in PUFA content, and further studies are currently being performed to define the mechanistic basis for this difference in tumor cell response. Such studies offer the promise of gaining new insight into the potential role of GLA in the therapeutic treatment of prostate cancer.

ACKNOWLEDGMENTS

These studies were supported by grants from Scotia Pharmaceuticals Ltd., U.K., and NCI CA ES69838-01

REFERENCES

1. D.P. Rose and J. M. Connolly, Dietary fat, fatty acids, and prostate cancer. *Lipids* 27:798 (1992).
2. D.P. Rose and J. M. Connolly, Effects of fatty acids and eicosanoid synthesis inhibitors on the growth of two human prostate cancer cell lines. *Prostate* 18:243 (1991).
3. S.K. Clinton, S. S. Palmer, C. E. Spriggs, and W. J. Visek, Growth of Dunning transplantable prostate adenocarcinoma in rats fed diets with various fat contents. *J Nutrit.* 118:908 (1988).
4. R.A. Karmali, P. Reichel, L. A. Cohen, T. Terano, A. Hirai, Y. Tamura, and S. Yoshida, The effect of dietary ω-3 fatty acids on the DU-145 transplantable human prostatic cancer. *Anticancer Res* 7:1173 (1987).
5. G.W. Ells, K. A. Chisholm, V. A. Simmons, and D. F. Horrobin, Vitamin E blocks the cytoxic effect of γ-linolenic acid when administered as late as the time of onset of cell death - insight into the mechanism of fatty acid induced cytotoxicity. *Cancer Lett* 98:207 (1996).
6. S. Vartak, M. E. C. Robbins, and A. A. Spector, Polyunsaturated fatty acids increase the sensitivity of 36B10 rat astrocytoma cells to radiation-induced cell kill. *Lipids* 32:283 (1997).
7. S.I. Grammatikos, P. V. Subbaiah, T. A. Victor, and W. M. Miller, Diversity in the ability of cultured cells to elongate and desaturate essential (*n*-6 and *n*-3) fatty acids. *Annals New York Acad Sci.* 745:92 (1994).
8. A.A. Spector and C. P. Burns, Biological and therapeutic potential of membrane lipid modification in tumors. *Cancer Res* 47:4529 (1987).
9. M.S. Manku, N. Morse-Fisher, and D. F. Horrobin, Changes in human plasma essential fatty acid levels as a result of administration of linoleic acid and gamma-linolenic acid. *Eur J Clin Nutr.* 42:55 (1988).
10. M.E. Begin, G. Ells, U. N. Das, and D. F. Horrobin, Differential killing of human carcinoma cells supplemented with n-3 and n-6 polyunsaturated fatty acids. *JNCI* 77:1053 (1986).
11. B.A. Wagner, G. R. Buettner, and C. P. Burns, Free radical-mediated lipid peroxidation in cells: Oxidizability is a function of cell lipid *bis*-allylic hydrogen content. *Biochemistry* 33:4449 (1994).
12. D.P. Rose, The mechanistic rationale in support of dietary cancer prevention. *Preventive Med* 25:34 (1996).
13. F.H. Faas, A. Q. Dang, M. Pollard, X. Hong, K. Fan, P. H. Luckert, and M. Schutz, Increased phospholipid fatty acid remodeling in human and rat prostatic adenocarcinoma tissues. *J Urol.* 156:243 (1996).

74

FIVE-LIPOXYGENASE INHIBITORS REDUCE PANC-1 SURVIVAL: SYNERGISM OF MK886 WITH GAMMA LINOLENIC ACID

Harris, JE, Alrefai, WA, Meng, J, Anderson, KM.

Section of Medical Oncology,

Department of Medicine and Department of Biochemistry,

Rush Medical College, Chicago, Il 60612

OVERVIEW

The five-lipoxygenase inhibitors ETYA, SC41661A and MK 886 reduced the proliferation and viability of Panc-1 human pancreatic cancer cells (1). The extent of inhibition depended upon drug concentration and with continued culture, cells detached and stained with trypan blue. Although results from flow cytometry were consistent with programmed cell death, despite repeated attempts, no DNA laddering characteristic of its later stages was detected, and studies with the TUNEL assay were negative. Light and electron microscopy of cells cultured with SC41661A provided morphologic evidence of an incompletely expressed type 1 programmed cell death. Cells cultured with MK886 exhibited an unusual cytoplasmic mode of cell death characterized by vacuolization and widely separated smooth internal membranes without diagnostic nuclear changes. This is in marked contrast to the extensive type 1 PCD induced by 5-lipoxygenase inhibitors cultured with human U937 monoblastoid cells (2). The incomplete destruction of chromatin and absence of a fully expressed type 1 response in SC41661A-treated Panc-1 cells implies a defective initiation and/or implementation of the enzymatic "program" for cellular suicide by these agents. The response of Panc-1 cells to MK886 suggests expression of a variant "cytoplasmic" (type 2?) form of non-necrotic cell death. In a European clinical trial, gamma

Eicosanoids and Other Bioactive Lipids in Cancer, Inflammation, and Radiation Injury, 4
Edited by Honn *et al.*, Kluwer Academic / Plenum Publishers, New York, 1999.

505

linolenic acid, a polyunsaturated fatty acid that generates free radicals has been combined with 5-fluorouracil as chemotherapy for pancreatic cancer. Panc-1 cell proliferation was insensitive to inhibition by several chemotherapeutic agents employed clinically, including 5-fluorouracil, cisplatin or gemcitabine and only somewhat sensitive to GLA. When gamma linolenic acid was combined with MK886, the more effective of the two 5-lipoxygenase inhibitors, a synergistic reduction in Panc-1 cell number and viability occurred.

INTRODUCTION

Most solid tumors resist chemotherapy and ionizing radiation. Pancreatic cancers treated with agents including 5-fluorouracil, cis-platin or gemcitabine in various combinations and sequences rarely yield a major clinical response. This is in distinct contrast with the response to therapy of many types of malignant hematopoietic diseases in which marked reduction in host tumor cell load occurs, almost invariably followed by a return toward pretreatment levels of cells more resistent to therapy. While differences in cell cycle transit times between hematopoietic and solid cancers may account for part of the comparative lack of response by many solid cancers to chemotherapy, other factors could contribute. Because of interest in this question, we compared the effect on U937 monoblastoid, PC-3 prostate and Panc-1 pancreatic cancer cell proliferation and cell death of several selective inhibitors of the arachidonic acid-metabolizing enzyme, 5-lipoxygenase. These cell lines exhibit comparable order-of-magnitude cell cycle times. In what follows, a brief overview and summary is provided, with references to work in press or submitted for the details.

METHODS

U937, Panc-1 and PB-3 cell lines were from the ATCC. The cells were cultured in RPMI-1640 with 5 or 10 percent FBS, 25 mM HEPES, 50 ug/ml streptomycin, 50 units/ml penicillin G and 10 percent CO_2. The lipoxygenase inhibitors, SC41661A (3-((3,5-bis(1,1-dimethyl)-4-hydroxy-phenyl) thiol)-N-methyl-N-(2-(2-phridyl-propanamide; Searle) and, MK886 (3-(t-(4-chlorobenzyl-3-butyl)thio-5-isppropyindal-22-yl)-2,3-dimethyl propanoic acid; Merck Frosst) and the polyunsaturated fatty acid, gamma linolenic acid (6,9,12-octadecaoctatrienoic acid)) were added in DMSO. Control and treated samples contained at most 0.3 percent of the vehicle. The relative number of viable cells was estimated using a hemocytometer or by a modified MTT assay with a commercial kit (Promega, Madison, WI). DNA was isolated with a DNA-RNA isolation kit (United States Biochemicals, Cleveland, OH). and electrophoresed on 1 per cent agarose gels. The Apo-Tag in situ apoptosis detection kit (Oncor, Gaithersburg, VA) was used to identify any oligosomic DNA. Samples were prepared for transmission electron microscopy and flow cytometry by conventional means.

RESULTS

Rather than present the individual data, we will summarize it, supplemented by a table with references (Table 1).

Table 1.Comparative Characteristics of and Responses to 5-LPOX Inhibitors by U937, Panc-1 and PC-3 Cells (1,2,5-10,12,13).

	U937		Panc-1		PC-3	
5-LPOx mRNA	+		0		+	
FLAP mRNA	+		+		+	
Bcl-2 protein	+		0		0	
Ca2+ response to MK-886	+		0		0	
Prolif. response to inhibitors	SC	MK	SC	MK	SC	MK
	<	<	<	<	<	<
Programmed cell death						
Type 1	a,b,c,d :	a,b,c,d				
Atypical type 1	--	--	a,b,d	--	a.b.c.d	--
Type 2	--	--	--	--	--	a,b,c,d±
Atypical Type 2	--	--	--	a,b,d	--	--

Supporting evidence: a, flow cytometry; b, TUNEL; c, DNA laddering; d, morphology.
+, present; 0, absent; <, reduced.

Proliferation of the three cell lines was reduced by the two 5-lipoxygenase (5-LPOX) inhibitors,SC41661A or MK886, requiring from 10-20 uM for a significant effect. Depending upon the percent of cells in cycle and the concentrations of serum and inhibitor, the extent and rate of change varied. With continued culture, attached cells (PC-3 and Panc-1) gradually detached, and many of these stained with trypan blue. That was due either to primary or secondary necrosis and/ or to programmed cell death (PCD), to be decided by further study. Whether the inverse relationship between the increased sensitivity to inhibition and reduced serum concentration was due to reduced binding or other inactivation of the inhibitor by serum and/or to a lesser concentration of growth factors, cytokines or hormones is not settled. If cells were quiescent, the inhibitors had no evident effect on their survival, at least at the concentrations employed over the 48 to 72 hr of study. At these concentrations, cells in cycle were susceptible, ongoing metabolic processes related to cell cycle progression were necessary, and non-specific cytotoxicity did not account for the results.

To examine the nature of cell death, whether programmed (by cellular suicide) or necrotic ("unprogrammed"), flow cytometry was employed. All treated cells exhibited subdiploid fractions whose percentage varied, depending upon conditions. The expectation was that this would correlate with DNA laddering. This was found to be true for U937 and PC-3 cells but not for Panc-1 cells, despite numerous attempts to demonstrate it. The APO-Tag "TUNEL" reaction, in which digoxigen-dUTP is added to 3'-OH DNA with deoxynucleotidyl transferase, followed by digoxigen-antibody peroxidase complex and incubation with diaminobenzidine substrate yielded oligosome-positive, inhibitor-treated U937 and PC-3 cells but unstained Panc-1 cells, consistent with their lack of DNA laddering.

Lacking generally agreed upon biochemical criteria for distinguishing the different forms of PCD e.g., types 1, 2, 3A and 3B (3,4) from one another and from necrosis, reliance is placed on ultrastructure. U937 cells cultured with either 5-LPOx inhibitors exhibited "classic:" type 1 PCD ultrastructure (2,5). PC-3 cells cultured with SC41661A exhibited an atypical type 1 PCD. MK886 induced an even more atypical ultrastructure, resulting predominantly in a vesiculated cytoplasm without changes in nuclear chromatin consistent with either type 1 or 2 PCD. The response of Panc-1 cells was even more unusual, and while a form of PCD did occur, based on the presence of numerous shrunken "dark" cells, the cytoplasmic changes were even less characteristic than MK886-treated PC-3 cells.

Messenger RNAs for 5-LPOX and 5-lipoxygenase activating protein (FLAP), determined by RT-PCR in the laboratory of Dr. Jim Mulshine was present in U937 and PC-3 (6-8). In Panc-1 cells, the situation is more complicated in that mRNA for 5-LPOX was missing while that for FLAP was present (8). U937 cells contain a high concentration of immuno-reactive Bcl-2, as determined by western blots while it was not detected in either Panc-1 or PC-3 cells (9). MK886 induced a rapid rise in U937 cytosol Ca2+ but none was detected in Panc-1 or PC-3 cells examined in suspension with fura II (10).

Gamma linolenic acid (GLA) is a polyunsaturated fatty acid that generates free radicals during its metabolism. It also induces PCD (11), as does 5,8,11,14-eicosatetraynoic acid (ETYA), the alkyne congener of arachidonic acid (2). In Europe, GLA has been used for the treatment of pancreatic cancer with reported modest increases in mean survival. Based on an argument that combining agents that modulate apparently disparate biochemical events may yield synergism (12), Panc-1 cells were cultured with GLA and MK886, with synergistic inhibition of proliferation (1). MK886 inhibits the expression of Oct-1 mRNA in U937 cells and in head and neck cancer cell (13), raising the interesting possibility that inhibition of this homeobox gene is also involved in the events we described. Interdiction of multiple interrelated pathways in malignantly transformed cells with a reduced number of redundant (parallel?) pathways compared to normal cells may promote synergism. While cancer cells utilize growth-promoting pathways in ways that largely escape physiologic controls, to the extent that these cells are aneuploid, the number of redundant pathways that likely permit nontransformed cells to accommodate the metabolic stress of therapy should be reduced.

CONCLUSIONS AND SOME SPECULATION

1. In U937 cells, the presence of mRNA for 5-LPOX and FLAP and of Bcl-2 protein yields a "classic" type 1 PCD response to the two LPOx inhibitors studied.

2. In PC-3 cells with undetectable (by western blots) Bcl-2, but with the two mRNAs of interest, SC41661A (which is an oxidation/reduction agent that may also increase intracellular free radicals (14) yields an atypical type 1 PCD response; Mk886 with a different mechanism of action, and may include additional effects on arachidonc acid metabolism, induces an atypical form of PCD perhaps best considered as a form of atypical tyle 2 cellular suicide.

3. In Panc-1 cells without detectable expression of 5-LPOX mRNA or Bcl-2 protein, the inhibitors induce what is believed to be a form of PCD which includes features of type 2 and 3A, but is otherwise even more atypical than MK886-treated PC-3 cells.

4. Bcl-2 protein may be required for the MK-886 acutely induced Ca2+ response (10), present in U937 but absent in PC-3 and Panc-1 cells.

5. To the extent that Bcl-2 is either absent or present in very low concentration, and

provided other members of this protein family (e.g., Bcl-2$_{LX}$ do not substitute for it, Bcl-2 expression may be an important ("essential"?) component for raising intracellular Ca2+ by a 5-LPOx-dependent mechanism, for the expression of "classic" type 1 PCD and for an extensive response to chemotherapy via a mitochondria - dependent. mechanism (15) Solid cancers lacking a Bcl-2-dependent mechanism with which to respond to potentially effective therapy utilize other non-Bcl-2-dependent pathways (e.g., Fas / Fas ligand?) to implement PCD, which evidently are less responsive to the available cyto-reductive therapies that depend upon the induction of programmed cell death for their major efficacy.

6. The diverse response in a population of malignant cells exhibiting generally unknown forms of structural and functional aneuploidy; likely with the partial or absent expression of at least some components (e.g, Bcl-2, caspases, etc) involved in cellular suicide should not be unexpected.

7. Since the observations in 1-5 certainly represent a gross oversimplification of the biochemical "circuitry" contributing to the different responses to therapy exhibited by hematologic and solid cancers, additional studies are needed to understand these events.

ACKNOWLEDGMENTS

We thank the Weinberg Foundation, the Seidel Family Trust and the Wadsworth Memorial Fund for their continued support

BIBLIOGRAPHY

1. Anderson, KM, Seed, T, Meng, J, Ou, D, Alrefrai, W, Harris, JE. Five lipoxygenase inhibitors reduce Panc-1 survival: The mode of cell death and synergism of MK886 with gamma linolenic acid., in press, Antican Res, 1998.
2. Anderson, KM, Seed, T, Plate, J, Jejah, A, Meng, J, Harris, JE. Selective inhibitors of 5-lipoxygenase reduce CML blast cell proliferation and induce limited differentiation and apoptosis. Leuk Res 19, 789-801, 1995.
3. Clark, P. Developmental cell death: morphological diversity and multiple mechanisms. Anat Embryol 181, 1956-213, 1990.
4. Charruayt-Marlangue, C, Aggoun-Zouaoui, D, Repressa, A, Ben-Ari, Y. lApopotic features of selective neuronal death in ischemia and gp 120 toxicity. Trends in Neurol Sci 19, 109-114, 1996.
5. Anderson, KM, Seed, TM, Jajeh, A, Dudeja, P, Byun, T, Meng, J, Ou, D, Bonomi, P, Harris, JE, An in vivo inhibitor of 5-lipoxygenase, MK886, at micromolar concentration induces apoptosis in U937 and CML cells. Antican Res 16, 2589-2600, 1996.
6. Anderson, KM, Ondrey, FG, Harris, JE. Modulation of cellular proliferation and induction of apoptosis in a human lymphoma cell line after treatment with selective lipoxygtenase inhibitors. 5th Int Conf on Eicosanoids, La Jolla, CAS, 1997, abst. #94.
7. Harris, JE, Seed, T, Vos, M, Mulshine, J, Meng, J, OU, D, Anderson, KM. 5-lipoxygenase inhibitors reduce PC-3 cell proliferation and initiate non-necrotic cell death., Am

Assoc Can Res, New Orleans, LA, 1998, in press.

8. Personal communication, Dr. Jim Mulshine.

9. Anderson, KM, Alrefrai, WA, Ou, D, Harris, JE. Does the apparent absence of Bcl-2 in PC-3 cells contribute to the atypical programmed cell death induced by selective inhibitors of 5-lipoxygenase?, submitted to the 1998 Int Biol of Prostate Growth, NIDDK.

10. Byun, T, Dudeja, P, Harris, JE, Ou, D, Seed, Y, Sawlani, D. Meng, J, Bonomi, P, Andnerson, KM. A 5-lipoxygenase inhibitor at micromolar concentrtion raises intracellular calcium in U937 cells prior to their physiologic cell death, Prost Leuko Essen Fatty Acids 56, 69-77, 1997.

11. Maimou-Fowler, T, Proctor, SJ, Dickson, AM. Gamma linolenic acid (GLA) induces apoptosis in B chronic lymphocytic leukemia (BCLL) cells. Blood 90, 532a, #2370, 1997.

12. Anderson, KM, Bonomi, P, Meng, J, Harris, JE. Inhibition of cancer cell proliferation by disruption of interdigitated / concatenated hierarchies of metabolic control / implementation processes: A proposal. Med Hypo 50, 119-123, 1998

13. Personal communication, Dr. Frank Ondrey.

14. Kocan, GP, Partis, RA, Mueller, RS, Smith, WG, Nakao, A. Contrasting effects of two arachidonate 5-lipoxygenase inhibitors on formyl-methionyl-leucyl-phenylalanine (fMLP) and complement fragment 5a-induced human neutrophil superoxide generation. Biochem Pharm 47, 1029-1037, 1994.

15. Kroemer, G, Zamzami, N, Susin, SA. Mitochondrial control of apoptosis. Immunol Today 18, 44-51, 1997.

ANANDAMIDE AND OTHER
BIOACTIVE LIPIDS

75

ENZYMOLOGICAL AND MOLECULAR BIOLOGICAL STUDIES ON ANANDAMIDE AMIDOHYDROLASE

Natsuo Ueda,[1] Kazuhisa Katayama,[2] Yuko Kurahashi,[1] Mitsujiro Suzuki,[1] Hiroshi Suzuki,[1] Shozo Yamamoto,[1] Itsuo Katoh,[2] Vincenzo Di Marzo,[3]   and Luciano De Petrocellis[4]

Departments of [1] Biochemistry and [2] Cardiovascular Surgery
Tokushima University School of Medicine, Tokushima 770-8503, Japan
[3] Istituto per la Chimica di Molecole di Interesse Biologico and [4] Istituto di Cibernetica CNR Napoli 80072, Italy

INTRODUCTION

Anandamide (arachidonylethanolamide) loses its cannabimimetic activities when it is enzymatically hydrolyzed to arachidonic acid and ethanolamine.[1] The enzyme responsible for the hydrolysis will be referred to as "anandamide amidohydrolase" or just "hydrolase" in this article. Previously we partially purified this enzyme from the microsomes of porcine brain, and showed that the same enzyme preparation catalyzed not only the anandamide hydrolysis but also its synthesis by the reverse reaction.[2] We examined various inhibitors (arachidonyl trifluoromethyl ketone, *p*-chloromercuribenzoic acid, phenylmethylsulfonyl fluoride and diisopropyl fluorophosphate) which inhibited the hydrolase and synthase activities in parallel. Together with several other lines of enzymological evidence, it was suggested that the two enzyme activities are attributable to a single enzyme protein.[2]

CHARACTERIZATION OF RECOMBINANT ANANDAMIDE AMIDOHYDROLASE

Previously, Cravatt *et al.* reported that oleamide, primary amide of oleic acid, was an endogenous sleep inducer, and showed that the compound was hydrolyzed enzymatically.[3] In 1996 they cloned cDNA for the responsible enzyme of rat liver. They found that the recombinant enzyme hydrolyzed not only oleamide but also anandamide, and referred to the enzyme as fatty-acid amide hydrolase.[4]

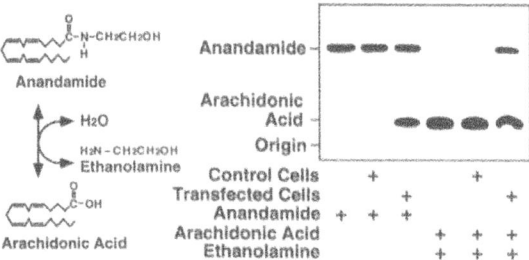

Figure 1. Anandamide hydrolase and synthase activities of the recombinant enzyme. The particulate fractions of COS-7 cells overexpressing rat anandamide amidohydrolase and the control cells were allowed to react with ^{14}C-labeled anandamide (100 μM) or arachidonic acid (200 μM). Products were analyzed by TLC with a solvent mixture of chloroform/methanol/ammonium hydroxide (80:20:2, v/v).

Since fatty-acid amide hydrolase was presumed to be identical to anandamide amidohydrolase, we prepared a cDNA for the enzyme by reverse transcriptase-polymerase chain reaction using rat liver mRNA as a template.[5] The cDNA was inserted to an eukaryotic expression vector pcDNA3.1 (Invitrogen), and COS-7 cells were transfected with the vector. When the particulate fraction of the transfected cells was allowed to react with [14C] anandamide, a band of arachidonic acid was detected on the TLC plate (Figure 1). Control cells, transfected with the insert-free vector, did not show the hydrolase activity. When we incubated [14C]arachidonic acid with the particulate fraction of COS-7 cells overexpressing the enzyme in the presence of ethanolamine, we observed the formation of anandamide, but not with the control cells (Figure 1).

The rate of anandamide hydrolysis increased depending on its concentration, and the Km value was 18 μM (Figure 2a). At the saturating concentration of anandamide, the specific enzyme activity was around 140 nmol/min/mg protein at 37 °C. As shown in Figures 2b and c, the anandamide synthesis was dependent on the concentrations of arachidonic acid and ethanolamine, and their Km values were 190 μM and 36 mM, respectively. The specific synthase activity was about 200 nmol/min/mg protein. These experimental results confirmed that the hydrolase and synthase activities were attributable to one enzyme. This finding was also reported by Deutsch's group.[6] However, due to the extremely high Km value for ethanolamine, the enzyme appears to act as the hydrolase rather than the synthase under the physiological conditions.

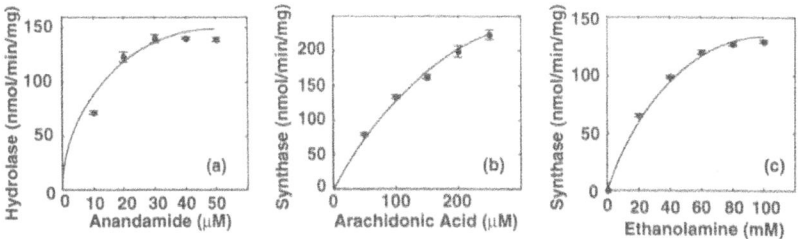

Figure 2. Substrate concentration-dependency of the anandamide hydrolysis and synthesis catalyzed by the recombinant enzyme. The particulate fraction of COS-7 cells overexpressing rat anandamide amidohydrolase was allowed to react with different concentrations of (a) anandamide, (b) arachidonic acid in the presence of 250 mM ethanolamine, or (c) ethanolamine in the presence of 200 μM arachidonic acid.

We also examined reactivity of the recombinant enzyme with other fatty acid derivatives (Figure 3).[5] Oleamide was hydrolyzed by the recombinant enzyme as reported by Cravatt *et al.* previously.[4] We found that the same enzyme preparation could form oleamide from oleic acid in the presence of as high as 1 M ammonium chloride. The oleamide synthesis did not occur with the control cells. Arachidonamide, primary amide of arachidonic acid, was also hydrolyzed with the same enzyme, and in its reverse reaction this compound was formed from arachidonic acid and ammonium chloride. Furthermore, methyl ester of arachidonic acid was hydrolyzed by this enzyme, but not by the control cells. In the reverse reaction, arachidonic acid was converted to its methyl ester in the presence of 2 M methanol. We noted that this ester formation preferred acidic pH around 5 although alkaline pH was optimum for the amide synthesis. The rates of these reactions are shown in Figure 3. At 100 μM the hydrolase activity was by far the most active with arachidonamide. Oleamide and methyl arachidonate were as active as anandamide. For the synthase reaction with 200 μM fatty acid as a substrate, the anandamide synthesis proceeded much faster than the syntheses of primary amides and ester. These results indicate a wide substrate specificity of the enzyme and the reversibility of the enzyme reactions.

ORGAN DISTRIBUTION OF ANANDAMIDE AMIDOHYDROLASE

We examined the distribution of the enzyme in various rat organs by measuring the anandamide hydrolase and synthase activities.[7] When the homogenate of each organ was tested, liver showed by far the highest activities of the hydrolase and synthase with specific activities of 4 - 5 nmol/min/mg protein at 37 °C (Figure 4a). Considerable activities were also detected in cerebrum, cerebellum, testis and parotid gland. In most organs tested, the synthase activity was comparable to the hydrolase activity. However, we noted that the hydrolase activity was much lower than the synthase activity in small intestine, stomach and colon. We presumed the presence of endogenous inhibitors for the hydrolase activity in gastrointestinal organs. Alternatively, a different enzyme may be present in these organs.

The anandamide hydrolase and synthase activities of the liver homogenate increased depending on the protein amounts. However, the enzyme activities did not increase linearly as the amount of the homogenate of small intestine was raised. When we added the homogenate of small intestine to the solubilized enzyme of rat liver microsome, the anandamide hydrolase activity was inhibited dose-dependently, suggesting the presence of

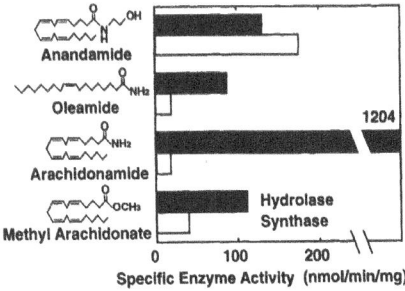

Figure 3. Substrate specificity of the recombinant enzyme. The particulate fraction of COS-7 cells overexpressing rat anandamide amidohydrolase was allowed to react with the indicated compounds (100 μM) for the hydrolytic activity. For the synthetic activity the enzyme was incubated with free fatty acids (200 μM) in the presence of 250 mM ethanolamine, 1 M NH4Cl or 2 M methanol.

endogenous inhibitory factors in the homogenate. The heat-treated homogenate showed the inhibitory effect, suggesting that the inhibitors were not proteins.

The homogenate was extracted with acetone, and the extract was tested for its inhibitory effect on the anandamide hydrolysis and synthesis. The acetone extract inhibited the hydrolase activity dose-dependently, but the synthase activity was inhibited to a lesser extent. In order to identify the inhibitory factors, we subjected the acetone extract to TLC, and scraped various lipid bands from silica gel. Then, we eluted the lipids from silica gel with methanol, and tested for the inhibitory effects. The bands corresponding to free fatty acids, monoacylglycerols and polar lipids inhibited the hydrolase and synthase activities. In agreement with this result, when pure free oleic acid, 2-arachidonoyl glycerol or phosphatidylcholine were included in the reaction mixture, the anandamide hydrolase and synthase activities of the liver microsome were reduced depending on the concentrations of these lipids. We should note that the hydrolase activity was inhibited more potently than the synthase activity.

We attempted to remove such endogenous lipid inhibitors from the enzyme preparation, and treated the homogenate of small intestine with 90% cold acetone. The resultant pellet was resuspended in Tris-HCl buffer. When we compared the enzyme activities in the homogenate of small intestine before and after the acetone precipitation, the specific hydrolase activity was increased about 10 fold by the acetone treatment. We, therefore, re-examined the organ distribution of the hydrolase and synthase activities with the acetone-treated homogenates. As shown in Figure 4b, small intestine now showed the hydrolase activity as high as 2.0 nmol/min/mg protein, second only to liver. Stomach and colon also showed considerable activities of the hydrolase. Previously, Desarnaud et al. reported that the hydrolase activity of rat was high in liver and brain, but low in small intestine and stomach.[8] The apparent low enzyme activity is presumably attributable to the endogenous lipid inhibitors.

We also examined distribution of mRNA of the enzyme in rat organs. Twenty-five μg of total RNA isolated from various organs was subjected to Northern blotting with ^{32}P-labeled cDNA of rat anandamide amidohydrolase as a probe. An intense radioactive band around 2.5 kb was detected in small intestine and stomach as well as liver. Faint bands were observed at the same position in cerebrum, testis, parotid gland, kidney, submaxillary gland, and spleen. The identity of the small intestine enzyme with the liver enzyme may be supported by this Northern blot analysis.

We applied this Northern blot analysis to various mammalian cell lines. Rat pheochromocytoma PC-12 and rat basophilic leukemia cells RBL-1 and RBL-2H3 gave mRNA at the same position as mRNA of rat liver. Although PC-12 cells differentiate to

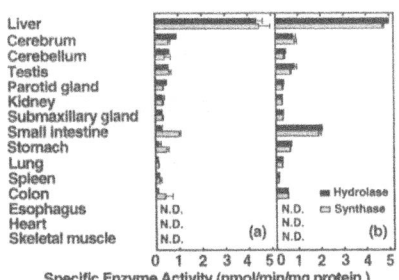

Figure 4. Organ distribution of anandamide amidohydrolase in rats. Homogenates of the indicated organs (a) and proteins precipitated with 90% acetone from the homogenates (b) were assayed for the anandamide hydrolase and synthase activities. N.D., not detectable.

neuron-like cells in response to nerve growth factor, the mRNA level was not changed by N18TG2 and human breast cancer cells EFM-19, but not in mouse macrophage J774 cells. These results were consistent with specific activities of anandamide amidohydrolase in these cell lines.[9,10]

We were also interested in the distribution of anandamide amidohydrolase in ocular tissues since anandamide was reported to reduce intraocular pressure in rabbit eyes.[11] When the homogenates of various tissues of porcine eyes were assayed, retina, choroid, iris, optic nerve and lacrimal gland showed high specific activities of the hydrolase and synthase comparable to those in porcine brain.[12] In contrast, lens did not show either the hydrolase or the synthase activity.

SUMMARY

Previously we suggested that anandamide amidohydrolase partially purified from porcine brain catalyzed the anandamide synthesis. The reversibility of the anandamide hydrolytic reaction was confirmed with a recombinant enzyme of rat liver. We also showed that the recombinant enzyme had a wide substrate specificity hydrolyzing primary amides and esters of fatty acids in addition to anandamide. When the organ distribution of anandamide amidohydrolase was examined with rats, a large amount of the enzyme was contained in small intestine as well as liver and brain. The intestinal hydrolase was masked by endogenous lipid inhibitors. The enzyme was also found in various eye tissues.

ACKNOWLEDGMENTS

This work was supported by grants-in-aid from the Ministry of Education, Science, Sports and Culture of Japan, the Human Frontiers in Science Program, the Japanese Foundation of Metabolism and Disease, and Ono Medical Research Foundation.

REFERENCES

1. D.G. Deutsch and S.A. Chin, Enzymatic synthesis and degradation of anandamide, a cannabinoid receptor agonist, *Biochem. Pharmacol.* 46:791 (1993).
2. N. Ueda, Y. Kurahashi, S. Yamamoto, and T. Tokunaga, Partial purification and characterization of the porcine brain enzyme hydrolyzing and synthesizing anandamide, *J. Biol. Chem.* 270:23823 (1995).
3. B.F. Cravatt, O. Prospero-Garcia, G. Siuzdak, N.B. Gilula, S.J. Henriksen, D.L. Boger, and R.A. Lerner, Chemical characterization of a family of brain lipids that induce sleep, *Science* 268:1506 (1995).
4. B.F. Cravatt, D.K. Giang, S.P. Mayfield, D.L. Boger, R.A. Lerner, and N.B. Gilula, Molecular characterization of an enzyme that degrades neuromodulatory fatty-acid amides, *Nature* 384:83 (1996).
5. Y. Kurahashi, N. Ueda, H. Suzuki, M. Suzuki, and S. Yamamoto, Reversible hydrolysis and synthesis of anandamide demonstrated by recombinant rat fatty-acid amide hydrolase, *Biochem. Biophys. Res. Commun.* 237:512 (1997).
6. G. Arreaza, W.A. Devane, R.L. Omeir, G. Sajnani, J. Kunz, B.F. Cravatt, and D.G. Deutsch, The cloned rat hydrolytic enzyme responsible for the breakdown of anandamide also catalyzes its formation via the condensation of arachidonic acid and ethanolamine, *Neurosci. Lett.* 234:59 (1997).
7. K. Katayama, N. Ueda, Y. Kurahashi, H. Suzuki, S. Yamamoto, and I. Kato, Distribution of anandamide

amidohydrolase in rat tissues with special reference to small intestine, *Biochim. Biophys. Acta* 1347:212 (1997).

8. F. Desarnaud, H. Cadas, and D. Piomelli, Anandamide amidohydrolase activity in rat brain microsomes Identification and partial characterization, *J. Biol. Chem.* 270:6030 (1995).

9. S. Maurelli, T. Bisogno, L. De Petrocellis, A. Di Luccia, G. Marino, and V. Di Marzo, Two novel classes of neuroactive fatty acid amides are substrates for mouse neuroblastoma 'anandamide amidohydrolase', *FEBS Lett.* 377:82 (1995).

10. T. Bisogno, S. Maurelli, D. Melck, L. De Petrocellis, and V. Di Marzo, Biosynthesis, uptake, and degradation of anandamide and palmitoylethanolamide in leukocytes, *J. Biol. Chem.* 272:3315 (1997).

11. Y. Mikawa, S. Matsuda, T. Kanagawa, T. Tajika, N. Ueda, and Y. Mimura, Ocular activity of topically administered anandamide in the rabbit, *Jpn. J. Ophthalmol.* 41:217 (1997).

12. S. Matsuda, N. Kanemitsu, A. Nakamura, Y. Mimura, N. Ueda, Y. Kurahashi, and S. Yamamoto, Metabolism of anandamide, an endogenous cannabinoid receptor ligand, in porcine ocular tissues, *Exp. Eye Res.* 64:707 (1997).

76

A PATHWAY FOR THE BIOSYNTHESIS OF FATTY ACID AMIDES

Kathleen A. Merkler,[1] Laura E. Baumgart,[1] Jodi L. DeBlassio,[1]
Uta Glufke[2], Lawrence King III,[1] Kimberly Ritenour-Rodgers,[1]
John C. Vederas,[2] Benjamin J. Wilcox,[1] and David J. Merkler[1,*]

[1]Department of Chemistry and Biochemistry
Duquesne University
Pittsburgh, PA 15282-1530

[2]Department of Chemistry
University of Alberta
Edmonton, Canada T6G 2G2

INTRODUCTION

Fatty acid primary amides are potent, bioactive molecules. Oleamide induces physiological sleep[1,2], potentiates the action of 5-hydroxytryptamine on rat brain 5-HT_{2A} and 5-HT_{2C} receptors[3], and blocks gap junction communication in glial cells[4]. Erucamide stimulates angiogenesis[5], arachidonamide is a tight-binding inhibitor of human synovial phospholipase A_2[6], and valpromide (n-propylpentanamide) is used clinically as an antiepileptic[7]. These studies indicate that the fatty acid primary amides are an exciting new class of mammalian, bioactive lipid. The biosynthetic pathway leading to the formation of the fatty acid primary amides is not understood and is information critical to the understanding and control of diseases related to their production.

The biosynthesis of another class of amidated hormones, the α-amidated neuropeptides, is well characterized. Amidated peptides arise from the post-translational processing of large precursor polypeptides. Typically, a prohormone with a C-terminal glycine-extension is flanked by pairs of basic amino acids[8] and is excised by the sequential actions of a Kex2p-like endoprotease (PC2, PC3, or furin) and an exopeptidase, carboxypeptidase E[9]. Oxidative cleavage of the glycine-extended precursor by peptidylglycine α-amidating monooxygenase (PAM) produces the active, α-amidated peptide hormone[10,11].

We have recently demonstrated that PAM will cleave N-myristoylglycine to myristamide[12]. We further proposed that the fatty acid primary amides are produced in mammals by the sequential actions of acyl-CoA:glycine N-acyltransferase (ACGNAT) and PAM. The N-fatty acylglycines are first made from fatty acyl-CoA thioesters and glycine by ACGNAT[13,14], and then amidated by PAM yielding the corresponding fatty acid amides (Figure 1). Tissue distribution studies of ACGNAT in the Japanese monkey (Macaca fuscata)[15] and of PAM in humans[16] co-localize the two enzymes in the brain suggesting that the brain is the site of fatty acid amide biosynthesis. Another possibility is the synthesis of N-fatty acylglycines by ACGNAT in the liver[13,14], their transport into the blood[17,18], and then their amidation by the PAM present in serum[16]. In this report, we extend our initial observations and demonstrate that purified rat PAM will amidate a diversity of N-fatty acylglycines from N-formylglycine to N-linolenoylglycine. We also show that the N-fatty

---

*To whom correspondence should be addressed.

Eicosanoids and Other Bioactive Lipids in Cancer, Inflammation, and Radiation Injury, 4
Edited by Honn et al., Kluwer Academic / Plenum Publishers, New York, 1999.

519

acylglycines inhibit the human serum PAM-dependent amidation of dansyl-Tyr-Val-Gly to dansyl-Tyr-Val-NH$_2$. These data provide additional evidence that PAM is involved in the biosynthesis of the fatty acid primary amides.

MATERIALS AND METHODS

Materials

Dansyl-Tyr-Val-Gly and D-Tyr-Val-Gly were from Sigma, N-acetylglycine and N-octanoylglycine were from Aldrich, N-decanoylglycine and N-lauroylglycine were from Novabiochem, N-myristoylglycine was from BACHEM, N-isovaleroylglycine was from TCI America, and N-formylglycine was from Fluka. N-Propionylglycine, N-butyrylglycine and N-hexanoylglycine were synthesized from the appropriate anhydride and glycine in aqueous sodium hydroxide solution. N-Oleoylglycine and N-linolenoylglycine were prepared by coupling the corresponding fatty acid and glycine methyl ester in dichloromethane using benzotriazol-1-yloxy-trispyrrolidinophosphonium hexafluorophosphate with subsequent hydrolysis of the methyl ester by lithium hydroxide in aqueous tetrahydrofuran. All other chemicals were of the highest quality available from commercial sources.

Human serum was obtained from whole blood drawn from a healthy 22 year old female volunteer by venipuncture. Blood was allowed to clot and serum collected by centrifugation. Serum was stored frozen at -70 °C until use.

Peptidylglycine α-Amidating Monooxygenase

Chinese hamster ovary cells, which secrete recombinant type A rat medullary thyroid carcinoma PAM into the culture medium[19,20], were grown in a Cellco Cellmax-100 hollow fiber bioreactor. PAM was purified as described[20] except that the final Sephacyl S-300 step was carried out using 50 mM HEPES/NaOH pH 7.8, 20 mM NaCl.

Enzyme Activity

Amidation activity was measured using HPLC separation of dansyl-Tyr-Val-NH$_2$ from dansyl-Tyr-Val-Gly as described by Jones et al.[21]. Oxygen consumption was measured using a Yellow Springs Instrument Model 53 oxygen monitor. Glyoxylate was quantitated colorimetrically as described by Katopodis and May[22].

RESULTS AND DISCUSSION

Amidation of N-Fatty Acylglycines

Merkler et al.[12] first reported that recombinant rat PAM catalyzed the amidation of N-myristoylglycine to myristamide. As presented in Tables 1 and 2, we have found that PAM will amidate a diversity of N-acylglycines. The long chain N-fatty acylglycines have limited aqueous solubility. In our experience, N-lauorylglycine is the N-acylglycine with the longest acyl chain that could be prepared in an aqueous solution at pH 6.0. The amidation of N-fatty acylglycines with acyl chains of ≥13 carbon atoms required the addition of an organic cosolvent, 2.5% (v/v) DMSO, to solubilize the long chain N-fatty acylglycines at concentrations ≤300 μM. The amidation of long chain N-fatty acylglycines in 2.5% (v/v) DMSO is shown in Table 2.

The data in Tables 1 and 2 demonstrate that PAM will amidate a variety of N-acylglycines. The V/K for acylglycine amidation increases as the acyl chain length increases reaching a plateau value of approximately 1.5×10^5 M^{-1} s^{-1} at N-decanoylglycine. The long chain N-fatty acylglycines are respectable substrates with V/K values ~4-fold higher than D-Tyr-Val-Gly. Two of the acylglycines amidated by PAM are of particular significance. N-Oleoylglycine would be the precursor for oleamide if PAM is demonstrated to have a role in fatty acid amide biosynthesis and N-isovaleroylglycine has been detected in the plasma of a patient suffering from isovaleric acidemia[17].

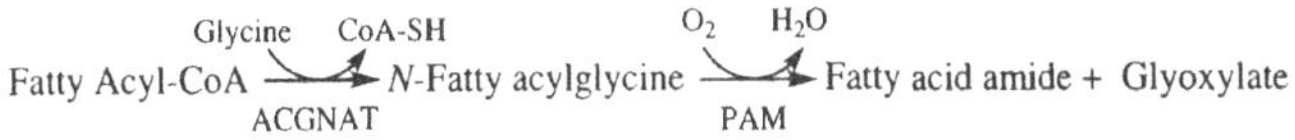

Glycine   CoA-SH O_2 H_2O

Fatty Acyl-CoA $\longrightarrow$ N-Fatty acylglycine $\longrightarrow$ Fatty acid amide + Glyoxylate

ACGNAT   PAM

Figure 1. The pathway proposed by Merkler et al.[12] for the biosynthesis of fatty acid primary amides

Table 1. The amidation of N-fatty acylglycines by purified rat PAM[1,2]

Substrate	$\underset{R}{\overset{O}{\|}}\!\!\!\!\!\overset{\displaystyle\|}{C}\!\!-\!\!\underset{H}{N}\!\!-\!\!COOH$	K_M (mM)	V_{MAX} (μmol/min/mg)	Relative V/K
D-Tyr-Val-Gly	R = D-Tyr-Val	0.22 ± 0.03	7.9 ± 0.8	52
N-Formylglycine	R = H	23 ± 0.9	8.2 ± 0.1	0.5
N-Acetylglycine	R = CH_3	9.3 ± 0.5	6.4 ± 0.2	1.0
N-Propionylglycine	R = CH_3CH_2	2.5 ± 0.2	8.6 ± 0.3	5.0
N-Butyrylglycine	R = $CH_3(CH_2)_2$	2.0 ± 0.1	8.6 ± 0.1	6.2
N-Isovaleroylglycine	R = $(CH_3)_2CHCH_2$	2.0 ± 0.1	8.8 ± 0.2	6.4
N-Hexanoylglycine	R = $CH_3(CH_2)_4$	0.42 ± 0.06	8.0 ± 0.3	28
N-Octanoylglycine	R = $CH_3(CH_2)_6$	0.20 ± 0.01	10.4 ± 0.2	76
N-Decanoylglycine	R = $CH_3(CH_2)_8$	0.10 ± 0.004	12.6 ± 0.2	177
N-Lauroylglycine	R = $CH_3(CH_2)_{10}$	0.061 ± 0.003	7.9 ± 0.2	190

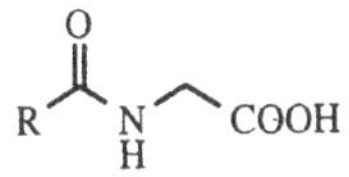

[1]Reactions were carried out at 37 °C in 100 mM MES/NaOH pH 6.0, 30 mM NaCl, 1.0% (v/v) ethanol, 0.001% (v/v) Triton X-100, 5.0 mM ascorbate, 1.0 μM $Cu(NO_3)_2$, 10 μg/ml catalase, and the indicated N-acylglycine (0.2K_M to 3.0K_M). Initial rates were measured by following the PAM-dependent consumption of O_2.
[2]Kinetic constants ± standard error are from computer fits of the initial rate data to the Michaelis-Menton equation.

Table 2. The amidation of long chain N-fatty acylglycines by purified rat PAM in DMSO[1,2]

Substrate		K_M (mM)	V_{MAX} (μmol/min/mg)	Relative V/K
N-Acetylglycine	R = CH$_3$	14.9 ± 1.4	7.5 ± 0.2	1.0[3]
N-Lauroylglycine	R = CH$_3$(CH$_2$)$_{10}$	0.11 ± 0.01	12.3 ± 0.8	200[3]
N-Myristoylglycine	R = CH$_3$(CH$_2$)$_{12}$	0.071 ± 0.008	9.2 ± 0.5	260
N-Oleoylglycine[4]	R = CH$_3$(CH$_2$)$_7$CH=CH(CH$_2$)$_7$	0.094 ± 0.034	11.0 ± 2.4	230
N-Linolenoylglycine[4]	R = CH$_3$(CH$_2$CH=CH)$_3$CH$_2$(CH$_2$)$_6$	0.069 ± 0.011	7.5 ± 0.8	220

The substrate structure is represented by R–C(=O)–N(H)–CH$_2$–COOH.

[1]Reactions were carried out at 37 °C in 100 mM MES/NaOH pH 6.0, 30 mM NaCl, 1.0% (v/v) ethanol, 0.001% (v/v) Triton X-100, 2.5% (v/v) DMSO, 5.0 mM ascorbate, 1.0 μM Cu(NO$_3$)$_2$, 10 μg/ml catalase, and the indicated N-acylglycine (0.2K_M to 1.5K_M). Initial rates were measured by following the PAM-dependent consumption of O$_2$.
[2]Kinetic constants ± standard error are from computer fits of the initial rate data to the Michaelis-Menton equation.
[3]DMSO has a small negative effect on the V/K for amidation. In the presence of 2.5% (v/v) DMSO, the V/K for N-acetylglycine amidation decreases 17% from 860 M^{-1}s^{-1} to 630 M^{-1}s^{-1} and the V/K for N-lauroylglycine amidation decreases 21% from 1.6 × 10^5 M^{-1}s^{-1} to 1.3 × 10^5 M^{-1}s^{-1} (compare relevant data in Tables 1 and 2).
[4]N-Oleoylglycine and N-linolenoylglycine show substrate inhibition at ≥150 μM (data not shown). The substrate inhibition is likely the result of the formation of micelles. The critical micellar concentration for N-oleoylglycine and N-linolenoylglycine is not known.

The PAM Catalyzed Production of Glyoxylate from N-Hexanoylglycine

The data in Tables 1 and 2 show only that there is an PAM-dependent consumption of O$_2$ in the presence of the N-fatty acylglycines. These data do not prove that the corresponding fatty acid amide is a product of the PAM catalyzed reaction. The PAM-dependent formation of glyoxylate provides a general method to demonstrate the oxidation cleavage of any N-fatty acylglycine (Figure 1). Glyoxylate production from N-hexanoylglycine is illustrated in Figure 2.

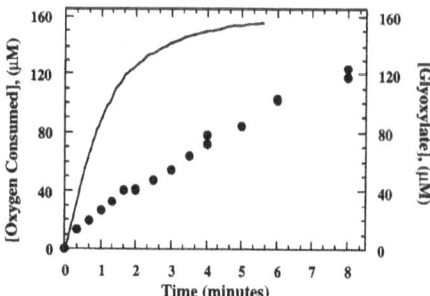

Figure 2. The PAM catalyzed consumption of O$_2$ (ff) and the production of glyoxylate (●) from N-hexanoylglycine. Reactions at 37 °C were initiated by the addition of 106 μg of purified rat PAM to 3.0 ml of 100 mM MES/NaOH pH 6.0, 30 mM NaCl, 1% (v/v) ethanol, 0.001% (v/v) Triton X-100, 10 μg/ml catalase, 1.0 μM Cu(NO$_3$)$_2$, 5.0 mM ascorbate, and 200 μM N-hexanoylglycine. In one sample, O$_2$ consumption was measured with an O$_2$ electrode. In a matched second sample, aliquots (100 μl) were removed at the indicated time and added to a vial containing 20 μl of 6% (v/v) trifluoroacetic acid to quench the reaction. The concentration of glyoxylate in each quenched samples was determined using the procedure of Katopodis and May[22].

Similarly, glyoxylate production was found upon the amidation of *N*-acetylglycine, *N*-butyrylglycine and *N*-(*tert*-butoxycarbonyl)glycine (data not shown). In each case, the rate of glyoxylate formation was slower than the rate of O_2 consumption. This pattern of results indicates that there must be an initial formation of an intermediate from the acylglycine before glyoxylate is produced. Analogous to the work on the amidation of glycine-extended peptides[23,24], this intermediate is postulated to be *N*-acyl-α-hydroxyglycine (Figure 3).

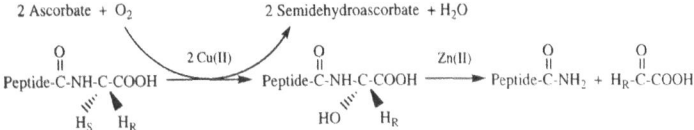

Figure 3. The peptide amidation reaction catalyzed by PAM. PAM is a bifunctional enzyme consisting of separate catalytic entities for peptide hydroxylation and carbinolamide dealkylation[24].

The Inhibition of Human Serum PAM by the N-Fatty Acylglycines

The HPLC separation of dansyl-Tyr-Val-Gly from dansyl-Tyr-Val-NH$_2$ provides a sensitive assay for peptide amidation activity[21]. Non-fluorescent substrates like the *N*-fatty acylglycines can be treated as competitive inhibitors because their binding to PAM will prevent the formation of dansyl-Tyr-Val-NH$_2$. As shown in Table 3, the N-fatty acylglycines inhibit crude human serum PAM (hsPAM). The binding of the N-fatty acylglycines to hsPAM provide strong support the pathway we have proposed for the biosynthesis of the the fatty acid primary amides (Figure 1).

Table 3. Comparison of inhibition constants (K_i values) measured using crude human serum PAM and Michealis constants (K_M values) measured using purified rat PAM[1]

N-Acylglycine	K_i	K_M
	$(mM)^{2,4}$	$(mM)^{3,4}$
N-Acetylglycine	7.0 ± 0.9	10.3 ± 0.5
N-Isovaleroylglycine	2.0 ± 0.2	3.8 ± 0.3
N-Lauroylglycine	0.18 ± 0.02	0.12 ± 0.01
N-Oleoylglycine	0.36 ± 0.04	~0.2[5]

[1]These experiments were carried out at the optimal reaction conditions described for crude human serum PAM[16]. The optimal reaction conditions for purified rat PAM[20] are different (see legends to Table 1 and Figure 2). As a consequence, the K_M values listed here differ slightly from the values shown in Table 1.
[2]Reactions were carried out at 37 °C in 50 mM HEPES/NaOH pH 7.4, 1.2 mM ascorbate, 10 μM Cu(NO$_3$)$_2$, 60 μg/ml catalase, 4.0 μM dansyl-Tyr-Val-Gly, 6.1 mg/ml human serum, and the indicated *N*-acylglycine ($0.4K_i$ to $3.0K_i$). Initial rates were measured by following the hsPAM-dependent amidation of dansyl-Tyr-Val-Gly by HPLC[22]. Inhibition constants ± standard errors were determined by Dixon analysis.
[3]Reactions were carried out at 37 °C in 50 mM HEPES/NaOH pH 7.4, 1.2 mM ascorbate, 1.0 μM Cu(NO$_3$)$_2$, 60 μg/ml catalase, 10-40 μg/ml rat PAM, and the and the indicated *N*-acylglycine ($0.4K_M$ to $10K_M$). Initial rates were measured by following the rat PAM-dependent consumption of O_2. Michealis constants ± standard error are from computer fits of the initial rate data to the Michaelis-Menton equation.
[4]The specific activity of purified rat PAM is ~400,000-fold higher than that of crude human serum PAM.
[5]Estimated from the data in Table 2. The ratio of the K_M values in Table 1 to those in Table 3 for *N*-isovaleroylglycine and *N*-lauroylglycine is ~2.0.

CONCLUSIONS

The data presented here show that PAM will amidate a diversity of N-fatty acylglycines *in vitro*. The V/K for N-acylglycine amidation increases ~400-fold as the length of the acyl chain increases from zero carbon atoms (N-formylglycine) to eleven carbon atoms (N-lauroylglycine). The V/K for long-chain N-fatty acylglycine amidation is approximately constant with a value that is ~200-fold higher than that for N-acetylglycine and ~4-fold higher than that for D-Tyr-Val-Gly (Tables 1 and 2). These data indicate that the long-chain N-fatty acylglycines are efficiently amidated by PAM, consistent with the pathway outlined in Figure 1. The glycine-conjugated precursor to oleamide (N-oleoylglycine) is a good substrate for PAM (Table 2) while the glycine-conjugated precursors to the other known fatty acid amides, erucamide (N-erucoylglycine) and arachidonamide (N-arachidonoylglycine), are anticipated to also be good substrates for PAM. Given the variety of N-acylglycines synthesized by ACGNAT[25] and recent results from our laboratory demonstrating that long-chain acyl-CoA thioesters are ACGNAT substrates (Baumgart, Merkler, and Merkler, unpublished), the sequential actions of ACGNAT and PAM could produce the fatty acid amides isolated from mammalian sources to date and suggests that other fatty acid amides await discovery.

The data in Table 3 indicate that the N-fatty acylglycines inhibit hsPAM. The agreement between the K_i values measured using hsPAM and the K_M values measured using purified rat PAM suggests that the inhibition of hsPAM is a consequence of their binding to the human enzyme and is not an artifact resulting from the use of serum. There are two general methods available to measure PAM activity: monitoring the consumption of O_2 or monitoring the production of glyoxylate (Figure 3). Neither method is sufficiently sensitive to measure the amidation an N-fatty acylglycine in serum (DeBlassio and Merkler, unpublished). The binding of N-fatty acylglycines to hsPAM suggests that an assay of sufficient sensitivity will measure N-fatty acylglycine amidation in serum and tissue extracts.

ACKNOWLEDGMENTS

This research was supported by grants from the Human Growth Foundation, the Hunkele Foundation, and the Duquesne University Faculty Development Fund to D.J.M., a Bayer Undergraduate Summer Research Fellowship to B.J.W., a grant from the Natural Sciences and Engineering Research Council of Canada to J.C.V., and a Deutsche Forschungsgemeindschaft Fellowship to U.G.

REFERENCES

1. R.A. Lerner, G. Siuzdak, O. Prospero-Garcia, S.J. Hendriksen, D.L. Boger, and B.F. Cravatt, Cerebrodiene: a brain lipid isolated from sleep-deprived cats, *Proc. Natl. Acad. Sci USA* 91:9505 (1994).
2. B.F. Cravatt, O. Prospero-Garcia, G. Siuzdak, N.B. Gilula, S.J. Henriksen, D.L. Boger, and R.A. Lerner, Chemical characterization of a family of brain lipids that induce sleep, *Science* 268:1506 (1995).
3. J.F. Hiudobro-Toro and R.A. Harris, Brain lipids that induce sleep are novel modulators of 5-hydroxytryptamine receptors, *Proc. Natl. Acad. Sci. USA* 93: 8078 (1996).
4. X. Guan, B.F. Cravatt, G.R. Ehring, J.E. Hall, D.L. Boger, R.A. Lerner, and N.B. Gilula, The sleep-inducing lipid oleamide deconvolutes gap junction communication and calcium wave transmission in glial ells, *J. Cell Biol.* 139:1785 (1997).
5. K. Wakamatsu, T. Masaki, F. Itoh, K. Kondo, and K. Sudo, Isolation of fatty acid amide as an angiogenic principle from bovine mesentery, *Biochem. Biophys. Res. Commun.* 168:423 (1990).
6. M.K. Jain, F. Ghomashchi, B.-Z. Yu, T. Bayburt, D. Murphy, D. Houck, J. Brownell, J.C. Reid, J.E. Solowiej, S.-M. Wong, U. Mocek, R. Jarrell, M. Sasser, and M.H. Gelb, Fatty acid amides: scooting mode-based discovery of tight-binding inhibitors of secreted phospholipase A_2, *J. Med. Chem.* 35:3584 (1992).

7. M. Bialer, Clinical pharmacology of valpromide, *Clin. Pharmacokin.* 20:114 (1991).
8. R.E. Mains, B.A. Eipper, C.C. Glembotski, and R.M. Dores, Strategies for the biosynthesis of bioactive peptides, *Trends Neurosci.* 6:229 (1983).
9. S.P. Smeekens, S.J. Chan, and D.F. Steiner, The biosynthesis and processing of neuroendocrine peptides, *Prog. Brain Res.* 92:235 (1992).
10. A.F. Bradbury, M.D.A. Finnie, and D.G. Smyth, Mechanism of C-terminal amide formation by pituitary enzymes, *Nature* 298:686 (1982).
11. B.A. Eipper, R.E. Mains, and C.C. Glembotski, Identification in pituitary tissue of a peptide α-amidation activity that acts on glycine-extended peptides and requires molecular oxygen, copper, and ascorbic acid, *Proc. Natl. Acad. Sci. USA* 80:5144 (1983).
12. D.J. Merkler, K.A. Merkler, W. Stern, and F.F. Fleming, Fatty acid amide biosynthesis: a possible new role for peptidylglycine α-amidating enzyme and acyl-coenzyme A:glycine N-acyltransferase, *Arch. Biochem. Biophys.* 330:430 (1996).
13. D. Schachter, and J.V. Taggart, Glycine N-acylase: purification and properities, *J. Biol. Chem.* 208:263 (1954).
14. N. Gregersen, S. Kølvraa, and P.B. Mortensen, Acyl-CoA:glycine N-acyltransferase: *in vitro* studies on the glycine conjugations of straight- and branched-chain acyl-CoA esters in human liver, *Biochem. Med. Metabol. Biol.* 35:210 (1986).
15. K. Asaoka, Enzymes that metabolize acyl-coenzyme A in the monkey - their distribution, properties, and roles in an alternative pathway for the excretion of nitrogen, *Int. J. Biochem.* 23:429 (1991).
16. G.S. Wand, R.L. Ney, S. Baylin, B. Eipper, and R.E. Mains, Characterization of a peptide α-amidation activity in human plasma and tissues, *Metabolism* 34:1044 (1985).
17. C. de Sousa, R.A. Chalmers, T.E. Stacey, B.M. Tracey, C.M. Weaver, and D. Bradley, The response to L-carnitine and glycine therapy in isovaleric acidaemia, *Eur. J. Pediatr.* 144:451 (1986).
18. Z. Gregus, T. Fekete, F. Varga, and C.D. Klaassen, Availability of glycine and coenzyme A limits glycine conjugation *in vivo*, *Drug Metab. Dispos.* 20:234 (1992).
19. A.H. Bertelsen, G.A. Beaudry, E.A. Galella, B.N. Jones, M.L. Ray, and N.M. Mehta, Cloning and characterization of two alternatively spliced rat α-amidating cDNAs from rat medullary thyroid carcinoma, *Arch. Biochem. Biophys.* 279:87 (1990).
20. D.A. Miller, K.U. Sayad, R. Kulathila, G.A. Beaudry, D.J. Merkler, and A.H. Bertelsen, Characterization of a bifunctional peptidylglycine α-amidating enzyme expressed in Chinese hamster ovary cells, *Arch. Biochem. Biophys.* 298:380 (1992).
21. B.N. Jones, P.P. Tamburini, A.P. Consalvo, S.D. Young, S.J. Lovato, J.P. Gilligan, A.Y. Jeng, and L.P. Wennogle, A fluorometric assay for peptidyl α-amidation activity using high-performance liquid chromatography, *Anal. Biochem.* 168:272 (1988).
22. A.G. Katopodis, and S.W. May, Novel substrates and inhibitors of peptidylglycine α-amidating monooxygenase, *Biochemistry* 29:4541 (1990).
23. S.D. Young and P.P. Tamburini, Enzymatic peptidyl α-amidation proceeds through formation of an α-hydroxyglycine intermediate, *J. Am. Chem. Soc.* 111:1933 (1989).
24. A.G. Katopodis, D. Ping, and S.W. May, A novel enzyme from bovine neurointermediate pituitary catalyzes dealkylation of α-hydroxyglycine derivatives, thereby functioning sequentially with peptidylglycine α-amidating monooxygenase in peptide amidation, *Biochemistry* 29:6115 (1990).
25. P.T. Ozand and G.G. Gascon, Organic acidurias: a review, *J. Child Neurol.* 6:288 (1991).

77

INVESTIGATION OF STRUCTURAL ANALOGS OF PROSTAGLANDIN AMIDES FOR BINDING TO AND ACTIVATION OF CB₁ AND CB₂ CANNABINOID RECEPTORS IN RAT BRAIN AND HUMAN TONSILS

Barbara A. Berglund[1], Daniel L. Boring[*], and Allyn C. Howlett[1]

[1]Department of Pharmacological and Physiological Science
Saint Louis University School of Medicine
1402 South Grand
St. Louis, MO 63104 USA

[*]Division of New Drug Chemistry III
Center for Drug Evaluation and Research
Food and Drug Administration
Rockville, MD 20857 USA

INTRODUCTION

Arachidonyl ethanolamide (anandamide, AEA) is an endogenously produced lipid amide which binds to and activates the CB_1 and CB_2 cannabinoid receptors. Further, administration of arachidonyl ethanolamide to animals produces many of the same physiological effects as classical and non-classical cannabinoids[1,2]. The CB_1 cannabinoid receptors are found in the brain and central nervous system; the CB_2 cannabinoid receptors are found in immune cells and tissue. Because arachidonyl ethanolamide is an arachidonic acid derivative, it is possible that it may be metabolized to a prostaglandin derivative. The carboxylic acid moiety is not necessary for binding to cyclooyxgenase, the enzyme which begins the arachidonic acid cascade[3,4], suggesting that an ethanol amide derivative might be possible as a substrate for this enzyme. This lends potential physiological relevance to a prostaglandin amide.

Previous studies have shown little ability of prostaglandins to bind to or activate the CB_1 cannabinoid receptor[5]. The binding properties of many arachidonyl ethanolamide derivatives have been characterized for the CB_1 and CB_2 cannabinoid receptors[5-11]. Based on these binding data as well as binding data for other classical and non-classical cannabinoids, a model was prepared to define the necessary pharmacophoric elements and conformations of cannabinoid ligands for binding to the cannabinoid receptor[12]. The prostaglandin amide series contains the necessary pharmacophoric elements but is conformationally restricted due to the cyclopentyl moiety. Previous prostaglandin ligands tested for cannabinoid activity were not α-alkylamide derivatives and did not effectively bind to the CB_1 cannabinoid receptor[6]. This study shows that α-alkylamide substitution enhances binding affinity compared with prostaglandins, but not to an affinity comparable to arachidonyl ethanolamide. Further the prostaglandin E_2 amides were able to activate G proteins, including Gs, the G protein which stimulates adenylyl cyclase. These signal transduction events are probably not mediated by cannabinoid receptors.

EXPERIMENTAL PROCEDURES

Chemistry

The prostaglandin amides were prepared by placing 1 equivalent of either PGE_2 or $PGF_{2\alpha}$ and 1.5 equivalents of triethylamine in dry methylene chloride, followed by cooling to -10°C under a blanket of argon for 15 min. With continuous stirring, 1.5 equivalents of ethylchloroformate were added all at once using a syringe. This was maintained at -10°C for 30 min, then 1.5 equivalents of the desired amine was added all at once using a syringe. This was allowed to warm to room temperature over a 15 min period. Saturated aqueous NH_4Cl was added to quench the reaction, then the mixture was vigorously shaken. This was extracted twice with ethyl acetate, the organic fractions combined and dried over anhydrous Na_2SO_4, and the solvent removed by rotary evaporation. The residue was chromatographed on silica gel using chloroform:methanol: ammonium hydroxide (90:10:1). The structures were verified by a combination of ^{1}H-NMR, chemical ionization mass spectrometry and Fourier transform IR spectroscopic techniques. The purity was demonstrated by a single spot on a TLC plate using two eluents. Compounds were stored as 100 mM solutions in ethanol under N_2 gas at -80°C. Multiple small aliquots of stock solutions were stored for a single use.

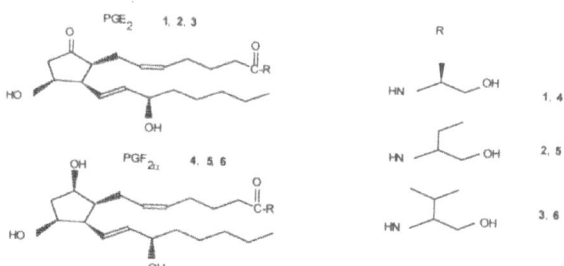

Figure 1. Structures of prostaglandin amides.

Cannabinoid receptor binding assays

Ligand binding to the CB_1 cannabinoid receptor was measured as described[6]. Ligand binding to the CB_2 cannabinoid receptor was measured in human tonsil membranes as described[7]. Rat brain and human tonsil membranes were pretreated with 100 μM aminoethylbenzene sulfonylfluoride (AEBSF), a serine amidase inhibitor, prior to assay. Data were analyzed by non-linear least squares regression analysis using Graphpad Inplot, and K_i values were determined from IC_{50} values for simple competitive inhibition using 350 pM as the K_d for [^{3}H]CP-55,940 for CB_1 and 8 nM as the K_d for [^{3}H]CP-55940 for CB_2.

GTPγS binding

Activation of G proteins in rat brain membranes using [^{35}S]GTPγS binding was measured as described[7]. Activitation of G proteins is expressed as pmol/mg.

Adenylyl cyclase determination in N18TG2 membranes

Activation or inhibition of adenylyl cyclase was measured as described[13] using 100 μM Ro20-1724 as the phosphodiesterase inhibitor. Results were calculated as the % inhibition or activation of forskolin-stimulated or basal activity at the indicated concentrations of compounds. Data from three individual experiments were averaged and non-linear regression analysis was used to determine the EC_{50} values using Graphpad Inplot.

RESULTS

None of the prostaglandin amides bound with high affinity to the CB_1 cannabinoid receptor (Fig 2A). Isopropyl derivatives of neither PGE_2 and $PGF_{2\alpha}$ (3 and 6) showed any binding even at 100 μM. The methylethanolamide derivative of PGE_2 (1) exhibited the highest affinity binding (K_i = 29 μM); the ethylethanolamide derivative of $PGF_{2\alpha}$ (5) was the best binder in that series (K_i = 40 μM).

Each of the prostaglandin amides bound with some affinity to the CB_2 cannabinoid receptor in human tonsils (Fig 2B). In this case, the compound with the highest affinity was the ethanolamide derivative of $PGF_{2\alpha}$ (4), which had a K_i of 733 nM. The highest affinity ligand of the PGE_2 series was also the ethanolamide derivative, 1. Of the $PGF_{2\alpha}$ derivatives, the lowest affinity binder was the isopropyl derivative (6) (K_i = 3.2 μM); for the PGE_2 series the isopropyl derivative was also the lowest affinity ligand (3), with a K_i of 7.1 μM.

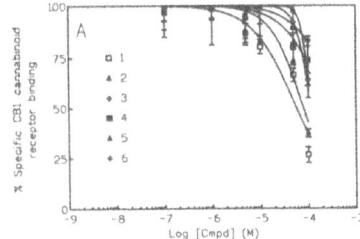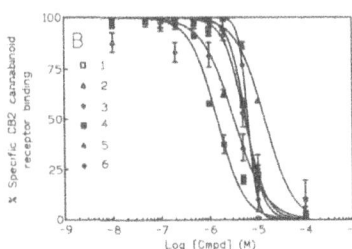

Figure 2. Prostaglandin amide affinity for CB_1 and CB_2 cannabinoid receptors.

The three PGE_2 amide derivatives were able to effectively activate G proteins in rat brain membranes (Table 1), with maximal effects seen at 1 μM. At these concentrations, there was little or no binding to the CB_1 cannabinoid receptor (in brain) (Fig 1). Not all of the $PGF_{2\alpha}$ amide derivatives were capable of G protein activation; only the ethanolamide derivative (4), at 1 μM, had any G protein activity. The action of the four compounds which were able to activate G proteins (1, 2, 3, and 4) were not inhibited by addition of the cannabinoid antagonist, SR-141716A (Table 1). This indicates that the nature of this interaction is not related to the CB_1 cannabinoid receptor.

Table 1. Activation of G proteins in rat brain membranes.

	GTPγS Binding			
Compound	100 nM	1 μM	1 μM + 10 nM SR	1 μM + 100 nM SR
Vehicle	0.79	0.85	0.80	0.81
PGE_1	3.66	3.72	3.70	3.71
PGE_2	2.11	2.31	2.08	2.33
1	1.34	2.24	2.28	2.17
2	2.26	2.81	2.84	2.82
3	1.81	2.16	2.00	2.11
4	1.02	1.31	1.32	1.30
DALN	-	2.87	0.95	0.90

The CB_1 cannabinoid receptor interacts with G_i, the G protein that inhibits adenylyl cyclase. In neuroblastoma cells, prostaglandins ($I_2 > E_1 > E_2$) stimulate cAMP production[13], and cannabinoid agonists inhibit cAMP produced by hormonal stimulation[6]. The prostaglandin amides did not inhibit forskolin-stimulated cAMP production, and in fact increased the production of cAMP (Table 2). PGE_2 and PGE_1 (10 μM) were also examined for comparison; these have been shown previously to not bind to the CB_1 cannabinoid receptor[5] but to activate adenylyl cyclase in neuroblastoma cells[13]. At 10 μM, analogs 1- 4 produced a stimulation of cAMP over forskolin-stimulated adenylyl cyclase ranging from 19% to 25%.

Table 2. Adenylyl cyclase activity in N18TG2 neuroblastoma cells.

		Adenylyl cyclase activity pmol/mg·min		% Stimulation of cAMP compared to vehicle	
Compound		(-)forskolin	(+)forskolin	(-)forskolin	(+)forskolin
vehicle		8.69	39.9	100	100
PGE_1		47.5	79.9	326	200
PGE_2		10.7	56.1	123	141
1		7.4	47.9	85	120
2		10.8	49.7	124	124
3		9.75	49.9	112	125
4		7.59	47.6	87	119
DALN		8.18	23	94	58

DISCUSSION

The prostaglandin amides bound more effectively to the CB_1 cannabinoid receptor than did the prostaglandins; previous work has shown that PGE_2 and $PGF_{2\alpha}$ do not bind to the CB_1 cannabinoid receptor[5]. This indicates that the alcohol amide substitution is necessary for CB_1 receptor binding in these constrained molecules. These compounds did not, however, bind with affinities comparable to arachidonyl ethanolamide. This indicates that perhaps the conformational restrictions of the prostaglandins prohibits high-affinity CB_1 cannabinoid receptor binding. This supports a conformational model[12] which reveals a region of steric hindrance when the prostaglandin amides are aligned with the pharmacophoric elements of hexahydrocannabinol, the prototypical cannabinoid used in the model.

The prostaglandin amides had at least one order of magnitude higher binding affinity for the CB_2 cannabinoid receptor than for the CB_1 cannabinoid receptor; however, the affinity was reduced by ten-fold compared to arachidonyl ethanolamide for the CB_2 cannabinoid receptor[7]. The structural requirements for the two cannabinoid receptors are not the same. This indicates that the prostaglandin amides contain the necessary pharmacophoric elements and perhaps some of the necessary conformational constraints which allow higher affinity binding to the receptor.

The PGE_2 prostaglandin derivatives and the ethanolamide $PGF_{2\alpha}$ derivative each activated G proteins. The PGE_2 derivatives increased the cAMP production by forskolin-stimuled neuroblastoma membranes. The IP_1 prostanoid receptor is present in the neuroblastoma cell line which was used in this study[14,15]; PGE_1 is an agonist for this receptor, and it produces an increase in cAMP[16,17]. In brain, other prostaglandin receptors include EP_3 receptor, which transduces its signal transduction through inhibition of cAMP and increase of PI turnover, and EP_4, which increases cAMP by increasing adenylyl cyclase activity[18]. The concentration at which the prostaglandin amides produce a stimulation correspond to a concentration at which competitive binding to the CB_1 receptor does not occur. It is possible that the prostaglandin derivatives are interacting with a different receptor to transduce the signal seen; perhaps the IP_1 prostanoid receptor is a candidate for this interaction. There is currently no IP_1 receptor antagonist; an assay using an antagonist to this receptor would help to prove the interaction was through this receptor.

The prostaglandin derivatives used in this study bind with relatively poor affinities to the CB_1 and CB_2 cannabinoid receptors, compared with arachidonyl ethanolamide, providing additional insight to conformational and pharmacophoric restrictions on binding to the cannabinoid receptors. However, these unique prostaglandin derivatives promote signal transduction via G protein; this may be due to an interaction with IP or EP receptors in the brain.

ACKNOWLEDGMENTS

DLB would like to acknowledge Mr. David Pate for the generous gift of the prostaglandins used as starting materials. BAB would like to acknowledge Mr. Gerald Wilken for performing the adenylyl cyclase assays.

REFERENCES

1. P.B. Smith, D.R. Compton, S.P. Welch, R.K. Razdan, R. Mechoulam, and B.R. Martin. The pharmacological activity of anandamide, a putative endogenous cannabinoid, in mice. *J. Pharmacol. Expt. Ther.* 270:219 (1994).

2. J. Wiley, R. Balster, and B. Martin. Discriminative stimulus effects of anandamide in rats. *Eur. J. Pharmacol.* 276:49 (1995).

3. I.K. Khanna, R.M. Weier, Y. Yu, P.W. Collins, J.M. Miyashiro, C.M. Koboldt, A.W. Veenhuizen, J.L. Currie, K. Siebert, and P.C. Isakson. 1,2-Diarylpyrroles as potent and selecitive inhibitors of cyclooxygenase-2. *J. Med. Chem.* 40:1619 (1997).

4. I.K. Khanna, R.M. Weier, Y. Yu, X.D. Xu, F.J. Koszyk, P.W. Collins, C.M. Koboldt, A.W. Veenhuizen, W.E. Perkins, J.J. Casler, J.L. Masferrer, Y.Y. Zhang, S.A. Gregory, K. Siebert, and P.C. Isakson. 1,2-Diarylimidazoles as potent, cyclooxygenase-2 selective, and orally active antiinflammatory agents. *J. Med. Chem.* 40:1634 (1997).

5. A.C. Howlett, D.M. Evans, and D.B. Houston. The cannabinoid receptor. In *Marihuana/Cannabinoids: Neurobiology and Neurophysiology.* L. Murphy, and A. Bartke, eds. (Boca Raton: CRC Press). 35. (1992).

6. J.C. Pinto, F. Potie, K.C. Rice, D. Boring, M.R. Johnson, D.M. Evans, G.H. Wilken, C.H. Cantrell, and A.C. Howlett. Cannabinoid receptor binding and agonist activity of amides and esters of arachidonic acid. *Mol. Pharmacol.* 46:516 (1994).

7. B.A. Berglund, D.L. Boring, G.H. Wilken, S. Lin, A. Makriyannis, and A.C. Howlett. Structural requirements for arachidonylethanolamide interaction with CB_1 and CB_2 cannabinoid receptors: Pharmacology of the carbonyl and ethanolamide groups. *Prostaglandins Leukot. Essent. Fatty Acids* (publ. submitted).

8. D.L. Boring, B.A. Berglund, and A.C. Howlett. Cerebrodiene, arachidonyl ethanolamide, and hybrid structures: potential for interaction with brain cannabinoid receptors. *Prostaglandins Leukot. Essent. Fatty Acids* 55:207 (1996).

9. I.B. Adams, W. Ryan, M. Singer, B.F. Thomas, D.R. Compton, R.K. Razdan, and B.R. Martin. Evaluation of cannabinoid receptor binding and in vivo activities for anandamide analogs. *J. Pharmacol. Expt. Ther.* 273:1172 (1995).

10. A.D. Khanolkar, V. Abadji, S. Lin, W.A.G. Hill, G. Taha, K. Abouzid, Z. Meng, P. Fan, and A. Makriyannis. Head group analogs of arachidonylethanolamide, the endogenous cannabinoid ligand. *J. Med. Chem.* 39:4515 (1996).

11. T. Sheskin, L. Hanus, J. Slager, Z. Vogel, and R. Mechoulam. Structural requirements for binding of anandamide-type compounds to the brain cannabinoid receptor. *J. Med. Chem.* 40:659 (1997).

12. W. Tong, E.R. Collantes, W.J. Welsh, B.A. Berglund, and A.C. Howlett. Derivation of a pharmacophore model for anandamide using constrained conformational searching and comparative molecular field analysis (CoMFA). *J. Med. Chem.* (in press).

13. A.C. Howlett. Stimulation of neuroblastoma adenylate cyclase by arachidonic acid metabolites. *Mol. Pharmacol.* 21:664 (1982).

14. H. Wise. Neuronal prostacyclin receptors. *Prog. Drug Res.* 49:123. (1997).

15. H. Wise, Y.M. Qian, and R.L. Jones. A study of prostacyclin mimetics distinguishes neuronal from neutrophil IP receptors. *Eur. J. Pharmacol.* 278:265. (1995).

16. T. Namba, H. Oida, Y. Sugimoto, A. Kakizuka, M. Negishi, A. Ichikawa, and S. Narumiya. CDNA cloning of a mouse prostacyclin receptor. Multiple signaling pathways and expression in thymic medulla. *J. Biol. Chem.* 269:9986. (1994).

17. D. Oliva and S. Nicosia. PGI2-receptors and molecular mechanisms in platelets and vasculature: state of the art. *Pharmcol. Res. Commun.* (1987).

18. R.A. Coleman et al. Classification of prostanoid receptors. In: *IUPHAR receptor compendium.* IUPHAR, eds. (1998).

78

HEPOXILIN A3 IS METABOLIZED INTO ITS ω-HYDROXY METABOLITE BY HUMAN NEUTROPHILS

Cecil R. Pace-Asciak,[1,2] Denis Reynaud,[1] Olga Rounova,[1]
Peter Demin,[1,3] and Kazimir K. Pivnitsky[3]

[1]Research Institute, Hospital for Sick Children,
555 University Avenue, Toronto, CANADA M5G 1X8,
[2]Department of Pharmacology, Faculty of Medicine,
University of Toronto, CANADA M5S 1A8, and
[3]N.D. Zelinsky Institute of Organic Chemistry, Russian
Academy of Sciences, 117913 Moscow, RUSSIAN FEDERATION

INTRODUCTION

Hepoxilins (Hx) elicit a variety of biological actions based on their ability to affect ion movements in the cell [Pace-Asciak 1994; Pace-Asciak et al., 1995a; Pace-Asciak et al., 1995b] . The most notable of these actions includes the release of insulin from pancreatic islets [Pace-Asciak and Martin 1984] as well as the regulation of cell volume in platelets [Margalit et al., 1993] . In the human neutrophil, HxA_3 releases calcium from intracellular stores and this release is Hx-receptor mediated [Reynaud et al., 1996; Reynaud et al., 1995] . During the course of our investigations in neutrophils, we noted that radiolabeled HxA_3 was metabolized into a single product which was identified as the corresponding ω-hydroxy product retaining the basic hydroxy-epoxide functional group of the parent hepoxilin intact [Reynaud et al., 1997] . Additional studies reported herein also indicate that the hepoxilin ω-hydroxylase is different from that which metabolizes LTB_4.

MATERIALS AND METHODS

Neutrophils were isolated from venous blood obtained from healthy drug-free human volunteers as reported previously [Reynaud et al., 1997] . The cells were finally adjusted to a concentration of 10^7 cells/mL in RPMI 1640 medium. Incubations utilized 100 x 10^6 cells/experiment in a siliconized glass tube to which had been added $[^3H_6]$-HxA_3 methyl ester (5 x 10^6 cpm)(prepared as in [Demin et al., 1994]) diluted with 30 μg of the unlabeled compound. The composition of the buffer consisted of the following in mM: NaCl 140, KCl 5, $MgCl_2$ 1,

Eicosanoids and Other Bioactive Lipids in Cancer, Inflammation, and Radiation Injury, 4
Edited by Honn *et al.*, Kluwer Academic / Plenum Publishers, New York, 1999.

535

CaCl$_2$ 1, HEPES sodium-free 10, and glucose 10, pH 7.3. The mixture was incubated for 60 m at 37°. The sample was extracted with ethyl acetate without acidification to isolate the inta compound should it have retained the hydroxy-epoxide functionality which is sensitive to acid. further experiments we treated the extract with acid to convert the hepoxilin structure to tl trihydroxy derivative which was analyzed by GCMS. Samples were analyzed by preparative TL (silica gel G, ethyl acetate/acetic acid 99/1, v/v), and appropriate radiolabeled zones were isolate and extracted for analysis of structure by GCMS. In separate experiments, time course and ce number requirements of the metabolism were investigated. As well, comparison of the metabolis of HxA$_3$ with that of LTB$_4$ was investigated. CCCP was useful in demonstrating that tl metabolism of HxA$_3$ was inhibited by this compound, while that of [^{3}H$_8$]-LTB$_4$ was unaffected all concentrations of drug (0 - 100 μM). Methyl esters (Me) or pentafluorobenzyl esters (PFB trimethylsilyl ether (TMSi) derivatives were prepared as reported previously [Reynaud et al., 199'

RESULTS AND DISCUSSION

When [^{3}H]-HxA$_3$ methyl ester is incubated with intact human neutrophils, a single

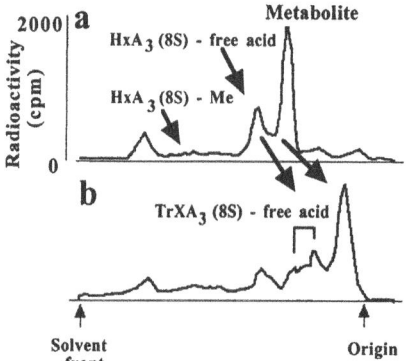

Figure 1. Radio thin layer chromatogram showing the profile of metabolites formed after incubation of [^{3}H$_6$] HxA$_3$ methyl ester with intact human neutrophils. See methods for details. Chromatograms are shown of extract untreated (a) and treated (b) with acid. Note that one major metabolite is formed which is acid-sensitive indicatin retention of the labile hydroxy-epoxide functionality. The metabolite was identified as ω–hydroxy HxA$_3$ (free acid)

major metabolite with increased polarity is observed. The formation of this metabolite is time- an cell number dependent (data not shown). It is also formed from an intermadiate which wa identified as the free acid of HxA$_3$. Hence, the methyl ester is taken up by the intact cells where i is first hydrolyzed into the free acid form which is subsequently converted into the majo metabolite shown in Fig. 1. Experiments in which the cells were broken (homogenization, freeze thawing etc) were devoid of metabolizing activity despite the presence of protease inhibitors in th buffer during cell disruption (data not shown). The main metabolite was isolated and subjected t GCMS analysis. NICI detection of PFB-TMSi derivative clearly showed the presence of a additional hydroxyl group in the hepoxilin structure due to a major fragment ion at m/z 49! representing M-PFB fragmentation (the corresponding fragment of the parent HxA$_3$ is at m/z 407) Important structural information was obtained from EI mass spectra of the Me-TMSi derivative The following fragment ion shifts due to the presence of the additional hydroxyl group in th metabolite were: m/z 281 (C8-C20) shifted to m/z 391 (i.e. 281 + 88) in the metabolite. A important fragment of structural significance locating the hydroxyl group at the ω-position was th ion at m/z 103 in the metabolite which was absent in the parent compound as it lacks this moiety

This fragment represented the terminal CH$_2$-OTMSi. Additional confirmation of the location of the hydroxyl group at the terminal end of the hepoxilin structure in the metabolite was obtained from the tetrahydroxy derivative of the metabolite formed through acidic workup. Fragment ion shifts at m/z 213 (C12-C20) to 301 (C12-C20) was observed while the common fragment ion in both the parent compound and the metabolite occurred at m/z 243 (C1-C8). Minor fragment ions occured in the spectrum of the metabolite at m/z 657 [M-15] and m/z 492 [M-(2x90)]. Final proof of structure was obtained through comparison of the mass spectrum of the metabolite with an authentic chemically synthesized product [Demin et al., 1996] .

Although LTB$_4$ is also metabolized into the ω-hydroxy metabolite by human

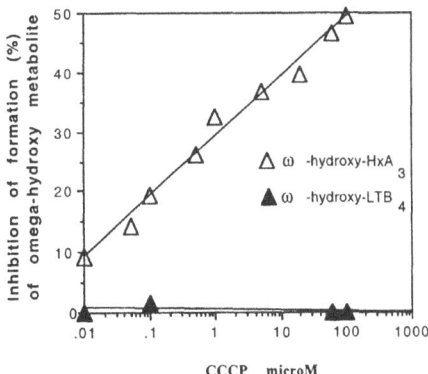

Figure 2. Dose-dependent inhibition by CCCP of ω-oxidation of HxA$_3$ but not LTB$_4$ by intact human neutrophils.

neutrophils, we discovered that its metabolism was unaffected by cell disruption, while that of HxA$_3$ was totally abolished despite the supplementation of the broken cell preparation with cofactors and protease inhibitors [Reynaud et al., 1997] . Additional evidence that the metabolism of HxA$_3$ took place by a different ω-hydroxylase was noted through a dose-dependent inhibition of HxA$_3$ metabolism by CCCP, a mitochondrial inhibitor (see Fig. 2).

The CCCP experiments suggested that while LTB$_4$ ω-hydroxylase is a microsomal enzyme [Powell 1984], HxA$_3$ oxidation takes place in the mitochondria. Unfortunately, all attempts to retain HxA$_3$ oxidizing activity in broken cell preparations were unsuccessful; this has prevented us from further investigating the mitochondrial nature of the Hx ω-oxidation system.

ACKNOWLEDGMENTS

Support of the Medical Research Council of Canada (grant MT#4181 to CRP-A) is gratefully acknowledged.

REFERENCES

Demin, P. M., Manukina, T. A., Pace-Asciak, C. R. and Pivnitsky, K. K., 1996, Total synthesis of 20-hydroxy hepoxilins, new metabolites of the hepoxilin family, *Mendeleev Commun.* **4:** 130-132

Demin, P. M., Pivnitsky, K. K., Vasiljeva, L. L. and Pace-Asciak, C. R., 1994, Synthesis of methyl [5,6,8,9,14,15-^{3}H$_6$]-hepoxilin B$_3$ and its conversion into methyl [5,6,8,9,14,15-^{3}H$_6$]-hepoxilin A$_3$, *J. Labelled Compounds and Radiopharmaceuticals.* 34: 221-230

Margalit, A., Sofer, Y., Grossman, S., Reynaud, D., Pace-Asciak, C. R. and Livne, A., 1993, Hepoxilin A$_3$ is the endogenous lipid mediator opposing hypotonic swelling of intact human platelets, *Proc. Natl. Acad. Sci. (USA).* 90: 2589-2592

Pace-Asciak, C. R., 1994, Hepoxilins: a review of their cellular actions, *Biochim. Biophys. Acta.* 1215: 1-8

Pace-Asciak, C. R. and Martin, J. M., 1984, Hepoxilin, a new family of insulin secretagogues formed by intact rat pancreatic islets, *Prostagl. Leukotrienes and Med.* 16: 173-180

Pace-Asciak, C. R., Reynaud, D. and Demin, P., 1995a, Mechanistic aspects of hepoxilin biosynthesis, *J. Lipid Mediat. Cell Signal.* 12: 307-311

Pace-Asciak, C. R., Reynaud, D. and Demin, P. M., 1995b, Hepoxilins: A review on their enzymatic formation, metabolism and chemical synthesis, *Lipids.* 30: 1-8

Powell, W. S., 1984, Properties Of Leukotriene B$_4$ 20 Hydroxylase From Polymorphonuclear Leukocytes, *J. Biol. Chem.* 259: 3082-3089

Reynaud, D., Demin, P. and Pace-Asciak, C. R., 1996, Hepoxilin A$_3$-specific binding in human neutrophils, *Biochem. J.* 313: 537-541

Reynaud, D., Demin, P. M. and Pace-Asciak, C. R., 1995, Hepoxilin binding in human neutrophils, *Biochem. Biophys. Res. Comm.* 207: 191-194

Reynaud, D., Rounova, O., Demin, P. M., Pivnitsky, K. K. and Pace-Asciak, C. R., 1997, Hepoxilin A$_3$ is oxidized by human neutrophils into its omega-hydroxy metabolite by an activity independent of LTB$_4$ omega-hydroxylase, *Biochim. Biophys. Acta Lipids Lipid Metab.* 1348: 287-298

79

OXIDATION OF ARACHIDONATE CONTAINING GLYCEROPHOSPHOLIPIDS IN
INTACT RED BLOOD CELLS AND RED BLOOD CELL MEMBRANES WITH
TERT-BUTYLHYDROPEROXIDE

Tatsuji Nakamura, Lisa Hall, and Robert C. Murphy

Division of Basic Sciences
National Jewish Medical and Research Center
1400 Jackson Street
Denver, CO 80206

INTRODUCTION

Intact red blood cells and red blood cell membranes (ghosts) offer two interesting yet different models to investigate chemical events taking place within the lipid bilayer following exposure of cells to reactive oxygen species. Lipid peroxidation and more specifically oxidation of the glycerol phospholipids that contain esterified polyunsaturated fatty acyl groups such as arachidonate and linoleate, can now be examined directly using techniques of electrospray ionization tandem mass spectrometry for both quantitative and qualitative analysis. Previously, we reported the quantitative analysis of oxidized fatty acids derived from phospholipids or red blood cells and that monohydroxyeicosatetraenoic acids (HETEs) increase significantly as esterified products following treatment with tert-butylhydroperoxide (tBuOOH). Interestingly, four separate epoxyeicosatrienoic acids (EETs) that were isobaric to these HETEs were also found in normal red blood cells and the abundance of these components significantly increased after treatment with tBuOOH (1). Chromatograms have been presented for the HETE and EET molecular species derived from glycerophos-phatidylethanolamine (GPE), but not for another major arachidonate containing glycerophospholipid, glycerphosphatidylcholine (GPC). Hydroperoxy containing fatty acyl groups, likely precursors of the esterified HETE were not detected in this model system.

Mass spectrometric investigation of intact glycerophospholipids isolated from control red blood cell membranes as well as those phospholipids isolated following treatment of the red blood cells membranes with tBuOOH resulted in the discovery of six isobaric HETE-containing phospholipids in the GPE class of phospholipids as well as three hydroperoxyeicosatetraenoates (5-, 12-, 15-HpETE) esterified to phospholipids (2). The hydroxyoctadecadienoates (9-, and 13-HODE) and hydroperoxy derivatives of linoleic acid (9- and 13-HpODE) esters were also observed. Analysis of the intact phospholipids revealed the distribution of these HETE and HODE as well as HpETE esterified products in not only the different phospholipid classes, but also in the subclasses such as the 1,2-diacyl-GPE and the plasmalogen containing GPE molecular species. Although arachidonic acid was most abundant in these plasmenyl glycerophospholipids, the major phospholipid molecular species which contain HETE and HpETE were identified as 1,2-diacyl glycerophospholipids (2). These two different models of lipid peroxidation provide an interesting glimpse into the stable products that can be isolated from glycerophospholipids and the effect of antioxidant mechanisms that can reduce reactive intermediates such as hydroperoxy fatty acyl groups.

EXPERIMENTAL

Materials

EET and HETE standards were purchased from Cayman Chemical (Ann Arbor, MI). $[^{18}O_2]12$-HETE was prepared from butyryl cholinesterase and $H_2^{18}O$ as previously described (3). *Tert*-butylhydroperoxide was purchased from Sigma Chemical Co. (St. Louis, MO). All solvents were HPLC grade and other reagents were of the highest grade commercially available.

RBC Preparation and Incubation

Human blood was obtained from healthy volunteers and RBCs isolated by a two-step centrifugation as previously described (1). After a final wash, cells were suspended to a final concentration of 2×10^8 RBC/mL in phosphate buffered saline (PBS). RBC ghosts were prepared by lysing the RBCs and in PBS diluted 5-fold containing 0.1 mM EGTA and 1 mM MgCl$_2$ as previously described (2). For those experiments with RBC ghosts, the resealed membranes were resuspended at 0.2 mg protein/mL buffer (PBS). Intact RBC (50 mL) and ghost preparations were incubated with tBuOOH (10 mM final concentration) at 37/C for 90 min. For both experimental protocols, resuspensions were centrifuged to obtain the cell/membrane pellet after incubation.

Phospholipid Extraction and Purification

Phospholipids were extracted from intact RBC cell pellets following the method of Rose and Oklander (4) with chloroform and isopropanol to maximize the yield of phospholipids yet minimize the influence of hemoglobin on the extraction process. Phospholipids from RBC ghosts were extracted by the method of Bligh and Dyer (5) substituting methylene chloride for chloroform. Phospholipids were purified and GPC was isolated from GPE and GPS classes with normal phase HPLC essentially following a previously published protocol (1). Briefly, normal phase HPLC was performed using a 5 ☐m silica (Licrosorb; Phenomenex, Torrance, CA) analytical column (4.6 x 250 mm) with a gradient of 62.5% solvent A (isopropanol/hexane/260 mM ammonium acetate, pH 7.0, 58/40/2) programmed to 100% solvent B (isopropanol/hexane/260 mM ammonium acetate, pH 7.0, 50/40/10) over a 20 min period. The solvent flow rate was 1 mL/min and 1 min fractions were collected.

Phospholipid Hydrolysis

Normal phase HPLC fractions from intact RBCs were saponified as previously described (1) after the addition of $[^{18}O_2]$12-HETE (10 ng) as internal standard. Free fatty acids were extracted twice with hexane prior to LC/MS/MS quantitation (1). Reverse phase HPLC was carried out using a Prodigy 5⬛ ODS 100⬥ (1 x 250 mm, Phenomenex, Rancho Palos Verdes, CA) with a linear gradient from 20% ammonium acetate (6.5 mM, pH 5.7) programmed to 100% methanol/acetonitrile (35/65) in 15 min at 50 ⬛L/min.

Mass spectrometric analyses were performed on a Sciex API III$^+$ triple quadrupole (PE-Sciex, Thornhill, Ontario, Canada) in the negative ion mode using multiple reaction monitoring (MRM) for quantitation of HETEs and EETs and monitoring the transitions m/z 319 ↝115 for 5-HETE, m/z 319 ↝ 151 for 9-HETE, m/z 319 ↝155 for 8-HETE and 8,9-EET, m/z 319 ↝ 167 for 11-HETE, m/z 319 ↝191 for 5,6-EET, m/z 319 ↝ 208 for 12-HETE and 11,12-EET, m/z 319 ↝219 for 15-HETE and 14,15-EET, and m/z 323 ↝ 183 for $[^{18}O_2]$12-HETE. Analysis of intact phospholipids was carried out with purified (normal phase HPLC) phospholipid classes which did not separate plasmalogen from 1,2-diacyl subclasses (2).

RESULTS AND DISCUSSION

Analysis of EETs and HETE-GPC in Human Red Blood Cells

Previous investigations (6) have reported the most abundant phospholipid class in the RBC was GPC (32.2%). Mass spectrometric analysis was carried out to determine whether HETE and EET phospholipid molecular species were also present in this phospholipid class from membranes of normal red blood cells. Furthermore, human RBCs were incubated with tBuOOH (10 mM) then analyzed for EET and HETE-containing GPC molecular species. GPC was extracted and purified by normal phase HPLC as described previously (1). Isolated GPC was saponified and the fatty acid components analyzed by LC/MS/MS using a MRM protocol specific for EETs and HETEs.

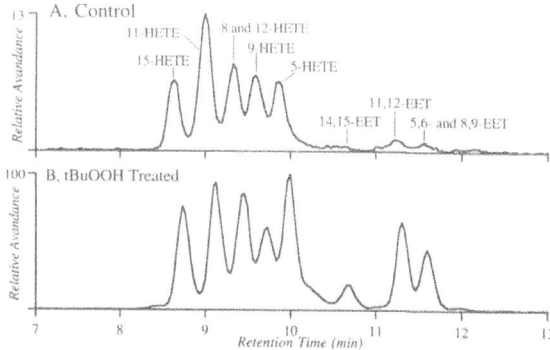

Figure 1. LC/MS/MS analysis (negative ion electrospray ionization) of HETEs and EETs derived from GPC in (A) untreated and (B) treated (tBuOOH, 10mM) human RBCs. GPC was extracted from RBCs then purified by normal phase HPLC. The isolated GPC fraction was saponified with 1 N NaOH, acidified, and then subjected to reverse phase LC/MS/MS. The mass spectrometer was operated in the MRM mode. Both chromatograms represent the summation of the total ion current derived from the transitions of m/z 319 to 115, 151, 155, 167, 191, 208 and 219.

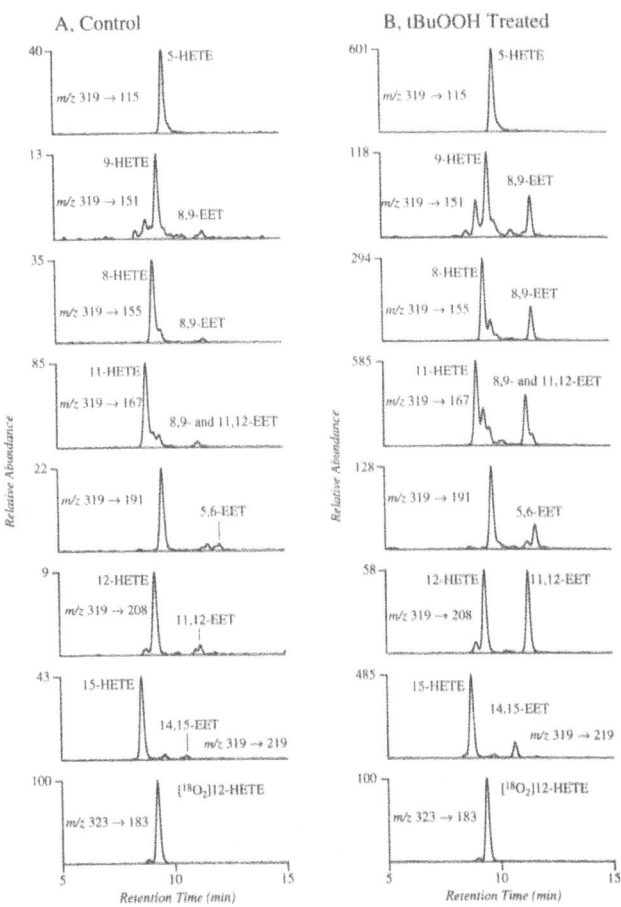

Figure 2 Multiple reaction monitoring profiles recorded during the analysis of HETEs and EETs derived from GPC of (A) untreated and (B) treated (tBuOOH, 10 mM) intact human RBCs. Samples were prepared and analyzed as described in Figure 1.

All six HETE isomers expected from autooxidation of esterified arachidonate were observed in approximately the same relative concentration (Fig. 1A). Although present, there was a substantially lower abundance of esterified EET regioisomers (Fig. 1A). Initiation of lipid peroxidation by tBuOOH and the intact RBCs resulted in a substantial increase in both HETEs and EETs esterified to GPC (Fig. 1B). The individual HETE and EET isomers derived from GPC were identified using MRM with unique ion transitions (Fig. 2). In some cases the ion transitions employed to detect the elution of HETEs were identical to the ion transitions used to detect the elution of the EETs. In these cases, the HPLC separation was used as a criteria to differentiate the two isobaric molecules. The combination of reverse phase HPLC and MRM enabled specific quantitation of each of the HETEs and the EETs derived from human red blood cell GPC using [^{18}O$_2$]12-HETE as internal standard. HETEs increased approximately 5-15 fold and all six HETE isomers were observed consistent with a direct oxidation of arachidonic acid present in the GPC of RBCs. The formation of the four isomeric EETs was strikingly enhanced (50-100 fold) by treatment with tBuOOH (Fig. 2B) as previously shown for GPE molecular species (1).

In experiments with RBC ghosts, mass spectrometric analysis of intact GPE lipids by electrospray tandem mass spectrometry permitted direct assessment of not only the oxidized fatty acyl groups, but also assessment of the sn-1 structural moiety. Plasmenyl phospholipids that have a vinyl ether structural motif at sn-1 contain a substantial fraction of total esterified arachidonate (2) as summarized in Figure 3A. Following treatment of RBC membranes with tBuOOH, plasmenyl GPE containing oxidized arachidonate was not proportionally higher than the 1,2-diacyl-GPE oxidized arachidonate (Fig. 3B). While this observation could result from a lower susceptibility of arachidonate-containing plasmenyl GPE to oxidation initiated by tBuOOH, reasonable evidence suggests that plasmenyl GPE is in fact more susceptible to oxidation (7,8). The results from this experiment are consistent with the suggestion that oxidized plasmenyl arachidonoyl GPE products may be unstable and more readily decompose to products not detected in this experiment.

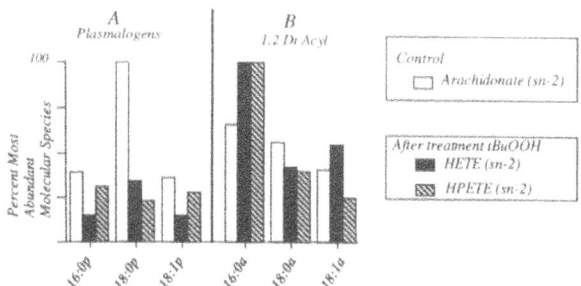

Figure 3 (A) Relative abundance of arachidonate-containing GPE phospholipids containing a vinyl ether moiety (plasmalogen) or ester moiety (1,2-diacyl) at the sn-1 position. Phospholipids were extracted from RBC ghosts and analyzed as intact species by electrospray tandem mass spectrometry (2). (B) Relative abundance of monohydroxyeicosatetraenoic acid containing phospholipids and RBC ghosts following treatment with tBuOOH (10 mM).

HpETE fatty acyl groups esterified predominately to 1,2-diacyl-GPC were also found to be quite abundant following treatment of RBC ghosts with tBuOOH and different isomers could be detected (9). These molecular species had a characteristic loss of water from the carboxylate anion generated during the collision induced dissociation of the molecular anion [M-H]⁻ (10). Following saponification and LC/MS/MS analysis of the free fatty acids, the 5-, 12-, and 15-HpETE isomers were separated from corresponding HETE isomers and the 5-, 12-, and 15-oxo-ETE oxidized products of esterified arachidonate.

Thus, oxidation of RBC membranes without antioxidant defense mechanisms results in abundant formation of hydroperoxy glycerophospholipids. Some reduction of these oxidized fatty acyl chains takes place to hydroxyglycerophospholipids, but there was no substantial formation of epoxy fatty acyl phospholipids. Analysis of the intact phospholipids enabled association of the observed oxidized fatty acyl group with a phospholipid subclass, *e.g.* vinyl ether versus acyl motif at sn-1. Interestingly, the stable oxidized arachidonate-containing phospholipids (GPE) were predominantly 1,2-diacyl-GPE even though substantially more arachidonate was esterified to plasmenyl GPE. A different picture of fatty acyl oxidation products emerged when intact human RBCs were treated with tBuOOH. Extant mechanisms result in the rapid reduction of the hydroperoxy moiety in acyl phospholipids to yield abundant hydroxy fatty acyl groups (11). Furthermore, hemoglobin likely participated in the epoxidation of esterified arachidonate yielding esterified epoxyeicosatetraenoate glycerophospholipids. Such products, either EETs, HETEs, or HpETEs could be released as free fatty acids following hydrolysis catalyzed by PLA₂ (12) resulting in the appearance of biologically active eicosanoids as products of lipid peroxidative events.

ACKNOWLEDGMENTS

This work was supported, in part, by a grant from the National Institutes of Health (HL34303).

REFERENCE

1. T. Nakamura, T., D.L. Bratton, and R.C. Murphy, Analysis of epoxyeicosatrienoic and monohydroxyeicosatetraenoic acids esterified to phospholipids in human red blood cells by electrospray tandem mass spectrometry, *J. Mass Spectrom.* 32:888-896 (1997).
2. L.M. Hall and R.C. Murphy, Analysis of oxidized molecular species of glycerophos-pholipids following treatment of red blood cell ghosts with t-butylhydroperoxide. *Anal. Biochem.*, in press.
3. R.C. Murphy and K.C. Clay, Preparation of labeled molecules by exchange with oxygen-18 water, *Methods Enzymol.*,193:338-348 (1990).
4. H.G. Rose and M. Oklander, Improved procedure for the extraction of lipids from human erythrocytes, *J. Lipid Res.* 6:428.
5. E.G. Bligh and W.J. Dyer, A rapid method of total lipid extraction and purification, *Can.J.Biochem.Physiol.* 37:911 (1959).
6. A. Diagne, J. Fauvel, M. Record, H. Chap, and L. Douste-Blazy, Studies on ether phospholipids. II. Comparative composition of various tissues from human, rat and guinea pig, *Biochim. Biophys. Acta* 793:221 (1984).
7. O.H. Morand, R.A. Zoeller, and C.R. Raetz, Disappearance of plasmalogens from membranes of animal cells subjected to photosensitized oxidation, *J. Biol. Chem.* 263:11597 (1988).

8. D. Reiss, K. Beyer, and B. Engelmann, Delayed oxidative degradation of polyun-saturated diacyl phospholipids in the presence of plasmalogen phospholipids *in vitro, Biochem. J.* 323:807 (1997).
9. L.M. Hall and R.C. Murphy, Electrospray mass spectrometric analysis of 5-hydro-peroxy and 5-hydroxyeicosatetraenoic acids generated by lipid peroxidation of red blood cell ghost phospholipids, *J. Am. Soc. Mass Spectrom.*, in press.
10. D.K. MacMillan and R.C. Murphy, Analysis of lipid hydroperoxides and long-chain conjugated keto acids by negative ion electrospray mass spectrometry, *J. Am. Soc. Mass Spectrom.* 6:1190 (1995).
11. H. Imai, K. Narashima, M. Arai, H. Sakamoto, N. Chiba, and Y. Nakagawa, Suppres-sion of leukotriene formation in RBL-2H3 cells that overexpressed phospholipid hydroperoxide glutathione peroxidase, *J. Biol. Chem.*273:1990 (1998).
12. M.G. Salgo, F.P. Corongiu, and A. Sevanian, Peroxidation and phospholipase A_2 hydrolytic susceptibility of liposomes consisting of mixed species of phosphatidylcholine and phosphatidylethanolamine, *Biochim. Biophys. Acta* 1127:131 (1992).

80

CANNABINOID MODULATION OF NEURONAL ACTIVITY IN ADULT RAT HIPPOCAMPUS

Paul Schweitzer, George R. Siggins, and Samuel G. Madamba

Neuropharmacology
The Scripps Research Institute
10550 North Torrey Pines Road
La Jolla, CA 92037

INTRODUCTION

Cannabinoids have powerful psychoactive properties and alter many physiological processes (Howlett, 1995). These effects are mediated through specific receptors found throughout the brain (Herkenham et al., 1990). The first isolated endogenous ligand for cannabinoid receptors was the lipid derivative N-arachidonylethanolamine, dubbed anandamide (Devane et al., 1992), which has been shown to be produced in neurons (Di Marzo et al., 1994). It is believed that there may be several additional cannabimimetic lipid mediators. Indeed, a second endogenous cannabinoid ligand, 2-arachidonylglycerol (2-AG), is found in brain (Stella et al., 1997). The levels of 2-AG markedly increase upon fiber tract stimulation in hippocampal slices, suggesting that 2-AG may act as a signaling molecule in hippocampus.

Cannabinoids substances are associated with memory impairment (Howlett, 1995). One of the highest densities of cannabinoid receptors (CB) is found in the hippocampus (Herkenham et al., 1990), a brain structure associated with learning and memory processes. Studies conducted on hippocampal slices *in vitro* have shown that cannabinoid agonists inhibit long-term potentiation (LTP) of synaptic transmission (Terranova et al., 1995), an electrophysiological model for learning and memory. This effect was prevented by the CB1 receptor antagonist SR141716, indicating a CB1 receptor-mediated mechanism.

Cannabinoids also alter postsynaptic properties of central neurons. The transient K+ A-current is increased (Deadwyler et al., 1995), while Ca^{2+} currents are decreased (Twitchell et al., 1997), in cultured hippocampal neurons exposed to cannabinoid agonists. However, no postsynaptic studies have been conducted on native adult brain preparations such as the hippocampal slice. Furthermore, cannabinoid effects on non-inactivating K+ conductances that control the resting membrane potential of neurons have not been investigated in brain. We studied the voltage-dependent M-current (I_M), a persistent K+ current that modulates neuronal excitability and is controlled by numerous neurotransmitters (Marrion, 1997).

In the present study, we used the adult rat slice preparation to investigate the effects of cannabinoids in hippocampus. Our Experiments indicate that cannabinoids affect neuronal activity in both inhibitory and excitatory directions. Synaptic potentiation is prevented by cannabinoid agonists, pointing at reduced neuronal activity. On the other hand, these agonists decrease the K+ M-current, raising the excitability of hippocampal neurons. We also show that cannabinoids and eicosanoids have opposite electrophysiological actions.

METHODS

Slice preparation. We used standard intracellular recording techniques in adult rat hippocampal slices (350 μm thick) as described previously (Schweitzer et al., 1993). The slices were submerged and superfused at a constant rate in gassed artificial cerebrospinal fluid (ACSF) of the following composition in mM: NaCl, 130; KCl, 3.5; NaH_2PO_4, 1.25; $MgSO_4$, 1.5; $CaCl_2$, 2.0; $NaHCO_3$, 24; glucose, 10. Lipids were dissolved in 0.1% DMSO.

Extracellular recordings. We recorded extracellular fields of excitatory postsynaptic potentials with a glass micropipette (3M NaCl) placed in CA1 stratum radiatum. To evoke synaptic activity, the Schaffer collaterals were stimulated with a bipolar tungsten electrode delivering constant voltage pulses that elicited a 40% maximal response. Long-term potentiation (LTP) was induced by applying 2 trains (20 sec apart) of high frequency stimulations (100 Hz, 1s) at the same intensity. Voltage records were acquired and analyzed with software.

Intracellular recordings. We performed single-electrode voltage-clamp studies using sharp micropipettes (3M KCl) to penetrate CA1 pyramidal neurons. Tetrodotoxin (1 μM) was added to the ACSF to block action potentials and synaptic transmission. Current and voltage records were acquired by D/A sampling and analyzed using software. To observe I_M, neurons were depolarized to potentials around –45 mV and hyperpolarizing commands were applied. The various problems (for example, space-clamp) associated with voltage-clamping of neurons with extended processes are discussed elsewhere (Halliwell and Adams 1982).

RESULTS AND DISCUSSION

Long-Term Potentiation of Synaptic Transmission

We performed extracellular recordings in hippocampal area CA1 to investigate the effect of the newly described brain endogenous cannabinoid 2-AG on LTP expression. Application of a high frequency stimulation (HFS) to the Schaffer collaterals, a fiber tract that projects from CA3 to CA1, induced a sustained potentiation of field excitatory postsynaptic potentials (fEPSPs) recorded in CA1 stratum radiatum. In control condition, the fEPSP remained at 156% of basal values 60 minutes after delivery of the tetanus (figure 1).

The addition of 2-AG in the superfusate had little effect on basal synaptic activity. However, when HFS was delivered in the continued presence of 2-AG, fEPSPs were only shortly potentiated and returned to basal values within 30 minutes of the tetanus (figure 1). There was no LTP phenomenon observed 60 minutes post-HFS, indicating that 2-AG completely prevented the establishment of LTP. The addition of the cannabinoid receptor (CB1) antagonist SR141716A prior to the addition of 2-AG completely prevented the effect of 2-AG, indicating that the suppression of LTP by the lipid occurred via activation of CB1 receptors.

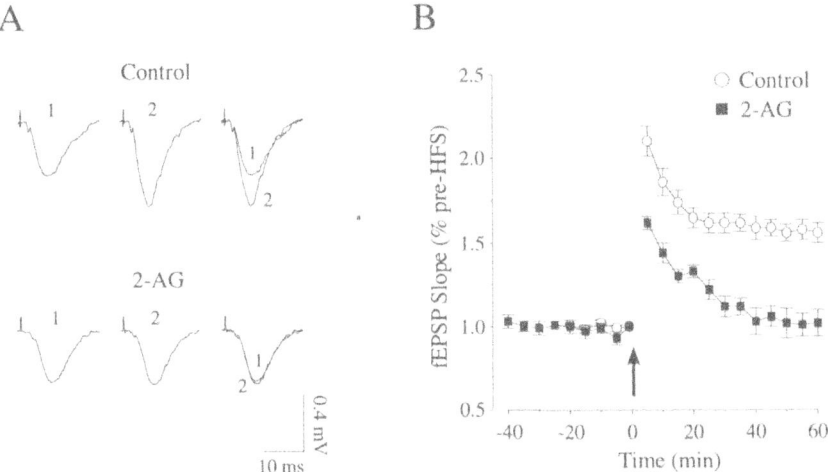

Figure 1. The cannabinoid 2-AG prevents long-term potentiation. **A.** Recordings of CA1 fEPSPs before (1) and 60 min after (2) HFS of the Schaffer collaterals. The top traces show the potentiation obtained in ACSF (control). The bottom traces depict the absence of potentiation observed in ACSF containing 2-AG. **B.** Time course of the fEPSP slope (normalized to pre-HFS) during the experiment (arrow indicates HFS).

These results show that the endogenous cannabinoid agonist 2-AG is physiologically active on brain neurons and prevents the establishment of LTP via CB1 receptors in adult rat hippocampus. Taken together, the marked inhibition of synaptic plasticity by 2-AG and its increased formation upon fiber tract stimulation implicate this lipidic substance as a neuromodulator in hippocampus.

Cannabinoids Decrease the Potassium M-Current

Cannabinoids affect transient K^+ and Ca^{2+} conductances in cultured hippocampal neurons. We investigated whether cannabinoids affect non-inactivating K^+ conductances in CA1 hippocampal pyramidal neurons (HPNs) using the slice preparation. We performed intracellular voltage-clamp recordings to study I_M, a persistent K^+ current that activates at slightly depolarized potentials to clamp the neuronal membrane near rest.

Because endogenous cannabinoids are quickly degraded in biological tissue, we used a non-degradable analog of anandamide, R1-methanandamide (MAEA). Addition of MAEA (5 µM) in the superfusate consistently reduced I_M amplitude by about 50-60% and elicited an inward steady-state current at holding potential (figure 2). All current values returned to control levels on washout of MAEA. To determine if this effect occurred via activation of CB1 receptors, we treated the slices with the CB1 receptor antagonist SR141716A (1 µM). A subsequent application of MAEA onto HPNs in the continued presence of the CB1 antagonist did not affect I_M, indicating that MAEA decreased the K^+ current by activating CB1 receptors.

These results indicate that cannabinoids decrease a non-inactivating K^+ conductance in adult hippocampal neurons. Such effect tends to raise neuronal excitability by affecting the ability of the membrane to clamp the potential near rest when the neurons depolarize. This excitatory action is opposite to the reported postsynaptic inhibitory effects of cannabinoids in various preparations (Howlett, 1995).

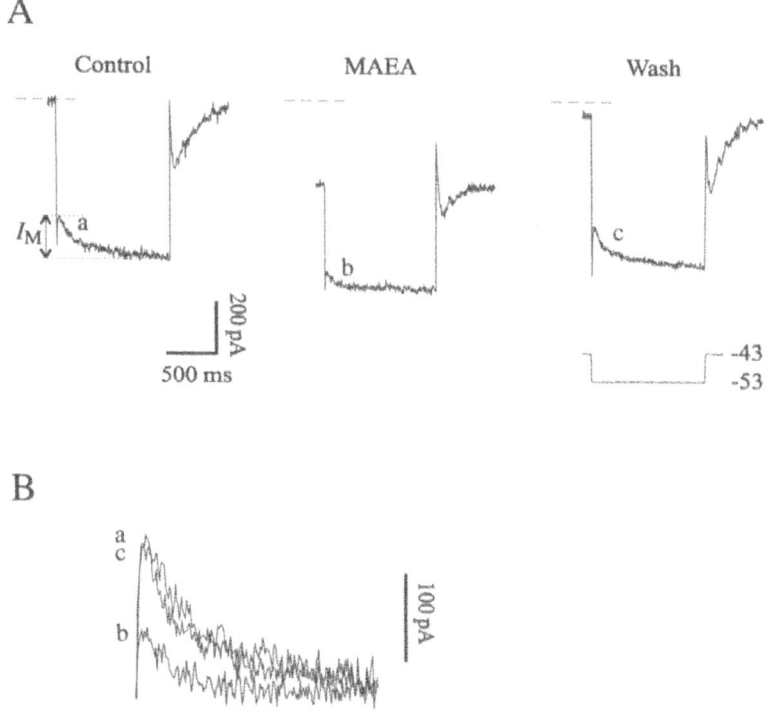

Figure 2. Methanandamide (MAEA) reduces I_M. **A.** Voltage-clamp recording (voltage protocol on lower right) of a HPN. The current traces show the I_M relaxations in control (a), during application of 5 µM MAEA (b), and after washout (c). Note the I_M decrease and the inward holding current (dashed line is control level) upon application of MAEA. **B.** Magnification of the relaxations, aligned and superimposed for comparison.

Arachidonic Acid and its Metabolites Augment the M-Current

Arachidonic acid and its metabolites, the eicosanoids, are lipidic messengers involved in several forms of neuromodulation (Piomelli, 1994), including regulation of ion channels and synaptic transmission. The fatty acid is also postulated to be a degradation product of anandamide and 2-AG, indicating that arachidonic acid may be formed and possibly released when endogenous cannabinoids are metabolized in biological tissue.

At the postsynaptic level, studies performed in our laboratory indicate that arachidonic acid augments I_M. This effect occurs via activation of the phospholipase A2 pathway by the neuropeptide somatostatin (Schweitzer et al., 1990) and generation of eicosanoids. The mechanism of arachidonic acid augmentation of I_M in HPNs involves the conversion of the fatty acid to 5-lipoxygenase metabolites: the blockade of this route with various specific inhibitors prevents the I_M increase, while blocking the 12-lipoxygenase or the formation of prostaglandins does not affect the arachidonic acid-induced augmentation of I_M. Indeed, superfusion of leukotriene C4 markedly augments I_M (figure 3), with recovery of control current measurements upon washout of the eicosanoid. A large outward holding current is elicited with the I_M increase, producing a pronouced inhibitory effect on neuronal activity in the activation range of I_M.

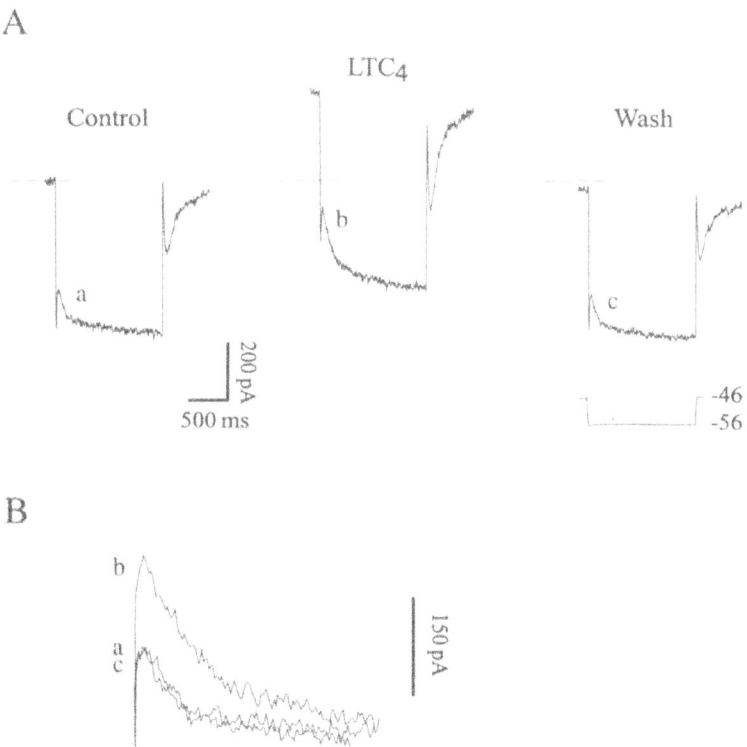

Figure 3. Leukotriene C4 (LTC4) increases I_M. **A.** Current recordings (voltage protocol on right) showing I_M in control (a), during 5 μM LTC4 (b), and washout (c). Note the I_M augmentation in LTC4, concomitant with an outward holding current (dashed line). **B.** Magnification of the relaxations.

The effect of arachidonic acid metabolites on I_M are opposed to those of cannabinoid agonists, indicating that eicosanoids are unlikely second messengers of cannabinoid effects. A tight compartmentalization of the different pools of arachidonic acid vraisemblably takes place into neurons, strictly regulating the levels of readily available arachidonate. However, a regulatory mechanism that would prevent the generation of arachidonic acid metabolites and re-route the fatty acid to another pathway is conceivable. For example, an indirect mechanism of action of arachidonic acid via activation of protein kinase C to decrease I_M has been reported in cultured hybrid cells (Schmitt and Meves, 1993).

CONCLUSION

These data show a marked effect of cannabinoid substances on I_M, a non-inactivating K+ conductance that plays an important role in HPN excitability. The diminution of this current points toward an excitatory role of cannabinoids at the postsynaptic level, an effect contrasting with our expectations. Indeed, the endogenous cannabinoids anandamide and 2-AG produce arachidonic acid upon degradation. However, the fatty acid and its metabolites augment I_M, while cannabinoids decrease this conductance. Interestingly, arachidonic acid also decreases the K+ A-current (Villarroel, 1993), while cannabinoids increase this current

(Deadwyler et al., 1995). Furthermore, the cannabinoids prevent the establishment of long-term potentiation, while arachidonic acid elicits this phenomenon (Williams et al., 1989).

Cannabinoids can postsynaptically increase hippocampal excitability by diminishing I_M, but may also decrease neuronal activity by inhibiting synaptic plasticity. Surprinsigly, the cannabinoids and the eicosanoids, two closely related families of lipid mediators, have opposite effects on various electrophysiological measurements.

Acknowledgments

Supported by NIH grants K01DA00291 and MH44346.

REFERENCES

Deadwyler, S.A., Hampson, R.E., Mu, J., Whyte, A., and Childers, S., 1995, Cannabinoids modulate voltage sensitive potassium A-current in hippocampal neurons *via* a cAMP-dependent process, *J.Pharmacol. Exp.Ther.* 273:734.

Devane, W.A., Hanus, L., Breuer, A., Pertwee, R.G., Stevenson, L.A., Griffin, G., Gibson, D., Mandelbaum, A., Etinger, A., and Mechoulam, R., 1992, Isolation and structure of a brain constituent that binds to the cannabinoid receptor, *Science* 258:1946.

Di Marzo, V., Fontana, A., Cadas, H., Schinelli, S., Cimino, G., Schwartz, J.-C., and Piomelli, D., 1994, Formation and inactivation of endogenous cannabinoid anandamide in central neurons, *Nature* 372:686.

Halliwell, J.V., and Adams, P.R., 1982, Voltage-clamp analysis of muscarinic excitation in hippocampal neurons, *Brain Res.* 250:71.

Herkenham, M., Lynn, A.B., Little, M.D., Johnson, M.R., Melvin, L.S., De Costa, B.R., and Rice, K.C., 1990, Cannabinoid receptor localization in brain, *Proc.Natl.Acad.Sci.USA* 87:1932.

Howlett, A.C., 1995, Pharmacology of cannabinoid receptors, *Annu.Rev.Pharmacol.Toxicol.* 35:607.

Marrion, N.V., 1997, Control of M-current, *Annual Review of Physiology* 59:483.

Piomelli, D., 1994, Eicosanoids in synaptic transmission, *Critical Rev.Neurobiol.* 8:65.

Schmitt, H., and Meves, H., 1993, Protein kinase C as mediator of arachidonic acid-induced decrease of neuronal M current, *Pflugers Archiv* 425:134.

Schweitzer, P., Madamba, S.G., and Siggins, G.R., 1990, Arachidonic acid metabolites as mediators of somatostatin- induced increase of neuronal M-current, *Nature* 346:464.

Schweitzer, P., Madamba, S.G., Champagnat, J., and Siggins, G.R., 1993, Somatostatin inhibition of hippo-campal pyramidal neurons: mediation by arachidonic acid and its metabolites, *J.Neurosci.* 13:2033.

Stella, N., Schweitzer, P., and Piomelli, D., 1997, A second endogenous cannabinoid that modulates long-term potentiation, *Nature* 388:773.

Terranova, J.P., Michaud, J.C., Le Fur, G., and Soubrié, P., 1995, Inhibition of long-term potentiation in rat hippocampal slices by anandamide and WIN55212-2: Reversal by SR141716 A, a selective antagonist of CB1 cannabinoid receptors, *Naunyn Schmiedebergs Arch.Pharmacol.* 352:576.

Twitchell, W., Brown, S., and Mackie, K., 1997, Cannabinoids inhibit N- and P/Q-type calcium channels in cultured rat hippocampal neurons, *J.Neurophysiol.* 78:43.

Villarroel, A., 1993, Suppression of neuronal potassium A-current by arachidonic acid, *FEBS Lett.* 335:184.

Williams, J.H., Errington, M.L., Lynch, M.A., and Bliss, T.V.P., 1989, Arachidonic acid induces a long-term activity-dependent enhancement of synaptic transmission in the hippocampus, *Nature* 341:739.

APOPTOSIS

81

CYCLOOXYGENASE-INDEPENDENT INDUCTION OF P21[WAF-1/CIP1], APOPTOSIS
AND DIFFERENTIATION BY L-745,337 AND SALICYLATE IN HT-29 COLON
CANCER CELLS

Paola Patrignani[1], Giovanna Santini[1], Maria G. Sciulli[1], Rosanna Marinacci[2], Ornella Fusco[2], Luciana Spoletini[2], Clara Natoli[2], Antonio Procopio[2] and Jacques Maclouf[3]

Departments of [1]Medicine and Aging, [2]Oncology and Neuroscience, University of Chieti "G. D'Annunzio", 66013 Chieti, Italy and [3]INSERM, Unité 348, 75475 Paris, France

INTRODUCTION

A large body of experimental evidence indicates that in animal models of bowel cancer, nonsteroidal antiinflammatory drugs (NSAIDs) can suppress carcinogenesis[1,2]. Moreover, both case-control and cohort studies have consistently found that regular use of aspirin is associated with approximately 50% reduction in the incidence and mortality of colorectal cancer, both in men and women[1,3]. Thirdly, sulindac reduces the number and size of colorectal adenomas in patients with familial adenomatous polyposis (FAP), though its effect is incomplete and reversible[1,2].

The mechanism(s) of the antineoplastic effects of NSAIDs is not completely understood but most hypotheses have focussed on their property to reduce the levels of prostanoids [prostaglandin(PG)s, prostacyclin and thromboxane(TX)A$_2$] in gastrointestinal tissues[4].

Multiple enzymatic steps are involved in the biosynthetic pathway of prostanoids. Prostaglandin endoperoperoxide synthase (PGHS), an enzyme which exhibits both cyclooxygenase and peroxidase activities, catalyzes the conversion of arachidonic acid (AA) to PGH$_2$, the upstream metabolite for PG and TX biosynthesis[5]. Two isoforms of PGHS have been identified in humans, referred to as PGHS-1 and PGHS-2 or COX-1 and COX-2. PGHS-1 mRNA and protein are present at relatively stable levels in many tissues. PGHS-2 is an inducible enzyme expressed in a restricted number of cell types in response to mitogenic and inflammatory stimuli [5].

Eicosanoids and Other Bioactive Lipids in Cancer, Inflammation, and Radiation Injury, 4
Edited by Honn *et al.*, Kluwer Academic / Plenum Publishers, New York, 1999.

555

Recent findings suggest that overexpression of PGHS-2 is an early, central event in colon carcinogenesis[6] and support the involvement of the shared inhibitory property of NSAIDs towards the cyclooxygenase activity of PGHS-1 and -2 in the prevention of colon cancer.

However, growth inhibition of colon adenocarcinoma cell lines *in vitro* by exposure to different NSAIDs has also been shown to involve PGHS-independent inhibition of cell proliferation and induction of apoptosis[7-9]. In fact, similar growth inhibiting effects were demonstrated with sulindac sulfoxide[7], sulfide[7] and sulfone[8], despite markedly different potency in inhibiting cyclooxygenase activity. Moreover, similar effects were reported in colon cancer cells regardless of PGHS expression[9].

Structurally different NSAIDs have been shown to reduce the levels of cell cycle-regulatory proteins[2,7] and sulindac and sulindac sulfide induce the tumor suppressor protein p21[WAF-1/cip17] an important signal linking apoptosis, differentiation and cell cycle alterations in response to exogenous stimuli[10-12].

In order to investigate the cyclooxygenase-dependence and the PGHS-isozyme specificity of growth inhibition by cyclooxygenase inhibitors, we have studied 3 compounds with varying potency and biochemical selectivity towards PGHS isozymes: aspirin, a relatively selective inhibitor of platelet PGHS-1[13], its metabolite sodium salicylate, a weak inhibitor of both PGHS isozymes[14], and the selective PGHS-2 inhibitor L-745,337[15,16], on prostanoid production, induction of the tumor suppressor protein p21[WAF-1/cip1], mutant p53 levels and morphological changes in the human colon adenocarcinoma cell line HT-29.

We used HT-29 cells because this human cell line of colon adenocarcinoma has been extensively studied for the antiproliferative effects of different NSAIDs *in vitro*[7-9] and PGHS-2 and PGHS-1 expression has been recently detected by reverse transcription-polymerase chain reaction (RT-PCR)[9].

PROSTANOID BIOSYNTHESIS AND PGHS ISOZYME EXPRESSION

Prostanoid biosynthesis

HT-29 cells cultured both in the absence and in the presence of foetal calf serum (FCS) produced very low quantities (<100 pg/10^6 cells) of prostanoids (PGE_2, $PGF_{2\alpha}$, 6-keto-$PGF_{1\alpha}$ and TXB_2). The treatment of HT-29 cells with the calcium ionophore A23187 (2 μM) or AA (10 μM) for 1 h caused a significant stimulation of prostanoid formation. PGE_2 and $PGF_{2\alpha}$, the predominant prostanoids, were increased by 6- and 14-fold in response to A23187 and AA, respectively.

In successive experiments, we selected the measurement of $PGF_{2\alpha}$ as a index of PGHS activity of HT-29 cells because PGE_2 is unstable in long-term cell cultures due to covalent binding to plasma proteins and enzymatic degradation.

PGHS-1 and PGHS-2 expression

To assess the contribution of PGHS-1 and PGHS-2 to the biosynthesis of prostanoids by HT-29 cells, we evaluated transcript levels in cells incubated for 2 and 24 h with 0.5 and 10% FCS by Northern blot. PGHS-1 and human glyceraldehyde phosphate dehydrogenase (GAPDH, used as control) mRNAs were detected in HT-29 cells both at 2 and 24 h of incubation. PGHS-1 cDNA probe hybridized to two transcripts of 2.7 and 5.2 kb. Similar results were found in the human histiocytic lymphoma cell line U-937 (used as control) treated for 72 h with PMA (5 nM). When the same filter was probed with the PGHS-2 cDNA, we did not detect any transcript in HT-29 cell extracts while a mRNA species of 5 kb was detected in U-937 cells.

As shown in figure 1 (panel A), Western blot analysis of HT-29 cell lysates cultured for 24 h in the presence of 0.5 or 10% FCS using a specific anti-PGHS-1 antibody, showed that HT-29 cells contained PGHS-1 protein. The use of rabbit polyclonal antibodies directed against a unique aminoacid sequence present in human PGHS-2 but not in PGHS-1 protein[17], evidenced a protein of approximately 70 kDa in lysates obtained by HT-29 cells both treated with 0.5 and 10% FCS (figure 1, panel A). The specificity of PGHS-2 like immunoreactivity detected in HT-29 cells was studied by incubating membranes with polyclonal rabbit anti-PGHS-2 serum or non-immune rabbit serum. As shown in figure 1 (panel B), a similar band was detected when Western blot was performed both in the absence and in the presence of specific anti-PGHS-2 antibodies. These results suggest that antibodies were detecting a non specific immunoreactive band. In contrast, in human monocytes stimulated for 24 h with LPS (10 μg/ml) (used as control), a protein of approximately 70 kDa was detected only when nitrocellulose membrane was incubated with specific anti-PGHS-2 antibodies.

Thus, HT-29 cells display low levels of PGHS-2 expression that can be detected only by RT-PCR[9].

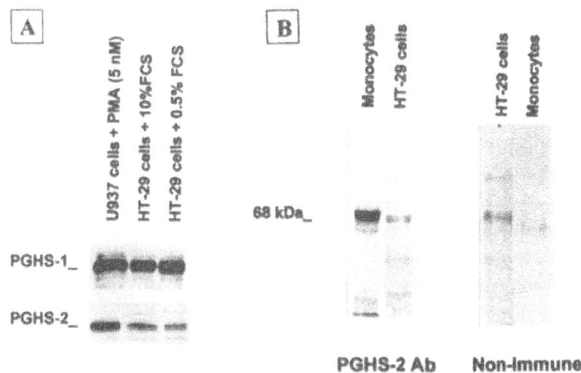

Figure 1. PGHS-1 and PGHS-2 protein levels in HT-29 cells. (A) HT-29 cells cultured for 24 h with 0.5 or 10% FCS and U-937 cells (used as control) cultured for 72 h with PMA (5 nM) were lysed and proteins were analysed by sodium dodecyl sulphate (SDS)-polyacrylamide gel electrophoresis and immunoblotting techniques using rabbit antibodies directed against PGHS-1 or the carboxyl-terminal of PGHS-2[17]. (B) Specificity of PGHS-2 like immunoreactivity detected by Western blot in HT-29 cells. HT-29 cells cultured for 24 h with 10% FCS and isolated human monocytes incubated for 24 h with LPS (10 μg/ml), used as a control, were lysed, proteins were electrophoresed and then transferred to nitrocellulose membranes. The membranes of nitrocellulose were incubated with polyclonal rabbit anti-PGHS-2 serum or non-immune rabbit serum. Equal amounts of proteins (20 μg) were loaded in all lanes. Immune complexes were visualized by incubating the membranes with biotin-conjugated anti-rabbit IgG and steptavidin-peroxidase.

Inhibition of HT-29 cell cyclooxygenase activity by L-745,337 and aspirin

L-745,337 is a novel antiinflammatory, antipyretic and analgesic compound that produces fewer gastrointestinal lesions in experimental animals than conventional NSAIDs[16]. In whole cell assays, this compound inhibits PGHS-2 with an IC_{50} of 23 nM, but is inactive on PGHS-1 at doses as high as 10 μM[16]. In HT-29 cells, L-745,337 inhibited $PGF_{2\alpha}$ production with an IC_{50} of 240$\pm$130 nM (mean$\pm$SEM, n=6) and almost complete suppression could be demonstrated at 10 μM. The potent and complete inhibitory effect of L-745,337 on PGHS activity of HT-29 cells suggests that the low expression of PGHS-2, only detectable by RT-PCR, can largely account for prostanoid production in HT-29 cells.

Aspirin dose-dependently inhibited the production of $PGF_{2\alpha}$ by HT-29 cells with an IC_{50} value of 27$\pm$10 μM (mean$\pm$SEM, n=3) while its metabolite sodium salicylate did not significantly affect $PGF_{2\alpha}$ production at concentrations up to 10 mM.

EFFECTS OF L-745,337, ASPIRIN AND SALICYLATE ON THE LEVELS OF P21[WAF-1/CIP1] AND MUTANT P53

HT-29 colon carcinoma cells harbor a G to A mutation at codon 273 in the p53 gene, resulting in the expression of a mutant p53 protein[18]. Mutant p53 and the tumor suppressor p21[WAF-1/cip1] proteins were constitutively expressed in HT-29 cells. Following 4 days of incubation of HT-29 cells in serum starving conditions (0.5% FCS) with aspirin and sodium salicylate up to 1 mM, the levels of mutant p53 and p21[WAF-1/cip1] were not significantly altered. At 10 mM, sodium salicylate caused a reduction by about 1/3 of mutant p53 and a 3-fold induction of p21[WAF-1/cip1].

Under the same experimental conditions, L-745,337 (5-500 μM) dose-dependently induced (1.7-6-folds) the tumor suppressor protein p21[WAF-1/cip1] (ED_{50}: 100 μM) while reducing the levels of mutant p53 protein (IC_{50}: 200 μM).

Expression of the tumor suppressor p21[WAF-1/cip1], a potent cyclin-dependent kinase (cdk) inhibitor[10], has been implicated in growth arrest in response to a variety of conditions, including DNA damage and terminal differentiation[11]. Induction of p21[WAF-1/cip1] in response to DNA damage requires the function of p53 tumor suppressor protein[11]. However, p21[WAF-1/cip1] expression was recently shown to be induced through a p53-independent mechanism[11]. In the present study, HT-29 cells constitutively expressed p21[WAF-1/cip1] that was increased by exposure to high concentrations of L-745,337 and salicylate in the absence of wild type p53.

MORPHOLOGICAL CHANGES INDUCED BY L-745,337 AND SALICYLATE

Apoptosis

To determine whether aspirin, sodium salicylate and L-745,337 induce cell death by apoptosis, HT-29 cell nuclei were stained with the chromatin staining Hoechst 33342 dye and examined with a fluorescence microscope. Cells with condensed, bright chromatin were scored as apoptotic. Approximately 60% of cell nuclei in the cultures treated for 4 days with 10 mM of sodium salicylate or 0.5 mM of L-745,337 in serum starving conditions showed typical signs of apoptosis, while cells treated with 1 mM of aspirin and DMSO vehicle did not.

Moreover, the effects of aspirin (1 mM), sodium salicylate (10 mM) and L-745,337 (5-500 μM) on the degradation of genomic DNA, a marker of apoptosis, were also analyzed by agarose gel electrophoresis and UV light after staining with ethidium bromide. The treatment of HT-29 cells with aspirin, sodium salicylate and L-745,337 for 4 days resulted in the degradation of genomic DNA of these cells, leading to small DNA fragments with double-strand break intervals of 180-200 bp below the predominant genomic DNA band.

Differentiation

HT-29 cells, a poorly differentiated, multipotent tumor cell type, can be induced to mature into both enterocyte-like goblet and absorptive cell phenotype by various substances such as sodium butyrate[19].

The morphological changes induced by exposure of HT-29 cells to aspirin (1 mM), sodium salicylate (10 mM) and L-745,337 (500 μM) for 4 days were evaluated by transmission electron microscopy. Untreated, DMSO- and aspirin-treated HT-29 cells showed monostrate growth of elongated cellular elements with large nuclei and finely dispersed chromatin. Lipofuscin-rich lamellar bodies, small mucus vacuoles, well condensed mitochondria and a well differentiated Golgi apparatus were observed in the scarce cytoplasm.

Exposure to L-745,337 (500 μM) was associated with the appearance of a differentiated phenotype. Frequently, cells aggregated forming lumens with well evident desmosomes and gap-junctions. In the lumen, fence-like microvilli were also seen. Abundant mucus granuli were observed in the cytoplasm. Similar morphological changes were observed in cells treated with 10 mM of sodium salicylate.

In HT-29 cells treated for 4 days with 10 mM of sodium salicylate and 500 μM of L-745,337, the number of lumens significantly increased by 5 and 15-fold, respectively, respect to untreated and DMSO treated cells. In contrast, the treatment with 1 mM of aspirin did not induce the formation of lumens.

At variance with our findings, it has been reported that HT-29 cell treatment with sulindac and sulindac sulfide for 1 day caused the up-regulation of p21[WAF-1/cip1] levels but this effect was not associated with intestinal cell differentiation evaluated by the expression of alkaline phosphatase, a marker of colon cancer cell differentiation[7]. It can be speculated that prolonged exposure to NSAIDs is required in order to up-regulate p21[WAF-1/cip1] to a critical level necessary for cell differentiation.

CONCLUSIONS

The main novel finding of our study is that the selective PGHS-2 inhibitor L-745,337 causes differentiation and apoptosis in human colon cancer cells displaying low levels of expression of PGHS-2 and low rates of prostanoid production. Similar responses were caused by high doses of sodium salicylate. These effects occurred through a mechanism(s) unrelated to cyclooxygenase inhibition that may involve p53-independent induction of the tumor suppressor p21[WAF-1/cip1]. Further studies are in progress in order to verify whether p21[WAF-1/cip1] accumulation may represent a functional response of cancer cells to NSAIDs mediating inhibition of growth, apoptosis and differentiation.

The relevance of these cyclooxygenase-independent effects remains to be investigated in appropriate pathophysiologic settings.

REFERENCES

1. M.J. Thun, Aspirin, NSAIDs, and digestive tract cancers, *Cancer Metastasis Rev.* 13: 269 (1994).
2. G.N. Levy, Prostaglandin H synthases, nonsteroidal anti-inflammatory drugs, and colon cancer, *FASEB J.* 11: 234 (1997).
3. E. Giovannucci, K.M. Egan, D.J. Hunter, et al., Aspirin and the risk of colorectal cancer in women, *N.Engl.J.Med.* 333: 609 (1995).
4. J.R. Vane, Inhibition of prostaglandins as a mechanism of action for aspirin-like drugs, *Nature New Biol.*, 23: 232 (1971).
5. W.L. Smith, R.M. Garavito, and D.L. DeWitt, Prostaglandin endoperoxide H synthases (cyclooxygenases)-1 and –2, *J. Biol. Chem.* 271: 33157 (1996).
6. M. Oshima, J.E. Dinchuk, S.L. Kargman, H. Oshima, B. Hancock, E. Kwong, J.M. Trzaskos, J.F. Evans, and M.M. Taketo, Suppression of intestinal polyposis in APC$^{\Delta 716}$ knockout mice by inhibition of cyclooxygenase 2 (PGHS-2), *Cell* 87: 803 (1996).
7. Y. Goldberg, I.I. Nassif, A. Pittas, L.L. Tsai, B.D. Dynlacht, B. Rigas, and S.J. Shiff, The anti-proliferative effect of sulindac and sulindac sulfide on HT-29 colon cancer cells: alteration in tumor suppressor and cell cycle-regulatory proteins, *Oncogene* 12: 893 (1996).
8. G.A. Piazza, A.J. Kulchak, M. Krutzsch, G. Sperl, N.S. Paranka, P.H. Gross, K. Brendel, R.W. Burt, D.S. Alberts, R. Pamukcu, and D.J. Ahnen, Antineoplastic drugs sulindac sulfide and sulfone inhibit cell growth by inducing apoptosis, *Cancer Res.* 55: 3110 (1995).
9. R. Hanif, A. Pittas, Y. Feng, M.I. Koutsos, L. Qiao, L. Staiano-Coico, S.I. Shiff, and B. Rigas, Effects of nonsteroidal anti-inflammatory drugs on proliferation and on induction of apoptosis in colon cancer cells by a prostaglandin-independent pathway, *Biochem. Pharmacol.* 52: 237 (1996).
10. J.W. Harper, G.R. Adami, N. Wei, K. Keyomarsi, and S.J. Elledge, The p21 Cdk-interacting protein Cip1 is a potent inhibitor of G$_1$ cyclin-dependent kinases, *Cell* 75: 805 (1993).
11. K.F. Macleod, N. Sherry, G. Hannon, D. Beach, T. Tokino, K. Kinzler, B. Vogelstein, and T. Jacks, P53-dependent and independent expression of p21 during cell growth, differentiation and DNA damage, *Genes Dev.* 9: 935 (1995).
12. J. Schwaller, H.P. Koeffler, G. Niklaus, P. Loetscher, S. Nagel, M.F. Fey, and A. Tobler, Posttranscriptional stabilization underlies p53-independent induction of p21$^{\text{WAF-1/CIP/SDI1}}$ in differentiating human leukemic cells, *J. Clin. Invest.* 95: 973 (1995).
13. F. Cipollone, P. Patrignani, A. Greco, M.R. Panara, R. Padovano, F. Cuccurullo, C. Patrono, A.G. Rebuzzi, G. Liuzzo, G. Quaranta, and A. Maseri, Differential suppression of thromboxane biosynthesis by indobufen and aspirin in patients with unstable angina, *Circulation* 96: 1109 (1997).
14. O. Laneuville, D.K. Breuer, D.L. DeWitt, T. Hla, C.D. Funk, and W.L. Smith, Differential inhibition of human prostaglandin endoperoxide H synthases-1 and –2 by nonsteroidal anti-inflammatory drugs, *J. Pharmacol. Exp. Ther.* 271: 927 (1994).

15. M.R. Panara, A. Greco, G. Santini, M.G. Sciulli, M.T. Rotondo, R. Padovano, M. di Giamberardino, F. Cipollone, F. Cuccurullo, C. Patrono, and P. Patrignani, Effects of the novel anti-inflammatory compounds, N-[2-(cyclohexyloxy)-4-nitrophenyl] methanesulphonamide (NS-398) and 5- methanesulphonamido-6-(2,4-difluoro-thiophenyl)-1-indanone (L-745,337), on the cyclo-oxygenase activity of human blood prostaglandin endoperoxide synthases, *Br. J. Pharmacol.* 116: 2429 (1995).

16. C.-C. Chan, S. Boyce, C. Brideau, A.W. Ford-Hutchinson, R. Gordon, D. Guay, R.G. Hill, C.-S. Li, J. Mancini, M. Penneton, P. Prasit, R. Rasori, D. Riendeau, P. Roy, P. Tagari, P. Vickers, E. Wong, and I.W. Rodger, Pharmacology of a selective cyclooxygenase-2 inhibitor, L-745,337: a novel nonsteroidal antiinflammatory agent with an ulcerogenic sparing effect in rat and nonhuman primate stomach, *J. Pharmacol. Exp. Ther.* 274: 1531 (1995).

17. A. Habib, C. Creminon, Y. Frobert, J. Grassi, P. Pradelles, and J. Maclouf, Demonstration of an inducible cyclooxygenase in human endothelial cells using antibodies raised against the carboxyl-terminal region of the cyclooxygenase-2, *J. Biol. Chem.* 268: 23448 (1993).

18. N.R. Rodrigues, A. Rowan, M.E.F. Smith, I.B. Kerr, W.F. Bodmer, J.V. Gannon, and D.P. Lane, p53 mutations in colorectal cancer, *Proc. Natl. Acad. Sci. U.S.A.* 87: 7555-7559 (1990).

19. C. Augeron, and C.L. Laboisse, Emergence of permanently differentiated cell clones in a human colonic cancer cell line in culture after treatment with sodium butyrate, *Cancer Res.* 44: 3961 (1984).

82

MODULATION OF CELLULAR PROLIFERATION AND INDUCTION OF APOPTOSIS
IN A HUMAN LYMPHOMA CELL LINE AFTER TREATMENT WITH SELECTIVE
LIPOXYGENASE INHIBITORS

Kenning M. Anderson , Frank G. Ondrey [1], and Jules E. Harris

Dept. of Medical Oncology
Rush Medical College
Chicago, IL 60612, USA

[1]Head and Neck Surgery Branch, Tumor Cell Biology Division
National Institute on Deafness and Other Communication Disorders
National Institutes of Health
Bethesda, MD 20892, USA

INTRODUCTION

It has long been recognized that eicosanoids have the ability to modulate cellular proliferation in many in vitro cell culture systems. It had also been recognized that both lipoxygenase and cyclooxygenase inhibitors have the ability to alter cellular proliferation in many cell types. Previously, we have shown that lipoxygenase inhibition is associated with decreased cellular proliferation in human lymphoma (1), leukemic blast- crisis cells (2), prostate adenocarcinoma (3), glioblastoma (4), and squamous cell carcinoma cells (5). In some cases, sub-populations of treated cells appear to undergo changes associated with cellullar differentiation (6, 7). Classic inhibitors of eicosanoid metabolism have been shown to have multiple physiologic effects, many of which are not attributable to their inhibition of eicosanoid synthesis (8, 9). In the present work, U937 cells were cultured with selective 5- lipoxygenase inhibitors as well as the less selective inhibitors, nordihydroguaeretic acid (NDGA) and eicosatetraynoic acid (ETYA), and examined for effects on cell proliferation and the induction of programmed cell death.

MATERIALS AND METHODS

Cell Culture Techniques: Mycoplasma- free U937 cells were grown in continuous culture in 10% FCS containing RPMI 1640 media, with penicillin, streptomycin, 25mM HEPE and 5mM L-glutamine. For several experiments, cells were conditioned to grow in 5%, 3% and 1.5% FCS- containing media without other additives than described. The rate of cellular proliferation was decreased compared to cells grown in 10% media but these low serum clones were able to be maintained in continuous culture and were subcultured weekly.

MTT assays: Log-phase U937 cells were gently pelleted at 125 x g, and placed into 96 well microtiter plates in 10, 5, 3, or 1.5, or 0 % FCS- containing media. The inhibitors, NDGA (Caymen Chemical, Ann Arbor, MI), A63162 (Abbott Labs, Abbott Park, IL), MK886 (Caymen Chemical, Ann Arbor, MI) were added at various concentrations by serial dilution. The carrier, DMSO, was added to control wells at appropriate concentrations as well. After incubation for 3 to 5 days, MTT reagent (Boehringer Mannheim, Indianapolis, IN) was added to all wells for 4 hours. Overnight incubation with the kit-supplied lysis buffer was performed. The assays were read the following day at 570 nM on an EIA Model 530 Plate Reader. Five replicates/data point are plotted.

Apoptosis assays: DNA laddering assays were performed by standard techniques. DNA was isolated with DNA/RNA isolation kit (US Biochemicals, Cleveland, OH) and treated with DNAse- free RNAse at 40^0 C for 1 hour. Two to five ug samples were loaded on 0.8 % Agarose gels with ethidium bromide and electrophoresed at 70-90 volts for 1 hour.

RT-PCR for Lipoxygenase pathway enzymes: This was performed as previously described (10). First strand cDNA, prepared from random hexamers was PCR-amplified from cDNA sequences of 5-LO and FLAP and selected with DNASTAR software. PCR was performed for 35 cycles with 94^0 C denaturation for 30 seconds, 60^0 C annealing for 15 seconds and 72^0 C extenson for 1 minute. RT-PCR products were resolved on 1.3% agarose gels and confirmed by southern blotting onto NC filters and hybridized to radiolabelled internal primer probes. Both 5-LO and FLAP products were of the expected size.

RESULTS

U937 cells were conditioned to grow under low serum conditions over several cell passages. Cells were kept in continuous culture at 10%, 5%, 3% and 1.5% FCS and grew without additional supplementation. There were no changes in gross cellular morphology with these culture adaptations. In previous work (1) we observed that eicosanoid inhibitors need to be at higher concentrations in the presence of 5 or 10% serum compared to serum- free conditions to observe significant decreases in thymidine incorporation. In the present series of experiments, this was also seen with the selective lipoxygenase inhibitor A63162 (Figure 1). Cells grown in 10 %, 5%, and 3% serum exhibited minimally decreased cellular proliferation compared to control cells, as judged by MTT assay (Figure 1 upper left, upper right, and lower left, respectively). Cells that were grown in 10% FCS-containing medium were also placed in media without serum (lower right panel). Cells grown in media without serum for the duration of the assay exhibited more significant decreases in proliferation (Figure 1. lower right). Baseline MTT -incorporation for cells plated at the start of the assay was not subtracted from the final O.D. in these representations.

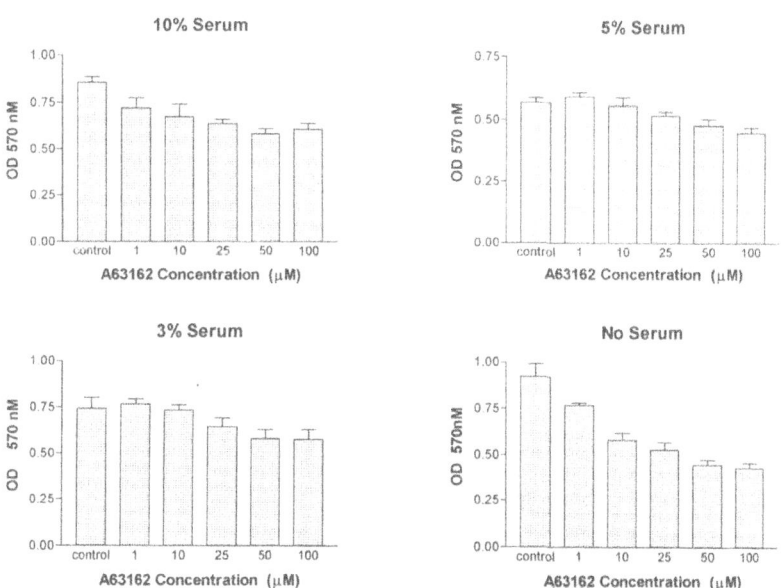

Figure 1 Concentration-dependent inhibition of cellular proliferation in subconfluent U937 cells in varying concentrations of fetal calf serum. Cells were cultured for 3-4 days in the presence of the inhibitors. 5000 cells/ well on a 96 well plate were initially plated at the time of addition of the inhibitors. Upper left, upper right, and lower left panels represent U937 cell line clones that were adapted to grow at decreasing concentrations of serum. The lower right panel represents the parental cell line (cells grown in 10% FCS) placed in media without serum at the time the inhibitors were added.

In the next experiments, U937 cells maintained in 1.5% fetal calf serum were grown for 96 hours in the presence of A63162, NDGA, or MK 886, a selective inhibitor of the 5 lipoxygenase activating protein, or FLAP. Exposure to these inhibitors resulted in a dose dependent decrease in cellular proliferation, as judged by MTT assay (Figure 2). Cells cultured with NDGA experienced significant decreases in MTT incorporation at concentrations above 5 uM (left panel). The reported IC_{50} for purified lipoxygenase enzyme in a cell free and serum free system is between 1 and 5 uM (11). In the center panel, inhibition of MTT incorporation was observed at concentrations of 10 uM and above. The reported IC_{50} for A63162 for purified

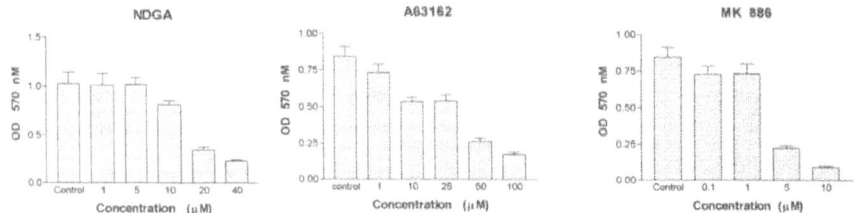

Figure 2 Inhibition if U937 cellular proliferation with NDGA, A63162, and MK 886, in 1.5% serum-containing media. Cells were grown in 1.5 fetal bovine serum-containing media. X-axis represents concentration of each inhibitor in uM and the y axis the mean optical density of five replicates +/- 1 S.E.M. One example experiment of three experiments is depicted.

lipoxygenase enzyme in a cell free and serum free system is between 3 and 5 uM (11). In the right panel, the inhibition of MTT incorporation by MK886 is depicted. Significant inhibition of cellular proliferation was observed at concentrations of MK 886 at 5 uM and above. The reported IC $_{50}$ of MK 886 for five lipoxygenase activating protein is reportedly as low as 3 nM in systems without serum or purified enzyme preparations. However the IC $_{50}$ has been reported to be as high as 2.1 uM in preparations containing serum (i.e. whole blood) (12).

U937 cells were next cultured with several concentrations of lipoxygenase inhibitors for 72 hours in the presence of 10% serum. Cell treatment for three lipoxygenase and one FLAP inhibitor result in classic, oligomeric DNA laddering indicative of programmed cell death (Figure 3). Other experiments were performed to identify 5 Lipoxygenase and FLAP. Messenger RNA for both of these proteins was found in our U937 cell clone as well as a second U937 cell clone obtained from ATCC by RT-PCR.

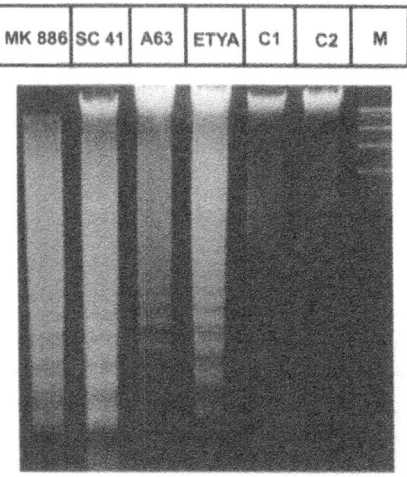

Figure 3 DNA laddering in U937 cells treated with lipoxygenase inhibitors. Cells were cultured with 40 uM MK886, 40 uM SC41661, 40 uM A63162, or 40 uM eicosatetraynoic Acid (ETYA) as denoted in the figure. C1 represents control cells incubated with the drug vehicle DMSO and C2 represent unincubated control cells.

566

DISCUSSION

In these experiments, we show that over several days in culture, both selective and non-selective lipoxygenase inhibitors decreased U937 cellular proliferation. The effects observed with A63162 were dependent on the concentration of serum, as the presence of 3% or greater FCS in the experiments resulted in insignificant inhibition of cellular proliferation. Previously we have observed that the presence of serum offered significant protection from the anti-proliferative effects of several lipoxygenase inhibitors (1). These were short term experiments , in which decreased thymidine incorporation was the endpoint for decreased cellular proliferation. In the case of MK 886, others have observed 1000-fold greater concentrations of the inhibitor to obtain an equivalent IC $_{50}$ for serum-containing cell systems. These factors would certainly be important for the potential in vivo use of these compounds as anti-proliferative agents parenterally in humans.

Both NDGA and A63162 reduced cell proliferation at concentrations similar to their IC $_{50}$ values for purified enzyme. From this one can conclude that these inhibitors may be functioning as anti-proliferative agents via their direct inhibition on lipoxygenase pathways. These effects were observed with both selective and non-selective inhibitors and agree with our previous findings in several cell types.

MK 886, an inhibitor of FLAP, in addition to inhibiting cellular proliferation, induced programmed cell death (PCD). This proved to be typical Type 1 PCD (13). MK886 also induced an abrupt rise in cytosolic Ca++ in other experiments (14), an event often associated with PCD in lymphohematopoietic cells. The growth inhibitory effects were dependent upon serum concentration, as decreased serum conditions were necessary to observe the maximal inhibitory effect. It is not known whether serum binds the inhibitors under these experimental conditions, thereby effectively decreasing their concentration or whether some other inactivation by the serum components is responsible. Alternatively or in addition, the concentration of growth factors, hormones, or cytokines in serum may provide a cytoprotective effect that prevented the decreases in proliferation or the induction of apoptosis by MK 886. It is not established that a direct effect of the lipoxygenase inhibitors on that pathway is responsible for these findings, or other cellular effects such as altered lipid peroxidation, arachidonic acid signaling, peroxide tone, cellular redox changes, inhibition of mitochondrial function (15), etc. may contribute. Clearly, the expression of PCD due to lipoxygenase inhibitors requires initial or concomitant reduction in cellular proliferation, as other experiments with less effective lipoxygenase inhibitors or quiescent cells do not result in PCD (13). This effect is not simply due to "non-specific cytotoxicity" but represents an active selection and implementation of specific metabolic events. Since this clone of U937 cells expresses mRNA for 5 LO and FLAP, they represent an appropriate model for further studies with these lipoxygenase inhibitors.

ACKNOWLEDGMENTS:

This work was supported by the Weinberg Foundation. We would like to thank Dr. J. Mulshine and Dr. Michelle Vos, BPRB, NCI for their assistance with the RT-PCR analysis. We would like to thank N. Falotico, Abbott Labs, Abbott Park, IL for the generous gift of A63162.

BIBLIOGRAPHY

1. Ondrey FG, Harris JE, Anderson KM: Inhibition of Eicosanoid and DNA Synthesis by 5,8,11,14-Eicosatetraynoic Acid, an Inhibitor of Arachidonic Acid Metabolism and its Partial Reversal by LT C4. Cancer Research 49:1138-1142, 1989.
2. Anderson KM, Seed T, JajehA, et al, An in vivo Inhibitor of 5-Lipoxygenase, MK 886, Induces Apoptosis in U937 and CML Cells, Anticancer Research, 16, 2589-2600, 1996.
3. Anderson KM, Wygodny J, Ondrey FG, Harris JE: Human PC-3 Prostate Cell Line DNA Synthesis is Suppressed by an In Vitro Inhibitor of Arachidonic Acid Metabolism, The Prostate, 12:1, 3-12, 1988.
4. Wilson DE, Digianfillipo A, Ondrey FG, Anderson KM, Harris JE: Effects of Nordihydroguaiaretic Acid on Cultured Rat and Human Glioma Cell Proliferation, Journal of Neurosurgery, 71:551-557, 1989.
5. Ondrey FG, Juhn SK, Adams GL: Inhibition of Head and Neck Tumor Cell Growth with Arachidonic Acid Metabolism Inhibition. Laryngoscope, 106(2), 129-134, 1996
6. Ondrey FG, Harris JE, Anderson KM: Differentiation of U937 Cells Induced by 5,8,11,14-Eicosatetraynoic Acid, a Competitive Inhibitor of Arachidonic Acid Metabolism, Experimental Cell Research, 179:477-487, 1988.
7. Anderson KM, Seed T, Ondrey FG, and Harris JE: The Selective 5-Lipoxygenase Inhibitor, A63162, Reduces Prostate (PC3) Cell Proliferation and Initiates Morphologic Changes Consistent with Secretion Anticancer Research, 14(5a):1951-60, 1994.
8. Anderson KM, Ondrey FG, Harris JE: ETYA, A Pleotropic Membrane- Active Arachidonic Acid Analogue Affects Multiple Signal Transduction Pathways in Cultured Transformed Mammalian Cells, Clinical Biochemistry 25:1-9, 1992.
9. Anderson KM, Ondrey FG, and Harris JE: Arachidonic Acid Analogues An Additional Class of Membrane- Active Agents with Potential Anticancer Activity, Prostaglandins, Leukotrienes, and Essential Fatty Acids: Reviews 35:231-241, 1989.
10. Avis A, Jett M, Boyle T, Vos M, Moody T, Treston A, Martinez A, Mulshine J: Growth Control of Lung Cancer by Interruption of 5-Lipoxygenase-mediated Growth factor Signaling, Journal of Clinical Investigation, 97(3): 806-813, 1996.
11. Marshall P, Griswold D, Breton J, et al. Pharmacology of the Pyrroloimidazole, SK & F 105809-I, Biochemical Pharmacology 42(4):813-24, 1991
12. Young R, Gillard J, Hutchinson J, et al: Discovery of Inhibitors of the 5-lipoxygenase activating Protein (FLAP), Journal of Lipid Mediators 6:233-238, 1993.
13. Anderson KM, Seed T, Jajeh A, et al: An in Vivo Inhibitor of 5-lipoxygenase, MK886, at Micromolar Concentration induces Apoptosis in U937 and CML Cells, Anticancer Research 16:2589-2600,1996
14. Byun T, Dudeja P, Harris J, et al : A 5-lipoxygenase inhibitor raises intracellular calcium in U937 Cells Prior to their Physiologic Cell Death, Prostaglandins, Leukotrienes, and Essential Fatty Acids 56:69-77, 1996
15. Kroemer G, Zamzani N and Susin S: Mitochondrial Control of Apotposis. Immunology Today 18:44-51, 1997.

83

MECHANISMS OF PEROXYNITRITE-INDUCED APOPTOSIS IN HL-60 CELLS

King-Teh Lin, Ji-Yan Xue, and Patrick Y-K Wong

Department of Cell Biology
School of Osteopathic Medicine
University of Medicine and Dentistry of New Jersey
Stratford, New Jersey 08084

INTRODUCTION

Apoptosis, aptly termed programmed cell death, is an active process critical for the homeostatsis of organisms, functioning to eliminate superfluous cells (1,2). It can be recognized by a characteristic series of morphological and biochemical alterations including cell shrinkage, blebbing, chromatin condensation, and DNA degradation into nucleosomal fragments (1,3). Enzymes of the caspase family are responsible for executing this process (4,5). Of importance, proteolytic cleavage of pro-caspase-3 (CPP32/Yama/apopain) to the active form (caspase-3) has been proposed to be a crucial step in initiating the execution process of apoptosis (5,6).

Peroxynitrite (ONOO2) is a biological product generated from the interaction of nitric oxide (ANO) and superoxide anion (O_2A2), present in a variety of mammalian cells including endothelial cells, neurons, neutrophils, and macrophages (7-10). Recently, we demonstrated that ONOO2 induces apoptosis in human leukemia HL-60 cells in a concentration- and time-dependent manner (11). As a strong oxidant and reactive species, peroxynitrite can oxidize protein and non-protein sulfhydryls (12), initiate lipid peroxidation (13), generate free radical intermediates (14), and cause DNA breakage (15). However, the exact mechanism involved in peroxynitrite-induced apoptosis is still unclear. In the present study, we elucidate the signal cascade elicited by peroxynitrite and provide insight toward the mechanism of peroxynitrite-induced apoptosis of HL-60 cells.

EXPERIMENTAL PROCEDURES

Peroxynitrite treatment: HL-60 cells were resuspended in 5 ml of Dulbecco=s phosphate-buffered saline (D-PBS) and treated with various concentrations of peroxynitrite as described in Lin et al. (11,16).

Quantitative assay of apoptotic DNA fragmentation: The extent of apoptotic DNA fragmentation was determined by a method adapted from that of Sellins and Cohen (17). The cell pellets were lysed with 0.3 ml hypotonic lysing buffer (10 mM Tris, pH 8.0, 1 mM EDTA, 0.5% Triton X-100) and the lysates centrifuged to separate intact and

Eicosanoids and Other Bioactive Lipids in Cancer, Inflammation, and Radiation Injury, 4
Edited by Honn *et al.*, Kluwer Academic / Plenum Publishers, New York, 1999.

569

fragmented chromatin. Both pellet and supernatant were precipitated with 12.5% trichloroacetic acid (TCA). The DNA precipitate was heated to 90EC for 10 min in 400 μl of 5% TCA, and quantitative analysis was carried out by reaction with diphenylamine (18). The percentage of DNA fragmentation was calculated from the ratio of DNA in the supernatant to the total DNA (supernatant plus pellet).

Measurement of reactive oxygen species (ROS): Generation of reactive oxygen species was measured by using an oxidation-sensitive fluorescent probe, 2′,7′-dichlorofluo-rescin diacetate (DCFH-DA), which oxidized form (2′,7′-dichlorofluorescein, DCF) is highly fluorescent (19). After washed with D-PBS twice, HL-60 cells were resuspended in 5 ml of D-PBS (1x10⁶/ml) and treated with indicated concentrations of ONOO2 for 5 min at 37EC. Then, the cells were centrifuged and resuspended in Krebs buffer (pH 7.4; NaCl, 118 mM; KCl, 4.7 mM; CaCl₂, 1.5 mM; NaHCO₃, 25 mM; MgSO₄, 1.1 mM; KH₂PO₄, 1.2 mM). DCFH-DA was added into cell suspension for measurement of ROS using Cytofluor9 2300 (PerSeptive Biosystems, Inc. MA) with an excitation wavelength of 485 nm and an emission wavelength of 530. Usually, 30 min interval was needed from peroxynitrite treatment to DCFH-DA addition.

Measurement of superoxide anion: Generation of O₂A2 was measured by chemiluminescence probe, bis-N-methylacridinium nitrite (lucigenin, Sigma) (20,21). HL-60 cells were resuspended in 5 ml of D-PBS (1x10⁶/ml) and treated with indicated concentrations of ONOO2 for 2 min at 37EC. The cells were centrifuged, resuspended in Krebs buffer. After the addition of lucigenin (final concentration: 0.25 mM), the cells were counted using a Mark 5303 scintillation counter (TM Analytic, IL) (22) at indicated time and continuously for an additional 90 min. Usually, 25 min interval was needed from peroxynitrite treatment to lucigenin addition. The zero time starts with the addition of lucigenin. In case of exogenous superoxide dismutase (SOD) treatment, SOD was added immediately after ONOO2 addition, and was also present in the Krebs buffer.

Western blot analysis: For detecting PARP cleavage, HL-60 cells were treated with 100 μM peroxynitrite and harvested at the indicated times. After washing once with PBS, cells were suspended in sample buffer (62.5 mM Tris/HCl, pH 6.8; 6 M urea; 10% glycerol; 2% SDS; 0.00125% bromophenol blue; 5% β-mercaptoethanol), then sonicated for 15 s and incubated at 65EC for 15 min. Equal amount of each cell extract was subjected to 7.5% SDS-PAGE for detecting PARP cleavage and 12% SDS-PAGE for detecting proteolysis of pro-caspase-3. After transfer of proteins to polyvinylidene difluoride membrane (PVDF, Bio-Rad), and blockade with 5% nonfat dry milk in PBS-0.05% Tween 20, the membranes were incubated with anti-PARP mouse monoclonal antibody (C-2-10, Biomol, PA) or anti-procaspase-3 mouse monoclonal antibody (Transduction Laboratories, KY) for 2 h. Antibody binding was detected using peroxidase-conjugated anti-mouse IgG (Sigma) and visualized by a standard chemiluminescence method (DuPont NEN, MA) performed according to the manufacturer's instructions.

Pretreatment of HL-60 cells with caspase-1 and caspase-3 inhibitors: The caspase-1 inhibitor, Ac-YVAD-CHO (N-acetyl-Tyr-Val-Ala-Asp-aldehyde, Biomol, PA) or caspase-3 inhibitor, Ac-DEVD-CHO (N-acetyl-Asp-Glu-Val-Asp-aldehyde, Biomol, PA), was directly added into the culture medium one hour before peroxynitrite treatment. Cells were harvested at selected intervals following ONOO2 addition and apoptotic DNA fragmentation was measured as described.

Measurement of caspase-3 activity using fluorogenic substrate DEVD-AMC: The activity of caspase-3 was measured as described by Nicholson et al. (5). The fluorescence emission of the 7-amino-4-methyl coumarin (AMC), released upon proteolytic cleavage of the fluorogenic substrate DEVD-AMC by active caspase-3, was measured using Cytofluo™ 2350 (excitation wavelength = 380 nm, emission wavelength = 460 nm).

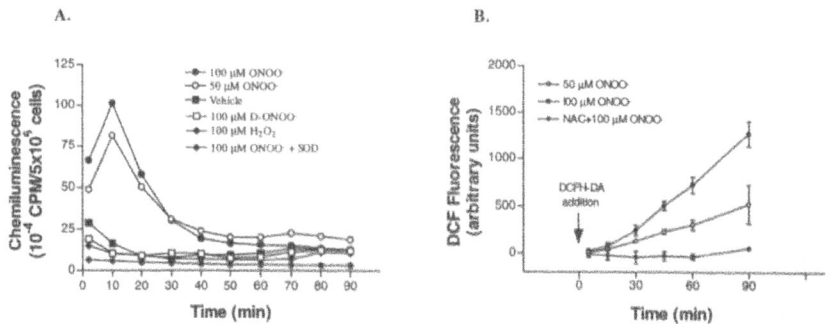

Figure 1. Peroxynitrite induces superoxide generation (A) (16) and ROS formation (B) after brief exposure of HL-60 cells to various concentrations of peroxynitrite and reagents.

RESULTS AND DISCUSSION

To elucidate the possible signals elicited by peroxynitrite that contribute to peroxynitrite-induced apoptosis in HL-60 cells, we employed highly specific probes, lucigenin (20) and DCFH-DA (19) to detect the formation of superoxide anion and other ROS after peroxynitrite treatment. Brief exposure of HL-60 cells to various concentrations of peroxynitrite induces elevation of lucigenin chemiluminescence, indicating generation of superoxide anion (Fig. 1A). Peroxynitrite-elicited superoxide anion generation was completely abolished by exogenous SOD, a scavenger of O_2A2, which was added immediately after peroxynitrite treatment. In contrast, decomposed peroxynitrite (D-ONOO2, 100 μM) failed to induce superoxide anion generation and the increase of superoxide was not observed after 100 μM H_2O_2 treatment under the similar experimental condition, indicating the effect is specific for peroxynitrite. Corresponding to the formation of O_2A2, the accumulation of ROS as measured by the fluorescent DCF was observed after brief exposure of cells to various concentrations of peroxynitrite (Fig 1B). To demonstrate the contribution of ROS elicited by ONOO2 to the ONOO2-induced apoptotic cell death, exogenous SOD (100 u/ml), catalase (200 u/ml) or both SOD and catalase (SOD/catalase) were added into the cell suspensions immediately after the addition of 100 μM ONOO2. The same amount of SOD, catalase or SOD/catalase were also resupplemented into the culture medium after the washout of post-treated ONOO2. As shown in Figure 2, peroxynitrite caused significant apoptotic DNA fragmentation, an indication of the apoptotic cell death, but decomposed peroxynitrite had no effect in HL-60 cells at 4 h after treatment. The addition of catalase or SOD/catalase significantly attenuated the peroxynitrite-induced apoptotic effect. In contrast, the addition of exogenous SOD did not block, instead, slightly exacerbated the peroxynitrite-induced apoptosis. These results indicated that H_2O_2 formation produced from the dismutation of ONOO2-elicited superoxide anion contributes to apoptosis in HL-60 cells. To further test this hypothesis, HL-60 cells were pretreated with cell permeable ROS scavenger, N-acetyl-cysteine (NAC, 15 mM, 3 h). NAC completely abolished the ONOO2-mediated DCF fluorescence (Fig. 1B). Associated with the effect of scavenging ROS, NAC attenuated the apoptotic response of peroxynitrite in HL-60 cells as determined by LDH release (Fig. 3). Taken together, these results indicated that the generation of ROS elicited by peroxynitrite contributes to the peroxynitrite-induced apoptosis in HL-60 cells.

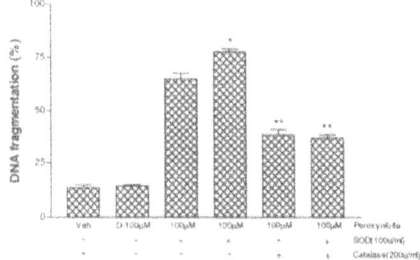

Figure 2. Effects of SOD or catalase on ONOO⁻-induced apoptosis of HL-60 cells as quantified by apoptotic DNA fragmentation (16).

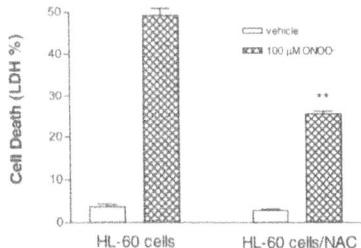

Figure 3. Effects of NAC on peroxynitate-induced apoptosis of HL-60 cells as measured by LDH Release at 24 h after peroxynitrate treatment (16).

Although reactive oxygen species (ROS) formation (16) elicited by peroxynitrite participate in peroxynitrite-induced apoptosis, the execution process of peroxynitrite-induced apoptosis has not been identified. To further elucidate the essential mechanisms critical for executing peroxynitrite-induced apoptosis, we examined the proteolytic cleavage of PARP during peroxynitrite-induced apoptosis in HL-60 cells. Proteolytic cleavage of PARP from a 116 KDa polypeptide to an 85-KDa fragment is a sensitive marker at the onset of apoptosis and also reflects caspase-3 activation (23). HL-60 cells were treated with 100 μM of peroxynitrite and harvested at the indicated times. Western blot analysis using anti-PARP antibody revealed that proteolytic cleavage of the 116 KDa PARP holoenzyme into the 85 KDa fragment (Fig. 4) was slightly visible 90 min after peroxynitrite treatment. The intense cleavage at 180 min was correlated with overt apoptotic DNA fragmentation (24). These results suggested participation of a caspase-3 family protease in executing peroxynitrite-induced apoptosis of HL-60 cells.

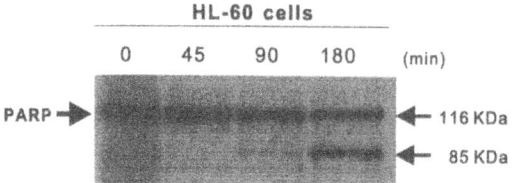

Figure 4. Time-dependent proteolytic cleavage of PARP during peroxynitrite-induced apoptosis of HL-60 cells.

To ascertain whether caspase-3 might be proteolytically activated by proteolysis of the inactive proenzyme during ONOO2-induced apoptosis, a monoclonal antibody against human caspase-3 proenzyme (32 KDa) was used to detect the levels of caspase-3 proenzyme during peroxynitrite-induced apoptosis by Western blot analysis. The results obtained from Western blot (Fig. 5) lend support to this hypothesis that the levels of caspase-3 proenzyme diminished in a time-dependent fashion due to proteolysis of the proenzyme. As shown in figure 5, in response to 100 μM peroxynitrite, the levels of caspase-3 proenzyme were considerably reduced at 3 h, and the proenzyme was hardly detected at 5 h after treatment. However, the latent 32-KDa proenzyme was clearly observed even 5 h after treatment in untreated and vehicle-treated HL-60 cells (Fig. 5). This Western blot analysis demonstrated that caspase-3 was indeed proteolytically activated at the onset of peroxynitrite-induced apoptosis.

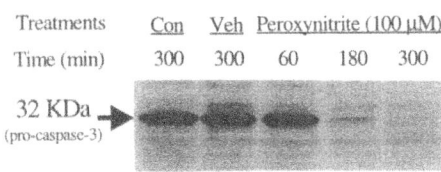

Figure 5. Effects of peroxynitrite on proteolysis of procaspase-3 (activation of caspase-3) during peroxynitrite-induced apoptotic cell death in HL-60 cells .

572

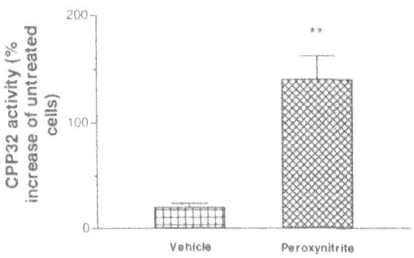

Figure 6. Effects of peroxynitrte on the activities of caspase-3 as measured by the emission of AMC fluorescence during peroxynitrite-induced apoptosis of HL-60 cells. The relative levels of AMC in peroxynitrite and vehicle groups were expressed as percent increase of untreated normal HL-60 cells.

The activity of caspase-3 was further investigated using a specific fluorogenic substrate of caspase-3 protease, Ac-DEVD-AMC (5). A marked increase of AMC fluorescence was observed at 3 h after peroxynitrite (Fig. 6), corroborating the results obtained by Western blot analysis (Fig 5). If caspase-3 is the major cysteine protease responsible for Acommitting≅ the HL-60 cells to undergo apoptosis in response to peroxynitrite, factors which specifically antagonize the action of caspase-3 should be able to prevent apoptosis. To test this hypothesis, HL-60 cells were pretreated with either Ac-DEVD-CHO, a potent and selective inhibitor of caspase-3, or Ac-YVAD-CHO, a known specific inhibitor of caspase-1, one hour before 100 μM of peroxynitrite treatment. As shown in Figure 7, pretreatment of HL-60 cells with 100 μM of Ac-DEVD-CHO significantly ($p < 0.01$) attenuated the apoptotic response to 100 μM peroxynitrite as determined by DNA fragmentation (Fig. 7). By contrast, Ac-YVAD-CHO failed to inhibit peroxynitrite-induced apoptosis under the same experimental conditions, indicating that caspase-1 activation is not involved in peroxynitrite-induced apoptosis of HL-60 cells. Taken together, these results strongly indicate that caspase-3 is the key cysteine protease engaged in peroxynitrite-induced apoptotic cell death in HL-60 cells.

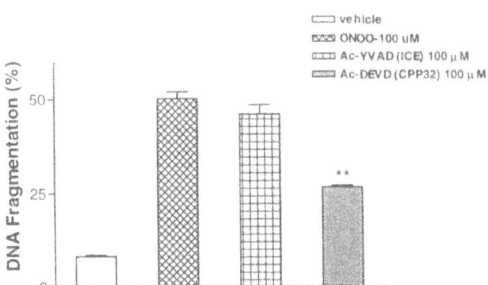

Figure 7. Effects of protease inhibitors on peroxynitrite-induced apoptosis of HL-60 cells. The caspase-1 (ICE) inhibitor Ac-YVAD and the caspase-3 (CPP32) inhibitor Ac-DEVD were directly added into the culture medium 1 h before peroxynitrite treatment. Cells were harvested 4 h after peroxynitrite treatment and apoptotic DNA fragmentation was quantified as described in A Materials and Methods≅.

573

It has been demonstrated that PARP, an enzyme implicated in DNA repair and genome surveillance, can be activated in response to DNA damage (25). Once recruited to the sites of DNA damage, the activated PARP will catalyze poly(ADP-ribosy)lation in the process of DNA repair at the expense of NAD$^+$, leading to depletion of the cellular ATP energy source (26). Exposure of cultured J774 macrophages to high concentrations of peroxynitrite (500 μM to 1 mM) can cause serious DNA strand breaks and consequent PARP activation (27). The consequent depletion of intracellular NAD$^+$ and ATP has been suggested to be responsible for the cytotoxicity in these cells. In our experimental system, during peroxynitrite-induced apoptotic cell death, PARP appears to be inactivated due to proteolytic cleavage of PARP by a caspase-3 family protease rather than activated. Although it appears that PARP cleavage may not be a direct cause of apoptosis (28), proteolytic cleavage of PARP by caspase-3 may function to conserve cellular energy required for the apoptotic process (2), thereby avoiding severe depletion of ATP stores that could cause irreversible structural damage and necrosis (29).

The result of the present studies demonstrate that activation of caspase-3 is critical for executing peroxynitrite-induced apoptotic cell death. Since the caspase-1 inhibitor Ac-YVAD-CHO did not appreciably affect peroxynitrite-induced apoptosis, caspase-1 is probably not involved in the upstream event leading to caspase-3 activation. However, we cannot rule out the possibility that other caspase-3 like family proteases, in addition to caspase-3, might be activated concurrently during peroxynitrite-induced apoptosis of HL-60 cells. How peroxynitrite causes activation of the caspase-3 family protease remains to be established. Since PARP cleavage was not apparent until after a considerable delay (90 min after peroxynitrite exposure), it is unlikely that caspase-3 activation is directly due to peroxynitrite *per se*, but rather due to either activation of upstream events such as release of cytochrome c from mitochondria (30) or generation of reactive oxygen species (16). It has been suggested that reactive oxygen intermediates (ROI) are the common mediators of PARP cleavage, DNA fragmentation and apoptosis in leukemia cells (31). Although brief exposure of HL-60 cells to peroxynitrite stimulates the release of ROS in a concentration- and time-manner (16) and the formation of ROS participate in peroxynitrite-induced apoptotic cell death. Whether this ROS generation is indeed primarily responsible for the caspase-3 activation needs further investigation.

Acknowledgments:

This work was supported by grants from NIHLBI: 25316-14; NIDDK: 41747 to Dr. P.Y-K W.

REFERENCE
1. Kerr, J. F. R., Wyllie, A. H., and Currie, A. R. (1972) *Br. J. Cancer* **26**, 239-257
2. Wyllie, A. H., Kerr, J. F. R., and Currie, A. R. (1980) *Int. Rev. Cytol.* **68**, 251-306
3. Wyllie, A. H., Morris, R. G., Smith, A. L., and Dunlop, D. (1984) *J. Pathol.* **142**, 67-77
4. Miura, M., Zhu, H., Rotello, R., Hartwieg, E. A., and Yuan, J. (1993) *Cell* **75**, 653-660
5. Nicholson, D. W., Ali, A., Thornberry, N. A., Vaillancourt, J. P., Ding, C. K., Gallant, M, Gareau, Y., Griffin, P. R., Labelle, M., Lazebnik, Y. A., and et al (1995) *Nature* **376**, 37-43
6. Nicholson, D. W. (1996) *Nature Biotechnol.* **14**, 297-301
7. Beckman, J. S., Beckman, T. W., Chen, J., Marshall, P. A., and Freeman, B. A. (1990) *Proc. Natl. Acad. Sci. U. S. A.* **87**, 1620-1624
8. Bonfoco, E., Krainc, D., Ankarcrona, M., Nicotera, P., and Lipton, S. A. (1995) *Proc. Natl. Acad. Sci. U. S. A.* **92**, 7162-7166
9. Fukuyama, N., Ichimori, K., Su, Z., Ishida, H., and Nakazawa, H. (1996) *Biochem. Biophys. Res. Commun.* **224**, 414-419

10. Ischiropoulos, H., Zhu, L., and Beckman, J. S. (1992) *Arch. Biochem. Biophys.* **298**, 446-451

11. Lin, K.-T., Xue, J.-Y., Nomen, M., Spur, B., and Wong, P. Y.-K. (1995) *J. Biol. Chem.* **270**, 16487-16490

12. Radi, R., Beckman, J. S., Bush, K. M., and Freeman, B. A. (1991) *J. Biol. Chem.* **266**, 4244-4250

13. Rubbo, H., Radi, R., Trujillo, M., Telleri, R., Kalyanaraman, B., Barnes, S., Kirk, M., and Freeman, B. A. (1994) *J. Biol. Chem.* **269**, 26066-26075

14. Karoui, H., Hogg, N., Frejaville, C., Tordo, P., and Kalyanaraman, B. (1996) *J. Biol. Chem.* **271**, 6000-6009

15. Inoue, S. and Kawanishi, S. (1995) *FEBS Lett.* **371**, 86-88

16. Lin, K.-T., Xue, J.-Y., Sun, F. F., and Wong, P. Y.-K. (1997) *Biochem. Biophys. Res. Commun.* **230**, 115-119

17. Sellins, K. S. and Cohen, J. J. (1987) *J. Immunol.* **139**, 3199-3206

18. Burton, K. (1956) *Biochem. J.* **62**, 315-323

19. LeBel, C. P., Ischiropoulos, H., and Bondy, S. C. (1992) *Chem. Res. Toxicol.* **5**, 227-231

20. Allen, R. C. (1986) *Methods Enzymol.* **133**, 449-493

21. Gyllenhammar, H. (1987) *J. Immunol. Methods* **97**, 209-213

22. Cherry, P. D., Omar, H. A., Farrell, K. A., Stuart, J. S., and Wolin, M. S. (1990) *Am. J. Physiol.* **259**,H1056-62

23. Kaufmann, S. H., Desnoyers, S., Ottaviano, Y., Davidson, N. E., and Poirier, G. G. (1993) *Cancer Res.* **53**, 3976-3985

24. Lin, K.-T., Xue, J.-Y., Lin, M. C., Spokas, E. G., Sun, F. F., and Wong, P. Y.-K. (1998) *Am. J. Physiol.* (in press)

25. Ohgushi, H., Yoshihara, K., and Kamiya, T. (1980) *J. Biol. Chem.* **255**, 6205-6211

26. Junod, A. F., Jornot, L., and Petersen, H. (1989) *J. Cell. Physiol.* **140**, 177-185

27. Szabo, C., Zingarelli, B., Oconnor, M., and Salzman, A. L. (1996) *Proc. Natl. Acad. Sci. U. S. A.* **93**, 1753-1758

28. Leist, M., Single, G., Kunstle, C., Volbracht, H., Hentze, H., and Nicotera, P. (1997) *Biochem. Biophys. Res. Commun.* **233**, 518-522

29. Wu, Y. K., Sun, F. F., Tong, D. M., and Taylor, B. M. (1996) *Biophysical Journal* **71**, 91-100

30. Liu, X., Kim, C. N., Yang, J., Jemmerson, R., and Wang, X. (1996) *Cell* **86**, 147-157

31. McGowan, A. J., Ruiz-Ruiz, M. C., Gorman, A. M., Lopez-Rivas, A., and Cotter, T. G. (1996) *FEBS Lett.* **392**, 299-303.

575

84

CENTRAL ROLE OF ARACHIDONATE 5-LIPOXYGENASE IN THE REGULATION OF CELL GROWTH AND APOPTOSIS IN HUMAN PROSTATE CANCER CELLS

Jagadananda Ghosh and Charles E. Myers

University of Virginia Cancer Center
Charlottesville, Virginia 22908

Dietary fat and the risk of prostate cancer: Prostate cancer has emerged as the most frequently diagnosed malignancy among men in the United States, taking thousands of lives every year (1). It is a multistep disease. In the United States, while there are more than 15 million men with localized prostate cancer, only 40-50,000 men will develop malignancy and die in a given year. Thus the development of the metastatic phenotype is an important rate limiting step in the natural history of prostate cancer. Through the incidence of the latent form of the diseases the same worldwide, there is significant geographic variation in the diagnosis of clinically evidence prostate cancer and mortality (2-4) suggesting involvement of environmental factors in this process. Moreover, people who migrate from low-incidence countries to high-incidence countries show increased risk of invasive prostate cancer (5-8) indicating a positive role for lifestyle factors, such as diet and nutrition.

Among the dietary factors, high intake of fat is frequently found to be associated with the diet in affluent nations, like the United States and Western Europe, where prostate cancer is also more common. Moreover, a number of epidemiological studies as well as experiments with animal models suggest a link between high dietary fat consumption and cancer of the prostate (9-14). Despite these reports, knowledge about the factors in fat associated with the risk of prostate cancer, such as high calories, a particular fat source or component fatty acid, is very limited. Based on this background information, our new project is focused on analysis of the effects of dietary fatty acids on the malignant behavior of human prostate cancer cells.

Androgen is a potent growth stimulator for prostate cancer cells. Fatty acids, e.g. arachidonic acid, stimulate production of androgen by the Leydig cells in testis and might have an indirect effect on the growth of cells of prostatic origin. Recent reports suggest a role for arachidonic acid, an omega-6 poly-unsaturated fatty acid, in the proliferation of prostate cancer cells (15-17). Moreover, we recently reported that arachidonic acid is a potent mitogen for both the androgen sensitive and androgen resistant human prostate cancer cells *in vitro* (18). Arachidonic acid also strongly stimulates *in vitro* invasion of human prostate cancer cells. Therefore, high intake of dietary fat rich in arachidonic acid might have a profound impact on the clinical progression of prostate cancer.

Eicosanoids and Other Bioactive Lipids in Cancer, Inflammation, and Radiation Injury, 4
Edited by Honn *et al.*, Kluwer Academic / Plenum Publishers, New York, 1999.

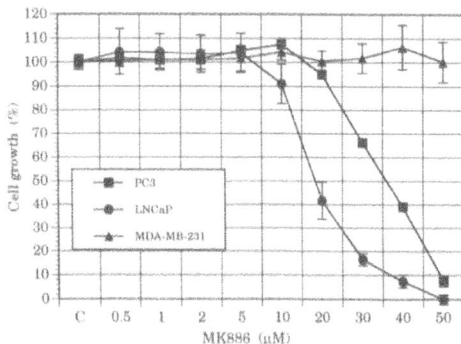

Figure 1. Inhibition of prostate cancer cell growth by 5-lipoxygenase specific inhibitor MK886. Cells (3000 per well) were plated overnight in RPMI medium supplemented with 10% FBS in 96 well tissue culture plates. On day-2, cells were treated with varying doses of MK886 and the plates were further incubated for 72 hours. Control cells were treated with the plating medium containing 0.02% DMSO. At the end of incubation period, cell growth was measured by MTS/PMS Cell Titer Assay (18). The results are presented as the mean ± standard error (n=8).

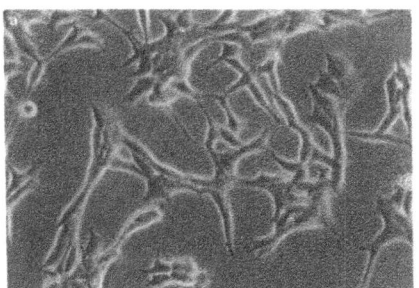
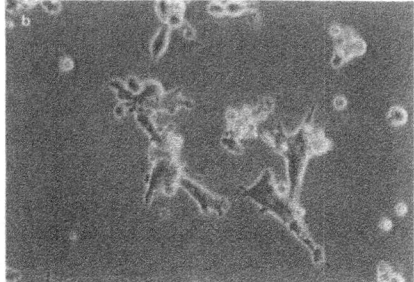
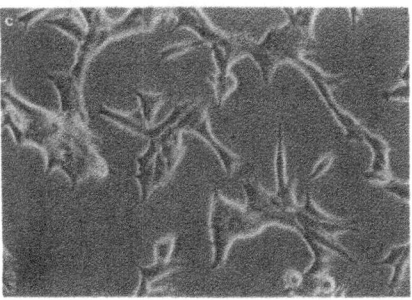

Figure 2. Photomicrographs showing morphology of LNCaP prostate cancer cells upon treatment with MK886. Cells (3 x 10^5) were plated in 60 mm diameter plates (Falcon) in RPMI medium supplemented with 10% FBS and grown for 48 hours. The old medium was then replaced with 2 ml fresh serum-free RPMI medium and the cells were treated with 10 μM MK886 for 6 hours with or without the addition of 5-lipoxygenase metabolite, 5-oxoETE (500 nM). Control cells were treated with serum-free medium containing 0.02% DMSO. Photographs were taken with a Zeiss inverted microscope at 20 X. (A) Control; (B) MK886 only; (C) MK886 plus 5-oxoETE.

Arachidonic acid metabolism, and prostate cancer cell growth an survival: Arachidonic acid and its precursor, linoleic acid, are essential fatty acids which we should obtain from dietary sources. Western style foods, particular rich in red meat and dairy products, are good source of these fatty acids. How these fatty acids modulate the malignant behavior of prostate cancer cells is an open question. After seeing the stimulatory effect of arachidonic acid on human prostate cancer cell growth, we were interested in the molecular mechanism involved in this process. Arachidonic acid can directly modulate the activity of a number of cellular proteins, like protein C and ras-GAP, known to have positive effects on malignant transformation (19-21). In addition to that, arachidonic acid can be metabolized to a wide range of eicosanoids (22, 23) which modulate diverse physiological and pathological functions including growth and invasion of tumor cells and suppression of immune surveillance (24, 25).

Using selective inhibitors of the various pathways of arachidonic acid metabolism, we found that metabolic conversion of arachidonic acid by 5-lipoxygenase is a critical requirement for its growth stimulation of prostate cancer cells. Inhibition of other major pathways of arachidonic acid metabolism (e.g. cyclooxygease, cytochrone P450, 12-lipoxygenase, etc.) did not show any appreciable effect. the hydroxydicosatetraenoid metabolites of 5-lipoxygenase (5-HETE, 5-HETE lactone and 5-oxoETE) exert mitogenic effects on prostate cancer cells and reverse the inhibitory effect of 5-lipoxygenase blockade. On the other hand, leukotrienes (e.g. LTB4, LTC4, etc.), another class of 5-lipoxygenase metabolites, are ineffective. These observations suggest that production of the 5-HETE series of eicosanoids is essential for arachidonic acid-stimulated growth of human prostate cancer cells. This proved to be the case: by specific radioimmunoassay measurement we demonstrated that prostate cancer cells treated with arachidonic acid produce 5-HETE and release this eicosanoid into the medium.

In other cell types, metabolism of arachidonic acid through cyclooxygenase and lipoxygenase pathways shows a strong correlation with growth factor-stimulated cell proliferation (26,27). Both EGF and androgen are well known mitogens for prostate cancer cells. Interestingly, inhibition of arachidonic 5-lipoxygenase blocks the mitogenic activity of these agents. Moreover, MK886, a specific inhibitor of 5-lipoxygenase activity, is a strong inhibitor of serum-stimulated growth of both androgen-responsive (LNCaP) and non-responsive (PC3) human prostate cancer cells, indicating an essential role for this pathway in the regulation of growth by these cells (Figure 1). Under the same experimental conditions, a hormone independent human breast cancer cell line. MDA-MD-231, is completely refractory to this inhibition.

Apoptosis, or programmed cell death, is a genetically based suicide mechanism preserved through evolutionary ages (28). It is the dominant mechanism of cell death in multicellular species. In prostate cancer cells, apoptosis is induced with treatment by drugs that bind with DNA or the cytoskeleton. Eicosanoids are now well known both as inducers and blockers of apoptotic cell death. The effect of a particular eicosanoid is cell type specific. Prostaglandin E_2 (PGE$_2$) is central to the apoptosis needed for egg release du ring ovulation (29), whereas it protects CD4+/CD8+ cells from activation-induced apoptotic cell death (30). Moreover, TNF and FAS-mediated cell deaths are associated with activation of phospholipase A2 and formation of lipoxygenase metabolites (31,32). Our previous observation of the dramatic inhibition of prostate cancer cell growth by selective inhibition of 5-lipoxygenase led us to explore whether operation of the 5-lipoxygenase pathway has any role in the survival of human prostate cancer cells.

We observed that inhibition of 5-lipoxygenase by MK886 induces a massive alteration of membrane morphology in prostate cancer cells, forming numerous blebs within hours of treatment. Inhibitors of other well characterized pathways of arachidonic acid metabolism (cyclooxygenase, cytochrone P450 or 12-lipoxygenase) did not cause any appreciable change in membrane morphology. Exogenous addition of 5-oxoETE, a metabolite of arachidonate 5-lipoxygenase, protects these cells from the severe membrane

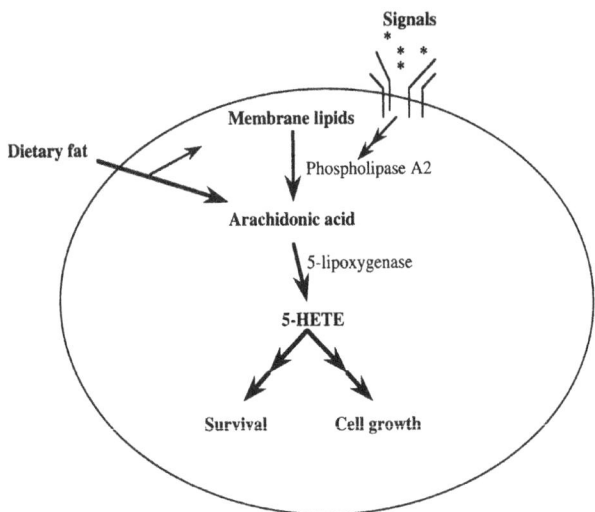

Figure 3. A model illustrating regulation of prostate cancer cell growth and survival by arachidonic acid.

changes induced by inhibition of 5-lipoxygenase, indicating a critical role for this pathway in the survival of prostate cancer cells (Figure 2). The extensive formation of membrane blebs suggest that it might be a case of apoptotic cell death. Later, we confirmed that the extensive formation of membrane blebs is associated with other standard features of apoptosis, e.g., mitochondrial permeability transition, externalization of phosphatidylserine and degradation of DNA to nucleosomal subunits.

The rapid induction of prostate cancer cell death upon inhibition of 5-lipoxygenase and its prevention by 5-HETE series of eicosanoids mean that these cells are constitutively making these metabolites. By direct immunoassay measurement, we demonstrated that both androgen sensitive (LNCaP) and androgen resistant (PC3) human prostate cancer cells consititutely produce 5-HETE in serum-free medium without any externally added stimulus. Moreover, this production of 5-HETE is dramatically increased upon treatment with exogenous arachidonic acid, indicating that prostate cancer cells can metabolize both endogenous and exogenous pools of arachidonic acid to produce a critical 5-lipoxygenase metabolite, 5-HETE.

Our experimental observations reveal that a continuous supply of 5-HETE is an absolute requirement for the survival and regulation of growth of both hormone responsive and non-responsive human prostate cancer cells. Based on our recent findings we want to propose a model showing the regulation of prostate cancer cell growth and survival by arachidonic acid metabolism (Figure 3).

ACKNOWLEDGMENTS

We express our sincere gratitude to the Lynch and Reynolds Foundation for their financial support. This work was also supported in part by the CapCure Foundation and an NCI Developmental Grant in Prostate Cancer, R21-CA 69848.

REFERENCES

1. Parker, S.L., Tong, T., Bolden, S., and Wingo, P.A. Cancer Statistics, 1997. *CA Cancer J. Clin.* 47:5-27 (1997).
2. Jackson, M.A., Ahluwalia, B.S., Herson, J., et al. Characterization of prostatic carcinoma among blacks: a continuation report. *Cancer Treat. Rep.* 61:167-172 (1977).
3. Brelow, N.E., Chan, C.W., Dhom, G., et al. Latent carcinoma of prostate at autopsy in seven areas. *Int. J. Cancer* 20:680-688 (1977).
4. Yatani, R., Chigusa, I., Akazaki, K., Stemmerman, G.N., Welsh, R.A., and Correa P. Geographic pathology of latent prostatic cancer. *Int. J. Çancer* 29:611-616 (1982).
5. Haenszel, W., and Kurihara, M. Studies of Japanese migrants. I. Mortality from cancer and other diseases among Japanese in the United States. *J. Nat. Cancer Inst.* 40:43-68 (1968).
6. Shimizu, H., Ross, R.K., Bernstein, L., Yatani, R., Henderson, B.E. and Mack, T.M. Cancers of the prostate and breast among Japanese and white immigrants in Los Angeles County. *Br. J. Cancer* 63:963-966 (1991).
7. Muir, C.S., Nectoux, J., and Straszewski, J. The epidemiology of prostatic cancer: geographical disribution and time-trends. *Acta Oncol.* 30:133-140 (1991).
8. Yu, H., Harris, R.E., Gao, Y.T., Gao, R., and Wynder, E.L. Comparative epidemiology of cacers of the colon, rectum, prostate and breast in Shanghai, China versus the United States. *Int. J. Epidemiol.* 20:76-81 (1991).
9. Whittemore, A.S., Kolonel, L.N., Wu, A.H., John, E.M. Gallagher, R.P., Howe, G.W., Burch, J.D., Hankin, J., Dreon, D.M., West, D.W., Tehm C., and Paffenbarger, R.S. Prostate cancer in relation to diet, physical activity and body size, in blacks, white and asians in the U.S. and Canada. *J. Natl. Cancer Inst.* 87:652-661 (1995).
10. Mettlin, C., Selenskas, s., Natarajan, N., and Huben, R. Beta-carotene and animal fats and their relationship to prostate cancer risk. *Cancer* 64:605-612 (1995).
11. Giovannucci, E., Rimm, E.B., Colditz, G.A., Stampfer, M.J., Ascherio, A., Chute, C.C., and Wilett, W.C. A prospective study of dietary fat and risk of prostate cancer. *J. Natl. Cancer Inst.* 85:1571-1579 (1993).
12. Le Marchand, L., Kolonel, L.N., Wilkins, L.N., Myers, B.C., and Hirohata, T. Animal fat consumption and cancer of the prostate: a prospective study in Hawaii. *Epidemiology* 5:276-282 (1994).
13. Gann, P.H., Hennekens, C.H., Sacks, F.M., Grodstein, f., Giovannucci, E., and Stampfer, M.J. Prospective study of plasma fatty acids and risk of prostate cancer. *J. Natl. Cancer Inst.* 86:281-286 (1993).
14. Wang, Y., Corr, J.G., Thaler, H.T., Tao, Y., Fair, W.R., and Heston, W.D. Decreased growth of established human prostate LNCaP tumors in nude mice fed a low-fat diet. *J. Natl. Cancer Inst.* 87:1456-1462 (1995).
15. Chaudry, A.A., Wahle, K.W., McClinton, S. and Moffat, L.E. Arachidonic acid meabolism in benign and malignant prostatic tissue in vitro: effects of fatty acids and cyclooxygenase inhibitors. *Int. J. Cancer* 57:176-180 (1994).
16. Dahiya, R., Yoon, W.H., Boyle, B., Schoenberg, S., Yen, t.B. and Narayan, P. biochemical, cytogenetic and morphological characteristics of human primary and metastatic prostate cancer cell lines. *Biochem. Int.* 27:567-577 (1992).
17. Anderson, K.M., Wygodny, J.B., Ondrey, F., and Harris J. Human PC3 prostate cell line DNA synthesis is suppressed by eicosatetraynoic acid, an in vitro inhibitor of arachidonic acid metabolism. *Prostate* 12:3-12, 1988).
18. Ghosh, J. and Myers, C.E. Arachidonic acid stimulates prostate cancer cell growth: critical role of 5-lipoxygenase. *Biochem. biophys. Res. Commun.* 235:418-423 (1995).

19. Khan, W.A., Blobe, G.C., and Hannun, Y.A. Arachidonic acid and free fatty acids as second messengers and the role of protein kinase C. *Cell. Signaling* 7:171-184 (1995).

20. Schachter, J.B., Lester, D.S., and Alkon, D.L. Synergistic activation of protein kinase C by arachidonic acid and diacyglycerols in vitro: generation of a stable membrane-bound, cofactor-independent state of protein kinase c activity. *Biochim. Biophys. Acta.* 1291: 167-176 (1996).

21. Han, J., McCormick, F., and Macara, I.G. Regulation of rasGAP and the neurofibromatosis-1 gene product by eicosanoids. *Science* 252:576-579 (1991).

22. Needleman, P., Turk, J., Jakschik, B.A., Morrison, A.R., and Lefkowith, J.B. Arachidonic acid metabolism. *Ann. Rev. Biochem.* 55:69-102 (1986).

23. Samuelsson, B., Dahlen, S.E., Lindgren, J.A., Rouzer, C.A., and Serhan, S.N. Leukotrienes and lipoxins: structures, biosynthesis and biological effects. *Science* 237:1171-1176 (1987).

24. Ara, G., and Teicher, B.A. Cyclooxygenase and lipoxygenase inhibitors in cancer therapy. *Prostaglandins, Leukotrienes and Essential Fatty Acids* 54:3-16 (1996).

25. Young, M.R. Eicosanoids and he immunology of cancer. *Cancer Mets. Rev.* 13:337-348 (1994).

26. Nolan, R.D., Danilowicz, R.M., and Eling, T.E. Role of arachidonic acid metabolism in the mitogenic reponse of BALB/c 3T3 fibroblasts to epidermal growth factor. *Mol. Pharmacol.* 33:650-656 (1988).

27. Eling, T.E. and Glasgow, W.C. Cellular proliferation and lipid metabolism: importance of lipoxygenases in modulating epidermal growth factor-dependent mitogenesis. *Cancer Met. Rev.* 13:397-410 (1994).

28. Steller, H. Mechanisms and genes of cellular suicide. *Science* (Washington, DC) 267:1445-1449 (1995).

29. Lim, H., Paria, B.C., Das, S.K., Dinchuk, J.E., Langenbach, R., Trzaskos, J.M., and Dey, S.K. Multiple female reproductive failures in cyclooxygenase 2-deficient mice. *Cell* 91:197-208 (1997).

30. Goetzel, E.J., An, S., and Zheng, L. Specific suppression of prostaglandin E_2 of activation-induced apoptosis of human CD4+8+ T lymphoblasts. *J. Immunol.* 154:1041-1047 (1995).

31. Jaattela, M., Benedict, M., Teweri, M., Shayman, J.A., and Dixit, V.M. Bcl-x and Bcl-2 inhibit TNF and FAS-induced apoptosis and activation of phospholipase A2 in breast carcinoma cells. *Oncogene* 10:2297-2305 (1995).

32. O'Donnell, V.B., Spycher, S., and Azzi, A. Involvement of oxidents and oxident-generating enzyme(s) in tumor-necrosis-factor-alpha-mediated apoptosis: role for lipoxygenase pathway but not mitochondrial respiratory chain. *Biochem. J.* 310:133-141 (1995).

85

ROLE OF AUTOCRINE MOTILITY FACTOR IN A 12-LIPOXYGENASE DEPENDENT ANTI-APOPTOTIC PATHWAY

Keqin Tang[1], Daotai Nie[1], Kenneth V. Honn[1,2,3]

Department of Radiation Oncology[1], Pathology[2] and Chemistry[3]
Wayne State University School of Medicine, Detroit, MI 48202

INTRODUCTION

Apoptosis or programmed cell death, plays an indispensable role in embryonic development, maturation of the immune system and maintenance of tissue and organ homeostasis (1-2). A wide spectrum of molecular entities including oncogenes, tumor suppressor genes, signal transducers, cell cycle proteins, free radicals, cations and proteases have been implicated in apoptosis. Recent work from our laboratory (3) demonstrated that 12-lipoxygenase (12-LOX) may play an essential role in regulating apoptosis in rat Walker carcinosarcoma (W256) cells. Down regulation of 12-LOX gene expression by antisense oligonucleotides triggered time and dose dependent apoptosis. The W256 cell death, triggered by antisense oligos could be partially blocked by exogenous 12(S)-HETE. The antisense oligo treatment also resulted in a significant decrease in the ratio of Bcl-2/Bax proteins which are critical determinants of apoptosis. As expected, over expression of Bcl-2 protein provided partial death sparing effects (3). It also was demonstrated that exogenous arachidonic acid (AA) protected against apoptosis (4). Exogenous AA in a time and dose dependent fashion suppressed NDGA induced W256 apoptosis as well as DNA fragmentation. Serum withdrawal also caused W256 cells to undergo typical apoptosis which was not rescued by several growth factors commonly found in serum. However, exogenous AA suppressed serum starvation induced W256 cell apoptosis (4). Further studies demonstrated that the factor responsible for AA induced protection against apoptosis was the 12-LOX metabolite of AA 12(S)-HETE. 12-LOX has been demonstrated by our laboratory to be expressed in numerous tumor cells from human, rat and mouse (5). Factors controlling the regulation of 12-LOX are largely unknown. Several growth factors have been shown to stimulate 12-LOX activity in tumor cells. For example, in A431, human epidermoid carcinoma cells platelet type 12-LOX is expressed (6). Serum starvation induces a time and dose dependent reduction in the expression of 12-LOX protein (8). Another cytokine which has been demonstrated to regulate 12-LOX levels in tumor cells is autocrine motility factor (AMF). AMF stimulates cell motility and invasion via receptor mediated signaling pathways. AMF is also a potent growth factor. Our laboratory demonstrated that AMF induced cell motility was mediated by the lipoxygenase generation of 12(S)-HETE (7). In a subsequent report, we also demonstrated that AMF was able to increase the cellular protein levels of 12-LOX in high but not in low metastatic clones (8).

Given the fact that AMF is a link to cellular proliferation, that it's immediate signaling mechanism is mediated through the generation of 12(S)-HETE, that it increases 12-LOX protein content in murine melanoma cells and that 12-LOX/12(S)-HETE have been shown to protect cells against the induction of apoptosis, we hypothesize that AMF may, through a 12-LOX dependent mechanism, increase cellular resistance to the induction of apoptosis. In this report we tested this hypothesis by determining whether AMF would increase message/protein levels of 12-LOX, whether it conferred a protective effect against the induction of apoptosis and preliminarily determined a possible mechanism of action.

METHODS AND MATERIALS

1.   Cells

The A431 human epidermoid carcinoma cell line was obtained from the American Type Culture Collection (Bethesda, MD).

2.   RT-PCR

Total RNA from different dose treatments of AMF was obtained by the acid-guanadinium isothiocyanate method using Tri-Reagent kit (Molecular Research Center, Inc., Cincinnati, OH). Total RNA (2mg) from each sample was reverse-transcribed using oligo-dT primer and MMLV reverse transcriptase supplied by Epicentre Technologies (Madison, WI), according to manufacturer's instructions. The following reagents were added: 5xRT buffer, 10mM dNTPs, 50mg/ml BSA, 30 units RNasin (Promega, Inc) and 200 units MMLV RNase in a final volume of 20 ml. The reaction was incubated at 42°C for 60 min and terminated by heating 95°C for 10 min. The RT mixture (1 ml) was amplified in 50 ml of PCR buffer containing 30 mM Tris-HCl (pH 8.5), 4 mM MgCl2, 100mg/ml BSA, 250 nM each of 12-LOX /actin sense and antisense primers, 0.25mM dNTPs, and 1 unit AmpliTaq (Perkin Elmer Cetus, Foster City, CA). PCR was performed in the GeneAmp PCR system 9600 (Perkin Elmer Cetus) by using the following cycles for 12-LOX /Actin: 94°C for 30", 70°C for 30", 72°C for 45" or 94°C for 20", 55°C for 30", 72°C for 1 min repeated for 35 cycles and followed by a 10-min extension at 72°C. The PCR product (15 ml) was separated on a 2% agarose-TAE gel and stained with ethidium bromide to visualize the amplified product. Omission of tumor cell RNA from the reaction mixture served as a negative control for RT-PCR.

3.   Western Blotting Assay

Aliquots of total cell lysate were mixed with 1 vol. of SDS sample buffer (85mM Tris-HCl, pH 6.8, containing 1.4 (w/v) SDS, 14% (v/v) glycerol, 5% (v/v) mercaptoethanol and a trace of bromophenol blue), boiled for 5 min, and subjected to SDS-PAGE on 8% acrylamide gel. Proteins were electrophoretically transferred to nitrocellulose membranes. After transfer, nonspecific sites were blocked with 5% (w/v) nonfat-dry milk in TTBS for 2h at 25°C with a 1:1000 final dilution of rabbit polyclonal antiserum to human platelet-type 12-LOX (Oxford Biomedical Research, Oxford, MI) in TTBS. After washing the blot in TTBS, the membranes were incubated for 1h at 25°C with a 1:4500 final dilution of horseradish peroxidase-conjugated donkey anti-rabbit IgG (Amersham, Arlington Heights, IL) washed again as above in TTBS, developed them using ECL (Amersham, Arlington Heights, IL) and exposed to Fuji Medical X-ray film at 25°C.

4.   Trypan blue dye exclusion assay

1x106 tumor cells were plated into 10cm culture dishes and cultured in DMEM with 10% FCS for two days. Cells were then treated with different doses of BHPP in the same media. At the end of treatment, dead cells were gently removed and surviving cells counted using the trypan blue dye exclusion assay. Trypan blue-stained cells and those cells with typical apoptotic morphology (i.e., membrane blebbing) were excluded from counting. The cell survival was expressed as the percentage of the control (i.e., the vehicle treatment). Each condition was run in triplicate or quadruplicate and the experiments repeated twice.

5.   DNA fragmentation

5x106 cells were treated with BHPP (0.01, 0.1, 1, 5, 10 µM). Harvested and fragmented DNA extracted with 200 ml of the lysis buffer for 5 min. Samples were then centrifuged at 500g for 5 min. The resultant supernatants were transferred to a clean set of eppendorf tubes and the pellets redissolved in 200 ml lysis buffer and extracted for 2 min. Samples were recentrifuged and the resultant supernatants combined with previous supernatants. Subsequently, SDS and DNase-free RNase were added to the pooled supernatants to the final concentration of 0.1% and 5 mg/ml, respectively, and samples incubated at 56°C for 2h. At the end of RNase treatment, proteinase K (2.5ml/ml) was added and samples further incubated for 2h at 37°C. Samples were extracted once with alkaline phenol/chloroform/isoamyl alcohol (25:24:1) and DNA precipitated with 0.3M NaAc (pH5.2). DNA from equal number of cells or equal amounts of DNA (20mg) was run on a 1.2% agarose gel and the DNA ladder formation visualized by ethidium bromide staining.

6.   Cell death detection ELISA

The ELISA kit was purchased from Boehringer Mannnheim. It is a photometric enzyme immunoassay for the qualitative and quantitative in vitro determination of cytoplasmic histone-associated DNA fragments (mono- and oligonucleosomes) present after induced cell death. After treatment cells are harvested, lysed and the cell pellet resuspended with 500ml incubation buffer, and incubated for 30 min at 4°C.The lysate is centrifuged at 20,000g for 10 min and 400 ml of the supernatant (=cytoplasmic fraction) carefully removed. 100ml of predilute sample solution (1:10) is added to each well of a microtiter plate coated with MTP-modules with anti-histone and incubated for 90 min at 25°C. After washing, 100 ml anti-DNA-peroxidase solution is added to each well and incubated for 90 min at 25°C. After another washing, 100 ml of substrate solution is added and incubated on a plate shaker at 250 rpm for color development (15 min). Measurements are recorded at 405 nm/490nm against solution as blank.

7.   Measurement of Caspase-3(CPP32)-like activity

The Caspase-3 activity was measured using a fluorogenic substrate DEVD-AFC(Biomol). Briefly A431 cells were treated with BHPP(10mM;36h) in the presence of increasing concentrations of AMF. At the end of the time period whole cell lysates were prepared in cell lysis buffer (10mM Tris-HCL, pH 7.5, NaH2P04/NaHP04,Ph 7.5, 130mM NaPL, 1% TritonX-100, 10mM PPi). After 100mg protein was coincubated with 15mM DEVD-AFC substrate in a total of 1 ml reaction buffer (20mM HEPES, pH 7.5, 10% glycerol, 2mM DTT) at 37oC for one hr. Subsequently, the activation of CPP32 was monitored by measuring the AFC liberated from the Ac-DEVD-AFC using Perkin-Elmer LS-5 fluorescence spectrophotometer at an excitation wavelength of 400 nm and emission wave length 505 nm. All samples were blanked against the reaction buffer and with the substrate alone. The results are expressed as a fold increase in activation.

RESULTS AND DISCUSSION

Previous work from our laboratory suggests that 12-LOX, but not 5-LOX, plays an essential role in regulating apoptosis in the rat Walker carcinsarcoma cells (W256). These cells express platelet type 12-LOX (6). Down regulation of the 12-lipoxygenase gene expression by antisense oligonucleotides triggered time and dose dependent apoptosis. The W256 cell death triggered by antisense oligos could be partially blocked by exogenous 12(S)-HETE but not by 5(S)-HETE. 15(S)-HETE was minimally effective (3). In addition, other tumor cells such as RBL-1 (rat basophilic leukemia cells), mMTLn-3 (rat mammary adenocarcinoma), and HEL (human erythroleukemia cell line) all undergo apoptosis following treatment with NDGA (a general lipoxygenase inhibitor) although the doses required are cell line dependent. In contrast, normal cells such as endothelial cells do not undergo apoptosis upon treatment with antisense oligos or LOX inhibitors suggesting the tumor cells differ significantly from normal cells with respect to their usage of 12-LOX in regulating cell survival and apoptosis. In our previous work (3) we artificially over expressed 12-LOX in order to demonstrate its role in apoptosis. In the present study we utilized a naturally occurring cytokine, autocrine motility factor, which has previously been demonstrated to increase 12-LOX protein levels in high metastatic murine cancer cells (8). The cell line used in this study, A431, has previously been shown by us to express the platelet type 12-LOX (6). To determine whether AMF will affect 12-LOX level in human cancer cells we used RT-PCR to amplify a predicted 12-LOX fragment as well as Western blotting to detect 12-LOX protein following treatment with AMF (1-100pg/ml). AMF treatment with the above mentioned dose range resulted in a concentration dependent increase in 12-LOX mRNA (Figure 1a). The maximum effect occurred at 18 hours with 100pg/ml of AMF. Similarly, AMF resulted in a dose dependent increase in 12-LOX protein concentration as detected by Western blot (Figure 1b).

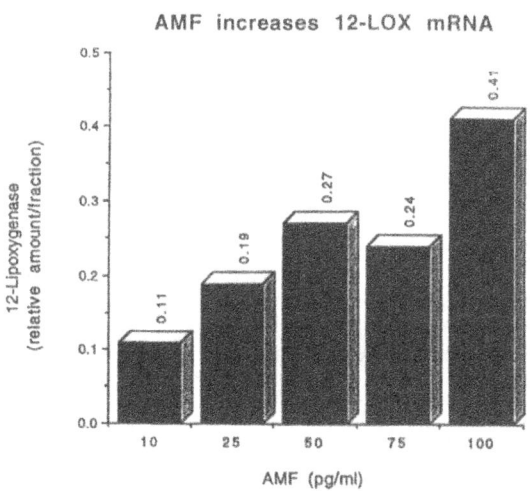

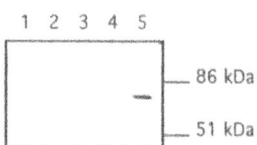

Figure 1A   **Figure 1B**

AMF enhances the expression of 12-LOX mRNA levels in A431 cells. A431 cells were grown in DMEM with 10% serum to 75% confluence. The cells were washed twice with PBS and the media replaced with serum-free DMEM. After 24 hours the cells were treated with AMF(1-100pg/ml) for 18 hours and total RNA isolated for RT-PCR as described above. b-actin was used as an internal control. 12-LOX mRNA (925bp) levels increased after AMF treatment in a dose dependent manner (A.) Similarly, 12-LOX protein levels also increased (B). lane 1: 1 pg/ml, lane 2: 25 pg/ml, lane 3: 50 pg/ml, lane 4: 75 pg/ml, lane 5: 100pg/ml.

We next examined whether BHPP, a select 12-lipoxygenase inhibitor (9), could affect survival of A431 cells. In the absence of AMF treatment BHPP resulted in a dose dependent decrease in cell survival (Figure 2).

586

BHPP Effect on A431 Cells Survival

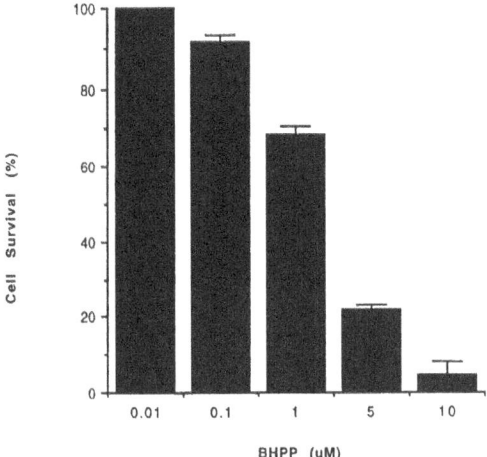

Figure 2
Cell survival as determined by Trypan blue dye exclusion assay. Maximum cell killing was observed at 10μM BHPP. Treatment was for 48 hours and results expressed as % of cell survival compared to the control in the absence of BHPP. Each condition was run in triplicate and results were derived from the mean + or - SE of 3 independent experiments.

BHPP-induced Apoptosis in A431 Cells

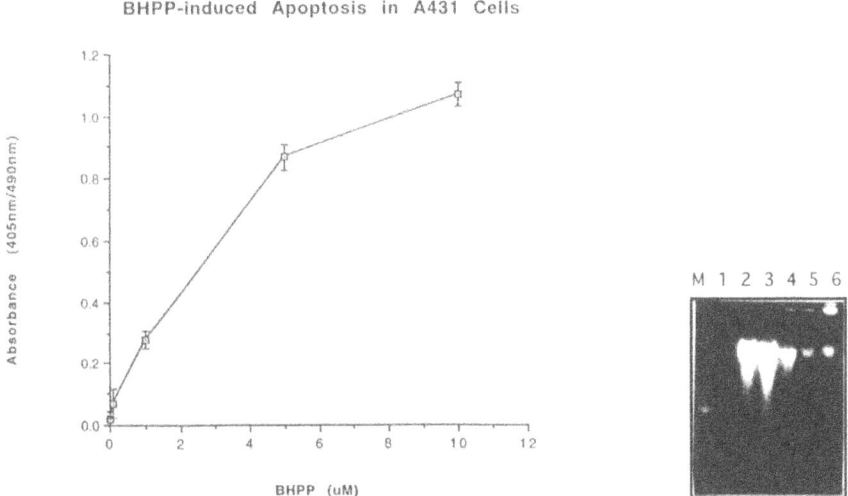

Figure 3A   **Figure 3B**
Detection of nucleosomes in the cytoplasm of A431 cells treated with BHPP for 48 hours by ELISA (Fig 3A). Fragmentation DNA isolated from equal starting cell number (5x106) was loaded on a 1.2 % agarose gel. Note the increase of ladder at 1μM, the maximun at 10 μM. lane 1: control, lane 2: 10μM, lane 3: 5μM, lane 4: 1μM, lane 5: 0.1μM, lane 6: 0.01μM.

In order to confirm that BHPP induced A431 cell death was a result of apoptosis a cell death ELISA assay and morphological observations (DNA fragmentation) were used. The cell death ELISA is based on a quantitative sandwich-immunoassay-principle using monoclonal antibodies directed against histone associated DNA and histones. This allows the specific determination of mono-, and oligonucleosomes in the cytoplasmic fraction of cell lysates following treatment with the apoptosis inducing agent, in this case, BHPP. When A431 cells were exposed to BHPP for 48 hours as described above, a dose dependent increase in cytoplasmic nucleosomes (Figure 3A) and DNA fragmentation were observed (Figure 3B).

Next we determined whether pre-treatment of cells with AMF could result in an amelioration of BHPP inducted apoptosis. A431 cells were treated in serum-free media with AMF (100pg/ml). After 24 hours the cells were treated with BHPP in a dose range of 0.01-10μM and nucleosomes detected as described above. AMF treatment resulted in a significant reduction of BHPP induced apoptosis (Figure 4).

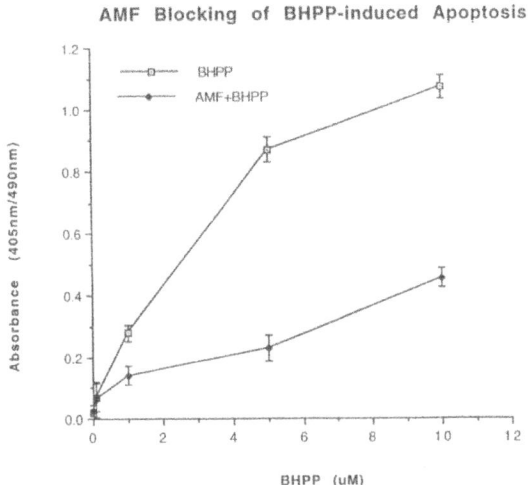

Figure 4 AMF prevents BHPP induced apoptosis

These results were confirmed by DNA laddering (data not shown). One of the distinguishing biochemical features of apoptosis is the DNA fragmentation which can be observed as a distinct ladder following electrophoresis. This laddered pattern is believed to result from an orderly fragmentation in DNA (from 100 kb-50 kb and finally to internucleosomal oligomers of approximately 180bp) which are cleaved preferentially at nuclear matrix attachment sites. This can be summed up in the hypothesis that activation of specific endonucleases prior to apoptosis may be a critical event. However, numerous observations suggest that apoptosis can be reconstituted in a nuclei free system suggesting the cytosolic enzymes may also be responsible for triggering apoptosis (10). Recently a family of proteinases known as ICE (interleukin-1b converting enzyme) and related proteinases were found to be essential triggers of apoptosis (11). Recent evidence suggests a cascade of proteinases in which pro enzymes are sequentially converted to active enzymes leading to an amplification of proteolytic activity within the cytoplasm. The terminal proteinase CPP32 (a 32kd cysteine protease; caspase-3) has emerged as the terminal effector of apoptosis induced by a diversity of agents which include glucocorticoids,

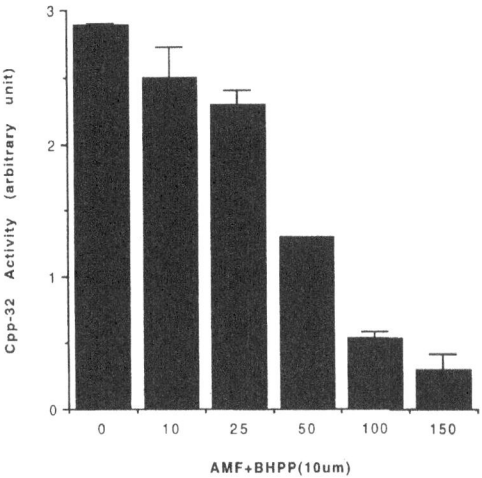

Figure 5 AMF results in decreased caspase activity.

chemotherapeutic drugs, TNF-α among others (11). This family of related cysteine proteinases has been called caspases and have been found to play an essential role in the execution of apoptosis (11). Caspases kill cells by specifically cleaving important cellular constituents such as PARP, U1-70 kPa protein, DNA-dependent protein kinase and actin (11). Therefore, we tested whether BHPP induced apoptosis lead to the activation of CPP32 (caspase-3) and whether AMF treatment could reduce caspase activation. Treatment of cells with BHPP (10uM) resulted in a distinct activation of caspase3 (Figure 5). Pretreatment of cells with AMF resulted in a dose dependent decrease in caspase activation (Figure 5) presumably due to the increased production of 12(S)-HETE.

In summary, we have demonstrated that the cytokine, AMF, through its action on its cell surface receptor (i.e. gp78) can induce transcription of the 12-LOX gene resulting in increased mRNA and 12-LOX protein. This increase in 12-LOX protein confers a resistance to the induction of apoptosis with a select (i.e. BHPP) and general (i.e. NDGA) 12-LOX inhibitors. In addition, AMF treatment prevents the induction of apoptosis due to the loss of trophic growth factors resulting from serum starvation (data not shown). The mechanism for this AMF induced protection may result from 12(S)-HETE down regulation of caspase3 activity. The mechanism whereby 12(S)-HETE down regulates caspase activity is currently under investigation.

REFERENCES

1. Kroemer, G., Petit, P., Zamzami, N., Vayssiere, J., and Mignottem, B. The biochemistry of programmed cell death. FASEB J. 9:1277-1287 (1995).
2. Manjo, G. and Joris I. Apoptosis, oncosis and necrosis. An overview of cell death. Am. J. Pathol. 146:3-15 (1995).
3. Tang, D.G., Chen, Y.Q., and Honn, K.V. Arachidonate lipoxygenases as essential regulators of cell survival and apoptosis. Proc. Natl. Acad. Sci. USA 93:5241-5246 (1996).
4. Tang, D.G., Guan, K. L., Li, L., Honn, K.V., Chen, Y.Q., Rice, R.L., Taylor, J.D., and Porter, A.T. Suppression of W256 carcinosarcoma cell apoptosis by arachidonic acid and other polyunsaturated fatty acids. Int. J. Cancer 72:1078-1087 (1997).

5. Tang, D.G. and Honn, K.V. Eicosanoids and Tumor Cell Metastasis in: Prostaglandin Inhibitors in tuimor Immunology and Immunotherapy (Harris, J.e Braun, D.P., and Anderson, K.U., eds.), pp. 73-108, CRC Press, Boca Raton. (1994).
6. Hagmann, W., Gao, X., Timar, J., Chen, Y.Q., Strohmaier, A.R., Fahrenkopf, C., Kagawa D., Lee, M., Zacharek, A., and Honn, K.V. 12-Lipoxygenase in A431 cells: Genetic identity, modulation of expression, and intracellular localization. Exp. Cell Res. 228:197-205 (1996).
7. Timar, J., Silletti, S., Bazaz, R., Raz, A., and Honn, K.V. Regulation of melanoma-cell motility by the lipoxygenase metabolite 12(S)-HETE. Int. J. Cancer 55:1003-1010 (1993).
8. Silletti, S., Timar, J., Honn, K.V., and Raz, A. Autocrine motility factor induces differential 12-Lipoxygenase expression and activity in high- and low-metastatic K1735 melanoma cell variants. Cancer Research 54:5752-5756 (1994).
9. Liu, B., Marnett, L.J., Chaudhary, A., Ji, C., Blair, I.A., Johnson, C.R., and Honn, K,V. Biosynthesis of 12(S)-hydroxyeicostatetraenoic acid by B16 amelanotic melanoma cells is a determinant of their metastatic potential. Lab. Invest. 70(3):314-23 (1994).
10. Martin, S.J., Newmeyer, D.D., Mathias, S., et al. Cell-free reconstitution of Fas-, UV irradiation- and ceramide-induced apoptosis. EMBO J. 14:5191-5200 (1995).
11. Kumar, S. ICE-like proteases in apoptosis. TIBS 20: 198-202 (1995).

CANCER PREVENTION AND TREATMENT

86

THE POSSIBLE INVOLVEMENT OF 15-LIPOXYGENASE/LEUKOCYTE TYPE 12-LIPOXYGENASE IN COLORECTAL CARCINOGENESIS

Hideki Kamitani, Mark Geller and Thomas Eling

Laboratory of Molecular Carcinogenesis,
NIEHS, NIH
Research Triangle Park, NC 27709

The initial genetic alteration in colorectal carcinoma is mutation of the *APC* gene. As a result, the expected rate of apoptosis in colorectal cells is attenuated leading to the expansion of the progeny of the *APC* mutant cells and hence tumor formation[1]. The up-regulation of cyclooxygenase-2 (Cox-2) occurs at this early stage of colorectal carcinoma and high expression is observed in multiple intestinal neoplasm (*Min*) mice, the model of familiar adenomatous polyposis[2,3], and in 85% of human colorectal carcinomas[4]. The over-expression of Cox-2 inhibits apoptosis in rat intestinal cells[5] and non-steroidal anti-inflammatory drugs (NSAIDs), such as sulindac sulfide, indomethacin and aspirin promote apoptosis in colorectal carcinoma cells. NSAIDs serve in chemoprevention of colorectal cancer in experimental animal models indicating Cox as a target for preventing colorectal tumors in man[6,7]. However, the functional role of the mutated *APC*, the overexpression Cox-2 and the inhibition of apoptosis is not clearly understood.

Our laboratory has shown that 15-lipoxygenase (LO) metabolites of arachidonic acid (AA) and linoleic acid (LA) alter signaling pathways which regulate cell proliferation and apoptosis in carcinogenic cells[8,9]. For example, 13(S)-hydroxyoctadecadienoic acid (HODE)/13(S)-hydroperoxyoctadecadienoic acid (HpODE) is formed from linoleic acid in response to EGF and contributes to EGF-dependent mitogenesis in hamster cells[8]. The 13(S)-HpODE augments cell growth by the up-regulation of the EGFR signaling pathway and appears to stimulate the EGF pathway by attenuation of receptor dephosphorylation[9]. Furthermore, 13(S)-HpODE inhibits starvation induced apoptosis (Eling et al unpublished observation) in these cells. Tang et al[10] showed that lipoxygenase metabolites, 12-

Eicosanoids and Other Bioactive Lipids in Cancer, Inflammation, and Radiation Injury, 4
Edited by Honn *et al.*, Kluwer Academic / Plenum Publishers, New York, 1999.

hydroxyeicosatetraenoic acid (HETE) and 15-HETE inhibit apoptosis in rat W256 cells. These findings suggest a potential involvement of lipoxygenases in the apoptotic process. In this study we wish to determine if lipoxygenase, in addition to Cox-2, also plays a role in the apoptotic process in colorectal cells.

The human colorectal carcinoma cell lines Caco-2, LS 123, HT-29, HCT-116 and SW-480 were obtained from ATCC. Wild type rat intestinal epithelial (RIE) cells and RIE-sense cells, in which Cox-2 cDNA is transfected[5], were provided by Dr. R.N. DuBois. Apoptosis was induced by treatment with 5 mM of NaBT in fetal bovine serum containing medium (FBS). The attached cells were used for Northern analysis, Western analysis and metabolism study of AA/LA. NaBT induces the colon/intestinal cells to undergo apoptosis, therefore the floating cells were assayed for apoptosis by DNA ladder analysis. Table 1 summarizes the results for apoptosis and for expressions of Cox-1, Cox-2 and 15-LO/12-LO by Northern analysis. The expression of Cox-1 was low in most cell lines but we did observe a significant expression of Cox-1 in HT-29 cells. Cox-1 expression was enhanced by NaBT treatment in HT-29, HCT-116 and SW-480 but the most significant expression was observed in HT-29 cells after treatment with NaBT. Thus, in three of five cell lines, the expression of Cox-1 was induced by treatment with NaBT. No expression of Cox-1 was observed in RIE cells including those treated with NaBT. Cox-2 expression was detected in Caco-2, HT-29, HCT-116 and SW-480 human cell lines and both RIE cell lines. The greatest expression of Cox-2 was observed in Caco-2 and HT-29 human cells and the RIE-sense cells. Modest Cox-2 expression was observed in the SW-480 cells. Treatment with NaBT did not enhance the expression of Cox-2 except in the SW-480 cells where a modest increase was detected. The expression of 15-LO/rat 12-LO was not detected in any of the cell lines cultured in FBS. However, treatment of the cells with NaBT resulted in the expression of 15-LO most readily observed in Caco-2, HT-29 and SW-480 human cells. NaBT also increased the expression of rat 12-LO in the RIE-wild cells. Note the human 15-LO and the rat 12-LO share high homology[12]. Interestingly, the human cells which express Cox-2 were easily induced to apoptosis by treatment with NaBT. The RIE-sense cells, which constitutively overexpress Cox-2 and are resistant to NaBT induced apoptosis, did not appear to express rat 12-LO. Thus, a correlation exists between the induction of apoptosis and the inducible 15-LO/rat 12-LO expression in colon/intestinal cells. Immunoblot analysis using antibodies specific for human 15-LO also support the conclusion for the induction of 15-LO in these human colorectal cells during NaBT induced apoptosis (data not shown).

We investigated metabolism of exogenous AA in Caco-2 cells treated with NaBT. Radiolabelled [^{14}C]AA (25 μM, 3 μCi) was incubated with the total cell lysates of Caco-2 cells treated with FBS and NaBT. PGE$_2$ was the main metabolite in FBS treated cells but 15-HETE was the major metabolite produced by NaBT treated cells as shown in Fig. 1. Furthermore, 15-HETE formation by NaBT treated cells was not inhibited by addition of 5 μM of indomethacin, suggesting Cox was not responsible for the formation of the 15-HETE. This result is consistent with data demonstrating the induction of a 15-LO in Caco-2 cells treated with NaBT. Thus NaBT treatment resulted in a dramatic shift in arachidonic and linoleic acid metabolism (data not shown) fom prostaglandin H synthase derived metabolites to lipoxygenase derived metabolites.

Table 1. Expression of COX-1, COX-2 and 15-LO/12LO and apoptosis in colorectal/intestinal derived cells

Cell line	Culture condition	COX-1	COX-2	15-LO/12-LO	apoptosis
Caco-2	FBS (72 h)	negative	overexpression	negative	no
	NaBT (72 h)	negative	moderate	high expression	yes
LS 123	FBS (72 h)	negative	negative	negative	no
	NaBT (72 h)	negative	negative	negative	yes
HT-29	FBS (72 h)	slight	overexpreesion	negative	no
	NaBT (72 h)	moderate	moderate	moderate	yes
HCT-116	FBS (72 h)	negative	slight	negative	no
	NaBT (72 h)	slight	slight	negative	yes
SW-480	FBS (72 h)	negative	slight	negative	no
	NaBT (72 h)	slight	moderate	slight	yes
RIE-wild	FBS (24 h)	negative	moderate	negative	no
	NaBT (24 h)	negative	moderate	moderate	yes
RIE-sense	FBS (24 h)	negative	overexpression	negative	no
	NaBT (24 h)	negative	overexpression	negative	no

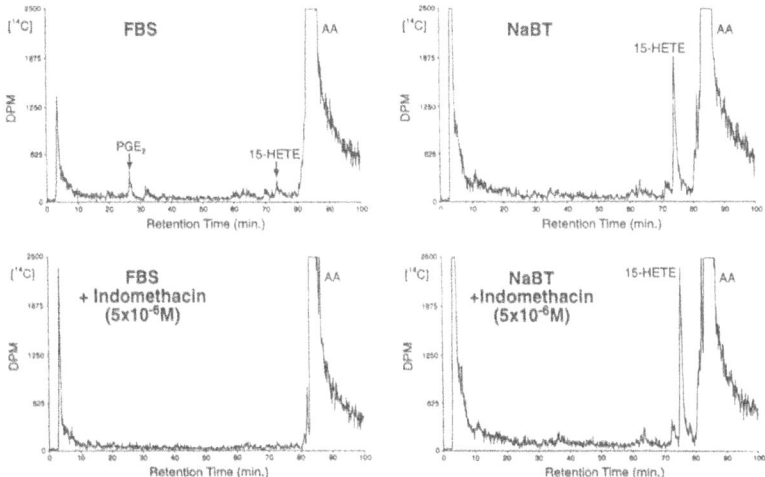

Figure 1. Analysis of metabolites of [^{14}C]arachidonic acid in NaBT treated Caco-2 cells. The (0.8 mg) protein from total cell lysates were reacted in 2 ml of reaction buffer containing 25 μM [^{14}C]arachidonic acid (3 μCi) and incubated for 30 min at 37 ᵦC. Radiolabeled products were separated by reverse-phase HPLC. The retention time of PGE$_2$ and 15-HETE were identified by co-elution with authentic standards.

This inducible 15-LO in Caco-2 cells during apoptosis was isolated by RT-PCR and characterized as the reticulocyte 15-LO that was demonstrated by Sigal et al[12] (data not shown)

While rat 12-LO, which was detected in RIE cells during apoptosis (Table 1), has 75% homology with human reticulocyte 15-LO in amino acid sequence[13], 73% homology can be seen between mouse leukocyte-type 12-LO[14] and human 15-LO. These three LO's are classified as the"leukocyte-type" category[15] and have similar enzymatic functions despite originating from different species. To clarify the possible involvement of leukocyte-type 12-LO of mice in intestinal tumorigenesis, we examined leukocyte-type 12-LO expression in normal tissue and polyps from *Min* mice by RT-PCR. The samples of *Min* mice were provided by Dr. R.N. DuBois. A pair of primers was employed based on the cDNA sequence of mouse leukocyte-type 12-LO. The primers of 12-LO that we used yielded 1253 bp. Mouse glyceraldehyde 3-phosphate dehydrogenase (G3PDH) was used as an internal control. Table 2 illustrates the summary of 12-LO expression by RT-PCR in various tissues of mice. Eleven out of 13 adenomas had clear expression of 12-LO, however 12-LO expression was not detected in 4 of 6 adjacent tissue samples from *Min* mice. This preliminary data indicats that there is a possible involvement of leukocyte-type 12-LO in *Min* mice adenoma.

The expression of human reticulocyte 15-LO has been reported in skin[16], lung[17] and blood cells[12], rat 12-LO is expressed in brain, lung, aorta, and blood cells[13], and mouse leukocyte-type 12-LO is expressed in brain, kidney and blood cells[14]. The clear expression of these LO's have not been shown in colorectal carcinoma cells. We have presented the first evidence of 15-LO/leukocyte 12-LO expression in colorectal tumors. Though the functional role of 15-LO/leukocyte 12-LO is still unclear, our data suggest that LO is possibly involved in apoptosis leading to tumorigenesis of colorectal cancer.

Table 2. Leukocyte-type 12-lipoxygenase expression in mice by RT-PCR

Organ	Number of samples	12-LO expression
adenomatous polyp (*min* mice)	13	11 ~ positive 2~ negative
adjacent tissue of adenoma (*min* mice)	6	2 ~ positive 4 ~ negative

Acknowledgments We thank Julie Angerman-Stewart for technical assistance and Dr. Raymond N. DuBois for providing RIE cell lines and *Min* mice samples, and for the helpful discussions throughout these studies.

REFERENCES

1. P.J. Morin, B. Vogelstein, and K.W. Kinzler, Apoptosis and APC in colorectal tumorigenesis, Proc. Natl. Acad. Sci. USA. 93: 7950 (1996).
2. M. Oshima, J.E. Dinchuk, S.L. Kargman, H. Oshima, B. Hancock, E. Kwong, J.M. Trzaskos, J.F. Evans, and M.M. Taketo, Suppression of intestinal polyposis in apc knockout mice by inhibition of cyclooxygenase 2 (COX-2), Cell 87: 803 (1996).
3. C.S.Williams, C. Luongo, A. Radhika, T. Zhang, L.W. Lamps, L.B. Nanney, R.D. Beauchamp, and R.N. DuBois, Elevated cyclooxygenase-2 levels in *min* mouse adenomas, Gastroenterology 111: 1134 (1996).
4. C.E. Eberhart, R.J. Coffey, A. Radhika, F.M. Giardiello, S. Ferrenbach, and R.N. DuBois, Upregulation of Cyclooxygenase-2 gene expression in human colorectal adenomas and adenomcarcinomas, Gastroentertology 107: 1183 (1994).
5. M.Tsujii, and R.N. DuBois, Alteration in cellular adhesion and apoptosis in epithelial cells overexpressing prostaglandin endoperoxide synthase 2, Cell 83: 493 (1995).
6. R. Hanif, A. Pittas, Y. Feng, M.I. Koutsos, L. Qiao, L. Staiano-Coico, S.I. Shiff, and B. Rigas, Effects of nonsteroidal anti-inflammatory drugs on proliferation and induction of apoptosis in colon cancer cells by a prostaglandin-independent pathway, Biochem Pharmacol. 52: 237 (1996).
   H. Sheng, J. Shao, S.C. Kirkland, P. Isakson, R.J. Coffey, J. Morrow, R.D. Beauchamp, and R.N. DuBois, Inhibition of human colon cancer cell growth by selective inhibitor of cyclooxygenase-2, J. Clin. Invest. 99: 2254 (1997).
7. W.C. Glasgow, C.A. Afshari, J.C. Barrett, and T.E. Eling, Modulation of the epidermal growth factor mitogenic response by metabolites of linoleic and arachidonic acid in Syrian hamster embryo fibroblasts, J. Biol. Chem. 267: 10771 (1992).
8. W.C. Glasgow, R. Hui, A.L. Everhart, S.P. Jayawickreme, J. Angerman-Stewart, B-B. Han, and T.E. Eling, The linoleic acid metabolite, (13S)-hydroperoxyoctadecadienoic acid, augments the epidermal growth factor receptor signaling pathway by attenuation of receptor dephosphorylation, J. Biol. Chem. 272: 19269 (1997).
9. D.G. Tang, Y.Q. Chen, and K.V. Honn, Arachidonic acid lipoxygenases as essential regulators of cell survival and apoptosis, Proc. Natl. Acad. Sci. USA. 93: 5241 (1996).
10. A. Hague, A.M. Manning, K.A. Hanlon, L.I. Huschtscha, D. Hart, and C. Paraskeva, Sodium butyrate induces apoptosis in human colonic tumour cell lines in a p53-independent pathway: implications for the possible role of dietary fibre in the prevention of large-bowel cancer, Int. J. Cancer. 55: 498 (1993).
11. E. Sigal, D. Grunberger, E. Highland, C. Gross, R.A.F. Dixon, and C.S. Craik, Expression of cloned human reticulocyte 15-lipoxygenase and immunological evidence that 15- lipoxygenase of different cell types are related, J. Biol. Chem. 265: 5113 (1990).
12. T. Watanabe, J.F. Medina, J.Z. Haeggstrom, O. Radmark, and B. Samuelsson, Molecular cloning of a 12-lipoxygenase cDNA from rat brain, Eur. J. Biochem. 212:605 (1993).

13. X-S. Chen, U. Kurre, N.A. Jenkins, N.G. Copeland, and C.D. Funk, cDNA cloning, expression, mutagenesis of C-terminal isoleucine, genomic structure, and chromosomal localizations of murine 12-lipoxygenases, J. Biol. Chem. 269: 13979 (1994).
14. S. Yamamoto, Mammalian lipoxygenases: Molecular structures and functions, Biochem. Biophys. Acta 1128:117 (1992).
15. H. Zhao, B. Richards-Smith, A.N. Baer, and F.A. Green, Lipoxygenase mRNA in cultured human epidermal and oral keratinocytes, J. Lipid. Res. 36: 2444 (1995).
16. E. Sigal, S. Dicharry, E. Highland, and W.E. Finkbeiner, Cloning of human airway 15-lipoxygenase: identity to the reticulocyte enzyme and expression in epithelium, Am. J. Physiol. 262: L392 (1992).

87

INDUCTION OF LEUKOTRIENE B4 METABOLISM BY CANCER CHEMOPREVENTIVE AGENTS

Thomas Primiano, Thomas W. Kensler, Michael A. Trush, and Thomas R. Sutter.

Department of Environmental Health Sciences, Johns Hopkins School of Hygiene and Public Health, 615 N. Wolfe St., Baltimore, MD 21205, USA.

INTRODUCTION

Dithiolethiones and other cancer chemopreventive agents inhibit the production of experimentally produced tumors by elevating the expression of several genes that encode for known cytoprotective enzymes[1]. Complementary DNA clones representing dithiolethione-inducible gene-1 (DIG-1*) were isolated from rat liver via differential hybridization screening[2]. The deduced amino acid sequence of DIG-1 was found to have 80% identity with the human liver enzyme leukotriene B_4 (LTB_4)-12-hydroxydehydrogenase (LTB_4 DH)[3,4]. DIG-1, purified >400-fold from the liver of rats dosed with 1,2-dithiole-3-dithiolethione (D3T), possessed an $NADP^+$-dependent activity to convert LTB_4 to 12-oxo-LTB_4 similar to LTB_4DH[3]. The formation of 12-oxo-LTB_4 by LTB_4 DH is the first step in the catabolism of LTB_4[4,5]. Subsequent conversion of 12-oxo-LTB_4 to 10,11-dihydro-12-oxo-LTB_4 (Met I) and 10,11,14,15-tetrahyro-12-oxo-LTB_4 (Met II) by as yet unidentified reductases has been described in liver and kidney cytosolic preparations as shown in Fig. 1[6]. These metabolic products of LTB_4 have diminished pro-inflammatory capacities, such as enhancing the mobilization of intracellular calcium[3], putatively via antagonism of the LTB_4 receptor[7]. Suppression of the pro-inflammatory actions of LTB_4, such as stimulation of neutrophil chemotaxis and superoxide anion generation, suggests that the catabolism of LTB_4 is a cytoprotective process.

*Abbreviations used: BNF, β-naphthoflavone, D3T, 1,2-dithiole-3-thione; DIG-1, dithiolethione-inducible gene-1; DS, disulfiram; EQ, ethoxyquin; 13-HODE, 13(S)-hydroxyoctadeco--9Z,13E-dienoic acid; I3C, indole-3-carbinol; LTB_4, leukotriene B_4; LTB_4DH, leukotriene B_4 12-hydroxydehydrogenase; OLT, oltipraz [5-(2-pyrazinyl)-4-methyl-1,2-dithiole-3-thione]; tBHQ, *tert*-butylhydroquinone.

In this report, it is shown that the hepatic levels of LTB4 metabolites produced by the actions of DIG-1 are elevated in rats dosed with D3T and other cancer chemopreventive agents. Hence, induction of DIG-1 activity by dithiolethiones and subsequent catabolism of LTB4 may contribute to additional mechanisms of cancer chemoprevention.

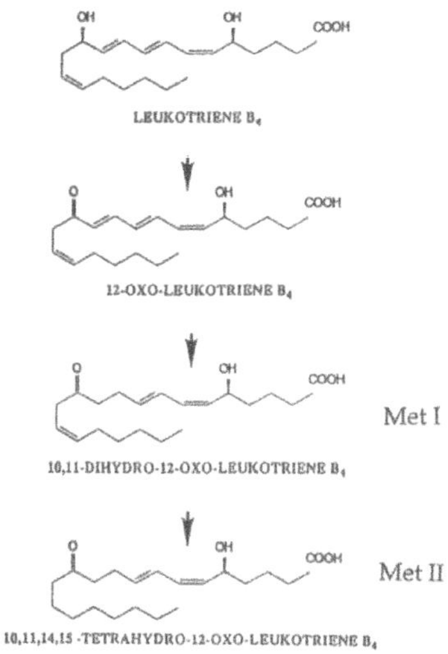

Fig. 1. Proposed metabolic pathway for the inactivation of LTB$_4$ in rat hepatic cytosol induced by dithiolethiones. LTB$_4$ is oxidized by DIG-1 to 12-oxo-LTB$_4$ which is subsequently converted to 10,11-dihydro-12-oxo- LTB$_4$ (Met I) and 10,11,14,15-tetrahydro-12-oxo- LTB$_4$ (Met II).

MATERIALS AND METHODS

Materials.

LTB$_4$ and 13(S)-hydroxyoctadeco-9Z,13E-dienoic acid (13-HODE) were purchased from Oxford Biomedical (Rochester Hills, MI). D3T was provided by Dr. Thomas J. Curphey (Dartmouth Medical School, Hanover, NH). Oltipraz (OLT) was provided by the Chemoprevention Branch of the National Cancer Institute (Bethesda, MD). Ethoxyquin (EQ), β-naphthoflavone (BNF), indole-3-carbinol (I3C), disulfiram (DS), and *tert*-butylhydroquinone (tBHQ) were purchased from Sigma (St. Louis, MO).

600

Animals and treatments.

Male F344 rats, 100 g, were obtained from Harlan (Indianapolis, IN) and housed under controlled conditions of temperature, humidity and lighting. Food and water were available *ad libitum*. Purified diet of the AIN-76A formulation without the recommended addition of 0.02% ethoxyquin was used. Rats were acclimated to this diet for 1 week before beginning the experiments. Rats were fed 0.4% EQ, 0.03% D3T, 0.075% OLT, 0.8% tBHQ, 0.5% I3C, and 1% disulfiram for 4 days. BNF (50 mg/kg /day,ip) was also administered for 4 days. Animals were sacrificed on the morning of day 5 and livers were excised and frozen in liquid nitrogen. Rat hepatic cytosolic preparations were made as previously described[8]. Total RNA was prepared and northern blots were performed as previously described[2].

Analysis of LTB$_4$ and LTB$_4$ metabolites.

Metabolism of LTB$_4$ was performed as described by Yokomizo *et al.*[4] with minor modifications[3]. The reaction was initiated by adding 3 nmol of LTB$_4$ in 2 µl to 86 µl 0.5x PBS buffer, 2 µl fresh 1 mM NADP$^+$, and 10 µl of rat hepatic cytosol containing 25 µg of protein. The reaction was incubated for1 h at 37°C and stopped by adding ethyl acetate containing 0.4 nmol 13(S)HODE. Following extraction LTB$_4$ and its metabolites were separated by reverse-phase HPLC at 235 nm for the 13-(S)HODE, LTB$_4$, Met I and Met II. Amounts of Met I and Met II were calculated from the extinction coefficient of 30,500 and 41,000, M^{-1}cm^{-1} , respectively[5]. To confirm the identities of Met I and Met II, LC/MS was performed utilizing negative ion electrospray with a mass scan range of 150-750 a.u., as previously described[2].

To determine if the levels of Met I and Met II were increased by D3T in rat hepatic tissue, differences in the amounts of each metabolite produced in cytosolic preparations from 12 separate animals dosed with D3T were compared to those dosed with vehicle by one-way ANOVA and the means judged for significance by a t-test ($p<0.05$).

RESULTS AND DISCUSSION

Induction of DIG-1 mRNA by cancer chemopreventive agents.

To determine the effectiveness of cancer chemopreventive agents in inducing the steady state levels of DIG-1 mRNA, rats were administered EQ, D3T, OLT, tBHQ, I3C, BNF, or DS using regimens known to induce carcinogen metabolizing enzymes[1]. Total RNA was isolated and northern blotted using DIG-1 or albumin (loading control) cDNA probes as shown in Fig. 2. Administration of EQ, tBHQ, I3C, BNF, or DS over 4 days yielded marginal (1.5- to 2.5-fold) increases in DIG-1 mRNA levels. In contrast, dosing rats with the dithiolethiones D3T and OLT resulted in elevations of 9.1- and 5.2-fold, respectively, in hepatic DIG-1 mRNA levels. Similar patterns in which the dithiolethiones were more efficacious than phenolic antioxidants and flavone-containing compounds at elevating mRNA levels and activities have been demonstrated with inducible carcinogen detoxication enzymes[1]. Furthermore, the parent compound, D3T, is consistently a more efficacious inducer of carcinogen metabolizing enzymes than $OLT^{1,9}$. Because of the LTB_4DH activity of DIG-1, its induction by D3T may translate to enhanced metabolism of LTB_4 *in vivo*.

Increased formation of LTB_4 metabolites in cytosolic fractions from rats dosed with D3T

To investigate effects of D3T on DIG-1/LTB_4DH activity, HPLC assays were perfromed to compare the concentrations of Met I and Met II following addition LTB_4 to hepatic cytosolic fractions from rats dosed with D3T or vehicle. A representative HPLC chromatogram showing the production of Met I and Met II by the activity of DIG-1 in cytosolic fractions is presented in Fig. 3. Peak retention times of 7.5 and 9.5 min were observed for Met I and Met II, respectively, by monitoring their absorbance at 235 nm. Molecular weights of 336 and 338 were determined for Met I and Met II, respectively, by negative-ion electrospray mass spectrometry, results consistent with those found by other investigators[4,5]. Negative ion electrospray mass spectra of Met I and Met II were primarily found as the water- or sodium-adducted ions. Masses $[M^{-1}]$ of 353 and 358 a.u. were found for the water-and sodium- adducted ions, respectively, for Met I. Similarly, masses

of $[M^{-1}]$ 355 and 360 a.u. were found for the water- and sodium-adducted ions, respectively Met II. Incubation of LTB_4 with hepatic cytosol prepared from rats dosed with either D3T or vehicle yielded a 2.2-fold increase in the amount of Met I (0.6 ± 0.11 versus 0.27 ± 0.05 nmol formed/mg cytosolic protein/60min), and a 2.0-fold increase in Met II (0.41 ± 0.05 versus 0.20 ± 0.05 nmol formed/mg protein/60 min) by treatment with D3T. The D3T-induced increase in Met I formation was statistically significant ($p<0.05$, $n=12$); whereas, the increase in Met II formation was not ($p=0.12$, $n=12$). However, the results indicate an overall increased catabolism of LTB_4 in the liver of rats receiving D3T ($p<0.05$).

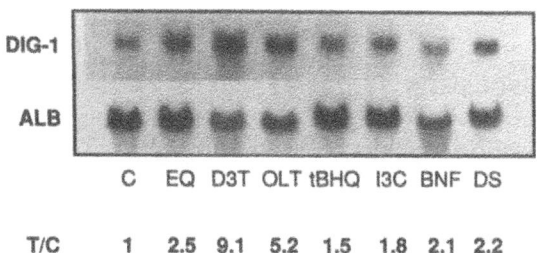

DIG-1

ALB

	C	EQ	D3T	OLT	tBHQ	I3C	BNF	DS
T/C	1	2.5	9.1	5.2	1.5	1.8	2.1	2.2

Fig. 2. Analysis of hepatic DIG-1 mRNA levels following treatment of rats with various cancer chemopreventive agents by northern blot hybridization.. Shown is a representative northern blot of total RNA isolated from the liver of rats fed 0.4 % ethoxyquin (EQ), 0.03% 1,2-dithiole-3-thione (D3T), 0.075% oltipraz (OLT), 0.8% *tert*-butylhydroquinone (tBHQ), 0.5% indole-3-carbinol (I3C), β-naphthoflavone (50 mg/kg/day,ip), or 1% disulfiram (DS) for 4 days. Loading of total RNA was corrected by normalizing autoradiographic signals to a blot probed with albumin (ALB). Fold induction (T/C) is defined as normalized signal from northern blots from D3T-treated rats with respect to vehicle-treated rats.

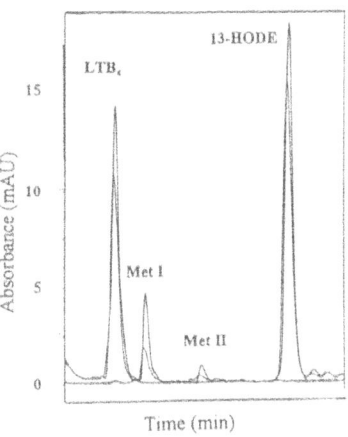

Time (min)

Fig. 3 HPLC profile of LTB₄ and its metabolites 10,11-dihydro-12-oxo-LTB₄ and 10,11,14,15-tetrahydro-12-oxo-LTB₄ generated by the activity of DIG-1. Shown is an HPLC chromatogram depicting the production of 10,11-dihydro-12-oxo-LTB₄ (Met I) and 10,11,14,15-tetrahydro-12-oxo-LTB₄ (Met II) generated by incubation with hepatic cytosol from rats dosed with D3T (bold tracing) or vehicle (thin tracing). LTB₄ , Met I, Met II, and the internal standard, 13-HODE, were detected by their absorbance at 235 nm (solid line). An aliquot (25 µg) of cytosolic protein was incubated with 30 µM LTB₄, 1 mM NADP⁺ in PBS at 37°C for 1 h

There is a lack of correspondence between the increased activity for the formation of Met I and Met II (≈ 2-fold) and increased levels of DIG-1 mRNA (≈ 9-fold) in the liver of D3T-treated rats. This discrepancy could be explained by the presence of an alternative pathway for the formation of Met I and Met II that functions constitutively, as evidenced by the catabolism of LTB₄ in the hepatic cytosol of vehicle-treated rats. Other pathways for LTB₄ catabolism have been defined in microsomal preparations of several tissues[6], however, cytosolic pathways other than those mediated by DIG-1 or LTB₄DH have not been defined. Operationally, other components that compete for the NADP⁺ cofactor may complicate the outcome of the assay by restricting available cofactor from binding to DIG-1. It is also possible that a 9-fold increase in DIG-1 mRNA leads to a 2-fold increase in DIG-1 protein. The production of a specific antibody will aid the analysis of DIG-1 protein levels, identify the location of DIG-1 and provide a specific inhibitor of DIG-1-associated turnover of LTB₄. In conclusion, increased catablism of LTB₄ is stimulated in rats by administration of D3T, mediated, in part, by the elevation of DIG-1 activity.

Epidemiological studies indicate that chronic inflammation associated with ulcerative colitis, gastritis, or hepatitis manifested during bacterial, parasitic, or viral infections has been linked to increased risks for the development of malignant disease. Increased levels of LTB₄ are found at sites of inflammation and tumor formation[10,11,12]. Therefore, LTB₄-stimulated neutrophil migration and activation may be important in the etiology of certain cancers. Neutrophils or other granulocytic cells present at sites of chronic infection are stimulated to release highly reactive oxidants, such as hydrogen peroxide, superoxide anion, nitric oxide and peroxynitrite[13,14,15]. Such oxidative events leading to mutagenesis and cytotoxicity are associated with various stages of tumorigenesis[16,17]. DIG-1 is the first enzyme in the degradative pathway of LTB$_4$found in tissues other than neutrophils. Elevating the expression of this enzyme may then increase inactivation of LTB$_4$. Consequently, induction of DIG-1 by dithiolethiones or other chemopreventive agents *in vivo* may afford protection from untoward tissue or organ damage by suppressing LTB$_4$-mediated inflammatory processes. We hypothesize that dithiolethiones, in addition to inducing genes enzymes that enhance the metabolism of carcinogens, can act to inhibit tumorigenesis by increasing the metabolism of LTB$_4$.

ACKNOWLEDGEMENTS

This work was funded by grants CA74136 and CA39416 from the National Cancer Institute, grants ES 03760 and ES 03818 from the National Institutes of Environmental Health Sciences, and by institutional grant IRG-11-36 from the American Cancer Society.

REFERENCES

1. Primiano, T., Sutter, T.R., and Kensler, T.W., Antioxidant-inducible genes, *Adv. Pharmacol.* **38,** 293-328 (1997).
2. Pri miano, T., Gastel, J.A., Kensler, T.W., and Sutter, T.R. Isolation of cDNAs representing dithiolethione-responsive genes, *Carcinogenesis* **17,** 2297-2303 (1996).
3. Primiano, T., Li, Y., Kensler, T.W., Trush, M.A., and Sutter, T.R. Identification of dithiolethione-inducible gene-1 as leukotriene B$_4$ 12-hydroxydehydrogenase: Implications for chemoprevention, Carcinogenesis, in press.
4. Yokomizo,T., Izumi,T., Takahashi,T., Kasama,T., Kobayashi,Y., Sato,F., Taketani,Y., and Shimizu,T. Enzymatic inactivation of leukotriene B$_4$ by a novel enzyme found in porcine kidney, *J. Biol. Chem.* **268,** 18128-18135 (1993).
5. Wainwright,S.L., and Powell,W.S. Mechanism of the formation of dihydro metabolites of 12-hydroxyeicosanoids, *J. Biol. Chem.* **266,** 20899-20906 (1991).
6. Powell, W.S., and Gravelle, F. Metabolism of leukotriene B$_4$ to dihydro and dihydro-oxo products by porcine leukocytes, *J. Biol. Chem.* **263,** 2170-2177 (1989).
7. Yokomizo,T., Izumi,T., Chang,K., Takuwa,Y., and Shimizu,T. A G-protein- coupled receptor for leukotriene B$_4$ that mediates chemotaxis, *Nature* **387,** 620-624 (1997).
8. Primiano,T., and Novak,R.F. Enhanced expression, purification, and characterization of a novel class alpha glutathione S-transferase isozyme appearing in rabbit hepatic cytosol following treatment with 4-picoline, *Toxicol. Appl. Pharmacol.* **112,** 291-299 (1992).
9. Kensler,T.W., Groopman, J.D., Roebuck, B.D., and Curphey,T.J. In Huang, M-T. Osawa,T., Ho, C-T., and Rosen, R.T. (eds.) *Food Phytochemicals for CancerPrevention I:Fruits and Vegetables.* American Chemical Society, Washington D.C., pp. 154-163 (1994).
10. Demopoulos, H., Pietronigro, D.D., Seligman, M.L. The development of secondary pathology with free radical reactions as a threshold mechanism *J. Amer.Coll. Toxicol.* **2,** 173-184 (1994).

11. Ekbom, A., Helmick, C., Zack, M., and Adami, H-O. Ulcerative colitis and colorectal cancer: a population-based study, *N. Engl. J. Med.* **323,** 1228- 1233 (1990).
12. Parsonnet,J., Friedman,G.D., Vandersteen,D.P. Helicobacter pylori infection and the risk of gastric carcinoma, *N. Engl. J. Med.* **325,** 1127-1131 (1991).
13. Hagen, T.M., Huang, S., Curnutte, J., Fowler, P., Martinez, V., Wehr, C.M., Ames, B.N., Chisari,F.V. Extensive oxidative DNA damage in hepatocyres of transgenic mice with chronic active hepatitis destined to develop hepatocellular carcinoma, *Proc.Natl. Acad. Sci. USA* **91,** 12808-12812 (1994).
14. Sharon, P., and Stenson, W.F. Enhanced synthesis of leukotirene B$_4$ by colonic mucosa in inflammatory bowel disease, *Gastroenterology* **86,** 453-460 (1984).
15. Li,Y., Ferrante, A., Poulos, A., and Harvey, D.P. Neutrophil oxygen radical generation: synergistic responses to tumor necrosis factor and mono/polyunsaturated fatty acids, *J.Clin. Invest.* **97,** 1605-1609 (1996).
16. Cerutti, P.A. Prooxidant states and tumor promotion. *Science* **227,** 375-381, (1985).
17. Trush, M.A., and Kensler, T.W. An overview of the relationship between oxidative stress and chemical carcinogenesis. *Free Radical Biol. Med.* **10,** 201-209 (1991).

88

INTESTINAL TUMOR LOAD IN THE *MIN/+* MOUSE MODEL IS NOT CORRELATED WITH EICOSANOID BIOSYNTHESIS

Jay Whelan,[1] Chun-Hung Chiu,[1] and Michael F. McEntee[2]

[1]Department of Nutrition
[2]Department of Pathology
University of Tennessee
Knoxville, TN 37996-1900

INTRODUCTION

Several lines of evidence demonstrate an inverse relationship between the use of nonsteroidal anti-inflammatory drugs (NSAIDs), such as aspirin and aspirin-like drugs, and intestinal cancer. NSAIDs have been shown to reduce the risk of intestinal cancer in humans by 50%.[1-4] It is well known that the anti-inflammatory properties of NSAIDs are related to their ability to inhibit prostaglandin biosynthesis. Cyclooxygenase (COX) catalyzes the committed step in prostaglandin formation. Two isoforms of cyclooxygenase exist, COX-1 and COX-2. NSAIDs can inhibit both. COX-1 is the constitutively expressed isoform, and COX-2 is the inducible form of the enzyme involved in inflammation.[5-7]

Heterozygous C57BL/6J-Min (*Min/+*) mice are highly susceptible to spontaneous development of intestinal adenomas due to a germline mutation in one allele of the murine adenomatous polyposis coli (*Apc*) gene with tumor initiation following loss of heterozygosity.[8-10] The *Min/+* mouse is an APC model for human intestinal tumorigenesis. Germline and somatic mutations of the APC tumor suppressor gene in humans are associated with a majority of colorectal cancers. Familial adenomatous polyposis (FAP) patients have a germline mutation in APC resulting in an autosomal dominant predisposition for intestinal adenomas.[11-13] A majority of individuals who develop sporadic intestinal tumors also develop an APC defect.[14] The use of sulindac, a non-selective COX inhibitor,[15] decreases the number and size of intestinal adenomas in FAP patients and *Min/+* mice.[16-19] It is generally accepted that sulindac and other NSAIDs mediate their effects via a reduction in prostaglandin biosynthesis; however, a growing body of

Eicosanoids and Other Bioactive Lipids in Cancer, Inflammation, and Radiation Injury, 4
Edited by Honn *et al.*, Kluwer Academic / Plenum Publishers, New York, 1999.

607

contradictory evidence suggests the possibility of a prostaglandin independent mechanism.[15,16,20-22] Therefore, this study investigated the relationship between NSAID use, intestinal tumor load and intestinal prostaglandin production using the *Min*/+ mouse model.

ANIMALS, MATERIALS AND METHODS

Animals

Male C57BL/6J mice, wild-type (WT, +/+) and *Min*/+ (*Apc*/+) were purchased from Jackson Laboratories (Bar Harbor, ME) following genotyping (35-40 days of age). Mice were randomly divided into treatment groups (4-11 animals per group). The animals were housed in suspended stainless steel cages in a temperature controlled room (23 ± 2^0C) with a 12-h light-dark cycle and given free access to food and water. Prior to sacrifice, all animals were fasted over night. All animal procedures were approved by the University of Tennessee Animal Care and Use Committee and were in accord with the NIH *Guide for the Care and Use of Laboratory Animals* (NRC 1985). Groups of animals (WT or *Min*/+) were sacrificed at various time points between 60 and 120 days of age. It is our experience that *Min*/+ mice may not live beyond 120 days of age. Body weight and food consumption for all animals were monitored. At the time of sacrifice, PGE_2 production, tumor number, size and location were determined. Average tumor size was calculated as the weighted average for each size classification.

Diets

The control diets were composed of purified AIN-93G powder diet (Dyets, Inc., Bethlehem, PA). Sulindac (30 mg/Kg/day), indomethacin (0.9 mg/kg/day) and aspirin (44 mg/Kg/day) were all purchased from Sigma Chemical Co (St. Louis, MO) and were administered by thoroughly mixing them in the diet each day. Indomethacin and aspirin were supplied to mice upon arrival for 40 days. Sulindac was supplied to mice, starting at the time of their arrival, for 40 or 80 days, or to 80 day old *Min*/+ mice for 2, 4 or 20 days. For those experiments using methyl esters of arachidonic acid and oleic acid, diets were supplemented with 1.5% (w/w) of oleic acid or arachidonic acid (Nu Chek Prep, Elysian, MN) upon the arrival of the mice (35 days of age) for 25 days (60 days of age). All diets were prepared weekly and stored at -20^{0}C. All animals were provided fresh food daily to prevent oxidation. Food consumption was monitored daily and the actual drug doses were calculated according to food intakes.

Eicosanoids Measurement

Following sacrifice, normal appearing intestinal sections (mid-jejunum) and pooled samples of individually excised small intestinal tumors were homogenized using a Polytron homogenizer in 0.1M cold Tris-HCl buffer (pH 7.4). *Ex vivo* production and basal levels of eicosanoids were determined as described previously.[16] An aliquot of the homogenate from each sample was used to determine protein concentration.[24]

Fatty Acid Analysis

The fatty acid methyl esters of the tissue phospholipids were prepared and analyzed as previously described.[23]

Statistical Analysis

Data are expressed as means ± standard error (SE). Differences between the groups were analyzed statistically by a Students t-test or analysis of variance (ANOVA) using Statistical Analysis Systems (SAS Institute, Inc, Cary, NC). Fisher's Least Significant difference multiple comparison method was used to determine differences among groups. Data was considered significant at $p < 0.05$.

RESULTS

No tumors were ever observed in the WT mice. The average number of tumors along the intestinal tract (small and large intestines) in Min/+ mice remained relatively constant between 60 and 115 days of age, with the average size of the tumors doubling over the same time period (Tables 1 and 5). When sulindac was supplemented to the diet for 80 days, tumor number and size were reduced by 93% and 55%, respectively, when compared to age-matched litter mates. The production of PGE_2 by the intestines was not different between WT and *Min/+* mice at any age (data not shown, see Chiu, et al[16]) and the production of PGE_2 was not reduced by 80 days of sulindac treatment (Table 1).

In order to determine if sulindac would cause regression of pre-existing tumors, animals were provided a sulindac-free diet for approximately 40 days following arrival and then supplemented with sulindac for either 2, 4 or 20 days. When sulindac was administered for two days, the number of tumors was 19% lower than in the age-matched *Min/+* control group which received no sulindac (29±3 vs 36±7 tumors; Table 2). Similar results were obtained when this experiment was repeated (data not shown). Within four days of sulindac treatment, tumor number (9±1) was 75% lower than the untreated group, with only a slight further reduction (7±2 tumors) after 20 days of sulindac treatment. Sulindac treatment had no effect on PGE_2 production (data not shown).

Table 1. Effect of age and sulindac on tumor number, size and PGE_2 production.[1]

	Age Groups[2]			
	77 (n=4)	101 (n=4)	115 (n=4)	115 + S (n=4)
Average tumor #/mouse	39.5 ± 4.6^b	45.3 ± 6.0^b	41.0 ± 4.6^b	3.0 ± 1.1^a
Average tumor size	1.20 ± 0.01^b	1.71 ± 0.03^c	1.82 ± 0.02^c	0.83 ± 0.07^a
PGE_2 (ng/mg protein)	2.76 ± 0.40^a	2.93 ± 0.28^a	2.92 ± 0.51^a	2.68 ± 0.26^a

[1] Published with permission of authors[16] and Cancer Research.
[2] *Min/+* mice were fed AIN93G powdered diet from time of arrival until sacrifice at 77, 101 and 115 days of age. One group (115+S) was given sulindac (30 mg/kg/day) from the time of arrival through 115 days of age.
[a] Values in each row with different superscript letter are significantly different at $p < 0.05$.

Table 2. Effects of sulindac supplementation for 2, 4, or 20 days on pre-existing tumors in the intestines of *Min/+* mice.[1]

	Treatment Groups[2]			
	-S (n=6)	+S 2 days (n=7)	+S 4 days (n=8)	+S 20 days (n=5)
Average tumor #/mouse	36.3 ± 7.3^b	29.4 ± 2.7^b	9.1 ± 0.8^a	6.8 ± 1.6^a
Average tumor size	1.14 ± 0.01^b	0.99 ± 0.01^a	0.92 ± 0.03^a	0.89 ± 0.02^a
PGE_2 (ng/mg protein)	3.71 ± 0.37^b	2.11 ± 0.44^a	2.64 ± 0.41^a	3.75 ± 0.6^b

[1] Part of data published with permission of authors[16] and Cancer Research.
[2] *Min/+* mice were provided a sulindac-free diet (-S) for an average of 37 days following arrival. Different groups of mice were then provided sulindac (+S; 30 mg/kg/day) for 2, 4, or 20 days prior to sacrifice
[a] Values in each row with different superscript letter are significantly different at $p < 0.05$.

To determine whether sulindac treatment was required to maintain a reduced tumor load, animals were supplemented with sulindac for 40 days followed by 40 more days of sulindac-free treatment (+S(40):-S(40)) (Table 3). When sulindac treatment stopped, tumor number increased almost 15 fold compared to the sulindac treated controls (2.0 tumors/mouse vs. 28.8 tumors/mouse).

When *Min/+* mice were treated with indomethacin, tumor number was reduced by 91% compared to untreated age-matched controls, with a concomitant reduction in PGE_2 production by 75% (Table 4). However, even though aspirin (ASA) treatment reduced PGE_2 production by 78%, there was not significant change in tumor number.

Arachidonic acid was provided in the diet to enrich tissue phospholipids with arachidonic acid and augment eicosanoid biosynthesis in order to evaluate what impact these changes would have on tumor frequency and/or size. Tissue arachidonic acid levels in the arachidonic acid-supplemented groups were 60% higher compared to the oleic acid-supplemented groups, independent of sulindac treatment (Table 5). The feeding of arachidonic acid resulted in significant increases in the production of PGE_2 but, again, sulindac did not modulate PGE_2 production. Arachidonic acid supplementation had no effect on tumor number compared to oleic acid (Table 5). Sulindac treatment reduced the number of tumors by 96% and 83% in the oleic acid and arachidonic acid groups, respectively. The average size of the tumors in the sulindac-treated animals were not different among the arachidonic acid and oleic acid-supplemented groups.

Table 3. Residual effect of sulindac treatment on intestinal tumor load in *Min/+* mice.[1]

	Treatment Groups[2]	
	+S 40	+S 40 : -S 40
Average tumor #/mouse	2.0 ± 0.8^a	28.8 ± 6.0^b
Average tumor size	0.94 ± 0.08^a	1.14 ± 0.03^b
PGE_2 (ng/mg protein)	3.12 ± 0.21^a	3.71 ± 0.52^a

[1] Part of data published with permission of authors[16] and Cancer Research.
[2] *Min/+* mice were treated with sulindac (30 mg/kg/day) for 40 days (+S 40). The other group (+S 40 : -S 40) was treated with sulindac for 40 days, then placed on a sulindac-free diet for an additional 40 days.
[a] Values in each row with different superscript letter are significantly different at $p < 0.05$.

Table 4. Effect of indomethacin and aspirin on tumor load and PGE_2 production.[1]

	Treatment Groups[2]		
	Control (n=8)	Indomethacin (n=4)	Aspirin (n=8)
Average tumor #/mouse	33.3 ± 5.8^b	4.5 ± 0.7^a	33.9 ± 4.0^b
Average tumor size	1.09 ± 0.08^a	1.16 ± 0.11^a	1.10 ± 0.04^a
PGE_2 (ng/mg protein)	6.74 ± 0.76^b	1.87 ± 0.66^a	1.68 ± 0.19^a

[1] Data, in part, has been submitted for publication and is pending review.
[2] Upon arrival, *Min/+* mice were treated with indomethacin (0.9 mg/kg/day) or aspirin (44 mg/kg/day) for 40 days prior to sacrifice. Control animals received no NSAID treatment.
[a] Values in each row with different superscript letter are significantly different at $p < 0.05$.

Table 5. Effect of dietary arachidonic acid supplementation on tumor load, PGE_2 production and arachidonic acid content of intestinal phospholipids.[1]

	Treatment Groups[2]			
	OA (n=5)	OA + S (n=5)	AA (n=5)	AA + S (n=5)
20:4 n-6 content (mol %)[3]	15.71 ± 0.29^a	15.73 ± 0.32^a	24.62 ± 0.78^b	25.66 ± 0.36^b
Average tumor #/mouse	40.2 ± 3.1^a	1.8 ± 0.6^b	34.2 ± 6.4^a	5.8 ± 0.4^c
Average tumor size	0.97 ± 0.01^b	$0.90 \pm 0.10^{a,b}$	0.82 ± 0.01^a	0.79 ± 0.05^a
PGE_2 (ng/mg protein)	3.00 ± 0.23^a	2.83 ± 0.44^a	4.12 ± 0.28^b	4.50 ± 0.63^b

[1] Published with permission of authors[16] and Cancer Research.
[2] Upon arrival *Min/+* mice were fed diets supplemented with 1.5% (w/w) ethyl esters of oleic acid (OA) or arachidonic acid (AA) with 30 mg/kg/day of sulindac (OA + S; AA + S) or without sulindac (OA; AA).
[3] Phospholipids were separated by TLC and methylated using ethereal diazomethane. Fatty acid methyl esters were then analyzed with gas chromatography. Data is expressed as mean ± standard error (SE).
[a] Values in each row with different superscript letter are significantly different at $p < 0.05$.

DISCUSSION

The commonly accepted hypothesis for the role of the anti-tumor effects of NSAIDs is their ability to inhibit prostaglandin biosynthesis, in particular PGE_2.[25-28] This study was designed to determine whether alterations of eicosanoid biosynthesis are correlated with the anti-tumor effects of NSAIDs. To accomplish this, various NSAIDs were administered to *Min/+* mice to inhibit prostaglandin biosynthesis. Sulindac and indomethacin are reversible inhibitors of cyclooxygenases, while aspirin irreversibly inhibits cyclooxygenases via acetylation. In addition, the ethyl ester of arachidonic acid was supplemented to diets of the mice to enhance prostaglandin production. Human and animal studies have clearly demonstrated that supplementing diets with arachidonic acid augments eicosanoid biosynthesis systemically and, as such, provides an effective non-invasive technique to elevate prostaglandins in vivo.[23,29-31]

Our results demonstrate that intestinal tumor initiation occurs prior to 60 days of age in this model as the average number of tumors did not increase between 60 and 115 days of age while the average tumor size doubled over the same time period. This shift in tumor size, but not tumor number, is consistent with the growth of pre-existing tumors with minimal induction of new tumors. As observed in other studies,[19,32] sulindac administration for 25, 40 or 80 days shortly after genotyping invariably reduced tumor number by 90%. Withdrawal of sulindac resulted in a 15 fold increase in tumor number within 40 days, indicating an obligate requirement of sulindac for a sustained antitumor effect. We also observed that sulindac effectively reduced the number of preexisting tumors by 75% within four days, probably through induction of apoptosis.[15,19,33,34] It has been previously reported that sulindac was ineffective on preexisting colonic tumors induced by 1,2-dimethylhydrozine in mice.[26] It is possible that tumor location could account for this apparent contradiction. Greater than 95% of the intestinal tumors in the *Min*/+ model occur in the small intestines and it was these tumors which were affected by sulindac treatment.

Sulindac is a pro-cyclooxygenase inhibitor and lacks activity until it is modified to its sulfide derivative in the liver and by colonic bacteria.[35,36] We were unable to link the effects of sulindac to prostaglandin formation. We did not observe a reduction in PGE_2 production in normal appearing intestines of *Min*/+ mice following sulindac treatment. We tested for inhibition of intestinal prostaglandin formation using sections of the mid-jejunum instead of tumors because sulindac eliminated virtually all of the tumors. In fact, when sulindac was used, intestinal prostaglandin biosynthesis was never inhibited in WT or *Min*/+ mice regardless of the length of administration. In addition, sulindac treatment was unable to reduce elevated prostaglandin production in mice supplemented with arachidonic acid. These results, while consistent, were nevertheless puzzling since exogenous administration of sulindac sulfide has been shown to inhibit cyclooxygenases in vitro.[37-39]

In an effort to further evaluate the effects of NSAIDs and prostaglandins, we investigated the effects of indomethacin, aspirin and dietary arachidonic acid on tumor load and prostaglandin production. Like sulindac, administration of indomethacin reduced tumor number by 90%; however, aspirin was ineffective in reducing tumor load. Both indomethacin and aspirin significantly reduced PGE_2 biosynthesis, indicating a lack of correlation between tumor load and the ability to produce prostaglandins. This conclusion was further supported by the fact that augmenting prostaglandin formation through arachidonic acid administration did not alter tumor number.

Our data suggests that tumor load is not correlated with eicosanoid production[16] and that the ability of sulindac to reduce tumor load in the *Min*/+ mouse model does not appear to be dependent upon the inhibition of prostaglandins, in particular, PGE_2. The mechanism by which sulindac and other NSAIDs cause tumor regression in this animal model has been linked to their ability to inhibit COX-2 activity.[40] Tumors over-express COX-2 in *Min*/+ mice and humans compared to adjacent normal appearing intestinal tissue and this may account, in part, for increased production of PGE_2.[16,41-43] More recently it has been suggested that the tumor suppressive effects of NSAIDs are not likely related to reductions in prostaglandins, but may be mediated by cellular elevations of free arachidonic acid stimulating ceramide biosynthesis as a result of COX inhibition.[20] However, this does not explain the discordant effects of aspirin and indomethacin, as both NSAIDs potently inhibited prostaglandin biosynthesis; nor does it explain the lack of efficacy by dietary arachidonic acid which doubled tissue arachidonic acid content (Table 5).

ACKNOWLEDGMENTS

Supported by a grant from the American Institute for Cancer Research, Tennessee Agricultural Experiment Station and private donors of the Arachidonic Acid Project.

REFERENCES

1. O. Suh, C. Mettlin, and N.J. Petrelli. Aspirin use, cancer, and polyps of the large bowel, *Cancer.* 72:1171 (1993).
2. P.H. Gann, J.E. Manson, R.J. Glynn, J.E. Buring, and C.H. Hennekens. Low-dose aspirin and incidence of colorectal tumors in a randomized trial, *J Natl Cancer Inst.* 85:1220 (1993).
3. I.I. Peleg, H.T. Maibach, S.H. Brown, and C.M. Wilcox. Aspirin and nonsteroidal anti-inflammatory drug use and the risk of subsequent colorectal cancer, *Arch Intern Med.* 154:394 (1994).
4. R.N. DuBois. Nonsteroidal anti-inflammatory drug use and sporadic colorectal adenomas, *Gastroenterol.* 108:1310 (1995).
5. C.S. Williams, R.N. DuBois. Prostaglandin endoperoxide synthase: why two isoforms?, *Am J Physiol.* 270:G393-G400 (1996).
6. G.P. O'Neill, HA Ford. Expression of mRNA for cyclooxygenase-1 and cyclooxygenase-2 in human tissues, *FEBS Lett.* 330:156 (1993).
7. S.H. Lee, E. Soyoola, P. Chanmugam, S. Hart, W. Sun, H. Zhong, S. Liou, D. Simmons, and D. Hwang. Selective expression of mitogen-inducible cyclooxygenase in macrophages stimulated with lipopolysaccharide, *J Biol Chem.* 267:25934 (1992).
8. A.R. Moser, W.F. Dove, K.A. Roth, and J.I. Gordon. The Min (multiple intestinal neoplasia) mutation: its effect on gut epithelial cell differentiation and interaction with a modifier system, *J Cell Biol.* 116:1517 (1992).
9. D.B. Levy, K.J. Smith, Y. Beazer-Barclay, S.R. Hamilton, B. Vogelstein, and K.W. Kinzler. Inactivation of both APC alleles in human and mouse tumors, *Cancer Res.* 54:5953 (1994).
10. C. Luongo, A.R. Moser, S. Gledhill, and W.F. Dove. Loss of Apc+ in intestinal adenomas from Min mice, *Cancer Res.* 54:5947 (1994).
11. Y. Miyoshi, H. Ando, H. Nagase, I. Nishisho, A. Horii, Y. Miki, T. Mori, J. Utsunomiya, S. Baba, G. Petersen, and et-al. Germ-line mutations of the APC gene in 53 familial adenomatous polyposis patients, *Proc Natl Acad Sci USA*. 89:4452 (1992).
12. K.W. Kinzler, M.C. Nilbert, L.K. Su, B. Vogelstein, T.M. Bryan, D.B. Levy, K.J. Smith, A.C. Preisinger, P. Hedge, D. McKechnie, and a et. Identification of FAP locus genes from chromosome 5q21, *Science.* 253:661 (1991).
13. J. Groden, A. Thliveris, W. Samowitz, M. Carlson, L. Gelbert, H. Albertsen, G. Joslyn, J. Stevens, L. Spirio, M. Robertson, and et-al. Identification and characterization of the familial adenomatous polyposis coli gene, *Cell.* 66:589 (1991).
14. S.M. Powell, N. Zilz, Y. Beazer-Barclay, T.M. Bryan, S.R. Hamilton, S.N. Thibodeau, B. Vogelstein, and K.W. Kinzler. APC mutations occur early during colorectal tumorigenesis, *Nature.* 359:235 (1992).
15. G.A. Piazza, A.L.K. Rahm, M. Krutzsch, G. Sperl, N.S. Paranka, P.H. Gross, K. Brendel, R.W. Burt, D.S. Alberts, R. Pamukcu, and D.J. Ahnen. Antineoplastic drugs sulindac sulfide and sulfone inhibit cell growth by inducing apoptosis, *Cancer Res.* 55:3110 (1995).

16. C.-H. Chiu, M.F. McEntee, and J. Whelan. Sulindac causes rapid regression of preexisting tumors in Min/+ mice independent of prostaglandin biosynthesis, *Cancer Res.* 57:4267 (1997).

17. W.R. Waddell, R.W. Loughry. Sulindac for polyposis of the colon, *J Surg.Oncol.* 24:83 (1983).

18. F. Tonelli, R. Valanzano, and P. Dolara. Sulindac therapy of colorectal polyps in familial adenomatous polyposis, *Dig.Dis.* 12:259 (1994).

19. S.K. Boolbol, A.J. Dannenberg, A .Chadburn, C. Martucci, X. Guo, J.T Ramonetti, M. Abreu-Goris, H.L. Newmark, M.L. Lipkin, J.J. DeCosse, and M.M. Bertagnolli. Cyclooxygenase-2 overexpression and tumor formation are blocked by Sulindac in a murine model of familial adenomatous polyposis, *Cancer Res.* 56:2556 (1996).

20. T.A. Chan, P.J. Morin, B. Vogelstein, and K.W. Kinzler. Mechanisms underlying nonsteroidal antiinflammatory drug-mediated apoptosis, *Proc.Natl.Acad.Sci.U.S.A.* 95:681 (1998).

21. R. Hanif, A. Pittas, Y. Feng, M.I. Koutsos, L. Qiao, L. Staiano-Coico, S.I. Shiff, and B. Rigas. Effects of nonsteroidal anti-inflammatory drugs on proliferation and on induction of apoptosis in colon cancer cells by a prostaglandin-independent pathway, *Biochem Pharmacol.* 52:237 (1996).

22. G.A. Piazza, A.K. Rahm, T.S. Finn, B.H. Fryer, H. Li, A.L. Stoumen, R. Pamukcu, and D.J. Ahnen. Apoptosis primarily accounts for the growth-inhibitory properties of sulindac metabolites and involves a mechanism that is independent of cyclooxygenase inhibition, cell cycle arrest, and p53 induction., *Cancer Res.* 57:2452 (1997).

23. J. Whelan, M.E. Surette, I. Hardardottir, G. Lu, K.A. Golemboski, E. Larsen, and J.E. Kinsella. Dietary arachidonate enhances tissue arachidonate levels and eicosanoid production in syrian hamsters, *J Nutr.* 123:2174 (1993).

24. M.A. Markwell, S.M. Haas, N.E. Tolbert, and L.L. Bieber. Protein determination in membrane and lipoprotein samples: manual and automated procedures, *Methods Enzymol.* 72:296 (1981).

25. J.R. Vane. Inhibition of prostaglandin synthesis as a mechanism of action for aspirin-like drugs, *Nat.New Biol.* 231:232 (1971).

26. M. Moorghen, P. Ince, K.J. Finney, J.P. Sunter , D.R. Appleton, and A.J. Watson. A protective effect of sulindac against chemically-induced primary colonic tumours in mice, *J Pathol.* 156:341 (1988).

27. R.F. Logan, J. Little, P.G. Hawtin, and J.D. Hardcastle. Effect of aspirin and non-steroidal anti-inflammatory drugs on colorectal adenomas: case-control study of subjects participating in the Nottingham faecal occult blood screening programme, *BMJ.* 307:285 (1993).

28. M. Pollard and P.H. Luckert. Effect of piroxicam on primary intestinal tumors induced in rats by N-methylnitrosourea, *Cancer Lett.* 25:117 (1984).

29. B. Li, C. Birdwell, and J. Whelan. Antithetic relationship of dietary arachidonic acid and eicosapentaenoic acid on eicosanoid production in vivo, *J Lipid Res.* 35:1869 (1994).

30. N.J. Mann, G.E. Warrick, K. O'Dea, H.R Knapp, and A.J. Sinclair. The effect of linoleic, arachidonic and eicosapentaenoic acid supplementation on prostacyclin production in rats, *Lipids.* 29:157 (1994).

31. A. Ferretti, G.J. Nelson, P.C. Schmidt, D.S. Kelley, G. Bartolini, and V.P. Flanagan. Increased dietary arachidonic acid enhances the synthesis of vasoactive eicosanoids in humans, *Lipids.* 32:435 (1997).

32. Y. Beazer-Barclay, D.B. Levy, A.R. Moser, W.F. Dove, S.R. Hamilton, B. Vogelstein , and K.W. Kinzler. Sulindac suppresses tumorigenesis in the Min mouse, *Carcinogen.* 17:1757 (1996).

33. S.J. Shiff, M.I. Koutsos, L. Qiao, and B. Rigas. Nonsteroidal antiinflammatory drugs inhibit the proliferation of colon adenocarcinoma cells: effects on cell cycle and apoptosis , *Exp Cell Res.* 222:179 (1996).

34. Y. Goldberg, I.I. Nassif, A. Pittas, L.L. Tsai , B.D. Dynlacht, B. Rigas, and S.J. Shiff. The anti-proliferative effect of sulindac and sulindac sulfide on HT-29 colon cancer cells: alterations in tumor suppressor and cell cycle-regulatory proteins, *Oncogene.* 12:893 (1996).

35. C.V. Rao, A. Rivenson, B. Simi, E. Zang, G. Kelloff, V. Steele, and B.S.. Reddy. Chemoprevention of colon carcinogenesis by sulindac, a nonsteroidal anti-inflammatory agent, *Cancer Res.* 55:1464 (1995).

36. D.E. Duggan, L.E. Hare, C.A. Ditzler, B.W. Lei, and K.C. Kwan. The disposition of sulindac, *Clin.Pharmacol.Ther.* 21:326 (1977).

37. E.A. Meade, W.L. Smith, and D.L. Dewitt. Differential inhibition of prostaglandin endoperoxide synthase (cyclooxygenase) isozymes by aspirin and other non-steroidal anti- inflammatory drugs, *J Biol Chem.* 268:6610 (1993).

38. W.L. Smith, E.A. Meade, and D.L. Dewitt. Interactions of PGH synthase isozymes-1 and -2 with NSAIDs, *Ann.N.Y.Acad.Sci.* 744:50 (1994).

39. O. Laneuville, D.K. Breuer, D.L. Dewitt, T. Hla, C.D. Funk, and W.L. Smith. Differential inhibition of human prostaglandin endoperoxide H synthases-1 and -2 by nonsteroidal anti-inflammatory drugs, *J Pharmacol Exp Ther.* 271:927 (1994).

40. S. Nakatsugi, M. Fukutake, M. Takahashi, K. Fukuda, T. Isoi, Y Taniguchi, T Sugimura, and K Wakabayashi. Suppression of intestinal polyp development by nimesulide, a selective cylooxygenase-2 inhibitor, in Min mice, *Jpn J Cancer Res.* 88:1117 (1997).

41. C.S. Williams, C. Luongo, A. Radhika, T. Zhang, L.W. Lamps, L.B. Nanney, R.D. Beauchamp, and R.N. DuBois. Elevated cyclooxygenase-2 levels in Min mouse adenomas, *Gastroenterol.* 111:1134 (1996).

42. S.L. Kargman, G.P. O'Neill, P.J. Vickers, J.F. Evans, J.A. Mancini, and S. Jothy. Expression of prostaglandin G/H synthase-1 and -2 protein in human colon cancer, *Cancer Res.* 55:2556 (1995).

43. H. Sano, Y. Kawahito, R.L. Wilder, A. Hashiramoto, S. Mukai, K. Asai, S. Kimura, H. Kato, M. Kondo, and T Hla. Expression of cyclooxygenase-1 and -2 in human colorectal cancer, *Cancer Res.* 55:3785 (1995).

89

12-LIPOXYGENASE EXPRESSION IN HUMAN MELANOMA CELL LINES

Tímár J.[1], Rásó E.[1], Honn K.V.[2], Hagmann W.[3]

[1]1st Institute of Pathology and Experimental Cancer Research, Semmelweis University of Medicine, Budapest, Hungary
[2]Department of Radiation Oncology, Wayne State University, Detroit, MI 48202
[3]German Cancer Research Center, Heidelberg, Germany

ABSTRACT

12-lipoxygenase (12-LOX) expression and function in the regulation of the metastatic phenotype was demonstrated in several murine melanoma lines before. Here we have provided novel evidences that, though at a low level (in max. 15% of the cell population), human melanoma lines (HT168, M1, HT199, HT18 and WM35) express the platelet-type isoform of 12-LOX both at mRNA and protein levels. 12-LOX expression was demonstrated in cultured tumor cells and in skin tumor xenografts. Comparison of the expression of 12-LOX in skin primary tumors and its lung metastases indicated a stable expression. The low level of 12-LOX expression in human melanoma cell lines suggests that other lipoxygenase(s) could also be responsible for the metabolism of arachidonic acid to 12-HETE breakdown products.

INTRODUCTION

Signaling pathways provide the biochemical link between the extracellular milieu and the intracellular machinery's responsible for the cellular activities. These pathways though unique at key points such as the receptor, transducer or effector levels, are connected to each other providing a complex network. Bioactive lipids are essential components of the signaling pathways where arachidonic acid (AA) metabolism breakdown products of the lipid membranes play important roles. Phospholipases are activated in various signaling pathways thereby producing AA which could be metabolized by COX or LOX enzymes. Recently it turned out that the 12(S)-HETE metabolite generated via 12-LOX or 15-LOX enzymes has important physiological functions[1]. 12(S)-HETE is involved in vascular functions (endothelial proliferation, retraction),

Eicosanoids and Other Bioactive Lipids in Cancer, Inflammation, and Radiation Injury, 4
Edited by Honn *et al.*, Kluwer Academic / Plenum Publishers, New York, 1999.

617

thrombogenesis (platelet activation) or inflammation (chemokinesis of leukocytes). 12-LOX is expressed in two forms, as platelet-type enzyme or as the leukocyte or epithelial type isoform[2].

It was revealed that 12-HETE can be produced by tumor cells suggesting that some tumors may express a 12-LOX enzyme. Expression of the platelet-type 12-LOX indeed was reported in murine melanomas (B16a[3], K1735 clones[4]), carcinomas (3LL[3,5] and AtT-20[6]), rat carcinosarcoma (W256)[3] and in human hematopoietic (HEL, HL-60)[7] and epithelial tumors (A431[8], clone-A[3]). The analysis of the role of 12-HETE in tumor cells indicated that it is involved in the generation of both the motogenic[9] and mitogenic[10] signals. Since 12-LOX function seemed to be critical in the regulation of the invasive/metastatic phenotype of the murine melanomas studied, we have questioned if human melanoma lines express 12-LOX.

MATERIALS AND METHODS

Cell and tumor lines

HT168 and M1 melanoma was derived from the human melanoma cell line A2058 and established as cell- and tumor line characterized by high (HT168-M1) and lower (HT168) liver metastatic potentials[11,12]. HT199 human melanoma was established from a primary and characterized by high liver and lung metastatic potentials[11]. HT-18 melanoma was established from a metastatic lymph node and is characterized by a very low metastatic potential while WM35 melanoma was maintained as cell line as well as xenograft characterized by the absence of the metastatic potential (Table 1).

Table 1. Metastatic potential of the human melanoma lines [11,12]

tumorcell line	metastatic potential
WM35	non metastatic
HT18	low metastatic
HT168	metastatic
HT168-M1	highly metastatic
HT-199	highly metastatic

Detection of 12-LOX expression at RNA level

Total RNA was obtained using guanidinium isothiocynate-CsCl method as described[8] and suspended at μg/μl in distilled water. One μg of RNA was reverse transcribed using oligo(dT)$_{18}$ primers and 2 μl of cDNA mixture was amplified by PCR with sense and antisense primers. DNase was used to prevent PCR product carry-over and genomic DNA contamination. The RT-PCR was performed in the GeneAmp PCR 9600 (Perkin-Elmer-Cetus) at 94°C, 30 sec, 70°C, 30 sec, 72°C, 30 sec for 35 cycles. The sequence of the platelet specific 12-LOX sense and antisense probes was described before[8]. Fifteen μl of PCR products were separated by electrophoresis on 2% agarose gels stained by ethidium bromide and photographed.

Detection of 12-LOX expression at protein level

Flow cytometry. Adherent cells were detached from tissue flasks by using low EDTA washing or isolated from solid tumors as described, then were centrifuged at low speed and fixed in absolute MetOH for 10 min. After washing in PBS cells were labeled for platelet-type 12-LOX by using a rabbit polyclonal antibody (Oxford Scientific), biotinylated goat anti-rabbit IgG (Amersham) and Streptavidin-FITC (Amersham). Negative controls were used where the primary antibody was omitted. Flow cytometry of the samples was done as described[12], where 90% of the fluorescence of the negative control cell population served as the gate. Both % of positive cells and the fluorescence intensity was determined. Samples were run in triplicate.

Western blotting[8]. Adherent tumor cells cultured in vitro were dislodged from the flask washed in PBS and homogenized as described before. Aliquots of the homogenate was subjected to sodium dodecyl sulphate-polyacrylamide gel electrophoresis (SDS-PAGE) on 8% gels. Proteins were electrophoretically transferred to nitrocellulose membranes, the nonspecific binding sites were blocked by non-fat milk and the proteins were probed with the rabbit anti-platelet type 12-LOX IgG (Oxford Scientific) and the binding of the secondary HRP-conjugated anti-rabbit IgG was detected using ECL technique (Amersham).

RESULTS

Detection of 12-LOX expression at RNA level in vitro

RT-PCR was performed on mRNA extracted from in vitro cultured human melanoma lines (WM35, HT168 and M1) where HEL leukemia served as positive control. Results indicated that though at a much lower level than in HEL cells, it was possible to detect the 404 bp platelet-type12-LOX cDNA in all human melanoma lines (data not shown).

Detection of 12-LOX at protein level in vitro

Imunocytochemistry indicated that only a small proportion of cultured tumor cells (max.15%) express the pl-12-LOX where the highest expression level was detected in the non-metastatic cell line WM35 (Fig.1).

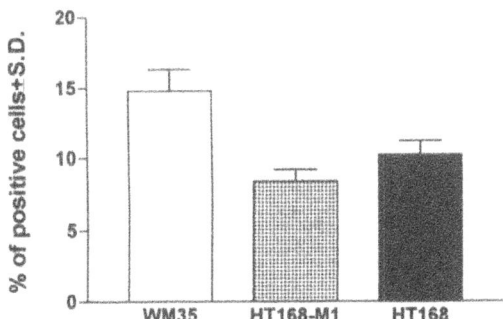

Fig.1. Expression of 12-LOX in human melanoma cell lines in vitro at protein level; flow cytometry. Data are expressed in % of positive cells (mean+S.D., n=3).

619

Western blotting of cellular proteins of HT168 and M1 melanoma cell lines indicated that they express the authentic 75 kD 12-LOX protein like HEL cells, though at a much lower level (Fig.2).

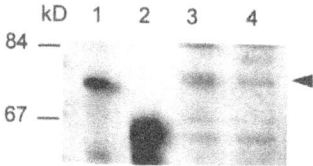

Fig.2. Western blotting of 12-LOX in cultured human melanoma cell lines. 1=HEL; 2=neg.;3=HT168;4=HT168-M1, arrow= 12-LOX

Detection of 12-LOX in melanomas in vivo

Tumor cells have been injected s.c. into newborn rats and tumor cells were re-isolated from the primary tumors after 30 days and the 12-LOX expression was measured by flow cytometry and immunocytochemistry. These studies indicated that pl-12-LOX expression was below 10% in the tumor cell populations and HT168 expressed at the highest level followed by HT199, M1 and HT18 which order almost follows the order of the metastatic potentials of the tumors (HT199>M1>HT168>HT18). In case of two melanomas (HT168 and M1) it was possible to analyze the 12-LOX expression in the lung metastases as well. These studies indicated that the % and the level of the expression of 12-LOX in the metastases follows the one found in the primary tumors indicating a stable phenotypic characteristics. (Table 2).

Table 2. Expression of 12-LOX in human melanoma lines in vivo (immunocytochemistry and flow cytometry)

melanoma line	% of positive cells	intensity
HT18	3.2$\pm$2.3	n.t.
HT168-M1	4.16$\pm$0.1	16.9$\pm$0.3
HT168-M1 lung met	4.19$\pm$1.3	20.3$\pm$1.3
HT199	4.44$\pm$1.97	28.8$\pm$1.6
HT168	9.52$\pm$4.9	163.1$\pm$4.1
HT168 lung met	5.67$\pm$0.1	149.4$\pm$0.8

Data are means of 3 parallel measurements $\pm$S.D. n.t.=not tested

DISCUSSION

This is the first report on the expression of 12-LOX both at mRNA and protein level in human melanomas. We have provided the evidence that human melanomas express the platelet-type isoform of 12-LOX both in vitro and in vivo. However, the expression is very low in the tumor cell populations (max.15%). These data are unexpected since in case of murine melanomas (B16a[3] and K1735 clones[4]) the expression of 12-LOX was comparable to HEL cells. Interestingly, in one of the studied melanoma lines, HT168, the micromilieu of the SCID mice skin can induce a huge increase in the 12-LOX expression (50-60%, data not shown), indicating that special local factors may change that low expression level in melanomas. These local factors could easily be the local immune effector cells which usually infiltrate and surround the primary skin melanomas. The SCID mice is lacking B cell activation while the newborn rats were immunosuppressed by an anti-thymocyte serum suggesting that the composition of the reactive infiltrate in and around the tumor must be different. That difference in the type of tumor-infiltrating lymphoid cells may influence the gene expressions, including the 12-LOX, in melanoma.

Since 12-HETE seems to be an important regulator of the generation of both the motogenic[9] and the mitogenic[10] signals in some tumor types, one can question the relevance of the low expression of 12-LOX in human melanoma. Since a particular cell type appears to express one isoform of 12-LOX[2] it is highly unlikely that another isoform is expressed in human melanoma (i.e. the epithelial-type). Since 15-LOX can also produce 12-HETE[1,2] we speculate that in human melanoma, instead of 12-LOX, the 15-LOX enzyme might also be expressed. To address this issue experiments are now underway in our laboratories.

ACKNOWLEDGMENTS

This work was supported by the Hungarian Science Found (OTKA T 21149, T 25155), Hungarian Ministry of Welfare (T 542/96) and NIH (CA-29997).

REFERENCES

1. Honn K.V., Tang D.G., Gao X., Butovich I.A., Liu B., Timár J., Hagmann W. 12-lipoxygenases and 12(S)-HETE: role in cancer metastasis. Cancer Metast Rev.13:365-396,1994.
2. Hagmann W. 12-lipoxygenase in human tumor cells. Pathol Oncol Res 3:83-88,1997.
3. Chen Y.Q., Duniec Z.M., Liu B., Hagmann W., Gao X., Shimoji K., Marnett L.J., Johnson C.R., Honn K.V. Endogenous 12(S)-HETE production by tumor cells and its role in metastasis. Canc Res 54:1574-1579,1994.
4. Silletti S., Timár J., Honn K.V., Raz A. Autocrine motility factor induces differential 12-lipoxygenase expression and activity in high- and low-metastatic K1735 melanoma cell variants. Canc Res 54:5752-5756,1994.
5. Hagmann W., Gao X., Zacharek A., Wojciechowski L.A., Honn K.V. 12-lipoxygenase in Lewis lung carcinoma cells: molecular identity, intracellular distribution of activity and protein, and Ca2+-dependent translocation from cytosol to membranes. Prostaglandins 49:49-62,1995.

6. Wang X.Q., Otsuka M., Takagi J., Kobayashi Y., Sato F., Saito Y. Inhibition of adenyl cyclase by 12(S)-hydroxyeicosatetraenoic acid. Biochem Biophys Res Commun 228:81-87,1996.

7. Hagmann W., Kagawa D., Renaud C., Honn K.V. Activity and protein distribution of 12-lipoxygenase in HEL cells: induction of membrane-association by phorbol ester TPA, modulation of activity by glutathione and 13-HPODE, and Ca2+-dependent translocation to membranes. Prostaglandins 46:471-477,1993.

8. Hagmann W., Gao X., Timár J., Chen Y.Q., Strohmayer A-R., Fahrenkopf C., Kagawa D., Lee M., Zacharek A., Honn K.V. 12-lipoxygenase in A431 cells: genetic identity, modulation of expression and intracellular localization. Exp Cell Res 228:197-205,1996.

9. Timár J., Silletti S., Bazaz R., Raz A., Honn K.V. Regulation of melanoma cell motility by lipoxygenase metabolite 12(S)-HETE. Int J Cancer 55:1003-1010,1993.

10. Tang D.G., Chen Y.Q., Honn K.V. Arachidonate lipoxygenases as essential regulators of cell survival and apoptosis. Proc Natl Acad Sci US 93:5241-5246,1996.

11. Ladányi A., Tímár J., Bocsi J., Tóvári J., Lapis K. Sex-dependent liver metastasis of human melanoma lines in SCID mice.Mel Res 5:83-86, 1995

12. Ladányi A., Tímár J., Paku S., Molnár G., Lapis K. Selection and characterization of human melanoma lines with different liver-colonizing capacity. Int J Cancer 46:456-461, 1990.

90

PLATELET-TYPE 12-LIPOXYGENASE REGULATES ANGIOGENESIS IN HUMAN PROSTATE CARCINOMA

Daotai Nie[1], Gilda G. Hillman[1], Timothy Geddes[1], Keqin Tang[1],
Christopher Pierson[2], David J. Grignon[2], and Kenneth V. Honn[1,2]

Departments of [1]Radiation Oncology and [2] Pathology
Wayne State University School of Medicine
Detroit, MI 48202

INTRODUCTION

The growth and metastasis of solid tumors are dependent on the ability of tumor cells to induce angiogenesis [1]. Angiogenesis, the formation of new blood vessels from pre-existing ones, involves endothelial cell proliferation, motility, and tubular differentiation. It is known that tumor cells can secrete a variety of angiogenic factors, such as basic fibroblast growth factor (bFGF) and vascular endothelial growth factor (VEGF), to stimulate angiogenesis. Tumor cells also produce angiogenesis inhibitors such as thrombospondin and angiostatin to control angiogenesis. The balance between angiogenesis stimulators and inhibitors determines the angiogenicity of tumor cells [2]. Acquisition of the ability to stimulate angiogenesis by tumor cells is an integral part of tumorigenesis. Activated oncogenes, such as ras, or inactivated tumor suppressor genes, such as p53, not only increase mitogenesis and prevent apoptosis in tumor cells, but also lead to the development of an angiogenic phenotype [for review, 3].

In human prostate adenocarcinoma (PCa), the level of vascularization in PCa has been demonstrated to be positively correlated with tumor stage [4-6]. Inhibition of angiogenesis by linomide or TNP-470 potently inhibits PCa growth and metastasis by promoting necrosis and apoptosis in the tumors [7-8]. Although various potential angiogenesis factors have been identified in prostate cancer [for review, 9], it still remains to be determined how PCa cells become angiogenic. Previously we detected the expression of platelet-type 12-LOX in human PCa and further demonstrated a correlation between 12-LOX mRNA expression and the stage of human PCa [10]. In the present study, we have examined the function of 12-LOX on PCa tumor growth. Our data show that 12-LOX has no detectable effect on PCa cell growth *in vitro* but stimulates PCa tumor growth *in vivo*. This effect of 12-LOX on tumor growth is closely related to increased angiogenesis. In vitro and in vivo angiogenesis assays suggest that PCa cells expressing high levels of 12-LOX are more angiogenic than those with no or low levels of 12-LOX. Our results provide a novel function for platelet-type 12-LOX in modulating PCa progression.

EXPERIMENTAL APPROACHES

Cell Culture: Rat angiogenic endothelial cell line RV-ECT (a gift from Dr. Clement Diglio) was cultured in DMEM-10%FBS [11]. PC3 cells were purchased from American Type Culture Collection (Manassas, VA) and maintained in RPMI 1640-10% FBS.

Stable Transfection of PC3 Cells and Characterization: Passage 28 PC3 cells were transfected with a pCMV-platelet 12-LOX construct (a gift from Dr. Collin Funk) [12] and pCMV-neo using a Lipofectin reagent (Gibco-BRL). PC3 cells transfected with pCMV-neo were used as controls. Transfectants were selected using 1 mg/ml geneticin (G418) in RPMI with 10% FBS and then cloned using a limiting dilution method in 96-well plates. The cloned transfectants were characterized regarding 12-LOX mRNA expression by Northern Blot and protein expression by Western Blot. The synthesis of 12(S)-HETE by 12-LOX transfectants was determined by using an RIA kit from Perspective Diagnostics (Cambridge, MA) according to the manufacturer's instruction.

Cell Proliferation Assay: To study cell proliferation in culture, 2,000 cells were seeded into each well of 96-well culture plate. The number of viable cells was assessed using an MTS cell proliferation assay kit (Promega Corp, Madison, MI). The A_{490nm} readings two to three hours after plating were used as baselines. The number of cells every 48 hours after plating was assessed using the exact same assay conditions, and were expressed as the percentage of increase by comparing with the A_{490nm} baselines.

Tumor Xenograft Model: 4×10^6 12-LOX transfectant cells or neo-control cells in 200 µl of HBSS were s.c. injected into the right flank of male nude mice (University of South Florida, Tampa, FL) of four to six weeks age (5~9 animals groups). Tumors were measured twice a week using a vernier caliper and tumor volume was calculated using the formula $(width)^2 \times length \times 0.52$ [13]. Six to seven weeks after injection, mice were sacrificed and the tumors were exposed, excised, photographed under an Olimpus SZ-4060 steremicroscope (Melville, NY) to record tumor vasculature and then fixed in 10% neutral buffered formalin for histochemical study.

Histochemical Studies: Each fixed tumor was cut in half, sections (5mm) were cut from each of two paraffin blocks representing one tumor and placed on gel coated slides. Hematoxylin and eosin staining was used to examine for the presence of necrosis and apoptotic bodies microscopically. Assessment of necrosis was done by a pathologist using a double-blind approach. To observe tumor cell apoptosis, slides were stained by *in situ* 3' end labeling ((terminal deoxynucleotidyltransferase-mediated dUTP nick end labeling; TUNEL) using ApopTag® *In Situ* Apoptosis Detection Kit-Peroxidase from Oncor (Gaithersburg, MD) according to the manufacturer's instruction. The apoptotic indices were expressed as the percentage of tumor cells with positive staining. A total of about one thousand tumor cells from each tumor were counted. To assess the vascularization status of tumors, immunohistochemical staining for CD 31 (DAKO Corp.; dilution 1:20) was performed using a standard ABC-immunoperoxidase procedure.

Cell Migration Assay: 500,000 RV-ECT endothelial cells in 0.5 ml RPMI-10%FBS were plated on the top chamber of a modified Boyden chamber (Becton Dickerson, Bedford, MA). After the cells settled, 1 ml of RPMI medium with 12(S)-HETE or the conditioned media was introduced into the lower chamber. After 4 hours of incubation, the cells on the top side of the transwell membrane were removed by cotton swabs. The membrane was then cut out, fixed in a quick-fix solution, double stained, and then mounted for observation and counting. Usually 12 high power fields representing two cross-lines of each membrane were counted. For each treatment, three chambers were used unless otherwise indicated.

In Vivo Angiogenesis Assay: Matrigel implantation assay for angiogenesis was essentially performed as described (14) by injecting 2 million PC3 12-LOX transfectants or neo-control cells s.c. into nude mice (4 mice/group) with 0.4 ml of Matrigel. Mice were sacrificed 12 days after injection and dissected to expose the implanted Matrigel for recording.

RESULTS AND DISCUSSION

To define the function of 12-LOX in PCa progression, over-expression of 12-LOX in PCa cells was achieved by transfecting PC3 cells with a platelet-type 12-LOX cDNA construct. Stable transfectants were cloned using the limiting dilution method. The expression of 12-LOX mRNA and protein in various clones was analyzed by Northern and Western blot analyses. As shown in **figure 1 A**, the levels of 12-LOX mRNA were increased in various transfectant clones when compared to the neo-controls or wild-type PC3 as revealed by Northern Blot. Western blot analysis demonstrated that 12-LOX transfectants also had higher levels of 12-LOX protein (**Fig. 1 B**). Among the various clones analyzed, PC3-nL2, nL8, nL11, and nL12 expressed 12-LOX at the highest levels. They also synthesized more 12(S)-HETE as measured by RIA: while PC3 parental cell line synthesized 14.6 pg 12(S)-HETE per million cells and neo-control, 30.1 pg, nL-2, nL-8, and nL-12 produced 197.2, 230.6, and 241 pg 12(S)-HETE per million cells. The results suggested that 12(S)-HETE biosynthesis was enhanced tremendously in 12-LOX transfectants even without exogenous arachidonic acid supplement.

Figure 1. Generation of PC3 transfectants synthesizing high levels of 12-LOX and 12(S)-HETE. The transfection of PC3 cells and the cloning of stable transfectants were performed as described in "Materials and Methods". **A.** Northern Blot analysis of 12-LOX mRNA levels in various clones of PC3 12-LOX transfectants. *Top panel* : Blot probed with 12-LOX cDNA. *Bottom panel* : Blot probed with actin cDNA as the loading control. **B.** Western Blot analysis of 12-LOX protein expression in various clones of PC3 12-LOX transfectants. The blot was probed with a 12-LOX polyclonal antibody and actin antibody.

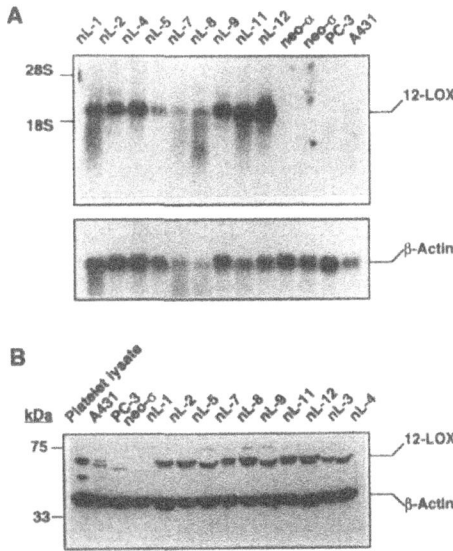

In vitro, the growth rates of several 12-LOX transfectant clones were similar to those of neo-controls and PC3 wild type (**Fig. 2 A**), with approximate doubling time of 36 hours, suggesting increased expression of 12-LOX in PC3 cells did not have detectable effect on cell proliferation. However, when s.c. injected into athymic mice, the tumors derived from 12-LOX transfectants (nL-12 and nL-2) grew faster and formed larger tumors than did the neo-control (Neo-σ and Neo-α) (**Fig. 2 B**). **Figure 2 C** shows the mice with tumors derived from 12-LOX transfectants and neo-controls. From the figure, 12-LOX transfectants clearly had an in vivo growth advantage over neo-control. We further tested one additional 12-LOX transfectant clone, nL-8, and obtained similar results (data not shown). The data strongly suggest that increased 12-LOX expression in PC3 cells stimulates tumor growth in vivo.

To understand the mechanism underlying the increased tumor growth by 12-LOX transfectants, histological analyses of tumors were performed. Hematoxylin and eosin staining revealed that, in the tumors derived from 12-LOX transfectant, tumor necrosis was significantly reduced. By staining DNA fragmentation products by *in situ* 3' end labeling, a slight but significant decrease in apoptotic index was observed in the tumors from 12-LOX transfectants (data not shown). These results suggest that the increased tumor growth by 12-LOX transfectants is mainly due to the reduction of necrosis and apoptosis in the tumors.

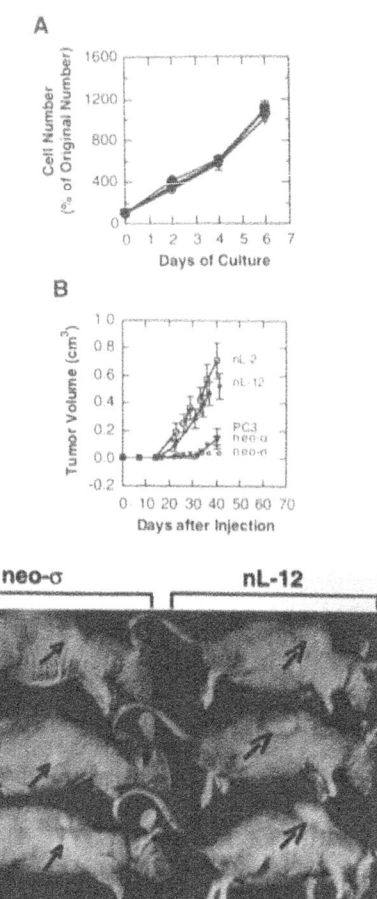

Figure 2. 12-LOX transfectants have an *in vivo* but not *in vitro* growth advantage. **A.** Growth kinetics of PC3 transfectants in culture. Cell proliferation of various transfectants was measured as described in Materials and Methods. Shown here are the growth curves of PC3 wild type (open circle), neo-σ (filled circle), nL-8 (open triangle), and nL-12 (filled triangle). Values are the mean of six determinations with S.E.M. as the error bars. Other clones such as nL-2 and neo-α also had similar growth kinetics (not shown). **B.** Growth kinetics of the tumors derived from 12-LOX transfectants and neo-controls. The values presented are the mean volume of eight tumors (n = 8) for nL-12 and neo-σ, five tumors (n = 5) for nL-2 and neo-α, and six tumors (n = 6) for PC3 wild type with S.E.M. as the error bars. **C.** Mice with tumors from 12-LOX transfectants or from neo-control. **Left panel**: three mice with tumors from neo-σ (arrowed). **Right panel**: three mice bearing tumors from 12-LOX transfected PC3 cells (nL-12, arrowed).

Since angiogenesis plays an important role in tumor growth by modulating tumor necrosis and apoptosis [2], we next examined whether the increased tumor growth by 12-LOX is a result of increased angiogenesis. Indeed, it was observed from the grossly exposed tumors that there was ample vascularization in tumors derived from 12-LOX transfectants while in those derived from neo-controls, little vessel penetration was evident (**Fig. 3 A**). We further did immunostaining of tumors with CD31 antibody which detects the presence of endothelial cells. As shown in **figure 3 B**, the vascular networks in tumors derived from nL-12 were sinusoidal in pattern and well developed in structure while in

626

those from neo-σ, endothelial cells were present but they were randomly distributed and did not form an organized vascular network. There were fewer vessels in neo-σ tumors than in nL-12 as suggested by microvessel density. Further, the vessels in the tumors derived from 12-LOX transfectants were mainly highly organized while in those from neo-σ, they were disorganized to intermediate. Similar increase in angiogenesis was also observed in the tumors derived from nL-2 and nL-8.

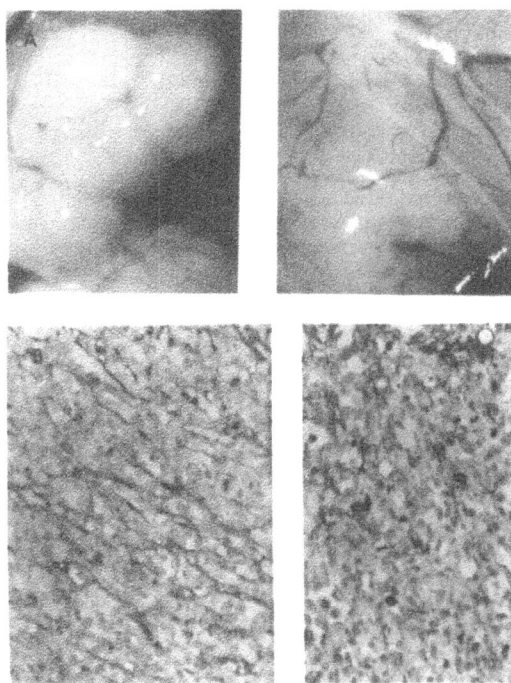

Figure 3. Increased angiogenesis in the tumors from 12-LOX transfected PC3 cells. **A.** Tumor morphology. Left panel: a tumor from neo-α. Right panel: a tumor from 12-LOX transfected PC3 cells, nL-2. (X 8). **B.** CD31 immunostaining. The brown color indicates the positive staining. **Left**: control tumor. Note the scattered vascular spaces which are randomly distributed and do not form a structured vascular network. **Right**: tumor from 12-LOX transfected PC3 cells. Note the numerous vascular channels showing a highly organized sinusoidal pattern surrounding small nests of tumor cells. (X 250).

The increased angiogenesis in the tumors from 12-LOX transfectants questions whether the observed increase in angiogenesis is the cause or a consequence of the increased tumorigenesis. To address this issue, we first assayed the conditioned media from those PC3 cell cultures for their ability to stimulate endothelial cell (RV-ECT) migration, a critical step in angiogenesis. As shown in **Figure 4 A,** the media from PC3 12-LOX transfectants induced more RV-ECT migration than those from neo-controls. Under similar assay condition, 12(S)-HETE itself also stimulated RV-ECT migration at nanomolar levels (**Figure 4 B**). The results suggest that 12-LOX transfectants secrete more motility factors, probably also including 12(S)-HETE, to stimulate endothelial cell migration. The increased angiogenicity of 12-LOX transfectants was further confirmed by chicken chrioallantoic membrane assay and the Matrigel implantation assay. As shown in **Figure 4 C**, the conditioned media from 12-LOX transfected cells stimulated angiogenesis while the media from neo-control had minimal effect in chicken chrioallantoic membrane, suggesting 12-LOX transfected PC3 cells secreted more angiogenic factors to stimulate angiogenesis. In Matrigel implantation assay as illustrated by **Figure 4 D**, 12-LOX transfectant nL-12 cells in Matrigel induced massive blood vessel penetration into the gel within twelve days while the neo-control did not. The results clearly illustrate that 12-LOX transfectants are more angiogenic than their neo-controls, suggesting that the increased tumor growth by 12-LOX transfectants is due to an increase in their ability to stimulate angiogenesis.

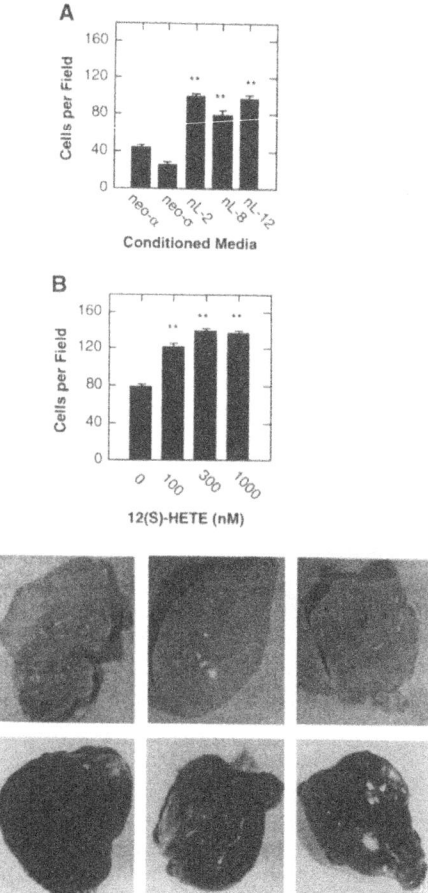

Figure 4. Increased angiogenicity of 12-LOX transfected PC3 cells. **A.** Stimulation of endothelial cell migration by the conditioned media from 12-LOX transfectants. The conditioned media were harvested after 24 hours of culture and used for migration assay as dscribed in "Materials and Methods." The values presented are the average number of cells migrated per field with S.E.M. as the error bars (**, p < 0.01 by Student's T test). **B.** 12(S)-HETE stimulates endothelial cell migration. The migration assay was performed esentially as in Figure 4 A except that media with various levels of 12(S)-HETE, instead of the conditioned media, was placed into the lower chamber. Values are presented as mean ± S.E.M. (**, p < 0.01 by Student's T test). **C.** Induction of angiogenesis in Matrigel by 12-LOX transfectants. ***Upper panel***: three Matrigel implants premixed with 2 X 10^6 Neo-σ cells. Note the vessel penetration into the gel was minimal, with little blood accumulated in the gel. In contrast, the Matrigel premixed with 2 X 10^6 12-LOX transfectant, nL-12, demonstrates considerable blood accumulation (***Lower panel***).

As the angiogenicity of tumor cells is controlled by the balance between stimulators and inhibitors of angiogenesis [2], it will be interesting to determine how 12-LOX upregulates the angiogenicity of PCa cells. First, 12-LOX or 12(S)-HETE may increase the angiogenicity of tumor cells by influencing the expression of angiogenic or angiostatic molecules such as VEGF and bFGF. Second, 12(S)-HETE may directly tip the balance in favor of angiogenic factors due to its pro-angiogenic nature. This is supported by the current finding that 12(S)-HETE stimulated endothelial cell migration at nanomolar levels and the previous documentation that 12(S)-HETE stimulated endothelial cell proliferation [15], retraction [16], and adhesion and that it increased the surface expression of integrin

$\alpha_v\beta_3$ in both macro- and micro-vascular endothelial cells [17]. It is noteworthy that integrin $\alpha_v\beta_3$ is predominantly associated with angiogenic blood vessels [18] and plays an essential role in human cancer angiogenesis [19]. Thus, 12(S)-HETE may directly increase the angiogenicity of PCa cells by stimulating angiogenesis or by eliciting several proangiogenic responses which are additive or synergistic to the effects from other angiogenic factors produced by PCa cells since different angiogenic factors have their own distinct effects on the process of angiogenesis [20]. Investigation is underway to determine whether increased 12-LOX expression in PCa cells influences the gene expression of angiogenic factors and whether 12(S)-HETE can stimulate angiogenesis alone or by its additive or synergistic interaction with other putative angiogenic factors.

To summarize, we found that increased expression of 12-LOX in human PCa cells stimulated prostate tumor growth by enhancing their angiogenicity. The findings have significant bearing on the regulation of PCa progression, and implicate that the increased expression of 12-LOX as PCa progresses to more advanced stages as observed *in vivo* [20] may play a contributory role in tumor progression by elaborating tumor vascularization. Thus, our work establishes a novel link between tumor angiogenesis, tumor growth, and the expression of platelet-type 12-LOX in PCa cells.

REFERENCES

1. Folkman, J., and Shing, Y. Angiogenesis. J. Biol. Chem., *267*: 10931-10934, 1992.
2. Hanahan, D., and Folkman, J. 1996. Patterns and emerging mechanisms of the angiogenic switch during tumorigenesis. Cell *86*: 353-364.
3. Vartanian, R. K., and Weidner, N. EC proliferation in prostatic carcinoma and prostatic hyperplasia: correlation with Gleason's score, microvessel density, and epithelial cell proliferation. Lab. Invest., *73*: 844-850, 1995.
4. Wakui, S., Furusato, M., Itoh, T., Sasaki, H., Akiyama, A., Kinoshita, I., Asano, K., Tokuda, T., Aizawa, S., and Ushigome, S. Tumor angiogenesis in prostatic carcinoma with and without bone marrow metastasis: a morphometric study. J. Pathol., *168*: 257-262, 1992.
5. Weidner, N., Carroll, P. R., Flax, J., Blumenfeld, W., and Folkman, J. Tumor angiogenesis correlates with metastasis in invasive prostate carcinoma. Am. J. Pathol., *143*: 401-409, 1993.
6. Vukanovic, J., and Isaacs, J. T. Human prostatic cancer cells are sensitive to programed (apoptotic) death induced by the antiangiogenic agent linomide. Cancer Res., *55*: 3517 - 3520, 1995.
7. Yamaoka, M., Yamamato, T., Ikeyama, S., Sudo, K., and Fujita, T. Angiogenesis inhibitor TNP-470 (AGM-1470) potently inhibits the tumor growth of hormone-independent human breast and prostate carcinoma cell lines. Cancer Res., *53*: 5233-5236, 1993.
8. Campbell, S. C. Advances in angiogenesis research: relevance to urological oncology. J. Urol., *158*: 1663-1674, 1997.
9. Gao, X., Grignon, D.J., Chbihi, T., Zacharek, A., Chen, Y.Q., Sakr, W., Porter, A.T., Crissman, J.D., Pontes, J.E., Powell, I.J., and Honn, K. V. Elevated 12-lipoxygenase mRNA expression correlates with advanced stage and poor differentiation of human prostate cancer. Urology *46*: 227-237, 1995.
10. Funk, C. D., Furci, L., and FitzGerald, G. A. Molecular cloning, primary structure, and expression of the human platelet/erythroleukemia cell 12-lipoxygenase. Proc. Natl. Acad. Sci. USA *87*: 5638-5642, 1990.
11. Diglio, C. A., Liu, W., Grammas, P., Giacomelli, F., and Wiener, J. Isolation and characterization of cerebral resistance vessel endothelium in culture. Tissue and Cell *25*: 833-846, 1993.
12. Hagmann, W., Gao, X., Timar, J., Chen, Y. Q., Strohmaier, A. R., Fahrenkopf, C., Kagawa, D., Lee, M., Zacharek, A., and Honn, K. V. 12-Lipoxygenase in A431 cells: genetic identity, modulation of expression, and intracellular localization. Exp. Cell Res., *228*: 197-205, 1996.

13. Yamaoka, M., Yamamoto, T., Ikeyama, S., Sudo, K., Fujita, T. Angiogenesis inhibitor TNP-470 (AGM-1470) potently inhibits the tumor growth of hormone-independent human breast and prostate carcinoma cell lines. Cancer Res., *53*: 5233-5236, 1993.

14. Ito, Y., Iwamoto, Y., Tanaka, K., Okuyama, K., and Sugioka, Y. A quantitative assay using basement membrane extracts to study tumor angiogenesis in vivo. Int. J. Cancer *67*:148-152, 1996.

15. Tang, D. G., Renaud, C., Stojakovic, S., Diglio, C. A., Porter, A., and Honn, K. V. 12(S)-HETE is a mitogenic factor for microvascular ECs: Its potential role in angiogenesis. Biochem. Biophys. Res. Commun., *211*: 462-468, 1995.

16. Honn, K. V., Tang, D. G., Grossi, I., Duniec, Z. M., Timar, J., Renaud, C., Leithauser, M., Blair, I., Johnson, C. R., Diglio, C. A., Kimler, V. A., Taylor, J. D., and Marnett, L. J. Tumor cell-derived 12(S)-hydroxyeicosatetraenoic acid induces microvascular endothelial cell retraction. Cancer Res., *54*: 565-574, 1994.

17. Tang, D. G., Chen, Y.Q., Diglio, C. A., and Honn, K. V. Transcriptional activation of EC integrin a_V by protein kinase C activator 12(S)-HETE. J. Cell Sci., *108*: 2629-2644, 1995.

18. Brooks, P. C., Clark, R. A. F., and Cheresh, D. A. Requirement of vascular integrin $\alpha_V\beta_3$ for angiogenesis. Science *264*: 569-571, 1994.

19. Brooks, P. C., Stromblad, S., Klemke, R., Visscher, D., Sarkar, F. H., and Cheresh, D. A. Antiintegrin $\alpha_V\beta_3$ blocks human breast cancer growth and angiogenesis in human skin. J. Clin. Invest., *96*:1815-1822. 1995.

20. Kumar, R., Yoneda, J., Bucana, C. D., and Fidler, I. J. Regulation of distinct steps of angiogenesis by different angiogenic molecules. Int. J. Oncol., *12*:749-757, 1998.

91

PROSTAGLANDIN H SYNTHASE AND LIPOXYGENASE MEDIATED ACTIVATION OF XENOBIOTICS IN PLATELETS

Kyoung-Min Lim, Joo-Young Lee, Jung-Sun Kim, and Jin-Ho Chung

College of Pharmacy
Seoul National University
Seoul 151-742 Korea

ABSTRACT

To investigate the involvement of prostaglandin H synthase (PHS) and lipoxygenase (LPO) in the activation of xenobiotics in platelets, platelet sonicates were preincubated with α-naphthol. Protein covalent binding of α-naphthol was measured following addition of arachidonic acid. Protein covalent binding was increased in a dose-dependent manner until it plateaued at 500 μM arachidonic acid. Pretreatment by two inhibitors of PHS, aspirin and indomethacin, resulted in a dose-dependent inhibition of α-naphthol-induced covalent binding, confirming PHS involvement. In addition, pretreatment by a LPO inhibitor, nordihydroguaiaretic acid (NDGA), also prevented covalent binding substantially, showing that LPO may be an alternative pathway for xenobiotic activation in platelets. Furthermore, combined treatment of aspirin and NDGA almost abolished the increase of α-naphthol-induced covalent binding, suggesting that PHS and LPO are both major pathways for xenobiotic activation in platelets.

Eicosanoids and Other Bioactive Lipids in Cancer, Inflammation, and Radiation Injury, 4
Edited by Honn *et al.*, Kluwer Academic / Plenum Publishers, New York, 1999.

INTRODUCTION

In many cases, xenobiotics must be activated to reactive intermediates by several enzymes to induce the adverse effects such as cytotoxicity, mutagenicity, carcinogenicity, etc (Marnett, 1990). These enzymes include cytochrome P-450 dependent monooxygenases (MFO), especially rich in hepatic tissues (Guengerich and Shimada, 1991), and prostaglandin H synthase (PHS) and lipoxygenase (LPO) in extrahepatic tissues which are generally low in MFO activity (Degen, 1993). PHS and LPO are found in most mammalian tissues such as platelets, reproductive tract, kidney medulla, endothelial cell, blood cell, and lung (Eling et al., 1990; Sigal, 1991; Yamamoto, 1992).

PHS and LPO are very similar in their metabolic mechanisms. Both of them dioxygenate arachidonic acid to organoperoxide producing prostaglandin G_2 (PGG_2) and hydroperoxyeicosatetraenoic acid (HPETE), respectively. These organoperoxides are reduced to prostaglandin H_2 (PGH_2) and hydroperoxyeicosatetraenoic acid (HETE). In the conversion of these peroxides, one electron oxidation of endogenous antioxidants such as tocopherol and glutathione can occur (Chan et al., 1991). However, in the presence of xenobiotics which can donate electrons, xenobiotics are often oxidized to reactive intermediates. Oxidized reactive intermediates can induce cytotoxicity through covalent binding, lipid peroxidation and oxidative stress. Roles of PHS and LPO were suggested in the activation of xenobiotics in tissues such as liver, lung, kidney medulla, and embryo (Yu and Wells, 1995; Liu and Wells, 1994; Kulkarni and Cook, 1988; Hastings, 1995). In addition, a large spectrum of chemicals was suggested to be activated by PHS and LPO, e.g. aromatic amines, benzidine, and aflatoxin (Eling et al., 1990; Hughes et al., 1988).

Platelets are important cells, functioning to prevent bleeding. Platelets are rich in PHS and LPO and these enzymes play a key role in the aggregation pathway, inducing formation of bioactive compounds such as thromboxane (TXA_2), 12-HPETE, and 12-HETE (Burch and Majerus, 1979; Lagarde, 1987). Therefore, platelets are a good model for PHS and LPO-mediated xenobiotic activation studies.

In the present study, we tried to demonstrate PHS and LPO-mediated xenobiotic activation in platelets using a phenolic compound, α-naphthol, which is known to be activated to reactive intermediates through various peroxidase-induced cooxidation (Doherty et al., 1986).

EXPERIMENTAL PROCEDURES

Chemicals and Animals

The following reagents were obtained from Sigma (St.Louis, USA): indomethacin, acetylsalicylic acid, nordihydroguaiaretic acid (NDGA), and dimethyl sulfoxide (DMSO). [1-C^{14}] α-naphthol was purchased from Amersham Co. (Arlington Heights, USA). All other reagents were commercial products of the highest available grade of purity. Female Sprague-

Dawley rats (180-220 g) were used for all experiments. The animals were allowed food and water *ad libitum*.

Preparation of Platelet Sonicates and Incubation Conditions

All procedures were conducted at room temperature, and the use of glass containers and pipettes was avoided. Blood was collected from the abdominal aorta of ether-anesthetized rats. For the preparation of platelet sonicates, acid-citrate-dextrose (ACD) (1:6; 85 mM trisodium citrate, 71 mM citric acid, 111 mM dextrose) was used as an anticoagulant. ACD-blood was centrifuged for 15 min at 150 g to obtain condensed platelet rich plasma (PRP). The PRP was pelleted by second centrifugation at 500 g for 10 min. The pelleted platelets were then washed in Tris-HCl buffer (25 mM Tris-HCl in 0.9% NaCl) and centrifuged a second time at 500 g for 10 min. Final platelet pellets were resuspended in the Tris-HCl buffer again and sonicated with a Brandson sonifier using a microtip (30% intensity, 10 sec period, 7 times with 30 sec interval, 1 ml volume). Platelet sonicates were incubated at 37°C in plastic flasks. α-Naphthol (25 μM) was added in an ethanol vehicle; control sonicates received of 2.5 μl vehicle alone. Aliquots were withdrawn for analysis after various time intervals.

Covalent Binding Assays

Covalent binding of α-naphthol in platelet sonicates was measured using radiolabelled α-naphthol. Platelet sonicates (490 μl) were incubated with α-naphthol and inhibitors at 37°C. Reaction was stopped by denaturing protein by methanol (1 ml) addition. Protein pellets were collected by centrifugation at 12,000 g for 3 min. Resulting pellets were extensively washed with methanol (1 ml) 9-10 times and digested with 1 N NaOH (1 ml) at 65°C for 1 hr. Aliquots of each sample (50 μl) was used for protein concentration assays by Biorad method and 0.5 ml aliquots were added to 4.5 ml of cocktail solution (Instagel®, Packard, USA) to measure radioactivity with a liquid scintillation counter (Wallace, USA). Covalent binding was assessed as nmoles/mg protein.

RESULTS

To investigate the involvement of prostaglandin H synthase (PHS) and lipoxygenase (LPO) in the activation of xenobiotics, protein covalent binding of α-naphthol induced by arachidonic acid, a common substrate for PHS and LPO, was measured as the indicator for activation of α-naphthol. Platelet sonicates were pretreated with 25 μM α-naphthol for 3 min

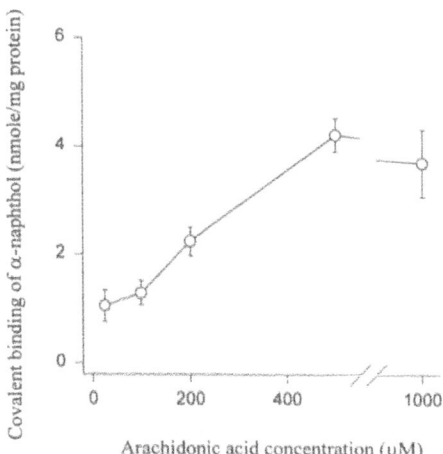

Figure 1. Covalent binding of α-naphthol to protein in platelet sonicates induced by arachidonic acid (AA). Platelet sonicates were incubated with 25 μM α-naphthol. Various concentrations of AA were added 3 min after α-naphthol and incubated for 15 min. Nonspecific binding (in the absence of arachadonic acid) was 2.1 ± 0.2 nmoles/mg protein. Values are means ± SEM (n=3-4).

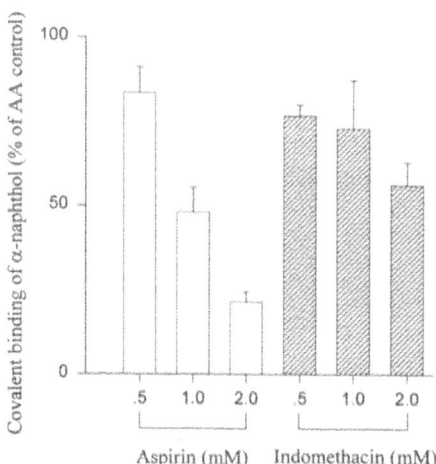

Figure 2. Inhibition of covalent binding of α-naphthol by aspirin or indomethacin in platelet sonicates. Platelet sonicates were incubated with aspirin or indomethacin at 37°C for 3 min prior to addition of 25 μM α-naphthol. Arachidonic acid (500 μM) was added 3 min after α-naphthol and the sonicates were incubated for an additional 15 min. Arachidonic acid control was 5.1 ± 0.2 nmoles/mg protein. Values are means ± SEM (n=3).

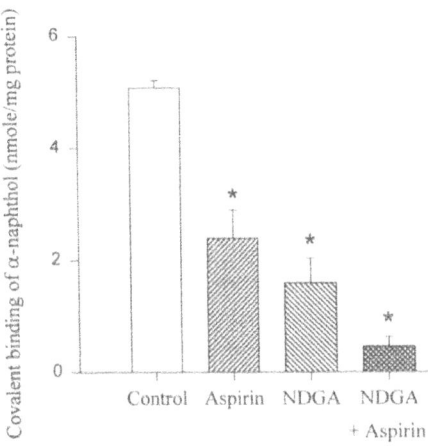

Figure 3. Inhibition of covalent binding of α-naphthol by aspirin and NDGA in platelet sonicates. Platelet sonicates were incubated with 2 mM aspirin and/or 50 μM NDGA at 37°C for 3 min prior to addition of 25 μM α-naphthol. Arachidonic acid (500 μM) was added 3 min after α-naphtol and the sonicates were incubated for an additional 15 min. Values are means ± SEM (n=3).

and then various concentrations of arachidonic acid were added as substrate for PHS and LPO. After incubation for 15 min, protein binding of α-naphthol was increased in a dose-dependent manner and this increase was plateaued at 500 μM arachidonic acid. On the other hand, in the absence of arachidonic acid, only minimal nonspecific binding was observed (Figure 1). These results suggest that xenobiotics like α-naphthol can be activated via an arachidonic acid-dependent pathway, such as by PHS and LPO activity.

In order to confirm the role of PHS in the activation of α-naphthol in platelets, inhibitors of PHS (aspirin or indomethacin) were preincubated with sonicates for 3 min prior to addition of α-naphthol (Figure 2). No inhibition of covalent binding by α-naphthol was observed in control incubations treated with vehicles alone (ethanol or DMSO, respectively). However, pretreatment with aspirin and indomethacin resulted in dose-dependent inhibition of α-naphthol-induced covalent binding, suggesting the contribution of PHS. Aspirin was a more potent inhibitor than indomethacin.

Involvement of LPO, another arachidonic acid-dependent pathway, was determined using nordihydroguaiaretic acid (NDGA), a LPO inhibitor (Figure 3). NDGA 50 μM treatment prior to α-naphthol addition prevented covalent binding substantially, suggesting that LPO is also involved in α-naphthol activation in platelets. Furthermore, when PHS inhibitor, aspirin, is combined with NDGA treatment, the increase of α-naphthol-induced covalent binding is almost totally abolished, suggesting that both PHS and LPO are major pathways for α-naphthol activation in platelets.

DISCUSSION

Prostaglandin H synthase (PHS) and lipoxygenase (LPO) are present in most mammalian tissues and relatively high levels are found in the reproductive tract, kidney medulla, endothelial cells, and platelets (Yamamoto, 1992). Since blood is readily exposed to various types of chemicals from both endogenous and xenobiotic sources, platelets are likely to play an important role in the biotransformation and therefore toxicity of many xenobiotics which are mediated by PHS and LPO. PHS and LPO of platelets are well known for their physiological importance in platelet aggregation. However, until this study, little was known regarding their toxicological role in bioactivation. In our study, xenobiotic activation mediated by PHS and LPO in platelets was investigated and demonstrated that a xenobiotic (α-naphthol) could be activated by PHS and LPO to induce covalent binding. This was accomplished by using platelet sonicates instead of microsomal fractions because our sonicates contained both microsomal PHS and cytosolic LPO.

α-Naphthol is a typical phenolic compound which is known to be activated by many peroxidases. According to the studies of Doherty et al.(1986), α-naphthol is one electron oxidized by peroxidases to reactive intermediates capable of inducing covalent bindng and oxidative stress. This process is very common to most phenolic compounds known to be activated by PHS and LPO. Our hypothesis that α-naphthol could be activated by PHS and LPO was confirmed by the observation that covalent binding of α-naphthol was increased by arachidonic acid (a common PHS and LPO substrate) (Figure 1) and this activity was inhibited by inhibitors of PHS and LPO (Figures 2 and 3). These data suggest that xenobiotics like α-naphthol can be activated in platelets by the arachidonic acid-dependent pathways PHS and LPO. This is the first study which conclusively reports that PHS and LPO in platelets are involved in the activation of a xenobiotic.

Recently, several metabolites of PHS and LPO were found to be involved in important physiological functions and pathophysiology such as inflammation, ischemic heart disease, and atherosclerosis. Since metabolic profiles by PHS and/or LPO may alter following activation of xenobiotics, characterization of the metabolites of PHS and LPO will be an important topic of future investigations.

ACKNOWLEDGMENTS

This work was supported by the Korea Science and Engineering Foundation (KOSEF) through the Research Center for New Drug Development at Seoul National University

REFERENCES

Burch, J.W. and Majerus, P.W., 1979, The role of prostaglandins in platelet function., *Semin. Hematol.* 16: 196-207.

Chan, A.C., Tran, K., Raynor, T., Ganz, P.R. and Chow, C.K., 1991, Regeneration of vitamin E in platelets., *J. Biol. Chem.* 266: 17290-17295.

Degen, G.H., 1993, Prostaglandin-H synthase containing cell lines as tools for studying metabolism and toxicity of xenobiotics., *Toxicology* 82: 243-256.

Doherty, M.D., Wilson, I., Patterson, L.H. and Cohen, G.M., 1986, Peroxidase activation of 1-naphthol to naphthoxy or naphthoxy-derived radicals and their reaction with glutathione., *Chem. Biol. Interact.* 58: 199-215.

Eling, T.E., Thompson, D.C., Foureman, G.L., Curtis, J.F. and Hoghes, M.F., 1990, Prostaglandin H synthase and xenobiotic oxidation, *Annu. Rev. Pharmacol. Toxicol.* 30: 1-45.

Guengerich, F.P. and Shimada, T., 1991, Oxidation of toxic and carcinogenic chemicals by human cytochrome P-450 enzymes., *Chem. Res. Toxicol.* 4: 391-407.

Hastings, T.G., 1995, Enzymatic oxidation of dopamine: The role of prostaglandin H synthase., *J. Neurochem.* 64: 919-924.

Hughes, M.F., Mason, R.P. and Eling, T.E., 1988, Prostaglandin hydroperoxidase-dependent oxidation of phenylbutazone: relationship to inhibition of prostaglandin cyclooxygenase., *Mol. Pharmacol.* 34: 186-193.

Kulkarni, A.P. and Cook, D.C., 1988, Hydroperoxidase activity of lipoxygenase: Hydrogen peroxide dependent oxidation of xenobiotics., *Biochem. Biophys. Res. Comm.* 155: 1075-1081.

Lagarde, M., 1987, Roles of cyclooxygenase and lipoxygenase metabolites in platelets, in: *Platelets in Biology and Pathology III.* D.E. White and J.L. Gordon, eds., Elsevier, New York.

Liu, L. and Wells, P.G., 1994, *In vivo* phenytoin-initiated oxidative damage to proteins and lipids in murine maternal hepatic and embryonic tissue organelles: potential molecular targets of chemical teratogenesis., *Toxicol. Appl. Pharmacol.* 124: 247-255.

Marnett, L., 1990, Prostaglandin synthase-mediated metabolism of carcinogens and a potential role for peroxyl radicals as reactive intermediates., *Environ. Health Perspect.* 88: 5-12.

Sigal, E., 1991, The molecular biology of mammalian arachidonic acid metabolism., *Am. J. Physiol.* 260: L13-L28.

Yamamoto, S., 1992, Mammalian lipoxygenases: molecular structures and functions., *Biochem. Biophys. Acta* 1128: 117-131.

Yu, W.K. and Wells, P.G., 1995, Evidence for lipoxygenase-catalyzed bioactivation of phenytoin to a teratogenic reactive intermediate: *In vitro* studies using linoleic acid-dependent soybean lipoxygenase and *in vivo* studies using pregnant CD-1 mice., *Toxicol. Appl. Pharmacol.* 131: 1-12.

92

CHARACTERIZATION OF TWO SPLICED VARIANTS OF HUMAN PHOSPHATIDIC
ACID PHOSPHATASE CDNAS THAT ARE DIFFERENTIALLY EXPRESSED IN
NORMAL AND TUMOR CELLS

David W. Leung,[1] Christopher K. Tompkins,[1] and Thayer White[2]

[1]Molecular Biology Department
[2]Lipid Biochemistry Department
 Cell Therapeutics, Inc.,
 Seattle, WA 98119.

INTRODUCTION

Phosphatidic acid (PA) and diacylglycerol (DG) are lipids involved in signal transduction and in
structural membrane lipid biosynthesis in cells. Phosphatidic acid phosphatase (PAP) (EC 3.1.3.4)
catalyzes the conversion of PA to DG. This enzyme is known to exist in at least two isoforms, one
of which (PAP1) is presumed to be cytosolic and membrane associated and the other (PAP2), an
integral membrane protein (Brindley and Waggoner, 1996). PAP1 has been implicated in
glycerolipid biosynthesis; whereas PAP2 is suggested to play a role in signal transduction. In
addition to dephosphorylating PA, purified PAP2 from rat liver has also been found to
dephosphorylate lysophosphatidic acid (LPA), ceramide-1-phosphate (C-1-P), sphingosine-1-
phosphate (S-1-P) (Waggoner et al., 1996), and DG pyrophosphate (Carman, 1997), suggesting
the involvement of PAP2 in modulating the balance of a broad spectrum of bioactive lipids
generated during cell signaling. A murine PAP2 cDNA (Kai et al., 1996) and two human cDNAs
isoforms, hPAP-2a and hPAP-2b, coding for PAP2 have been identified (Kai et al., 1997).
Homology search of the GenBank database using the murine PAP2 sequence probe has also
enabled us to isolate several putative human isoenzymes. This paper reports the isolation and
expression of two alternatively spliced variants termed PAP2-α1 and PAP2-α2 encoded by the
human PAP-2a gene and two other isoforms of PAP2 derived from separate genes.

Eicosanoids and Other Bioactive Lipids in Cancer, Inflammation, and Radiation Injury, 4
Edited by Honn *et al.*, Kluwer Academic / Plenum Publishers, New York, 1999.

CLONING AND SEQUENCE ANALYSIS OF HUMAN PAP2 cDNAs

Homology search of the Genbank dbEST (Boguski et al., 1994) uncovered several putative peptide sequences homologous to the mouse PAP (Kai et al., 1996) protein sequence. (Leung et al., 1998). DNA sequence analysis of potential human PAP2 clones obtained revealed at least two different classes of clones with sequences that diverged within a putative exon, suggesting the presence of two alternatively spliced forms of PAP2. These two alternatively spliced forms of PAP2 are designated here as PAP2-α1 and PAP2-α2. The coding sequence of PAP2-α1 is identical to that of hPAP-2a (Kai et al., 1997), whereas PAP2-α2 represents a novel alternatively spliced form of PAP2-α. In addition, we have also expressed two other isoforms of human PAP2, termed PAP2-β and PAP2-γ. The coding sequence of PAP2-β is identical to that of hPAP-2b, whereas the sequence of PAP2-γ is distinct but similar to all the other human PAP2 isoforms.

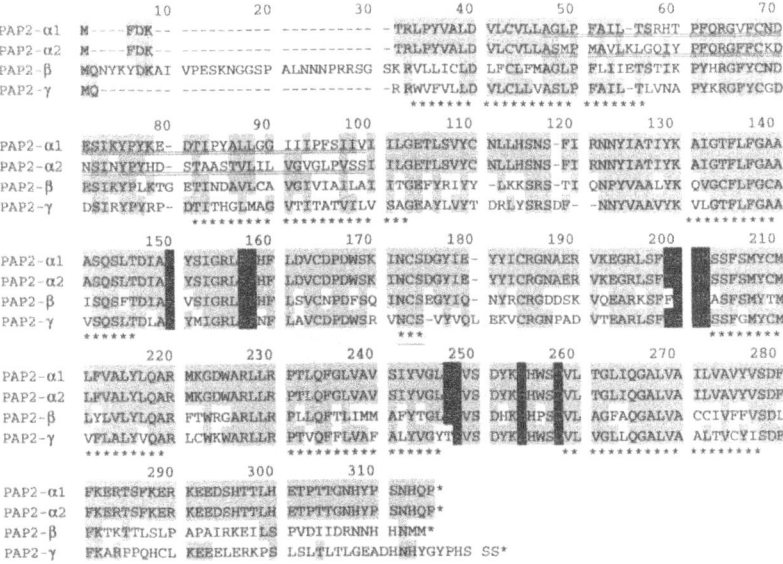

Figure 1. PAP2 amino acid sequence alignment from four human isoforms. The coding regions of the two alternatively spliced exons for human PAP2-α1 and PAP2-α2 are doubly underlined. Amino acids identical in at least two sequences are highlighted. The putative glycosylation sites are indicated by underlined asterisks below. Residues that are characteristic of a phosphatase sequence motif (Stukey and Carman, 1997) are inversely highlighted. Possible transmembrane domains are indicated by asterisks below. The nucleotide sequences reported have been submitted to GenBank under accession numbers AF014402, AF014403, AF043329 and AF035959.

Amino acid sequence comparisons of the individual human PAP2s (Fig. 2) show that the PAP2-α isoforms have overall amino acid matches of 35% and 44% with the PAP2-β and the PAP2-γ sequences, respectively. The coding sequences of the two alternatively spliced exons within the two human PAP2-α isoforms have an overall amino acid match of 40%. The human PAP2-α isoforms have an overall amino acid match of 79%, 25%, 3%, and 6%, respectively, with the murine (Kai et al., 1996), drosophila (Zhang et al., 1997), yeast (Mandala et al., 1998; Carman, 1997) and *E. coli* homologs (Carman, 1997). The proteins encoded by human PAP2 isoforms have a hydrophobicity profile containing six transmembrane stretches similar to that of the mouse PAP2 (Kai et al., 1996), the rat Dri42 (Barila et al., 1996), a homolog of PAP2-β, and the putative PAP2 homologue in *Drosophila*, *Wunen* (Zhang et al., 1997), suggesting that PAP2s are integral membrane proteins. These proteins also contain a conserved phosphatase sequence motif (Neuwald., 1997; Stukey and Carman, 1997), KXXXXXXXRP---(P/Y)SGH---SRXXXXXHXXXD, found in certain lipid phosphatases, the mammalian glucose-6-phosphatases, certain bacterial nonspecific acid phosphatases, and a fungal chloroperoxidase. Structure-function analysis of human glucose-6-phosphatases (Lei et al., 1995) and X-ray structure of the chloroperoxidaes showed the key residues within this motif were involved in substrate interaction (Messerschmidt and Wever, 1996). Based on similarity to the membrane topology of Dri42 elucidated by glycosylation patterns of various Dri42 fusion proteins after co-translational insertion into membranes (Barila et al., 1996), all the essential residues in the phosphatase motif in PAP2 are predicted to be found in non-cytoplasmic loop regions, consistent with the notion that PAP2 enzymes are integral membrane proteins with the catalytic sites exposed on cell surface or organellar lumena (English et al., 1997; Perry et al., 1993).

EXPRESSION OF HUMAN PAP2-α1 AND PAP-α2 IN MAMMALIAN CELLS

Plasmid DNAs for PAP2 expression were transfected stably into human vascular endothelial cell line ECV304 or transiently into human embryonic kidney cell line 293-EBNA to determine if they would increase PAP2 activity in cell extracts. Fig. 2 (left panels) compares the endogenous PAP activity in ECV304 cells to that in ECV304 lines stably transfected with a PAP2-α1 expression vector, with a PAP2-α2 expression vector, and with its control vector (pCE9), using a TLC assay that measures the conversion of $[^{14}C]PA$ to $[^{14}C]DG$; and (right panels) the endogenous PAP activity in 293-EBNA cells to that in 293-EBNA cells transiently transfected with a PAP2-β expression vector, with a PAP2-γ expression vector, and with the control vector (pCE9). Cells transfected with either PAP2-α1, PAP2-α2, PAP2-β, or PAP2-γ cDNA demonstrate significantly greater PAP activity than those of control cells transfected with a control vector or untransfected cells as evidenced by the increased conversion of $[^{14}C]PA$ to $[^{14}C]DG$ (Fig. 2, upper panels). Quantification of the DG bands (bottom panels) with a phosphorescence imager (Molecular Dynamics, Sunnyvale, CA) consistently shows an 8 fold, a 2 fold, a 2.5 fold and a 6 fold increase in intensity in, respectively, the PAP2-α1, PAP2-α2, PAP2-β, and PAP2-γ transfected cells versus control or untransfected ECV304 cells or 293-EBNA cells.

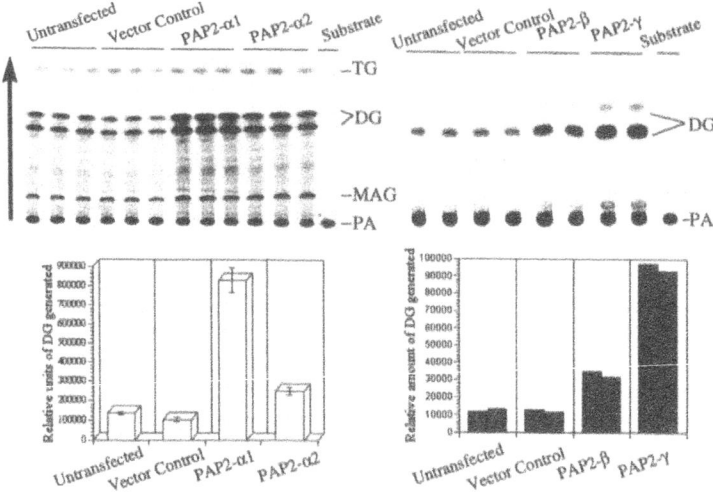

Figure 2. TLC analysis of PAP2 activity in ECV304 and in 293-EBNA cells transfected with PAP2-α1, PAP2-α2, PAP2-β, or PAP2-γ expression plasmids. PAP activity was assessed in triplicate in ECV304 cell extracts and in duplicate in 293-EBNA cell extracts using [¹⁴C]PA as substrate. Equal amounts of proteins were used in each reaction. The substrate and products were separated by TLC and visualized using a Phosphorimager. The identities of labeled bands were determined by migration of lipid standards visualized by fluorescence after primulin staining. The lower panels show the quantification by phosphorimager analysis of the DG products from triplicate samples (Mean± SD) and duplicate samples.

To determine whether overexpression of PAP2 cDNA would alter the steady state level of various lipids in cells, ECV304 cells overexpressing PAP2-α1or PAP2-α2 were grown to quantities where individual lipid levels could be measured. FF-TLC analysis (Fig. 3, left panel), a system optimized for resolution of PA from all other lipids, shows cells transfected with either PAP2-α1or PAP2-α2 cDNA had a greater than 50% reduction in PA content when compared to cells transfected with a control vector or untransfected cells; this is consistent with an increase in PA-hydrolyzing activity, as detected in the transfected cells. MOD-TLC analysis, a system designed for the separation of neutral lipids, shows no concurrent rise in DG levels, after normalization of the total lipids loaded per lane, in cells transfected with either PAP2-α1or PAP2-α2 when compared to control cells (Fig. 3, right panel). This suggests that the DG generated from PA hydrolysis was further metabolized to other lipid products. All other lipids examined, cholesterol-ester (CE), cholesterol (Chol), TG, PE, and PC, showed little variation (<10%) among the four cell lines, with the exception of an apparent 30% elevation of TG in cells transfected with PAP2-α1.

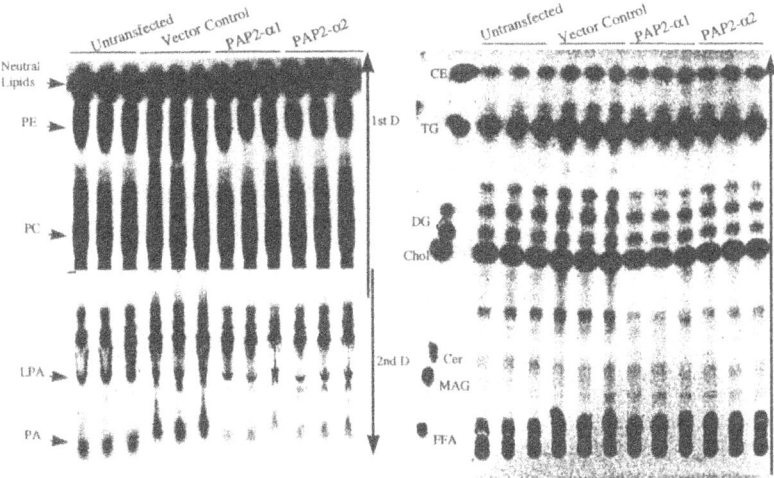

Figure 3. TLC analysis of of the steady state level of various lipids in ECV304 endothelial cells transfected with PAP2-α1 or PAP2-α2 expression plasmids. Lipid samples were prepared in triplicates from untransfected ECV304 cells and from ECV304 cells transfected with pCE9 control vector, with PAP2-α1, or with PAP2-α2 expression plasmids. The charged lipids (left panel) were separated by ffTLC (West et al., 1997). The neutral lipids (right panel) were resolved by multi-one-D TLC (Leung et al., 1998) and the lipid bands were visualized by fluorescence after primulin staining. The identities of labeled bands were determined based on the migration of lipid standards.

Detection of PAP2-α mRNA in human tissues

In agreement with published results (Kai et al., 1997; Ulrix et al., 1998), Northern blot analysis (Leung et al., 1998) shows that PAP2-α mRNA can be detected in all tissues tested. The similarity in size and sequence between PAP2-α1and PAP2-α2 mRNAs would not allow one to distinguish the relative abundance of the two isoforms by Northern analysis. Accordingly, PCR primers corresponding to regions specific to each individual exon and to a common region downstream of the two alternative exons were used to amplify defined fragments of 397 bp (specific to PAP2-α1)and 338 bp (specific to PAP2-α2) from various cDNA libraries. PCR analysis of the fragments generated (Fig. 4) shows both isoforms can be readily detected in all six of the tissues examined. PAP2-α1 was found to be the predominant isoform in kidney, lung, placenta, and liver; PAP2-α2 was the predominant isoform in heart and pancreas.

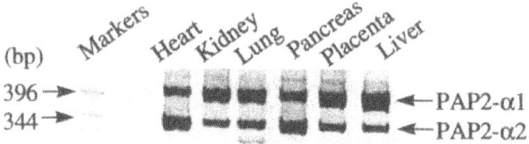

Figure 4. Estimation of the relative abundance of PAP2-α1 and PAP2-α2 isoforms by PCR analysis. PCR primers corresponding to regions specific to each of the alternatively spliced exons and to a common region downstream were used to amplify defined fragments of 397 bp and 338 bp expected from the PAP2-α1 and PAP2-α2 sequences, respectively, from various cDNA libraries. The lengths of standard DNA markers are indicated on the left-hand margin.

DIFFERENTIAL EXPRESSION OF PAP2-α mRNA IN NORMAL VS. TUMOR COLON TISSUES

PA has been implicated in mitogenesis of several cell lines (English et al., 1996). PA level has been found to be increased, whereas both PAP1 and PAP2 activities have been found to be decreased in either *ras* or *fps* transformed cell lines compared to the parental Rat2 fibroblast cell line (Martin et al., 1997; Martin et al., 1993). On the other hand, PAP2-α was found to be up-regulated in a prostate cell line after induction with androgen (Ulrix et al., 1998). To test the hypothesis that PAP2-α expression may be suppressed in certain tumor cells, we examined the expression of PAP2 mRNA in human tumor panel blots (Invitrogen, Carlsbab, CA) that contained tumor RNAs isolated from various malignant tissues and RNAs from the normal tissues in the surgical margins, run in adjacent lanes. Of a total of eight different tissues examined, PAP2-α mRNA expression was found to be suppressed in five tumor tissues (colon, rectum, breast, fallopian tube, and ovary) when compared to its expression in the corresponding normal tissues (Leung et al., 1998). To determine whether the suppression of PAP2-α expression in selected tumor tissues could be generalized to similar tissues derived from several donors, mRNAs from four pairs each of malignant and normal breast and colon tissues were obtained. Of the four breast tissue donors examined, only one showed suppression of PAP2-α expression in the tumor tissue (Leung et al., 1998). Of greater interest, all four colon adenocarcinomas showed suppression of PAP2-α mRNA expression, as compared to the mRNA from the corresponding normal tissues (Fig. 5), whereas PAP2-β expression was suppressed in two (donors 1 and 2) out of four cases in tumor versus normal tissues. PAP-γ level was too low to be detected by Northern analysis.

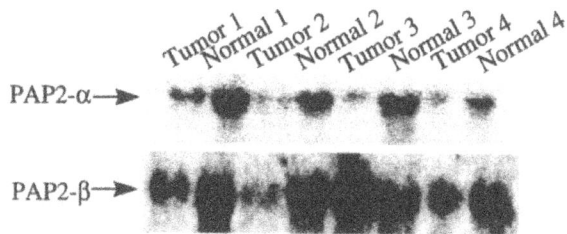

Figure 5. Northern Analysis of PAP2-α and PAP2-β mRNA in colon tissues. Each tissue set (Invitrogen, Carlsbab, CA) contains total RNA from human tumor tissues and donor-matched normal tissues.

644

The downregulation of PAP2-α expression may be a mechanistic contributor to development of this neoplasm. Human PAP2-α cDNAs could be candidates for gene therapy of selected cancers. As a multifunctional lipid-specific phosphatase located on cellular membranes, PAP2-α can dephosphorylate and hence modulate the balance of lipid messengers between the phosphorylated substrates (PA, LPA, S-1-P, and C-1-P) and the dephosphorylated products (DG, sphingosine, and ceramide) in cells (Brindley and Waggoner, 1996). Changes in a multicellular microenvironment have been found to drastically affect carcinogenesis in three-dimensional cell culture models (Weaver et al., 1996). The availability of various cDNAs for PAP2 isoforms may thus also provide the tools for understanding the roles of their lipid substrates and products in regulating cell activation and signaling.

ACKNOWLEDGMENTS

We thank Noel Balantac for isolating transfected cell lines expressing PAP2-α1 or PAP2-α2 cDNAs and Dr. Robert Lewis for critically reviewing this work.

REFERENCES

Barila, D., Platerot, M., Nobili, F., Muda, A.O., Xie, Y., Morimoto, T., and Perozzi, G., 1996, The *Dri 42* gene, whose expression is up-regulated during epithelial differentiation, encodes a novel endoplasmic reticulum resident transmembrane protein. *J Biol Chem.* 271:29928.

Boguski, M.S., Tolstoshev, C.M., and Bassett, J., 1994, Gene discovery in dbEST. *Science.* 265:1981.

Brindley, D.N., and Waggoner, D.W., 1996, Phosphatidate phosphohydrolase and signal transduction. *Chem Phys Lipids.* 80:45.

Carman, G.M. 1997, Phosphatidate phosphatases and diacylglycerol pyrophosphate phosphatases in Saccharomyces cerevisiae and Escherichia coli. *Biochim Biophys Acta.* 1348:45.

English, D., Cui, Y., and Siddiqui, R.A., 1996, Messenger functions of phosphatidic acid. *Chem Phys Lipids.* 80:117.

English, D., Martin, M., Harvey, K.A., Akard, L.P., Allen, R., Widlanski, T.S., Garcia, J.G.N., and Siddiqui, R.A., 1997, Characterization and purification of neutrophil *ecto*-phosphatidic acid phosphohydrolase. *Biochem J.* 324:941.

Kai, M., Wada, I., Imai Si, Sakane, F., and Kanoh, H., 1997, Cloning and characterization of two human isozymes of Mg2+-independent phosphatidic acid phosphatase. *J Biol Chem.* 272:24572.

Kai, M., Wada, I., Imai, S., Sakane, F., and Kanoh, H., 1996, Identification and cDNA cloning of 35-kDa phosphatidic acid phosphatase (Type 2) bound to plasma membranes - Polymerase chain reaction amplification of mouse H₂O₂-inducible *hic*53 clone yielded the cDNA encoding phosphatidic acid phosphatase. *J Biol Chem.* 271:18931.

Lei, K.J., Pan, C.J., Liu, J.L., Shelly, L.L., and Chou, J.Y., 1995, Structure-function analysis of human glucose-6-phosphatase, the enzyme deficient in glycogen storage disease type 1A. *J Biol Chem.* 270:11882.

Leung, D.W., Tompkins, C., and White, T., 1998, Molecular cloning of two alternatively spliced forms of human phosphatidic acid phosphatase cDNAs that are differentially expressed in normal and tumor cells. *DNA Cell Biol.* (in press).

Mandala, S., Thornton, R., Tu, Z., Kurtz, M.B., Nickels, J., Broach, J., Menzeleev, R., and Spiegel, S., 1998, Sphingoid base 1-phosphate phosphatase: A key regulator of sphingolipid metabolism and stress response. *Proc Natl Acad Sci USA.* 95:150.

Martin, A., Duffy, P.A., Liossis, C., Gomez-Munoz, A., O'Brien, L., Stone, J.C., and Brindley, D.N., 1997, Increased concentrations of phosphatidate, diacylglycerol and ceramide in *ras*- and tyrosine kinase (*fps*)-transformed fibroblasts. *Oncogene.* 14:1571.

Martin, A., Gomez-Munoz, A., Waggoner, D.W., Stone, J.C., and Brindley, D.N., 1993, Decreased activities of phosphatidate phosphohydrolase and phospholipase D in *ras*- and tyrosine kinase (*fps*)-transformed fibroblasts. *J Biol Chem*. 268:23924.

Messerschmidt, A., and Wever, R., 1996, X-ray structure of a vanadium-containing enzyme: chloroperoxidase from the fungus *Curvularia inaequalis*. *Proc Natl Acad Sci USA*. 93:392.

Neuwald, A.F., 1997, An unexpected structural relationship between integral membrane phosphatases and soluble haloperoxidases. *Protein Sci*. 6:1764.

Perry, D.K., Stevens, V.L., Widlanski, T.S., and Lambeth, J.D., 1993, A novel ecto-phosphatidic acid phosphohydrolase activity mediates activation of neutrophil superoxide generation by exogenous phosphatidic acid. *J Biol Chem*. 286:25310.

Stukey, J., and Carman, G.M., 1997, Identification of a novel phosphatase sequence motif. *Protein Sci*. 6:469.

Ulrix, W., Swinnen, J.V.; Heyns, W., and Verhoeven, G., 1998, Identification of the phosphatidic acid phosphatase type 2a isozyme as an androgen-regulated gene in the human prostatic adenocarcinoma cell line LNCaP. *J Biol Chem*. 273:4660.

Waggoner, D.W., Gómez-Muñoz, A., Dewald, J., and Brindley, D.N., 1996, Phosphatidate phosphohydrolase catalyzes the hydrolysis of ceramide 1-phosphate, lysophosphatidate, and sphingosine 1-phosphate. *J Biol Chem*. 271:16506.

Weaver, V.M., Fischer, A.H., Peterson, O.W., and Bissell, M.J., 1996, The importance of the microenvironment in breast cancer progression: recapitulation of mammary tumorigenesis using a unique human mammary cell model and a three-dimensional culture assay. *Biochem Cell Biol*. 74:833.

West, J., Tompkins, C.K., Balantac, N., Nudelman, E., Meengs, B., White, T., Bursten, S., Coleman, J., Kumar, A., Singer, J.W., and Leung, D.W., 1997, Cloning and expression of two human lysophosphatidic acid acyltransferase cDNAs that enhance cytokine-induced signaling responses in cells. *DNA Cell Biol*. 16:691.

Zhang, N., Zhang, J., Purcell, K.J., Cheng, Y., and Howard, K., 1997, The Drosophila protein Wunen repels migrating germ cells. *Nature*. 385:64.

93

EICOSAPENTAENOIC ACID ALTERS MANGANESE SUPEROXIDE DISMUTASE IMMUNOREACTIVE PROTEIN LEVELS IN NORMAL BUT NOT MALIGANT CENTRAL NERVOUS SYSTEM DERIVED CELLS

[1]Geoffrey D. Girnun, [1]Larry W. Oberley, [2]Steven A. Moore, and [1]Mike E.C. Robbins

[1]Radiation Research Laboratory,
[2]Department of Pathology
University of Iowa
Iowa City, IA 52252

INTRODUCTION

The dose of radiation that safely can be administered to patients with malignant gliomas is limited by the potential for severe CNS morbidity. Manipulations that might increase the radiation sensitivity of gliomas and/or decrease injury to normal components of the CNS, e.g., the microvascular endothelium, one of the cell types believed to play a major role in radiation myelopathy, would provide a therapeutic benefit. One such approach may be the use of polyunsaturated fatty acids (PUFAs). PUFAs can inhibit tumor cell growth both *in vitro*[1] and *in vivo*[2]. Moreover, Vartak et al.[3], have recently shown that PUFAs are not only selectively cytotoxic to glioma cells, but also enhance the radiosensitivity of glioma cells.

PUFAs have also been shown to reduce the severity of radiation-induced injury to normal tissue such as skin[4] and CNS[5]. The damaging effects of ionizing radiation are due mainly to the generation of reactive oxygen species (ROS) which alter cell proteins, lipids and DNA. Therefore, we hypothesize that PUFAs may modify the response of cells to ionizing radiation by altering antioxidant enzymes that scavenge ROS. One such antioxidant enzyme, manganese superoxide dismutase (Mn-SOD), a member of the superoxide dismutase family of enzymes which scavenge superoxide radicals, has been shown[6,7] to protect a variety of cells from ionizing radiation. In addition, several studies have shown[8,9] that PUFAs can alter Mn-SOD message, protein and activity. Therefore, we determined if one such PUFA, eicosapentaenoic acid (EPA;

Eicosanoids and Other Bioactive Lipids in Cancer, Inflammation, and Radiation Injury, 4
Edited by Honn *et al.*, Kluwer Academic / Plenum Publishers, New York, 1999.

647

20:5n-3), could alter Mn-SOD in a normal CNS-derived microvascular endothelial cell and a malignant CNS-derived glioma cell line.

MATERIALS AND METHODS

Cell culture

Rat brain microvascular endothelial (RBMEn) cells isolated[10] from newborn rat pups were used between 10-15 passages after isolation. RBMEn cells were grown in minimum essential medium containing 1% L-glutamine, 0.1% gentamycin, 10% heat-inactivated fetal bovine serum (Gibco, Grand Island, NY), 6 g/L glucose (Fisher, Fairlawn, NJ),. The

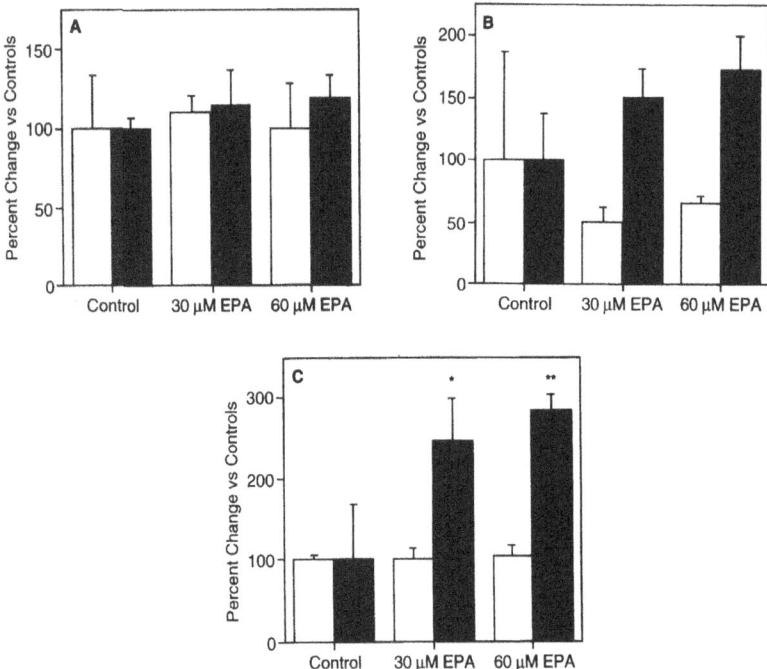

Figure 1. Percent change for Cu/Zn-SOD (open columns) or Mn-SOD (solid columns) immunoreactive protein in RBMEn cells following supplementation with EPA for 24 (A), 48 (B) or 72 (C) hr. Values are percent compared to controls and represent mean ± SD. *p < 0.05 and **p < 0.01 compared to controls.

media was changed every 3 days. For a malignant cell, the ethylnitrosurea-induced rat astrocytoma cell line[11] (36B10) was used. 36B10 cells were grown in high glucose (4.5 g/L) Dulbecco's modified Eagle's medium containing 1% L-glutamine, 0.1% gentamycin, 10% heat-inactivated fetal bovine serum (Gibco, Grand Island, NY). The media was changed every other day and cells passaged twice per week.

Fatty acid supplementation

Cells were serum deprived for 20 hr to synchronize cell cycle and down regulate transcription and translation. The RBMEn cells were then supplemented with 0, 30, or 60 μM eicosapentaenoic acid (EPA, Cayman Chemical, Ann Arbor, MI) for 24-72 hr in serum containing media for SOD analysis or 72 hr for fatty acid analysis. The 36B10 cells were supplemented for 72 hr for both SOD and fatty acid analysis. All experiments were carried out in triplicate.

Fatty acid analysis

Total cell lipids were extracted by the method of Folch et al.[12] and transesterified[13]. The percent fatty acid composition was determined by gas-liquid chromatography as previously reported[3].

Immunoblotting

Proteins from cell lysates were seperated using denaturing SDS-PAGE according to the method of Laemmeli[14]. Samples then were transferred to nitrocellulose by standard procedures. Membranes were probed using antibodies to Mn-SOD and Cu/Zn-SOD. Protein bands were visualized using chemiluminescence. Bands were imaged and digitized by the Image Analysis Facility at the University of Iowa. Densitometry was determined as a function of optical density and band size (integrated densitometric value), and treated samples expressed as a percent change compared to controls. ANOVA was performed on densitometric values and followed by Student's t-test as a post-hoc test where applicable. A $p < 0.05$ was considered significant.

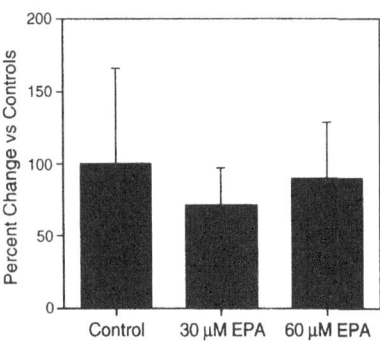

Figure 2. Percent change for Mn-SOD immunoreactive protein in 36B10 cells following supplementation with EPA for 72 hr. Values are percent compared to controls and represent mean ± SD.

RESULTS:

Fatty acid analysis

Table 1 shows the percent of EPA and its elongation product docosapentaenoic acid (DPA) in RBMEn cells and 36B10 cells following supplementation with 0-60 μM EPA for 72 hr. There is a dose dependent increase in both EPA and DPA in the RBMEn cells following EPA supplementation. The percent of EPA increased 5- and 8-fold over controls, following 30 and 60 μM EPA supplementation, respectively. Likewise, the percent of DPA increased 3.5- and 4-fold under the same conditions. The 36B10 cells displayed a similar trend with EPA increasing 12- and 15-fold and DPA increasing 8- and 14-fold over controls after supplementation with 30 and 60 μM EPA for 72 hr, respectively. The composition of docasahexaenoic acid (DHA) did not appear to increase substantially in either cell line following EPA supplementation (data not shown), indicating an inability of both of these cell lines to convert EPA to DHA.

Table 1. EPA and DPA composition of RBMEn and 36B10 cells following supplementation with EPA for 72 hr.

	Fatty acid	Control	30 μM	60 μM
RBMEn	EPA	1.55 ± 0.63	8.09 ± 0.69^2	12.48 ± 0.89^2
	DPA	3.28 ± 0.40	11.25 ± 1.62^1	14.05 ± 0.56^2
36B10	EPA	0.63 ± 0.06	7.67 ± 0.57^2	9.23 ± 0.65^2
	DPA	1.27 ± 0.15	10.2 ± 0.2^2	17.47 ± 0.83^2

Percent of total fatty acids. Values do not add up to 100% since not all fatty acids are shown. Values are mean $\pm$ SD (n=3).
[1] $p < 0.005$; [2] $p < 0.0005$, compared to control values.

SOD Analysis

Figure 1 shows densitometric changes from a representitive immunoblot for Mn-SOD and Cu/Zn-SOD in RBMEn cells following 0-60 μM EPA supplementation for 24-72 hr. At 24 and 48 hr there appears to be an increase in Mn-SOD immunoreactive protein, however, it is not significantly different from unsupplemented controls (Fig. 1A and 1B). In contrast, by 72 hr, there is a significant 2.5 and 3.0-fold increase in Mn-SOD immunoreactive protein following 30 and 60 μM EPA supplementation respectively (Fig.1C; $p < 0.05$ and 0.01, respectively). The

Cu/Zn-SOD content of the RBMEn cells was not altered by EPA supplementation at any time point.

In contrast to the RBMEn cells, there does not appear to be a change in Mn-SOD immunoreactive protein in 36B10 cells, after supplementation with 30 and 60 μM EPA (Figure 2). Cu/Zn-SOD levels were also unaltered in the 36B10 cells after supplementation with EPA for 72 hr (data not shown).

DISCUSSION:

Mn-SOD is a tetrameric enzyme with a total MW of 88 kDa found in the mitochondria of eukaryotic cells. CuZn-SOD is a dimer with a total MW of 32 kDa, found primarily in the cytosol. The main function of these enzymes is the dismutation of superoxide radicals into hydrogen peroxide and oxygen.

Our results indicate that EPA did not alter Cu/Zn-SOD levels in either cell line. These findings support previous studies[15,16] which indicate that Cu/Zn-SOD, in general, is not inducible. However, we did observe an increase in Mn-SOD immunoreactive protein levels in the RBMEn cells following supplemenation with 30 and 60 μM EPA for 72 hr. Although there appeared to be an increase by 24 and 48 hr, it did not reach statistical significance until 72 hr. In contrast, Mn-SOD levels were not altered in a malignant rat glioma cell line treated with similar doses of EPA over the same time course. This inability of tumor cells to induce Mn-SOD has been demonstrated previously[17], and appears to be a general characteristic of cancer cells. Differences in uptake and elongation of EPA do not appear to be the reason for selective increase in Mn-SOD, since both cell lines were able to take up EPA and elongate it to DPA.

Studies have shown[18] that ROS can increase Mn-SOD expression in normal cells, which might, in part, explain the observed increase in Mn-SOD protein in the RBMEn cells. EPA has been shown[19] to stimulate ß-oxidation in peroxisomes, which leads to an increase hydrogen peroxide production. In addition, we have observed an increase in mitochondrial activity in the RBMEn cells following supplementation with EPA(data not shown). This increased mitochondrial activity may lead to an increased leakage of superoxide radicals off the electron transport chain[20] (ETC). Hence, the increased generation of hydrogen peroxide and superoxide radicals may in part be responsible for the observed increase in Mn-SOD. This might also explain the lack of an increase in Mn-SOD protein in the 36B10 cells. In general tumor cells operate with increased glycolysis[21], rather than oxidative phosphorylation. Thus with less oxidative phosphorylation occuring, fewer superoxide radicals would leak off the ETC. Indeed, results from our lab indicate that in contrast to the RBMEn cells, mitochondrial activity in the 36B10 cells is not significantly different from controls following EPA supplementation (data not shown). Hence, there would be a smaller ROS signal for the induction of Mn-SOD in the tumor cells.

Further studies are needed to determine if other fatty acids produce similar effects. In addition, it remains to be determined if this increase in Mn-SOD in the RBMEn cells leads to protection of these cells from radiation or other oxidative stress. Neverthless, these studies indicate that EPA is able to selectively increase Mn-SOD immunoreactive protein in normal but not malignant CNS-derived cells, and suggests that EPA may be of potential therapeutic benefit in anti-glioma therapy.

ACKNOWLEDGMENTS

This work was supported in part by Scotia Pharmaceuticals, U.K., NIH grant NS-24621 and American Heart Association Grant-In-Aid 96-50661N

REFERENCES

1. Gonzalez, M.J., Schemmel, R.A., Dugan, Jr.,L., Gray, J.I. and Welsch, C.W. Dietary fish oil inhibits human breast carcinoma growth: a function of increased lipid peroxidation. *Lipids* 28:827-832 (1993).
2. Karmali, R.A., Reichel, P., Cohen, L.A., Terano, T., Hirai, A., Tamura, Y. and Yoshida, S. The effect of dietary n-3 fatty acids on the DU-145 transplantable human prostatic cancer. *Anticancer Res.* 7:1173-1180 (1987).
3. Vartak, S., Robbins, M.E.C. and Spector, A.A. Polyunsaturated fatty acids increase the sensitivity of 36B10 rat astrocytoma cells to radiation-induced cell kill. *Lipids* 32:283-292 (1997).
4. Hopewell, J.W., van den Aardweg, G.J.M.J., Morris, G.M., Rezvani, M., Robbins, M.E.C., Ross, G.A., Whitehouse, E.M., Scott, C.A. and Horrobin, D.F. Amerlioration of both early and ;ate radiation-induced damage to pig skin by essential fatty acids. *Int.J.Radiat.Oncol.Biol.Phys.* 30:1119 (1994).
5. Hopewell, J.W., van den Aardweg, G.J.M.J., Morris, G.M., Rezvani, M., Robbins, M.E.C., Ross, G.A., Whitehouse, E.M. In: *New Approaches to Cancer Treatment: Unsaturated Lipids and Photodynamic Therapy,* D.F. Horrobined., Churchill Comm. Europe, London (1994).
6. Wong, G.H. Protective roles of cytokines against radiation: induction of mitochondrial Mn-SOD. *Biochim.Biophys.Acta.* 1271:205 (1995).
7. Juan, S., Yuan, C., Mei, Z., Mingtao, L. and Zhongliang, G. Mitochondrial alterations in CHO cells exposed to X-ray after transfecting with human MnSOD cDNA. *Med.Sci.Res.* 25:81 (1997).
8. Venkatraman, J.T., Chandrasekar, B., Kim, J.D. and Fernandes, G. Effects of n-3 and n-6 fatty acids on the activities and expresssion of hepatic antioxidant enzymes in autoimmune-prone NZBxNZW F1 mice. *Lipids* 29:561 (1994).
9. Phylactos, A.C., Harbige, L.S. and Crawford, M.A. Essential fatty acids alter the activity of manganese-superoxide dismutase in rat heart. *Lipids* 29:111 (1994).
10. Debault, L.E., Henriquez, E., Hart, M.N. and Cancilla, P.A. Cerebral microvessels and derived cells in tissue culture II. Establishment, identification, and preliminary characterization of an endothelial cell line. *In Vitrc* 17:480 (1981).
11. Spence, A.M., Coates,P.W. Scanning electron microscopy of cloned astrocytic lines derived from ethyl-nitrosourea-induced rat glioma. *Virchows Arch.* B: 28, 27 (1978).
12. Folch, J., Lees, M.B. and Sloane Stanley, G.H. A simple method for the isolation and purification of total lipids from animal tissues. *J.Biol.Chem.* 226:497 (1957).
13. Morrison, W.R., Smith, M.L. Preparation of fatty acid methyl esters and dimethylacetals from lipids with boron fluoride-methanol. *J.Lipid Res.* 5:600 (1964).
14. Laemmli, U.K. Cleavage of structural proteins during the assembly of the head of bacteriophage T4. *Nature.* 227:680 (1970).
15. Monkuno, K., Ohtani, K., Suzumura, A., Kiyosawa, K., Hirose, Y., Kawai, K. and Kato, K. Induction of manganese superoxide dismutase by cytokines and lipopolysaccharide in cultured mouse astrocytes. *J.Neurochem.* 63:612 (1995).
16. Visner, G.A., Chesrown, S.E., Monnier, J., Ryan, U.S. and Nick, H.S. Regulation of manganese superoxide dismutase: Il-1 and TNF induction in pulmonary artery and microvascular endothelial. Biochem. *Biophys.Res.Commun.* 188:453 (1992).
17. Bravard, A., Sabatier, L., Hoffschir, F., Ricoul, M., Luccioni, C., Dutrillax, B. SOD2: a new type of tumor-suppressor gene? *Int.J.Cancer.* 51:476 (1992).

18. Yoshioka, T., Homma, T., Meyrick, B., Takeda, M., Moore-Jarret, T., Kon, V. and Ichikawa, I. Oxidants induce transcriptional activation of manganese superoxide dismutase in glomerular cells. *Kidney Int.* 46:405 (1994).

19. Willumsen, N., Vaagenes, H., Lie, Ø., Rustan, A. and Berge, R.K. Eicosapentaenoic acid, but not docosahexaenoic acid, increases mitochondrial fatty acid oxidation and upregulates 2,4-dienoyl-CoA reductase gene expression in rats. *Lipids* 31:579 (1996).

20. Benzi, G., Moretti,A. Age- and peroxidative stress-related moficications of the cerebral enzymatic activities linked to mitochondria and the glutathione system. *Free Radic.Biol.Med.* 19:77 (1995).

21. Brand, K.A. and Hermfisse, U. Aerobic glycolysis by proliferating cells: a protective strategy against reactive oxygen species. *FASEB J.* 11:388 (1997).

94

GAMMA RADIATION AND RELEASE OF NOREPINEPHRINE IN THE HIPPOCAMPUS

Sathasiva B. Kandasamy

Radiation Pathophysiology and Toxicology Department
Armed Forces Radiobiology Research Institute
Bethesda, MD 20889-5603

This article reviews research work being done in this laboratory.[1-7]

Although the central nervous system (CNS) is generally considered to be relatively resistant to direct effects of ionizing radiation, exposure to ionizing radiation can have a complex effect on the CNS that is dependent on the dose and time elapsed after radiation exposure.[8] The biochemical basis for radiation-induced behavioral/physiological changes mediated by the CNS is unknown.

The hippocampus is important in the critical functions of learning, memory, and motor performance, and these functions are impaired after exposure to ionizing radiation.[8] Histofluorescence and immunohistochemical techniques have shown noradrenergic pathways in the hippocampus.[9] Norepinephrine may act as a neuromodulator because it enhances the magnitude, duration, and probability of induction of long-term potentiation (LTP) at mossy fiber synapses in the formation of the hippocampus, and LTP is widely believed to be a cellular mechanism for aspects of learning and memory.[9]

An essential step in neurotransmission is the influx of Ca^{2+} into presynaptic nerve terminals through voltage-gated calcium channels. The resulting elevation of the intracellular Ca^{2+} concentration triggers the release of neurotransmitters. Calcium channels, therefore, play a pivotal role in excitation-secretion coupling.[10] Calcium channels modulated norepinephrine (NE) release from several neuronal preparations.[10,11] The L-type calcium channel activator Bay K 8644 [methyl-1,4-dihydro-2,6-dimethyl-3-nitro-4-2(trifluoromethylphenyl) pyridine-5-carboxylate], a dihydropyridine with convulsant actions, increases KCl-evoked NE release from brain slices, and the effect is antagonized by the L-type calcium channel antagonists nifedipine and other dihydropyridines.[12] On the other hand, the evoked NE release is reduced by blockade of N-type calcium channels by ω-conotoxin GVIA, indicating a central role for this type of calcium channel in transmitter release processes.[13]

Protein kinase C (PKC) is a family of Ca^{2+}- and phospholipid-dependent enzymes, which

Eicosanoids and Other Bioactive Lipids in Cancer, Inflammation, and Radiation Injury, 4
Edited by Honn *et al.*, Kluwer Academic / Plenum Publishers, New York, 1999.

655

are activated by the second messenger 1,2-diacylglycerol and its analogue, phorbol 12-O-tetradecanoate 13-acetate (TPA).[14] A wide spectrum of neuronal processes, including the transduction system, ion channels, and synaptic plasticity, has been demonstrated to be sensitive to PKC. This enzyme also plays a critical role in LTP in the hippocampus.[15] PKC is highly concentrated in the rat hippocampus.[16]

Earlier studies determined the effects of γ irradiation (^{60}Co) on ^{45}Ca^{2+} uptake in rat brain synaptosomes.[1-3] γ irradiation reduced KCl-stimulated voltage-dependent uptake of ^{45}Ca^{2+} in rat whole-brain, cortical, and striatal synaptosomes.[2-3] The enhancement of KCl-stimulated ^{45}Ca^{2+} influx in whole-brain synaptosomes by the calcium L-type agonist Bay K 8644 was also reduced by radiation exposure. Radiation-induced inhibition of calcium uptake enhanced by Bay K 8644 was not a result of an altered L-type calcium channel recognition receptor, because nimodipine (calcium channel L-type receptor antagonist) binding to L-type calcium channel receptors was not altered following radiation exposure.[2] Prostaglandins, inositol 1, 4,5-trisphosphate, and phorbol esters were tested for their ability to inhibit radiation-induced decreases in calcium influx in rat whole brain synaptosomes.[3] If treatment with these drugs inhibited the radiation-induced reduction in calcium influx, it would indicate that radiation interferes with the formation and/or function of second messengers that are necessary for calcium influx. Pharmacological studies with TPA have provided indirect evidence to suggest that radiation-induced decreases in calcium uptake are due to impairment of PKC acitivity,[3] which is responsible for the opening and closing of ion channels and the mobilization of calcium by these drugs, because none of the compounds tested alone completely prevented radiation-induced decreases in Ca^{2+} uptake. Some drug combinations, however, did inhibit the decreases.[3] Similar synergistic effects of TPA with Bay K 8644 or calcimycin, which increases intracellular Ca^{2+}, have been reported.[17]

Experiments were done to determine whether exposure to γ radiation decreases hippocampal PKC levels in rats. Chen et al.,[6] demonstrated that exposure to 10 Gy or 30 Gy (dose rate of 10 Gy/min) of γ radiation (^{60}Co) either *in vitro* or *in vivo* induces a significant decrease in PKC levels in rat hippocampus.

Gamma radiation (10 Gy at 10 Gy/min from a ^{60}Co source) induces a decrease in hippocampal NE release 48 h after exposure.[4,5,7] This review, in addition to reviewing the effects of ionizing radiation on calcium channels[1-3] summarizes the research work on the effect of Bay K 8644, calcimycin 6S-[6α(2S*,3S*),8β(R*),9β,11α]-5-(methylamino)-2-[[3,9, 11-trimethyl-8-[1-methyl-2-oxo-2(1H-pyrrol-2-yl)ethyl]-1,7-dioxaspiro[5,5]undec-2-yl] methyl]-4-benzoxazolecarboxylic acid (a calcium-ionophore that increases intracellular Ca^{2+}) and TPA alone or in combination on decreases in hippocampal NE release after radiation exposure.

As a Ca^{2+}-dependent second messenger system, PKC is intimately involved in intracellular signal transduction and plays an important role in neurotransmitter release. Phorbol ester-mediated activation of PKC enhanced the release of various neurotransmitters,[18] whereas inhibition of PKC decreased it. Therefore, any experimental conditions that interfere with the influx of Ca^{2+} and/or impair PKC activity might result in decreased neurotransmitter release.

This study has demonstrated that exposure to 10 Gy of γ radiation induces a decrease in NE release in the hippocampus after 48 h. The present study determined the effects of two drugs (Bay K 8644 and calcimycin) that increase intracellular Ca^{2+} by different mechanisms (Bay K 8644 acts by stimulating L- and N-type calcium channels, whereas calcimycin acts as an ionophore) with and without TPA on the release of NE in the hippocampus of sham-irradiated and irradiated animals. The present study also determined the effects of TPA, which stimulates

PKC, with and without Bay K 8644 or calcimycin. Neither Bay K 8644 (1-100 nM), calcimycin, nor TPA alone prevented the radiation-induced decreases in the release of NE in the hippocampus. However, 10 nM of Bay K 8644 or calcimycin in combination with1-100 nM of TPA or 10 nM of TPA in combination with 1-100 nM of Bay K 8644 or calcimycin did prevent the radiation-induced decreases in the release of NE (Fig. 1).

Although the mechanisms involved in radiation-decreased NE release in the hippo- campus are unknown, it is suggested that ionizing radiation might affect the activity of PKC and the mobilization of Ca^{2+} by these drugs because it decreases PKC levels in the hippocampus[6] and Ca^{2+} influx in brain synaptosomes.[2,3] In addition, the stimulation of PKC by TPA and the mobilization of Ca^{2+} together by Bay K 8644 or calcimycin are necessary for the restoration of radiation-decreased NE release in the hippocampus. In other words, it suggests that stimulation of PKC by TPA alone, mobilization of Ca^{2+} by Bay K 8644, or calcimycin alone is not enough to restore radiation-decreased NE release in the hippocampus.

In conclusion, these results suggest that stimulation of PKC by TPA with mobilization/ influx of Ca^{2+} by Bay K 8644 or calcimycin is necessary to restore radiation-decreased NE release in the hippocampus.

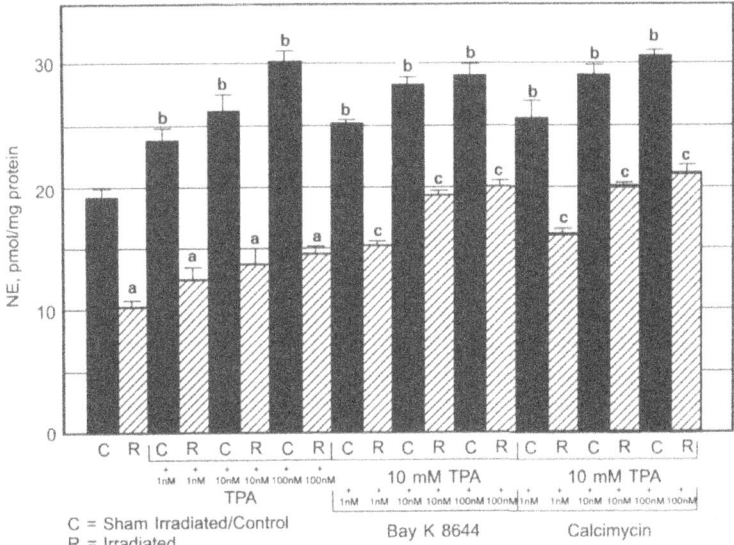

Figure 1. Effect of *in vitro* administration of various doses of TPA alone, 10 μM of TPA plus various doses of Bay K 8644, or 10 μM of TPA plus various doses of calcimycin on the release of norepinephrine in the hippocampus of sham-irradiated and irradiated animals. Values are expressed as the mean of releases for eight rats ± SEM. a, Significance between values for sham-irradiated rats without drug treatment and irradiated rats with or without drug treatment; $P < 0.05$; b, significance between values for sham-irradiated rats without drug treatment and sham-irradiated rats with drug treatment, $P < 0.05$; c, significance between values for irradiated rats and irradiated rats with drug treatment, $P < 0.05$. (From: Kandasamy SB. Effect of Bay K 8644, calcimycin and phorbol ester on radiation-induced decreases in the release of norepinephrine in the hippo- campus in rats. Radiat. Res., 149:277-283, 1998. With permission.)

REFERENCES

1. S.B. Kandasamy, and W.A. Hunt, Arachidonic acid and prostaglandins enhance potassium-stimulated calcium influx into rat brain synaptosomes, *Neuro-pharmacology* 29:825 (1990).
2. S.B. Kandasamy, T.C. Howerton, and W.A. Hunt, Reductions in calcium uptake induced in rat brain synaptosomes by ionizing radiation, *Radiat. Res.* 125:158-162 (1991).
3. S.B. Kandasamy, and A.H. Harris, Effect of prostaglandins, inositol 1, 4, 5- trisphosphate, and phorbol esters on radiation-induced decreases in calcium influx in rat brain synaptosomes, *Radiat. Res.* 131:43-46 (1992).
4. S.B. Kandasamy, S.A. Stevens-Blakely, T.K. Dalton, and A.H. Harris, Implication of nitric oxide synthase in radiation-induced decrease in hippocampal noradrenaline release in rats, in: *The Biology of Nitric Oxide*, S. Moncada, M.A. Marletta, J.B. Hibbs, Jr., and E.A. Higgs, eds., Portland Press, London (1992).
5. S.B. Kandasamy, Role of nitric oxide synthase, superoxide dismutase, and glutathione peroxidase in radiation-induced decrease in norepinephrine release, in: *The Neurobiology of NO and ·OH* (C.C. Chiueh, D.L. Gilbert, and C.A. Colton, eds., Ann. N.Y. Acad. Sci., 738, New York, (1994).
6. H.T. Chen, G. Chen, S.B. Kandasamy, and A.H. Harris, Ionizing radiation decreases hippocampal protein kinase C (PKC) levels in rats, *FASEB* 7:A272 (1993).
7. S.B. Kandasamy, Effect of Bay K 8644, calcimycin and phorbol ester on radiation- induced decreases in the release of norepinephrine in the hippocampus in rats, *Radiat. Res.*, 149:277-283 (1998).
8. D.J. Kimeldorf and E.L. Hunt, Neurophysiological effects of ionizing radiation, in: *Ionizing Radiation: Neural Function and Behavior* (D.J. Kimeldorf and E.L. Hunt, eds., Academic Press, New York (1965).
9. D.J. Heal, The effects of drugs on behavioural models of central noradrenergic function, in: *The Pharmacology of Noradrenaline in the Central Nervous System* (D.J. Heal, and C.A. Marsden, eds., Oxford University Press, New York, (1990).
10. G. Augustine, M.P. Charleton, and S.J. Smith, Calcium action in synaptic transmitter release, *Annu. Rev. Neurosci.* 10:633-693 (1987).
11. R.J. Miller, Multiple calcium channels and neuronal function. *Science* 235:46-47 (1987).
12. D.N. Midlemiss, The calcium channel activator, Bay K 8644, enhances K+-evoked efflux of acetylcholine and noradrenaline from rat brain slices. Naunyn-Schmiedebergs. *Arch. Pharmacol.* 331:114-116 (1985).
13. D.J. Dooley, A. Lupp, and G. Hertting, Inhibition of central neurotransmitter release by ω-conotoxin GVIA, a peptide modulator of the N-type voltage-sensitive calcium channel. Naunyn-Schmiedebergs, *Arch. Pharmacol.* 336:467-470 (1987).
14. Y. Nishizuka, Studies and perspectives of protein kinase C, *Science* (Wash. DC), 233: 305-312 (1986).
15. R.F. Akers, D.M. Lovinger, P.A. Colley, D.J. Linden, and A. Routenberg, Translocation of protein kinase C activity may mediate hippocampal long-term potentiation. *Science* 231:587-589 (1986).
16. U. Kikkawa, Y. Takai, R. Minakuchi, S. Inohara, and Y. Nishizuka, Calcium-activated, phospholipid-dependent protein kinase from rat brain, *J. Biol. Chem.* 257:13341- 13348 (1982).

17. R.Z. Litten, E.A. Suba, and B.L. Roth, Effects of a phorbol ester on rat aortic concentration and calcium influx in the presence and absence of BAY K 8644, *Eur. J. Pharmacol.* 144:185-191 (1987).
18. H.Y. Huang, C. Allgaier, G. Hertting, and R. Jackisch, Phorbol ester-mediated enhancement of hippocampal noradrenaline release: Which ion channels are involved? *Eur. J. Pharmacol.* 153:175-184 (1988).

BIOACTIVE LIPIDS IN
NON-MAMMALIAN CELLS

95

(3R)-HYDROXY-OXYLIPINS—A NOVEL FAMILY OF OXYGENATED POLYENOIC FATTY ACIDS OF FUNGAL ORIGIN

Santosh Nigam[1*], Tankred Schewe[1], and J. Lodewyk F. Kock[2]

[1]Eicosanoid Research Division, Department of Gynaecology
University Medical Centre Benjamin Franklin, Free University
Berlin, D-12200 Berlin, Germany

[2]Department of Microbiology and Biochemistry, University of
the Orange Free State, P.O. Box 339, Bloemfontein 9300,
South Africa

INTRODUCTION

Certain fungi and yeasts are capable of producing hydroxy fatty acids or other oxygenated fatty acid derivatives which are collectively named oxylipinds, and which apparently play a predominant role in the vegetative growth and sexual reproduction of the yeast (Kurtzman et al., 1974; Herman and Herman, 1985; Kerwin et al., 1986; Losel, 1988; Rattray, 1988; Van Dyk et al., 1991; Brodowski and Oliw, 1992; Jensen et al., 1992; Van der Berg, 1993; Van Dyk et al, 1994; Venter et al., 1997; Kock et al., 1998; Herman, 1998).

Although a number of fungi and related organisms are capable of synthesizing sizable amounts of arachidonic acid or other eicosanoid precursors (Shinmen et al, 1989; Ghandhi and Weete, 1991; Van der Westhuizen et al., 1994), little is known concerning the eicosanoid formation by fungi. With this background at the outset of our research few hundred strains of yeasts of the family *Lipomycetaceae* were screened for the formation of eicosanoids. While screening it was found that the yeast *Dipodascopsis uninucleata* transformed exogenous arachidonic acid to a biologically active 3-hydroxy-fatty acid, 3-hydroxy-5, 8, 11, 14-eicosatetraenoic acid (Van Dyk et al, 1991). This compound turned out to be not only a novel eicosanoid but also he prototype of a new family of fungal oxylipins, the (3R)-hydroxy-oxylipins. This chapter will present a short overview on the recent progress of the research around this new compound class.

---

*Author for correspondence
E-mail: Nigam@zedat.fu-berlin.de

CHEMICAL STRUCTURE

Structural elucidation of the metabolite formed from arachidonic acid by *Dipodascopsis uninucleata* (Van Dyk et al, 1991) revealed the presence of a 3-hydroxy group as judged from a base peak of *m/z* 175 in the mass spectrum of the carboxymethyl-trimethylsilyl derivative (Figure 1A). this fragment later proved to be a characteristic one for any member of the (3R)-hydroxy-oxylipin family. The comparison of the M⁺ peaks of the native and the hydrogenated metabolites in the mass spectrum uncovered the presence of 4 double bonds in the molecule. In the FT-IR spectroscopy the absorption bands - characteristic for *cis* double bonds - were observed. Also, the presence of a hydroxy-and a free carboxyl group was evident from the FT-IR spectrum. Moreover, a UV absorption maximum at 193 nm ruled out the presence of any conjugated diene in this molecule. These observations and the fact that in ¹H-NMR spectroscopy the olefinich protons behaved identically in arachidonic acid and is 3-hydroxy derivative allowed an assignment of the structure to 3-hydroxy-(5Z, 8Z, 11Z, 14Z)-eicosatetraenoic acid (3-HETE) (Van Dyk et al., 1991). Unlike the hydroxy-eicosatetraenoic

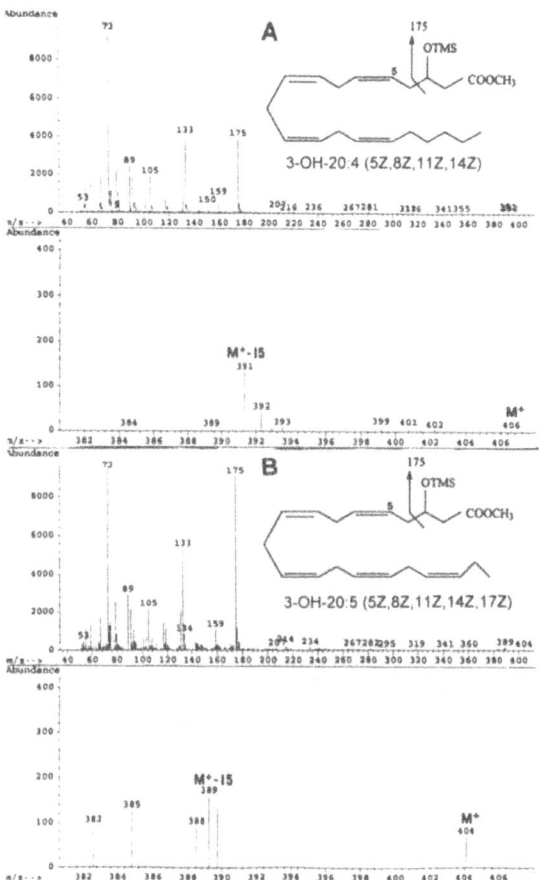

Figure 1.   Electron impact-mass spectra of methyl-trimethylsilylated 3-HETE (A) and 3-HEPE (B)
[*publ. from Lipids, 32(12),1997 with the permission of AOCS Press, IL*]

acids formed in mammalian cells *via* the various lipoxygenase or cyclooxygenase pathways which possess altered double bond systems as compared to arachidonic acid, the 3-HETE can be regarded as a genuine hydroxy-arachidonic acid.

For the analysis of the enantiomeric composition of the natural product, the (3R) and the (3S) enantimoers were chemically synthesized (Bhatt et al., 1998) and used as references for chiral-phase HPLC separation. A (3R)/(3S) of 95:5 was found in native 3-HETE indicating high stereospecificity of the bioformation of 3-HETE (Venter et al, 1997).

Upon feeding *Dipodascopsis uninucleata* with (5Z, 8Z, 11Z, 14Z, 17Z)-eicosapentaenoic or (5Z, 8Z, 11Z)-eicosatrienoic acids instead of arachidonic acid, the corresponding (3R)-hydroxy-eicosapolyeonic fatty acids were identified (Figures 1B & 2). In contrast, when the yeast was fed with linoleic acid, a 3-hydroxylated metabolite was isolated, the structure of which was established as (3R)-hydroxy-(5Z, 8Z)-tetradecadienoic acid (Figure 3), suggesting the retroconversion of linoleic acid to a C-14 fatty acid. Correspondingly, (3R)-hydroxy-(5Z, 8Z)-tetradecadienoic acid was formed from (11Z, 14Z, 17Z)-eicosatrienoic acid (Venter et al, 1997). No 3-hydroxylated metabolite could be so far detected upon feeding a number of saturated and other unsaturated fatty acids such as oleic acid, linolelaidic acid and γ-linolenic acid. Taken together, these data document the existence of a new family of oxylipins which display two common characteristics in all members, a (5Z, 8Zz) diene group and a as (3R)-hydroxy group.

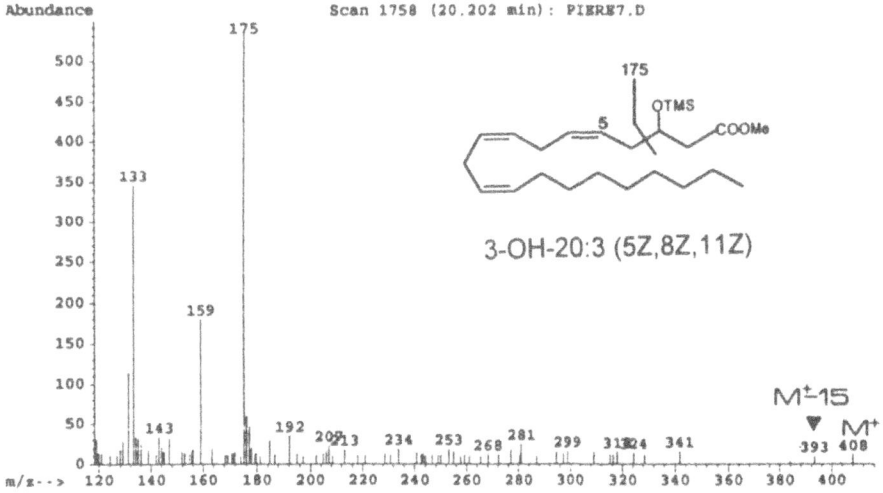

Figure 2.   Electron impact-mass spectrum of methyl-rimethylsilylated 3-hydroxy 20:3 (5Z, 8Z, 11Z), [*published from Lipids, 32(12),1997 with permission of AOCS Press, Champaign, IL*]

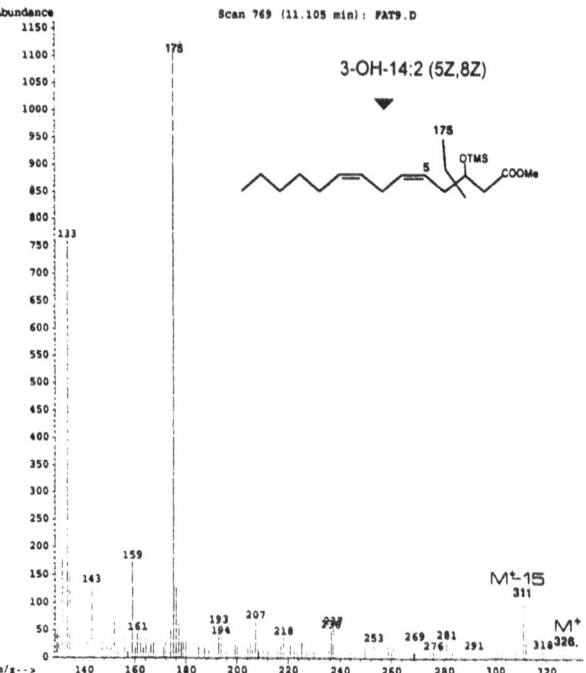

Figure 3. Electron impact-mass spectrum of methyl-trimethylsilylated 3-hydroxy 14:2 (5Z, 8Z), [published from Lipids, 32(12), 1997 with the permission of AOCS Press, Champaign, IL]

BIOSYNTHESIS AND DISTRIBUTIONS

The observation that the formation of (3R)-hydroxy-oxylipins in part requires retroconversion of the parent fatty acid suggests, that it occurs at the level of the free fatty acid or one of its activated species (*e.g.* CoA esters) rather than at the level of cellular lipids. At first sight, it is tempting to presume that these compounds may be formed by uncompleted β-oxidation, which, however, by closer examination appears to be very unlikely, since in that case a (3S)-hydroxy fatty acid has to be formed. Thus, a direct monooxygenation of the precursor fatty acid seems to be more probably way of ist biosynthesis. This hypothesis must, however, be confirmed by $^{18}O_2$ experiments.

The putative " (3R)-hydroxylase" may be a cytochrome P-450 enzyme or a flavo-protein. Notably, the formation of (3R)-hydroxylase-oxylipins is inhibited by millimolar concentrations of acetylsalicylic acid or salicylic acid.

An *in situ* mapping of the occurrence of (3R)-hydroxylase-oxylipins in *Dipodascopsis uninucleata* reveled that they occur only at defined stages of development of the yeast (see also Kock et al. in this volume). this selected formation of these oxylipinds at certain developmental sages may be an impediment for its demonstration o in various fungal species. Under certain circumstances, those developmental states which are competent for oxylipin connection, it is worth noticeable that the (3R)-hydroxylase-oxylipins formation by *Dipodascopsis uninuclea* is not unique, since a conversion by us in Mucor genevensis, a fungus which is not related to the former species (Pohl et al, 1998).

666

BIOLOGICAL ACTIONS

In *Dipodascopsis uninuclea* the occurrence of (3*R*)-hydroxylase-oxylipins is limited to the sexual phase of the life cycle of the yeast (See Kock et al, this volume), which is suppressed by acetylsalicylic acid, an inhibitor of (3*R*)-hydroxylase-oxylipins formation (Kock et al, 1998). Therefore, it is reasonable to assume that the (3*R*)-hydroxylase-oxylipins are growth regulators in fungi. The inhibition of cellular life cycle by salicylic acid (which is possibly also the inhibitor *per se*, when acetylsalicylic acid is used in long-term incubations of the yeast) may be biologically relevant in the context, that salicylic acid is a widely distributed plant hormone, and is implicated in pathogen defense (Delaney et al., 1994; Slusarenko, 1996). The suppression of (3*R*)-hydroxylase-oxylipins formation may protect the host plant from the invasion of pathogenic fungi.

Dipodascopsis uninuclea is unable to produce arachidonic acid *de novo*, but upon feeding this fatty acid, 3-HETE is formed. This reaction may be a paradigm for human-pathogenic fungi possessing similar metabolic and growth-regulatory patterns. The parasite may use the arachidonic acid released from host cells to convert to 3*R*-HETE, which, in turn, may affect host-parasite interactions in yet unknown ways. Work is currently in progress to elucidate the formation of 3*R*-HETE or other (3*R*)-hydroxy-oxylipins in human-pathogenic fungi which play an important clinical role in gynaecology and dermatology.

We also studied extensively the effects of 3*R*-HETE on human blood cells, in particular human neutrophils. 3*R*-HETE was found to be a strong chemotactic agent, the potency of which is comparable with those of LTB_4 or fMet-Leu-Phe. Unlike the latter mediators, however, 3*R*-HETE does not exert chemokinesis and exocytosis to a sizeable extent. Thus, 3*R*-HETE appears to be a comparatively specific stimulant of inflammatory cells. It augmented the release of arachidonic acid and of platelet-activating factor (PAF) *via* activation of phospholipase(s) A_2. The actions of 3*R*-HETE on neutrophils were observed to be directly correlated to an increase in the cytosolic calcium concentration, apparently resulting from the influx of calcium from extracellular medium. The cell signaling cascade triggered by 3*R*-HETE appears to imply G-protein-dependent processes. The actions of 3*R*-HETE on neutrophils will be a detailed subject of a separate communication. In summary, our data characterized the 3*R*-HETE as a unique cell stimulant, the spectrum of actions of which are different from all other eicosanoids studied so far.

ACKNOWLEDGMENTS

Authors wish to thank AICR (Ni 81025), FNK (Ni-97), FUB (Ni-FO) and FRD, RSA (JLFK) for supporting this study.

REFERENCES

Bhatt, R.K., Falck, J.R., and Nigam, S., 1998. Enantiospecific total chemical synthesis of a novel arachidonic acid metabolite 3-hydroxyeicosatetraenoic acid, *Tetrahedron Lett.* 39: 249.

Brodowski, I.D. and Oliw., E.H., 1992. Metabolism of 18:2(n-6), 18:3(n-3), 20:4(n-6) and 20:5(n-3) by the fungus *Gaeumannomyces graminis*: identification of metabolites formed by 8-hydroxylation and ω2 and ω3 oxygenation. *Biochim. Biophys. Acta* 1124:59.

Delaney, T.P., Ukness, S., Vernooij, B., Friedrich, L., Weymann, K., Negrotto, D., Gaffney, T., Gut-Rella, M., Kessmann, H., Ward, E., and Ryals, J., 1994. A central role of salicylic acid in plant disease resistance. *Science* 266:1247.

Gandhi, S.R. and Weete, J.D., 1991. Production of the polyunsaturated fatty acids arachidonic acid and eicosapentaenoic acid by the fungus *Pythium ultimum*. *J. Gen. Microbiol.* 137:1825.

Herman, P., 1998. Oxylipin production and action in fungi and related organisms. In: *Eicosanoids and Related Compounds in Plants and Animals*. A.S. Rowley, T. Schewe, and H. Kuhn, eds., Portland Press, London, UK (in press).

Herman, R.P. and Herman, C.A., 1985. Prostaglandins or prostaglandin like substances are implicated in normal growth and development in Oomycetes. *Prostaglandins* 29:819.

Jensen, E.C., Ogg, C. and Nickerson, K.W., 1992. Lipoxygenase inhibitors shift the yeast/mycelium dimorphism in *Ceratocystis ulmi. Appl. Environ. Microbiol.* 58:2505.

Kerwin, J.L., Simmons, C.A., and Washino, R.K., 1986. Eicosanoid regulation of oosporogenesis by *Lagenidium giganteum. Prostaglandins, Leukotrienes and Medicine* 23:173.

Kock, J.L.F., Jansen van Vuuren, D., Botha, A., Van Dyk, M.S., Coetzee, D.J., Botes, P.J., Shaw, N., Friend, J., Ratledge, C., Roberts, A.D., and Nigam, S. 1997. The production of biologically active 3-hydroxy-5,8,11,14-eicosatetraenoic acid (3-HETE) and linoleic acid metabolites by *Dipodascopsis, System. Appl. Microbiol.* 20:39.

Kock, J.L.F., Venter, P., Linke, D., Schewe, T., and Nigam, S. 1998. biological dynamics and distribution of 3-hydroxy fatty acids in the yeast *Dipodascopsis uninucleata* as investigated by immunofluorescence microscopy. Evidence for a putative regulatory role in the sexual reproductive cycle. *FEBS Lett.,* (in press).

Kurtzman, C.P., Versonder, R.F., and Smiley, M.J., 1974. Formation of extracellular 3-D-hydroxypalmitic acid by *Saccharmycopsis malanga. Mycologia* 66:582.

Losel, D.M. 1988. Fungal lipids. In: *Microbial Lipids Vol. 1,* C. Ratledge and S.G. Wilkinson, eds., Academic Press, London.

Pohl, C.H., Botha, A., Kock, J.L.F., Coetzee, D.J., Botes, P.J. and Nigam, S. 1998. Oxylipin formation in fungi. Retroconversion of arachidonic acid to linoleic acid and biotransformation to 3-hydroxy-5,8-tetradecadienoic acid by *Mucor genevensis,* (submitted).

Rattray, J.B.M. 1988. Yeasts. In: *Microgial Lipids, Vol. 1,* C. Ratledge and S.G. Wilkinson, eds., Academic Press, London.

Shinmen, Y., Shimizu, S., Akimoto, K., Kawashima, K. and Yamada, H. 1989. Production of arachidonic acid by *Mortierella* fungi. *Appl. Microbiol. Biotechnol.* 31:11.

Slusarenko, A.J., 1996. the role of lipoxygenase in plant resistance to infection. In: *Lipoxygenase and Lipoxygenase Pathway Enzymes,* G.J. Piazza, et al., eds. AOCS Press, Champaign, Illinois.

Van Der Berg, L. 1993. *Evidence for the Production of Arachidonic Acid Metabolites by Saccharomyces cerevisiae,* M.Sc Thesis, Department of Microbiology and Biochemistry, University of Orange Free State, Bloemfontein, South Africa.

Van Der Westhuizen, J.P.J., Kock, J.L.F., Botha, A. and Botes, P.J. 1994. The distribution of the ω3- ω6-series of cellular long-chain fatty acids in fungi. *System. Appl. Microbiol.* 17:327.

Van Dyk, M.S., Kock, J.L.F., Coetzee, D.J., Augustyn, O.P.H., and Nigam, S. 1991. Isolation of a novel arachidonic acid metabolite 3-hydroxy-5, 8, 11, 14-eicosatetraenoic acid (3-HETE) from the yeast *Dipodascopsis uninucleata* UOFS-Y128. *FEBS Lett.* 283:195.

Venter, P., Kock, J.L.F., Sravan Kumar, G., Botha, A., Coetzee, F.J., Botes, P.J., Bhatt, R.K., Falck, J.R., Schewe, T., and Nigam, S. 1997. Production of 3R-hydroxy-polyenoic fatty acids by the yeast *Dipodascopsis uninucleata. Lipids* 32:1277.

96

EICOSANOIDS IN THE BRAIN OF WARM- AND COLD-ACCLIMATED BULLFROGS

Ceil A. Herman and Georgia Luczy-Bachman

Department of Biology
New Mexico State University
Las Cruces, NM 88003

INTRODUCTION

Leukotrienes play important roles in numerous biological systems including the nervous system. They are synthesized from arachidonic acid, a 20-carbon fatty acid component of cell membrane phospholipids. Peptide leukotrienes (LT)C_4, LTD_4, and LTE_4, have been identified as the slow-reacting substance of anaphylaxis (SRS-A) and are mediators of asthma in mammals (Samuelsson et al., 1980). While LTC_4 shows high affinity binding in rat whole brain, LTB_4, LTD_4 and LTE_4 show very weak or no binding (Schalling et al., 1986). LTC_4 and LTD_4 induced prolonged excitation of Purkinje neurons in rat cerebellum (Palmer et al., 1980). LTC_4 is a potent agent for stimulating the dose-dependant release of luteinizing hormone (LH) and prolactin from rat pituitary cells and may also participate in a negative feed-back system for luteinizing hormone-releasing hormone (LHRH) (Hulting et al., 1985). LTC_4 binding sites in brain tissue have been characterized in guinea pig (Cheng & Townley, 1984), rat (Lindgren et al, 1984; Schalling et al., 1986), and mouse (Goffinet & Nguyen, 1987). Immunohistochemical localization of LTC_4 binding in rat brain identifies only the preoptic, hypothalamic and median eminence regions (Hulting et al., 1985).

Leukotriene C_4 is produced by bullfrog brain of both warm- and cold-acclimated animals (Martinez et al., 1994) and binding sites in bullfrog brain membranes have been characterized (Herman et al, 1992). Leukotriene C_4 receptors in bullfrog lung are affected by exposure to cold temperature (Herman et al, 1995). As in mammals, leukotrienes in amphibians may participate in the hypothalamus/pituitary hormone releasing mechanisms. The initiation of chorusing and establishment of territories by male bullfrogs is correlated with increased levels of LH (Mendonca et al., 1985). As mating activity is seasonal, LTC_4 receptor up- or down-regulation in response to temperature changes may be important in the reproductive cycle.

Eicosanoids and Other Bioactive Lipids in Cancer, Inflammation, and Radiation Injury, 4
Edited by Honn *et al.*, Kluwer Academic / Plenum Publishers, New York, 1999.

669

MATERIALS AND METHODS

Animals

American bullfrogs, *Rana catesbeiana,* (345.5 ± 13.7 g, mean ± S.E.M), of both sexes, were used in these studies. The frogs were collected locally or obtained from Charles Sullivan Company, Nashville, TN. Warm-acclimated frogs were fed beef liver three times a week and kept at room temperature (22°C) in tanks containing water. The cold-acclimated bullfrogs were fed once prior to being put into a refrigerated storage unit (5°C) in plastic boxes containing water for 4 wks. The difference in feeding is necessary due to the slow digestion of food at cold temperatures. The cold water was changed every 3 days.

Brain sectioning

Bullfrogs were anesthetized with MS-222 (0.3%) and the brains removed and immediately submerged in liquid nitrogen-cooled isopentane for approximately 1 min. The brains were removed and placed into liquid nitrogen-cooled vials and stored at -80°C until frozen sectioning began. 20 μm sagittal, horizontal and frontal sections were obtained using a frozen cryostat. Optimum slicing temperature was -17°C. Serial sections were mounted using Histobond (1%) for tissue adhesion and slides were allowed to dry for 3 hr at room temperature.

Incubation buffer was (mM)Tris-HCl, 20; $CaCl_2$, 3; $MgCl_2$, 5; NaCl, 1; DTT,1; pH 7.4. Serine borate (5mM) was added to the leukotriene C_4 (LTC_4) incubations to inhibit the conversion of LTC_4 to LTD_4 (Chiono et al., 1992). Tissues were incubated with [^{3}H]LTC_4, (1nM, .00013 μCi) and LTC_4 (1μM) was used as the competitor to [^{3}H]LTC_4. All chemical solutions and buffers were kept at 4° before and during the experiment. Only the radioligand incubation (30 min) was performed at room temperature. Sigma incubation chambers (200 μl) were used to contain treatment solutions on the slides

Prior to [^{3} H]LTC_4 radioligand incubation, all brain sections were rinsed in the buffer solution for 5 min to remove Histobond residue. Glass incubation chambers were positioned face up, filled with buffer and radioligand solution or buffer, radioligand and competitor. The slides were then placed on the filled chambers and automatically sealed with a thin rubber gasket around the edges of the chamber. After incubation, the slides were placed for 1 min in ice-cold buffer to remove any residue and then dipped in ice-cold double distilled water and let dry for1hr.

Autoradiography

Cardboard autoradiography cassettes containing Amersham Hyperfilm housed the tissue sections for 10 wks after which film was developed. Film was scanned with a Cohu high performance CCD video camera, and analyzed by NIH Image version 1.59. All tissue sections were subsequently stained with thionin and counter stained with eosin to identify neural bundle location on the films.

An autoradiographic standard curve with [³H]LTC₄ was use to calculate the concentration/cm² of tissue. Student's group t-test was used to test significance of the differences between the warm- and cold-acclimated animals. A difference of P<0.05 was considered significant.

RESULTS

Autoradiograpic studies and quantitation of binding of [³H]LTC₄ demonstrated that specific areas of the bullfrog brain showed dense binding. Results of a representative experiment are shown in Figure 1.

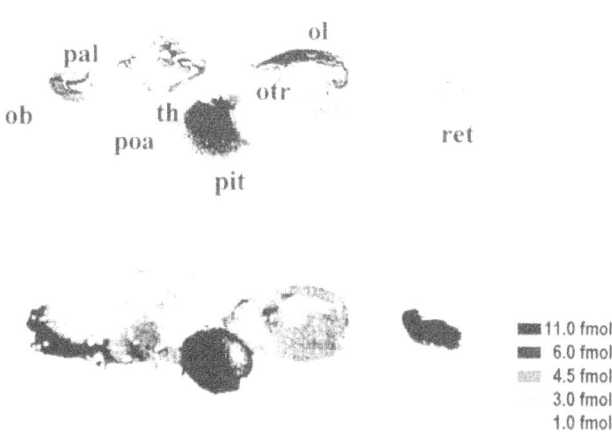

The top panel shows the sagittal histological tissue section (20 μm) on which the incubation was done. ob: olfactory bulb, pal: pallium, th: thalamus, poa: hypothalamic preoptic area, otr: optic tract, ot: optic tectum, pit: pituitary; ret: reticular formation. The bottom panel shows the sagittal section with dense binding of [³H]LTC₄ in the olfactory bulb, pituitary, and reticular formation. Values are expressed in fmol/cm².

Table 1 shows quantitation of [³H]LTC₄ receptor binding in warm and cold-acclimated bullfrog brains. Warm-acclimated anterior pituitary regions were significantly different from the same region in cold-acclimated frogs with higher maximum binding (19.5 ± 0.4 fmol/cm) than the cold-acclimated animals (16.3 ±0.2 fmol/cm). No significant differences were found in mean density of olfactory components, medulla or whole pituitary.

Table 1
Leukotriene C_4 Binding Using [^{3}H]-LTC$_4$ Binding
in Warm- and Cold-Acclimated Bullfrog Brain

Brain Area Binding	Warm-Acclimated	n	Cold-Acclimated	n
whole pituitary - mean	14.0 ± 0.3	3	13.1 ± 0.3	3
whole pituitary - maximum	22.7 ± 0.3	3	19.2 ± 0.3	3
anterior pituitary - maximum	19.5 ± 0.4	3	$16.3 \pm 0.2a$	3
olfactory bulb - mean	18.2 ± 0.3	3	16.1 ± 0.2	3
olfactory bulb - maximum	22.5 ± 0.2	3	21.3 ± 0.8	3
medulla - mean	10.2 ± 0.9	3	11.4 ± 1.2	3
medulla - maximum	16.8 ± 2.5	3	20.1 ± 0.3	3

Values represent mean $\pm$S.E.M. in fmol/cm^2
a= significantly different $p<.05$ by group T-test from corresponding warm-acclimated

DISCUSSION

This study found localization of the LTC$_4$ receptors in the bullfrog brain in areas which include olfactory components such as the nervous terminalis and olfactory bulb, the septal region, anterior pituitary gland and the brainstem regions. These areas have been shown to be involved in the gonadotropin-releasing hormone system of two anuran amphibians, *Rana pipiens and Hyla regilla*, suggesting that these two receptor systems are intricately linked to one another (Muske and Moore, 1988).

The localization of LTC$_4$ binding in bullfrog anterior pituitary is in agreement with reports that LTC$_4$ receptors are found in rat pituitary (Eberhardt et al, 1991) and that LTC$_4$ is a potent stimulator of luteinizing hormone (LH) release from rat pituitary cells in concentrations as low as 10^{-14} M (Lindgren et al., 1984; Hulting et al, 1985) There is good correlation between synthesis of leukotrienes and LH release following the administration of GnRH as well as the blockade of LH release using high doses of GnRH in rat pituitary cells (Hulting et al., 1985). Evidence suggests that arachidonic acid or its metabolites are involved in GnRH action on LH release in rat pituitary cells (Naor et al., 1985).

LTC$_4$ synthesis has been found in other regions of the bullfrog brain including cerebrum (including olfactory and nervus terminalis components), optic lobe, and medulla, which correlates to the moderate to dense binding of LTC$_4$ receptors shown in this study (Martinez et al., 1994). This endocrine system has its embryonic origins in both the olfactory (terminal nerve-septo-hypothalamic preoptic system) and brainstem regions (Muske, 1993). The presence of GnRH in the olfactory system suggests that the terminal nerve (TN) may respond to sexual pheromones and consequently regulate reproductive behavior. Additionally, regulation of the pituitary by GnRH from this region is achieved by TN secretion via its close contact with ventral surface blood vessels, and transported via circulation to the gonadotropic cells of the anterior pituitary, not from the hypothalamus by the hypophyseal-portal system (Muske, 1993).

Other LTC_4 binding areas have also been implicated in reproductive physiology. The septal region may have behavioral involvement, as well as being the possible relay station to the pineal gland with associations between GnRH activity and transmission of seasonal information (Maurel et al., 1992). Significance of the LTC_4 binding in the optic lobe, a sensory structure, is unknown. The brainstem is proposed to have been the earliest developing GnRH system in amphibians. Antibodies characterize brainstem GnRH as cGnRH II (Muske and Moore, 1987; 1988), and possibly these neurons serve a non-secretory function because their fibers also project to the midbrain. Amphibians may receive dual regulation by both direct involvement with the LH neurons in the pituitary and by associating closely with the GnRH neuronal system in the forebrain and hindbrain.

There were no acclimation temperature-related changes in LTC_4 receptor distribution of the olfactory components, whole pituitary, and medulla in bullfrog brain. However, distribution of the most dense binding areas in the anterior pituitary show significantly greater binding in warm-acclimated bullfrog pituitaries. Localization of LH in anterior pituitary has been shown in *Rana pipiens* (Gracia-Navarro and Licht, 1987) and the magnitude of response of LH release to GnRH stimulation was negatively affected in a dose dependant manner to lower temperatures (Porter and Licht, 1986). With receptor density diminished in the anterior pituitary of the cold-acclimated bullfrog, responsiveness may also decrease. More investigation to link frog LH release from the pituitary due to LTC_4 stimulation is needed.

ACKNOWLEDGMENTS

This work was supported by NSF Grant IBN-9318047

REFERENCES

Cheng, J.B. and Townley, R.G., 1984, Effect of serine borate complex on the relative ability of leukotriene C_4, D_4, and E_4 to inhibit lung and brain tritium-labeled leukotriene D_4 and tritium labelled leukotriene C_4 binding: demonstration of the agonist potency order for leukotriene D_4 and leukotriene C_4 receptors, Biochem. Biophys. Res. Commun. 119:612.

Chiono, M., Andazola, J.J., Torres, O.A., and Herman, C.A., 1992, Characterization and localization of leukotriene C_4 receptors in bullfrog lung. J. Pharmacol. Exp. Ther. 262:1248.

Eberhardt, I., Kiesel, L., Rosenberg, K., Klinga, K., and Runnebaum, B., 1991, Characterization of leukotriene C_4 binding in anterior pituitary membrane preparations, Prostaglandins 41:185.

Gracia-Navarro, F., 1989, Immunocytochemical and ultrastructural study of the frog (Rana pipiens) pars distalis with special reference to folliculo-stellate cell function during in vitro superfusion. Cell Tissue Res. 256:623.

Goffinet, A.M. and Nguyet, A., 1987, Leukotriene C_4 binding sites in mouse brain:pharmacological characteristics. Eur. J. Pharmacol. 140:343.

Herman, C.A., Charlton, G.A., and Chiono, M., 1992. Characterization of leukotriene C_4 binding sites in the brain of the American bullfrog, *Rana catesbeiana*. J. Exp. Zool. 262:1

Herman, C.A., Skarda, S., Romero, M.A., Chapunoff, D., Schulmeister, K., and Torres, O.A., 1995, Leukotriene C_4 stimulated contractions in bullfrog lung are affected by cold-acclimation and calcium antagonists. Prostaglandins 49:117.

Herman, C.A., Luczy, G., Wikberg, J.E.S. and Uhlén, S., 1996, Characterization of adrenoceptor types and subtypes in American bullfrogs acclimated to warm or cold temperature. Gen. Comp. Endocrinol. 104:168.

Hulting, A., Lindgren, J.Å., Hökfelt, T., Eneroth, P., Werner, S., Patrono, C., and Samuelsson, B., 1986. Leukotriene C_4 as a mediator of leuteinizing hormone release from rat pituitary cells, Proc. Natl. Acad. Sci, USA. 82:3834.

Lindgren, J.Å., Hökfelt, T., Dahlen, S.E., Patrono, C., and Samuelsson, B., 1984, Leukotrienes in the rat central nervous system. Proc. Natl. Acad. Sci. USA., 81:6212.

Martinez, J.M., Chapunoff, D., Romero, M.A., and Herman, C.A., 1994. Eicosanoid synthesis by warm- and cold-acclimated American bullfrog (*Rana catesbeiana*) brain, J. Exp. Zool. 269:298.

Maurel, D., Boissin-Agasse, L., Roch, G., and Boissin, J., 1992, Short-day stimulation of testicular activity and immunoreactivity of the hypothalamic GnRH system in mink following deafferentation of the pineal body by bilateral superior cervical ganglionectomy and melatonin replacement. Brain Res. 578:99.

Mendonca, M.T., Licht, P., Ryan, M.J., and Barnes, R., 1985, Changes in hormone levels in relation to breeding behavior in male bullfrogs (*Rana catesbeiana*) at the individual and population levels. Gen. Comp. Endocrinol. 58:270.

Muske, L.E. 1993, Evolution of gonadotropin-releasing hormone (GnRH) neuronal systems. Brain Behav. Evol. 42:215.

Muske, L.E. and Moore, F.L., 1987, Luteinizing hormone-release hormone-immunoreactive neurons in the amphibian brain are distributed along the course of the nervus terminalis. Ann. N.Y. Acad. Sci. 519:433.

Muske, L.E. and Moore, F.L., 1988, The amphibian nervus terminalis: anatomy, chemistry and relationship with the hypothalamic LHRH system. Brain Behav. Evol.. 32:141.

Naor, Z., Kiesel, L., Vangerhoek, J.Y., and Catt, K.J., 1985, Mechanism of action of gonadotropin releasing hormone: role of lipoxygenase products of arachidonic acid in luteinizing hormone release. J. Steroid Biochem. 23:711.

Palmer, M.R., Mathews, W.R., Murphy, R.C., and Hoffer, B.J., 1980, Leukotriene C elicits a prolonged excitation of cerebellar Purkinje neurons. Neurosci. Lett. 18:173.

Porter, D.A. and Licht, P., 1986, Effects of temperature and mode of delivery on responses to gonadotropin-releasing hormone by superfused frog (*Rana pipiens*) pituitaries. Gen. Comp. Endocrinol. 63:236.

Samuelsson, B., Hammarstrom, S., Murphy, R.C., and Borgeat, P., 1980, Leukotrienes and SRS-A. Allergy 35:375.

Schalling, M., Neil, A., Terenius, L., Lindgren, J.Å, Miamoto, T., Hökfelt, T., and Samuelsson, B., 1986, Leukotriene C_4 binding sites in the rat central nervous system. Eur. J. Pharmacol. 122:251.

97

PRODUCTION OF 3-HYDROXY FATTY ACIDS BY THE YEAST *DIPODASCOPSIS UNINUCLEATA*. BIOLOGICAL IMPLICATIONS

J. Lodewyk F. Kock,[1] Pierre Venter,[1] Alfred Botha,[1] Dennis J. Coetzee,[1] Pieter W.J. van Wyk,[2] Dandré P. Smith,[1] Tankred Schewe,[3] and Santosh Nigam[3]

[1] Department of Microbiology and Biochemistry
[2] Department of Botany and Genetics
   University of the Orange Free State
   P.O. Box 339, Bloemfontein, 9300,
   South Africa
[3] Eicosanoid Research Division
   Department of Gynaecology
   University Medical Centre Benjamin Franklin
   Free University Berlin, D-12200 Berlin, Germany

INTRODUCTION

Hydroxy fatty acids (OH-FAs) including 3-OH-FA are widely distributed in micro-organisms (Schweizer, 1989). Thus, the production of 3-OH-FA by fungi has been reported in various studies (Van Dyk et al., 1994). Using thin-layer chromatography, autoradiography, ^{1}H-NMR as well as gas chromatography-mass spectrometry, we isolated and identified the novel 3-hydroxyeicosanoid (3R)-hydroxy-(5Z,8Z,11Z,14Z)-eicosatetraenoic acid (3-HETE) from the yeast *Dipodascopsis uninucleata* following incubation with arachidonic acid (Van Dyk et al., 1991; Venter et al., 1997). This compound now turned out to be the prototype of the new family of fungal unsaturated bioactive (3R)-hydroxy-oxylipins (see Nigam et al., this volume).

Since *D. uninucleata* is not capable of producing arachidonic acid *de* novo, the endogenous oxylipin formed by this yeast is most likely (3*R*)-hydroxy-(5*Z*,8*Z*)-tetradecadienoic acid (3-HTDE) which was shown to be formed from linoleic acid (Venter et al., 1997). Recent investigations using immunofluorescence microscopy provided ample evidence that (3*R*)-hydroxy-oxylipins are produced during the ascosporogenesis stage of the sexual reproductive cycle of *D. uninucleata* (Kock et al., 1998) which substantiates earlier suggestions implicated for 3-OH-FAs in this process (Botha and Kock, 1993).

BIOTECHNOLOGICAL PRODUCTION OF (3*R*)-HYDROXY-OXYLIPINS

3-HETE has been found to exert potent biological activities in mammalian cells (Nigam et al., 1997; Nigam et al., this volume). The detailed elucidation of its mode of action requires sufficient quantities of this compound and related (3*R*)-hydroxy-oxylipins. Therefore, the production of 3-HETE by the yeast *D. uninucleata* has been optimised in such a way that approximately 4% of the arachidonic acid added is transformed to 3-HETE (Kock et al., 1997). By feeding other precursor polyenoic fatty acids to the yeast, other homologues can be obtained (Venter et al., 1997).

MAPPING (3*R*)-HYDROXY-OXYLIPINS *IN SITU*

It was of interest to know the exact location of the formation of (3*R*)-hydroxy-oxylipins during the course of the life cycle of *D. uninucleata*. For this purpose, an antibody was raised against authentic (3*R*)-HETE, which fortunately revealed comparable immunoreactivity against 3-HTDE and other (3*R*)-OH-FAs. This group-reactivity allowed us to localise by mapping all 3R-OH FAs during the life cycle of *D. uninucleata* with the help of immunofluorescence microscopy using fixed cells with and without cell walls (Kock et al., 1998). As a result of this investigation, (3*R*)-OH-FAs, probably 3-HTDE, were found to be associated with the sexual reproductive cycle and were visualised in the gametangia, asci and between the liberated ascospores, *i.e.* interspore matrix (Figure 1A). A scanning electron micrograph of the interspore matrix is shown in Figure 1B.

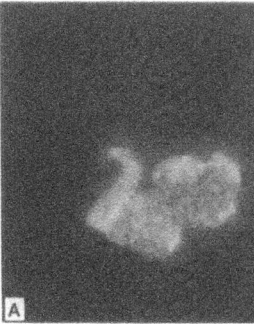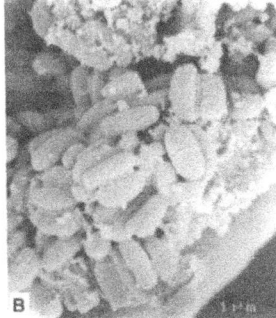

Figure 1. A: Cluster of ascospores released from ascus of *Dipodascopsis uninucleata* and fluorescing interspore matrix visualised through immunofluorescence mapping (10 mm = 10 μm cell size). B: Scanning electron micrograph (SEM) of ascospore cluster demonstrating interspore matrix.

These observations may be useful to further optimise the production of 3-HETE and related compounds as well as to elucidate the metabolism and function of endogenously formed (3R)-hydroxy-oxylipins in D. uninucleata. It is conceivable that these oxylipins might have a function in gametangiogamy, ascosporogenesis and in attaching these ascospores in probably „sticking clusters" for the effective distribution, for instance, by insects. Since the (3R)-hydroxy-oxylipins do not seem to be associated with the cell surface (i.e. cell wall and cell membrane) or the cytoplasm of the hyphae, but are rather localised in the interspore matrix, it is tempting to speculate that the depletion of these oxylipins is causative for the separation of spores from each other prior to germination.

It should be emphasised that a similar role in the life cycle of fungi has also been reported for other oxylipins, such as psi (precocious sexual inducers) factors (Herman, 1998).

ACKNOWLEDGMENTS

The authors wish to thank the Association for International Cancer Research, UK (Ni-81025), Free University Berlin and the Foundation for Research Development, South Africa.

REFERENCES

Botha, A., and Kock, J.L.F., 1993, The distribution and taxonomic value of fatty acids and eicosanoids in the Lipomycetaceae and Dipodascaceae. *Ant. v. Leeuwenhoek* 63: 111-123.

Herman, P., 1998, Oxylipin production and action in fungi and related organisms. In: *Eicosanoids and Related Compounds in Plants and Animals*, A.S. Rowley, T. Schewe and H. Kühn, eds, Portland Press, London, UK, in press.

Kock, J.L.F., Jansen van Vuuren, D., Botha, A., Van Dyk, M.S., Coetzee, D.J., Botes, P.J., Shaw, N., Friend, J., Ratledge, C., Roberts, A.D., and Nigam, S., 1997, The production of biologically active 3-hydroxy-5,8,11,14-eicosatetraenoic acid (3-HETE) and linoleic acid metabolites by *Dipodascopsis. System. Appl. Microbiol.* 20: 39-49.

Kock, J.L.F., Venter, P., Linke, D., Schewe, T., and S. Nigam, 1998, Biological dynamics and distribution of 3-hydroxy fatty acids in the yeast *Dipodascopsis uninucleata* as investigated by immunofluorescence microscopy. Evidence for a putative regulatory role in the sexual reproductive cycle. *FEBS Lett.*, in press.

Nigam, S., Goparaju, S.K., Fox, S., Friend, J., and Kock, J.L.F., 1997, Signal transducing effects of 3-HETE, a novel compound from the yeast *Dipodascopsis uninucleata*, on human neutrophils, 5th International Conference on Eicosanoids and Other Bioactive Lipids in Cancer, Inflammation and Related Diseases, La Jolla, CA, Sept. 17-20, 1997 (Abstract #203).

Schweizer, E., 1989, Biosynthesis of fatty acids and related compounds. in: *Microbial Lipids Vol. 2*, C. Ratledge, and S.G. Wilkinson, eds, pp. 3-50, Academic Press, San Diego.

Van Dyk, M.S., Kock, J.L.F., and Botha, A., 1994, Hydroxy long-chain fatty acids in fungi. *World J. Microbiol. Biotech.* 10: 495-504.

Van Dyk, M.S., Kock, J.L.F., Coetzee, D.J., Augustyn, O.P.H., and Nigam, S., 1991, Isolation of a novel arachidonic acid metabolite 3-hydroxy-5,8,11,14-eicosatetraenoic acid (3-HETE) from the yeast *Dipodascopsis uninucleata. FEBS Lett.* 283: 195-198.

Venter, P., Kock, J.L.F., Sravan Kumar, G., Botha, A., Coetzee, D.J., Botes, P.J., Bhatt, R.K., Falck, J.R., Schewe, T., and Nigam, S., 1997, Production of 3R-hydroxy-polyenoic fatty acids by the yeast *Dipodascopsis uninucleata. Lipids* 32: 1277-1283.

98

CATALYTIC PROPERTIES OF LINOLEATE DIOL SYNTHASE OF THE FUNGUS
GAEUMANNOMYCES GRAMINIS: A COMPARISON WITH PGH SYNTHASES

Ernst H. Oliw, Chao Su, and Margareta Sahlin[1]

Department of Pharmaceutical Biosciences, Uppsala
Biomedical Center, Uppsala, Sweden, and [1]Department
of Molecular Biology, Arrhenius Laboratories,
Stockholm University, Stockholm, Sweden

INTRODUCTION

The fungus *Gaeumannomyces graminis* causes "take-all", a devastating root disease of wheat. Our interest in *G. graminis* is due to its biosynthesis of oxylipins, which might be important for its reproduction. The fungus can sequentially metabolize linoleic acid (18:2n-6) to 8R-hydroperoxylinoleic acid (8-HPODE) via an 8R-dioxygenase and to 7S,8S-dihydroxylinoleic acid (7,8-DiHODE) via a hydroperoxide isomerase[1]. A highly purified protein with both enzymes activities was recently found to be a hemoprotein[2]. The protein had an apparent molecular size of 130 kDa on SDA-PAGE and it appeared to be a tetramer on gel filtration. The holoprotein will be referred to as linoleate diol synthase (LDS).

Polyunsaturated fatty acid can be oxygenated by two major classes of dioxygenases: lipoxygenases with non-heme metal centers and hemoproteins (PGH synthases, LDS). Dioxygenases can introduce oxygen into their substrates in two ways. Lipoxygenases and PGH synthases activate the fatty acid by proton abstraction and the carbon centered radical reacts with oxygen[3]. PGH synthases also have peroxidase activity, which initiates the cyclooxygenase activity by generating a Tyr radical, which abstracts the proton. Other oxygenases, e.g., the ferrous hemoprotein L-tryptophan 2,3-dioxygenase[4], activate oxygen, which then reacts with the substrate.

The main objective of the present study was to determine whether LDS contained ferrous or ferric iron, to study the effects of hydroperoxides and other agents on LDS and to determine whether incubation of LDS with 18:2n-6 or 8-HPODE generated protein radicals.

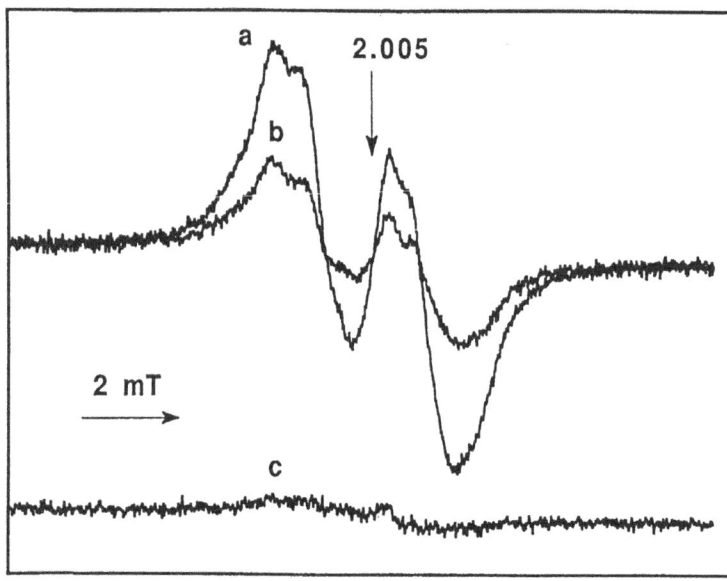

Figure 1. EPR spectra of linoleate diol synthase. The spectra were recorded at 77 K and the samples were frozen following incubation of enzyme with substrate for 3 s or for 45 s on ice. The trace (a) marks the 3 s sample, trace (b) the 45 s sample, and trace (c) enzyme without added substrate. The vertical arrow marks g=2.005. Recording conditions: microwave frequency 9.3 MHz, microwave power 2 mW, modulation 0.16 mT, receiver gain 8e5, 16 scans. In the 3 s sample, the diol synthase concentration was about 4.9 µM (holoenzyme), the 18:2n-6 concentration 0.1 mM, and the radical concentration was estimated to be about 3.2 µM.

EXPERIMENTAL

LDS was purified from mycelia of *G. graminis* as described[1,2]. Spectroscopy was performed with a dual beam spectrophotometer (Shimadzu UV-2101PC). High and low spin derivatives were prepared by addition of a few grains of NaF, NaN3, or KCN directly to the sample cuvette[5]. A TLC assay[1,2] or an oxygen electrode was used to measure enzyme activity. EPR spectra were recorded with an EPR spectrometer (ESP 300, Bruker) at 10 K (with an Oxford-Instrument cryostat) or at 77 K (with a cold finger Dewar). The radical content was measured by comparing double integrals with that from a 1mM Cu-10 mM EDTA standard or by using a secondary standard of protein R2 ribonucleotide reductase from *Escherichia coli* with known tyrosyl radical content. Power of half saturation ($P_{1/2}$) was evaluated from microwave saturation curves as described[6]. About 0.4-0.9 mg of LDS (specific activity 1-2 μmol min-1 mg-1 protein) was incubated with 0.3 mM 18:2n-6 or 0.1 mM 8-HPODE in 140 μl for 3 s or for 45 s on ice, immediately frozen (in pre-cooled n-pentane at -105°C) and analyzed by EPR.

RESULTS

Light absorption and EPR

LDS absorption light with maxima at about 630 nm, 566 nm (α), 528 nm (β), and 406 nm (γ)[2]. The absorption maxima at 630 nm and at 406 nm suggested a ferric hemoprotein in a high spin state[5]. Ferric iron of hemoproteins can be transformed into low and high spin states by ions and pH changes[5]. CN- and N_3^- formed low spin derivatives of LDS with γ bands at 421 nm and 411 nm, respectively. F- can induce a high spin state, but F- had little effect on LDS. High spin is also favoured at low pH and low spin under alkaline conditions. At pH 6.5 (0.1 M KHPO₄ buffer) the spectrum was essentially unchanged but the γ band was reduced in intensity and shifted to 412 nm at pH 9.5 (in 0.1 M glycine buffer). These observations suggested that LDS was ferric and in a high spin state.

EPR at 10 K confirmed the presence of mainly high spin ferric iron in LDS from the resonance at g=6.3 (Fig. 1). EPR studies of the 3 s and 45 s samples incubated with 18:2n-6 showed that a radical was formed. Analysis of the 3 s sample at 77 K revealed an EPR doublet hyperfine pattern with characteristics of a major splitting at 2.3 mT and minor, poorly resolved splittings of ca 0.6 mT (data not shown). The signal was centered at g=2.005. EPR analysis of the 45 s sample displayed the doublet but to a lower content. However, an underlying contribution from a second species was obvious from the change of line-shape in the spectrum. The radical thus appeared to be transient. Work in progress suggests that similar protein radicals were formed during 3 s and 45 s incubation of LDS with 8-HPODE.

$P_{1/2}$ is a measure of the relaxation behaviour of the unpaired electron and provides an indication of the interaction between a free radical and a metal center is given by $P_{1/2}$. We measured $P_{1/2}$=18 mW for LDS, assuming inhomogenous broadening (b=1). This is in the same order of magnitude as the tyrosyl radicals of PGH synthase-1 by Lassmann et al.[7] and the tyrosyl radical in the protein R2 of ribonucleotide reductase from *E. coli*, which was determined in a parallel measurement. The tyrosyl radical is located within 10 Å from the heme iron center of PGH synthase-1[8] and about 6 Å from the dinuclear ferric iron center of protein R2[9]. The results suggest that the heme iron of LDS is within 10Å from the protein radical.

Inhibition and Activation of LDS

Biosynthesis of 8-HPODE by the purified protein occurred with a time lag of about 10s (oxygen electrode). Addition of 8-HPODE reduced the time lag but 13-HPODE was without effect, whereas large concentrations of GSH peroxidase and GSH increased the lag time and could totally block the dioxygenase and isomerase activities.

Tetranitromethane (TNM), which reacts with Tyr radicals, is a potent time-dependent inhibitor of the cyclooxygenase reaction[3]. We investigated the effect of TNM on LDS. TNM at 0.4 mM caused about 50% inhibition of LDS activity. Diethylpyrocarbonate (DEPC) is supposed to react with histidine ligands and can inhibit cyclooxygenase[10]. DEPC also reduced the LDS activity. His and Tyr residues might thus be important for the catalytic activity of LDS. Some detergents augmented the enzyme activity significantly. The effects of Tween-20 (0.04%) and CHAPS (0.5 mM) were most striking. Addition of heme was without effect on LDS activity.

DISCUSSION

Spectroscopy of LDS suggested that it contained ferric heme in a high spin state. The first step of LDS catalysis is likely substrate activation by hydrogen abstraction in analogy with PGH synthases and lipoxygenases. PGH synthases and lipoxygenases are activated by hydroperoxides[3], e.g., by 13S-hydroperoxylinoleic acid (13-HPODE). We found that 13-HPODE did not reduce the lag time of LDS. 13-HPODE is not substrate for the hydroperoxide isomerase, while 8-HPODE is. The finding that 8-HPODE can reduce the lag time, indicating a mechanistic connection between the 8R-dioxygenase and hydroperoxide isomerase activities.

EPR demonstrated that a protein radical was formed during LDS catalysis. EPR analysis at 77 K showed a doublet hyperfine pattern at 3 s, which was reduced in intensity at 45 s. The relaxation properties of the protein radical indicated that the radical was located near the heme group, presumably within 10 Å. However, a radical at g=2.00 could conceivably also be a substrate radical, but we excluded this possibility from EPR of linoleate radicals[11].

Protein radicals have been described in a handful of biological systems. Tyrosyl radicals have been studied in greatest detail. They are found in aerobic ribonucleotide reductase[9], PGH synthase-1[3, 12], photosystem II[13], and bovine liver catalase[14]. Glycyl and tryptophan radicals have also been described, e.g., transient tryptophan radicals in cytochrome c peroxidase[15] and glycyl radicals in anaerobic ribonucleotide reductases[16]. These radicals can be distinguished from tyrosyl radicals by EPR[16, 17].

The main splitting in the EPR spectrum of a protein radical seems to be due to coupling of the unpaired electron to one or both of the β-methylene hydrogens of the radicalized amino acid. In Tyr the spin density will be delocalized over the ring system. A large portion is found at C-1, which will couple to the β-protons. The ring protons at C-3 and C-5 of Tyr are almost equivalent and will give rise to hyperfine splitting of ca 0.6 mT[18]. The LDS doublet intermediate seems most similar to tyrosyl radicals. Both the g-values, the magnitude of the hyperfine splitting and also the poorly resolved 0.6 mT splitting, which could be ascribed to the C-3 and C-5 ring protons, are similar in tyrosyl radicals of ribonucleotide reductase. A tyrosyl radical is therefore a likely candidate for the intermediate of LDS. However, a conclusive identification by EPR will require isotope labeling of the Tyr residues of recombinant LDS.

PGH synthase-1 forms protein radicals with very similar EPR characteristics as tyrosyl radicals[19]. Lassmann et al.[7] compared the PGH doublet to the tyrosyl radical in aerobic ribonucleotide reductase and concluded that PGH synthase-1 formed a tyrosyl radical. In addition, a radical at Tyr[385] is assumed to be catalytically important[20]. This assignment is based on the 3D structure of PGH synthase-1 and other observations[12]. TNM is, for example, a selective agent for the nitration of tyrosyl residues and an inhibitor of the cyclooxygenase reaction[3, 21].

TNM reduced the LDS activity in concentrations used to inhibit PGH synthase-1. This result suggests that tyrosyl residues may be important during LDS catalysis as well.

How is the tyrosine radicals of LDS generated? It seems likely that the formation is related to oxidation of the ferric heme iron during the hydroperoxide isomerase reaction. A series of observations suggest that LDS could be mechanistically related PGH synthase-1 in this respect. First, exogenous 8-HPODE both stimulates LDS activity, is metabolized by the hydroperoxide isomerase, and generates a protein radical. In addition, PGH synthase-1 and LDS are both inhibited by GSH peroxidase and both enzymes are also inactivated during catalysis by suicide inactivation. Finally, work in progress shows that a partial amino acid sequence of LDS has significant homology with a peroxidase (Hörnsten, Su, Hellman, Osbourn, and Oliw, unpublished observation) and stopped-flow experiments indicate formation of ferryl oxygen intermediates during catalysis (unpublished observations).

PGH synthase-1 will oxygenate 20:3n-9 to its main metabolite, 13-hydroxyeicosatrienoic acid, in an "abortive" cyclooxygenase reaction[21] in analogy with dioxygenation of 18:2n-6 by LDS. After these initial oxygenations of the monoallylic carbons, however, there is an obvious difference. LDS has hydroperoxide isomerase and PGH synthase-1 has peroxidase activity. The ferryl oxygen of PGH synthase-1 is usually reduced to water and ferric heme iron, but the ferryl oxygen of LDS is not. The ferryl oxygen of LDS is likely reduced to ferric heme iron during biosynthesis of 7,8-DiHODE.

In summary, our observations suggest that the two enzyme activities of LDS could be described by the following catalytic cycle:

1. Reduction of 8-HPODE to 8R-HODE and the formation of ferryl oxygen and the tyrosyl radical.
2. The ferryl oxygen inserts oxygen at C-7 of 8-HODE so that 7,8-DiHODE and ferric heme iron are formed.
3. The tyrosyl radical initiates a new cycle by eliminating the 8-*pro*-S hydrogen from 18:2n-6, the carbon-centered radical reacts with molecular oxygen and forms a peroxyl radical.
4. 8-HPODE is formed from the peroxyl radical and a tyrosyl radical regenerated.

ACKNOWLEDGMENTS

Supported by the Swedish Medical Research Council (06523), MBS (to E.O. and S.U., and to M.S.) and the Swedish Cancer Society (to B-M. Sjöberg for M.S.).

REFERENCES

1. ID. Brodowsky, M. Hamberg, and E.H. Oliw, A linoleic acid (8R)-dioxygenase and hydroperoxide isomerase of the fungus *Gaeumannomyces graminis*, *J. Biol. Chem.* 267:14738 (1992).
2. C. Su, and E.H. Oliw, Purification and characterization of linoleate 8-dioxygenase from the fungus *Gaeumannomyces graminis* as a novel hemoprotein, *J. Biol. Chem.* 271:14112 (1996).
3. W.L. Smith, and L.J. Marnett, Prostaglandin endoperoxide synthase: structure and catalysis. *Biochim. Biophys. Acta* 1083:1 (1991).

4. J.M Leeds, P.J. Brown, G.M. McGeehan, F.K., Brown, and J.S. Wiseman, Isotope effects and alternative substrate reactivities for tryptophan 2,3-dioxygenase, *J. Biol. Chem.* 268:17781 (1993).

5. E. Antonini, and M. Brunori. *Hemoglobin and Myoglobin in Their Reactions with Ligands.* North-Holland Publishing Co., Amsterdam (1971).

6. M. Sahlin, L. Peterson, A. Gräslund, A. Ehrenberg, B.M. Sjoberg, and L.Thelander, Magnetic interaction between the tyrosyl free radical and the antiferromagnetically coupled ion center in ribonucleotide reductase, *Biochemistry* 26:5541 (1987).

7. G. Lassmann, G., R. Odenwaller, J.F. Curtis, J.A. DeGray, R.P Mason, L.J. Marnett, and T.E. Eling, Electron spin resonance investigation of tyrosyl radicals of prostaglandin H synthase. Relation to enzyme catalysis, *J. Biol. Chem.* 266:20045 (1991).

8. D. Picot, P.J. Loll, and R.M. Garavito, The X-ray crystal structure of the membrane protein prostaglandin H2 synthase-1, *Nature* 367:243 (1994).

9. B.M. Sjöberg, The ribonucleotide reductase jigsaw puzzle: A large piece falls into place, *Structure* 2, 793 (1994).

10. A. Tsai, L.C. Hsi, R.J. Kulmacz, G. Palmer, and W.L. Smith, Characterization of the tyrosyl radicals in ovine prostaglandin H synthase-1 by isotope replacement and site directed mutagenesis, *J. Biol. Chem.* 269:5085 (1994).

11. W. Chamilutrat, and R.P. Mason, Lipid peroxyl radical intermediates in the peroxidation of polyunsaturated fatty acids by lipoxygenase. Direct electron spin resonance investigations, *J. Biol. Chem.* 264:20968 (1989).

12. W.L. Smith, R-M. Garavito, and D.L. DeWitt, Prostaglandin endoperoxide H synthases (cyclooxygenases)-1 and -2, *J. Biol. Chem.* 271:33157 (1996).

13. C.W. Hoganson, and G.T. Babcock, Protein-tyrosyl radical interactions in photosystem II studied by electron spin resonance and electron nuclear double resonance spectroscopy: Comparison with ribonucleotide reductase and in vitro tyrosine, *Biochemistry* 31:11874 (1992).

14. A. Ivancich, H.M. Jouve, and J. Gaillard, EPR evidence for a tyrosyl radical intermediate in bovine liver catalase, *J. Am. Chem. Soc.* 118:12852 (1996).

15. M. Sivaraja, D.B. Goodin, M. Smith, and B.M. Hoffman, Identification by ENDOR of Trp191 as the free-radical site in cytochrome c peroxidase compound ES, *Science* 245:738 (1989).

16. P.Young, J. Andersson, M. Sahlin, and B-M. Sjöberg, Bacteriophage T4 anaerobic ribonucleotide reductase contains a stable glycyl radical at position 580, *J. Biol. Chem.* 271:20770 (1996).

17. M. Sahlin, G. Lassmann, S. Pötsch, A. Slaby, B-M. Sjöberg, and A. Gräslund, Tryptophan radicals formed by iron/oxygen reaction with Escherichia coli ribonucleotide reductase protein R2 mutant Y122F, *J. Biol. Chem.* 269:11699 (1994).

18. J.A. DeGray, G. Lassmann, J.F. Curtis, T.A. Kennedy, L.J. Marnett, T.E. Eling, and R.P. Mason, Spectral analysis of the protein-derived tyrosyl radicals from prostaglandin H synthase, *J. Biol. Chem.* 267:23583 (1992).

19. R. Karthein, R. Dietz, W. Nastainczyk, and H.H. Ruf, Higher oxidation states of prostaglandin H synthase. EPR study of a transient tyrosyl radical in the enzyme during the peroxidase reaction, *Eur. J. Biochem.* 171:313(1988).

20. A-L. Tsai, R.J. Kulmacz, and G. Palmer, Spectroscopic evidence for reaction of prostaglandin H synthase-1 tyrosyl radical with arachidonic acid, *J. Biol. Chem.* 270: 10503 (1995).
21. E.H. Oliw, L. Hörnsten, H. Sprecher, and M. Hamberg, Oxygenation of 5,8,11-eicosatrienoic acid by prostaglandin endoperoxide synthase and by cytochrome P450 monooxygenase: structure and mechanism of formation of major metabolites, *Arch. Biochem. Biophys.* 305:288 (1993).

CONTRIBUTORS

Martina Andberg
Department of Medical Biochemistry and Biophysics
Karolinska Institutet, S-171
77 Stockholm, Sweden

Kenning M. Anderson
Dept. of Medical Oncology
Rush Medical College
Chicago, IL 60612

Alaa F. Badawi
Laboratory of Pharmacogenetics
Clinical Pharmacology
Altantic Veterinary College
550 University Avenue
Charlottetown, Prince Edward Island
Canada C1A 4P3

Carsten T. Beuckmann
Department of Molecular Behavioral Biology
Osaka Bioscience Institute
6-2-4 Furuedai, Suita,
Osaka, 565-0874, Japan

Barbara A. Berglund
Department of Pharmacological and Physiological Science
Saint Louis University School of Medicine
1402 South Grand
St. Louis, MO 63104

S. Borngraber
Institute of Biochemistry
University Clinics Charite
Humboldt University
Hessische Str. 3-4
10115 Berlin, Germany

Alan R. Brash
Department of Pharmacology
Vanderbilt University Medical Center
Nashville, TN 37232-6602

Michael R. Buchanan
Department of Pathology
McMaster University
Hamilton, Ontario L8N 3Z5

Valerie Capra
Laboratory of Molecular Pharmacology
Institute of Pharmacological Sciences
University of Milan
Via Balzaretti 9
20133 Milan, Italy

Antonio Celardo
Istituto di Ricerche Farmacologiche "Mario Negri"
Department of Vascular Medicine and Pharmacology
"G. Bizzozero" Laboratory of Platelet and Leukocyte Pharmacology
Consorzio Mario Negri Sud
66030 S. Maria Imbaro, Italy

Jerold Chun
Department of Pharmacology
Neurosciences and Biomedical Sciences Program
School of Medicine
University of California, San Diego
9500 Gilman Drive
La Jolla, CA 92093-0636

Marina Dadaian
Department of Woman and Child Health
Division of Reproductive Endocrinology
Karolinska Hospital, Bldg. L5
S-171 76 Stockholm, Sweden

Pallavi R. Devchand
Institut de Biologie Animale
Universite de Lausanne
Batiment de Biologie
CH 1015, Lausanne
Switzerland

P. Dieter
TU Dresden
Institut fur Physiologische Chemie
Dresden, Germany

David M. Dirig
Department of Pharmacology
University of California, San Diego
La Jolla, CA 92093

Christine Eberhardt
ICOS Corporation
Bothell, WA 98021

Yang-Yi Fan
Faculty of Nutrition and Molecular and Cell
Biology Group
Texas A&M University
College Station, TX 77843

Jean Pierre Gascard
CNRS ERS 566
Centre Chirurgical Marie Lannelongue
133 av. de la Resistance
92350 Le Plessis Robinson
France

Jagadananda Ghosh
University of Virginia Cancer Center
Charlottesville, VA 22908

Geoffrey D. Girnun
Radiation Research Laboratory
Department of Radiology
University of Iowa
Iowa City, IA 52252

Wayne C. Glasgow
Division of Basic Medical Sciences
Mercer University School of Medicine
Macon, Georgia 31207

Nuria Godessart
Research Center
Almirall Prodesfarma
Cardener 68-74
08024 Barcelona, Spain

Edward J. Goetzl
Departments of Medicine and Microbiology-
Immunology
University of California Medical Center
San Francisco, CA 94143-0711

Rachel Goldman
Department of Biological Chemistry
The Weizmann Institute of Science
Rehovot 76100, Israel

Jessica F. Greene
Departments of Entomology and
Environmental Toxicology
University of California at Davis
Davis, CA 95616

Zhonghong Guan
Departments of Medicine and Molecular
Biology and Pharmacology
Washington University
School of Medicine
St. Louis, Missouri 63110

Cecilia Guastadisegni
Laboratory of Environmental Hygiene
Instituto Superiore di Sanita
Viale Regina Elena 299
00161 Rome, Italy

Namrata Gupta
Department of Pharmacology and
Therapeutics
McGill University
3655 Drummond, Montreal
Quebec, Canada H3G 1Y6

Geetha M. Habib
Department of Pathology and Cell Biology
Baylor College of Medicine
Houston, TX 77030

Damaris Harle
Institute of Pharmaceutical Chemistry
Johann Wolfgang Goethe University
Frankfurt, Marie-Curie-Str. 9
D-60439 Frankfurt, Germany

J.E. Harris
Section of Medical Oncology
Department of Medicine
Rush Medical College
Chicago, IL 60612

Catherine C. Hedrick
Division of Cardiology
University of California Los Angeles
Los Angeles, CA 90095

Helena Herbertsson
Department of Biomedicine and Surgery
Division of Cell Biology
Linkoping University
S-58185 Linkoping,
Sweden

Ceil A. Herman
Department of Biology
New Mexico State University
Las Cruces, NM 88003

Harvey R. Herschman
Depaprtments of Biological Chemistry and
Molecular and Medical Pharmacology
UCLA School of Medicine
Molecular Biology Institute
611 Circle Drive East
Los Angeles, CA 90025-1570

Charles S. T. Hii
Department of Immunopathology
Women's and Children's Hospital
North Adelaide, South Australia 5006

Christian W. Honemann
Klinik and Poliklinik fur Anasthesiologie and
Operative Intensivmedizin
Westfalische Wilhelms Universitat Munster
Albert-Schweister-Strasse 33,
D-48149 Munster, Germany

Takashi Izumi
The Department of Biochemistry and
Molecular Biology
Faculty of Medicine
The University of Tokyo
Hongo 7-3-1
Bunkyo-ku, Tokyo 113,
Japan

Wolfgang Jira
Department of Organic Chemistry I
University of Bayreuth
D-95440 Bayreuth

Amit S. Kalgutkar
A.B. Hancock, Jr., Memorial Laboratory for
Cancer Research
Department of Biochemistry
Center in Molecular Toxicology
Vanderbilt University School of Medicine
Nashville, TN 37232-0146

Hidecki Kamitani
Laboratory of Molecular Carcinogenesis
NIEHS, NIH
Research Triangle Park, NC 27709

Sathasiva B. Kandasamy
Radiation Patholphysiology and Toxicology
Department
Armed Forces Radiobiology Research
Institute
Bethesda, MD 20889-5603

Kunio Kato
Department of Physiology
Exploratory Research for Advanced
Technology (ERATO)
Japan Science and Technology Corporation
(JST)
2-9-3, Shimo-Meguro
Meguro-ku, Tokyo 152,
Japan

Kuljeet Kaur
University of Rochester Cancer Center
601 Elmwood Avenue
Rochester, NY 14642

Uddhav Kelavkar
Renal Division
Emory University
Atlanta, GA 30322

J. Lodewyk F. Kock
Department of Microbiology and
Biochemistry
University of the Orange Free State
P.O. Box 339
Bloemfontein, 9300
South Africa

Ichiro Kudo
Department of Health Chemistry
School of Pharmaceutical Sciences
Showa University
1-5-8 Hatanodai, Shinagawa-ku
Tokyo 142, Japan

Hartmut Kuhn
Institute of Biochemistry
University Clinics Charite
Humboldt University
Hessische Str. 3-4
10115 Berlin, Germany

Hiroshi Kuwata
Department of Health Chemistry
School of Pharmaceutical Sciences
Showa University 1-5-8 Hatanodai
Shinagawa-ku, Tokyo 142
Japan

Edward F. LaBelle
Department of Physiology
Allegheny University of the Health Sciences
2900 Queen Lane
Philadelphia, PA 19129

Joung H. Lee
Department of Neurosurgery
Cleveland Clinic Foundation
Cleveland, OH

Soo H. Lee
Department of Physiology
School of Medicine
Ajou University
Suwon 442-749, Korea

David W. Leung
Molecular Biology Department
Cell Therapeutics, Inc.
Seattle, WA 98119

Michael W. Lieberman
Department of Pathology
Baylor College of Medicine
Houston, TX 77030

Kyoung-Min Lim
College of Pharmacy
Seoul National University
Seoul 151-742 Korea

King-Teh Lin
Department of Cell Biology
School of Osteopathic Medicine
University of Medicine and Dentistry of New
Jersey
Stratford, NJ 08084

Alon Margalit
The Institute for Applied Biochemistry
Ben-Gurion University of the Negev
P.O. Box 653
Beer-Sheva 84105
Israel

Lisa A. Marshall
Department of Innumopharmacology
SmithKline Beecham Pharmaceuticals
709 Swedeland Road
King of Prussia, PA 19406

Michael Matejka
Department of Oral Surgery
School of Dentistry
University of Vienna

Kathleen A. Merkler
Department of Chemistry and Biochemistry
Duquesne University
Pittsburgh, PA 15282-1530

Steven A. Moore
Department of Pathology
The University of Iowa
Iowa City, IA 52242

Angel Montero
Center for Glomerulonephritis
Renal Division
Emory University
Decatur, GA 30033

Jason D. Morrow
Department of Medicine
Vanderbilt University School of Medicine
Nashville, TN 37232-6602

Makoto Murakami
Department of Health Chemistry
School of Pharmaceutical Sciences
Showa University
1-5-8 Hatanodai, Shinagawa-ku
Tokyo 142, Japan

Tatsuji Nakamura
Division of Basic Sciences
National Jewish Medical and Research Center
1400 Jackson Street
Denver, CO 80206

Daotai Nie
Department of Radiation Oncology
Wayne State University
School of Medicine
Detroit, MI 48202

Santosh Nigam
Eicosanoid Research Division
Department of Gynaecology
University Medical Centre Benjamin Franklin
Free University Berlin
D-12200 Berlin, Germany

Makoto Nishiyama
Division of Neurosurgery
Tottori University
Faculty of Medicine
Yonago 683, Japan

690

Mark G. Obukowicz
Discovery Pharmacology
G.D. Searle
800 N. Lindbergh Blvd.
St. Louis, MO 63167

Sachiko Oh-ishi
Department of Pharmacology
School of Pharmaceutical Sciences
Kitasato University
Minato-ku, Tokyo 108-8641

Ernst H. Oliw
Department of Pharmaceutical Biosciences
Uppsala Biomedical Center
Uppsala, Sweden

Richard L. Ornberg
Analytical Sciences Center
Science and Technology
Monsanto Company
St. Louis, MO 63198

Cecil R. Pace-Asciak
Research Institute
Hospital for Sick Children
555 University Avenue
Toronto, Canada M5G 1X8

Paola Patrignani
Department of Medicine and Aging
University of Chieti "G. D'Annunzio"
66013 Chieti, Italy

Michael Potter
Laboratory of Genetics
National Cancer Institute
National Institutes of Health
Bethesda, MD 20892

Thomas Primiano
Department of Environmental Health
Sciences
Johns Hopkins School of Hygiene and
Public Health
615 N. Wolfe Street
Baltimore, MD 21205

Mike Robbins
Radiation Research Laboratory
Department of Radiology
University of Iowa
Iowa City, IA 52242

Dwight Robinson
Arthritis Unit
Massachusetts General Hospital
Boston, MA 02114

L. Jackson Roberts, II
Departments of Pharmacology and Medicine
Vanderbilt University
Nashville, TN 37232

Shlomo Sasson
Department of Pharmacology
School of Pharmacy
Faculty of Medicine
Hebrew University
Jerusalem, IL-91120
Israel

Kerstin Schnurr
Institute of Biochemistry
University Clinics Charite
Humboldt University
Hessische Str. 3-4
10115 Berlin, Germany

Paul Schweitzer
Neuropharmacology
The Scripps Research Institute
10550 North Torrey Pines Road
La Jolla, CA 92037

Charles N. Serhan
Center for Experimental Therapeutics and
Reperfusion Injury
Department of Anesthesia
Brigham and Women's Hospital
Harvard Medical School
Boston, MA 02115

Satoko Shimbara
Department of Health Chemistry
School of Pharmaceutical Sciences
Showa University
1-5-8 Hatanodai, Shinagawa-ku
Tokyo 142, Japan

Keqin Tang
Department of Radiation Oncology
Wayne State University
School of Medicine
Detroit, MI 48202

Bernd-J. Thiele
Institute of Biochemistry
University Clinics Charite
Humboldt-University
Berlin, Hessische Str. 3-4
D-10115 Berlin, Germany

Ulrich Tibes
Boehringer Mannheim GmbH
Dept. of Preclinical Research
D-68298 Mannheim

G. Tigyi
Dept. of Physiology and Biophysic
The University of Tennessee
894 Union Avenue
Memphis, TN 38163

Jozsef Timar
1st Institute of Pathology and Experimental
Cancer Research
Semmelweis University of Medicine
Budapest, Hungary

Natsuo Ueda
Department of Biochemistry
Tokushima University School of Medicine
Tokushima 770-8503, Japan

Yoshihiro Urade
Department of Molecular Behavioral Biology
Osaka Bioscience Institute, Suita
Osaka 565-0874, Japan

Sujata Vartak
Radiation Research Laboratory
Department of Radiology
University of Iowa
Iowa City, IA 52242

Jay Whelan
Department of Nutrition
University of Tennessee
Knoxville, TN 37996-1900

Guishan Xiao
Division of Hematology
Department of Internal Medicine
University of Texas Health Science Center at
Houston
Houston, TX 77030

Shozo Yamamoto
Department of Biochemistry
Tokushima University
School of Medicine
Tokushima 770-8503, Japan

Yorihiro Yamamoto
Research Center for Advanced Science and
Technology
University of Tokyo
Komaba, Meguro-ku,
Tokyo 153

692

INDEX

Airway morphology, cysteinyl leukotrienes and, 313
Albumin
  eicosanoid metabolism and, 23–27
  12-HETE production and, 24–25
Alzheimer's disease
  neuroprostanes and, 344–346
  reactive oxygen species and, 413
Anandamide (N-arichidonylethandamine), 547, 550
Anandamide amidohydrolase
  characterization, 513–515
  organ distribution, 515–517
  Anesthesia, thrombaxane A_2 and, 269–273, 277–281
APC gene, in colorectal carcinoma, 593
APHS (2-acetoxyphenylhept-2-ynyl sulfide), 139–142
Apoptosis
  arachidonate 5-lipoxygenase and, 579–580
  arachidonic acid-induced protection, 579–580, 583–589, 593–596
  caspase and, 569, 572–574, 588–589
  in colon cancer cells, 555–560
  in colorectal carcinoma cells, 593–596
  execution process, 569
  in leukemia HL-60 cells, 569–574
  in lymphoma cells, 563–567
  lysophosphatidic acid and, 245, 259–260
  peroxynitrite-induced, 569–574
  sphingosine 1-phosphate inhibition of, 259–260
Arachidonic acid
  apoptosis and, 579–580, 583–589, 593–596
  biological actions of, 365
  cell proliferation and, 463, 579–580
  CNS injury and, 125–129
  diabetes vascular disease and, 378–381
  metabolism
    M-current augmentation, 550–551
    platelet enzyme in, 23
    $sPLA_2$-IIA and, 183, 213
    vascular glucose transport and, 378–381
  neuromodulation, 550–552
  neutrophil superoxide production and, 365–370
  prostaglandin conversion, 409
  reactive oxygen species formation and, 415

Arachidonic acid (*cont.*)
  release
    Ca^{2+}-dependent $sPLA_2$ and $cPLA_2$ in, 200
    from human monocyte phospholipids, 215–218
    nitric oxide and, 170–171
Arachidonyl ethanolamide, 527
Arthritis
  collagen-induced model, 145
  COX-2 inhibitor SC-046 and, 145–149
  linoleic acid peroxidation and, 479–483
Asthma, cysteinyl-containing leukotrienes and, 301, 313
Atherosclerosis
  development, 455
  growth regulatory factors, 485
  linoleic acid peroxidation and, 479–483
  platelet aggregation in, 53

Breast cancer, hormones, dietary fatty acids and, 119–123

Cannabinoid receptors
  arachidonyl ethanolamide and, 527
  neuronal hippocampal activity and, 547–552
  prostaglandin derivatives and, 527–531
Carcinogenesis
  chronic inflammation and, 435
  NSAIDs and, 555
CD spectroscopy, 57, 58
Central nervous system (CNS)
  cerebrovascular PG production and regulation, 125–129
  ionizing radiation effects, 647–652, 655–657
  lysophospholipid signaling and, 357–361
Ceramide, reactive oxygen species formation and, 413–416
Chemiluminescence, lipid-stimulated, 414–417
Clinical pain syndromes, spinal NSAIDs and, 405–406
Colon cancer
  L-745,337 PGHS inhibitor and, 558–559
  NSAIDs and, 555–560
  PGHS overexpression in, 556
  salicylate and, 558–559

Colorectal carcinoma
  APC gene in, 593
  apotosis in, 593–596
Cyclooxygenase (COX), 113–174
  activity measurement, 131–136
  catalytic activities, 132
  colorectal carcinoma apoptosis and, 593–596
  expression and activity, 131
  function and regulation, 3–7
  isozymes, 402–403
    cytokine regulation of, 125–129
    expression, 169–171
    nitric oxide activation of, 169–171, 422–423
  NSAIDs inhibition and, 607
  in periodontitis, 419, 422–423
  spinal
    inhibition of, 402
    in tissue-injury and hyperalgesia, 402–406
Cyclooxygenase (COX-1)
  function and expression, 9
  induction into human megakaryoblastic cell line,
    17–19
Cyclooxygenase (COX-2)
  cerebrovascular, 125–129
  expression
    MAPK signaling pathway, 9–11
    microglia, 169–171
  inhibitors, 131
    indomethacin and, 151–155
    L-745,337 and NS-398, 157–162
    NSAIDs, 402
    SC-046, 145–149
    selective vocalent modifer APHS, 139–142
  MEKK1 activation of, 11
Cyclopentenone isoprostane formation, 337
Cytokine
  interleukin-4 and –13
    biological role of, 80
  regulation
    of cerebrovascular COX-2 expression, 125–
      129
    of leukotriene (LTA₄) hydrolase, 449–453
    of 15-LOX and PH-GPx, 75–80
  synthesis and release agents, 443
  from Th1/Th2 cells, 410, 411
Cytokine-suppressive anti-inflammatory drugs
  (CSAIDs), 9

Diabetes
  atherosclerosis and, 455–458
  vascular disease, 378–381
1,2-diacylglycerol hydroperoxide, 431–435
Diethiolethiones, tumorigenesis and, 599–604
Dipodascopsis uninucleata, 3-hydroxy fatty acids
  and, 675–677
Dithiolethione, leukotriene B₄ metabolism and, 599–
  600

Eicosanoid
  albumin-modified metabolism, 23–27
  apoptotic cell death and, 579

Eicosanoid *(cont.)*
  biosynthesis
    intestinal cancer and, 607–613
    NSAIDs anti-tumor effect and, 607–613
    release agents and, 443
  enzyme regulation, 3–27
  fungal oxylipins, 663–667
  LPS and MTP-PE induced formation of, 443–447
  neuromodulation and, 550–552
Eicosapentaenoic acid (EPA), CNS-derived cells and,
  647–652
Epidermal growth factor (EGF) signal tranduction,
  371–375
Erucamide, 519
Extracellular signal-regulated protein kinase (ERK),
  365–370

Fatty acids, *see also* (3R)-hydroxy-oxylipins
  COX induction into human megakaryoblastic cell
    line, 17–21
  dioxygenases and, 679
  polyunsaturated, 365
    breast cancer and, 120, 122–123
    coronary disease and, 485
    cytotoxic action, 493–497
    lipid peroxidation, 479–480
    prostate cancer, 499, 577–580
    radiosensitivity of glioma cells and, 647–652
    Ras-specific negative regulator and, 391–397
    primary amides, biosynthetic pathway for, 519–
      524
Flourescence quenching assay, 56, 57–58
Fluorescence microscopy, COX activity and, 131–
  136
Fungal (3R)-hydroxy-oxylipins, 663–667
  biology, 666, 667
  biotechnological production and mapping, 676–
    677
  chemical structure, 664–665

Gaeumannomyces graminis, catalytic properties of,
  679–683
Gamma radiation
  γ-glutamyl leukotrienase
    function of, 305
    leukotriene C₄ and D₄ and, 301–305
    hippocampal norepinephrine release and, 655–
      657
Gamma-linolenic acid (GLA)
  anti-atherogenic effects, 485–490
  macrophage-smooth muscle cell interactions,
    487
  in pancreatic cancer treatment, 505–509
  prostaglandin E₁ role in, 487–489
  in prostate cancer cytotoxicity, 493–504
  radiation sensitivity and, 493–497
  smooth muscle cell proliferation, 481–487
    in vivo effect, 489–490
Gastric ulcers, indomethacin-induced, 157–162
Glomerulonephritis, cytokin regulation of, 449–
  453

Glucose
  metabolism
    protein kinase C (PKC) and, 425–430
    monocyte endothelial interactions and, 456–458
  transport
    lipoxygenase regulation of, 377–381
    transporter-1 (GLUT-1), 378, 381
Glutathione, 476
  peroxidases, 75
  prostaglandin biosynthesis regulation by, 165–167
Glutathione S-transferases (GSTs), xenobiotic metabolism and, 327
Glycerophospholipids, lipid peroxidation of, 539–544
Gout, NSAIDs and, 165
G-protein coupled receptors (GPCRs)
  for lysophoshpolipids, 259–263
  for spingosine 1-phosphate (S1P), 260–261, 263
  ventricular zone gene-1 (vzg-1) and, 357–361

Hepoxilin
  biological actions in human neutrophils, 535
  ω-hydroxy metabolite
HETE: see Hydroxyeicosatetraenoic acid
HODE: see Hydroxyoctadecadienoic acid
Hydroxy fatty acids: see Fungal (3R)-hydroxy-oxylipins
13(S) hydroperoxyoctadecadienoic acid (HpODE)
  cell proliferation/apoptosis regulation by, 593
  in epidermal growth factor signal transduction, 371–375
Hydroxyeicosatetraenoic acids (HETEs)
  apoptosis and, 594
  cell–cell interactions and, 464–465
  inflammation and, 464
  metabolites in glucose transport system, 378–381
  metastasis and, 463–464
  thrombosis and, 464
  tumor growth and secondary metastasis and, 465–468
  12-HETE
    albumin-modified production of, 24–25
    in tumor cells, 618, 621
    12(S) binding complex, 253–257
Hydroxyoctadecadienoic acid (HODE)
  9-HODE, 479–483
  linoleic acid and, 479–483
13-hydroxyoctadecadienoic acid (13-HODE)
  cell–cell interactions and, 464–465
  inflammation and, 464
  thrombosis and, 464
  in tumor growth and secondary metastasis, 465–468
Hyperalgesia, spinally-mediated
  COX inhibitors and, 402–403
  sensory input processing in, 401
  spinal prostaglandin release and, 404–405
Hyperglycemia
  atherogenesis and, 455, 477
  12-lipoxygenase activity and, 458

Indomethacin (INDO)
  gastric ulcers and, 157–162
  L-745,337 and NS-398 COX-2 inhibitors and, 157–162
  pristane-induced plasmacytomagenesis and, 152–155
Inducible nitric oxide synthase (iNOS)
  endotoxin-induced expression in murine macrophases, 425–430
  expression, 169
Inflammation
  arachidonic-acid lipid mediators of, 3
  carcinogenesis and, 435
  COX-2 inhibitors and, 131
  glomerular LTA$_4$ hydrolase expression and, 449–453
  HETEs and 13-HODE in, 464
  leukotrienes in, 437–440, 599–604
  linoleic acid oxidation products in, 295, 479–483
  15-lipoxygenase as therapeutic strategy in, 73
  lymphocyte-mediated, 409–411
  nitric oxide and, 419–423, 425–430
  prostglandin production in, 183–187, 209
  spinal synthesis and prostanoid release following, 401–406
Intestinal cancer, NSAIDs and, 607–613
Isoleukotoxin
  leukotoxin and, 473–474
  metabolite toxicity, 475–476
  toxic effects of, 471–473
Isolevuglandin formation, 337–339
Isoprostane
  F$_4$–isoprostane, 335
  reactive products formation and, 335–350
Isoprostane-like compounds, docosahexaenoic acid and, 343–346

Jun N-terminal kinase (JNK) activation, 9, 365–370

Kupffer cells activation, 443–447

L-699,333, 327–328
L-745,337, 555–560
Leukemia, and peroxynitrite-induced apoptosis, 569–574
Leukotoxin
  isoleukotoxin synthesis, 473–474
  metabolite toxicity, 471–473, 475–476
Leukotriene (LT), 3, 287–331
  biosynthesis, 61
  cysteinyl
    antagonists, 313–317
    asthma and, 301, 313
    biology, 327
    metabolism, 295
    pulmonary hypertension and, 307–310
  in inflammation, 437–440, 599–604
  metabolism in activated mast cells, 37–41
  peptide, 669
Leukotriene A$_4$ hydrolase, 295
  cytokine regulation of, 449–453
  enzymatic transformation into LTB$_4$, 319–324

Leukotriene B$_4$
dithiolethione-induced, 599–604
γ-glutamyl leukotrienase cleavage of, 301–305
in glomerular injury, 449–453
kinetic evaluation of, 437–440
molecular cloning and characterization, 237–242
receptor (BLT), 238–239
in tissue injury, 287
Leukotriene C$_4$
γ-glutamyl leukotrienase, and, 301–304
hypothalamus/pituitary hormone release in amphibians and, 669–673
metabolism, 295, 301
Leukotriene C$_4$ synthase
inhibitors, 327–328
random rapid equilibrium mechanism for, 327–331
Leukotriene D$_4$, 295
dipedtidases and, 295–299
γ-glutamyl leukotrienase and, 305
Leukotriene E$_4$, 295
congenital heart lesions and, 307–310
kinetic evaluation of, 437–440
Levuglandin formation, 335–336
Linoleate diol synthase (LDS)
catalytic properties of, 679–683
G. graminis and, 679
prostaglandin H synthase and, 679, 682–683
Linoleic acid
metabolites
excessive cell growth and, 463–468
toxicity, 471–477
oxidation products in inflammation, 479–483
Lipid mediators
cannabimimetic, 547
human keratinocyte, 413
LPA-like receptor subtype specificity in, 245–250
Ros formation and, 413–417
Lipid perioxides/peroxidation (LPO)
of arachidonate-containing glycerophospholipids, 539–544
in cellular metabolism of hydroperoxy ester lipids, 75–80
linoleic acid-derived, 479–483
neuronal plasticity and, 43–47
in pathogenesis, 75
Lipid phosphoric acids, see also Lysophosphatidic acid
Lipocalin-type prostaglandin D synthase Beta-trace
non-substrate lipophilic ligands and, 55–59
secreted from heart to plasma, 49–53
Lipopolysaccharide (LPS)
biological effects on macrophages, 443–447
PGE$_2$ release and, 169–171
Lipoxins, aspirin-triggered
neutrophil-mediated changes in vascular permeability and, 290–292
polymorphonuclear neutrophil infiltration and, 288–292
stable analogs of, 287–288

Lipoxygenase (LOX)
classification, 61, 62–64
colorectal cancer apoptosis and, 593–596
human expression, 83–87
inhibitors, 563–567
iron ligand sphere, 99–103
low density lipoprotein oxidation, 455–456
mammalian expression, 61–66
neuronal plasticity metabolites, 43–47
plant and animal, 99–103
platelet xenobiotics and, 631–636
positional specificity of, 91–96
prostate carcinoma and, 623–629
structure, 99–100
U937 cellular proliferation and, 563–567
5-lipoxygenase
in cell proliferation/apoptosis, 563–567, 577–580
mRNA expression, 105–110
pancreatic cancer and, 505–509
12-lipoxygenase
in AA-induced protection against apoptosis, 583–589
albumin effect on, 25–27
human melanoma cell line expression, 617–621
monoctye endothelial interactions and, 455–458
platelet-type, 623–629
15-lipoxygenase
cytokine induced regulation of, 75–80
expression and gene induction role, 67–73
leukocyte type 12-lipoxygenase, 593–596
15S-lipoxygenase
in epithelial tissues, 83–87
Liver macrophage activation, 443–447
Long-term potentiation (LTP) of synaptic transmission, 43, 45
Low density lipoprotein oxidation
linoleic acid hydroperoxides in, 479–480
lipoxygenases in, 455–456
Luteinizing hormone (LH), and leukotriene C$_4$, 669–673
Lymphoma, cellular proliferation modulation in, 563–567
Lysophosphatidic acid (LPA)
analogs
receptor subtype specificity in, 245–250
apoptosis inhibition, 259–260
biochemistry and biology, 245, 259–260, 351
function and characterization, 259–260
growth effects, 245
homology clusters and evolutionary tree, 262
receptors
cannabanoid receptor CB1 and, 360
G protein-coupled and, 260–263, 277
PSP24 gene, 249
ventricular zone gene-1 (vzg-1), 357–360
Lysophosphatidic acid acyltransferase (LPAAT)
cDNA cloning of, 351–355
in LPA biology, 351
phosphatidic acid formation and, 351, 354–355
Lyso-phosphatidylcholine (LPC), reactive oxygen species formation and, 413–416

Lysophospholipids: *see* Lysophosphatidic acid
Lysosphingolipids: *see* Sphingosine 1-phosphate

Manganese superoxide dismutase (Mn-SOD), radiation and, 647–652
MAPEG (membrane associated proteins in eicosanoid and glutathione metabolism), 327, 331
Mast cells
  biosynthetic pathway regulation in, 37–41
  PGD$_2$ synthesis in, 5–6
Melanoma, 12-LOX and 15-LOX expression in, 617–621
Memory impairment, cannabinoids and, 547
Mitogen-activated protein (MAP) kinase, 444, 446
  biology, 365
  p38 MAP kinase
    arachidonic acid-stimulated, 365–370
    Pak activation/potentiation of, 387
  signaling pathway
    COX regulation of, 9–14

α-Naphthol, 636
Neurofibromin, dietary fatty acid inhibition of, 391–397
Neurological disorders, role of free radicals in, 346
Neuronal plasticity, lipid peroxides and, 43–47
Neuropeptides, α-amidated, biosynthesis of, 519
Neuroprostanes, 344, 346
Neutrophils
  hepoxilin A$_3$ metabolized in, 535–537
  3*R*-HETE effects, 667
  p21-activated protein kinases in, 385–388
  p38 MAP kinase, 365–370
  superoxide production in, 365–370
Nitric oxide
  COX activation by, 422–423
  LPS and MTP-PE induced, 443–447
  microglial COX-2 expression and, 169–171
  in murine macrophages, 425–430
  periodontitis and, 419–423
  polymorphonuclear
    and lipoxin stable analogs, 287–292
  release
    agents, 443
    arachidonic acid-induced, 170–171
    endotoxin-induced, 425–430
  synthesis
    in periodontitis, 419–423
    prostaglandin E$_2$, 169–174
Nonsteroidal anti-inflammatory drugs (NSAIDs)
  carcinogenesis and, 555
  collagen-induced arthritis model and, 145
  colorectal cancer apoptosis and, 593
  cyclooxygenase and, 9
  gout and, 165
  intestinal cancer and, 607–613
  plasmacytomagenesis and, 153–155
  prostaglandin synthesis and, 4, 119–123, 607–613
  spinal, 402–406

Norepinephrine
  arachidonate release and, 177–181
  long-term potentiation in hippocampus and, 655–657
Nuclear hormone receptors: *see* Peroxisome proliferator-activated receptors (PPARs)

Oleamide, 519
Oxygen radicals, in carcinogenesis, 435

p120 GTPase activating protein (p120 GAP), fatty acids and, 391–397
p21-activated protein kinases (Paks)
  neutrophil responses and, 385–388
  phosphatidylinositol 3-kinase regulation of, 387
  phospholipids and sphingolipids effects on, 388
Pancreatic cancer
  gamma-linolenic acid in, 505–509
  5-lipoxygenase inhibitors and, 505–509
Peptidylglycine α-amidating monooxygenase (PAM), 519–524
Periodontitis
  COX activation in, 419, 422–423
  increased PGE$_2$ and PGI$_2$ production in, 410
  nitric oxide synthesis in, 419–423
Peroxidase
  PGH synthases of fungus *G. graminis*, 679–683
  PGH$_2$ formation, 115
Peroxisome proliferator-activated receptors (PPARs)
  8S-HETE and polyunsaturated fatty acid and, 231–235
  isotypes
    ligand specificity of, 232
Peroxynitrite, in programmed cell death, 569–574
Phosphatidic acid
  LPAAT and, 351, 354–355
  phosphatase
    expressed in mammalian cells, 641–643
    in glycerolipid biosynthesis and signal transduction, 639
    in normal v. tumor cells, 644–645
    PAP2 cDNA cloning and sequence analysis and, 640
Phospholipase A$_2$, *see also* Secretory phospholipase A$_2$s (sPLA$_2$s)
  human leukocyte chemotaxis and, 189–195
  in human monocyte eicosanoid formation, 215–218
  released from vascular smooth muscle cells, 177–181
Phospholipid
  growth factors, 245
  human monocyte release from 215–218
  p21-activated protein kinase (Paks) and, 388
Phospholipid hydroperoxide glutathione peroxidase, cytokine induced regulation of, 75–80
Plasmacytomagenesis, pristane
  induction system, 151
  indomethacin inhibition of, 152–155
    prostaglandin synthesis and, 154
    ROS formation and, 154–155

Platelet(s)
PGD$_2$/PGI$_2$ and, 53
PHS and LPO-mediated xenobiotic activation and, 631–636
TXA$_2$ receptor-mediated, 269–273
Platelet derived growth factor (PDGF), induction pathways, 5
Platelet-activating factor (PAF)
hippocampal long-term potentiation and, 221–227
mast cell antiflammatory action and, 37–41
mast cell metabolism and, 41
reactive oxygen species formation and, 413–416
Polymorphonuclear leukocytes (PMNs), activation of, 431–435
Prostaglandin (PG), see also Cyclooxygenase (COX)
amides
cannabinoid receptors binding with, 527–531
biosynthesis
in activated mast cells, 5–6
functional linkage of enzymes in, 29–33
glutathione regulation of, 165–167
hormones and fatty acid effects in, 119–123
ligand-induced, 4–5
NSAIDs and, 4, 119–123, 607–613
p38 MAPK signaling pathway and, 9–14
in periodontitis, 419–423
phases of, 183
cerebrovascular production/regulation, 125–129
in human megakaryoblastic cell line CMK, 19–21
roles of, 169
Prostaglandin D$_2$
breast cancer and, 119
formation and function, 49
spinal mechanism, 404–405
Prostaglandin D synthase: see Lipocalin-type prostaglandin D synthase Beta-trace
Prostaglandin E$_1$
gamma-linolenic acid and, 487–489
Prostaglandin E$_2$
B lymphocytes immunoregulation of, 409–410
biosynthesis
nitric oxide and, 169–174
MEKK1 activation and, 11
human disease and, 411
Th1/Th2 cytokine modulation by, 410
Prostaglandin endoperoperoxide synthase
cyclooxygenase activity, 17–19, 555
NSAIDs and, 556
human isoforms, 555
inhibitor L-745,337, 555–560
isozyme expression, 139, 556–557
in thromboxane and PG biosynthesis, 17
Prostaglandin G$_2$, 132
Prostaglandin H$_2$, 132
Prostaglandin H synthase
comparison of PGHS-1 and PGHS-2, 115–117
linoleate diol synthase and, 679, 682–683
xenobiotic activation in platelets and, 631–636
Prostaglandin I$_2$
antimetastic role, 463
as nociceptive mediator of writhing reaction, 265–268

Prostanoids, see also Prostaglandins; Thromboxanes
biosynthesis, 555, 556
in activated mast cells, 5–6
COX isozymes and, 402–403
ligand-induced, 4–5
prostaglandin synthase in, 3
spinal, post-injury, 401–406
Prostate cancer
arachidonate 5-lipoxygenase in, 579–580
cell growth and apoptosis in, 577–580
GLA-mediated cytotoxicity and, 493–504
platelet-type 12-lipoxygenase in, 623–629
polyunsaturated fatty acids and, 499–504, 577
Protein kinase, mitogen-activated: see Mitogen-activated protein kinase
Protein kinase C (PKC)
arachidonic acid-stimulated, 365–370
glucose metabolism and, 425–430
hippocampal long-term potentiation and, 655–656
iNOS expression and, 425, 430
receptor-mediated signal transduction role, 431
PSP24 gene, as lysophosphatidic acid receptor, 249
Pulmonary hypertension, cysteinyl-leukotrienes in, 307–310

Radiation sensitivity
gamma-linolenic acid cytotoxic effects on, 493–497
polyunsaturated fatty acids and, 647–652
RAS proteins
dietary fatty acids inhibition of, 391–397
neurofibromin and p120 GAP regulation of, 391–397
in tumorigenesis process, 392
Reactive oxygen species formation (ROS)
lipid mediator-induced, 413–417
in peroxynitrite-induced apoptosis, 571, 574
physiological and pathological roles, 413
platelet-activating factors, 413–416
pristane plasmacytomagenesis and, 154–155
radiation injury and, 647

Salicylate, colon cancer and, 555–560
Secretory phospholipase A$_2$ (sPLA$_2$)
mRNA
antisense oligonucleotides and, 199–206
low molecular weight sPLA$_2$ inhibitors and, 205–206
prostaglandin synthesis and, 6–7
type IIA
arachidonic acid metabolism and, 183, 213
compared to types IIC and V, 203–213
Seizure-induced synaptic potentiation (S1P), and lipid peroxides, 43–47
Signal transduction
arachidonic acid and, 365–370
dietary fatty acids and RAS proteins in, 391–397
epidermal growth factor and, 371–375
in glucose transport system, 377–381
LTB$_4$ receptor activation of, 242
p21-activated protein kinases in, 385–388

Signal transduction (*cont.*)
  protein kinase C (PKC) in, 431
  Ras proteins in, 392, 393
Sphingosine 1-phosphate (S1P)
  biochemistry and biology, 259–260
  G protein-coupled cellular receptors for, 260–261,
    263
  homology clusters and evolutionary tree, 262
  p21-activated protein kinase (Paks) and, 388
Spinal prostanoid synthesis, post-injury inflammation
  and, 401–406
Superoxide anion, generation and release of, 431–
  435

Th1/Th2 cytokine
  human disease role of, 411
  prostaglandin modulation of, 410
Thrombosis, 13-HODE and, 464
Thromboxane A_2
  anesthesia effects on
    local effects, 269–273
    receptor signaling as target of, 277–281
  formation, 23

Thromboxane A_2 (*cont.*)
  halothane and, 281
Thromboxane $B_{2, 19}$
Toxicity mechanism, 476–477
Tumor necrosis factor (TNF), formation of, 443–447
Tumorigenesis
  dithiolethiones and, 599–604
  Ras protein inhibition by dietary fatty acids in,
    391–397

Valpromide, 519
Vascular cells
  glucose transport activity in, 377–381
  hyperglycemia and, 377
Vascular disease, diabetes, 378–381
Ventricular zone
  cell generation from, 357–361
  gene-1 (vzg-1), 357–361

Xenobiotic epoxide hydrolase, 324
Xenobiotics:
  glutathione S-transferases and, 327
  platelet activation of, 631–636

The manufacturer's authorised representative in the EU is Springer
Nature Customer Service Centre GmbH, Europaplatz 3, 69115 Heidelberg,
Germany. If you have any concerns regarding our products, please
contact ProductSafety@springernature.com

Printed and bound by CPI Group (UK) Ltd, Croydon, CR0 4YY

23/04/2026

02095623-0008